ORGANIC CHEMISTRY

A Concise Approach

SECOND EDITION

FREDRIC M. MENGER
DAVID J. GOLDSMITH
LEON MANDELL

Department of Chemistry
Emory University

ORGANIC CHEMISTRY

A Concise Approach

SECOND EDITION

W. A. BENJAMIN, INC.
Menlo Park, California · Reading, Massachusetts
London · Amsterdam · Don Mills, Ontario · Sydney

Printed in the United States of America. Published simultaneously in Canada.

Library of Congress Catalog Card No. 74-7527.

ISBN 0-8053-3281-2
ABCDEFGHIJ-HA-798765

PREFACE

The objective of this second edition of *Organic Chemistry—A Concise Approach* is the same as that of the first edition: to present the principles of synthetic and physical organic chemistry in a manageable number of pages. Although we have added sufficient new material to make the text suitable for a full length course, conciseness has not been sacrificed. Nor has brevity been allowed to interfere with rigor. We believe that any student who masters this book will have the background required, both for intermediate and advanced courses in organic chemistry and for biochemistry.

By the end of the third chapter of *Organic Chemistry—A Concise Approach,* the student is handling simple syntheses and has even been exposed to the philosophy of organic synthesis. Compare this, for example, to recent texts in which 500 pages of material are covered before the chemistry of alcohols is first mentioned. By the time page 500 of *Organic Chemistry—A Concise Approach* is reached, syntheses of natural products as well as classic examples of physical organic thought have been presented.

The most important new material added in this second edition deals with

1) the planning of complex syntheses,
2) electrocyclic reactions,
3) stereochemistry,
4) biochemical transformations and
5) spectroscopy.

The first edition contained many problems which were largely thought-provoking and often (frankly) difficult. We have retained these problems (they are necessary to stimulate creativity), but we have also added more of the "bread and butter" variety. The student should solve the easier problems first, and then use the time saved in not having to read an extra 1200 pages to think about and discuss the remaining questions. Answers to *all* problems are given at the end of the book.

Undoubtedly, there will be topics which the individual professor, because of a particular preference or expertise, will wish to cover in more detail than they are covered in this book. These subjects can be readily discussed in lectures. What is even more important, these lectures *must* provide what no text can give—the personal touch that inspires a student to learn and think.

The authors would like to thank Professors A. G. Anderson, D. E. Applequist, R. B. Bennett, R. Breslow, T. Cohen, C. S. Foote, T. L. Jacobs, J. E. McMurry, J. O. Olson, and D. J. Woodman who made numerous valuable suggestions which were incorporated in this book.

Atlanta, Georgia
November 1974

F.M.M.
D.J.G.
L.M.

CONTENTS

1
BONDING

1.1 ENERGY LEVELS AND ATOMIC ORBITALS

A chemical reaction must involve the breaking and the forming of chemical bonds whether the reaction takes place in a test tube or as part of some physiological system. Thus, in order to understand the principles that govern chemical processes, it is necessary first to understand the nature of chemical bonds. We therefore begin our discussion of organic chemistry by examining the forces that hold the atoms of a molecule together.

Table 1.1 Electronic configuration of atoms

Atomic no.	Element	First electronic shell	Second electronic shell			
		1s	2s	$2p_x$	$2p_y$	$2p_z$
1	H	1e				
2	He	2e				
3	Li	2e	1e			
4	Be	2e	2e			
5	B	2e	2e	1e		
6	C	2e	2e	1e	1e	
7	N	2e	2e	1e	1e	1e
8	O	2e	2e	2e	1e	1e
9	F	2e	2e	2e	2e	1e
10	Ne	2e	2e	2e	2e	2e

Table 1.1 shows the electron configuration of the first-row elements. The electronic shells denote different energy states that may be assumed by the electrons surrounding the positive nucleus. Electrons in the second shell are, on the average, farther away from the nucleus and, therefore, of higher energy than the electrons in the first shell. Within the second shell there are two sublevels (2s and 2p), with the 2p electrons of higher energy than the 2s electrons. The shapes of the s and p orbitals are shown in Fig. 1.1. The 1s and 2s orbitals are spherical, while the p orbital is dumbbell shaped.

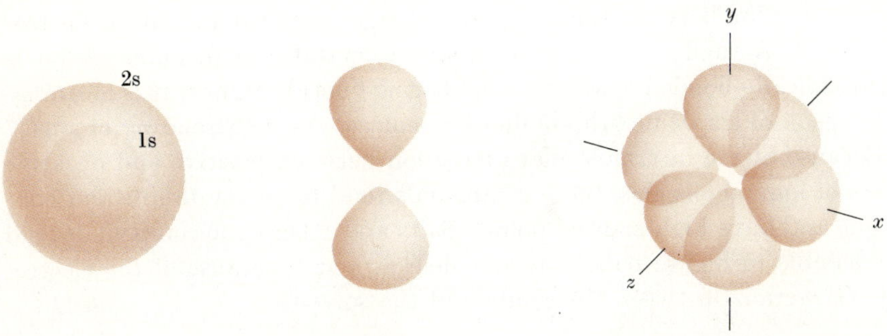

(a) (b) (c)

Fig. 1.1 Shapes of s and p atomic orbitals: a) s orbitals, b) p orbital, and c) three equivalent p orbitals

Note (Fig. 1.1) that in the second electronic shell there are three equivalent p orbitals (p_x, p_y, p_z) which are mutually perpendicular. Since each orbital may contain a maximum of two electrons (the Pauli Exclusion Principle), the nucleus of a first-row element (which has electrons only in the first two shells) may be surrounded by up to ten electrons. The first electronic shell, possessing only an s orbital, may contain two electrons, while the second electronic shell, having an s orbital and three p orbitals, may be occupied by up to eight electrons. Electrons normally fill available orbitals, with the lowest energy orbitals being occupied first. Moreover, electrons do not pair up in the p orbitals if equivalent vacant p orbitals are available (Hund's Rule).

The outer shell electrons are important in bonding because their tendency to reach an inert gas configuration (all orbitals filled) provides the impetus for bond formation. Carbon, nitrogen, and oxygen atoms, therefore, bond in such a manner as to achieve a stable electron configuration with a full octet of electrons in the second shell. An atom may acquire the inert gas configuration by either the loss or gain of electrons or by the sharing of electrons. As we will see in the next two sections, the nature of the bond formed depends on which of these routes is followed.

1.2 IONIC OR ELECTROSTATIC BONDS

An *ionic* or *electrostatic bond* results, for example, from the reaction between a lithium atom and a fluorine atom. Both atoms achieve the inert gas configuration when the 2s electron of lithium is donated to the fluorine, thereby filling the fluorine's last 2p orbital.

$$\mathrm{Li}\cdot + \cdot\ddot{\underset{\cdot\cdot}{\mathrm{F}}}: \longrightarrow \mathrm{Li}^{+} + :\ddot{\underset{\cdot\cdot}{\mathrm{F}}}:^{-}$$

This reaction leaves lithium as a cation and fluorine as an anion. The electrostatic interaction between these opposite charges is called an ionic or electrostatic bond. Actually, in a lithium fluoride crystal each lithium cation is electrostatically bonded to several neighboring fluoride anions. It is meaningless to speak of a specific lithium fluoride molecule in such an array of ions.

Because of the intermolecular attraction between positive and negative centers of ionic molecules, these compounds tend to be crystalline and nonvolatile, and have high melting points. Salts and other ionic compounds tend to be soluble in highly polar solvents such as water because of the electrostatic attractions between the solute and the solvent.

1.3 COVALENT BONDS

In ionic bond formation atoms fill their outer shells by losing or gaining electrons. *Covalent bonds* result when atoms achieve the stable octet configuration by sharing electrons. Consider compound formation by a carbon atom (with four electrons in its outer shell) and four hydrogens (each with one electron in its outer shell).

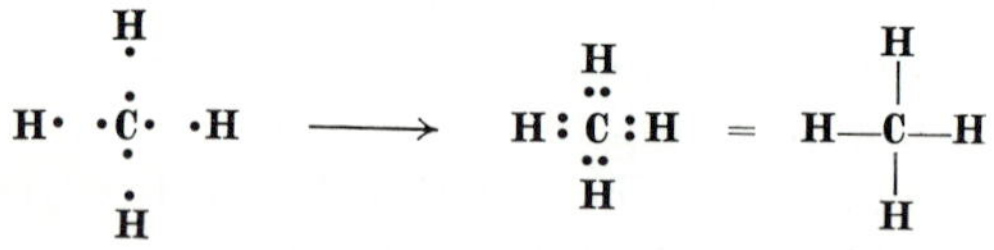

methane

Note that in the product, methane, each atom has filled orbitals by virtue of electron sharing. The carbon has a full octet of electrons and each hydrogen has an electron pair. The four carbon-hydrogen bonds are said to be covalent.

At this point it becomes necessary to define *electronegativity.* Electronegativity (an unfortunate name) refers to the degree of attraction an atom exerts on electrons. The more electronegative an atom, the greater its ability to attract electrons. The order of electronegativity of some atoms commonly found in organic chemistry can be expressed by

$$\mathrm{F} > \mathrm{O} > \mathrm{N} \approx \mathrm{Cl} > \mathrm{Br} > \mathrm{C} \approx \mathrm{H}.$$

Ionic bond formation in lithium fluoride, described above, results when an electron is completely transferred from the lithium to the fluorine. This is because the nonmetal is vastly more electronegative than the metal. On the other hand, carbon and hydrogen are both of low electronegativity. Neither atom is able to withdraw an electron from the other; instead, they share electrons and form a covalent bond. Covalent bonds are only mildly polar because the electrons are almost equally shared; and molecules having covalent bonds are said to be nonpolar. They resist solution in water and other polar solvents. Nonpolar compounds possess only weak intermolecular attractions and therefore have relatively low melting points and high vapor pressures compared to ionic materials.

The high polarity of water can now be understood in terms of the electronegativity sequence given above. Oxygen is much more electronegative than hydrogen, although not enough to effect complete electron transfer. The oxygen atom forms covalent bonds with the two hydrogen atoms via electron sharing. However, since the oxygen is much more electronegative, the electrons are closer to the oxygen, on the average, than to the hydrogens. Consequently, the oxygen has a partial negative charge and the hydrogens a partial positive charge. Both water and methane have covalent bonds, but the covalent bonds of water are polar, while those of methane are nonpolar.

Carbon's unique chemistry can be attributed to its ability to form strong carbon-carbon covalent bonds. This is typified in the series of *hydrocarbons* or *alkanes* shown below:

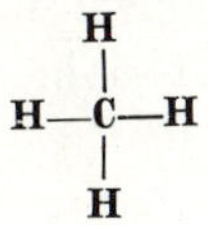

methane

(CH_4)

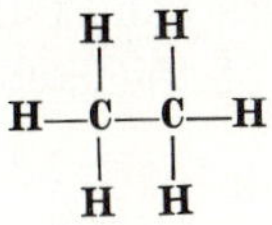

ethane

($H_3C—CH_3$)

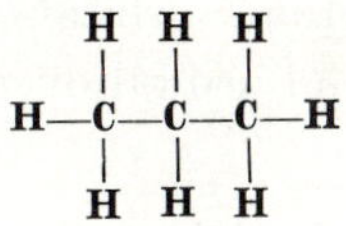

propane

($H_3C—CH_2—CH_3$)

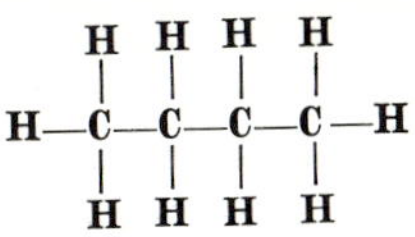

n-butane

($H_3C—CH_2—CH_2—CH_3$)

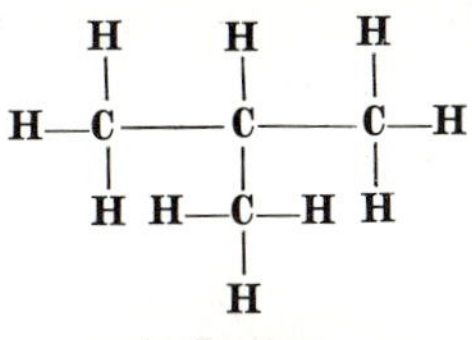

isobutane

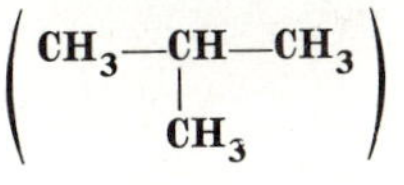

The two compounds isobutane and *n*-butane (see Appendix for nomenclature) illustrate a common phenomenon in carbon chemistry called structural *isomerism*. Isobutane and *n*-butane are said to be isomeric. Isomers are materials which have the same molecular formula (C_4H_{10} for isobutane and *n*-butane) but a different arrangement of atoms. Because of this difference in their *structures*, isomers have different physical properties (boiling points, refractive indexes, vapor pressures, heats of combustions, solubilities, etc.). The ability of carbon to form carbon-carbon bonds makes possible a tremendous number of compounds. Organic chemistry deals with the chemical and physical properties of these compounds.

Two other properties associated with covalent bonds are direction, which will be discussed in detail later, and characteristic bond distances. Thus, the carbon-hydrogen bond distances in all the compounds shown above are about the same (1.09 Å), as are the carbon-carbon bond distances (1.54 Å).*

1.4 MOLECULAR ORBITAL THEORY

The nature of the covalent bond has been rationalized most satisfactorily in terms of molecular orbital theory, which treats molecular orbitals in terms of combinations of atomic orbitals. Thus, the hydrogen-hydrogen bond of the hydrogen molecule can be pictured as the combination of two 1s atomic orbitals of the two hydrogens, yielding a molecular orbital containing two electrons. (Actually, the combination of the *two* atomic orbitals produces *two* molecular orbitals. One of these is a low-energy "bonding orbital", filled with the two electrons; the other is an empty high-energy "antibonding" orbital.) The overlap of the atomic orbitals which causes bond formation is shown in Fig. 1.2. The extent of orbital overlap is one of the factors that determine bond strength.

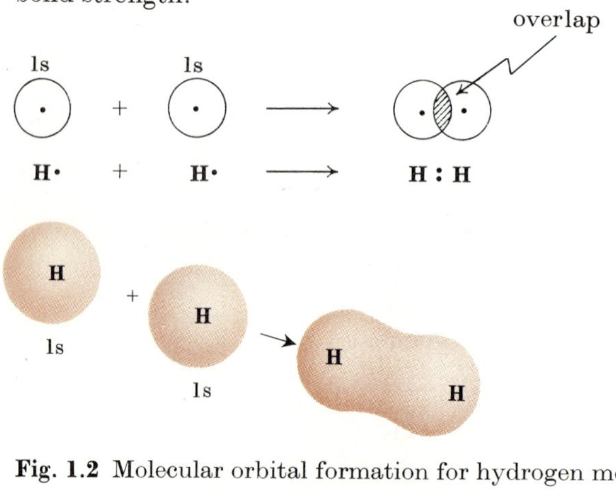

Fig. 1.2 Molecular orbital formation for hydrogen molecule

We can see the usefulness of this approach if we examine the water molecule. Consider first a water molecule formed by the overlap of the 1s orbitals of the two hydrogen atoms with each of the two unfilled 2p orbitals of oxygen (Fig. 1.3). This picture of water predicts a hydrogen-oxygen-hydrogen bond angle of 90°. However, the known bond angle for water is considerably greater than this (105°). The following section will explain this discrepancy.

* Å represents the Angstrom unit of 1×10^{-8} cm.

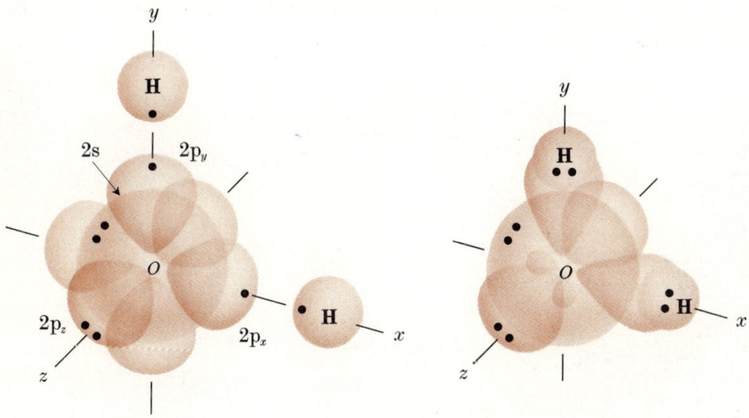

Fig. 1.3 Molecular orbital formation for water molecule

1.5 ORBITAL HYBRIDIZATION

We can account for the actual geometry of water by postulating that the various orbitals of oxygen can mix or "hybridize". For example, if we combine the single 2s orbital with the three 2p orbitals, we will produce four new orbitals. These are called sp^3 hybrid orbitals because they are composed of one s and three p orbitals. Combining a given number of orbitals always produces the same number of hybrid orbitals. The four sp^3 orbitals are directed toward the vertices of a regular tetrahedron, with the nucleus of the oxygen atom at its center. The shape is shown in Fig. 1.4.

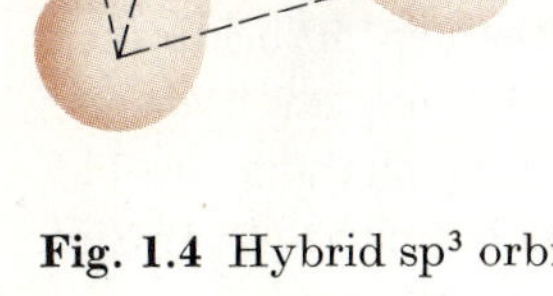

Fig. 1.4 Hybrid sp^3 orbitals

The six outer electrons must be accommodated by these four hybrid orbitals, none of which can contain more than two electrons. Thus, two sp^3 orbitals contain an electron pair each, whereas the other two orbitals have but a single electron each. The water molecule can be constructed by allowing overlap between the two half-filled sp^3 orbitals of the oxygen atom and the 1s orbitals of two hydrogen atoms (Fig. 1.5).

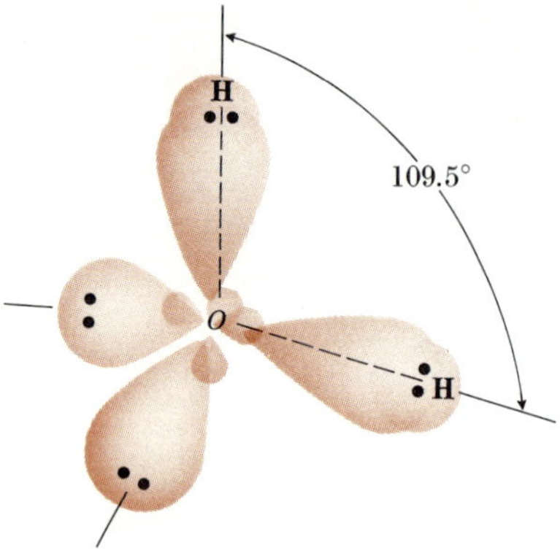

Fig. 1.5 Hybridized molecular orbital picture of water

This representation of the hybridized molecular orbitals predicts a bond angle of 109.5°, since this is the value of the angle defined by two vertices of a regular tetrahedron and its center. The value of 109.5° is considerably closer to the actual value of 105° than the 90° predicted by the unhybridized model.

It must be emphasized that the hybridization theory was developed to explain the facts; the structure of water is not a necessary consequence of a rigorous theoretical analysis. The 4.5° discrepancy between the predicted and the actual bond angles shows that, in fact, the sp^3 model is not entirely satisfactory. The smaller-than-expected hydrogen-oxygen-hydrogen bond angle has been attributed to the "desire" of the two unshared pairs of electrons to be in hybrid orbitals with more than just the 25% s character found in a pure sp^3 orbital. The electron pairs prefer greater s character because the s orbital is closer to the positively charged nucleus than the p orbitals. Consequently, the true hybridization of the oxygen may be such as to give the hybrid orbitals containing the unshared pair of electrons a greater share of the s orbital. The two orbitals of oxygen engaged in bonding to the hydrogens would in turn receive a greater share of the p orbitals than sp^3 hybridization would indicate. According to this model, the hydrogen-oxygen-hydrogen bond angle should be somewhat smaller than 109.5° (total p character would, of course, result in a 90° angle). This detailed discussion is a good illustration of how theory may be modified to accommodate indisputable facts.

The molecular orbital picture of carbon compounds involves hybridization similar to that of oxygen. Thus, an sp^3 hybridized carbon atom has a tetrahedral geometry, with each of the four outer shell electrons occupying one of the sp^3 orbitals. Methane is formed by overlapping the 1s orbital of a hydrogen atom with each of the half-filled sp^3 orbitals (Fig. 1.6).

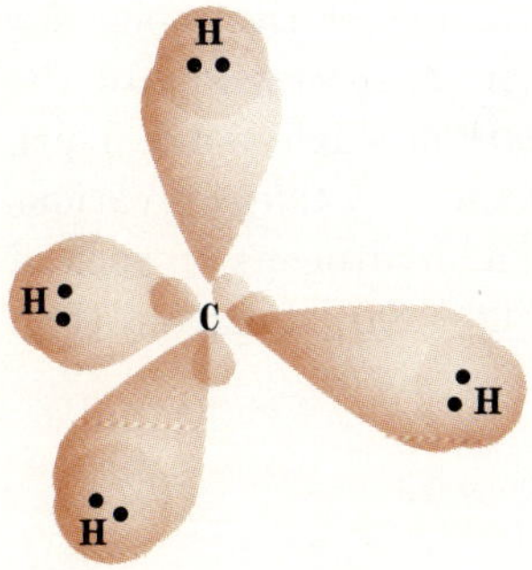

Fig. 1.6 Molecular orbital picture of methane

This gives methane a tetrahedral geometry with bond angles of 109.5°, which correspond to the experimental value. It is interesting that we arrive at the same geometry for carbon if we picture the four outer electrons as placed on the surface of a sphere and allow the electrons to position themselves so as to form the most stable system. The electrons would station themselves at the vertices of a regular tetrahedron inscribed in the sphere, for they would then be as far as possible from each other, thereby minimizing repulsive interactions between the negative charges.

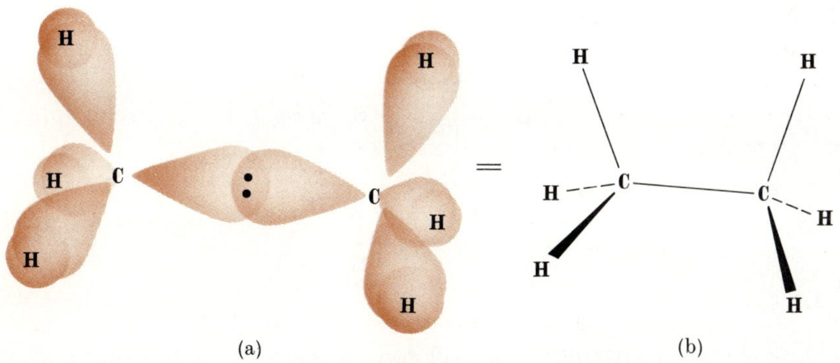

Fig. 1.7 a) Orbital and b) "stick" representations of ethane

Carbon-carbon bonds are formulated in a manner analogous to that shown for the carbon-hydrogen bonds in methane. Overlap of two carbon sp^3 orbitals, containing one electron each, generates a carbon-carbon bond, as in the formation of ethane (Fig. 1.7a). Extension of this procedure yields the larger hydrocarbons such as propane and butane. These compounds are known under the general name of *alkanes*.

Another, more convenient, representation of ethane is the stick drawing shown in Fig. 1.7(b) which does not depict the volume of the molecular orbitals but does show the three dimensionality or *stereochemistry* of the molecule. There is free rotation about the carbon-carbon single bond. Thus, the two methyl tetrahedra (or methyl groups) of ethane may take up various rotational dispositions with respect to each other. The hydrogens on each of the carbon atoms may be opposed to each other (Fig. 1.8a), staggered (Fig. 1.8b), or at some intermediate position.

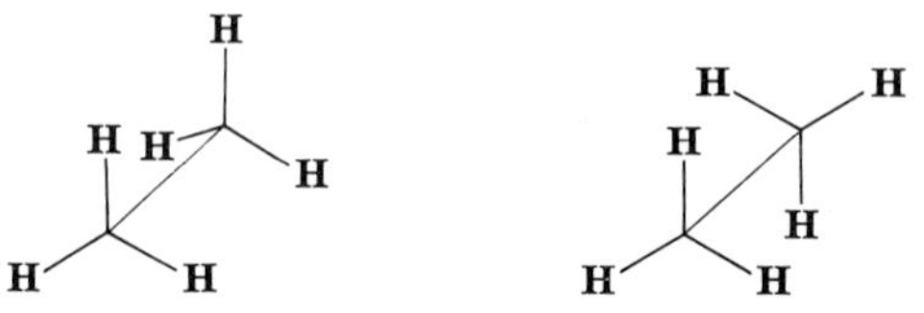

(a) (b)

Fig. 1.8 Conformers of ethane

The structures in Fig. 1.8 are known as *conformers* or *conformational* isomers. The conformers interconvert extremely rapidly, and it would be impossible to separate one from the others.

The tetrahedral nature of the carbon atom also permits the formation of cyclic molecules. For example, the compound cyclohexane can be formed from tetrahedral carbon atoms with no deviations from the normal bond angles (Fig. 1.9).

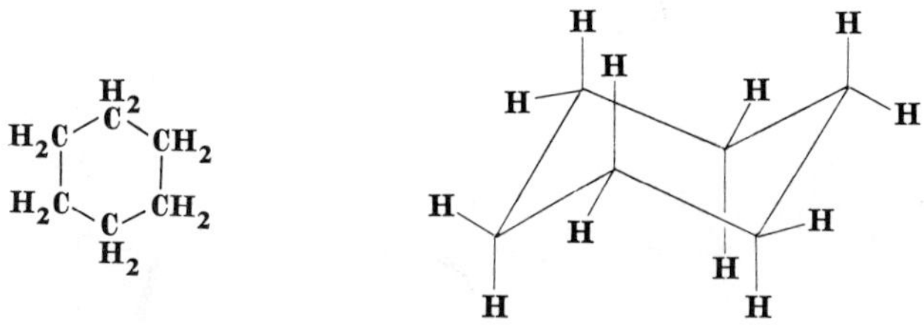

Fig. 1.9 Cyclohexane

The three-dimensional picture of cyclohexane shows that the ring must be somewhat puckered in order that the carbons retain an undistorted tetrahedral configuration. Smaller ring systems are also known; cyclopropane is an interesting small-ring compound (Fig. 1.10) whose bonding we shall consider in greater detail later. Figure 1.10 also shows a representation of cyclopropane in which the carbons and hydrogens are implied but the symbols C and H are not shown. This is common notation in ring systems.

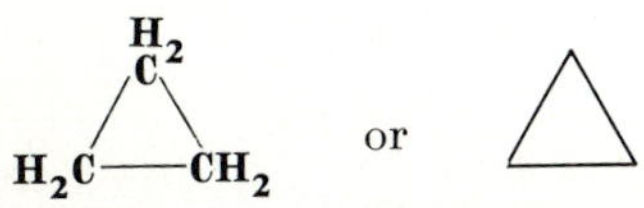

Fig. 1.10 Two representations of cyclopropane.

Pi bonds. Hybridizations other than sp^3 are possible for carbon and arise from various combinations of the one 2s and the three 2p orbitals. Consider combining the 2s orbital with *two* of the 2p orbitals. This generates a carbon with three sp^2 orbitals (each orbital is one part s and two parts p) and one unhybridized p orbital. Such an sp^2 hybridized carbon atom is shown in Fig. 1.11.

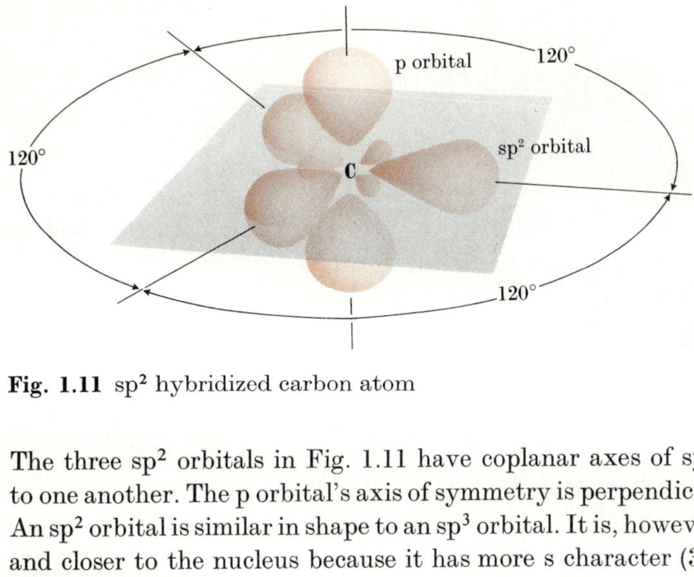

Fig. 1.11 sp^2 hybridized carbon atom

The three sp^2 orbitals in Fig. 1.11 have coplanar axes of symmetry at 120° to one another. The p orbital's axis of symmetry is perpendicular to this plane. An sp^2 orbital is similar in shape to an sp^3 orbital. It is, however, more compact and closer to the nucleus because it has more s character ($33\frac{1}{3}\%$ in sp^2 compared to 25% in sp^3). Carbon's four outer shell electrons may be placed one in each of the three sp^2 orbitals and one in the p orbital. Compound formation from such a species is pictured in the following way. Two sp^2 hybridized carbons are arranged so that all six sp^2 orbitals are coplanar; in addition, an sp^2 orbital of one carbon is collinear with an sp^2 orbital of the other carbon (Fig. 1.12a). The four other sp^2 orbitals are bonded to hydrogens, and the two carbons are joined by means of sp^2-sp^2 overlap. This geometry leads to parallel p orbitals, each containing one electron. These orbitals overlap to form a second bond between the two carbons (Fig. 1.12b). Both electrons in this bond are shown in the upper lobe of the molecular orbital, but there is equal probability that one or both of the electrons will be in the bottom lobe.

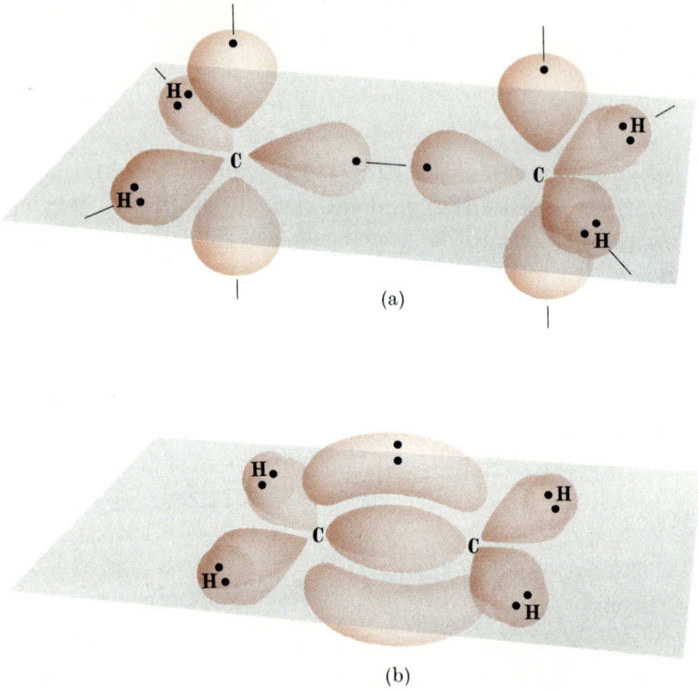

Fig. 1.12 Compound formations with two sp^2 hybridized carbons

The compound we have just described is called ethylene, and it possesses what is known as a double bond. A more classical structure of ethylene is

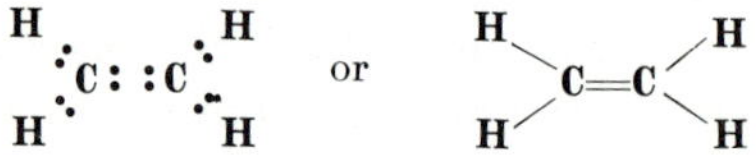

ethylene

Materials containing a double bond are called *olefins* or *alkenes*. Some other olefins are shown in Fig. 1.13.

The properties of the bond formed by the overlap of the two p orbitals are very different from the properties of single bonds such as those in ethane. The double bond in ethylene is characterized by *sideways* overlap between two orbitals. A bond formed in this manner is called a π (pi) bond. The bonds in ethane are all formed by *head-on* overlap between two orbitals, and are called σ (sigma) bonds. Thus the two carbon-carbon bonds of ethylene are different; one is a π bond and the other is a σ bond. Free rotation about the π bond of ethylene is not possible as it is about the carbon-carbon σ bond of

$CH_3—CH═CH_2$

propene

$CH_2═CH—CH_2—CH_3$

1-butene

$CH_3—CH═CH—CH_3$

2-butene

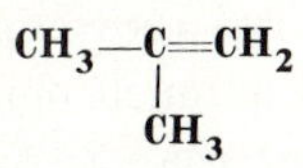

isobutylene or
2-methylpropene

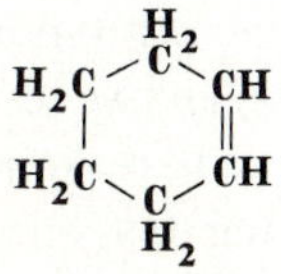

cyclohexene

$CH_2═CH—CH═CH_2$

1,3-butadiene

Fig. 1.13 Some alkenes

ethane. This fact becomes apparent when we realize that if the two carbons were to rotate about the central double bond, the overlap of the two p orbitals would be destroyed, and thus the π bond would be broken. Furthermore, the four groups on the carbons bearing the double bond must be held in a coplanar relation to one another, since only then will the two p orbitals experience maximum overlap. One consequence of the above factors is a type of isomerism called geometrical isomerism. In 2-butene for example, there are two possible dispositions or configurations of the substituents about the double bond, as shown in structures 1 and 2:

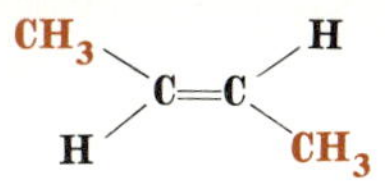

1

trans-2-butene

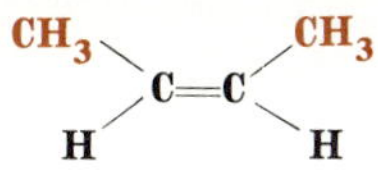

2

cis-2-butene

The two isomers have different physical and chemical properties and are interconvertible only via a chemical reaction. Compound 1 is called *trans*-2-butene and compound 2 is called *cis*-2-butene. The prefix *trans* indicates the two substituents (methyl groups) are on opposite sides of the double bond, while *cis* means that the substituents are on the same side.

Geometric isomerism is found wherever free rotation about a bond is not possible. For example, in cyclic compounds such as cyclobutane (Fig. 1.14),

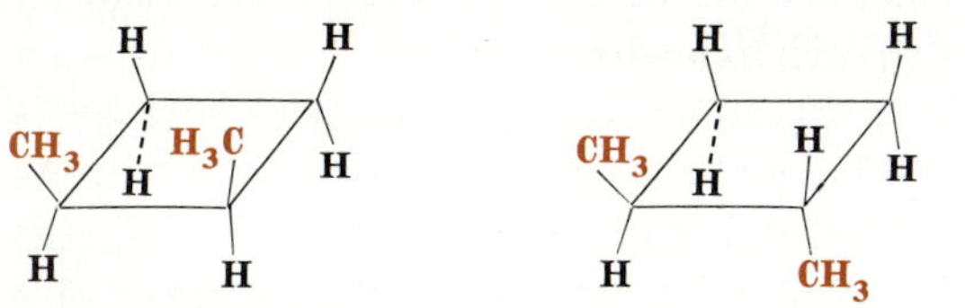

cis *trans*

Fig. 1.14 *cis* and *trans* 1,2-dimethylcyclobutane

free rotation about the σ bonds comprising the ring is not possible as long as the ring remains intact. As a result, two geometrical isomers of 1,2-dimethylcyclobutane are possible.

The electrons of the π bond of ethylene are not localized directly between the carbon atoms. Instead, the electrons exist in a diffuse orbital above and below the plane of the molecule. For this reason the π bond is much more accessible to attack by electron-seeking reagents like protons than is a σ bond.

So far we have described two hybridizations for carbon: sp^3 (the combination of one 2s orbital with three 2p orbitals) and sp^2 (the combination of one 2s and two 2p orbitals). Now consider generating an sp hybridized carbon by combining one 2s orbital with one 2p orbital, yielding a carbon surrounded by two sp (one part s and one part p) hybrid orbitals and two unhybridized 2p orbitals. Such a hybridized carbon has two collinear sp orbitals whose axis is perpendicular to both of the unhybridized 2p orbitals (Fig. 1.15). We may place one of the four valence electrons of carbon in each of these orbitals.

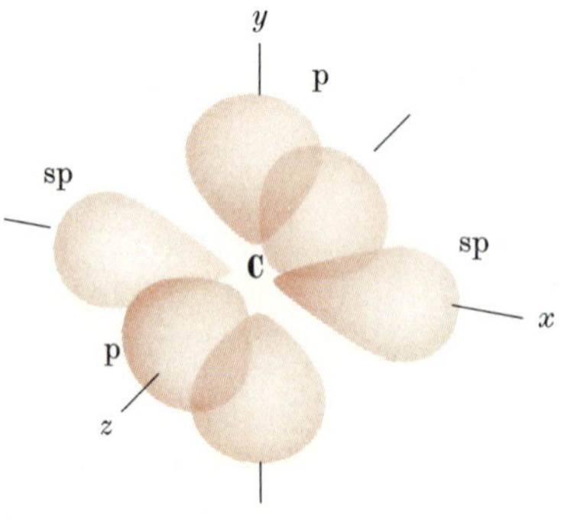

Fig. 1.15 sp hybridized carbon atom

The sp orbitals are similar to sp^3 and sp^2 orbitals, except that they are more compact. Since an sp orbital has 50% s character, it is drawn in closer to the nucleus than sp^2 and sp^3 orbitals (containing $33\frac{1}{3}$% and 25% s character, respectively). To form the molecular orbitals of a compound incorporating an sp hybridized carbon, we allow two such hybridized carbons to approach each other, oriented so the axes of their sp orbitals are collinear. We place hydrogens in the two sp orbitals that are directed away from each other (Fig. 1.16a). As in the case of alkene formation, p orbital overlap is simultaneous with the formation of the σ bond. In this instance, however, there are two pairs of p orbitals that may undergo π bond formation (Fig. 1.16b), so that we are forming a total of three carbon-carbon bonds. The molecule generated by the above process is linear; that is, the two hydrogen nuclei and the two carbon nuclei lie on a straight line. Thus, as we go from sp^3 to sp^2 to sp hybridization, we go from tetrahedral, to planar, to linear stereochemistry, and the bond angle changes from 109.5°, to 120°, to 180°.

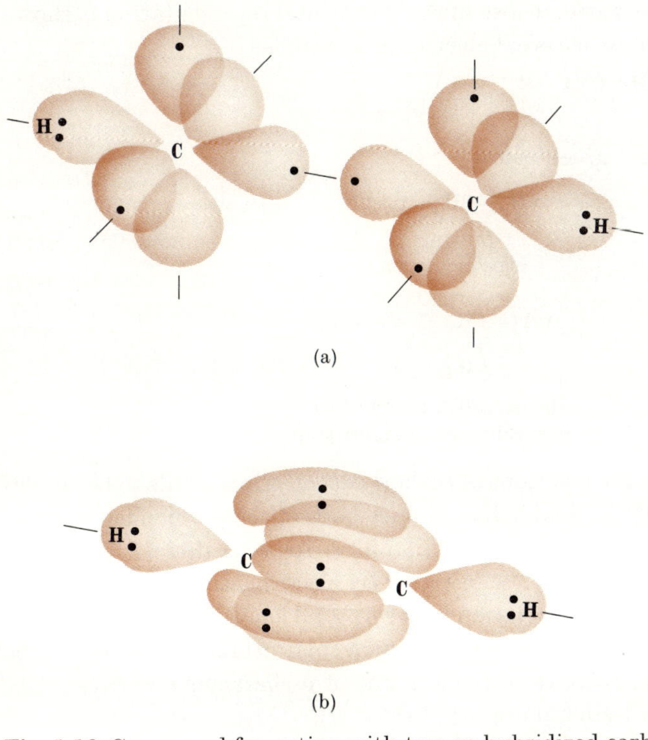

Fig. 1.16 Compound formation with two sp hybridized carbons

If we were to look down the axis of the molecule we have just formed, we might expect the ends of the π bond to appear as in Fig. 1.17(a).

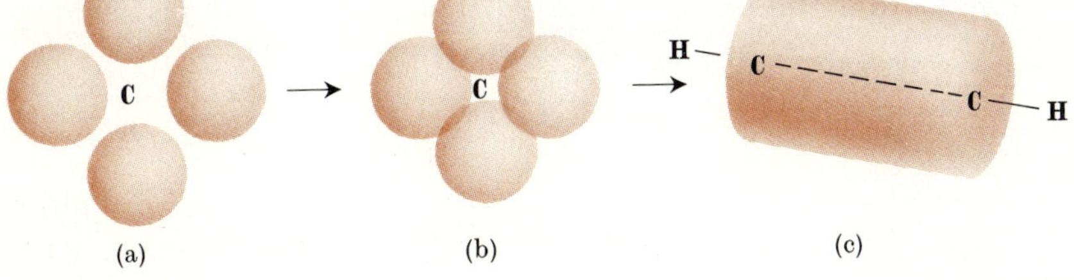

Fig. 1.17 π electron system in acetylene

In fact, the lobes of each of the π bonds overlap one another along the axis of the molecule (Fig. 1.17b) to form a cylinder of electrons surrounding the carbon-carbon σ bond axis. The π bonds thus lose their individuality and merge, giving a π system with cylindrical symmetry.

The sp hybridized system we have just described is called a carbon-carbon triple bond, and compounds containing a triple bond are called *alkynes*. The simplest alkyne is acetylene, whose molecular orbital representation is shown in Fig. 1.17(c) and whose classical electronic picture is

H : C : : : C : H or H—C≡C—H

acetylene

Some other alkynes are given below:

CH_3—C≡C—H

propyne or
methylacetylene

CH_3—CH_2—C≡CH

1-butyne or
ethylacetylene

CH_3—C≡C—CH_3

2-butyne or
dimethylacetylene

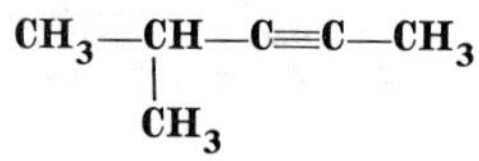

4-methyl-2-pentyne or
methylisopropylacetylene

We would expect the reactions of carbon-carbon triple bonds to be similar to those of carbon-carbon double bonds since both of these π bonded systems constitute centers of high electron availability. As with double bonds, the reactions of triple bonds should be initiated by electron-seeking species.

There is an interesting correlation between bond length and hybridization. As we would expect, the bond length of a molecular orbital decreases, as the amount of s character relative to the amount of p character increases, since an s orbital is smaller than a p orbital (Table 1.2).

Table 1.2 Correlation of bond distances with hybridization

Carbon orbital hybridization	Fractional s character, %	System	C—H bond length, Å	C—C bond length, Å
sp^3	25	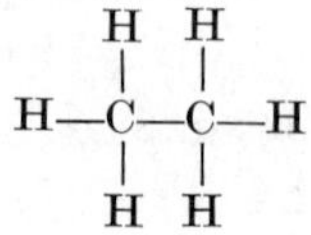	1.093	1.54
sp^2	33.3	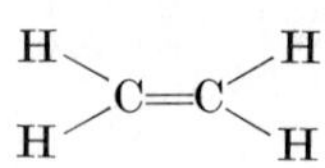	1.087	1.34
sp	50	H—C≡C—H	1.057	1.20

The concept of sp, sp^2, and sp^3 hybridization is convenient for visualizing organic structures, but it must be remembered that there is actually a continuum of orbital hybridizations. An atom involved in a particular bond could be of some intermediate hybridization which, of course, is more difficult to describe precisely than the three cases where orbitals mix in a simple manner. In general, when carbon is bonded directly to two, three, and four atoms, the hybridization resembles the sp, sp^2, and sp^3 situations, respectively.

Cyclopropane. To complete our discussion of the molecular orbital approach to covalent bonds we will return to our discussion of cyclopropane (Fig. 1.10). It is obvious that cyclopropane molecules should have unusual properties, for the arrangement of the three equivalent carbon atoms would require bond angles of 60°, and yet we know that the normal bond angle for sp^3 hybridized carbons is 109.5°. This deviation from the "normal" bond angle has been used as a measure of the "strain" introduced into a molecule by the formation of such a system. It is a destabilizing factor and leads to enhanced reactivity, usually of a type that tends to relieve the strain. This concept has been called the *Baeyer Strain Theory*.

The abnormal reactivity of cyclopropane bonds can also be described in terms of molecular orbitals. Consider two carbon atoms that form a carbon-carbon bond by the overlap of two sp^3 orbitals. It is apparent that at a given distance the overlap of the two sp^3 orbitals will be at a maximum if they are collinear (Fig. 1.18a) rather than at an angle to each other (Fig. 1.18b).

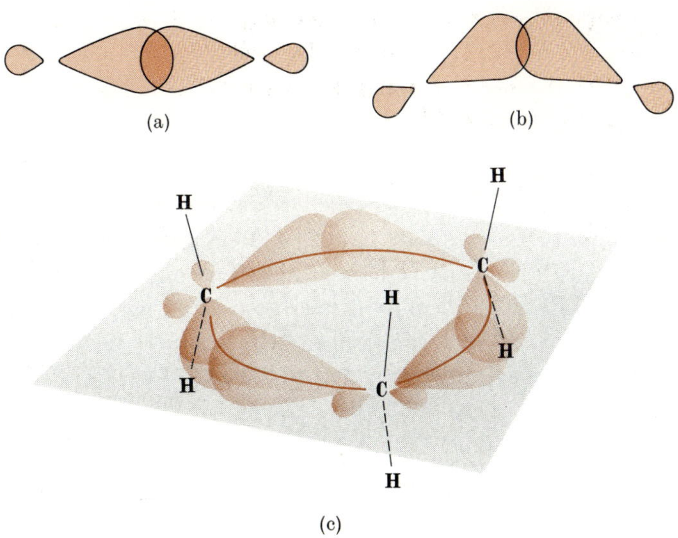

Fig. 1.18 Molecular orbital picture of cyclopropane

These two situations are illustrated by the carbon-carbon bonds of ethane and cyclopropane, respectively. We can construct the molecular orbital picture for cyclopropane by arranging three sp^3 hybridized carbons at normal carbon-carbon bond distances and by placing two of the sp^3 orbitals of each carbon in the plane formed by the three atoms (Fig. 1.18c). Notice that the carbon-carbon bonds so generated are not simple σ bonds since the orbitals do not overlap in a direct head-on manner. These bonds are actually banana shaped and are sometimes referred to as banana bonds. The enhanced reactivity of the cyclopropane bonds may be attributed to the diminished overlap of the orbitals relative to ordinary σ bonds. Moreover, the electrons in the banana bonds are more accessible to electron-seeking reagents since the orbital clouds project away from the carbon-carbon axes. The bonding situation in cyclopropane has also been explained by a change in the hybridization state of the cyclopropane carbons.

1.6 COORDINATE-COVALENT BONDS

So far we have mentioned ionic and covalent bonds. In the former, an electron is transferred completely from one atom to another. In the latter, two atoms each contribute an electron and share the pair of electrons between them. There is a third type of bond, called a *coordinate-covalent bond*, in which both atoms also share a pair of electrons. However, in such a bond *both* electrons are contributed by *one* of the coordinate-covalently bonded atoms. Coordinate-covalent bonds are typified in the compound formed from trimethylboron and trimethylamine.

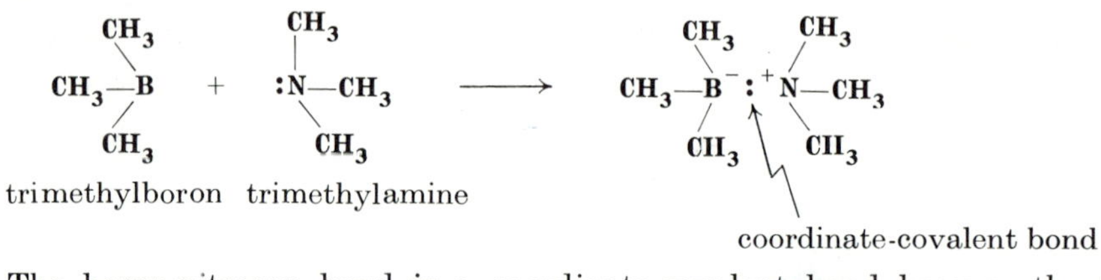

The boron-nitrogen bond is a coordinate-covalent bond because the two atoms share what was once an unshared pair of electrons on the nitrogen. Aside from this, the molecule has all the properties associated with covalent bonding. Note that the boron and nitrogen atoms of the product are charged. In the next chapter we will learn how to calculate atomic charge.

ADDITIONAL READING

E. Griswold, *Chemical Bonding and Structure*, Lexington, Mass.: Raytheon, 1968. Paperback.

J. March, *Advanced Organic Chemistry: Reactions, Mechanisms, and Structure*, New York: McGraw-Hill, 1968, Chapters 1–3.

PROBLEMS

1. Is lithium metal (atomic number = 3) electronegative? Explain.
2. Two hydrogen atoms (H·) prefer to share electrons and form H_2 rather than form H^+ and $H:^-$ by transfering an electron. Explain this behavior.
3. Define covalent bond. How is a covalent bond usually indicated in organic chemistry?
4. What is the maximum number of electrons an orbital can hold?
5. Carbon, nitrogen, and oxygen have atomic numbers of 6, 7, and 8, respectively.
   a) How many electrons are in the outer bonding shell of each atom?
   b) Draw sp^3 hybridized carbon, nitrogen, and oxygen atoms and fill the orbitals with the proper number of electrons.
   c) How many "unshared pairs" of electrons does each atom have?
6. Draw all the isomers of non-cyclic C_6H_{14}. Consult the Appendix for nomenclature, and name the isomers.
7. If cyclohexane were planar, what would be the carbon angles at each of the six vertices? How many degrees deviation would this be from the normal tetrahedral angle?
8. Draw a three dimensional structure for $CH_3C{\equiv}CH$.
9. Only one isomer of dichloromethane, CH_2Cl_2, is known to exist. Explain why this fact is consistent with a tetrahedral configuration of the carbon atom, and inconsistent with a planar configuration in which the chlorines and hydrogens occupy corners of a square about the central carbon.
10. Explain why cyclopentyne, drawn below, would be difficult to synthesize.

Note: In the above polygon notation, each vertex represents a carbon atom; hydrogens are omitted.

11. Show by means of a three-dimensional stick drawing (as in Fig. 1.8) that the structures drawn below are planar representations of the same compound.

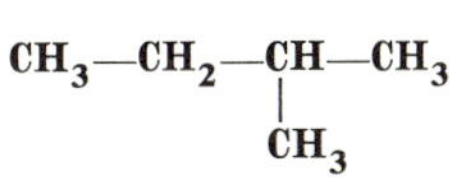

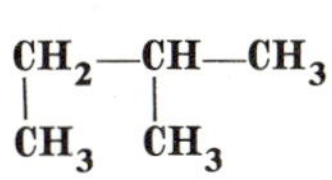

12. How many σ bonds and how many π bonds are in each of the following compounds?

a) $CH_3CH_2CH_3$ b) $HC{\equiv}C{-}CH{=}CH_2$ c) 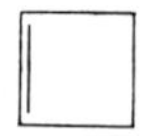

13. Explain why the carbon-chlorine bond distance decreases in the series ethyl chloride, vinyl chloride, chloroacetylene.

$CH_3CH_2Cl > H_2C{=}CHCl > HC{\equiv}CCl$

14. Are the following pairs of structures isomers or the same compounds?

a) $CH_3C{\equiv}CH$ $HC{\equiv}CCH_3$

b) 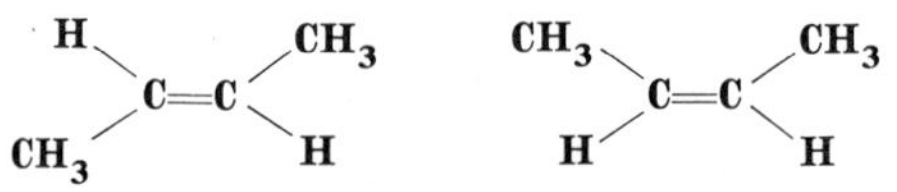

c) 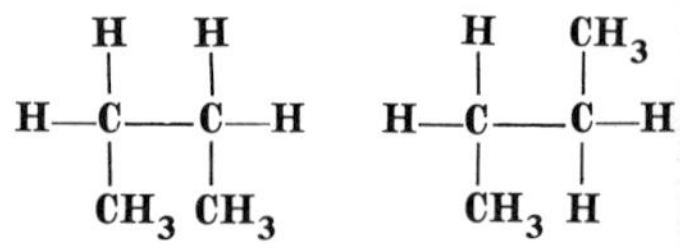

d) 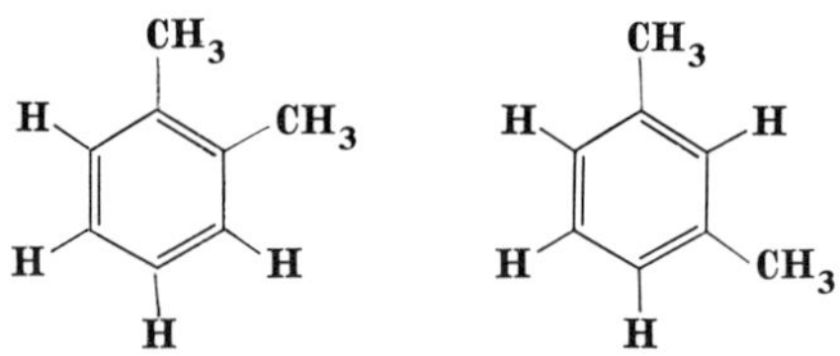

15. Write structures for all isomeric noncyclic pentenes (five carbons and one double bond).

16. Devise an algebraic relationship which will give the number of hydrogens in a noncyclic alkane when the number of carbons in the hydrocarbon is specified. Repeat for alkenes and alkynes.

17. Alkenes are susceptible to reactions with electron-seeking species such as HBr. The reaction of HBr with ethylene is

$$CH_2{=}CH_2 + HBr \longrightarrow CH_3{-}CH_2Br.$$

How many isomeric bromoalkanes would be expected from the addition of HBr to

a) 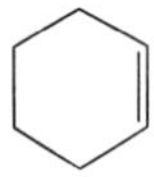 b) $CH_3CH{=}CHCH_3$ c) $CH_3CH{=}CHCH_2CH_3$

18. Rationalize the fact that it is impossible to remove a proton (H^+) from ethane, while a very strong base will remove a proton from acetylene according to the equation

$$HC{\equiv}CH + B^- \longrightarrow HC{\equiv}C^- + BH$$

19. Assume that carbon (with four electrons in the 2-shell) engaged in compound formation without hybridizing. How many covalent bonds would the carbon atom be able to form?

20. Draw three-dimensional pictures of the following known compounds. Include all the hydrogens.

a) 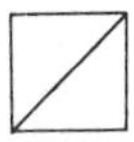

b) 

21. Draw an orbital picture of allene, showing the hybridization of all the carbon atoms and the relative arrangements of the hydrogens.

$CH_2{=}C{=}CH_2$

22. Carbene (CH_2:) is a divalent carbon with two unshared electrons. It is a very unstable species, formed only transiently during the course of certain reactions. Draw molecular orbital pictures of carbene assuming sp, sp^2, and sp^3 hybridization of the carbon.

23. Draw an orbital picture of ammonia (NH_3), assuming that the nitrogen is sp^3 hybridized. Why might the nitrogen-hydrogen bond formed in the following reaction be considered a coordinate-covalent bond?

$$NH_3 + H^+ \longrightarrow NH_4^+$$

24. Draw a picture of two sp^2 hybridized carbon atoms in which the two p orbitals are engaged in σ-type overlap.

25. When *trans*-2-butene is irradiated with ultraviolet light *cis*-2-butene is formed. Propose a mechanism for this isomerization; that is, describe what events might take place as the *trans* material, which absorbs the light energy, is converted into a product.

2

INDUCTIVE AND RESONANCE EFFECTS

2.1 NUCLEAR CHARGE

Since organic species, like inorganic ions, can bear charge, it is important to be able to determine whether the atoms in an organic compound are positive, negative, or neutral. In methane, for example, there are eight electrons in the outer shell of the carbon, but the carbon *shares* these electrons with four hydrogen atoms. The carbon therefore has a net number of four electrons, namely one per bond. Since the methane carbon has the same net number of electrons as a neutral atomic carbon, it bears no charge. However, when carbon has an average of three electrons, it bears a positive charge, and when it has a net number of five electrons, it bears a negative charge. Consider the case of the methyl cation (Fig. 2.1a). This cation, called methyl carbonium ion, is a highly unstable and transient species, not an isolable compound. The point here is that there must be a single positive charge on the carbon atom because there is a net number of three electrons around the carbon (one per bond). The methyl carbanion (Fig. 2.1b), also an unstable species, has a negative charge because there are five electrons around the carbon (two from the unshared pair of electrons, plus a net of three from the three bonds).

(a) (b)

Fig. 2.1 a) Methyl carbonium ion and b) methyl carbanion

Let us now calculate the charge on the three species in the equation below, keeping in mind that a neutral oxygen atom has six electrons in its outer 2-shell.

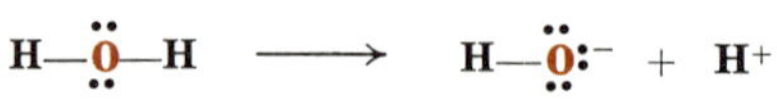

The oxygen of the water molecule is neutral because it has a net of six electrons (four from the two unshared pairs and one each from the two bonds). The

hydroxide ion is negatively charged because it has a net of seven electrons (two each from three unshared pairs and one from the bond). The proton bears a positive charge because it has one electron less than a hydrogen atom (a proton has no electrons in its shells). Formal charge may be calculated from the following equation:

$$\text{Charge} = A - (B/2 + C)$$

where A = the number of electrons in the outer shell of the neutral atom; B = the number of shared electrons; and C = the number of unshared electrons. For example, the hydroxide ion bears a negative charge because $A = 6$, $B = 2$, and $C = 6$. Further examples are shown below:

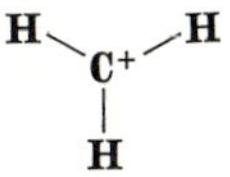
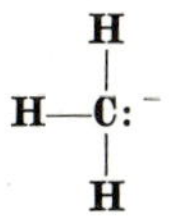
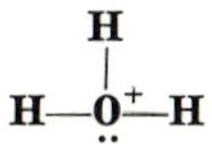

$A = 4$	$A = 4$	$A = 6$
$B = 6$	$B = 6$	$B = 6$
$C = 0$	$C = 2$	$C = 2$
Charge = +1	Charge = −1	Charge = +1

2.2 INDUCTIVE EFFECTS

In Chapter 1 we defined electronegativity as the ability of an atom to attract electrons. When two atoms of different electronegativity are covalently bonded to each other, the electrons will, on the average, be closer to the atom of greater electronegativity. In such cases, the bond is said to be polarized. The bond in hydrogen chloride is polarized because chlorine is more electronegative than hydrogen ($F > O > N \approx Cl > Br > C \approx H$). Displacement of electrons in the hydrogen chloride bond, as a result of the electronegativity difference between the atoms, produces a partial negative charge on the chlorine and a partial positive charge on the hydrogen. Such charges are usually indicated by a delta (δ), where delta is some fractional number between zero and one.

$$\overset{+\delta}{\text{H}}\text{—}\overset{-\delta}{\text{Cl}}$$

The hydrogen chloride molecule, therefore, is called a dipole. Associated with this dipole is a *dipole moment* μ, defined as the magnitude of the charge e multiplied by the distance of charge separation d:

$$\underset{\text{in Debye units (D)}}{\mu} = \underset{\text{in electrostatic units (esu)}}{e} \times \underset{\text{in centimeters}}{d} \times 10^{18}$$

The amount of ionic character displayed by the hydrogen chloride covalent bond may be estimated by means of dipole moment measurements. Assume that hydrogen chloride is completely ionic (that is, $\delta = 1$). Since the hydrogen-chlorine bond distance is 1.28×10^{-8} cm and the charge on an electron is 4.8×10^{-10} esu, the dipole moment calculated from the above equation would be 6.15 Debye units (D). The experimental value is only 1.03 D because hydrogen chloride is, in fact, not completely ionic. The ionic character of the covalent bond is only $(1.03/6.15) \times 100$, or 17%.

The preceding calculation applies to hydrogen chloride in the gaseous state. The hydrogen-chlorine bond consists of an electron pair which is shared by the two atoms, albeit unequally. When hydrogen chloride is dissolved in water, the molecule dissociates to form a proton (or, more properly, hydronium ion) and chloride ion. The bond breaks, with the electronegative chlorine atom taking both electrons. Hydrogen chloride dissociates in water because the water dipole solvates and stabilizes the resulting ionic species.

If a molecule has several polarized bonds, the *net* dipole moment of the molecule is the vector sum of the individual moments. Thus, carbon tetrachloride (CCl_4) has a zero dipole moment because the four chlorines are symmetrically disposed about the carbon in a tetrahedral arrangement, and the individual moments cancel. Each of the carbon-chlorine bonds is, of course, polarized in the direction of the chlorines (chlorine is more electronegative than carbon). Methyl chloride (CH_3Cl) has a dipole moment of 1.86 D since the molecule is unsymmetrical, and the polarity of the carbon-chlorine bond is not canceled.

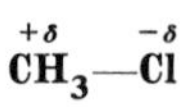

$\mu = 1.86$ D

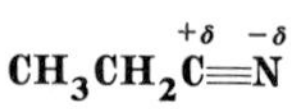

$\mu = 4.0$ D

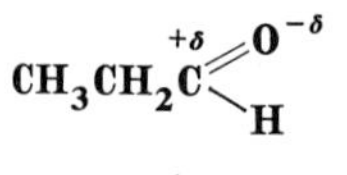

$\mu = 2.7$ D

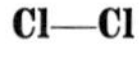

$\mu = 0$

Polarized electrons in a covalent bond give rise to an *inductive effect*. For example, the dipole moment of methyl chloride (CH_3Cl) is the result of an electron-withdrawing inductive effect exerted by the chlorine. Inductive effects, together with resonance and steric effects, are the three most important factors used to rationalize relative reactivities of organic compounds. The reactivities of the two olefins in Fig. 2.2 are another good example of an inductive effect.

Figure 2.2

The double bond of the fluorinated compound is much less susceptible to attack by an electron-seeking reagent (such as a proton) than is the non-fluorinated olefin. The three fluorines inductively withdraw electrons, thereby reducing the electron density of the π system and the reactivity toward electron-seeking reagents.

One important property of inductive effects arising from electronegativity differences is that these effects diminish rapidly with distance. Thus, in *n*-propyl chloride the carbon bonded to the chlorine bears a considerable positive charge; the charge on the next carbon is very small; and the charge on the methyl group is virtually non-existent.

$$CH_3—CH_2—\overset{+\delta}{CH_2}—\overset{-\delta}{Cl}$$

2.3 RESONANCE

Resonance is an extremely important and useful concept in organic chemistry. To introduce the concept, we will use the classic example of benzene (C_6H_6).

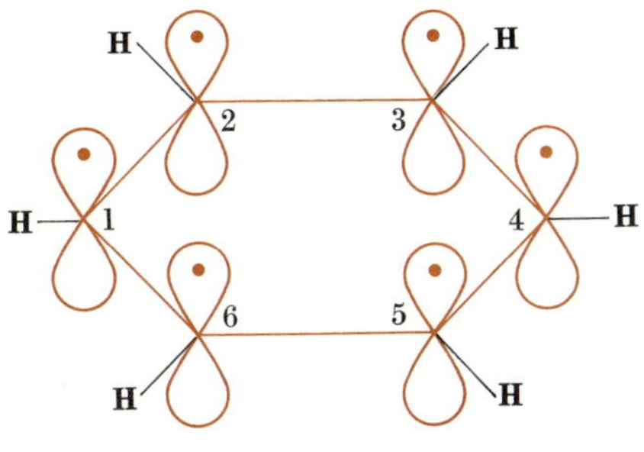

Fig. 2.3 Molecular orbital picture of benzene without p orbital overlap

Benzene is a cyclic compound consisting of six coplanar sp^2 hybridized carbons (Fig. 2.3) each of which has a p orbital holding one electron. Three double bonds can be formed by overlapping the p orbitals of C-1 and C-2, C-3 and C-4, and C-5 and C-6, or by overlapping the p orbitals of C-2 and C-3, C-4 and C-5, and C-6 and C-1 (Figs. 2.4a and 2.4b). For simplicity, hydrogens have been omitted in these diagrams.

(a) (b)

Figure 2.4

Which of these two structures is the correct formulation for benzene? In fact, *neither* structure adequately describes the properties of benzene. All carbon-carbon bond distances in benzene are known to be *equal*. Both structure 2.4(a) and structure 2.4(b) predict that the three single carbon-carbon bonds should be 1.54 Å, while the three double bonds should be 1.34 Å. X-ray and electron diffraction data affirm, to the contrary, that all six bond distances are 1.40 Å.

According to the resonance concept, benzene is a *resonance* hybrid of the two structures 2.4(a) and 2.4(b). This means that the true structure of benzene, although difficult to depict precisely, is intermediate between structures 2.4(a) and 2.4(b). Resonance theory simply says that benzene resembles structures 2.4(a) and 2.4(b), neither of which, in itself, has any physical reality. Since 2.4(a) and 2.4(b) are equivalent, they "contribute" equally to the structure of benzene. This explains why the carbon-carbon bonds of benzene are identical and intermediate in length between a normal single and double bond.

Benzene is *not* a mixture of resonance contributors 2.4(a) and 2.4(b). We emphasize again that these two structures do not exist; they are fictitious (but useful) representations of the actual benzene molecule. Resonance contributors are always indicated by a double arrow (Fig. 2.5).

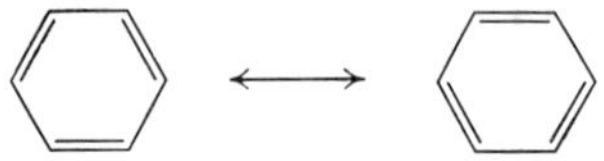

Figure 2.5

This symbol is not to be confused with the two arrows used for equilibria ($\rightleftarrows$). Resonance and equilibrium bear no similarity. In an equilibrium the species involved have physical existence and are interconvertible.

The resonance concept can also be illustrated by means of 1,2-dimethylbenzene (called *o*-xylene). This compound is a resonance hybrid of two nonsuperimposable contributors. (One has a single bond between the methyl groups; the other a double bond.)

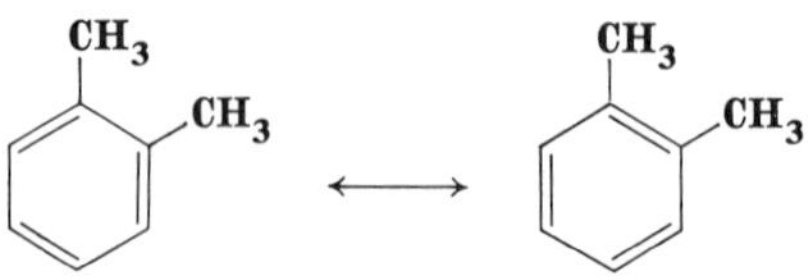

If *o*-xylene were not a hybrid of the two structures, then chemists would be able to find two isomers of the compound, each having different chemical and physical properties. However, only one *o*-xylene is known.

Calculations reveal that benzene, like all resonance hybrids, is more stable (of less energy) than the contributors would be if they existed. The stability associated with delocalization of the electrons in a π system is called the *resonance energy*. Benzene has a particularly great resonance stability. For example, benzene is inert to certain electron-seeking reagents that readily attack isolated double bonds. Benzene and related compounds display an inertness which has come to be known as *aromaticity*, and compounds which contain the benzene ring are said to be *aromatic*. (Aromaticity is discussed in detail in Section 10.2.)

Benzene can also be viewed in terms of a molecular orbital picture. The six unhybridized p orbitals (Fig. 2.2) combine to form six molecular orbitals, three of which are filled with two electrons each. The nature of these molecular orbitals is complicated; suffice it to say that benzene has a doughnut-shaped cloud of electrons above and below the plane of the molecule (Fig. 2.6).

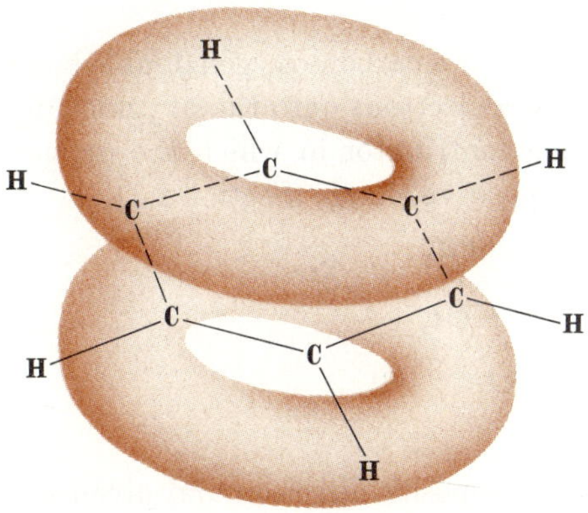

Fig. 2.6 Molecular orbital picture of benzene with p orbital overlap

Resonance "extremes" need not always contribute equally as they do in benzene. A resonance structure may contribute very little to the overall structure of the hybrid. Listed below are some rules which should be helpful in assessing the importance of resonance contributors.

1. *All resonance contributors must have precisely the same molecular structure and geometry.*

Benzene's two resonance contributors (Fig. 2.7) have the same bond angles and distances.

Figure 2.7

Only the electron density changes. Note that one contributor may be obtained from the other by "pushing electrons" (curved arrows). Each curved arrow indicates a two-electron movement. Pushing electrons is of great assistance in finding reasonable contributors to a molecule. Since all contributors must have the same spatial arrangement of atoms, it is obvious that *cis*- and *trans*-2-butene (Fig. 2.8) are different compounds, not structures contributing to some hybrid.

Fig. 2.8 *cis* and *trans* 2-butene

2. *Reasonable contributors should reflect the electronegativities of the atoms involved in the π system.*

The compound acetone (Fig. 2.9a) possesses a carbon-oxygen double bond (called a carbonyl group). By pushing the π electrons onto the oxygen (Fig. 2.9b) one obtains a reasonable resonance contributor in which the oxygen bears a negative charge and the carbon a positive charge (Section 2.1).

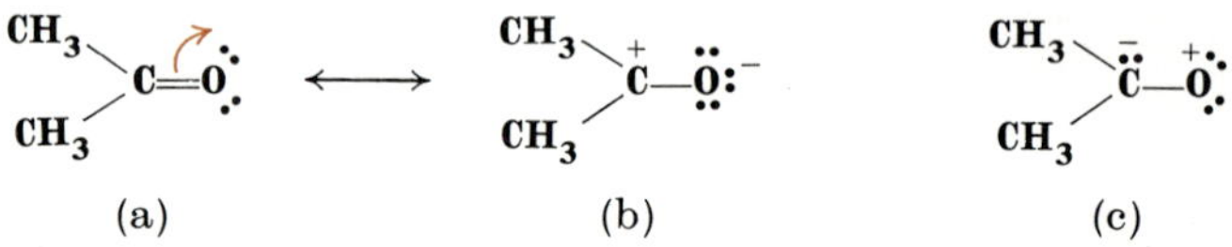

Figure 2.9

If we had pushed the electron pair in the opposite direction and given the electrons to carbon, then carbon would bear the negative charge and oxygen the positive charge (Fig. 2.9c). Since oxygen is more electronegative than carbon, structure 2.9(c) contributes little, if at all, to the hybrid structure of acetone. In contrast to benzene, the two reasonable contributors of acetone (Figs. 2.9a and 2.9b) are not identical and therefore would not be expected to contribute equally. This leads us to the next rule.

3. *The creation of atoms with charge or with less than a full octet of outer electrons detracts from the importance of a contributor.*

In principle, we could draw a resonance contributor of benzene in which one carbon bears a negative charge while a neighboring carbon bears a positive charge (Fig. 2.10b). However, since the two carbons are of equal electronegativity, there is no good reason why the electrons should be polarized in this manner. Moreover, one of the carbons (the cationic carbon) has only six outer electrons. Both these factors lead us to regard this contributor as unimportant relative to the two major contributors of benzene. In the case of acetone, charge is indeed created in contributor 2.9(b), but this is not quite so undesirable since oxygen is an electronegative atom. Nevertheless, structure 2.9(a), with all its atoms having complete outer shells, is undoubtedly more important to the overall structure of acetone than is contributor 2.9(b).

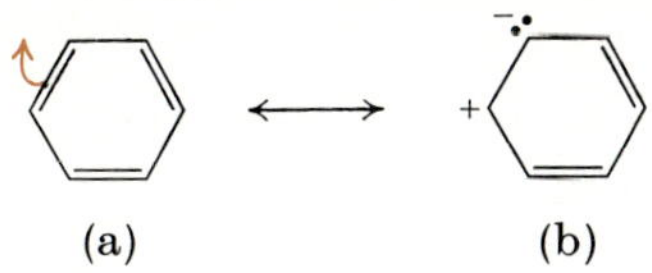

Figure 2.10

In Fig. 2.10 we have, for simplicity, omitted the hydrogens. This is a common procedure; however, if this omission gives rise to any confusion, the hydrogens should be inserted.

4. *No carbon, oxygen, nitrogen, or other first row atom may contain more than eight electrons in its outer shell.*

This rule will be illustrated using a substituted benzene derivative, aniline (Fig. 2.11).

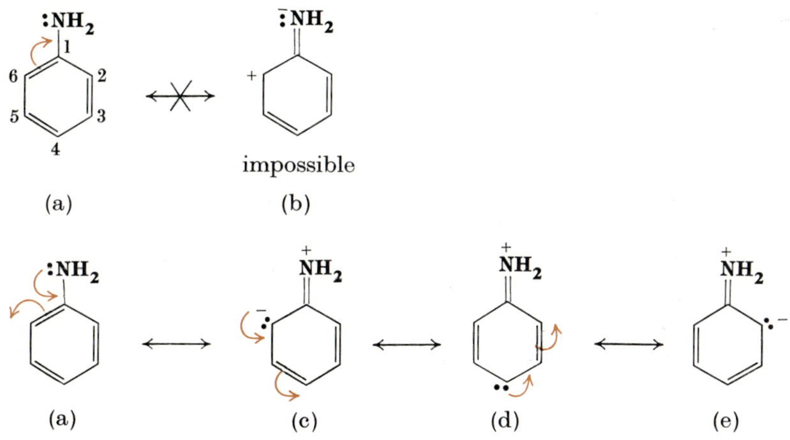

Fig. 2.11 Resonance in aniline

The aniline nitrogen is bonded to three groups, and it possesses an unshared pair of electrons (like the nitrogen of ammonia). Pushing electrons in the direction of the nitrogen (Fig. 2.11b) gives the nitrogen ten electrons in its outer bonding shell, two more than the maximum limit of eight. Therefore, Fig. 2.11(b) is an impossible structure. Reasonable resonance contributors are shown in Fig. 2.11(c), 2.11(d), and 2.11(e). These contributors involve charge separation, and for this reason they are not nearly as important as 2.11(a). Nevertheless, they contribute sufficiently to render the 2, 4, and 6 ring carbons of aniline more electron rich than the other three carbons. Note that the nitrogen of the contributors is positively charged; this is an acceptable state of nitrogen, as the stability of ammonium ion indicates.

5. *Maximum resonance stabilization is achieved when there are two or more equivalent low-energy contributors.*

This rule has already been illustrated by the benzene system. Another example is the resonance-stabilized acetate anion (Fig. 2.12).

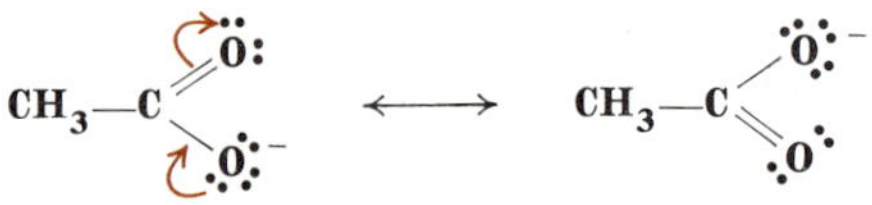

Fig. 2.12 Resonance in acetate anion

The two resonance structures contribute equally. Consequently, the most satisfactory representation of the acetate anion hybrid is one in which each oxygen bears exactly one-half of a negative charge. Frequently, for the sake of convenience, unshared pairs of electrons in organic structures are omitted (Fig. 2.13).

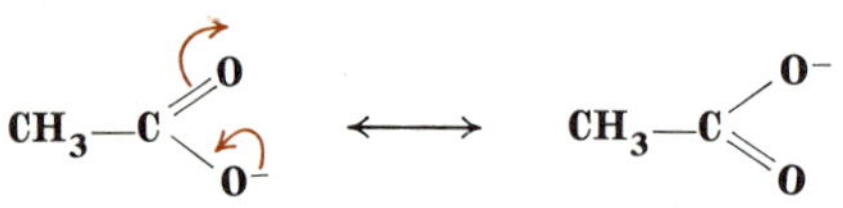

Fig. 2.13 Resonance in acetate anion

This is especially true for unshared pairs of electrons on carbon (Fig. 2.14).

Figure 2.14

PROBLEMS

1. Draw the hydroxide ion (HO^-) showing all the unshared pairs of electrons on the oxygen. Then calculate the charge by supplying the correct values of B and C (see Section 2.1).

   $$\text{Charge} = -1 = 6 - (B/2 + C)$$

2. Repeat the above for ammonia, NH_3.

   $$\text{Charge} = 0 = 5 - (B/2 + C)$$

3. Place a $+\delta$ or a $-\delta$ over the underlined atoms.

   $\underline{H}$—$\underline{F}$ $\qquad$ $H_3\underline{C}$—$\underline{Cl}$ $\qquad$ $H\underline{O}$—$\underline{Cl}$

4. Assess the importance of the charged resonance contributors.

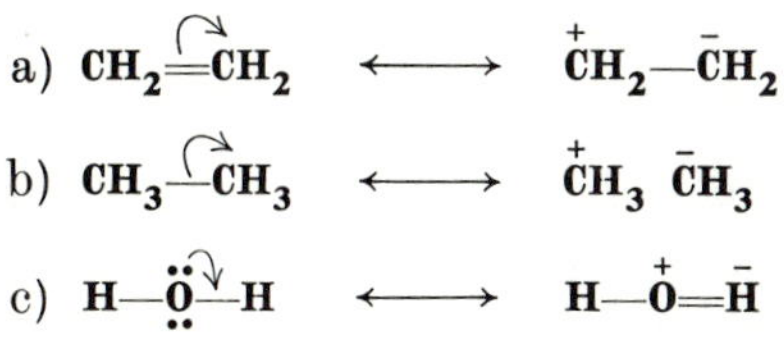

5. The central carbon of acetone is attacked by electron-rich species such as hydroxide ion. Suggest two explanations.

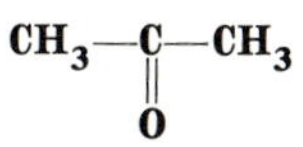

6. The following are reasonable resonance contributors. Supply arrows showing how one may be converted into the other by "electron pushing."

a) 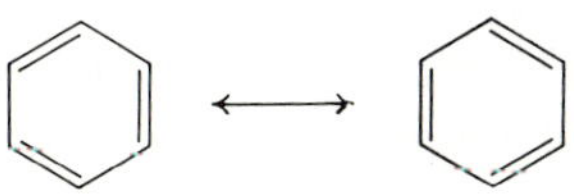

b) 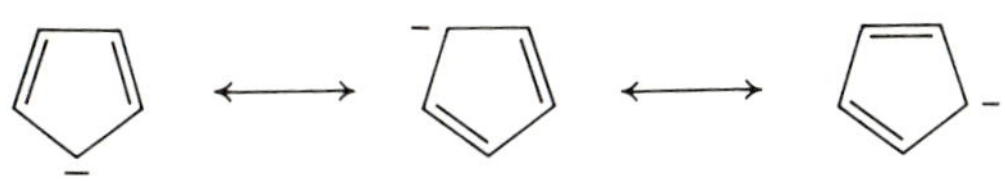

c) 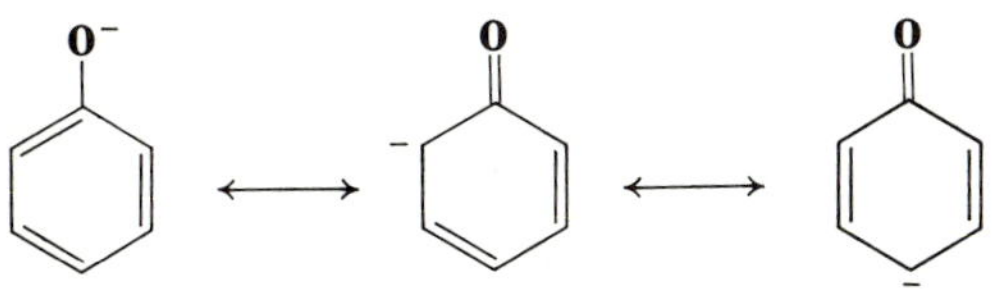

7. Are the indicated atoms of the following species positive, negative, or neutral?

a) nitrogen

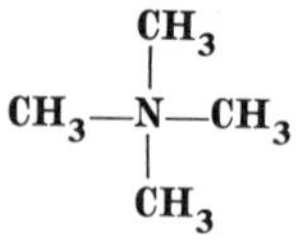

b) oxygen

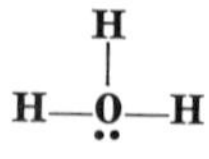

c) carbon

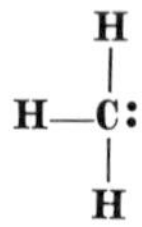

d) carbon

e) oxygen

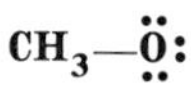

f) nitrogen

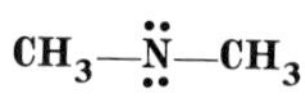

g) oxygen

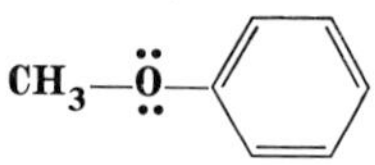

8. Write a reasonable charged resonance contributor for each of the following organic compounds. Unshared pairs of electrons are not included in the structures of the compounds.

a) 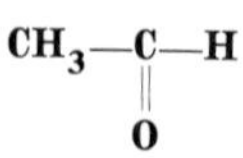

b) 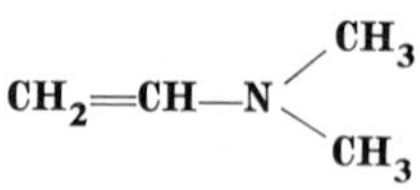

c) 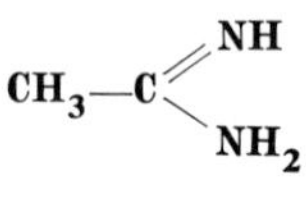

d) 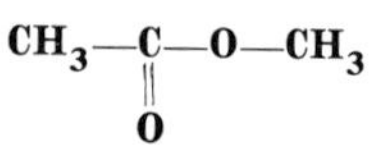

e) CH_3—NH—C≡N

9. The dipole moment of ethyl chloride is 2.05 D, whereas the dipole moment of vinyl chloride is only 1.44 D. Explain the diminished dipole moment of vinyl chloride in terms of a resonance effect.

CH_3CH_2Cl	CH_2=CHCl
2.05 D	1.44 D

10. Sulfur, with a total of 16 electrons, has its bonding electrons in the 3-shell. Determine whether the sulfur in dimethylsulfide is charged or neutral. How many unshared pairs of electrons does the sulfur have?

CH_3—S—CH_3

11. How many unshared pairs of electrons do the following atoms possess? The atoms are shown with their correct nuclear charge.

a) oxygen

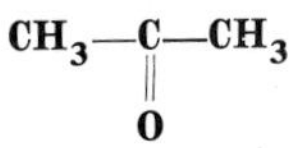

b) oxygen

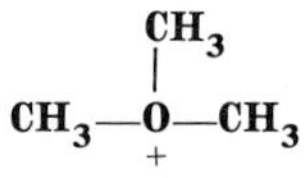

c) nitrogen

$CH_3—C≡N$

d) nitrogen

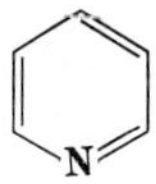

e) carbon

H—C—H

f) carbon

$H—C≡C^-$

g) oxygen

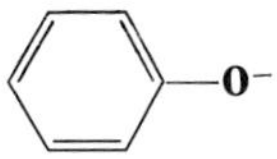

h) nitrogen

$CH_3—N{=}O$

12. There are two isomeric compounds of molecular formula $CH_3(NCO)$. Their structures may be represented as $CH_3—O—C≡N$ and $CH_3—N{=}C{=}O$. In contrast, there is only one ionic Na(NCO). Explain.

13. What charge does each carbon bear in the resonance hybrid of cyclopentadienyl anion?

14. Are the following pairs of structures resonance contributors or different species?

a) $CH_2{=}CH—CH{=}CH_2$ and $CH_3—CH{=}CH—CH_3$

b) 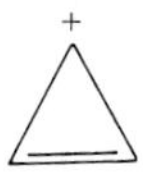 and 

c) CH_3OH and CH_3O^-

d) 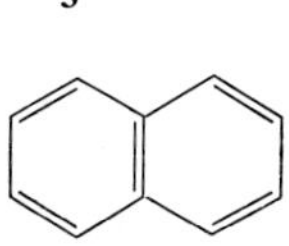 and 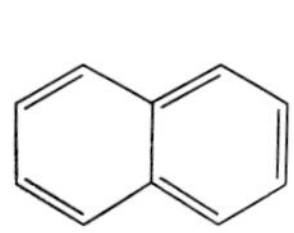

e) 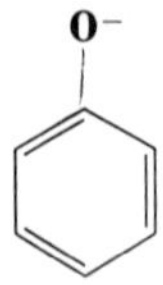 and 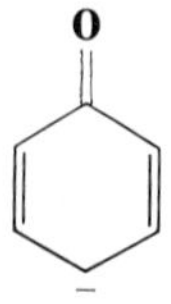

f) 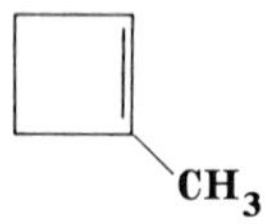 and 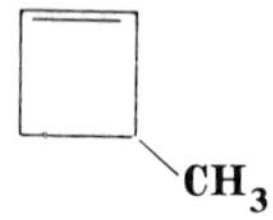

15. A carbonium ion is an unstable, positively charged carbon. Explain why it is easier to generate a carbonium ion bonded to a benzene ring than it is to make methyl carbonium ion.

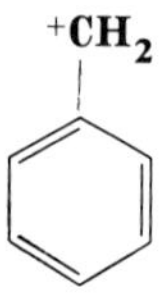 versus $^{+}CH_3$

16. A delta (δ) is used to indicate partial charges on atoms. In the following compounds place a +δ or −δ above the appropriate atoms.

a) $CH_2{=}CH{-}CH{=}O$

b) 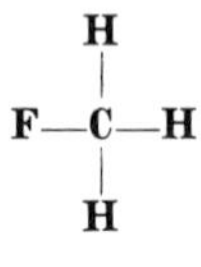

c) 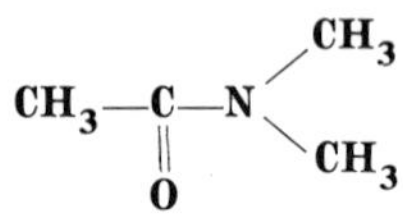

d) 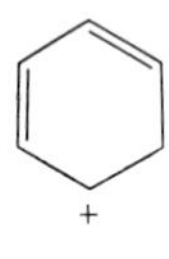

e) 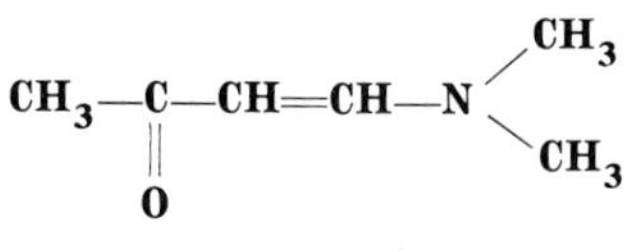

17. In the compound drawn below, called dimethylformamide, we do not find the free rotation about the central nitrogen-carbon bond that we would expect from a single bond. Explain.

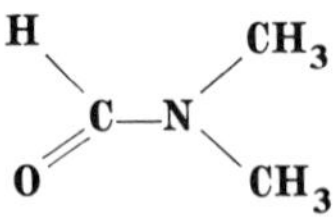

18. Account for the fact that pyrrole is much less basic than pyrrolidine. (Basicity is a measure of how well an atom with a free pair of electrons bonds in a coordinate-covalent manner to a proton.)

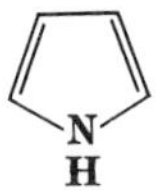

pyrrole

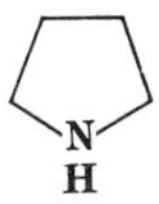

pyrrolidine

19. Explain why the following resonance contributor of cyclobutene is energetically unreasonable.

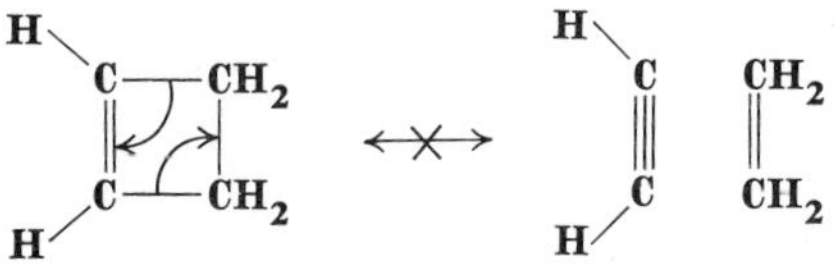

20. The single carbon-carbon bond of propylene is polarized as shown, giving rise to a small dipole moment. Explain the cause of this bond polarity.

$$CH_2{=}\overset{-\delta}{C}H{-}\overset{+\delta}{C}H_3$$

21. Draw the major resonance contributors of the following species.

a) 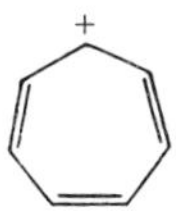

b) 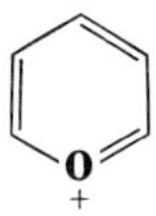

c) 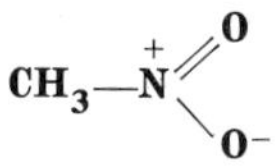

d) 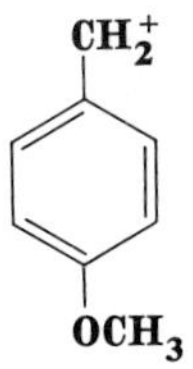

e) 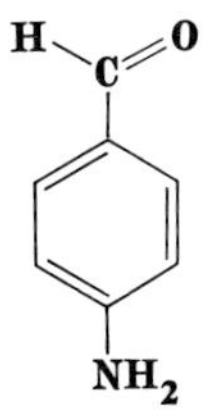

3

FUNCTIONALITY

3.1 CLASSES OF ORGANIC COMPOUNDS AND FUNCTIONALITY

The chemistry of organic materials is conveniently described by categorizing compounds according to their *functional groups.* A functional group is a particular combination of atoms in a molecule (such as the atoms comprising a carbon-carbon triple bond) which undergoes a characteristic set of reactions no matter what other atoms may be near the reactive center. The surroundings may affect the rates of reaction, but the essentials of the reaction remain the same. For example, the hydroxyl or alcohol group (—OH), exhibits similar reactions whether it is attached to a methyl (CH_3—), ethyl (CH_3CH_2—), or a hydrogen. This is illustrated by comparing the reactions of water and ethanol with sodium:

$$2\ \mathrm{H{-}OH} + 2\ \mathrm{Na} \longrightarrow 2\ \mathrm{HO^-Na^+} + \mathrm{H_2}$$

$$2\ \mathrm{CH_3CH_2{-}OH} + 2\ \mathrm{Na} \longrightarrow 2\ \mathrm{CH_3CH_2O^-\ Na^+} + \mathrm{H_2}$$

Although sodium reacts faster with water than with ethanol, the actual reaction processes are analogous. The purpose of this chapter is to expose the student to the more important functional groups and to one or two representative reactions of each. The student should not become alarmed that in this chapter many of the details of the reactions (*e.g. how* sodium reacts with water to form sodium hydroxide) are not explained. This understanding will be developed in later chapters. For the moment, however, it is only necessary to learn a few organic reactions and to appreciate how they may be used to construct a large variety of organic compounds.

3.2 ALKANES

Alkanes were introduced in our discussion of covalent bonds (Section 1.3). They are also known as *saturated* hydrocarbons, indicating the absence of multiple bonding. The open-chain compounds are called *aliphatic* hydrocarbons, and the cyclic saturated materials are said to be *alicyclic.*

Alkanes are comprised of σ bonds which are unpolarized and therefore relatively unreactive. In fact, so much energy is required to achieve bond

Table 3.1 Some saturated hydrocarbons

Name	Formula, C_nH_{2n+2}	Melting point, °C	Boiling point, °C	Specific gravity (as liquids)
Liquids				
n-Pentane*	C_5H_{12}	−129.7	36.1	0.6264
n-Hexane	C_6H_{14}	− 94.0	68.7	0.6594
n-Heptane	C_7H_{16}	− 90.5	98.4	0.6837
n-Octane	C_8H_{18}	− 56.8	125.6	0.7028
n-Nonane	C_9H_{20}	− 53.7	150.7	0.7179
n-Decane	$C_{10}H_{22}$	− 29.7	174.0	0.7298
n-Undecane	$C_{11}H_{24}$	− 25.6	195.8	0.7404
Solids				
n-Octadecane	$C_{18}H_{38}$	28.0	308.0	0.7767
n-Nonadecane	$C_{19}H_{40}$	32.0	330.0	0.7776
n-Eicosane	$C_{20}H_{42}$	36.4		0.7777
n-Heneicosane	$C_{21}H_{44}$	40.4		0.7782

*The symbol *n* (normal) means that the hydrocarbon is a straight chain (no branching). See section on nomenclature at the end of this text.

Table 3.2 Cycloparaffins

Name	Formula, C_nH_{2n}	Structure	Melting point, °C	Boiling point, °C	Specific gravity (as liquids)
Cyclopropane	C_3H_6	CH_2—CH_2 \ / CH_2	−127.0	−32.9	0.688
Cyclobutane	C_4H_8	CH_2—CH_2 | | CH_2—CH_2	−80.0	11.0	0.7038
Cyclopentane	C_5H_{10}	CH_2—CH_2 | | CH_2 CH_2 \ / CH_2	−94.0	49.5	0.7460

cleavage in alkanes that it is difficult to control the reactions, and mixtures of products result. The reaction of methane with chlorine can be initiated if energy (in the form of light) is provided:

$$CH_4 + Cl_2 \xrightarrow{\text{light}} CH_3Cl + CH_2Cl_2 + CHCl_3 + CCl_4 + HCl$$

The initial products are methyl chloride (CH_3Cl) and hydrogen chloride. But the reaction conditions are such that further substitution of the hydrogens of methyl chloride occurs until all the hydrogens are replaced and a complex mixture of products is obtained. A reaction of this type would be of little synthetic value in the laboratory where it is impossible to control what is produced. In industry, however, where efficient large-scale separation systems are available, these processes can be of great importance.

Many reactions can be utilized for the synthesis of alkanes. We shall illustrate their preparation with just two of these methods.

Synthesis of alkanes

$$\underset{\text{ethyl bromide}}{CH_3{-}CH_2{-}Br} + \underset{\text{lithium aluminum hydride}}{LiAlH_4} \longrightarrow \underset{\text{ethane}}{CH_3{-}CH_2{-}H}$$

or, in general,

$$R{-}X + LiAlH_4 \longrightarrow R{-}H$$

In the first procedure an alkyl halide (R—X, where R is an unreactive organic fragment and X is a halogen) is treated with lithium aluminum hydride. The reaction leads to replacement of the halogen by a hydrogen. Lithium aluminum hydride may therefore be regarded as a source of hydride ion (a hydrogen atom with the 1s-shell filled with two electrons):

$$\underset{\text{hydride ion}}{H{:}^-}$$

The hydride ion displaces the bromine of ethyl bromide:

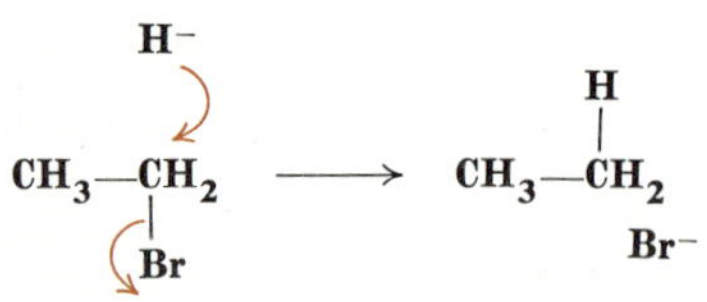

The above scheme is an example of a *reaction mechanism*. A mechanism depicts bond breakage and bond formation as reactants are converted into products. Of course, the mechanism is a gross over-simplification in that it ignores the roles of the lithium and aluminum portions of $LiAlH_4$. Nevertheless, the mechanism helps visualize how the alkane is formed.

A second way to form alkanes from alkyl halides is to treat the alkyl halide with magnesium metal and then to expose the product to water:

$$CH_3—CH_2Br + Mg \longrightarrow \underset{\text{Grignard reagent}}{CH_3—CH_2—MgBr} \xrightarrow{H_2O} CH_3—CH_2—H + MgBr(OH)$$

The product formed from an alkyl halide and magnesium is the very useful Grignard reagent, named after the chemist who won the Nobel Prize for developing it. The reactions of Grignard reagents can be rationalized by considering them to be polarized in the following sense:

$$[CH_3CH_2MgX \longleftrightarrow CH_3CH_2^- \;\; {}^+MgX]$$

The carbon bears a large negative charge. Since carbon is not electronegative, the presence of the high electron density creates an unstable situation, and the Grignard reagent reacts quickly with positive centers to remove the charge. This tendency is manifested in the reaction of a Grignard reagent with water to yield an alkane:

$$XMg^+ \;\; R^- \;\; H—OH \longrightarrow R—H + {}^-OH + {}^+MgX$$

The alkyl anion or *carbanion*, R^-, is so basic that it readily removes a proton from water to form hydroxide ion and the alkane.

3.3 ALKENES OR OLEFINS

It has already been pointed out (Section 1.5) that the π electron system of a double bond is more reactive than its σ system. For example, if ethylene gas is bubbled through a solution of bromine in carbon tetrachloride (CCl_4), one obtains 1,2-dibromoethane:

$$CH_2{=}CH_2 + Br \longrightarrow Br \quad —CH_2Br$$

In this reaction the π bond is destroyed and two new σ carbon-bromine bonds are generated. The σ electron framework is left intact; only the π bond of the olefin is involved. The addition of bromine to a π bond constitutes a good qualitative test for an olefin; excess olefin will decolorize a red solution of Br_2 in CCl_4.

The reaction between ethylene and bromine is called an *addition reaction*. Addition reactions are commonplace among *unsaturated* compounds (i.e., compounds containing multiple bonds). The mechanisms and synthetic utility of addition reactions will be discussed in Chapter 9.

There are a number of ways to achieve unsaturation in a molecule. Two of the more common methods are the dehydration of alcohols and the dehydrohalogenation of alkyl halides.

Dehydration of alcohols

$$CH_3CH_2OH \xrightarrow[\text{heat}]{H_2SO_4} CH_2{=}CH_2 + H_2O$$

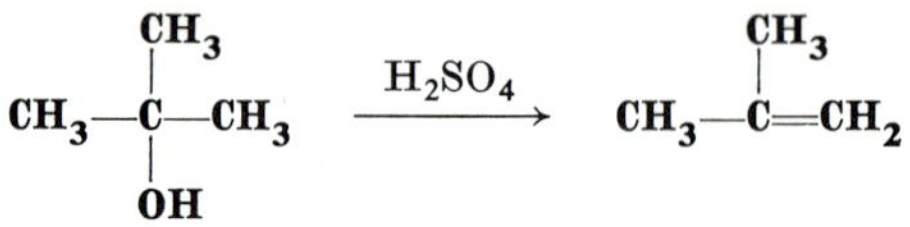

Dehydrohalogenation of alkyl halides:

$$CH_3CH_2X \xrightarrow[\text{(alcohol solvent)}]{KOH} CH_2{=}CH_2 + HX$$

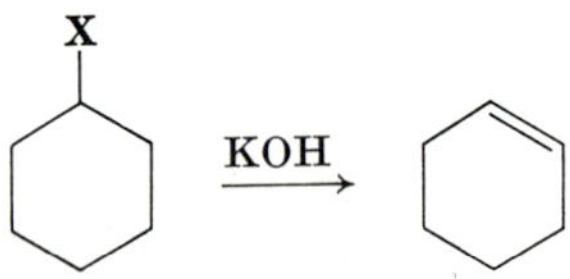

In the first reaction, an alcohol is heated in the presence of sulfuric acid, a powerful dehydrating agent. Water is eliminated, and the olefin is formed. The reaction could be carried out, for example, by heating ethanol at 180°C* in the presence of sulfuric acid. A smooth flow of gaseous ethylene would evolve from this mixture.

The mechanism for the dehydration reaction shows more clearly the nature of the transformation:

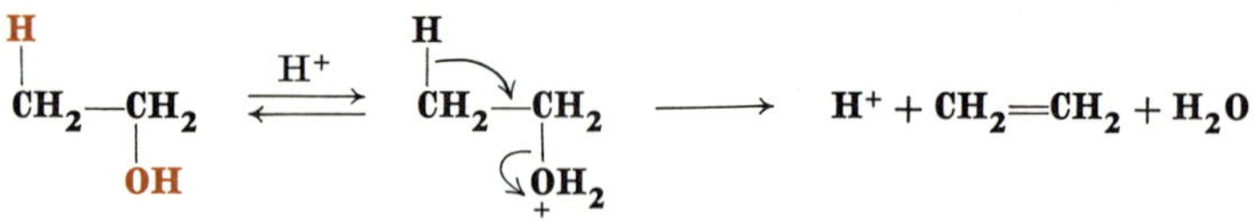

In the first step, the alcohol group is protonated. The resulting species then loses a proton and water to form ethylene. Note that there is no *net* consumption of a proton in the dehydration, yet the reaction would not proceed in the absence of an acid. In other words, the dehydration is *catalyzed* by the acid. One final point should be made clear. Although throughout this book a proton

* Throughout this text, all temperatures may be assumed to be Celsius degrees.

is represented as **H**$^+$, the proton is really a highly solvated species. For example a proton in water is coordinated to at least one solvent molecule and is called hydronium ion or H_3O^+.

The second way to prepare alkenes is to eliminate HX from an alkyl halide by heating a solution of the alkyl halide and potassium hydroxide in ethanol solution. (The inorganic by-products in this reaction are, of course, H_2O and KX).

In both olefin preparations, small fragments on adjacent carbon atoms (H_2O in the first case and HX in the second) are eliminated from the molecule. These reactions are essentially the reverse of the addition reaction and are called *eliminations*. This is a general reaction type and will be discussed more fully in Chapter 8.

3.4 ALKYNES

Since the triple bond of an alkyne or acetylene involves a π electron system similar to that of the double bond of an alkene, we might expect the reactions of alkynes and alkenes to be related. This is, in fact, the case. Alkynes, for example, undergo addition reactions. This is illustrated in the reaction between acetylene and bromine in which two moles of bromine add to one mole of acetylene:

$$\mathrm{H{-}C{\equiv}C{-}H + 2Br_2 \longrightarrow Br_2CH{-}CHBr_2}$$

Since an alkyne is more unsaturated than an alkene, acetylene adds two moles of bromine rather than one.

One way to prepare an alkyne is to use elimination. Thus, 1,1-dichloroethane treated with alcoholic potassium hydroxide affords acetylene by the elimination of two moles of HCl:

$$\mathrm{Cl_2CH{-}CH_3 \xrightarrow{KOH} HC{\equiv}CH}$$

3.5 AROMATIC COMPOUNDS

Aromatic compounds are substances that contain a benzene ring or similar "aromatic" system (Chapter 2). They are unique among unsaturated materials because they do not undergo the addition reactions described above for alkenes and alkynes. Typical aromatic compounds are shown in Fig. 3.1.

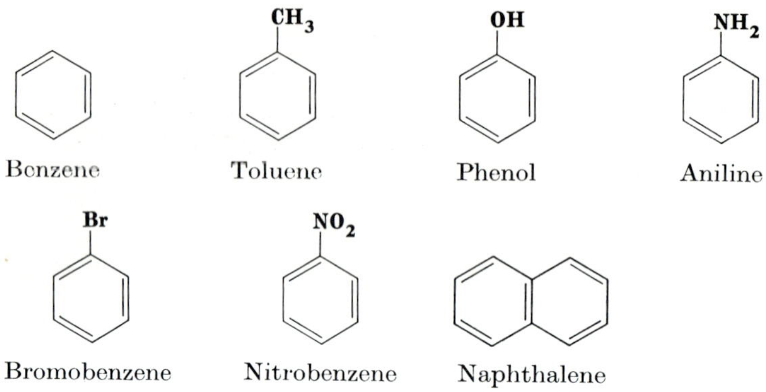

Fig. 3.1 Examples of aromatic compounds (aromatic hydrogens are omitted)

Aromatic compounds characteristically undergo *substitution* reactions. For example

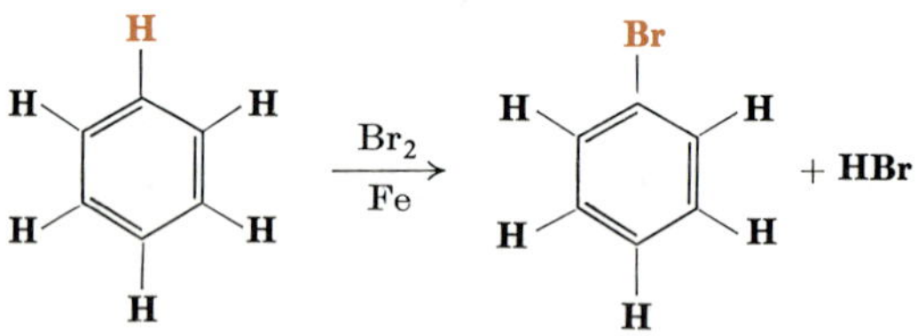

In this case a bromine is substituted for a hydrogen to yield bromobenzene. The reaction is produced by warming benzene with bromine in the presence of iron. As we will see in Chapter 10, the mechanism of this reaction involves a combination of addition and elimination reactions.

Aromatic compounds are usually synthesized by modifying structures that already have an aromatic ring, rather than by generating the aromatic ring from a nonaromatic fragment, although this can also be done. Thus, if cyclohexane is heated in the presence of platinum, *dehydrogenation* occurs and produces benzene and hydrogen:

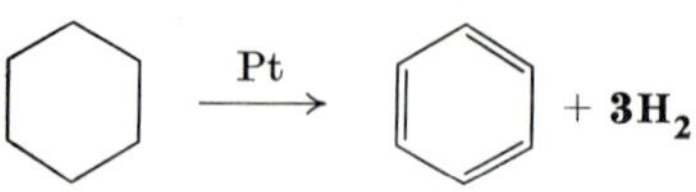

The great resonance stability of the aromatic product provides the "driving force" for the conversion.

3.6 CARBONYL GROUP

The carbonyl group, containing a carbon-oxygen double bond, is one of the most important functional groups in organic chemistry. Two classes of compounds which contain the carbonyl group are aldehydes and ketones.

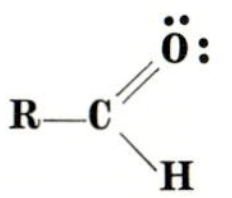
aldehyde

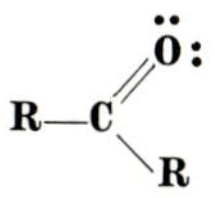
ketone

The oxygen atoms are shown here with both pairs of unshared electrons, although these are usually omitted for the sake of convenience.

The carbon atom of a carbonyl group possesses a partial positive charge because (a) oxygen is more electronegative than carbon and (b) the carbonyl has a dipolar resonance contributor. Therefore the carbon of a carbonyl group is subject to attack by electron-rich reagents (*nucleophiles*).

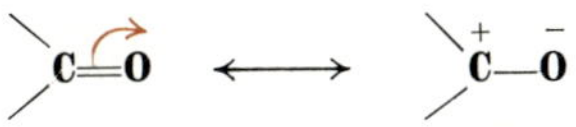

Nucleophilic addition to a carbonyl group can be illustrated by the reversible addition of hydroxide ion to acetaldehyde:

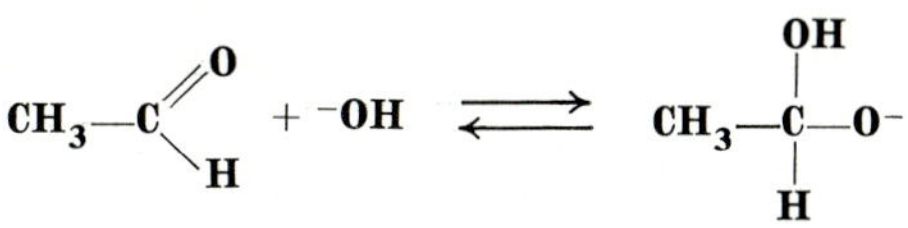

While electron-rich nucleophiles add to the carbonyl double bond, carbon-carbon double bonds of olefins are subject to attack only by electron-deficient species (*electrophiles*). Hydroxide ion will not attack an isolated carbon-carbon double bond; the unpolarized π system resists the approach of a species with a high electron density.

Enolate formation is another characteristic reaction of carbonyl compounds. In the presence of a strong base, compounds which have a proton on a carbon adjacent to a carbonyl will reversibly lose the proton:

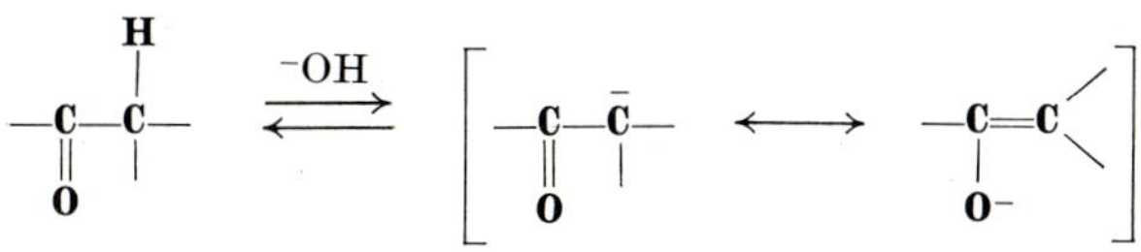
enolate anion

Ketones and aldehydes are not very acidic; even in a strong base only a tiny percentage of the carbonyl compound is in the enolate form. The only reason the enolate forms at all is that the species is resonance stabilized, with the electronegative oxygen bearing a large portion of the negative charge. The enolate anion is highly reactive, and leads to many useful synthetic reactions. The low concentration of enolate is not prohibitive in this regard; as enolate is consumed, more is produced by means of the equilibrium with the carbonyl compound. The bromination of acetophenone, an aromatic ketone is an example of a reaction involving an enolate:

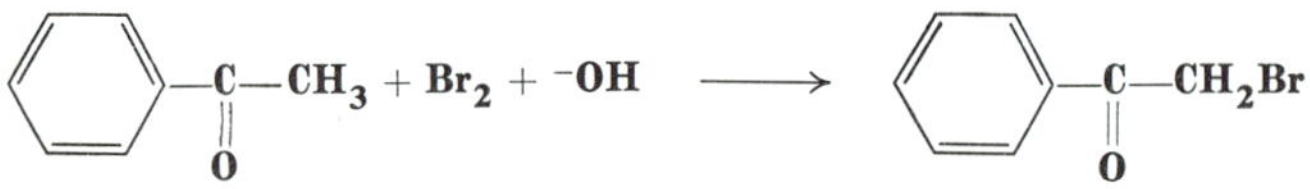

The mechanism shows more clearly the participation of the enolate:

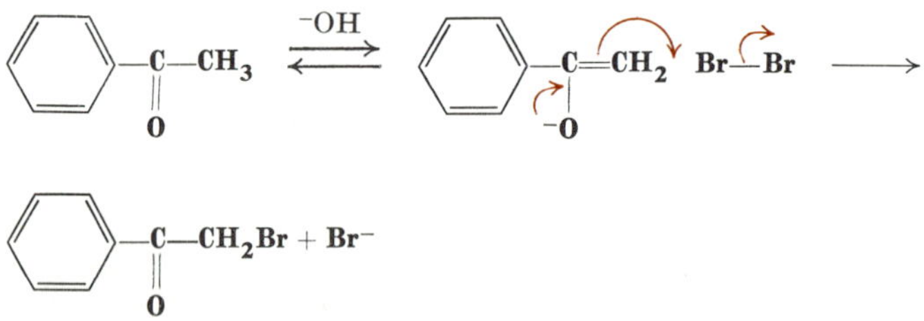

One method commonly used to prepare ketones and aldehydes is the *oxidation* of alcohols. (For the moment we will simply define oxidation as the loss of hydrogen). The oxidation of isopropyl alcohol by chromic acid produces

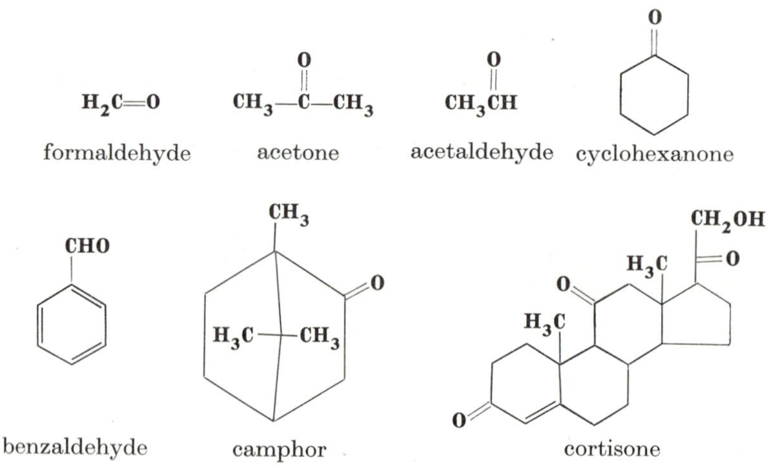

Fig. 3.2. Examples of aldehydes and ketones

acetone. We will not consider the mechanism of this reaction now, although its usefulness should be appreciated.

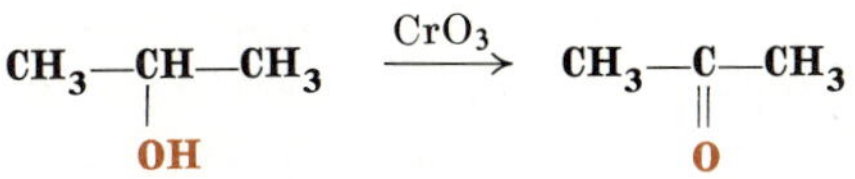

Some typical aldehydes and ketones are shown in Fig. 3.2.

3.7 HYDROXYL GROUP

The hydroxyl group (—OH) is widely used in organic synthesis because there are excellent ways to convert a hydroxyl group into other common functional groups. Before discussing the reactivity of alcohols, we should mention that hydroxyl groups associate with one another as a result of electrostatic attraction between the partially positive hydrogen and the partially negative oxygen:

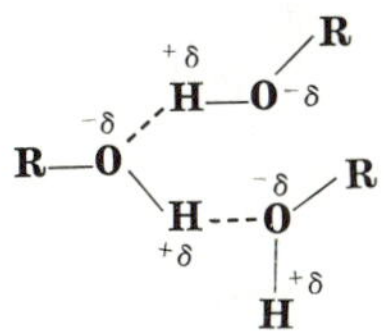

This dipole-dipole interaction, called *hydrogen bonding*, gives rise to relatively high boiling points for alcohols.

It is convenient to classify alcohols according to their substitution, for their reactivities toward many reagents may be correlated on this basis. A tertiary alcohol has a hydroxyl group on a carbon bonded to three substituent carbons; a secondary alcohol has a hydroxyl group located on a carbon atom bonded to two substituent carbons; and a primary alcohol has a hydroxyl group located on a carbon bonded to only one substituent carbon (Fig. 3.3). In those reactions involving cleavage of the O—H bond (e.g. the reaction of an alcohol with sodium), primary alcohols are the most reactive and tertiary the least.

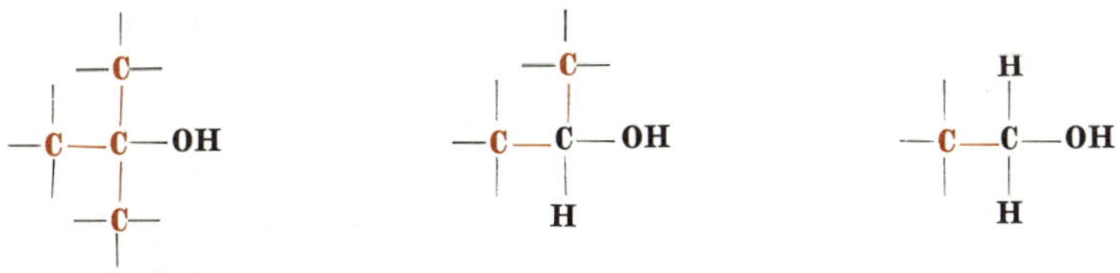

tertiary alcohol secondary alcohol primary alcohol

Tertiary alcohols react relatively slowly with sodium because the carbons interfere with solvation of the ionic product (RO^-). Interaction between the ions and the solvent dipoles stabilizes the charges. The bulky carbons of a tertiary alcohol reduce the ability of the solvent to approach the alcohol anion. This is an example of a *steric effect* in which reactivity is affected by atomic size. We have now discussed the three principal factors used to rationalize reactivity of organic molecules: inductive effects, resonance effects, and steric effects.

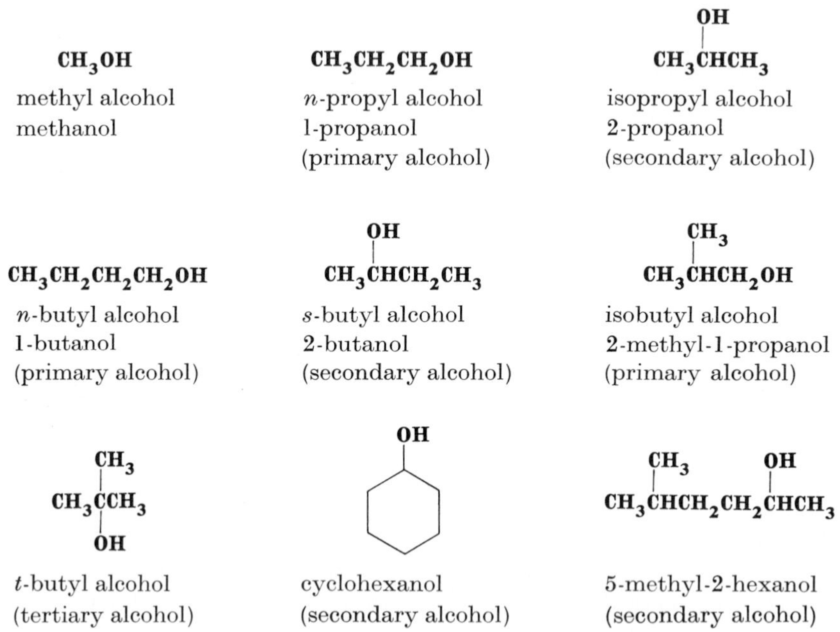

Fig. 3.3 Examples of alcohols (see Appendix for nomenclature)

Treatment of an alcohol with hydrogen bromide results in cleavage of the carbon-oxygen bond and formation of an alkyl bromide:

$$\mathrm{ROH + HBr \longrightarrow R{-}Br + H_2O}$$

The rate of this substitution reaction also depends on the type of alcohol treated. This will be discussed in detail later.

One method commonly used to generate alcohols is the reduction of carbonyl groups using lithium aluminum hydride ($LiAlH_4$), a reagent mentioned previously in connection with the synthesis of alkanes (Section 3.2). An aldehyde is reduced to a primary alcohol, and a ketone to a secondary one:

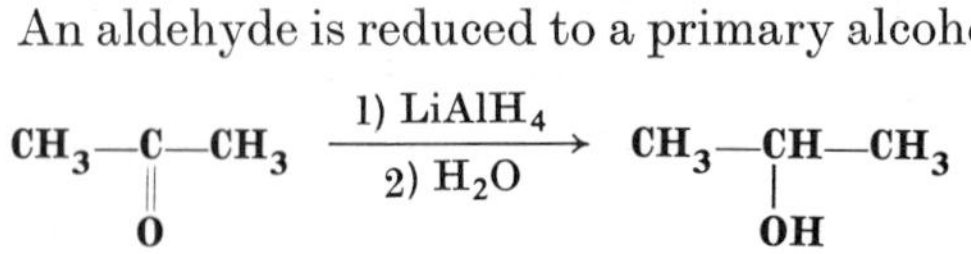

Recall that an alcohol can be oxidized back to the carbonyl group using chromic acid.

The Grignard reaction provides another useful method for synthesizing alcohols:

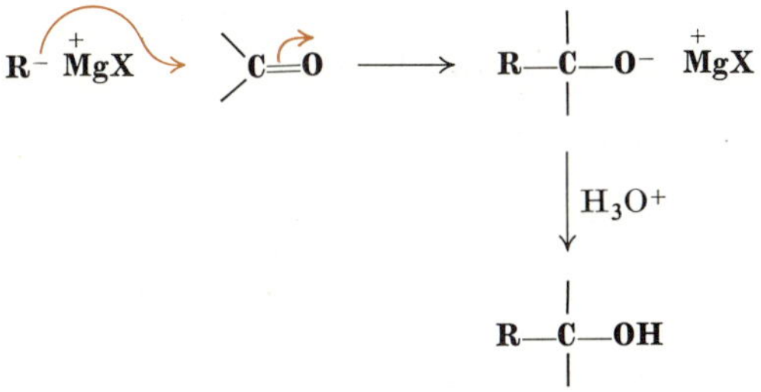

In this sequence, the carbonyl carbon provides a positive site for attack by the alkyl anion of the Grignard reagent. The resulting addition yields a salt which, upon addition of dilute acid, yields the corresponding alcohol. The reaction can be used for the preparation of primary, secondary, or tertiary alcohols:

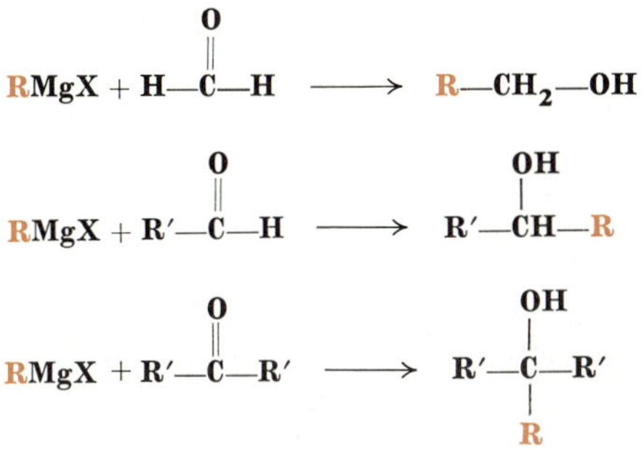

Thus, primary alcohols are derived from a Grignard reagent and formaldehyde; secondary alcohols from a Grignard reagent and aldehydes other than formaldehyde; and tertiary alcohols from a Grignard reagent and ketones.

The usefulness of this set of reactions becomes apparent when it is realized that Grignard reagents, aldehydes, and ketones can all be derived from alcohols (aldehydes by the oxidation of primary alcohols, and ketones by the oxidation of secondary alcohols). Thus, it is possible to devise synthetic schemes for fairly complex materials by building them up sequentially from smaller, more available substances (refer to Section 3.11).

3.8 CARBOXYL GROUP

The carboxyl group (—COOH) is a weakly acidic organic functionality. For example, acetic acid has a dissociation constant in water of about 2×10^{-5}. The acidity of this material (i.e., its ability to lose a proton) is associated with an equilibrium:

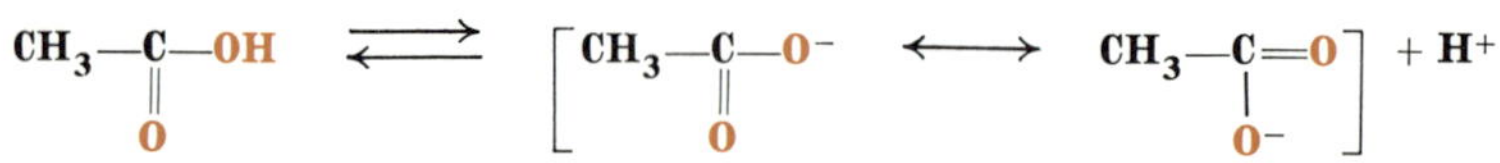

One of the chief reasons acetic acid is so much more acidic than ethyl alcohol is that the acetate anion is resonance stabilized, whereas ethoxide ($CH_3CH_2O^-$) is not.

The carboxyl function may be introduced into a molecule by oxidizing a primary alcohol with chromic acid. The initial product is the corresponding aldehyde, which may be either removed from the oxidizing medium by distillation or allowed to react further to yield the carboxylic acid.

$$RCH_2\text{—}OH \xrightarrow{CrO_3} RCHO \xrightarrow{CrO_3} RCOOH$$

A second method of synthesis is *carbonation* of a Grignard reagent. This reaction is a variation of the Grignard addition to a carbonyl group; in this case, the carbonyl compound is carbon dioxide:

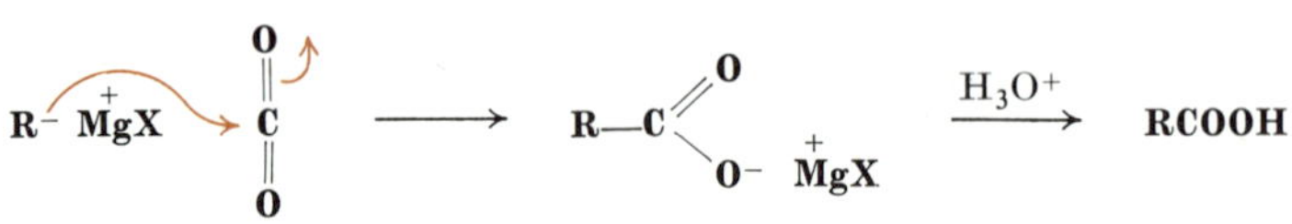

Many functional groups may be considered carboxyl group derivatives. These are listed below.

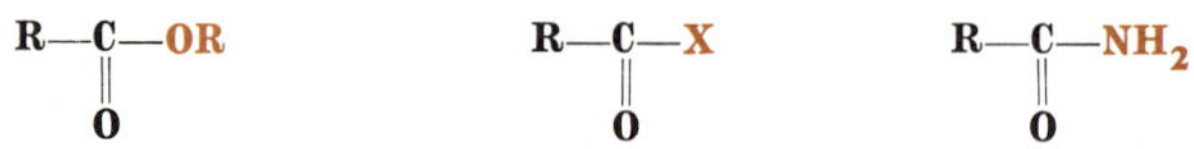

ester — acid halide (X = Cl, Br, I) — amide

Specific examples are given in Fig. 3.4.

An acid chloride may be prepared from an acid by treatment with thionyl chloride ($SOCl_2$):

$$RCOOH \xrightarrow{SOCl_2} RCOCl$$

CH_3COOH
acetic acid

$CH_3CH_2CH_2COOH$
butyric acid

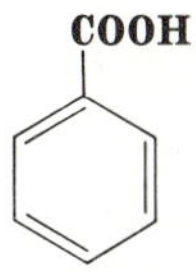

benzoic acid

CH_3COOCH_3
methyl acetate
(an ester)

$CH_3CH_2CH_2CONH_2$
butyramide
(an amide)

CH_3COCl
acetyl chloride
(an acid chloride)

$CH_3CH_2COOCH_2CH_3$
ethyl propionate
(an ester)

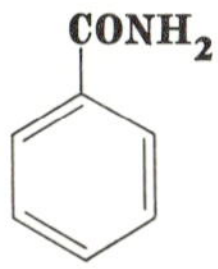

benzamide
(an amide)

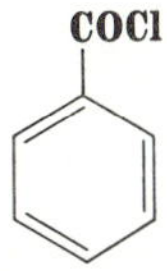

benzoyl chloride
(an acid chloride)

Fig. 3.4 Examples of carboxylic acids and their derivatives

Acid chlorides are useful in that the chloride is easily displaced by an alcohol or an amine to give an ester or amide, respectively:

$$RCOCl \xrightarrow{CH_3OH} RCOOCH_3$$

$$RCOCl \xrightarrow{NH_3} RCONH_2$$

Esters and amides can be converted back to the parent acid by a *hydrolysis* reaction (reaction with water catalyzed by acid or base):

$$R—\underset{\underset{O}{\|}}{C}—OR' \xrightarrow{H_3O^+} R—\underset{\underset{O}{\|}}{C}—OH + HOR'$$

Much more will be said about these important reactions later, when the chemistry of the carbonyl group is discussed.

3.9 AMINO GROUP

The amino group (RNH_2, R_2NH, and R_3N) may be viewed as a derivative of ammonia in which organic groups replace one or more of the hydrogens. RNH_2, R_2NH, and R_3N are called primary, secondary, and tertiary amines, respectively. The chemistry of amines parallels that of ammonia, the most striking feature being their basicity (ability to accept a proton) which arises from the unshared pair of electrons. Compare (below) the acceptance of a proton by an amine, and the acceptance by ammonia:

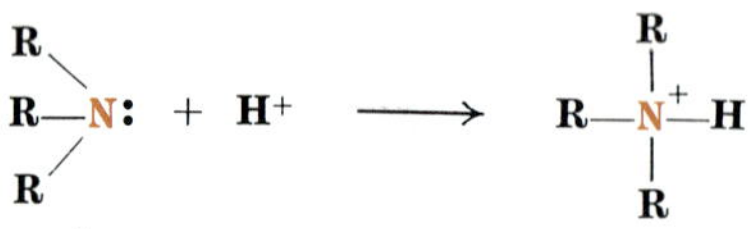

amine

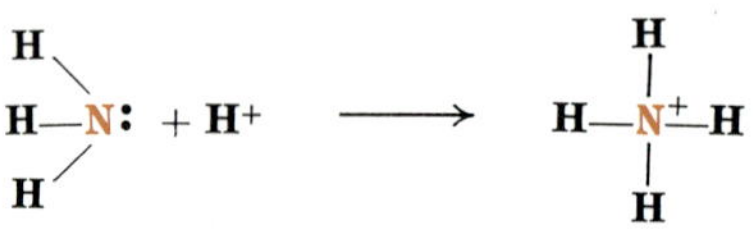

Amines may be prepared by a substitution process called *alkylation* in which an alkyl group becomes bonded to a nitrogen:

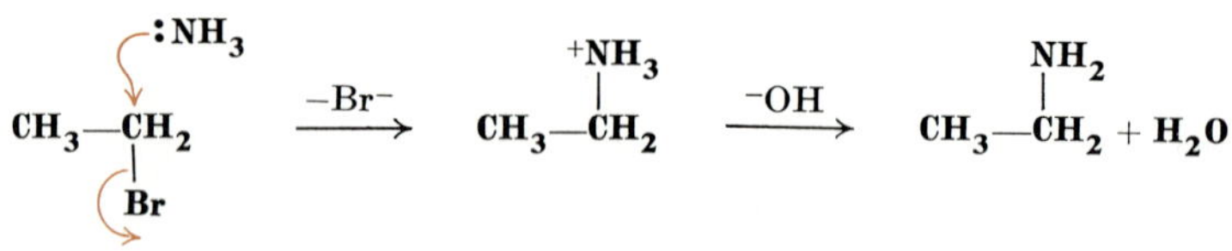

Note the similarity of the first step to the substitution reaction of hydride ion on ethyl bromide presented earlier.

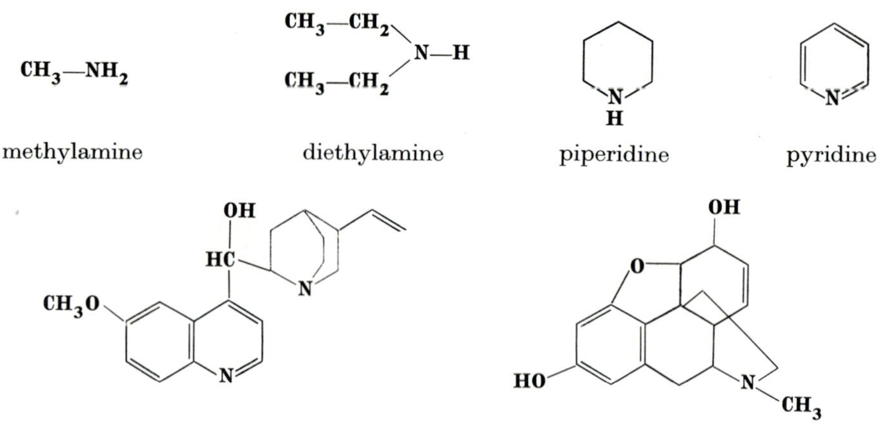

quinine morphine

Fig. 3.5 Typical amines

3.10 ETHER GROUP

The ether group (R—O—R) is a relatively inert function which may be regarded as a water derivative where alkyl groups replace both hydrogens. Since ethers have no reactive π system or removable proton, they lack reactivity and are useful solvents for organic reactions (such as the Grignard reaction).

Ethers, like water, have weak basicity:

$$H{-}\ddot{\underset{\cdot\cdot}{O}}{-}H + H^+ \longrightarrow H{-}\overset{H}{\overset{|}{O^+}}{-}H$$

$$R{-}\ddot{\underset{\cdot\cdot}{O}}{-}R + H^+ \longrightarrow R{-}\overset{H}{\overset{|}{O^+}}{-}R$$

Ethers are much weaker bases than amines because the electronegative oxygen is less "willing" to bear a positive charge than is nitrogen.

Ethers are synthesized most often by a substitution reaction:

$$ROH + Na \longrightarrow RO^- Na^+$$

$$RO^- Na^+ + CH_3CH_2Br \longrightarrow CH_3CH_2OR + Br^-$$

An alcohol is first converted into the corresponding *alkoxide anion*. The anion, with its negative charge and three unshared pairs of electrons, is much more reactive than the alcohol as a nucleophile. Next the alkoxide displaces the bromine from ethyl bromide. The mechanism of this second step is much like that for alkylation of ammonia to give an amine. Two examples of ether formation are shown below:

1) $$CH_3CH_2OH + Na \longrightarrow CH_3CH_2\bar{O}Na^+$$

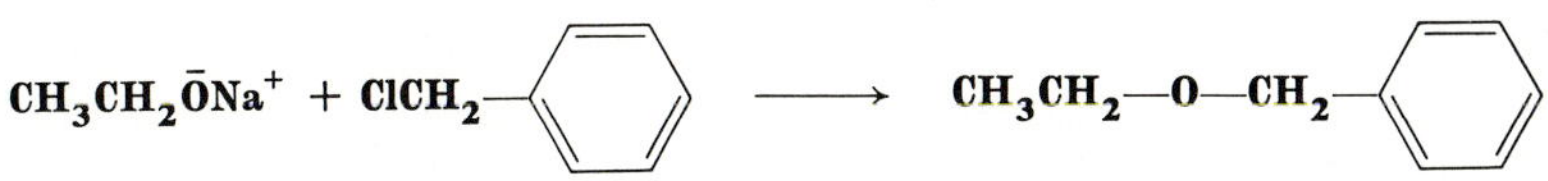

2)

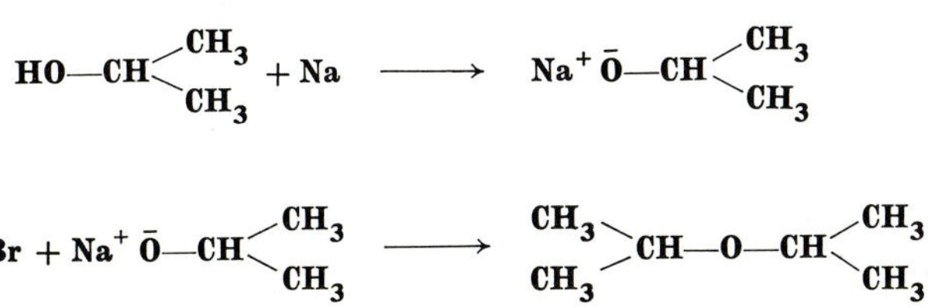

3.11 SYNTHESIS

The ability of the carbon atom to form chains, rings, and multiple bonds makes possible an extraordinarily large number of compounds from the combination of a relatively small number of atoms. For example, there are 366,319 possible structures for hydrocarbons containing only twenty carbon atoms. The variation of properties with structure is sufficiently great that a large part of the thrust of organic chemistry in this century has been directed toward the synthesis of new and different kinds of compounds. Despite this history, however, only a small fraction of the myriad kinds of materials possible have been synthesized. Thus it is likely that organic synthesis will continue its lively course in the forseeable future.

For the student beginning the study of organic chemistry, synthesis has a special value. It provides a base for the application of a given body of factual knowledge to the solution of problems. As we proceed in our study, both the extent of our knowledge and the complexity of the problems will increase. At every stage, however, the solution to a synthetic problem must be based on the tools at hand. The reaction of an organic molecule with a given reagent is not fortuitous. A molecule reacts in a specific fashion dependent upon its structure and the conditions to which it is subjected. Thus, before we attempt to solve a synthetic problem we must know all of the reactions which we have previously encountered.

3.12 GENERAL PRINCIPLES

A number of general principles may be formulated for finding a solution to a synthetic problem. We shall first list these principles, and then follow their use in a specific example.

1. *Begin at the end!*

Work backwards from the final structure, the target molecule, until readily available starting materials (compounds which do not require prior synthesis) are reached. If in a given synthetic problem the starting compound is specified, we still work back from the final structure. Such a specification only limits the number of possible answers to the problem, not the general method of solution.

2. *Identification and recognition of the target molecule*

In examining the structure of the target molecule, the following questions should be asked. What kind of compound is it? What functional group(s) does it contain? What is the nature of the carbon skeleton? Does the molecule have a normal alkyl chain, or are there branches in that chain? Are there any rings present, and are they cycloalkyl or aromatic rings?

3. *The functional group*

What general methods do we have available for its preparation? Which of these methods is applicable to the specific functional group in the target molecule?

4. *The carbon skeleton*

The major problem in most organic syntheses is the construction of the carbon skeleton. The interchange of functional groups is often easy to do; for example, ketone to alcohol, aldehyde to acid, or alcohol to bromide. The questions we ask about making carbon–carbon bonds are related to what we have already asked with regard to the functional group. Do any of the methods available for forming functional groups also provide a means for making carbon-to-carbon bonds? If so, is that method compatible with the specific carbon skeleton of the target molecule? If not, is there a method for carbon skeleton formation which produces a functional group convertible into the required one?

5. *Precursor molecules*

Analysis of the structure of the target molecule and consideration of the questions posed in steps 1 through 4 will generate two possible types of precursor molecules. One of these types is a functional group equivalent of the final structure. The other is a set of compounds having fewer carbon atoms than the target molecule. When these latter precursors are put together, the final carbon skeleton is achieved. The generation of either type of precursor molecule should result in a simplification of the problem. In general, if a projected route leads to precursors which are more difficult to synthesize than the target molecule itself, another route should be sought.

6. *Total synthetic plan*

Each precursor molecule is subjected to the same analysis (steps 1 through 5) originally applied to the target molecule. Eventually that analysis will lead back to acceptable starting materials, and we will have constructed a total synthetic plan, or synthetic "tree." A general picture of such a synthetic tree is shown in Fig. 3.6. Compound A is the target molecule, and A′ is a functional-group equivalent. Compounds B and C are precursor molecules, each of which has fewer carbon atoms than A or A′. Compounds F and G are precursors to B, and again have a simpler carbon skeleton. Compound C′ is a functional-group equivalent of C, and D and E are its precursors. Finally, F, G, D, and E fit the criterion of readily available starting materials.

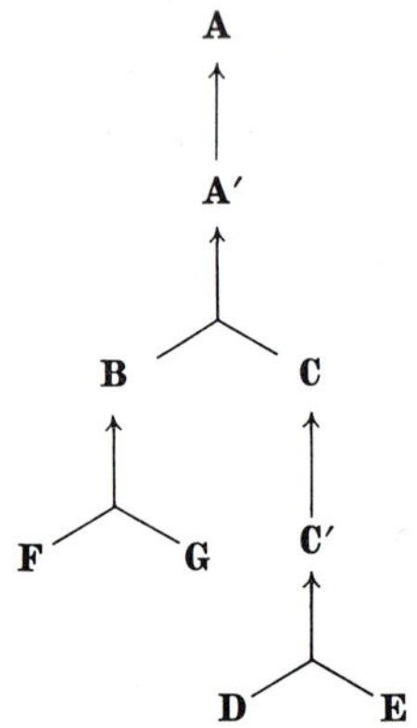

Fig. 3.6 A synthetic tree

3.13 A SYNTHETIC PROBLEM

Our first example of a synthetic problem will be the monofunctional compound 2,3-dimethyl-3-pentanol. How can this compound be synthesized from starting materials containing no more than two carbon atoms; e.g., methanol, ethyl bromide, formaldehyde, acetic acid, etc.?

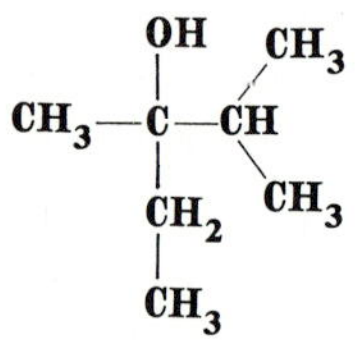

2,3-dimethyl-3-pentanol

First we must identify the kind of compound that we wish to synthesize and the specific aspects of its structure that our plan must accommodate. In examining the structure, we recognize that the compound is an alcohol. Specifically, it is a tertiary alcohol and one of the alkyl groups attached to the carbinol (hydroxyl-bearing) carbon is branched. The recognition of these features immediately raises two key questions:

1) What methods can be used to prepare an alcohol?
2) Which of these methods may be applicable to the preparation of a tertiary alcohol?

Our identification of the target molecule and our initial planning questions have focused on the functional group—the hydroxyl group. Implicit in the term *functional group* is the idea that this molecular feature is the prime center of reactivity. Thus, a functional group is the "handle" by which we turn one substance into another. A functional group is the means by which we construct carbon chains and rings—the means by which we make carbon-to-carbon bonds.

We have already encountered two methods of preparing an alcohol. These are:

1) the hydride reduction of a carbonyl group:

2) the addition of a Grignard reagent to a carbonyl group:

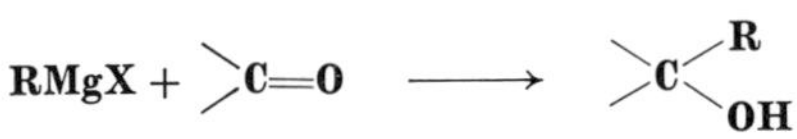

Two other methods for producing alcohols are:
3) the addition of the elements of water to a double bond (olefin hydration):

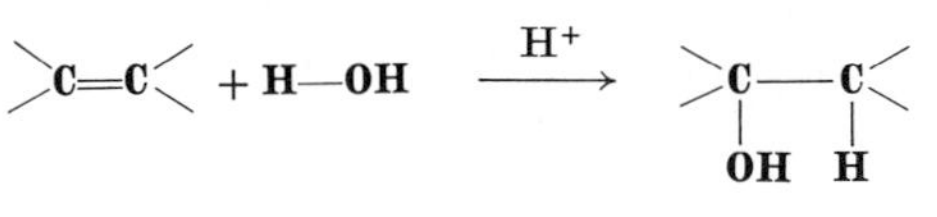

4) the basic hydrolysis of an alkyl halide:

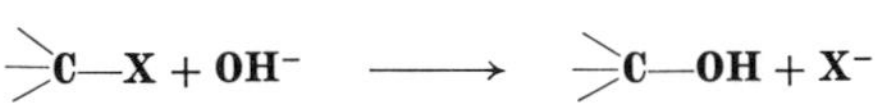

We must consider each of these methods in terms of the structure we wish to synthesize. Are any of them applicable to the preparation of a tertiary alcohol? More important, are any of them useful in solving the problem of constructing the carbon chain of 2,3-dimethyl-2-pentanol? On the basis of these two questions we may dismiss methods 1 and 4 from further consideration. Hydride reduction of a ketone or an aldehyde cannot yield a tertiary alcohol; the product of such a reaction must necessarily have at least one hydrogen atom attached to the carbinol carbon and therefore be either a primary or secondary alcohol. The second inappropriate method is the hydrolysis of a halide. The halide in our projected synthesis would have to be, for example, 2,3-dimethyl-3-bromopentane (shown below).

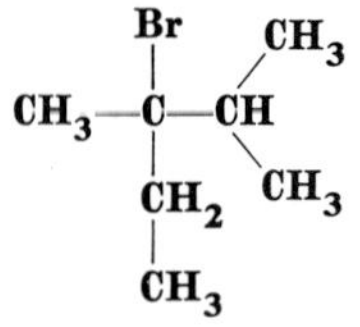

2,3-dimethyl-3-bromopentane

One of the halide reactions we have encountered in this chapter is conversion into an olefin by elimination of hydrogen halide:

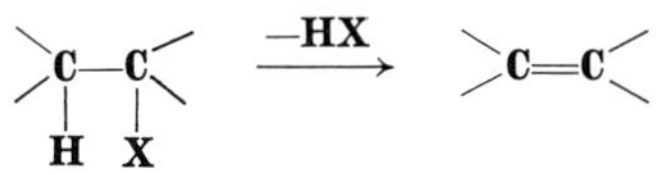

Since this is virtually the sole mode of reaction of tertiary substrates, halide hydrolysis is inapplicable to the synthesis of our desired alcohol. In addition, alkyl halides are normally prepared from the corresponding alcohol:

We are left then with either the addition of water to a double bond or the addition of a Grignard reagent to a suitable carbonyl substrate. For the former process, the question arises, addition of water to what olefin? There are three possibilities:

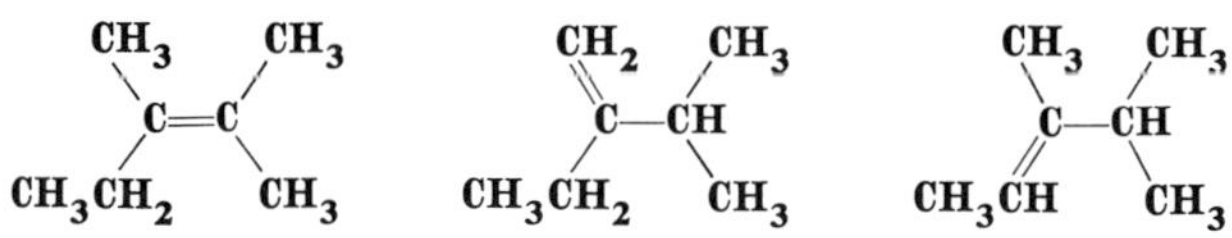

Even without any knowledge of the mechanism of olefin hydration we see that each of these olefins might afford *two* different alcohols, only one of which in each case would be the right one. It is known, however, that in the addition of water to double bonds the hydroxyl (OH) group becomes attached to the carbon atom of the double bond bearing the largest number of alkyl groups. Thus, the second and third olefins would give the desired product. The first one, having the *same number* of alkyl groups attached to each of the carbon atoms of the double bond, would be expected to yield two products; this would diminish the yield of the desired alcohol. Although we have at this point found a method for preparing the right alcohol, and two possible precursors on which to apply that method, we are still faced with our initial dilemma: how do we put together a 2,3-dimethylpentane system? We have not yet faced the problem of a carbon-carbon bond construction.

In this first synthetic problem one of the methods for introducing the functional group also provides a means for making carbon-to-carbon bonds. (This will not always be the case.) The method is, of course, the Grignard reaction. Referring to Section 3.7 we see that addition of a Grignard reagent to a *ketone* (not an aldehyde) yields a tertiary alcohol.

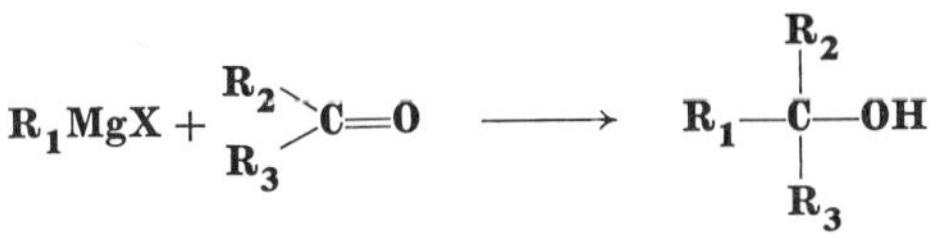

If the Grignard reaction is the key to both carbon–carbon bond formation and functional group preparation in this synthesis, what should be the specific structure of each of the R groups in our general reaction? The choice may seem to be arbitrary. That is, R_1 could be either a methyl (CH_3) group, an ethyl (CH_3CH_2) group, or an isopropyl ($(CH_3)_2CH$) group. We shall see however that, in terms of economy of effort (the minimum number of steps required to complete the total synthesis of our target molecule from one or two carbon precursors), the assignment of either an ethyl or an isopropyl group to the structure of the Grignard reagent is best. The complete synthesis is outlined in Fig. 3.7.

We have so far dealt only with step 1 in this plan. How was the rest of the synthetic strategy devised? The answer lies in applying the same kind of analysis used for the final structure for each of the precursor molecules. Thus for the Grignard reagent employed in step 1 we ask, how do we make isopropyl magnesium bromide? This step requires only the reaction of the appropriate bromide with magnesium metal in ether, step 2. The bromide in turn may be made from the corresponding alcohol, step 3. Note that steps 2 and 3 only involve the interchange of functional groups. The Grignard reagent and the bromide used in these transformations are functional group equivalents of the alcohol. Since the alcohol has three carbon atoms, it must be synthesized. Analysis of its structure again suggests the use of the Grignard reaction. The compound, isopropyl alcohol, is a secondary alcohol. Such alcohols can be prepared from Grignard reagents and aldehydes:

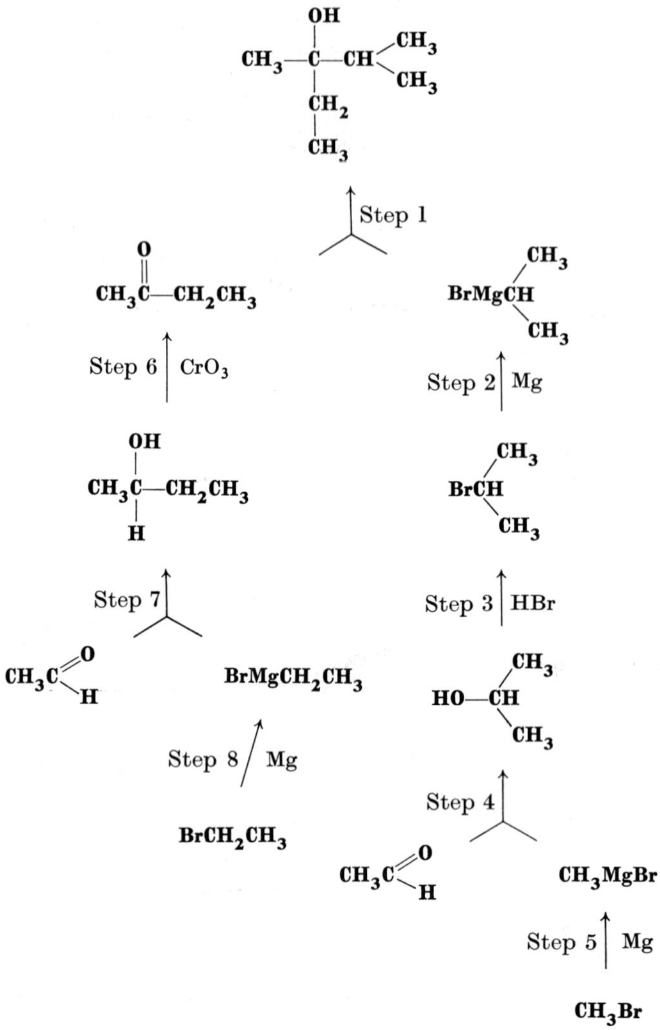

Fig. 3.7 A total synthesis of 2,3-Dimethyl-3-pentanol

Thus the combination of methyl magnesium bromide and acetaldehyde (step 4) can be used.

Returning to step 1 we see that we require a ketone, 2-butanone. Since the ketone has more than two carbon atoms we must synthesize it. We have no direct method for making a four-carbon ketone from smaller fragments. There is, however, a functional equivalent of this ketone which we can prepare. This is the alcohol used in step 6. Only a functional-group interconversion, oxidation of an alcohol, is needed to obtain 2-butanone. The analysis used for planning the preparation of isopropyl alcohol can be employed for the step 6 alcohol also. The compound is a secondary alcohol, and we may again use the reaction of a Grignard reagent with an aldehyde. We arrive then at a complete synthetic plan since the components for step 7 are both two-carbon compounds. Steps 8 and 5 are also listed since Grignard reagents are not easily storable substances and are usually prepared just prior to use.

The synthetic strategy of Fig. 3.7 is only one way of preparing 2,3-dimethyl-3-pentanol. Other schemes exist. For example, as we have noted earlier the Grignard synthesis of step 1 may be accomplished using either methyl or ethyl magnesium bromide and an appropriate ketone. The reader should outline these other approaches and verify that the use of methyl magnesium bromide requires two more steps than does either of the alternatives.

3.14 2,3-DIMETHYLPENTANE AND 2,3-DIMETHYL-2-PENTENE

The solution to our first synthetic problem was based on the realization that both the functional group and part of the carbon skeleton could be constructed *via* the same reaction, a Grignard addition. What do we do, however, if the target molecule contains *no* functional group, or if the one present cannot be prepared at the same time as a carbon–carbon bond? For example, how would we make 2,3-dimethylpentane or 2,3-dimethyl-2-pentene:

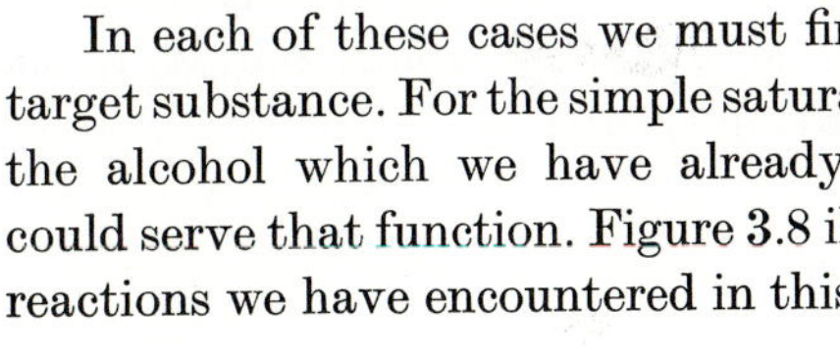

2,3-dimethylpentane 2,3-dimethyl-2-pentene

In each of these cases we must find a functional group equivalent of the target substance. For the simple saturated hydrocarbon, 2,3-dimethylpentane, the alcohol which we have already synthesized, 2,3-dimethyl-3-pentanol, could serve that function. Figure 3.8 illustrates the necessary conversion using reactions we have encountered in this chapter.

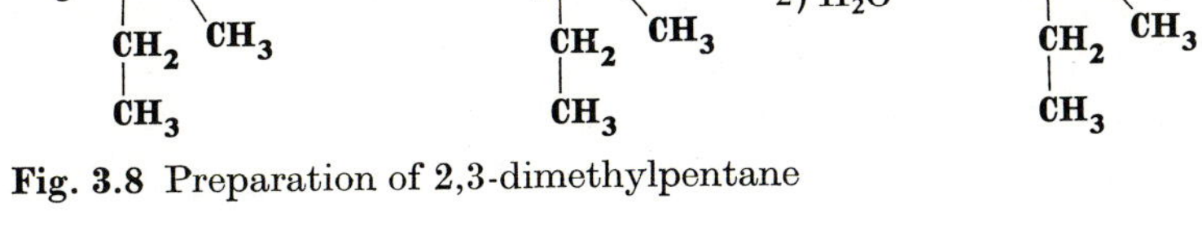

Fig. 3.8 Preparation of 2,3-dimethylpentane

The preparation of 2,3-dimethyl-2-pentene requires one additional fact. The dehydration of an alcohol (Section 3.3) generally yields the most substituted olefin as its major product when several olefins are possible. (The most-substituted olefin is that one with the largest number of alkyl groups attached to the double-bond carbon atoms.) Figure 3.9 shows how this information is used again to relate the final olefin product to 2,3-dimethyl-3-pentanol.

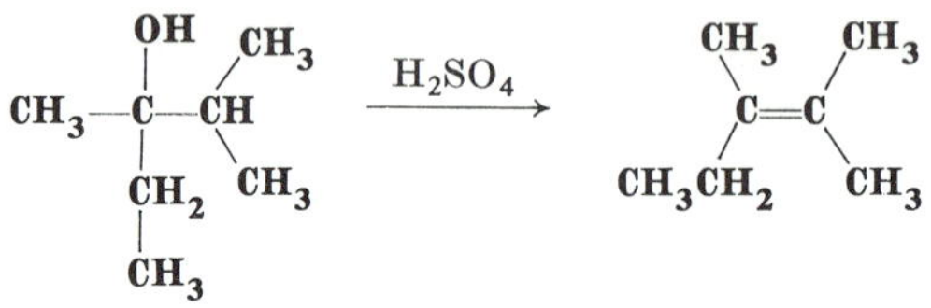

Fig. 3.9 The preparation of 2,3-dimethyl-2-pentene

3.15 A WORD TO THE WISE

The syntheses that we have previously examined and the problems which follow are based on only one method for the formation of carbon-carbon single bonds—the Grignard reaction. In addition, all our target and precursor molecules are monofunctional compounds. The Grignard synthesis is a very useful one. It is not, however, the only method known. In the following chapters we shall encounter a variety of other ways of making carbon–carbon bonds. These methods often provide more efficient, more economical, and more "elegant" ways of synthesis than does the Grignard reaction. We must be careful, therefore, not to use the latter in a rote and hackneyed manner when more powerful methods are also at our command.

Synthesizing multifunctional compounds requires a knowledge of *selectivity*. How do we carry out a change at one functional group without also effecting changes in others? Are there reactions which are inherently selective, or must one find a way of *protecting* certain functional groups in order to use another? These questions will be answered in later chapters.

PROBLEMS

1. Complete the following reactions by supplying the necessary inorganic reagent, starting material, or product.

a) 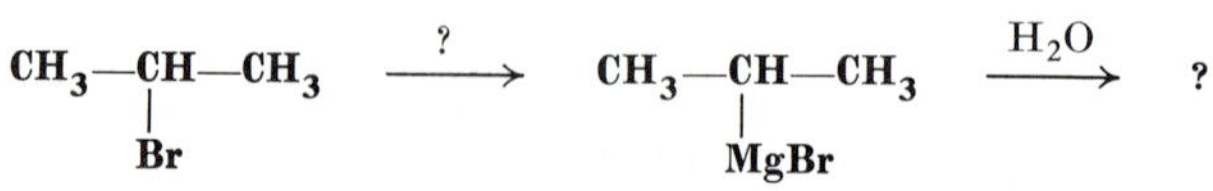

b) 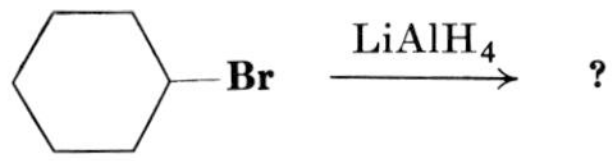

c) $? + Br_2 \longrightarrow BrCH_2\underset{\displaystyle CH_3}{\underset{|}{C}}HBr$

d) 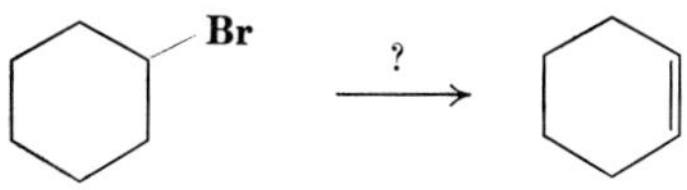

e) 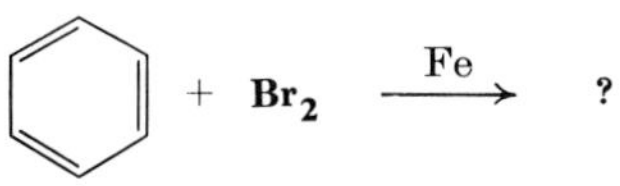

f) 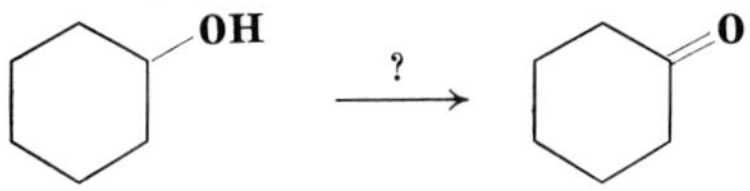

g) $CH_3MgBr + ? \longrightarrow CH_3CH_2OH$

h) 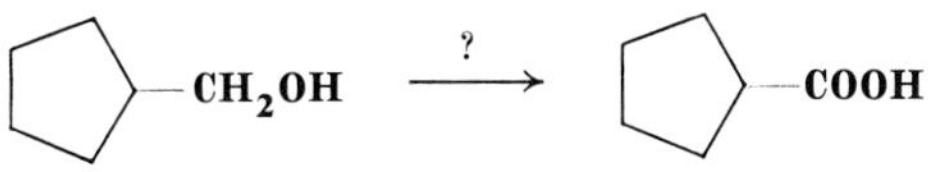

i) 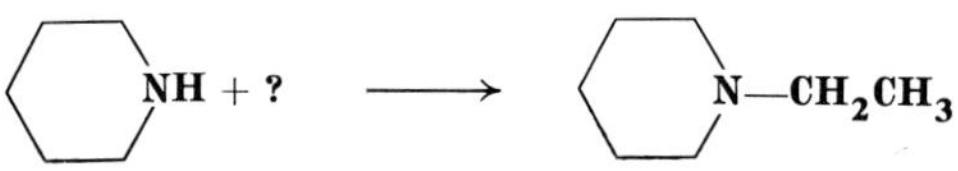

j) $CH_3CH_2Br + ? \longrightarrow CH_3-\overset{\displaystyle CH_3}{\overset{|}{\underset{\displaystyle CH_3}{\underset{|}{C}}}}-O-CH_2CH_3$

k) $(CH_3)_2CHMgBr + ? \longrightarrow (CH_3)_2CHCOOH$

l) 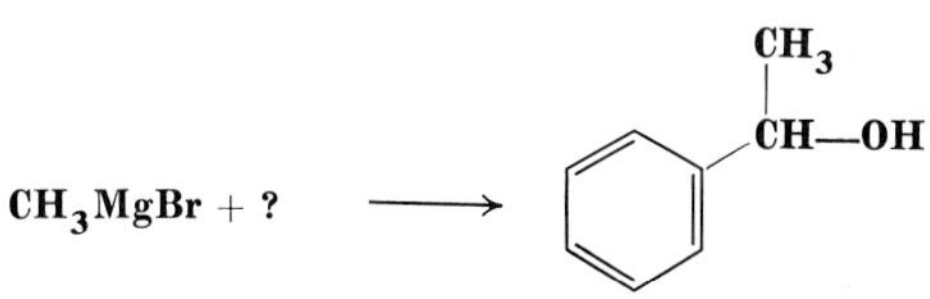

m) $CH_3CH_2CH_2\underset{\displaystyle OH}{\underset{|}{C}}HCH_3 + HBr \longrightarrow ?$

n) $CH_3CH_2\underset{\displaystyle OH}{\underset{|}{C}}HCH_2CH_3 \xrightarrow{H_2SO_4}$?

o) 

2. Draw the following compounds:
   a) a benzene ring bearing an aldehyde group,
   b) a cyclohexane ring bearing a hydroxyl group,
   c) a secondary alcohol,
   d) a five-membered cyclic ketone,
   e) diethylamine,
   f) carbon dioxide,
   g) acetic acid,
   h) an ether with six carbons,
   i) the lowest-molecular-weight primary amine,
   j) an ester of acetic acid.
3. Supply the necessary reagents for the following 2-step syntheses.

a) 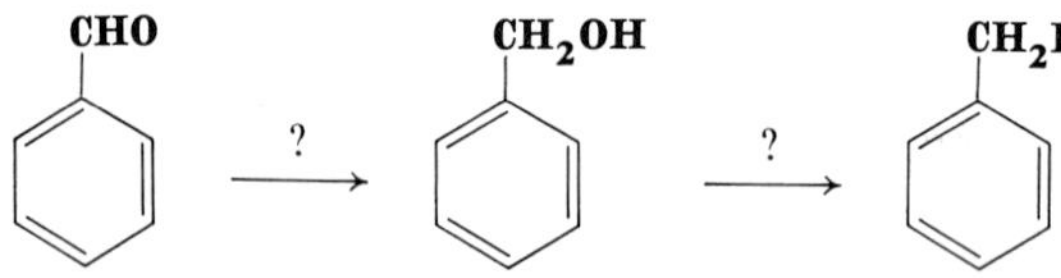

b) 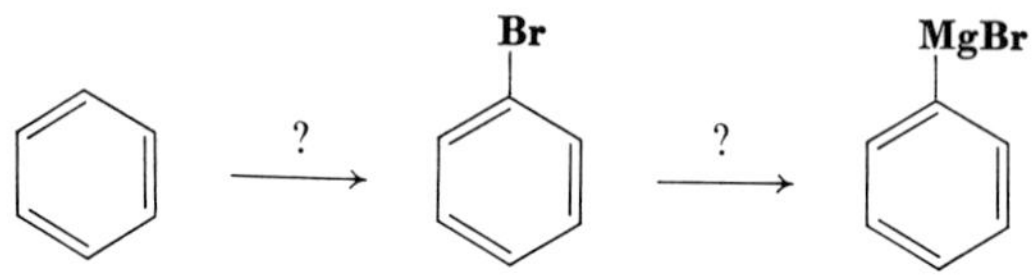

c) 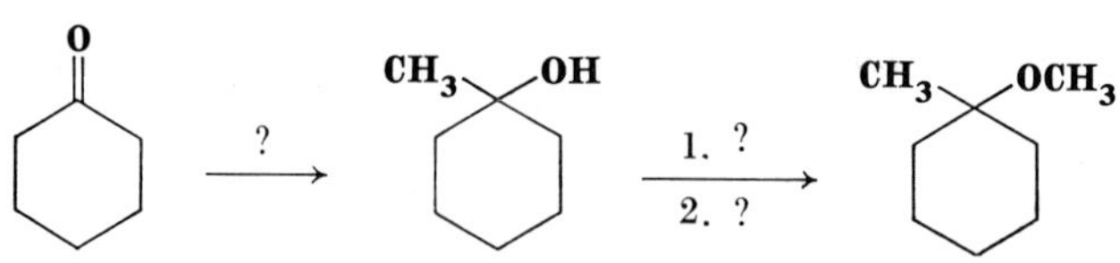

4. Supply the missing structures.

a) $CH_3CHO \xrightarrow{CH_3MgBr} \mathbf{A} \xrightarrow{CrO_3} \mathbf{B} \xrightarrow{CH_3MgBr} \mathbf{C} \xrightarrow[2)\ CH_3Br]{1)\ Na} \mathbf{D}$

b) 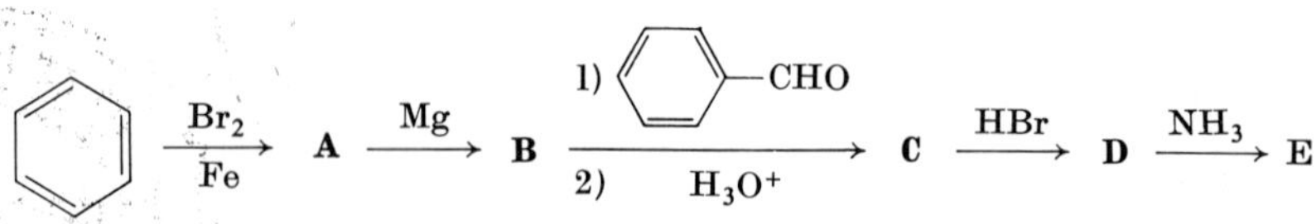

5. Cyclopentanone is a five-membered cyclic ketone that can be purchased for about \$6.00 per 100 grams.

cyclopentanone

Show what steps would be necessary to convert cyclopentanone into each of the following compounds. Indicate all the required reagents.

a) 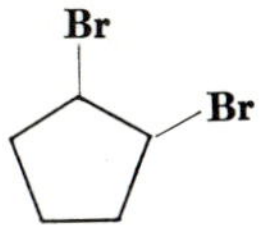

1,2-dibromocyclopentane

b) 

cyclopentyl methyl ether

c) 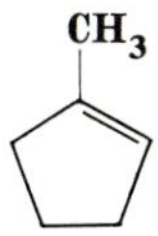

1-methylcyclopentene

d) 

Cyclopentane

e) 

cyclopentane carboxylic acid

6. Antistine, a useful antihistaminic drug, has been prepared by reacting an amine with a chlorine-containing compound.

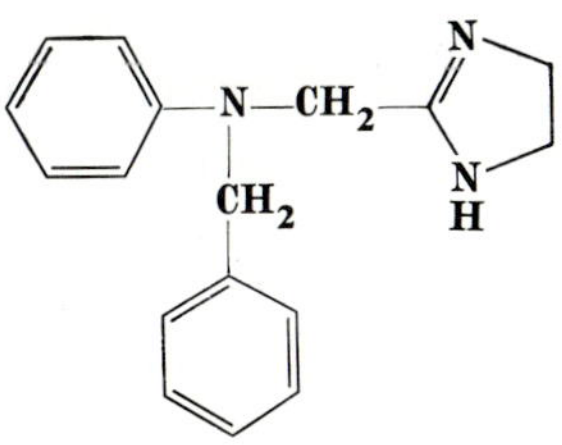

antistine

Draw the structures of these two reactants.

7. 4-Methylazulene, a compound with a beautiful blue color, was synthesized from a bicyclic unsaturated ketone:

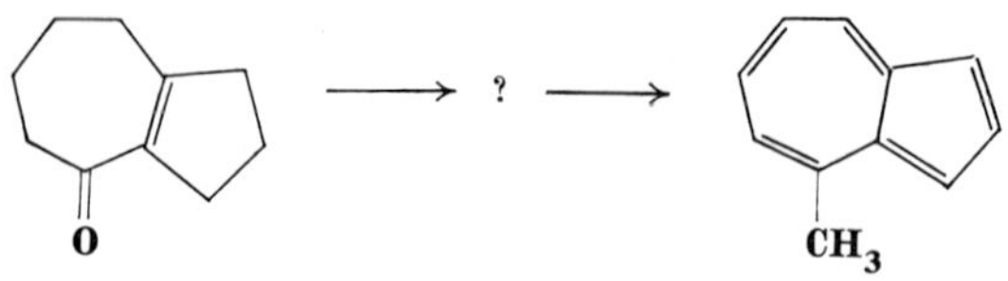

4-methylazulene

How would you carry out this transformation? Draw an important resonance contributor for the substituted azulene.

8. Estrone is a female hormone that is isolated from the urine of pregnant mares:

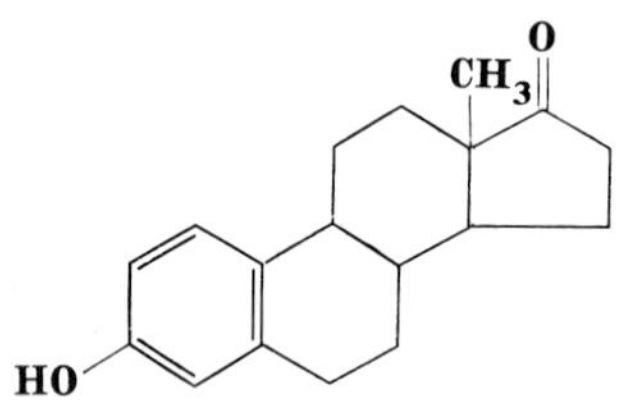

estrone

During the course of the total synthesis of estrone, the following reaction is carried out:

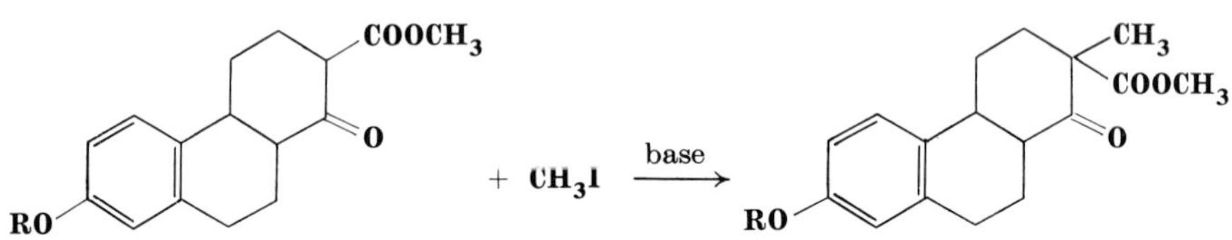

Propose a mechanism for this reaction.

9. Amphetamine is a central nervous system stimulant which is used as an antidepressant. Synthesize this drug starting with phenylacetone.

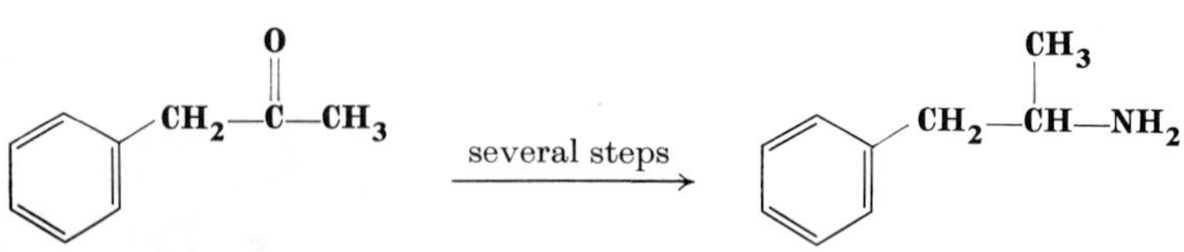

phenylacetone amphetamine

10. A weak nucleophile such as water does not generally add to a ketone or aldehyde at pH = 7. Trichloroacetaldehyde is an exception; in a neutral aqueous solution, this compound exists almost entirely as the addition product. The substance so formed is called chloral hydrate, a well-known hypnotic drug. Explain why water adds so readily to the chlorine-substituted aldehyde.

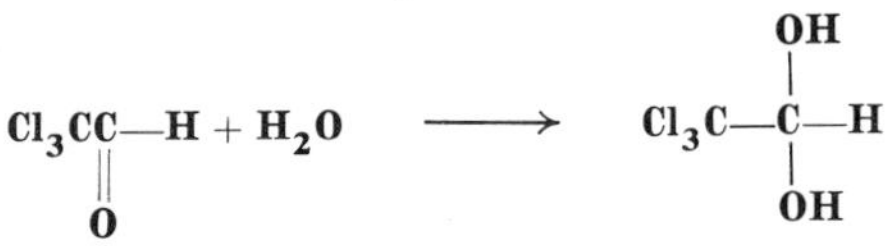

11. Starting with any alkyl halide of three carbons or less, formaldehyde, and any inorganic reagents, synthesize:

a) $CH_3CH_2CH_2CH_2OCH_3$

b) 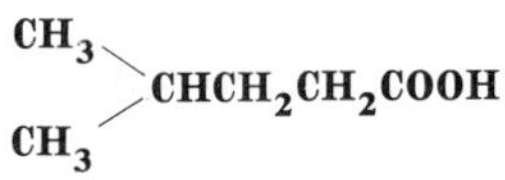

12. By-products are compounds which result from side reactions and which detract from the yield of the desired products. Suggest possible by-products for the following reactions.

a) 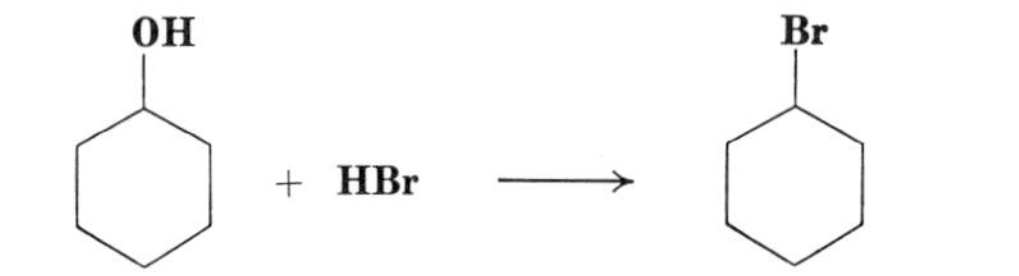

cyclohexanol   cyclohexyl bromide

b) 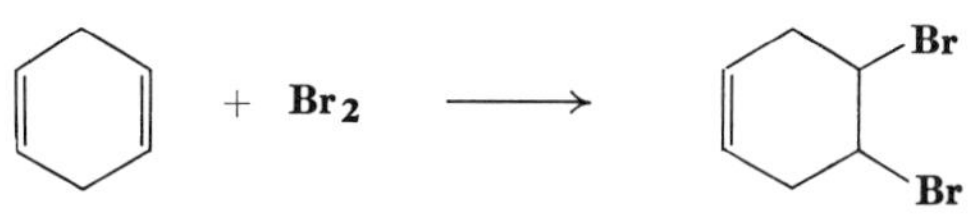

1,4-cyclohexadiene   4,5-dibromocyclohexene

c) 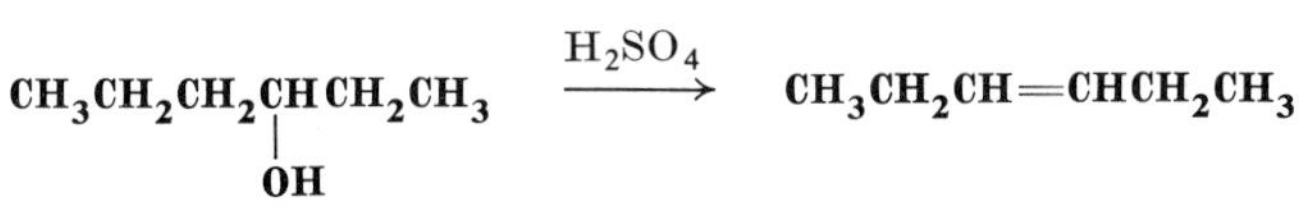

3-hexanol   3-hexene

d) 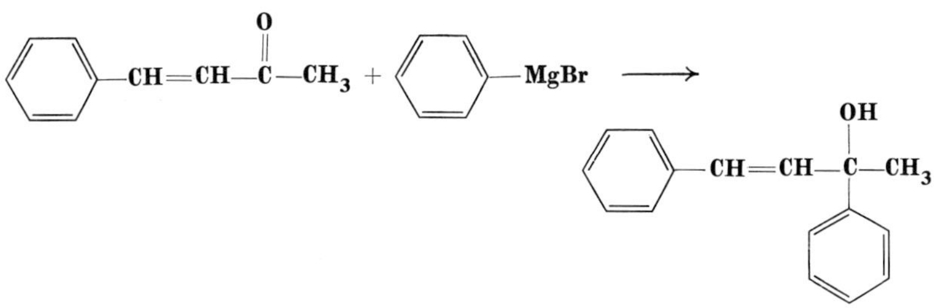

13. Ethyl ether can be cleaved by heating with hydroiodic acid. Suggest a mechanism for this reaction.

$$\underset{\text{ethyl ether}}{CH_3CH_2OCH_2CH_3} + HI \longrightarrow \underset{\text{ethyl iodide}}{CH_3CH_2I} + \underset{\text{ethanol}}{CH_3CH_2OH}$$

14. You have in your laboratory only organic compounds of two carbons or less. Using these compounds in conjunction with any inorganic reagents propose syntheses for:

a) 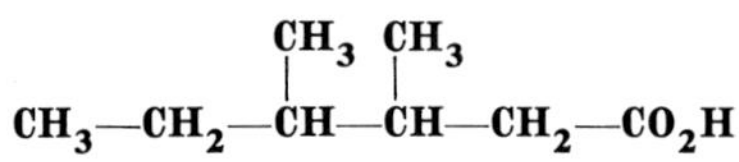

b) 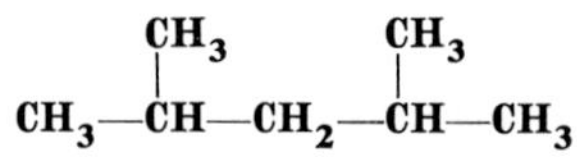

c) $CH_3—CH_2—CH{=}CH—CH_3$

4
EQUILIBRIA AND REACTION RATES

4.1 EQUILIBRIA AND REACTION RATES

Organic compounds undergo a wide range of reactions. Bond breakage and formation in these reactions are usually explained within the framework of a number of physical chemical principles. The principles that we shall be particularly concerned with are those that govern equilibria and reaction rates (kinetics).

4.2 EQUILIBRIUM

The term *equilibrium* is used in connection with reversible reactions:

$$\mathrm{A} \underset{k_2}{\overset{k_1}{\rightleftharpoons}} \mathrm{B}$$

The rate at which A goes to B is proportional to the concentration of A:

$$\text{rate (A to B)} = k_1[\mathrm{A}]$$

The proportionality constant, k_1, is called a *rate constant*. Similarly, the rate at which B goes to A is given by

$$\text{rate (B to A)} = k_2[\mathrm{B}]$$

When the rate at which A forms B equals the rate at which B forms A, the system is said to be in equilibrium. At equilibrium the concentration of A and B does not change with time; molecules of A still react to form B, and vice versa, but there is no net loss or gain in the concentration of either species. Since at equilibrium the forward and reverse rates are equal,

$$k_1[\mathrm{A}]_{\mathrm{eq}} = k_2[\mathrm{B}]_{\mathrm{eq}}$$

The equilibrium constant (K) is defined as the quotient of the concentrations of products and reactants at equilibrium and is equal to k_1 divided by k_2:

$$K = \frac{[\mathrm{B}]_{\mathrm{eq}}}{[\mathrm{A}]_{\mathrm{eq}}} = \frac{k_1}{k_2}$$

Whenever the rate constants for the forward and reverse reactions are very small, it takes a long time to secure equilibrium concentrations (starting with pure A). In this chapter, however, we are primarily concerned with acid-base equilibria which involve transfers of labile protons. Since proton transfers are often very fast reactions, these equilibria are established immediately. Thus, the equilibrium concentrations of acetic acid, acetate anion, and hydronium ion are formed as soon as acetic acid is mixed with water:

$$CH_3COOH \rightleftharpoons CH_3COO^- + H_3O^+$$

In the case of the acetic acid equilibrium, most of the material is in the form of acetic acid rather than acetate (that is, K is a very small number). Although both the forward and reverse proton transfer reactions are very fast, the reverse reaction is even faster than the forward one, and the equilibrium lies far to the left.

If A and B are in equilibrium, then the concentration of each depends on their relative stabilities. That is, the more stable (less energetic) material will be present in the higher concentration.

We may define relative stability as the difference in the *standard free energy* (ΔG°) between the product and reactant molecules. This free energy difference is related to the equilibrium constant K by the expression

$$\Delta G^\circ = -RT \ln K$$

where R is the gas constant and T is the absolute temperature. Thus, the greater the difference in the negative standard free energy of the reactants and the products, the larger the value of K, and the further the reaction proceeds to the right, or to "completion." Table 4.1 illustrates the quantita-

Table 4.1 The relationship between the percentage of the more stable isomer at equilibrium, the equilibrium constant, K, and the free-energy difference at 25°C for the equilibrium A$\rightleftharpoons$B

More stable Isomer, %	K	ΔG_{25°, kcal/mole
50.00	1.00	0.00
60.00	1.50	0.24
70.00	2.33	0.50
80.00	4.00	0.82
90.00	9.00	1.30
95.00	19.00	1.74
98.00	49.00	2.31
99.00	99.00	2.72
99.90	999.00	4.09
99.99	9999.00	5.46

tive relationship between ΔG° and the equilibrium constant K. When the energy difference between A and B is zero, $K = 1.00$ and there are equal amounts of A and B at equilibrium. If B is more stable than A by 1.30 kcal/mole, about 90% of the material will be in the form of B.

The *energy diagram* for the reaction A $\rightleftharpoons$ B in Fig. 4.1 provides a useful means of representing the equilibrium relationship. The ordinate represents the standard free energy of the system; the abscissa represents the progress of the reaction. Energy diagrams will be discussed in more detail later in the chapter, but for now we need only recognize that B predominates at equilibrium because B is of lower energy than A. Relative stability alone dictates the composition of a system at equilibrium.

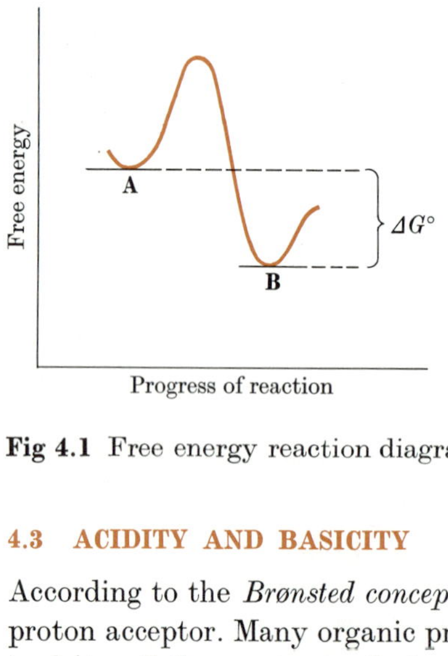

Fig 4.1 Free energy reaction diagram

4.3 ACIDITY AND BASICITY

According to the *Brønsted concept*, an acid is a proton donor and a base is a proton acceptor. Many organic processes occur as the result of the acidity or basicity of the reactants. It is useful, therefore, to consider the types of structures which give rise to acid-base properties.

4.4 ACIDITY

The *acidity* of a substance is measured by the position of an equilibrium:

$$\mathbf{HA} + \mathbf{B} \rightleftharpoons \mathbf{HB} + \mathbf{A}$$

acid 1 base 1 acid 2 base 2

In water, base 1 is H_2O and acid 2 is H_3O^+. To estimate relative acid strengths in water, therefore, we must consider the equilibrium:

$$\mathbf{HA} + \mathbf{H_2O} \rightleftharpoons \mathbf{H_3O^+} + \mathbf{A^-}$$

The equilibrium constant (K) for this process is a measure of the acidity of HA, since the larger the value of K, the larger the concentration of hydronium ion. When a strong acid like HCl is added to water, all the HCl is converted into hydronium ion and chloride. Hydronium ion is the strongest possible acidic species in an aqueous solution. Organic acids, such as acetic acid, are weak proton donors compared with HCl, as indicated by a K of $1.8 \times 10^{-5}\ M$ for acetic acid. Since the values of K for organic acids are usually small and unwieldy numbers, equilibrium constants are often used in the form of their negative logarithms (pK_a).

$$pK_a = -\log(K)$$

The pK_a for acetic acid, for example, is 4.76. The *smaller* the pK_a value, the *stronger* the acid. It is convenient to think of the pK_a as that pH at which an acid is half ionized, half un-ionized. At pH = 4.76, acetic acid is 50% acetate anion. At pH = 3.76 there is only about 10% acetate, and when the pH is raised to 5.76 by the addition of a strong base, there is 90% acetate. At pH = 10, essentially all of the acetic acid is in the form of acetate anion.

4.5 ORGANIC ACIDS

The most common class of organic acids, carboxylic acids, are characterized by the presence of the carboxyl group:

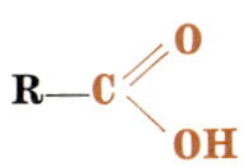

(also written **R—CO$_2$H** or **RCOOH**)

Some examples of carboxylic acids are shown below, and typical pK_a values are listed in Table 4.2.

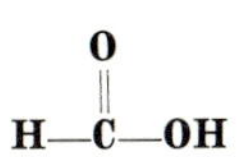

formic acid

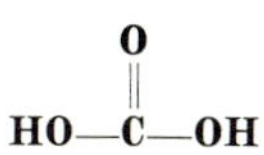

carbonic acid

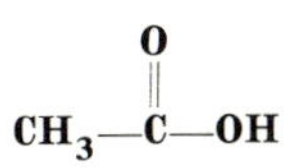

acetic acid

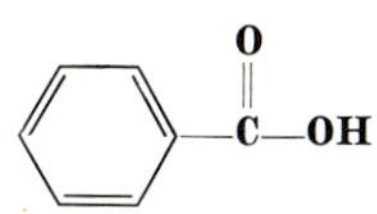

benzoic acid

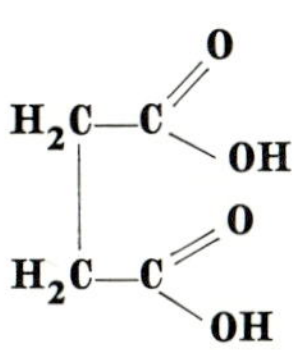

succinic acid

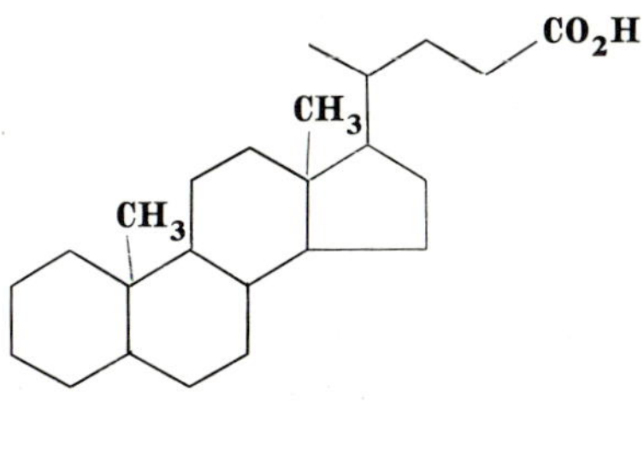

cholanic acid

Table 4.2 **pK_a values for carboxylic acids**

Aliphatic acids	pK_a	Aromatic acids	pK_a
HCO_2H	3.75	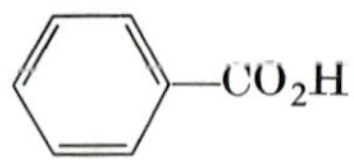	4.17
CH_3CO_2H	4.76	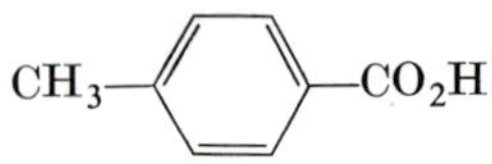	4.37
$ClCH_2CO_2H$	2.87	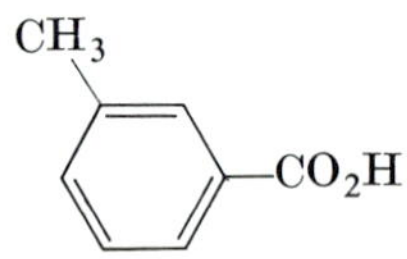	4.27
Cl_2CHCO_2H	1.25	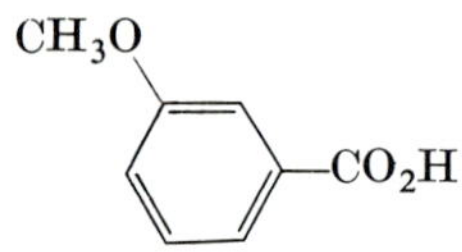	4.09
Cl_3CCO_2H	0.66	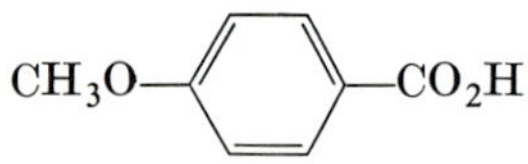	4.47
$O_2NCH_2CO_2H$	1.68	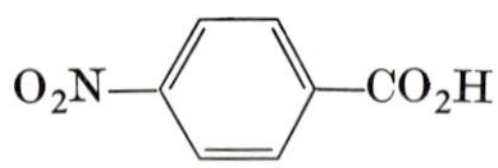	3.43
$CH_3C(=O)—CH_2CO_2H$	3.58	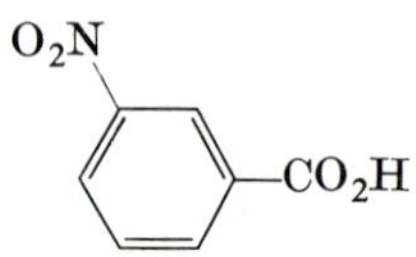	3.49
$CH_3CH_2CO_2H$	4.87	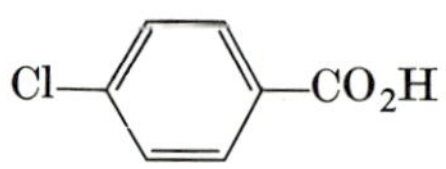	3.98
$CH_3(CH_2)_6CO_2H$	4.89	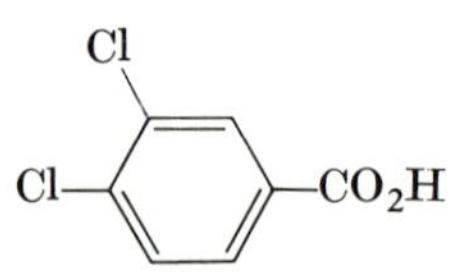	3.83

4.6 RELATIVE ACIDITY

The relative acidity of organic acids may be qualitatively predicted by examining their structures and their conjugate bases. Let us compare, for example, a carboxylic acid, acetic acid, with another potentially acidic molecule, ethanol. Both of these molecules may donate a proton to water:

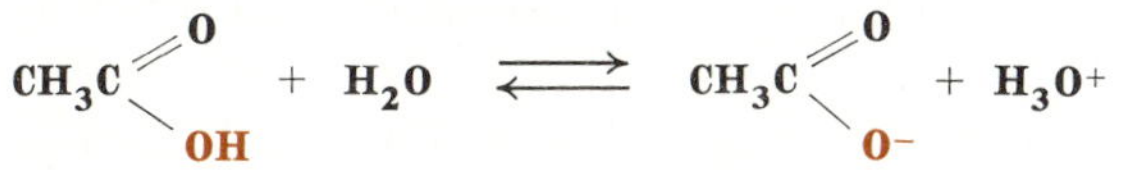

$$CH_3CH_2OH + H_2O \rightleftharpoons CH_3CH_2O^- + H_3O^+$$

The extent to which these equilibria lie to the right depends on the relative differences in energy between the acids and their corresponding bases. Acetic acid is a resonance-stabilized molecule; that is, acetic acid is a resonance hybrid whose two extremes are shown below:

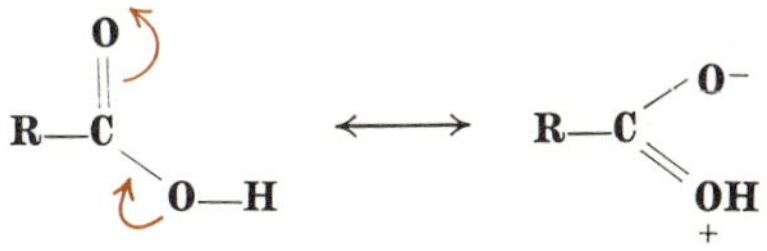

The resonance contribution of the second structure will be small, however, since the resonance interaction requires the separation of charge (Section 2.3). Acetate anion, on the other hand, is stabilized to a considerable extent by resonance since there are two equivalent low-energy contributors:

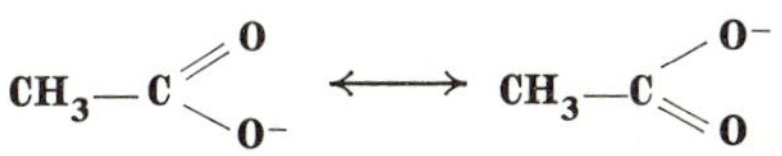

Since acetate anion is appreciably stabilized, the energy difference between acetic acid and acetate is much smaller than that between ethanol and ethoxide (Fig. 4.2) where similar stabilization is not found. We may conclude that

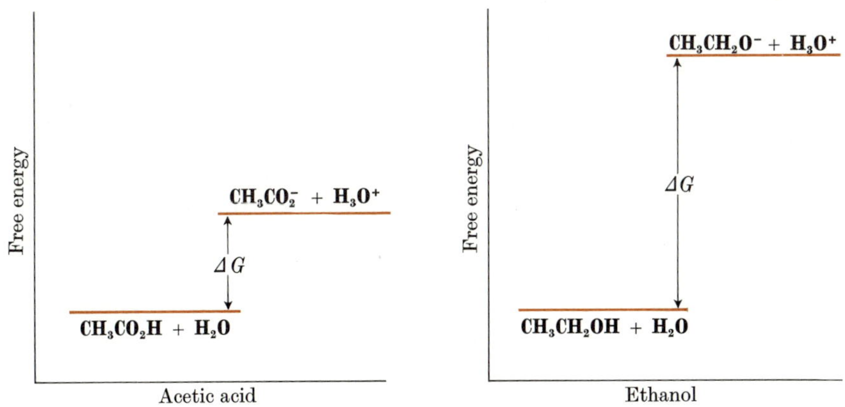

Note: free energy levels of acetic and ethanol are arbitrarily made the same.

Fig. 4.2 Free energy differences in two equilibria

acetic acid should be a much stronger acid than ethanol, and this is found to be the case. Acetic acid has a pK_a of 4.76, whereas the pK_a of ethanol is about 16. This means that the equilibrium constants for the two ionizations differ by more than eleven powers of 10.

The *conjugate base* of HA is A^- and the *conjugate acid* of A^- is HA. There is an inverse relationship between the acid strength of HA and the base strength of A^-. The stronger the acid HA, the weaker the base A^-. Thus, the conjugate base of HCl, namely chloride ion, is an extremely weak base. Since acetic acid is a much stronger acid than ethanol, ethoxide is a much stronger base than acetate.

4.7 SUBSTITUENT EFFECTS

The acid strength of RCOOH depends greatly on the nature of the R group. In general, any R group that withdraws electrons has an acid-strengthening effect (it lowers the pK_a). For example, chloroacetic acid is stronger than acetic acid (see Table 4.2). The electronegative chlorine withdraws electrons, making it easier for the carboxylate anion to accommodate a negative charge.

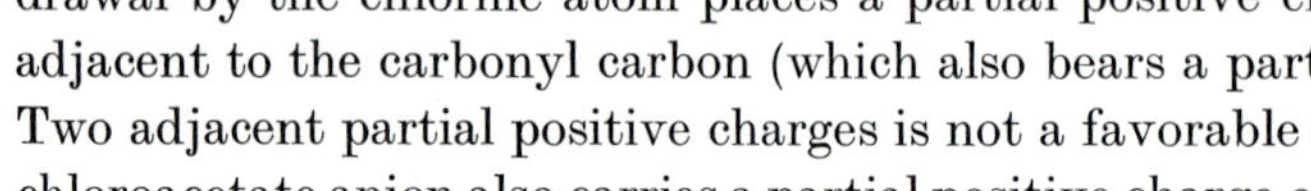

Actually, this explanation is an oversimplification, because in an equilibrium process it is necessary to consider how the substituent (or any other factor) affects *both* sides of the equilibrium. In chloroacetic acid, inductive withdrawal by the chlorine atom places a partial positive charge on the carbon adjacent to the carbonyl carbon (which also bears a partial positive charge). Two adjacent partial positive charges is not a favorable situation. While the chloroacetate anion also carries a partial positive charge on the carbon bonded to the chlorine, the carbonyl carbon bears almost no positive charge because a carboxylate is negatively charged and electron-rich. The unfavorable electrostatic interaction between the two carbons of the conjugate acid is thus relieved upon ionization of the acid. Chloroacetic acid is therefore stronger than acetic acid where inductive destabilization of the acid form is not present.

Table 4.2 shows that other electronegative substituents attached to a carboxyl group (such as Cl_2CH—, Cl_3C—, CH_3COCH_2—) have similar acid-strengthening effects.

Inductive effects of substituents fall off rapidly with distance. Thus, γ-chlorobutyric acid ($ClCH_2CH_2CH_2COOH$) is only twice as strong as butyric acid, while the α-chloro compound ($CH_3CH_2CHClCOOH$) is about 100 times as strong.

4.8 AROMATIC ACIDS

Table 4.2 lists pK_a values for a number of aromatic carboxylic acids. The simplest member of this series, benzoic acid (C_6H_5COOH), is stronger than acetic acid. This is because an sp^2 carbon exerts greater attraction on electrons than an sp^3 carbon. An sp^2 orbital, with $33\frac{1}{3}$% s character, is smaller (closer to the positive nucleus) than an sp^3 orbital with only 25% s character.

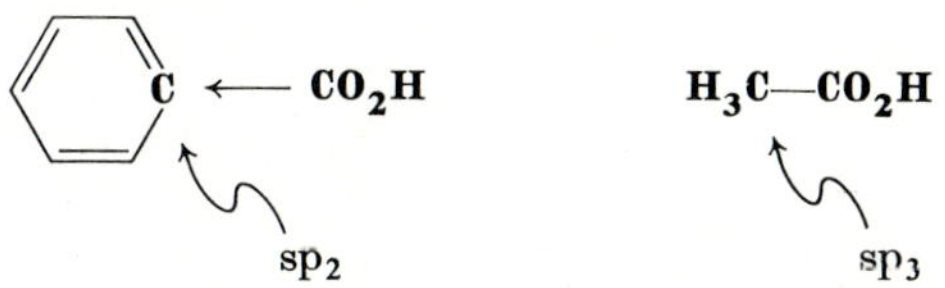

Before discussing the properties of substituted aromatic acids, we must become familiar with the terminology used to describe substituted aromatic rings. Figure 4.3 shows how the positions of substituents on a benzene ring are related to a reference position. These positions are either assigned numbers (with the reference substituent at position 1) or given the names *ortho* (*o*-), *meta* (*m*-), or *para* (*p*-).

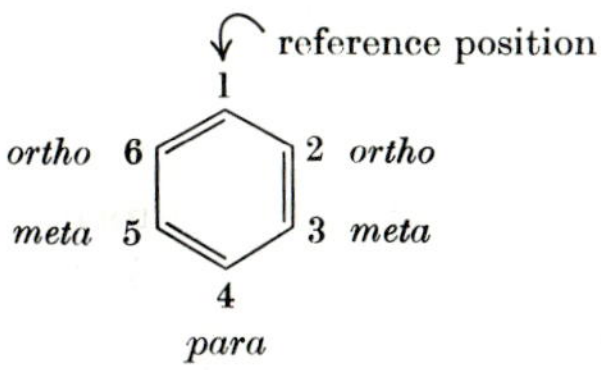

Figure 4.3

Therefore the three isomeric aromatic acids shown in Fig. 4.4 are called *o*-methoxybenzoic acid (2-methoxybenzoic acid), *m*-methoxybenzoic acid (3-methoxybenzoic acid), and *p*-methoxybenzoic acid (4-methoxybenzoic acid).

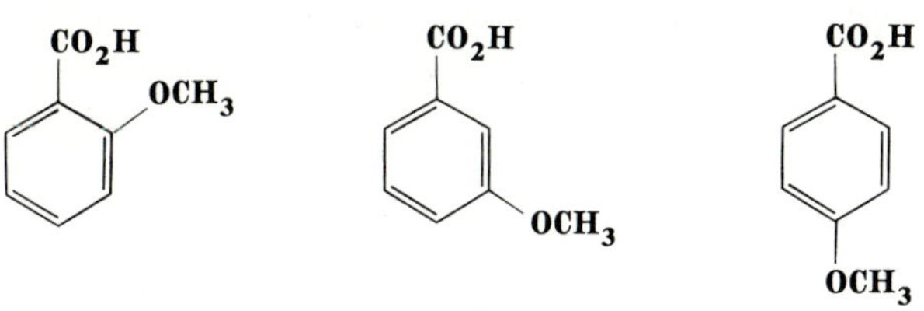

Figure 4.4

To compare the acidities of substituted aromatic acids, resonance and inductive effects have to be taken into account. For example, *p*-methoxybenzoic acid is weaker than benzoic acid primarily because of resonance.

Electron donation by the oxygen of the methoxy group (Fig. 4.5) stabilizes the acid more than its conjugate base, because an un-ionized carboxyl group has a greater capacity to attract electrons than a carboxylate anion.

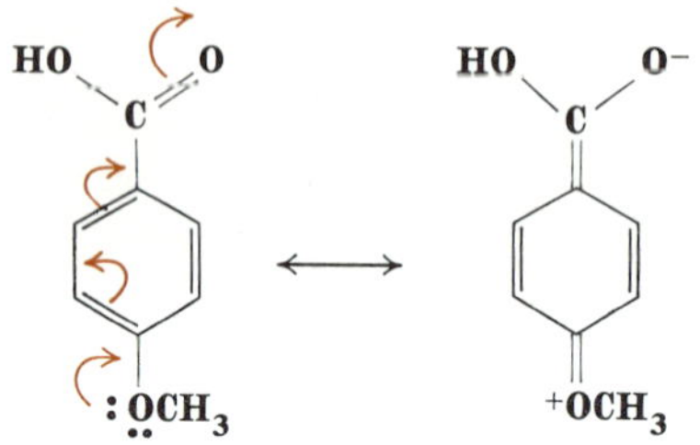

Fig. 4.5 Resonance in *p*-methoxybenzoic acid

Resonance stabilization of the acid shifts the ionization equilibrium to the left. In contrast, *m*-methoxybenzoic acid is stronger than benzoic acid because of the inductive withdrawal by the electronegative methoxyl oxygen. (Fig. 4.6).

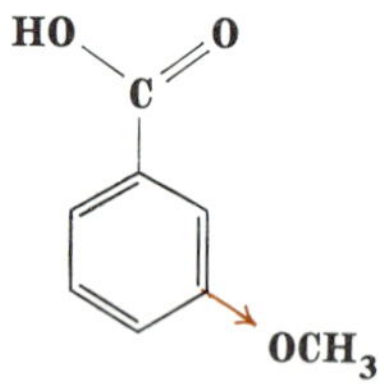

Figure 4.6

In the *meta* compound there can be no direct donation of electrons to the carboxyl group and the effect of resonance in this case is negligible (Fig. 4.7).

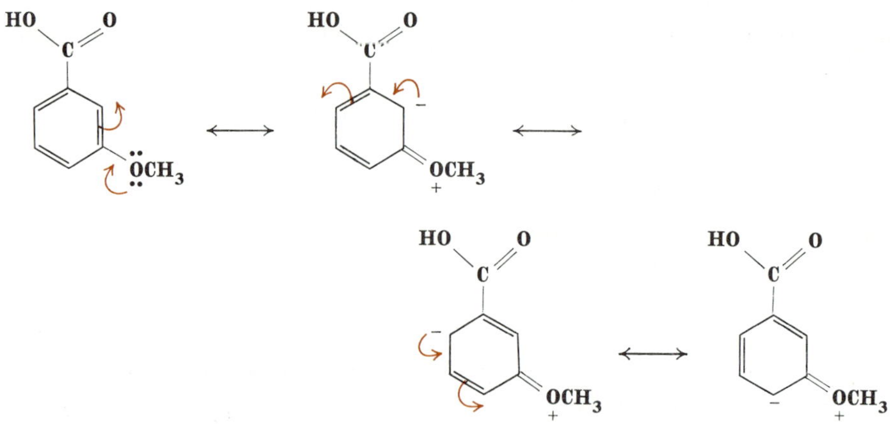

Figure 4.7

It is important to realize that the above rationale was established after the fact. In the *para*-substituted compound, resonance factors tend to decrease the acidity, whereas inductive effects increase it. Since it is known experimentally that *p*-methoxybenzoic acid is weaker than benzoic acid, we conclude that the resonance effect must predominate.

With *ortho*-substituted benzoic acids, a third factor is involved. An *ortho* substituent is so near the reactive site that it perturbs the manner in which the carboxyl and carboxylate groups are solvated by water. Solvation is an important, but poorly understood, aspect of chemical reactivity, and it is often difficult to rationalize the behavior of *ortho*-substituted compounds. The situation is further complicated by steric interactions between *ortho* substituents.

4.9 PHENOLS

A number of other classes of organic compounds also exhibit acidic properties. One such class is the phenols. The parent compound, phenol, has a pK_a of 9.98; at pH = 9.98 phenol is half ionized (Fig. 4.8).

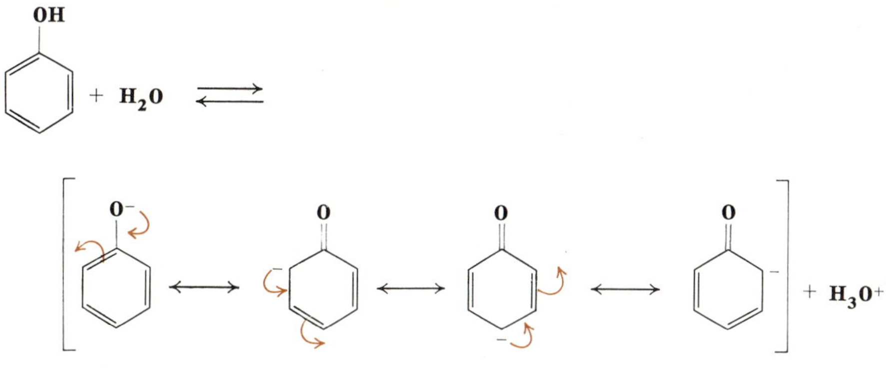

Figure 4.8

We noted earlier that ethanol (pK_a = 16) and other aliphatic alcohols are very weak acids. We can see that phenol, an aromatic alcohol, ought to be a stronger acid than an aliphatic alcohol since the conjugate base (phenoxide ion) is resonance stabilized. Of course, it is the *difference* in free energy between the acid and its conjugate base that determines acidity, and the phenol is also resonance stabilized. However, the stabilization of the phenol should be small compared to that of the anion because the former requires charge separation (Fig. 4.9).

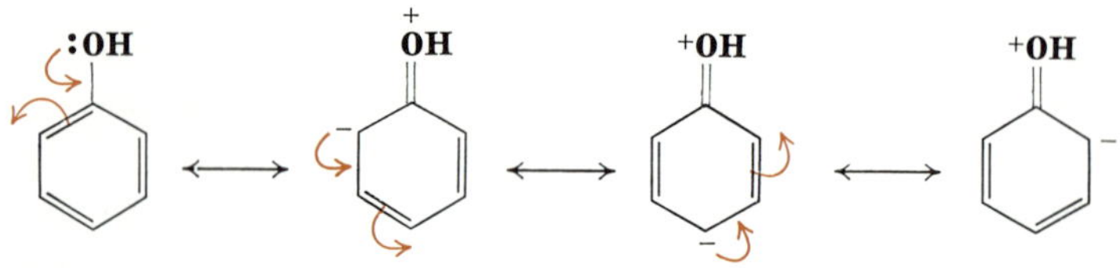

Figure 4.9

A comparison of phenol and ethanol in an energy diagram illustrates nicely that the energy difference between phenol and phenoxide is less than that between ethanol and ethoxide (Fig. 4.10).

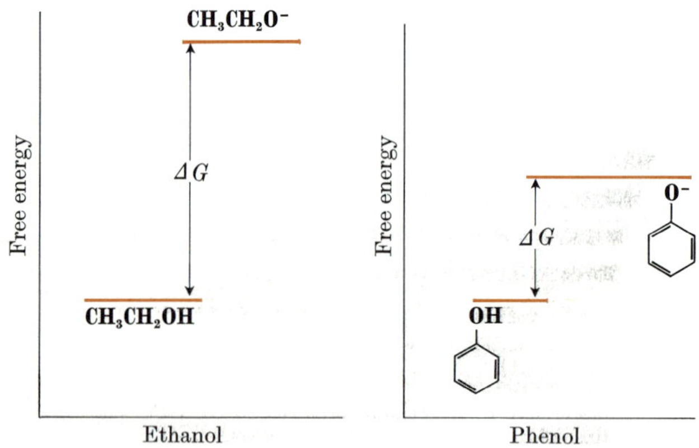

Note: free energy levels of ethanol and phenol are arbitrarily made the same.

Fig. 4.10 Free energy differences in two equilibria

The effect of substituents on the acidity of phenols is similar to their effect on carboxylic acids. The pK_a values of some typical phenols are given in Table 4.3. We might note particularly the acid strength of 2,4,6-trinitrophenol (picric acid). Because of the ability of the nitro groups to delocalize anionic charge, this compound is a strong acid (Fig. 4.11). Figure 4.11

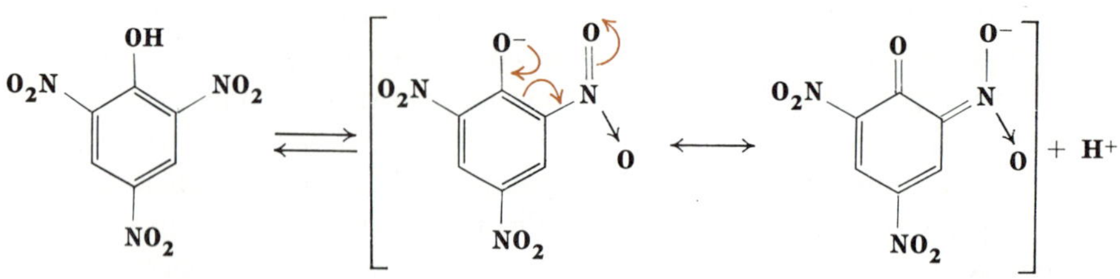

Fig. 4.11 Resonance in picrate anion

Table 4.3 pK_a values for phenols

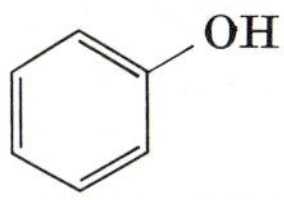	9.98
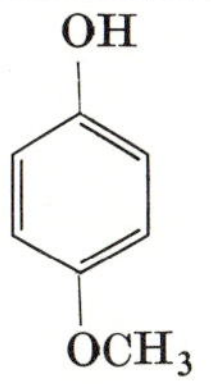	10.21
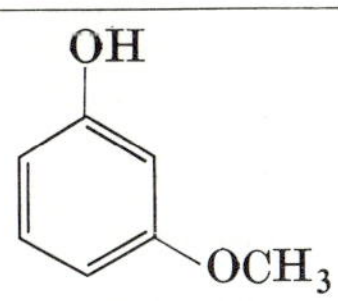	9.65
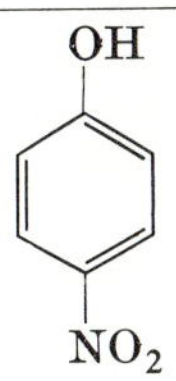	7.15
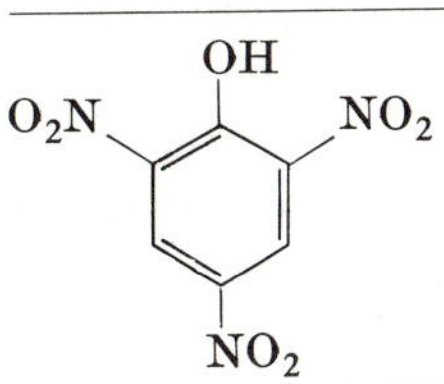	0.71

illustrates only two of the many possible resonance structures for the conjugate base of picric acid.

4.10 CARBON ACIDS

The two types of acids discussed thus far can be classified as oxygen acids since a hydroxyl group donates the proton. There are, in addition, a number of acids in which a carbon-hydrogen bond is broken. Many of these are of great importance in synthetic organic chemistry. In general, these are extremely weak acids, but in the presence of a strong base they may be partially converted into their conjugate bases. Proton transfers involving carbon acids, unlike those involving oxygen acids, are often slow reactions.

4.11 KETONES AND ALDEHYDES

Ketones and aldehydes, with a hydrogen-bearing carbon adjacent to the carbonyl group, are very weak carbon acids. Acetone, for example, ionizes in an aqueous base with an estimated pK_a of 20 (Fig. 4.12).

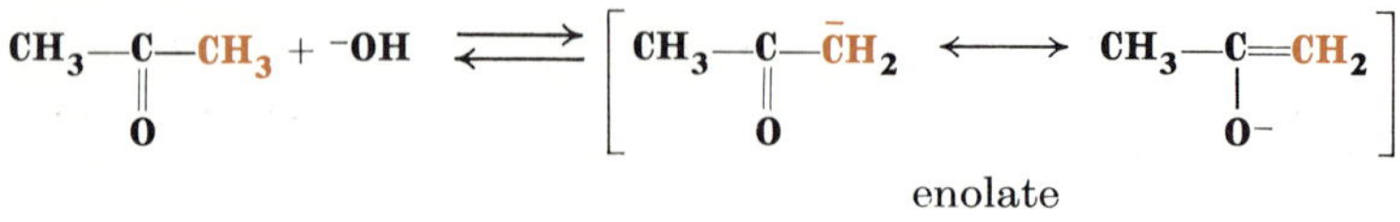

Figure 4.12

Even at very high pH values, only a small fraction of the acetone is in the conjugate base form. It is, of course, impossible to make an aqueous solution of pH = 20 in which acetone would be half ionized. At pH = 14, only one molecule in a million is anionic. The conjugate base of acetone, called an *enolate anion*, is a resonance hybrid. Once again, the possibility of resonance delocalization of the anionic charge, resulting from the loss of a proton, gives these compounds their acidic character. The major contributor to the hybrid is undoubtedly the form that has the negative charge on the oxygen, since oxygen is more electronegative than carbon and is better able to support the charge.

Since the acidity of the α-proton of a ketone (that is, the proton on a carbon atom adjacent to the carbonyl) is largely a function of the resonance stabilization of the anion, any additional delocalizing groups should increase the acidity. Thus, for example, ethyl acetoacetate (pK_a = 10.7) readily forms an enolate in the presence of a base such as ethoxide ion (Fig. 4.13). This carbon acid is almost as strong as phenol.

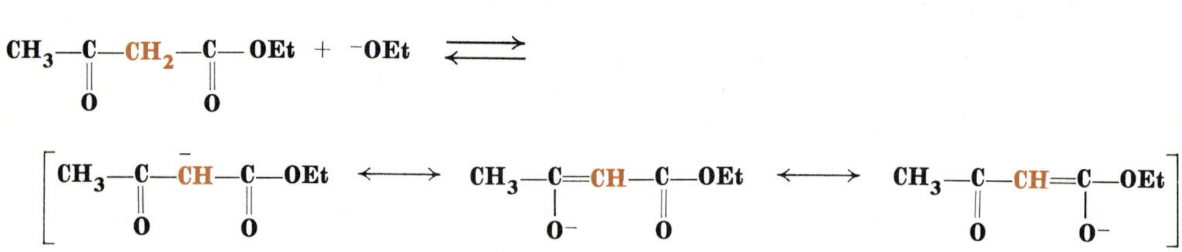

Fig. 4.13 Resonance in acetoacetate anion

4.12 TAUTOMERISM

In acidic solutions ketones and aldehydes engage in another type of equilibrium involving a species called an *enol* (Fig. 4.14):

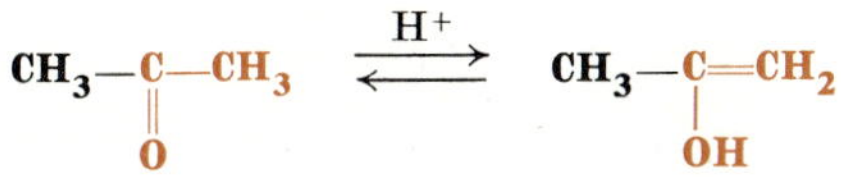

enol

Fig. 4.14 Keto-enol equilibrium

A ketone and its corresponding enol are called *tautomers*, a name given to equilibrating isomers which differ only in the position of a mobile proton. The position of a tautomeric equilibrium depends on the relative energies of the participating species. The enol form generally predominates with phenols, whereas the keto form is much more stable in the case of ketones (Fig. 4.15).

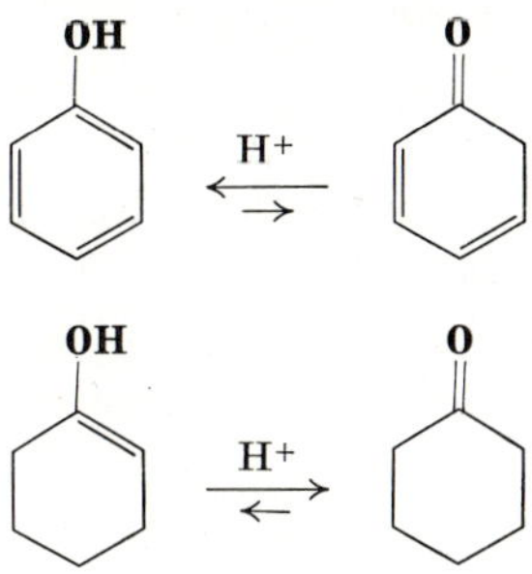

Fig. 4.15 Keto-enol equilibria

4.13 ACETYLENE

Acetylene (HC≡CH) is another carbon acid. It is such a weak acid ($pK_a = 25$) that only an extremely strong base such as *amide anion* (NH_2^-) can successfully remove a proton:

$$\underset{}{HC{\equiv}CH} + \underset{\text{amide anion}}{NH_2^-} \longrightarrow \underset{\text{acetylide anion}}{H{-}C{\equiv}C^-} + NH_3$$

The acidity of acetylene cannot be ascribed to resonance stabilization of acetylide anion. A reasonable explanation for the acidity is that the sp orbital of the acetylide anion has 50% s character, which means that the electrons are closer to the nucleus and hence more stable than they would be in a larger sp^2 or sp^3 orbital.

4.14 LEWIS ACIDS

The acids we have discussed thus far can be classified as Brønsted acids (proton donors). Acids can be defined more generally as species which accept electrons.

This is the Lewis Theory, and any electron acceptor is called a *Lewis acid.* A proton is a Lewis acid, for in the reaction of this cation with a base such as water, the proton accepts a pair of electrons to form a coordinate-covalent bond:

$$H-\ddot{O}(-H): + H^+ \longrightarrow H-\ddot{O}^{+}(-H)-H$$

Lewis acids are those species that contain an atom with an unfilled valence shell. Ferric chloride and aluminum chloride are typical examples of Lewis acids which are not Brønsted acids. The reactions of ferric chloride and aluminum chloride with an electron donor (water) are shown below:

$$H_2O + FeCl_3 \longrightarrow H_2\overset{+}{O}-\overset{-}{Fe}Cl_3$$

$$H_2O + AlCl_3 \longrightarrow H_2\overset{+}{O}-\overset{-}{Al}Cl_3$$

Lewis acids such as aluminum chloride are important catalysts in organic chemistry.

4.15 BASES

Organic bases may be classified as Brønsted bases (proton acceptors) or, more generally, as Lewis bases (electron donors). Thus a base B has an unshared pair of electrons which it may donate to an acid A with an empty orbital (Fig. 4.16).

Fig. 4.16 Acid-base equilibrium

4.16 AMINES

The most common organic bases are the amines. These compounds are ammonia derivatives in which one or more of the hydrogen atoms have been replaced by a carbon or other substituent. In the simplest amine, methylamine, one hydrogen is replaced by a methyl group (Fig. 4.17).

Figure 4.17

Amines owe their basicity, as does ammonia itself, to the presence of an unshared pair of electrons. These electrons may be donated to the empty orbital of a proton to form an ammonium salt. Methylamine reacts with a proton in water to form methylammonium ion (Fig. 4.18).

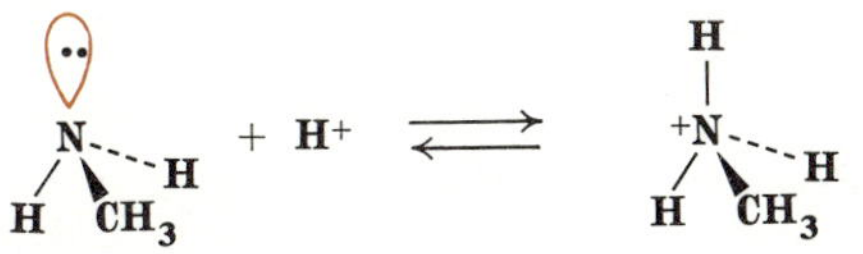

Figure 4.18

Methylamine is a stronger base than ammonia because alkyl groups are weak electron donors, and therefore help stabilize positive charge on a neighboring atom:

$$CH_3 \longrightarrow NH_3^+$$

The strength of organic bases is best discussed in terms of an equilibrium between the conjugate acid and the free base:

$$HB^+ \rightleftharpoons H^+ + B$$

Protonated amines, HB^+, are weak acids with characteristic pK_a values (Table 4.4, page 72).
Since a *large* pK_a value for HB^+ means that HB^+ is a *weak* acid, and since the conjugate base of a weak acid is a strong base, a large pK_a value indicates a strong base. Thus, dimethylamine (a secondary amine) is a slightly stronger base than methylamine (a primary amine).

Aniline, $C_6H_5NH_2$, is the simplest aromatic amine (p$K_a = 4.58$) and is a much weaker base than methylamine (p$K_a = 10.62$). The aniline molecule is a resonance hybrid to which the structures shown in Fig. 4.19 make a small but finite contribution.

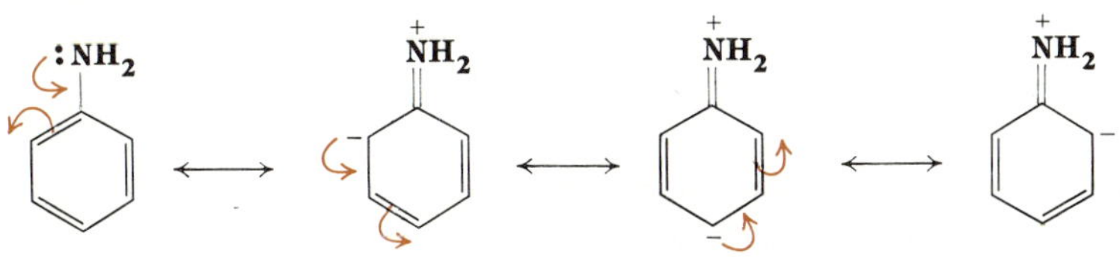

Figure 4.19

The formation of anilinium ion, by reaction of the base with a proton, results in the loss of resonance energy, since the nitrogen atom's pair of unshared electrons can no longer overlap with the aromatic π system. Methylamine, on the other hand, neither gains nor loses resonance stabilization in going from the free base to the conjugate acid.

Table 4.4 pK_a **values for conjugate acids of organic bases**

	pK_a		pK_a
CH_3NH_2	10.62		5.12
$(CH_3)_2NH$	10.77		1.02
$(CH_3)_3N$	9.80	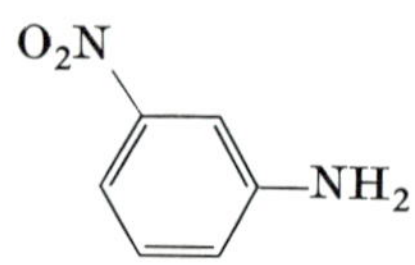	2.50
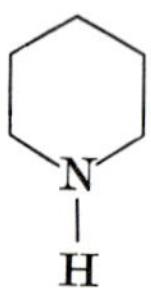	11.22	CH_3—C(=:O:)—CH_3	−7.20
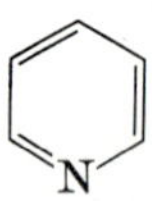	5.23	CH_3CH_2—Ö—CH_2CH_3	−3.59
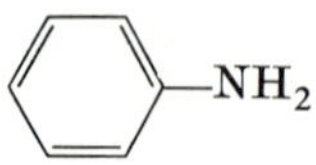	4.58		
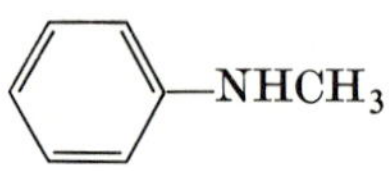	4.85		

4.17 OXYGEN BASES

In the following chapters we will examine a variety of reactions which result from oxygen basicity. The simplest oxygen base is water and, like nitrogen bases, it reacts by donating a pair of electrons to a proton.

$$H_2O + H^+ \longrightarrow H_3O^+$$

Any compound which has a neutral oxygen atom with an unshared pair of electrons is also a base, albeit a very weak one in most cases. Representative organic oxygen bases are ethers, alcohols, ketones, and carboxylic acids; their basic reactions are shown in Fig. 4.20. Note that the latter three classes of bases are *amphoteric* compounds (they have both acidic and basic properties). In ketones, ethers, and the other very weak oxygen bases, protonation takes place in only a very small percentage of the molecules, even in a strong acid such as 1*N* HCl. Nevertheless, as we shall see, the conjugate acids of the bases are important in terms of their chemical reactivity.

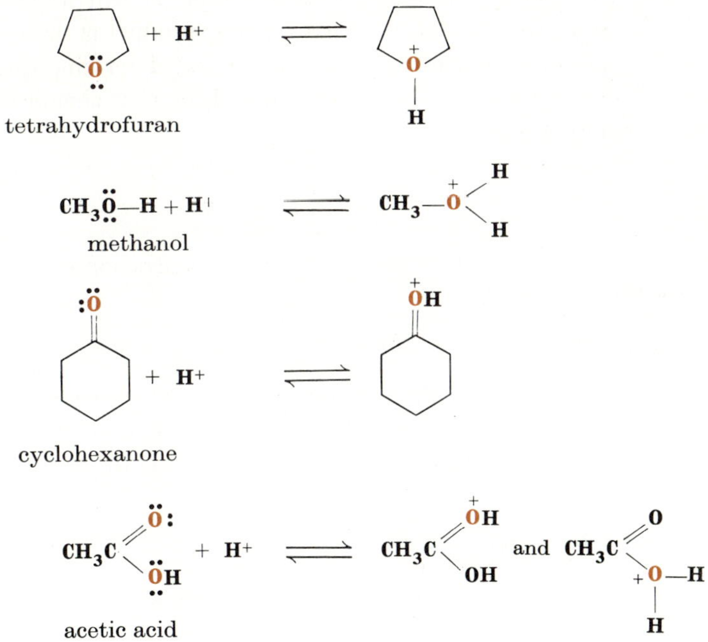

Fig. 4.20 Conjugate acids of ethers, alcohols, ketones, and acids

4.18 REACTION RATES

When we discussed equilibrium processes, we were concerned only with the energy differences of reactants and products. Our analysis of reversible reactions depended on some knowledge of the relative stability of the reactants and products, but was independent of the exact pathway by which they were interconverted. In theory, all reactions are reversible, but in fact many reactions are best treated as nonequilibrium, or irreversible, processes. Apparent irreversibility may result from a number of situations. For example, in the potentially reversible system

$$A + B \rightleftharpoons C + D$$

the reverse reaction may be so slow that even a high concentration of products C and D will not result in the re-forming of reactants A and B within a reasonable time period. The forward reaction, therefore, may be considered irreversible. Many reactions are irreversible by virtue of an insoluble or gaseous product. Since one of the products is removed from the system, the reverse reaction is precluded. Now, suppose that a molecule may follow either of two alternative pathways when treated with a certain reagent, and that these two pathways are irreversible. Clearly, the major reaction product will be the one formed the *fastest*. Since the reactions are irreversible, product stability is of no relevance. The products depend only on factors that determine the *rate* of conversion. The remainder of the chapter will deal with chemical kinetics, the study of reaction rates.

4.19 KINETICS

The rate of a chemical reaction is proportional to the concentration of the reacting species. For the reaction A $\rightarrow$ B,

rate of appearance of B $= k_1[\text{A}]$

where k_1 is a proportionality constant called the rate constant. This particular process is described as a *first-order* reaction because the formation of the product depends on the concentration of a single reactant raised to the first power. In the reaction A + B $\rightarrow$ C,

rate of appearance of C $= k_2[\text{A}][\text{B}]$

This is an example of an overall *second-order* reaction, first-order in A and first-order in B.

If the energy diagram for the reaction between A and B to give C were that shown in Fig. 4.21, the reaction would be instantaneous. The reactants would react as fast as they could diffuse together. Most organic reactions are

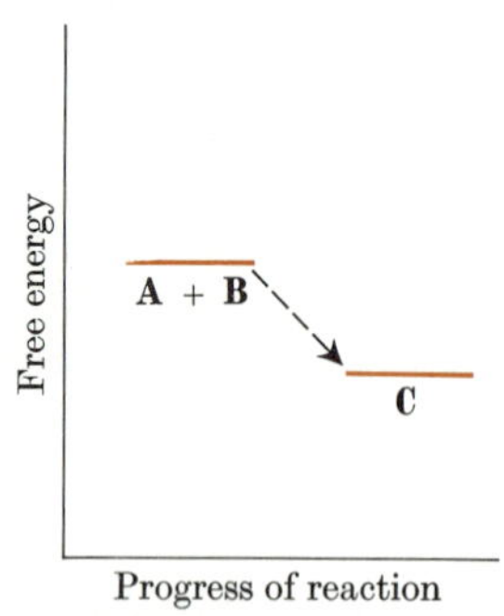

Fig. 4.21 Free energy diagram for reaction A + B $\rightarrow$ C

much slower than this because the reactants must pass an energy barrier before forming the product (Fig. 4.22). The abscissa of Fig. 4.22 represents the progress of the reaction (as measured, for example, by the distance between A and B which are forming a bond).

The rate at which A and B form C depends only on the energy difference between the reactants and the highest energy point of the energy diagram. The energy maximum is called the *transition state*, and the energy difference between the transition state and reactants (ΔG^*) is called the *activation energy*. The greater the activation energy, the slower the reaction (i.e., the smaller the rate constant).

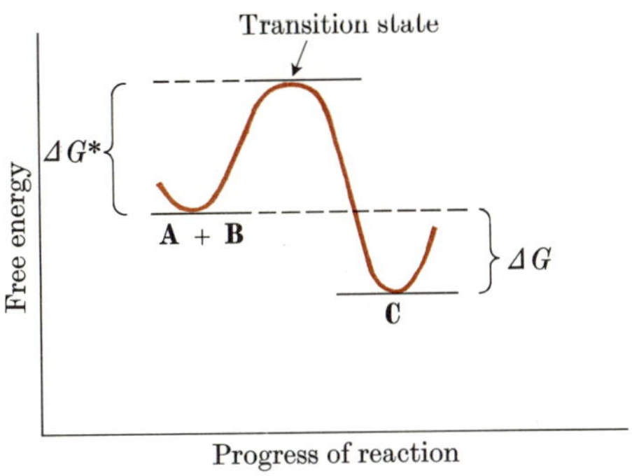

Fig. 4.22 Free energy changes during reaction A + B → C

If the energy of the products is less than the energy of the reactants, the reaction is *exothermic* (heat is released). If the reverse is true, the reaction is *endothermic* (heat is absorbed). The rate of reaction does not depend on the energy difference between reactants and products (ΔG in Fig. 4.22).

Every organic reaction involves the making and/or the breaking of chemical bonds. The transition state, which lies directly along the reaction path, is characterized by partially formed or broken bonds. In the reaction between X and Y to form X—Y, the transition state has a partial bond between X and Y. In other words, orbital overlap between X and Y is incomplete. Partial bonds are usually indicated by dashed lines:

X + Y	⟶	[X ---Y]	⟶	X—Y
reactants		transition state		products

A reaction in which one bond is being broken, while another is being made, could be depicted as:

A + B—X ⟶ **[A --- B ---X]** ⟶ **A—B + X**

Since a transition state is so unstable, and since it has such an extremely short lifetime, there is no way to determine its precise structure. The best we can do is guess what the transition state might look like. Consider again the reaction between A and B—X shown above. If the transition state looks like products (that is, if the transition state is near the products on the abscissa of the energy diagram), then there is appreciable bond formation between A and B, and appreciable bond breakage between B and X.

The extent to which bonds are made or broken in the transition state can be qualitatively correlated with the overall energy change of a reaction. A highly exothermic reaction usually has low activation energy with a transition state that lies only a short distance along the reaction path. Highly endothermic reactions are usually characterized by transition states that resemble the products.

4.20 INTERMEDIATES

Most organic reactions are not simple one-step, one-transition state processes. The reactions usually involve some type of *intermediate*. An intermediate is a species, often of high energy, which is formed during the course of a reaction, but which has all of its bonds intact. An intermediate is, therefore, different from a transition state which contains at least one partial bond. A process that involves a single intermediate may be viewed as a two-step reaction:

A → [intermediate] → B

Each step has a transition state (Fig. 4.23). The intermediate shown in Fig. 4.23 is of high energy, and it would have only a fleeting existence. An energy diagram with a somewhat more stable intermediate is shown in Fig. 4.24. Occasionally an intermediate is so stable that it can be isolated. On the other

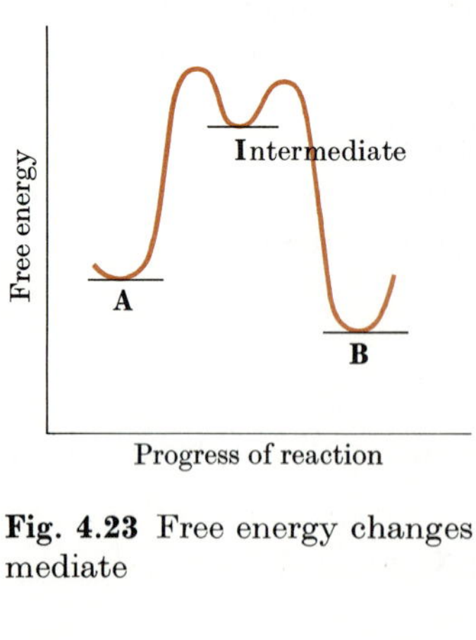

Fig. 4.23 Free energy changes for a reaction passing through an unstable intermediate

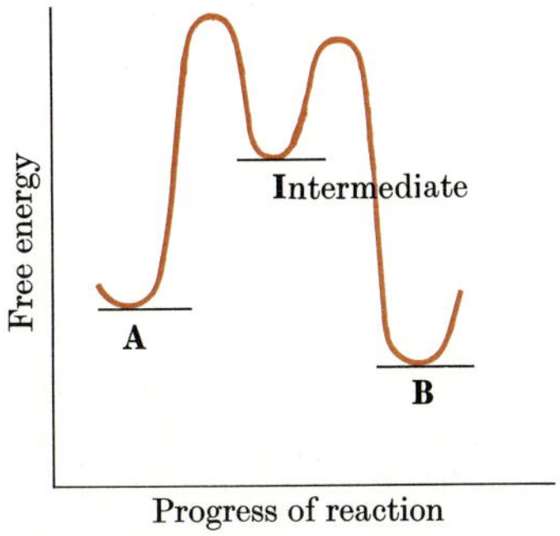

Fig. 4.24 Free energy changes for a reaction passing through a stable intermediate

hand, if the energy of the intermediate is very close to that of the two transition states, it may be difficult to prove the existence of the intermediate.

The halogenation of ketones is a classic example of a reaction involving an intermediate:

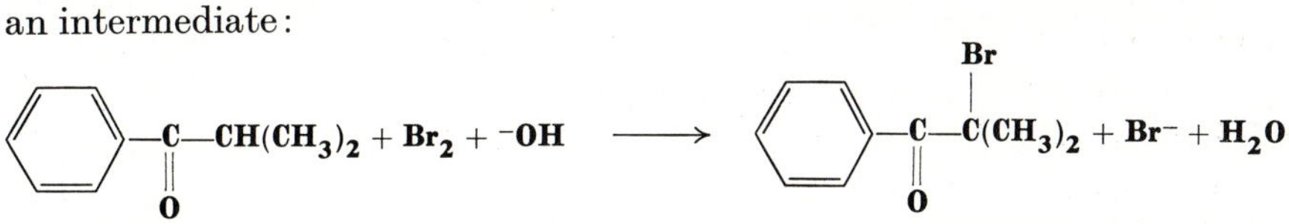

It was found that the rate of this reaction was dependent on the concentration of ketone and hydroxide ion, but was *independent* of the bromine concentration. The halogenation of a ketone takes place at the same rate, for example, whether the reaction is run in $1M$ or $2M$ Br_2.

rate of product formation $= k\,[\text{ketone}][\text{hydroxide}]$

The rate expression is best explained in terms of a mechanism whose first step is the reversible formation of an *enolate anion intermediate:*

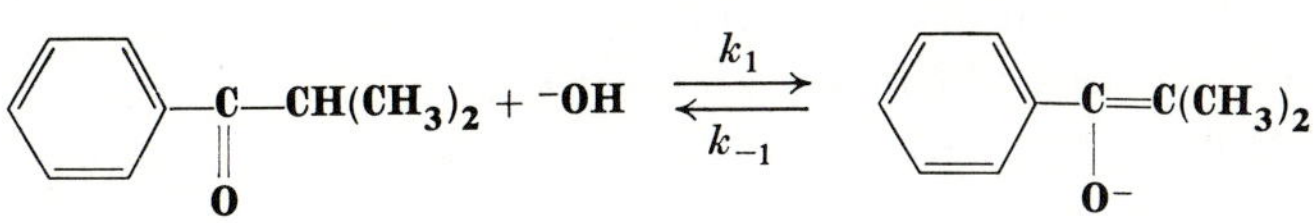

The enolate then reacts with Br_2 to form the products:

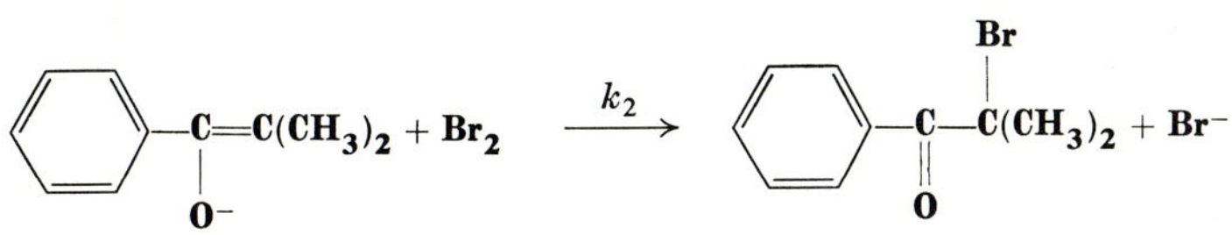

The rate of the reaction is independent of the bromine concentration as long as the second step is much faster than the reverse of the first step. This can be demonstrated by the following equations, where K, H, and E are ketone, hydroxide ion, and enolate, respectively:

$$\text{rate of product formation} = k_2[\mathrm{E}][\mathrm{Br}_2]$$

$$\text{rate of concentration change of E} = k_1[K][\mathrm{H}] - k_2[\mathrm{E}][\mathrm{Br}_2] - k_{-1}[\mathrm{E}]$$

Note that the second equation has three terms to the right of the equal sign. The first term represents the rate of formation of E, the other two terms represent the rate of decomposition of E. The *net* change in the concentration of E is the first term minus the decomposition terms. If E is unstable, then it will always be in very low concentration, and we can assume that the *change* in concentration of E with time is negligible. This is an example of the so-called *steady state assumption*. When the rate of concentration change of E is set equal to zero, then

$$k_1[K][\mathrm{H}] = k_2[\mathrm{E}][\mathrm{Br}_2] + k_{-1}[\mathrm{E}] \qquad \text{or} \qquad [\mathrm{E}] = \frac{(k_1[K][\mathrm{H}])}{(k_{-1} + k_2[\mathrm{Br}_2])}$$

If the second step is fast relative to the reverse of the first step, then

$$k_2[\mathrm{Br}_2] \gg k_{-1} \qquad \text{and} \qquad [\mathrm{E}] = \frac{(k_1[K][\mathrm{H}])}{k_2[\mathrm{Br}_2]}$$

Substituting this last equation into the equation for the rate of product formation, we find that

$$\text{rate of product formation} = k_1[K][\mathrm{H}]$$

The rate, in terms of starting materials, is seen to be *independent* of the bromine concentration, and this is consistent with experimental observation.

The formation of enolate in the bromination of a ketone in base is said to be the *rate determining step*. This means that the product cannot be formed faster than the rate of conversion of ketone into enolate. The rate of enolate formation "determines" the rate of the overall reaction.

A large number of organic reactions proceed via intermediates, and we shall now be concerned with three principal types.

Free radicals. Chemical species which contain an unpaired electron but which are electrically neutral are known as *free radicals*, or simply *radicals*. The hydrogen atom is a familiar free radical. Radicals are usually produced by *homolytic* cleavage of a single bond:

homolytic fission

Each species formed by this type of bond breakage retains one of the two bonding electrons. Chlorine radical, for example, can be produced by the photochemical splitting of a chlorine molecule:

$$Cl{-}Cl \xrightarrow{\text{light}} 2Cl\cdot$$

Carbon free radicals may be formed by homolytic cleavage of carbon-hydrogen or carbon-carbon bonds. For example, chlorine radical reacts with methane to produce methyl radical:

$$Cl\cdot + H{-}CH_3 \longrightarrow Cl{-}H + \cdot CH_3$$

The methyl radical is a reactive intermediate that can attack a chlorine molecule to produce methyl chloride:

$$CH_3\cdot + Cl_2 \longrightarrow CH_3Cl + \cdot Cl$$

A mixture of methane and chlorine in the presence of light will, therefore, produce methyl chloride via a free radical mechanism.

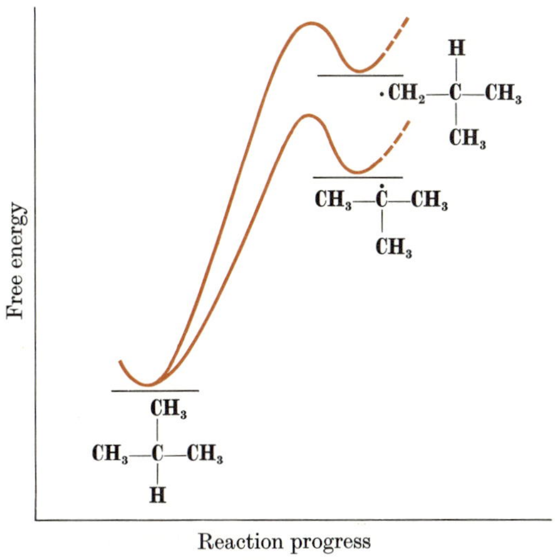

Fig. 4.25 Free energy changes for reactions involving free radical intermediates

The relative stability of free radicals depends upon their substitution. It has been found experimentally that the more substituted a radical, the more stable it is, and the faster it is formed. Tertiary radicals ($R_3C\cdot$) are formed the most easily, and methyl radicals the least easily, in the same type of radical-producing reaction.

$R_3C\cdot$	>	$R_2CH\cdot$	>	$RCH_2\cdot$	>	$CH_3\cdot$
tertiary		secondary		primary		methyl
3°		2°		1°		

In the reaction of isobutane with chlorine atom, we would predict that the product would arise principally from the most highly substituted radical, and this is found to be the case. Figure 4.25 shows an energy comparison of the two possible modes of radical formation for this reaction. The tertiary site is the most stable and has the lowest free energy of activation.

Carbonium ions. A number of reactive intermediates arise from *heterolytic* bond cleavage.

$$A\,\vdots\!:B \longrightarrow A^+ + B^-$$

heterolytic fission

One of the species produced by bond breakage retains both unshared electrons and is negatively charged, while the other has an empty orbital and is positively charged An electron-deficient carbon is called a carbonium ion. There are several ways to produce carbonium ions, three of which are shown below.

1) $C{-}X \longrightarrow C^+ + X^-$

Example: 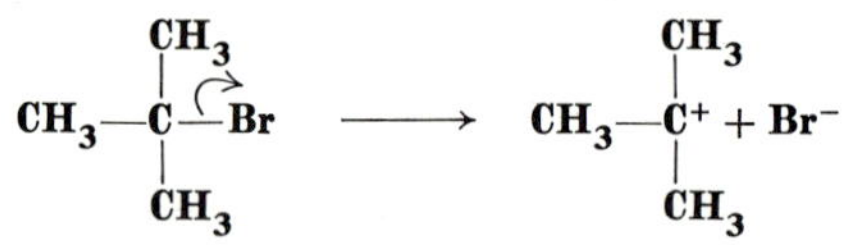

2) $C{=}C + H^+ \longrightarrow {}^+C{-}C{-}X$

Example: $CH_2{=}CH_2 + H^+ \longrightarrow \overset{+}{C}H_2{-}CH_3$

3) $C{=}O + X^+ \longrightarrow [C{=}\overset{+}{O}X \longleftrightarrow \overset{+}{C}{-}OX]$

Example: 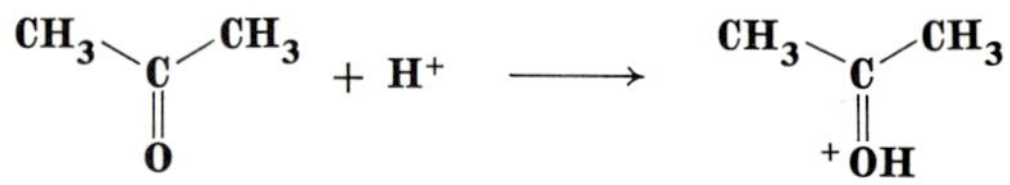

It should be emphasized that all the above carbonium ions are unstable intermediates and have only fleeting existences.

Carbonium ion stability follows the same order as radical stability.

R_3C^+	>	R_2CH^+	>	RCH_2^+	>	CH_3^+
tertiary		secondary		primary		methyl

This order of stability is due primarily to the electron-releasing inductive effect of the alkyl groups attached to the cationic, sp^2 hybridized carbon atom. The more alkyl groups there are, the more easily the positive charge is accommodated. Resonance stabilization of the positive charge can also enhance the ease of formation and the stability of carbonium ion intermediates. Thus, allyl carbonium ions are more stable than simple alkyl carbonium ions.

$$CH_2{=}CH{-}\overset{+}{C}H_2 \longleftrightarrow \overset{+}{C}H_2{-}CH{=}CH_2$$

Such resonance-stabilized ions have, in fact, sufficiently long lifetimes that they can sometimes be directly observed by spectroscopic methods.

Carbanions. Heterolytic bond cleavage leads to retention of the electron pair on a carbon atom and yields species called carbanions. We have already seen examples of such anions in our discussion of bases. The conjugate base of acetone is a typical carbanion. For convenience, the unshared pair of electrons on carbanions is usually omitted.

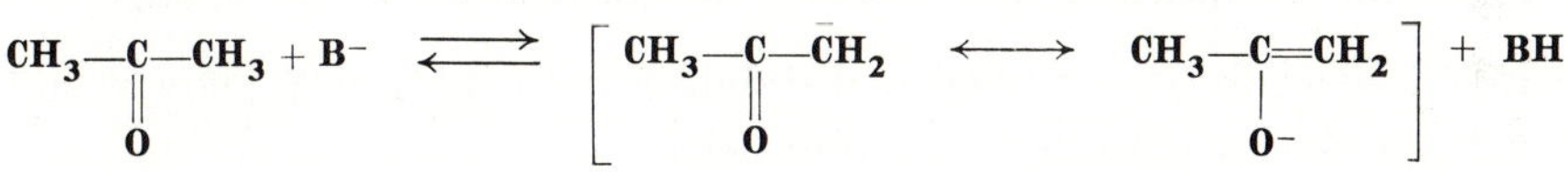

Simple alkyl carbanions can also be prepared. Butyl bromide reacts with lithium metal to yield butyl lithium (a compound resembling a Grignard reagent).

$$CH_3CH_2CH_2CH_2Br + 2Li \longrightarrow CH_3CH_2CH_2CH_2{}^{-}Li^{+} + Li^{+}Br^{-}$$

This anionic compound is relatively stable becuause of the considerable covalent character of the carbon-lithium bond, and it may be stored in an oxygen- and water-free atmosphere for several months. Alkyl carbanions show the opposite order of stability found with carbonium ions and radicals:

$^{-}CH_3$	>	$^{-}CH_2R$	>	$^{-}CHR_2$	>	$^{-}CR_3$
		1°		2°		3°
methyl		primary		secondary		tertiary

The intermediacy of these species will be illustrated in a variety of reactions in the following chapters.

ADDITIONAL READING

R. Stewart, *The Investigations of Organic Reactions*, Englewood Cliffs, N.J.: Prentice-Hall, 1966. Paperback.

R. Breslow, *Organic Reaction Mechanisms*, Menlo Park, Calif.: Benjamin, 1966. Paperback.

P. Sykes, *A Guidebook to Mechanism in Organic Chemistry*, New York: Wiley, 1961. Paperback.

C. A. Vanderwerf, *Acids, Bases, and the Chemistry of the Covalent Bond*, New York: Van Nostrand Reinhold, 1961. Paperback.
R. W. Alder, R. Baker, and J. M. Brown, *Mechanism in Organic Chemistry*, New York: Wiley, 1971.

PROBLEMS

1. The pK_a values of two acids are 4.8 and 5.8. Which is the stronger acid? By how much do the acid dissociation constants K differ?
2. In the equilibrium $\mathbf{HA + H_2O \rightleftarrows A^- + H_3O^+}$:

   a) would stabilization of HA alone make the acid weaker or stronger?
   b) would stabilization of A^- alone make the acid weaker or stronger?

3. If adding a substituent to an acid stabilizes both HA and A^-, how would the equilibrium in Problem 2 be affected?
4. Explain why acetic acid is a stronger acid than phenol.
5. Draw the resonance contributors of *o*-methoxybenzoic acid:

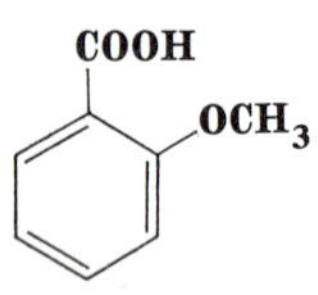

6. The conjugate base of a strong acid is a weak base. Explain.
7. 9.41 grams of phenol (p$K_a = 10$, molecular weight $= 94.1$) were dissolved in water, the pH adjusted to 11.0 and the volume brought to one liter. What is the concentration of phenolate ion (the conjugate base of phenol)?
8. Draw the major resonance contributors of phenolate ion. Which one is the most important contributor? Why?
9. The rate of an endothermic reaction can be faster than the rate of an exothermic reaction. Explain.
10. True or false?
    a) A reaction with one intermediate has two transition states.
    b) The larger the rate constant the faster the reaction.
    c) The larger the activation energy the slower the reaction.
    d) The chlorine radical is an intermediate, not a transition state.
    e) A tertiary carbonium ion is more stable than a primary carbonium ion.
    f) Methylamine is a much stronger base than methanol.
    g) The steady state assumption can be applied to reactions involving an unstable intermediate whose concentration is always extremely small.
    h) Homolytic fission of Cl_2 produces the chlorine radical.

11. Arrange the following compounds in order of increasing acidity (decreasing pK_a) and explain the difference in strength between them.

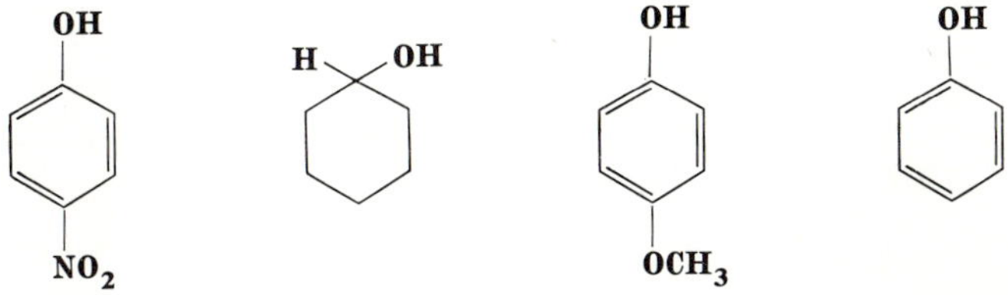

12. Explain the fact that the aromatic amine shown below protonates on nitrogen, whereas the vinyl amine protonates on carbon.

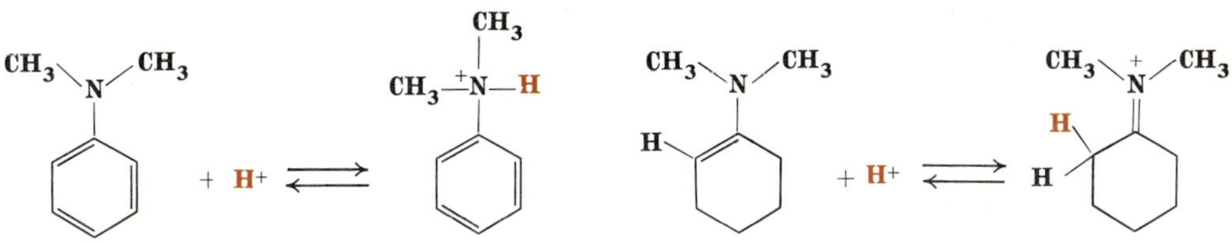

13. The pH of the gastric juice of the stomach is between 1 and 2, while the pH of the upper small intestines is about 8. The pK_a of aspirin (acetylsalicylic acid) is 3.5. Is aspirin ionized or un-ionized in the two regions?

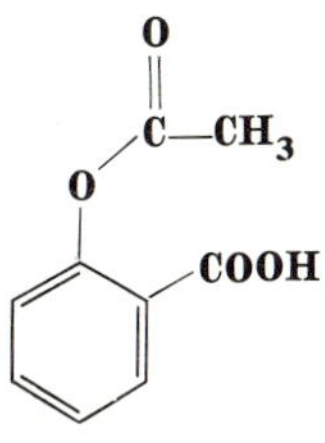

aspirin

14. Phthalic acid is an aromatic dicarboxylic acid. The first acid group ionizes with a pK_a = 2.98. The second one subsequently ionizes with a pK_a = 5.28. Explain the difference between the two pK_a values.

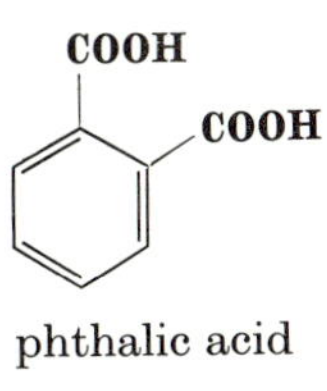

phthalic acid

15. Would you expect the pK_a of acetic acid to be less or greater in benzene than in water? Explain.

16. Derive the following relationship for $HA \rightleftarrows H^+ + A^-$.

$$pK_a = pH - \log \frac{[A^-]}{[HA]}$$

17. A protein molecule consists of long chains of small organic substances, called amino acids, which are linked up as shown below.

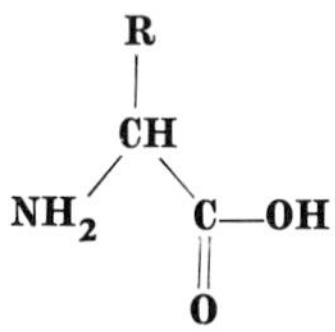

amino acid

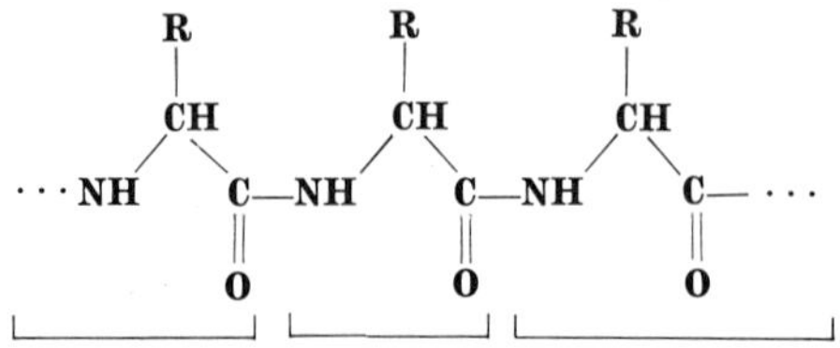

protein segment

There are about two dozen naturally occurring amino acids, with varying R groups, which combine in various permutations to form proteins (much as letters of the alphabet comprise written words). A few of the two dozen amino acids are listed in the table below.

Amino acid	R	pK_a
Histidine	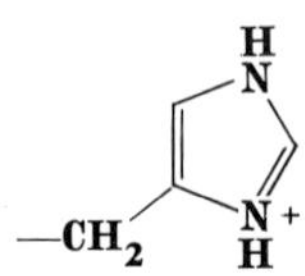	6.0
Serine	$-CH_2OH$	13.8
Lysine	$-(CH_2)_4NH_3^+$	10.5
Aspartic acid	$-CH_2COOH$	3.9
Arginine	$-(CH_2)_3NH-C(=\overset{+}{N}H_2)-NH_2$	12.5

Answer the following questions.

a) A protein is dissolved in a buffer of pH = 8.0. What charge, if any, would there be on each of the amino acid side chains listed in the table?

b) Explain why the pK_a of the arginine side chain is two pK_a units larger than that of the lysine side chain.

c) Which nitrogen of the five-membered imidazole ring of histidine would protonate in acidic solutions? Explain.

d) Which is a stronger base, the conjugate base of the serine hydroxyl or the unprotonated arginine?

18. Sulfonamides have an acidic proton as a result of resonance stabilization of the anionic nitrogen:

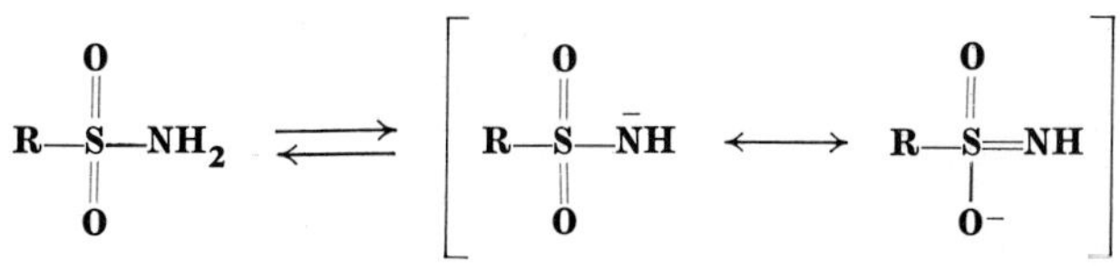

Note that the octet rule does not apply to sulfur because its bonding electrons are in the 3-shell (which can hold up to 18 electrons). There is an important class of sulfonamides, called *sulfa drugs*, whose biological activity depends on the pK_a of the compounds (see table below). The term C_r represents the minimum molar concentration necessary to cause

Sulfa drug*	pK_a	C_r
$R—SO_2—NH_2$	10.43	20.0
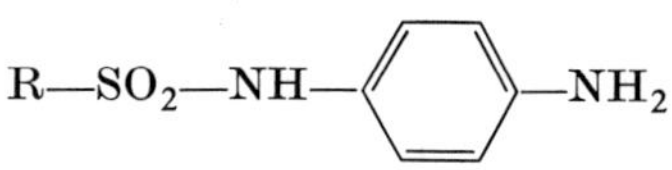	10.22	5.0
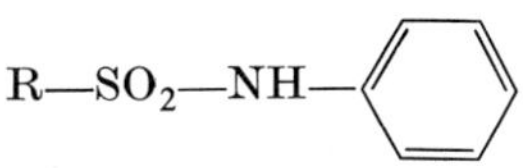	9.60	3.0
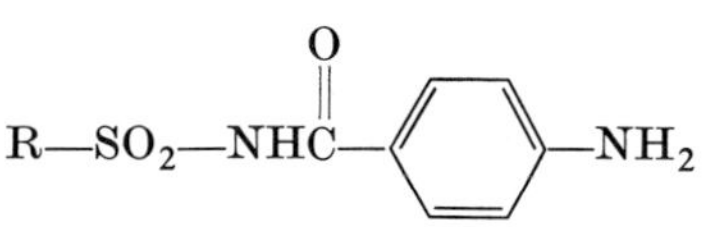	5.20	0.5
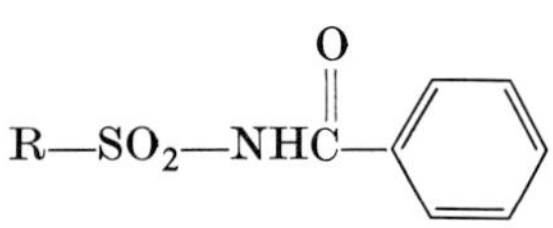	4.57	0.3

* R represents 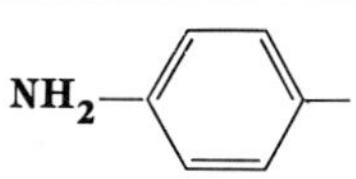

bacteriostasis of *E. coli* under standard conditions. The smaller the value of C_r, the greater the activity of the sulfa drug. From the data in the table, would you conclude that the active form of the drugs is ionic or nonionic?

19. Thiamine (vitamin B_1) is necessary in the human diet in order to prevent beriberi. It is also an important catalyst in many biological reactions.

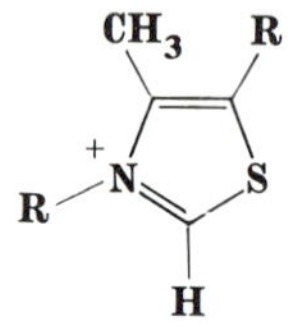

thiamine

The aromatic hydrogen at position 2 of the thiazolium ring of thiamine was found to be amazingly acidic. If thiamine is dissolved in heavy water (D_2O) at neutrality, the aromatic hydrogen will readily exchange for a deuterium:

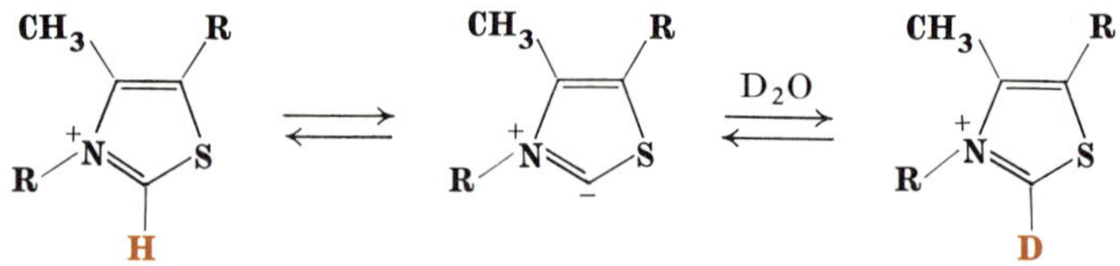

Cite two factors which might explain the acidity of the ring hydrogen.

20. Depending on which of the nitrogen atoms is protonated, 4-aminopyridine can be considered a stronger or weaker base than 3-aminopyridine. Rationalize this statement.

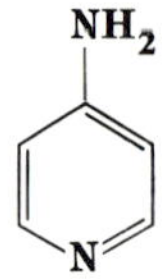

4-aminopyridine

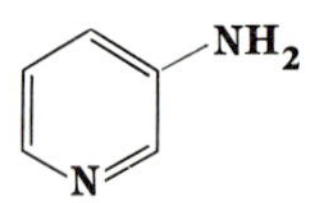

3-aminopyridine

21. The addition of hydrogen bromide to ethylene is a two-stage process in which the first step is a reversible formation of a carbonium ion intermediate. The second step, destruction of the intermediate, is a fast step compared to the reverse reaction of the first step. Draw an energy diagram for the addition reaction.

$$CH_2{=}CH_2 + HBr \rightleftharpoons CH_3{-}CH_2^+ + Br^- \longrightarrow CH_3CH_2Br$$

22. Draw, on a single graph, the energy profiles for the reactions A→B and A→C such that the first reaction is slower than the second, and B is more stable than C.

23. Draw an energy diagram for a reaction mechanism in which A forms an intermediate (I) that subsequently decomposes into two products, B and C. Assume that compound C is the major reaction product. The reactions forming B and C are known as *competitive reactions.*

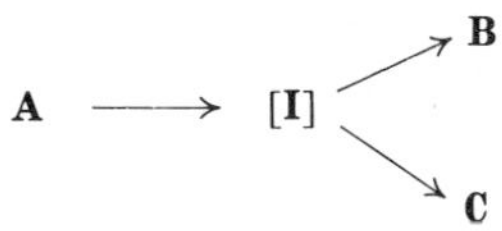

24. When acetyl peroxide is heated, acetoxy radical is formed via homolytic fission of the oxygen-oxygen bond.

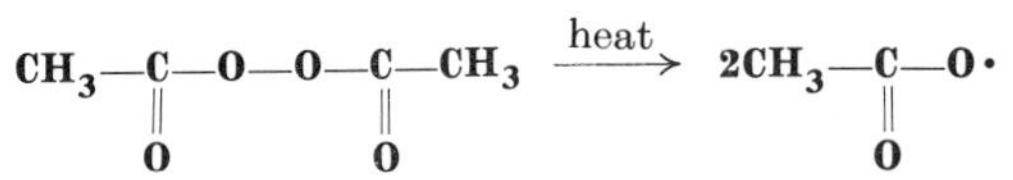

On the other hand, ionization of acetic acid involves heterolytic cleavage of the oxygen-hydrogen bond.

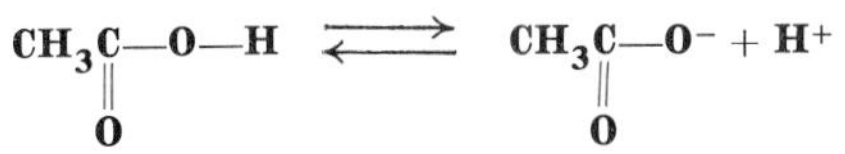

Why should one bond cleave homolytically and the other heterolytically?

25. In the presence of base, an α-hydroxyaldehyde is in equilibrium with an α-hydroxyketone.

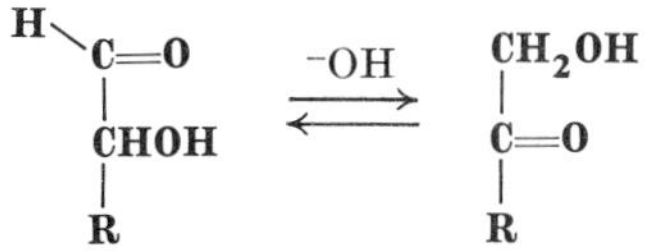

This reaction is called the Lobry de Bruyn–Alberda van Ekenstein transformation in honor of the discoverer and his mistress Alberda. Propose a mechanism for the reaction.

26. Chemical reactions within the living cell are catalyzed by proteins called enzymes. Many enzymes have been purified and crystallized. Enzymes can often catalyze in a test tube the same reactions they catalyze within the cell. Elucidation of the mechanism of the enzymatic reactions (that is, how enzymes lower activation energies) is an important problem now facing biologists and chemists. An example of an enzyme-catalyzed

reaction is the migration of the double bond in the steroid system drawn below.

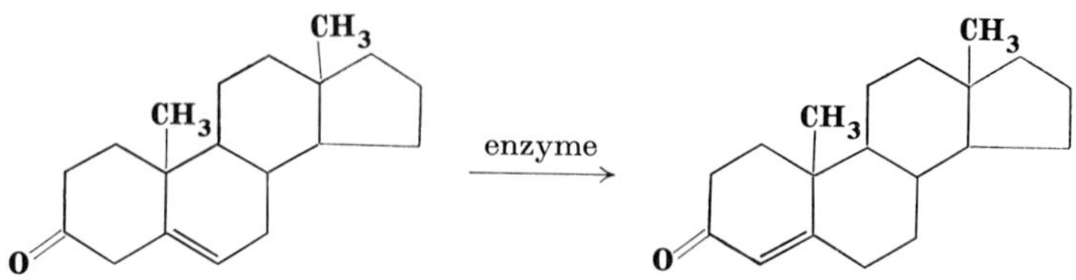

This reaction is also catalyzed by a proton, although much more slowly than by the enzyme. Propose a mechanism for the acid-catalyzed reaction, remembering that a proton can promote enol formation.

27. The reactions given below were described in Chapter 3. Draw their transition states showing all partial charges.

a) $CH_3CH_2X + {}^-OH \longrightarrow CH_2{=}CH_2 + X^- + H_2O$

b) $CH_3CH_2Br + NH_3 \longrightarrow CH_3CH_2\overset{+}{N}H_3 + Br^-$

28. The limit of solubility in water for a neutral monofunctional organic compound lies between C_4 and C_6. Thus, butanol (C_4H_9OH) is soluble in water whereas octanol ($C_8H_{17}OH$) is not. If a compound contains an ionic group, such as a carboxylate anion, it is usually soluble in water no matter how large the nonpolar portion of the molecule. Most neutral monofunctional compounds are soluble in benzene, a solvent which is immiscible with water. Propose a scheme for separating a mixture of *p*-methylphenol, decanol, and 1-naphthylamine into its components.

$CH_3(CH_2)_9OH$

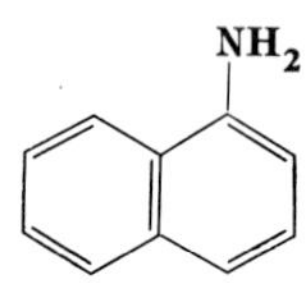

p-methylphenol	decanol	1-naphthylamine
$pK_a = 10$	$pK_a = 17$	$pK_a = 4$

5

STEREOCHEMISTRY

5.1 SYMMETRY AND DISSYMMETRY

Stereochemistry (three-dimensional chemistry) deals with those properties of molecules that result from the spatial relationships of their constituent groups. Like all three-dimensional objects, molecules are either *symmetric* or *dissymmetric*. When an object can be superimposed on its mirror image, it is said to be symmetric. A cube, for example, can be superimposed on its mirror image (Fig. 5.1).

Fig. 5.1 Cube reflected in a mirror—cube and its image superimposable

A dissymmetric object cannot be superimposed on its mirror image. The spot on the rectangular box in Fig. 5.2 is placed in such a position that the object is dissymmetric. Although the box itself and its mirror image can be superimposed, there is no way of doing this that will place one spot directly over the other.

Dissymmetric objects can be characterized by their *chirality*, or *handedness*. The most obvious example of objects which have handedness are hands themselves. Both of our hands have the same structure, and are nonsuperimposable mirror images; they are described as either right or left depending upon the spatial relationships of the fingers when viewed from a particular

Fig. 5.2 Spotted box reflected in a mirror—box and its image nonsuperimposable

side. Many other common objects also have chirality: for example, a glove (but not a flat mitten), the threaded base of a light bulb, or a golf club.

The shapes of many organic molecules are also dissymmetric and have chirality. Before we define a structural requirement for dissymmetry, let us consider the symmetry of some simple molecules.

5.2 ELEMENTS OF SYMMETRY

The simplest structural feature which makes a molecule symmetric is a *plane of symmetry*, or *mirror plane*. A plane of symmetry bisects a molecule into

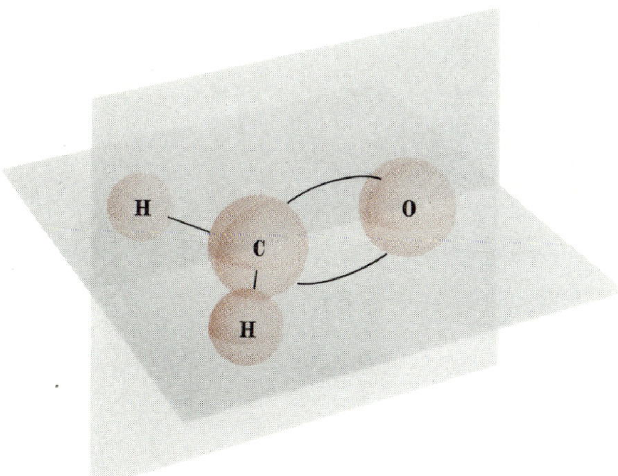

Fig. 5.3 Planes of symmetry in formaldehyde

halves which are exact mirror reflections of each other. Formaldehyde, for example, has two mirror planes (Fig. 5.3).

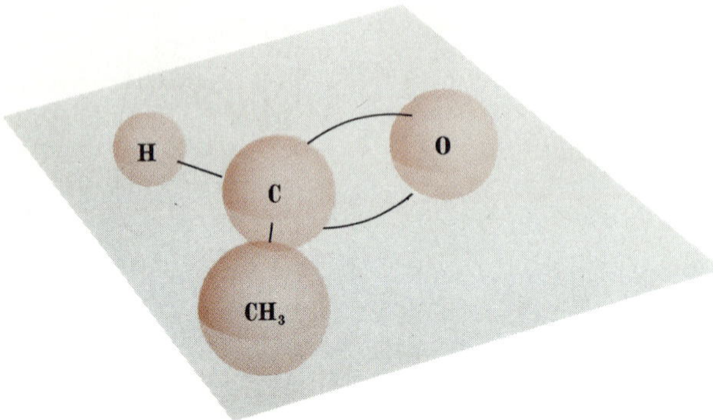

Fig. 5.4 Plane of symmetry in acetaldehyde

Acetaldehyde, on the other hand, has only one plane of symmetry (Fig. 5.4) since a plane cutting the molecule in a vertical direction would not yield two superimposable mirror images. The symmetry of these molecules is a function of their sp^2 hybridized atoms. The bond angles of such atoms are 120°, and the substituents attached to trigonal atoms must all lie in the same plane. Therefore, carbon-carbon or carbon-oxygen double bond functionalities (olefins, ketones, aldehydes, etc.) cannot impart dissymmetry to a molecule. The same is true of triple bonds. Figure 5.5 illustrates two of the planes of symmetry of dimethylacetylene.

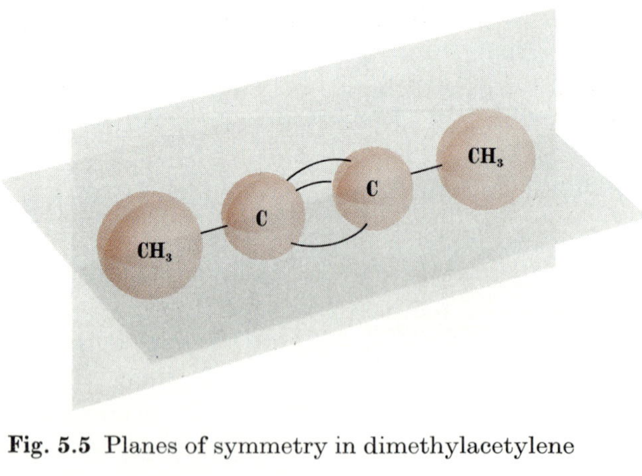

Fig. 5.5 Planes of symmetry in dimethylacetylene

Another common element of molecular symmetry is the *center of symmetry*. *Trans*-1,3-dichlorocyclobutane is typical of a compound which has a center of symmetry. When a line is extended from the central point of the ring, any substituent it intersects is the reflection of the group, met by a line extended equally in the opposite direction (Fig. 5.6). Note that this molecule also has a plane of symmetry.

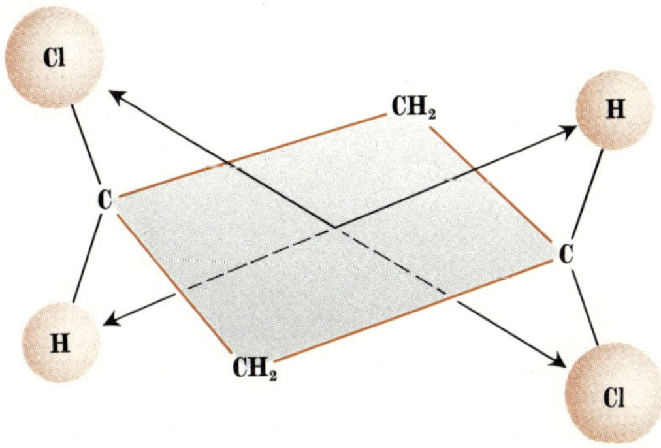

Fig. 5.6 Center of symmetry in *trans*-1,3-dichlorocyclobutane

5.3 THE ASYMMETRIC CARBON

The most common form of dissymmetry found in organic compounds is associated with the *asymmetric carbon atom* or *asymmetric center*. A carbon atom is asymmetric when it has four nonequivalent substituents. To understand why this structural unit may make a molecule dissymmetric, let us examine the symmetry properties of tetrahedral carbon atoms with varying kinds of substitution.

If we look at a carbon atom with four equivalent substituents, for example methane (Fig. 5.7), we find that the molecule has several planes of symmetry.

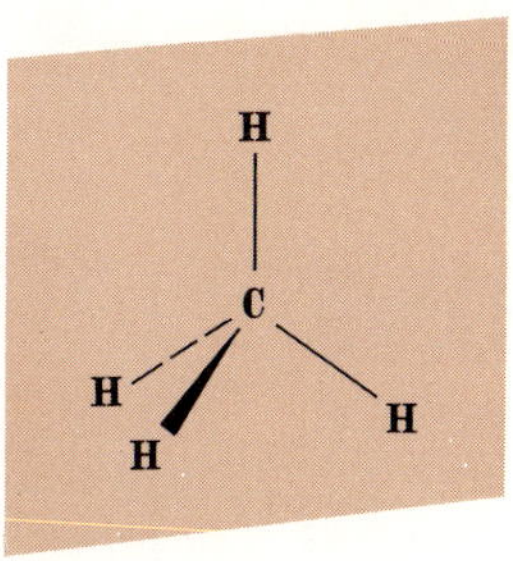

Fig. 5.7 Plane of symmetry in methane

Each of these planes bisects the carbon atom and two of the adjoining hydrogen atoms. The remaining two hydrogen atoms and the bonds connecting them to the carbon atom in each case are mirror reflections of each other.

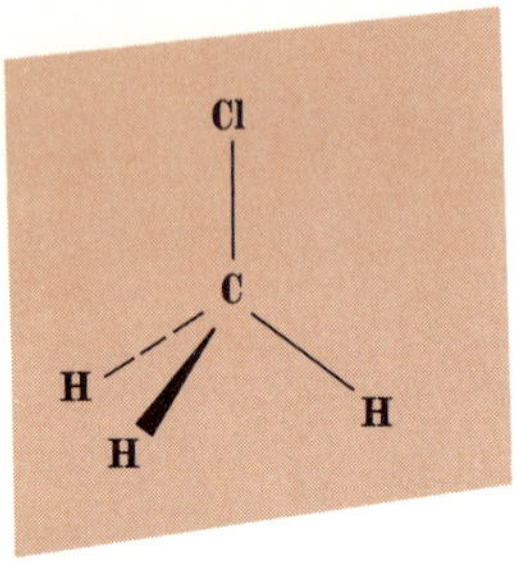

Fig. 5.8 Plane of symmetry in chloromethane

If we now substitute a chlorine atom for one of the hydrogens (Fig. 5.8), we find that the molecule is still symmetric. A plane bisecting the chlorine atom, the carbon atom, and any of the hydrogen atoms divides the molecule into mirror image halves.

When we substitute two of the hydrogen atoms, as in bromochloromethane (Fig. 5.9), the resulting molecule still has a plane of symmetry. In contrast, a molecule such as bromochlorofluoromethane (Fig. 5.10) has *four different substituents* attached to one carbon atom and no element of symmetry whatsoever; it is, therefore, dissymmetric. Consequently, there is possible a mirror image isomer of this compound which cannot be superimposed on it. Mirror image substances of this type which contain an asymmetric center are called *enantiomers* or *optical antipodes*. These stereoisomers are examples of *optical isomerism*.

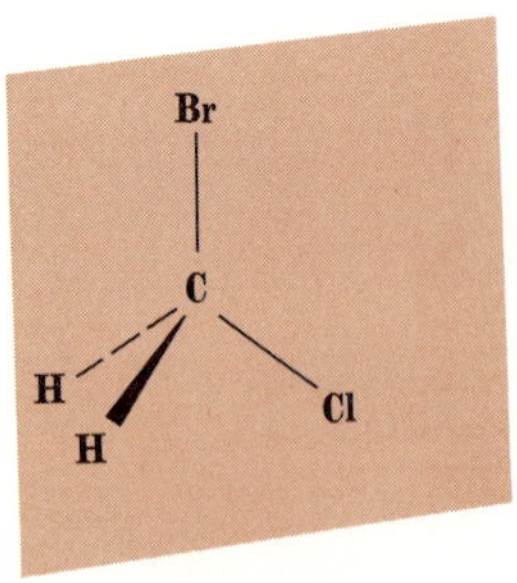

Fig. 5.9 Plane of symmetry in bromochloromethane

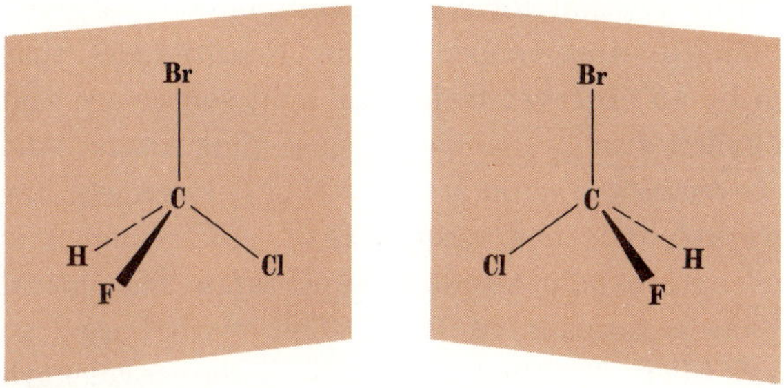

Fig. 5.10 Nonsuperimposable mirror images of a dissymmetric molecule

5.4 OPTICAL ISOMERISM

The interaction of enantiomers with a symmetric chemical or physical environment is identical in all respects. Thus optical antipodes will have identical boiling points, melting points, identical solubilities in symmetric solvents like water or benzene, and identical reactivities with symmetric molecules like hydrogen chloride or carbon dioxide. A flat mitten (a symmetric object) is a common analogy of this kind of behavior: it will fit a left or right hand equally well. A left or right hand will not fit with equal ease, however, into a left-handed glove (a dissymmetric object). In the same way, a left-handed molecule will interact differently from its mirror image with a dissymmetric environment. For example, enzymes are dissymmetric substances, and a living organism has only one optical isomer of each of these biological catalysts. Therefore, only one of the chiral forms of a dissymmetric substance such as a sugar can be metabolized by a given organism, since only this "hand" can fit the enzyme "glove".

The usual way to distinguish mirror image isomers is by means of their differing interactions with plane-polarized light. Ordinary light may be thought of as vibrating in all possible planes perpendicular to the direction from which the light beam is propagated. When this light is passed through a Nicol prism, or other polarizer, only light vibrating in one of these planes is transmitted. If this plane-polarized light is then passed through a tube of known length containing a known concentration of one enantiomer of a mirror image set of compounds, the plane of polarization rotates a certain number of degrees. In contrast, when a solution containing the same concentration of the other optical antipode is substituted, the plane of polarization rotates the same number of degrees in the opposite direction. Substances that have this effect on plane-polarized light are said to be *optically active*, and the number of degrees through which the plane of polarization rotates is a measure of their optical activity. This measurement is made on an instrument called a polarimeter and is constant for a given compound at a given wavelength of light, tube length, and concentration. Enantiomers that rotate light clockwise as seen by an observer facing the light source are said to be *dextrorotatory* and are called *d* or (+) isomers; those that rotate light counterclockwise are said to be *levorotatory* and are called *l* or (−) isomers. For example, the d-form of glyceraldehyde has a rotation of +8.7° when it is measured at the 589 mμ wavelength band of a sodium vapor lamp. The optical antipode, l-glyceraldehyde, has a rotation of −8.7°. The symbol used to indicate optical rotation is $[\alpha]_\lambda^T$, where T is the temperature, λ the wavelength of the light, and $[\alpha]$ the observed rotation divided by the length of tube in decimeters times the concentration in grams per milliliter. This expression of rotatory power is called the *specific rotation*.

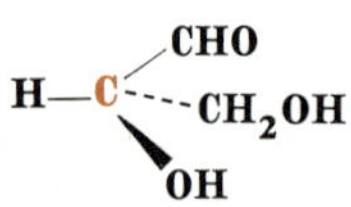

(+)-glyceraldehyde
$[\alpha]_D^{20°} = +8.7°$

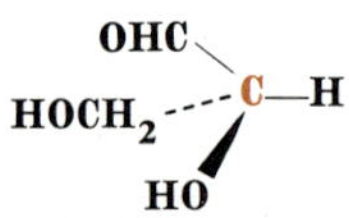

(−)-glyceraldehyde
$[\alpha]_D^{20°} = -8.7°$

An equimolar mixture of two enantiomers has no discernible effect on plane-polarized light. This must be so because the two isomers rotate light in equal but opposite directions. This kind of 1 : 1 mixture of optical antipodes is called a *racemic mixture* or *racemate*. When a compound containing an asymmetric center is synthesized from symmetric materials, equal amounts of the two enantiomers are always obtained.

5.5 CONFIGURATION

Since a molecule containing an asymmetric carbon atom has chirality, we must have some way to determine whether a given enantiomer is right- or left-handed. In order to do this we need a convention for describing the spatial arrangement of groups in the molecule when it is viewed from a given orientation. Using the analogy of human hands again, we could describe the chirality of any hand by specifying that all hands must be viewed from the back. Viewed from this orientation, when the thumb is on the left and the small finger on the right, we say we are looking at a right hand. If we observe the opposite arrangement, we say we are looking at a left hand. The particular spatial arrangement for any hand (or any dissymmetric molecule) when it is viewed according to a set convention is called its *absolute configuration*.

A number of conventions have been devised for assigning absolute configuration to optically active compounds. The most widely used convention is called the R (*rectus*) and S (*sinister*) *system*. According to this system we arrange the four groups attached to the asymmetric carbon of a dissymmetric molecule in order of decreasing atomic numbers. Thus, for our previous example of bromochlorofluoromethane the arrangement of the groups would be bromine (4), chlorine (3), fluorine (2), and hydrogen (1), where (4) indicates the highest atomic-number group and (1) indicates the lowest. When two or more of the atoms directly appended to the asymmetric carbon have the same atomic number, as in 2-methylbutyraldehyde,

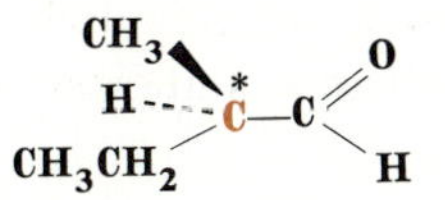

one enantiomer of 2-methylbutyraldehyde

the ranking order is determined by examining the atoms attached to each of these similar substituents. In 2-methylbutyraldehyde, the hydrogen atom attached to the asymmetric carbon (C*) has an atomic number of 1, but the other three substituents are all carbon groups and we cannot distinguish between them on the basis of the atomic numbers of the atoms directly attached to the asymmetric center. If we look at the next set of atoms we find that the aldehyde carbon is attached to oxygen (atomic number 8), the methylene is attached to a methyl carbon (atomic number 6), and the methyl is attached to only hydrogen (atomic number 1). The sequence of the groups of this compound must then be:

—CHO(4), $—CH_2CH_3$(3), $—CH_3$(2), —H(1)

The third, fourth, or *n*th set of atoms on the groups attached to an asymmetric carbon can be analyzed in this way whenever it is required. Finally, one additional sequence rule must be taken into account. Any group containing a multiple bond is treated as if it were a collection of the equivalent number of single bonds. Thus the groups $—CH{=}CH_2$ and $—COCH_3$ are viewed respectively as

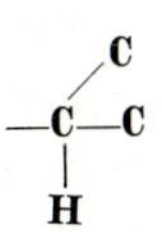

and

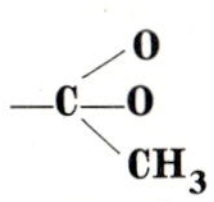

In order to assign the R or S configuration to one of the enantiomers of a dissymmetric compound, we must view that enantiomer from a particular orientation. Using the optical antipodes of 2-methylbutyraldehyde as an example (Fig. 5.11), we view the three-dimensional models of both compounds along the axis of the bond connecting the asymmetric carbon to the atom with the lowest atomic number, and from the side of the molecule opposite that atom. The remaining three groups will then appear like the spokes of a wheel. If these groups are arranged in such a way that we look in a *clockwise* direction from the group with the highest atomic number to the group with the lowest number, we say that the compound has an R *configuration*. If the decreasing order occurs in a *counterclockwise* direction, then the molecule has an S *configuration*.

The R and S scheme of naming absolute configuration is merely a nomenclature device and must not be thought to relate to the sign and magnitude of optical activity. For example, a compound with an R configuration may be either levorotatory or dextrorotatory.

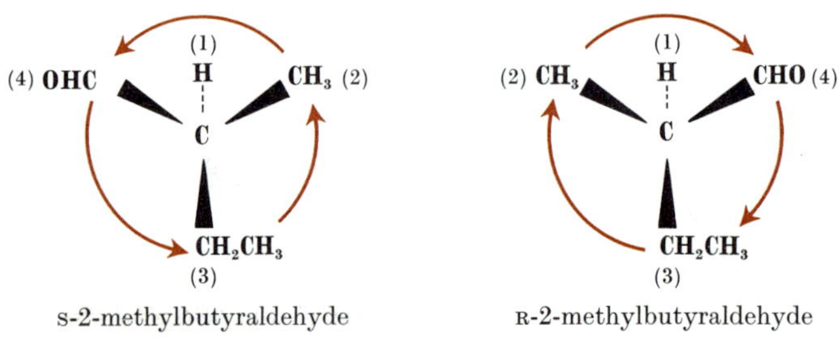

Fig. 5.11 Examples of R and S nomenclature

5.6 FISCHER PROJECTIONS

One problem which arises when we attempt to depict configurations or to assign chirality to them is the difficulty of representing three-dimensional structures on a two-dimensional surface. To overcome this difficulty one uses the so-called *Fischer projection:* the structure of an asymmetric carbon atom is drawn in a prescribed orientation and then projected onto a planar surface. In the prescribed orientation, the asymmetric carbon is placed so that two substituents are horizontal and project out toward the viewer, and the other two substituents are vertical and project away from the viewer. Thus we may draw the R and S configurations of 2-methylbutyraldehyde in the manner shown in the left halves of (a) and (b) of Fig. 5.12. If we now project these figures onto the plane of the paper we obtain the two-dimensional representations shown in the right halves of Fig. 5.12.

To go from the planar representations back to three-dimensional structures we picture pulling the two horizontal groups forward and pushing the two vertical ones backward. We can also represent a given enantiomer by other Fischer projections. Thus R-2-methylbutyraldehyde can be

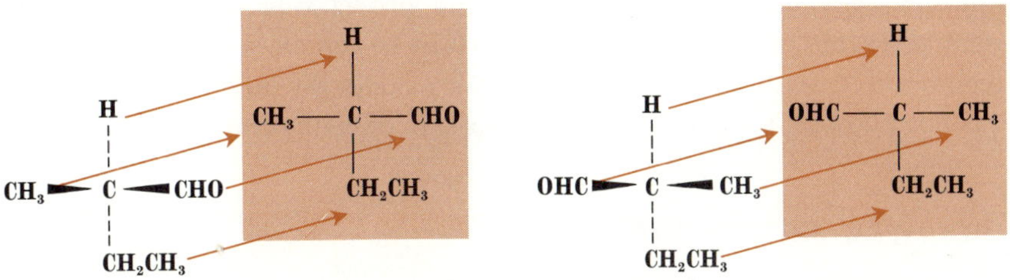

(a) R-configuration (b) S-configuration

Fig. 5.12 Fischer projection formulas

oriented in the way shown on the left in Fig. 5.13 which would lead to the Fischer projection shown on the right.

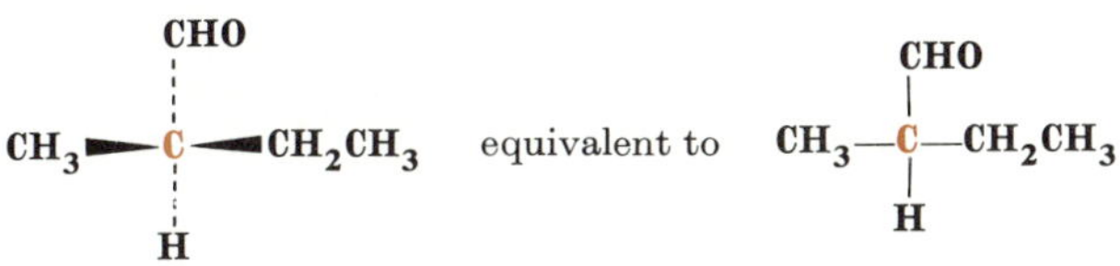

Fig. 5.13 Fischer projection conventions

The problem now arises of trying to decide by inspection whether the two Fischer projections of R-2-methylbutyraldehyde shown in Figs. 5.12 and 5.13 really represent the same enantiomer. This may be done by exchanging pairs of substituents step by step to convert one representation into the other. If the number of exchanges is even, the two projections represent the same configuration.

Fig. 5.14 180° rotation of Fischer projection formulas

For example (Fig. 5.14), if we exchange the methyl and ethyl groups of one projection, and then exchange the hydrogen and aldehyde groups, we obtain the second projection. Since two steps were necessary, the two projections must represent the same enantiomer. An odd number of exchanges (for example, if we switched only the methyl and hydrogen groups), converts one enantiomer into the other.

Another way to compare Fischer projections in certain cases is to rotate them 180° in the plane of the paper, and then to place one over the other to see whether they coincide. If they do, the projections must necessarily represent the same enantiomer. It is important to remember, however, that this technique is limited and that a rotation of only 90° in the plane of the paper is not allowed, as it results in an inversion of configuration.

5.7 MULTIPLE ASYMMETRIC CENTERS

Molecules may have more than one asymmetric center. As the number of such carbon atoms increases, the number of possible optical isomers goes up exponentially. Thus, a compound with n different asymmetric centers has 2^n optical isomers. The following pictorial representation will clarify this relationship for the case where $n = 2$:

(1) (2) (3) (4)

dl pair *dl* pair

It can be seen that there are four permutations. Optical isomers 1 and 2 are mirror images of each other as are 3 and 4. In other words, 1 and 2 (and 3 and 4) are enantiomers and have identical chemical and physical properties other than their effect on polarized light. But note that 1 and 3, for example, are *not* mirror images. They are, in fact, compounds with different melting points, solubilities, chemical reactivities, etc. Two optical isomers with multiple asymmetric centers which are not enantiomeric (not mirror images) are called *diastereomers*. Since diastereomers have different physical properties, they can be separated by techniques such as crystallization or chromatography. For example, a mixture of the four optical isomers above could be separated into two pairs (1 and 2) and (3 and 4).

The concept of diastereomers can be best illustrated by means of cyclic compounds. There are four optical isomers (two *dl* pairs) of 1-ethyl-2-methylcyclobutane:

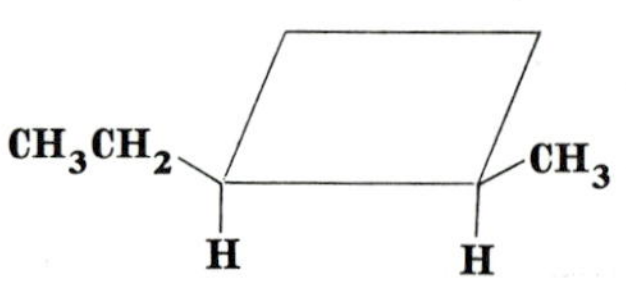

(1)

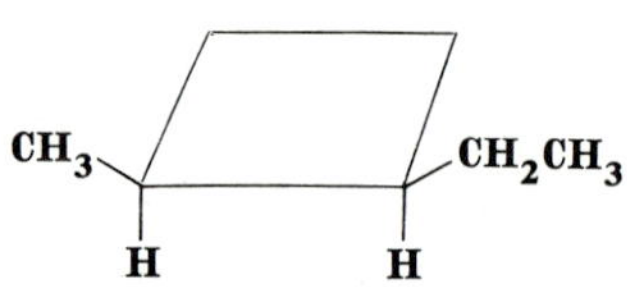

(2)

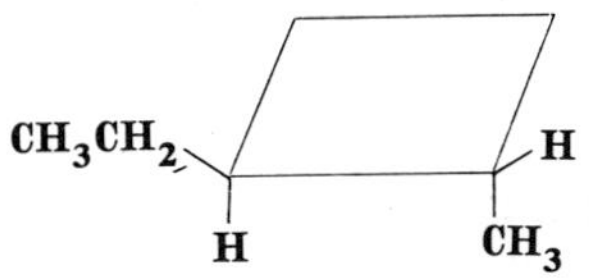

(3)

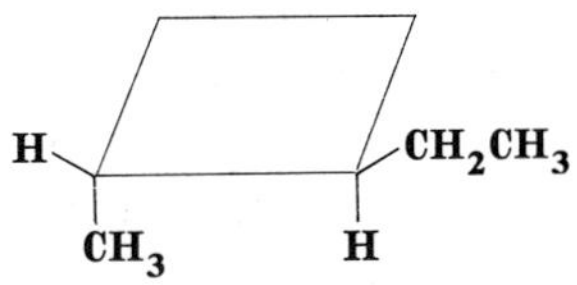

(4)

Clearly, structures 1 and 2 are nonsuperimposable mirror images (enantiomers). The same is true for 3 and 4. Compounds 1 and 2 are called *cis* to distinguish them from their diastereomers, *trans* 3 and 4. The properties of *dl cis* and *dl trans* 1-ethyl-2-methylcyclobutane are different and the two *dl* pairs can be separated.

Let us now examine an acyclic system, a sugar, with two different

$$HOCH_2\text{—}\overset{*}{C}HOH\text{—}\overset{*}{C}HOH\text{—}CHO$$

asymmetric centers. (The word "different" means that the two centers do not have identical substituents. As described in the next section, if the centers do bear identical groups, the 2^n law does not hold.) Figure 5.15 presents both three-dimensional and Fischer representations of the four optical isomers of the sugar.

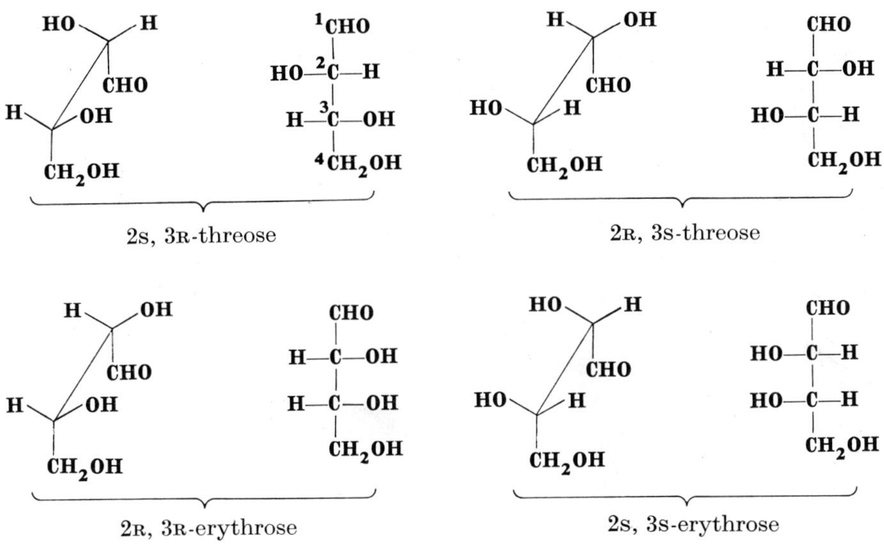

Fig. 5.15 Optical isomers of $HOCH_2$-CHOH-CHOH-CHO

One pair of mirror images is made up of S, R and R, S isomers, whereas the other is comprised of R, R and S, S compounds. The two sugars of the first pair are known as threose, and those of the second pair as erythrose. Each of the erythrose sugars (+ or −) is a diastereomer of each threose (+ or −). The optical rotations of the R, R and S, S enantiomers of erythrose are, of course, equal but opposite in sign; likewise are those for the S,R and R,S enantiomers of threose. However, the magnitudes of rotation of any two diastereomers (R,R-erythrose and S,R-threose, for example) are different as are other chemical and physical properties.

In cyclic compounds such as 1-ethyl-2-methylcyclobutane it is easy to name the two diastereomeric pairs. They are called *cis* and *trans*. The situation is a little more complex in acyclic systems. The following convention is used for naming diastereomeric pairs of compounds which contain two asymmetric centers differing by only *one* substituent, as in the above sugar with

$\overset{*}{C}(H)(OH)(CHO)$ and $\overset{*}{C}(H)(OH)(CH_2OH)$:

1) Imagine that the two different groups (—CHO and —CH_2OH in the sugar, for example) are identical.
2) If the three substituents on one asymmetric center can be made to correspond perfectly to the three groups on the other asymmetric center, the compound is called *erythro*. (The name is derived from erythrose which has this property—see Fig. 5.15.)
3) If it is impossible to line up substituents on the two asymmetric centers, no matter how the groups are rotated relative to one another, the compound is called *threo*.

The examples below will clarify this useful nomenclature which allows chemists to refer unambiguously to a particular diastereomer pair. The groups in color are those which are imagined to be the "same."

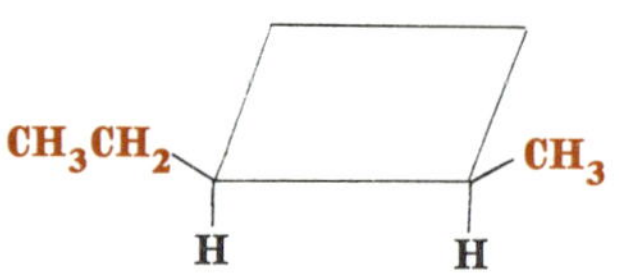

cis (*erythro*)

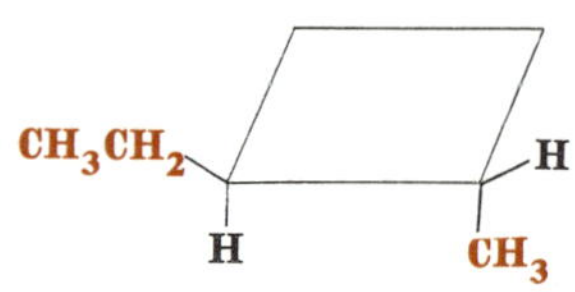

trans (*threo*)

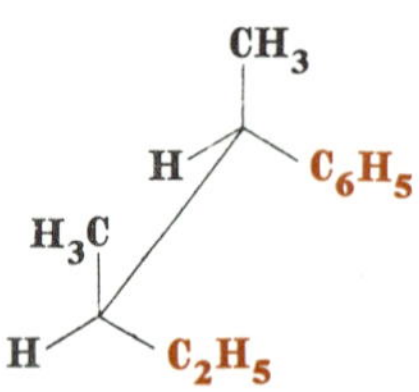

erythro

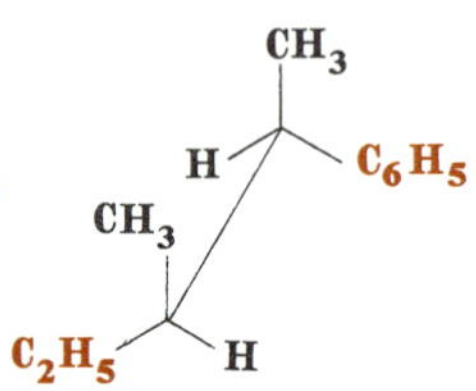

threo

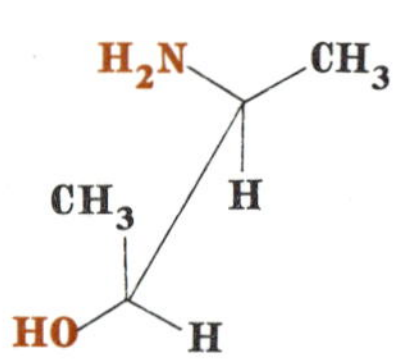

erythro

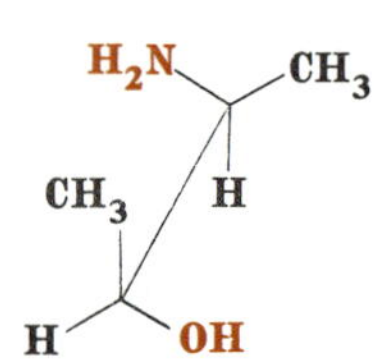

threo

In the third example, a 1:1 correspondence is achieved by imagining the —OH and —NH_2 to be identical and by rotating the front asymmetric carbon clockwise by 60°.

This section should be reread after reading the final sections of this chapter on conformational analysis.

5.8 MESO COMPOUNDS

Tartaric acid is an example of a special case of optical isomerism in which a compound has two asymmetric centers, each of which bears the *same* substituents:

$$HO_2C\text{—}\overset{*}{C}HOH\text{—}\overset{*}{C}HOH\text{—}CO_2H$$

When we draw Fischer projections of the possible optical isomers of tartaric acid (Fig. 5.16), we observe that two of the compounds form a pair of nonsuperimposable mirror images. These have the R,R and S,S configurations. The other pair of isomers, with R,S and S,R configurations, are superimposable mirror images; that is, they are identical.

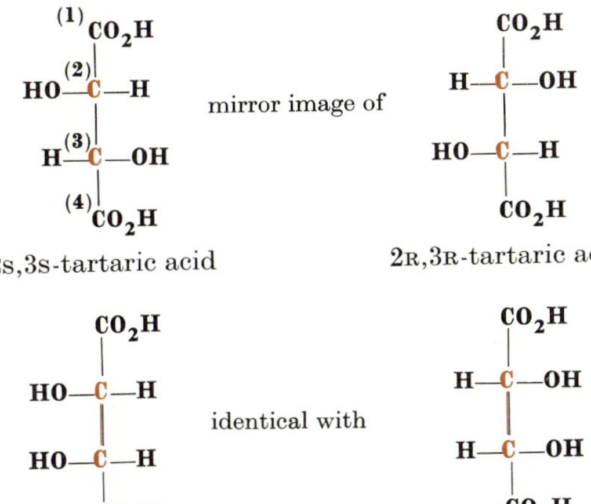

2S,3S-tartaric acid 2R,3R-tartaric acid

2S,3R-tartaric acid 2R,3S-tartaric acid

Fig. 5.16 Optical isomers of tartaric acid

Further examination of either one of these latter structures reveals that the molecule contains a mirror plane and is therefore symmetric. This may seem remarkable considering that each of the tetrahedral carbon atoms is itself an asymmetric carbon. However, each asymmetric center in a molecule rotates plane-polarized light a certain number of degrees in a given direction, depending on the structure of the particular asymmetric carbon. Thus, when a molecule has two asymmetric centers of the same substitution pattern but with opposite configurations, one center will rotate light $+x°$ and the other center will rotate the light $-x°$. The net rotation will be zero, and we can consider the molecule as a kind of internal racemate. The *single* isomer of tartaric acid represented by the R,S and S,R Fischer projections in Fig. 5.16 is called a *meso* compound and it is optically inactive.

5.9 CONFORMATION

The form of stereoisomerism we have been discussing so far—optical or configurational isomerism—is a property generally associated with fixed chemical bonds; that is, to convert one configuration into another covalent bonds must be broken and re-formed. However, there is a second type of stereoisomerism which results from the ability of groups to rotate with respect to one another about the axis of a single bond. This is called *conformational isomerism*, and the isomers are called conformational isomers, or *conformers*. Since the interconversion of two conformers of a compound requires only rotations about single bonds, we must examine conformational isomerism in terms of the energy characteristics of equilibrium reactions. We shall begin our discussion of conformational isomerism with an example of the simplest molecule which exhibits this property: ethane.

5.10 ETHANE

If one carbon atom of ethane is rotated 360° about the carbon-carbon bond relative to the second carbon atom, the molecule assumes an infinite number of conformations, each differing from the others by the geometrical relationship of the hydrogens on the adjacent carbon atoms. These conformations do not have equal energy. Three of them represent the minimum energy of the ethane molecule, and three represent the maximum energy. All of these conformers may be represented graphically as in Fig. 5.17, where the abscissa represents degrees of rotation about the carbon-carbon bond.

The maxima in this plot are those conformers in which the hydrogens are directly opposed to each other, or *eclipsed*. The minima are the three conformers whose hydrogens are in what is called a *staggered* relationship. The angle of separation of any two specified hydrogens on the two carbon atoms is called the *dihedral angle* ϕ, and in the eclipsed forms of ethane, ϕ has

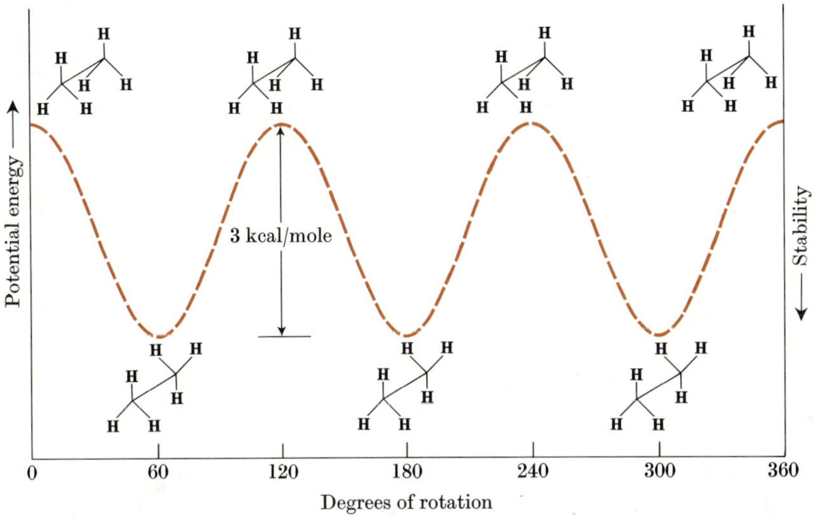

Fig. 5.17 Energies and conformers of ethane

values of 0°, 120°, 240°, and 360°. In the staggered conformations ϕ is equal to 60°, 180°, and 300°. Groups which have a dihedral angle relationship of 180° are said to be anti with respect to each other. Figure 5.18 shows the two principal conformations of ethane in the "Newman projection." In this projection we face the molecule directly along the carbon-carbon bond. The front carbon atom is represented as a circle with three bonds radiating from its center. The fourth bond and the carbon atom to which it is attached cannot be seen, but the three remaining bonds of the rear carbon atom project from the circumference of the forward atom. This projection is particularly useful for examining the dihedral angle relationships of groups attached to adjacent atoms.

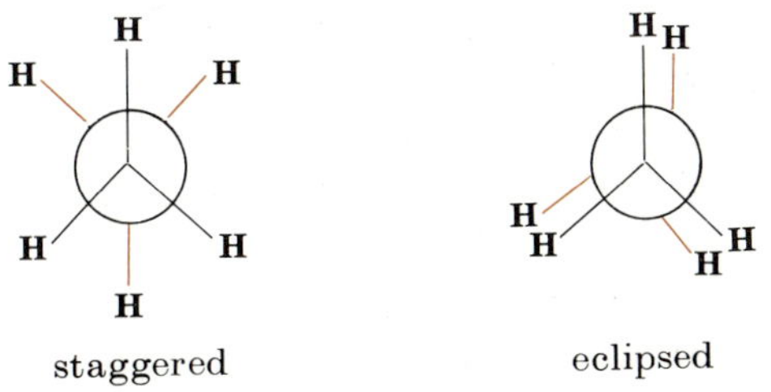

Fig. 5.18 Conformers of ethane

It has been found that an eclipsed conformation has greater energy than a staggered conformation. We might assume that the hydrogens of the eclipsed form interfere with each other, but the distance between eclipsed hydrogens is too great for the electron clouds of the hydrogens to interact and repel each other significantly. We cannot discuss here the factors that determine this energy difference, but staggered relationships tend to be stable, while eclipsed relationships tend to be unstable. The difference in energy between these arrangements appears to be a function of the bond itself: in ethane this energy difference is 3.0 kcal and is called a "barrier to rotation." In Fig. 5.17 the smooth curve which connects the two *extreme* kinds of conformations represents the energy of all other possible dihedral angle relationships, or conformers, of the ethane molecule. It should be noted that an energy difference of 3.0 kcal is very small; various conformers interconvert rapidly at room temperature.

5.11 BUTANE

The hydrocarbon *n*-butane has a more complex conformational picture:

$CH_3CH_2CH_2CH_3$

We may, however, consider butane as a disubstituted ethane molecule. By so doing we see that the curve of potential energy vs. rotation for butane (Fig. 5.19) shows two kinds of maxima and minima.

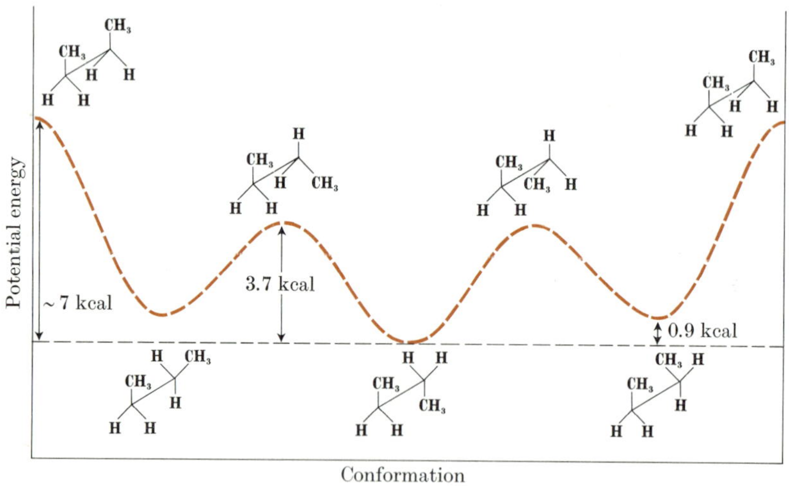

Fig. 5.19 Energies and conformers of *n*-butane

A molecule will spend most of its time at an energy minimum where it can essentially roll back and forth in the potential energy trough. In the case of ethane there is only a single form of minimum energy, the staggered conformer. With butane, however, we expect to find a distribution of molecules between two low-energy forms: one where the methyl groups have a dihedral angle of 180° (called the *anti* conformer) and a second with a dihedral angle of 60° (called the *gauche* conformer). The anti form is more stable than the gauche form by 0.9 kcal/mole. Thus, if we arbitrarily assign an energy of 0.0 kcal/mole to the anti, the energy of the gauche is 0.9 kcal/mole. Butane, then, will exist as an equilibrium mixture of anti and gauche conformers with the more stable anti being slightly in excess of the gauche.

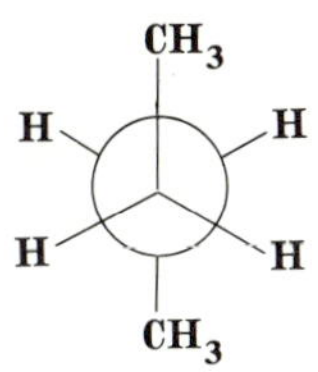

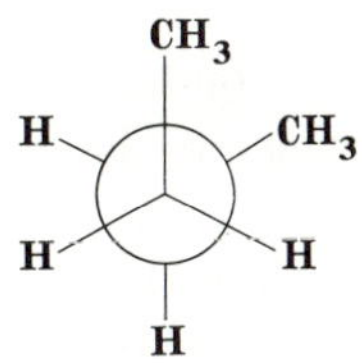

Conformer:	anti	gauche
Dihedral angle:	180°	60°
Relative energy:	0.0 kcal/mole	0.9 kcal/mole

It is important to note that the two eclipsed conformers of butane (the maxima in Fig. 5.19) will not be present in any significant concentration at equilibrium. Since they lie at maxima on the potential energy curve, they have only fleeting existence as transitional forms between the conformers found in the energy troughs.

Neither the anti nor the gauche conformers have eclipsed groups. Why then do they differ in energy by 0.9 kcal/mole? The answer lies in the fact that the methyl groups in the gauche conformer are relatively close to one another, and this gives rise to *nonbonded interactions* between the two. In other words, the electron clouds of the two methyl groups repel each other. The effect is nonexistent in the anti conformer because the methyl groups are too far away (the dihedral angle equals 180°).

5.12 CYCLOHEXANE

Cyclohexane is not planar. If it were planar, then the bond angles of the ring would be 120° which is 11° more than the normal tetrahedral angle of 109°. Instead, cyclohexane assumes a *chair conformation* which allows each carbon to possess an unstrained tetrahedral geometry. (These assertions are best seen with molecular models. It requires force to position all six carbon atoms of a six-membered ring into a plane.) We see from Fig. 5.20 that there are two different

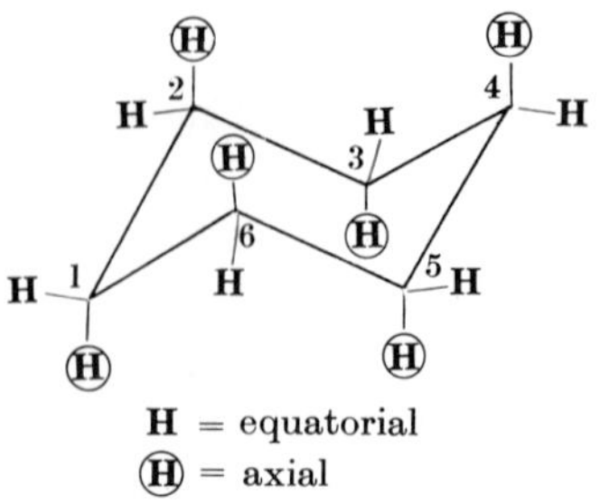

Fig. 5.20 Chair conformation of cyclohexane

kinds of hydrogen bonded to the chair conformation. Above and below the "plane" of the ring are six hydrogens which are more-or-less vertical (the circled hydrogens in Fig. 5.20). These are called, for obvious reasons, *axial* hydrogens. Note that there are three axial hydrogens above the ring on alternate carbons and three axial hydrogens below the ring on alternate carbons. In addition to the six axial hydrogens, we find six other hydrogens which lie in what might be called the "equator" of the ring and which project radially from the ring. These are called *equatorial* hydrogens (uncircled hydrogens in Fig. 5.20). The three equatorial hydrogens on carbons 1, 3, and 5 project slightly above the ring; the other three on carbons 2, 4, and 6, project slightly downward.

An important property of cyclohexane must now be described. The ring system is flexible, and one chair conformation may flip into another by rotation about the carbon-carbon bonds. This process is depicted in Fig. 5.21. (Actually, flipping involves the intermediacy of a so-called *boat* conformer, but this is a detail. It is highly recommended that the student duplicate the flipping process in Fig. 5.21 using molecular models if they are available.) Inspection of Fig. 5.21 reveals a property characteristic to cyclohexane stereochemistry: *when one chair flips into another, each axial bond becomes equatorial and each equatorial bond becomes axial.* For example, the circled axial hydrogen in the first chair conformer in Fig. 5.21 becomes equatorial in the second one.

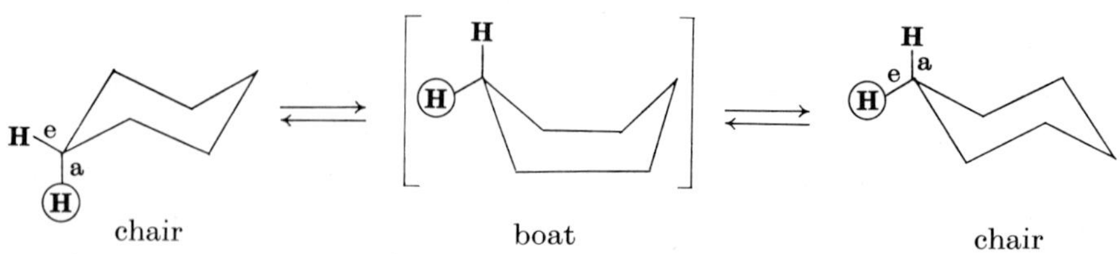

Fig. 5.21 Chair-chair conformational equilibration in cyclohexane

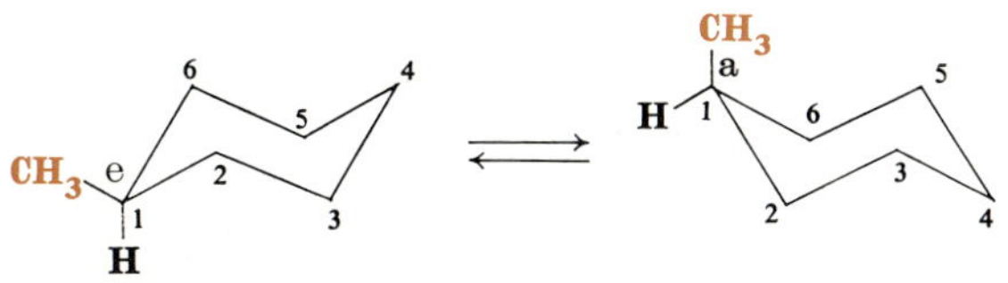

Fig. 5.22 Chair-chair conformational equilibration for methylcyclohexane

In the case of cyclohexane, all chair conformations are identical. This is not true for a compound like methylcyclohexane. The methyl substituent in this compound can be either equatorial or axial (Fig. 5.22). Although the two conformers interconvert rapidly, they are not identical in energy. The conformer with an equatorial methyl group is in fact more stable than the axial one by 1.8 kcal/mole. Thus at room temperature over 90% of the molecules exist in the equatorial conformation. The energy difference of 1.8 kcal/mole can be estimated by comparing the conformations in Fig. 5.22. In the axial conformation, the methyl group and carbons 1,2, and 3 form a butane-like system with a dihedral angle between the methyl group and carbon 3 of 60° (as in the gauche conformer of butane). This dihedral angle is seen readily in a molecular model by looking down the 1-2 bond. A gauche relationship also exists between the methyl group and the carbon 5 along the 1-6 bond. Thus there are two gauche interactions worth 0.9 kcal/mole each for a total of 1.8 kcal/mole.

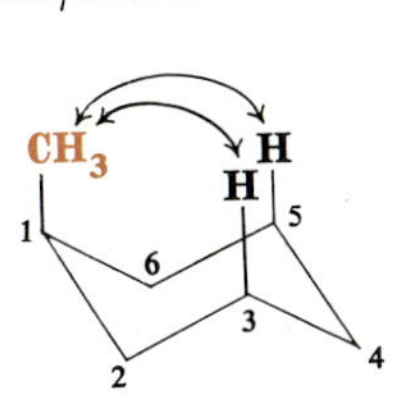

2 gauche interactions

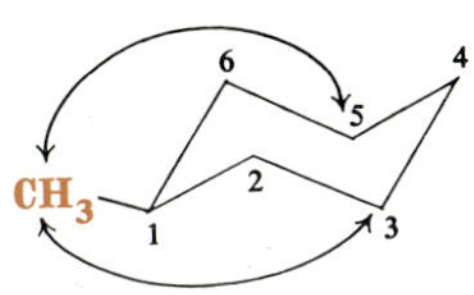

2 anti interactions

In the equatorial conformer of methylcyclohexane, the methyl group is anti to the carbons 3 and 5, so that the methyl group does not contribute to the nonbonded interactions in the molecule. The energy difference, then, between axial and equatorial conformers is 1.8 kcal/mole.

One can estimate the percentage of a conformer using the following rule. For each 1.4 kcal/mole difference in energy between two conformers, the ratio of the conformers differs by a factor of ten. Thus, if conformer A is 2.8 kcal/mole more energetic than conformer B, conformer B will be 100 times more prevalent than conformer A. The ratio will be 1000:1 if the energy difference is 4.2 kcal/mole (3 × 1.4 kcal/mole).

5.13 DIMETHYLCYCLOHEXANE

The two methyl groups of 1,2-dimethylcyclohexane can be either *cis* or *trans* to each other depending on whether they are on the same or opposite side of the ring. The question now arises as to how the *cis* and *trans* relationships can be described more precisely in terms of equatorial and axial bonds. It is obvious from Fig. 5.23 that when two methyl groups in the 1,2-positions are both axial, they are on opposite sides of the ring and therefore *trans*. Since the di-equatorial conformer can be obtained from the di-axial conformer by a mere flip of the ring, the di-equatorial arrangement must also be *trans*. Indeed, it is seen that one equatorial bond projects slightly upward and the other downward. Thus *trans*-1,2-dimethylcyclohexane is a mixture of two conformers: di-axial and di-equatorial. From what has already been said about methylcyclohexane, one might guess that the di-equatorial is the more stable. The energy difference between the two can be readily assessed using the value of 0.9 kcal/mole for a gauche interaction. The di-axial conformer has a total of four gauche interactions (two for each axial methyl) so that we can assign a value of 3.6 kcal/mole to the di-axial conformer.

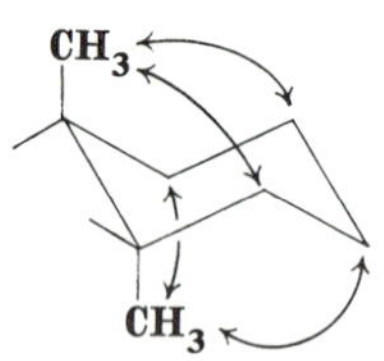

E = 3.6 kcal/mole

The di-equatorial conformer has only one gauche interaction, namely that between the two equatorial methyl groups. Therefore the di-equatorial conformer is more stable than the di-axial by 2.7 kcal/mole. This means that only about 1% of the molecules of *trans*-1,2-dimethylcyclohexane are di-axial.

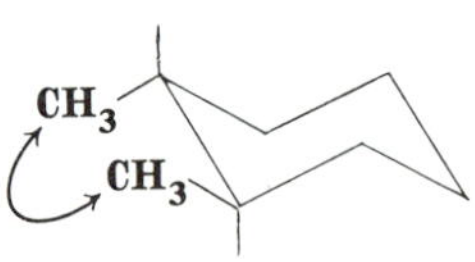

E = 0.9 kcal/mole

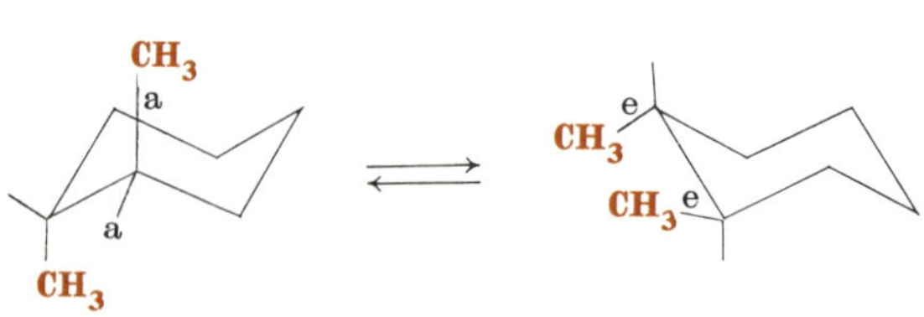

Fig. 5.23 Conformational equilibria for *trans*-1,2-dimethylcyclohexane

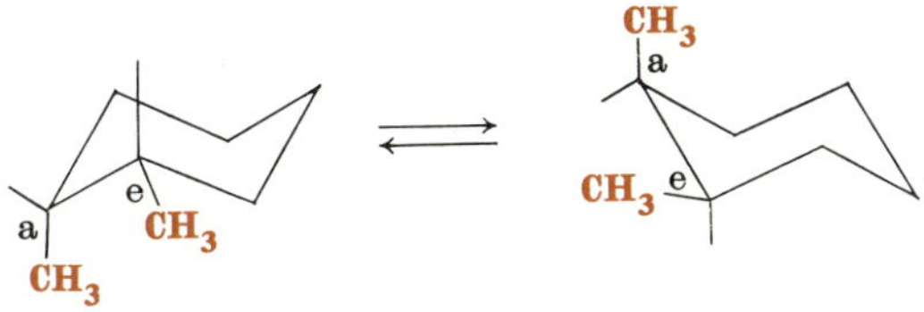

Fig. 5.24 Conformational equilibria for *cis*-1,2-dimethylcyclohexane

Let us now consider the *cis*-1,2-dimethylcyclohexane isomer. In this compound the methyl groups are on the same side of the ring which means that one is axial and the other equatorial (Fig. 5.24). When the ring flips, the axial methyl becomes equatorial and the equatorial methyl becomes axial. The two conformers are therefore identical in energy. It is now important to inquire which isomer, the *cis* or the *trans*, is the more stable. The *trans* compound can be assigned a value of 0.9 kcal/mole for its di-equatorial conformer (the di-axial conformer is of little consequence). The *cis* isomer, on the other hand, has three gauche interactions: one between the two methyl groups and two between the axial methyl and the ring. Thus, the *cis* has an energy of 2.7 kcal/mole or 1.8 kcal/mole more than the *trans*. We conclude that the *trans* is more stable than the *cis*.

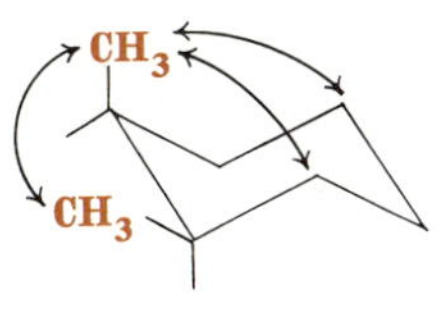

E = 2.7 kcal/mole

It is sometimes convenient to represent *cis* and *trans*-1,2-dimethylcyclohexane in a planar manner as shown below.

It is tempting to conclude from such drawings that the *trans* is more stable than the *cis* because the *trans* has its methyl groups further apart and therefore there is less steric interaction. Although the *trans* is indeed more stable than the *cis*, the rationale is *incorrect* because the two methyl groups in *both* the *cis* and the *trans* are 60° apart as seen in Fig. 5.25. Thus, despite the suggestion of the planar drawings to the contrary, the distance between the methyl groups in the *cis* compound equals the distance between the methyl groups in the *trans* isomer. Construction of molecular models would help the reader visualize this fact.

Fig. 5.25 Pictures of *trans* and *cis*-1,2-dimethylcyclohexane looking down the carbon-carbon bond bearing the methyl groups. The dotted line represents the six-membered ring.

Inspection of three-dimensional drawings (Fig. 5.26) or molecular models of *cis*-1,3-dimethylcyclohexane reveals that the two methyl groups must be either both axial or both equatorial in order to be on the same side of the ring. This contrasts with the 1,2-dimethylcyclohexane case in which the *cis* bonds are axial/equatorial. Clearly, the di-equatorial *cis* 1,3-conformer is more stable than the di-axial 1,3-conformer (by how much?).

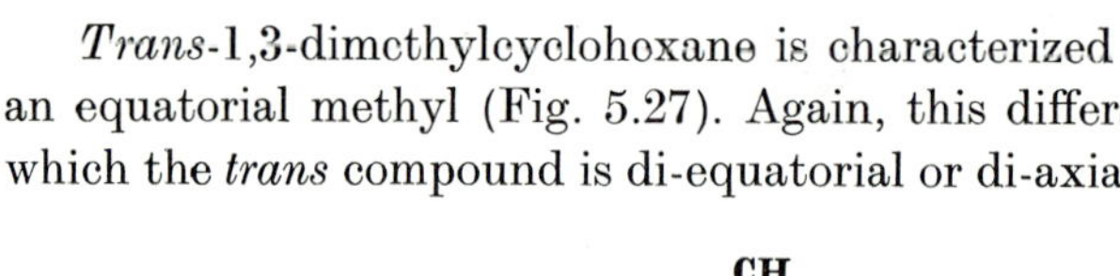

Fig. 5.26 Conformation equilibria for *cis*-1,3-dimethylcyclohexane

Trans-1,3-dimethylcyclohexane is characterized by an axial methyl and an equatorial methyl (Fig. 5.27). Again, this differs from the 1,2-isomer in which the *trans* compound is di-equatorial or di-axial.

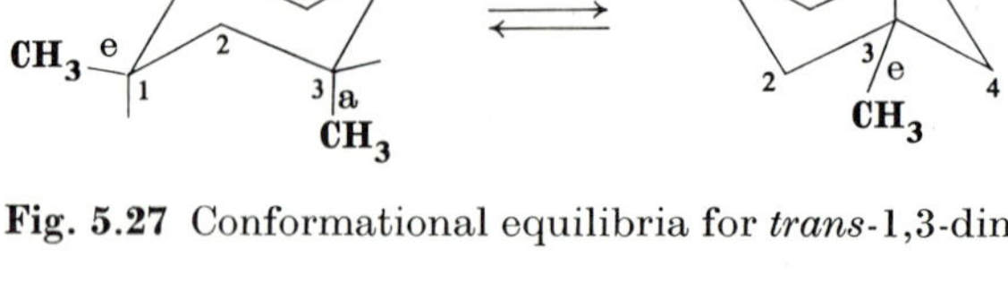

Fig. 5.27 Conformational equilibria for *trans*-1,3-dimethylcyclohexane

We leave it as an exercise for the student to determine the axial/equatorial relationships in *cis* and *trans*-1,4-dimethylcyclohexane.

ADDITIONAL READING

T. A. Geissman, *Workbook in Organic Chemistry*, San Francisco: W. H. Freeman, 1972. Chapter 3. Paperback.

M. S. Newman, Ed., *Steric Effects in Organic Chemistry*, New York: Wiley, 1956. Chapter 1.

C. D. Whittaker, *Stereochemistry and Mechanism*, Clarendon, Texas: Clarendon Press, 1973. Paperback.

PROBLEMS

1. One can tell whether a compound is optically active by determining whether or not the compound is superimposable on its mirror image. On this basis determine if the following are optically active:

a)

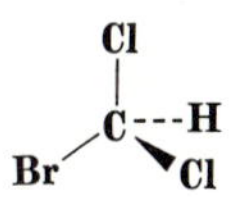

b)

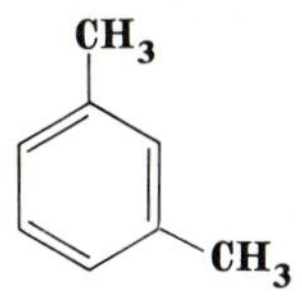

c)

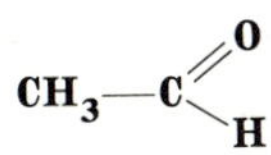

d)

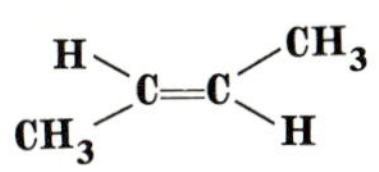

e)

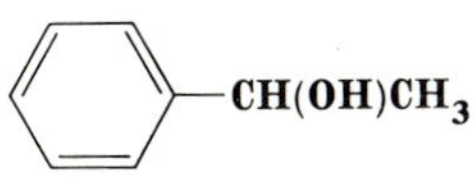

f)

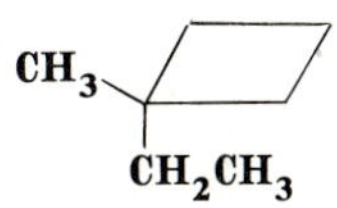

2. When a sample of 1-phenylethanol, purchased from a chemical supply house, was examined in a polarimeter, the plane of polarized light was not rotated. Explain the apparent lack of optical activity.

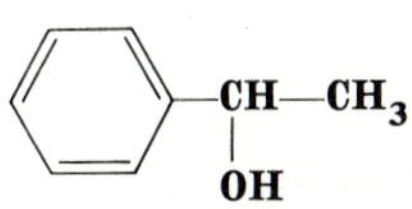

3. Does *s*-2-methylbutyraldehyde rotate polarized light in a levorotatory or dextrorotatory manner?
4. Draw an eclipsed conformation of *meso*-tartaric acid:

   HOOC–CHOH–CHOH–COOH.

5. Draw the highest and lowest energy conformations of *n*-butane. Explain the source of the energy difference.

6. True or false?
   a) If the *s* absolute configuration of a compound is dextrorotatory, then the R configuration must be levorotatory.
   b) Trichloromethane has a plane of symmetry.
   c) The reaction of two optically inactive compounds must necessarily give optically inactive products.
   d) In the Fischer projection, vertical bonds always project away from the viewer.
   e) Two enantiomers have identical physical and chemical properties other than their effect on polarized light.
   f) A compound with three different asymmetric centers has eight optical isomers (or four ± pairs).
   g) *Trans*-1,2-dimethylcyclohexane is more stable than *cis*-1,2-dimethylcyclohexane.
   h) *Trans*-1,2-dimethylcyclohexane and *cis*-1,2-dimethylcyclohexane are diastereomers.
   i) The staggered and eclipsed conformation of ethane interconvert rapidly and cannot be separated.
7. Draw the four optical isomers of 1-methyl-2-ethylcyclobutane. Indicate which are diastereomers and which are enantiomers.
8. For each of the following compounds indicate whether it is symmetric or dissymmetric, how many asymmetric centers it contains, and the configuration (R or S) of each of the asymmetric centers within it.

a) 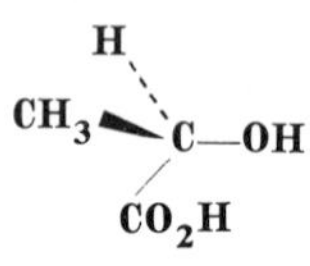

b) 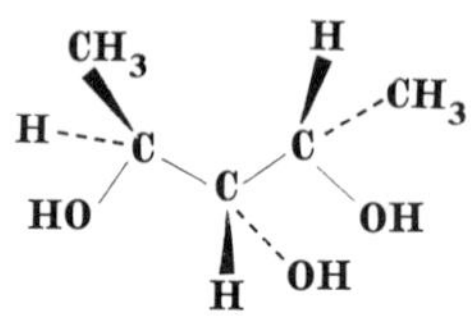

c) 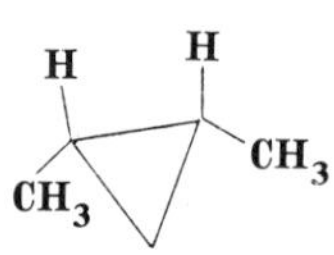

d) 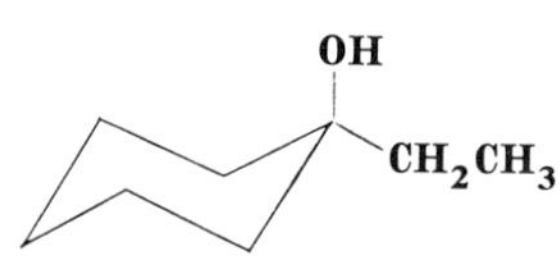

e) 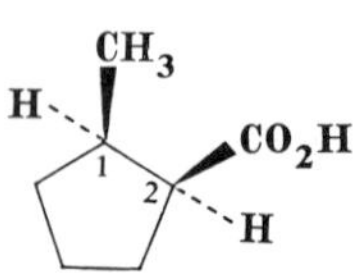

f) 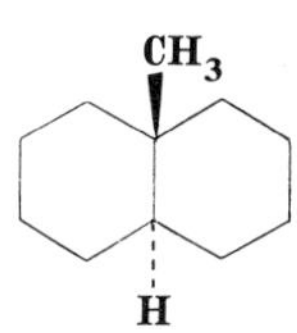

g) 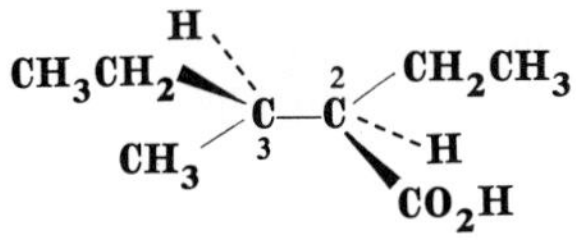

h) 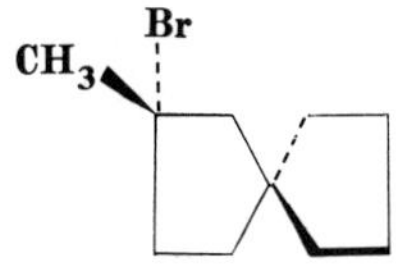

i) 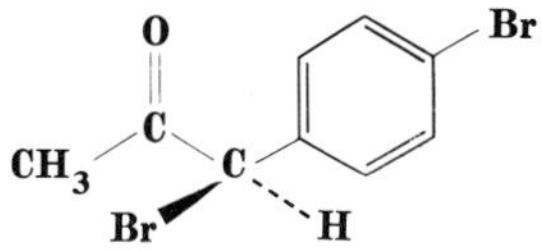

j) 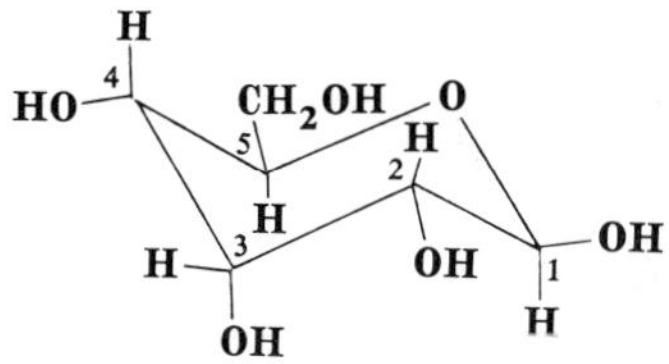

9. Give the number of optical isomers possible for the following compounds.

a) 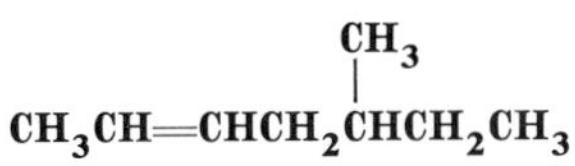

b) $CH_3CH_2CHClCH_2CH_3$

c) 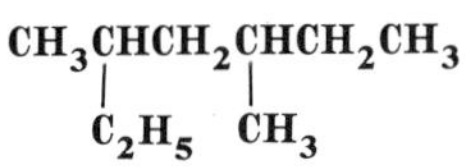

d) 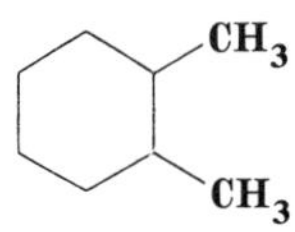

cis

e) 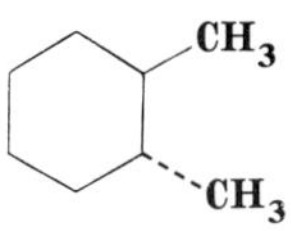

trans

f) 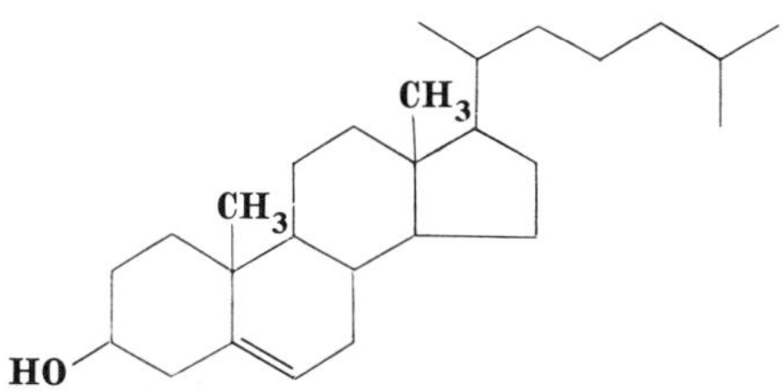

10. Draw Fischer projections of the following compounds and assign configurations (R or S) to each asymmetric center.

a) 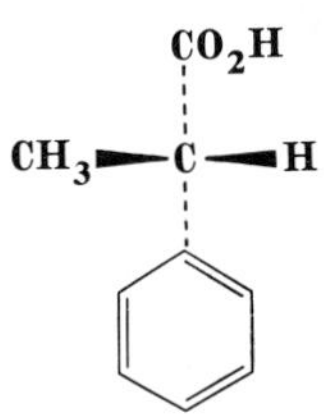

b) 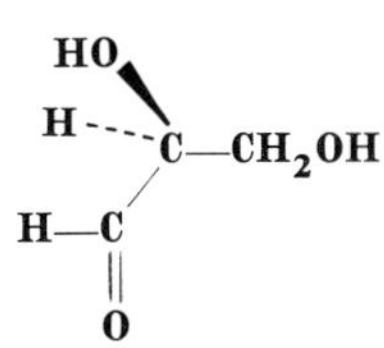

c) 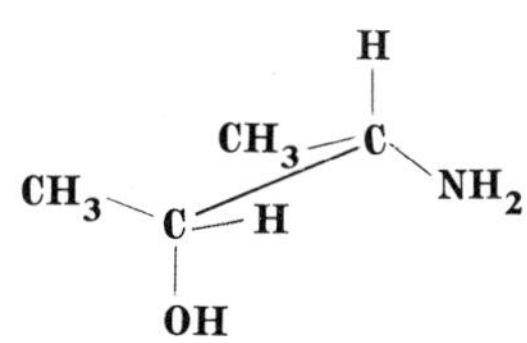

11. For each pair of Fischer projection formulas below indicate whether the two compounds are enantiomers, diastereoisomers, or identical.

a) 

b) 

c) 

12. Draw three-dimensional "sawhorse"

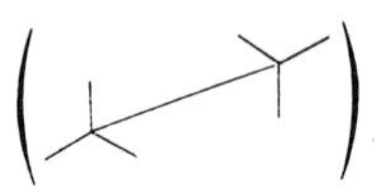

structures for the following compounds:

a) 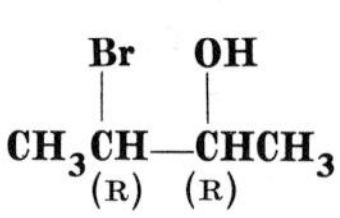
b) 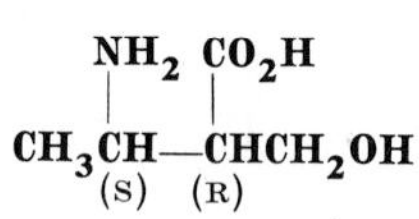
c) 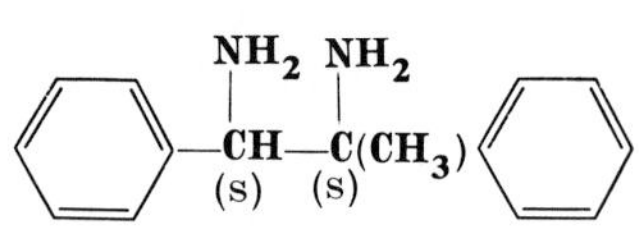

13. α-Pinene has two asymmetric carbon atoms. Only two optical isomers are known, however. Explain.

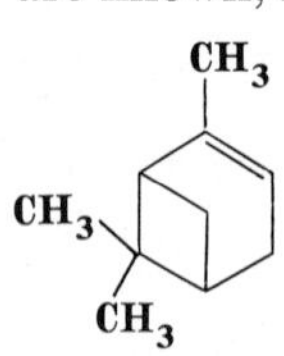

α-pinene

14. Draw the chair conformations of *cis*-1,2-dichlorocyclohexane.

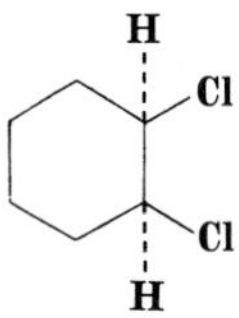

   a) Are these conformers superimposable?
   b) What is their stereochemical relationship?
   c) Explain why *cis*-1,2-dichlorocyclohexane cannot be obtained in an optically active form.

15. Draw the possible chair conformations of the following compounds.

a) 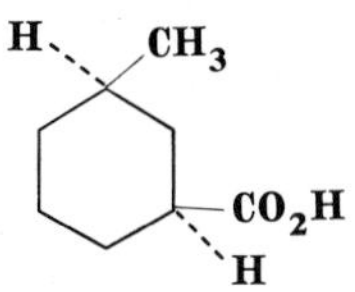

b) 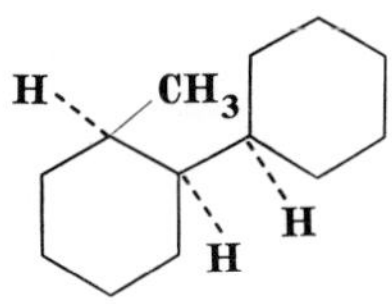

c) 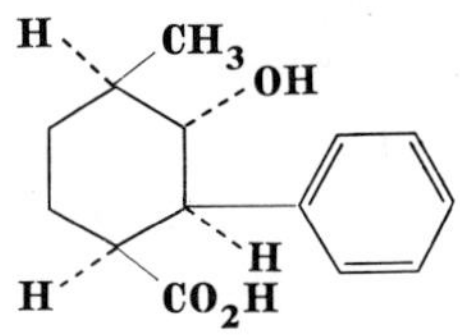

16. The preferred conformation of *cis* and *trans*-decalin are shown below:

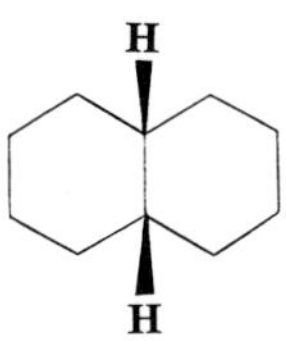

cis-decalin

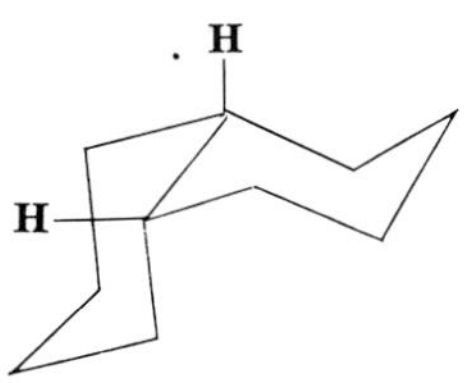

cis-decalin

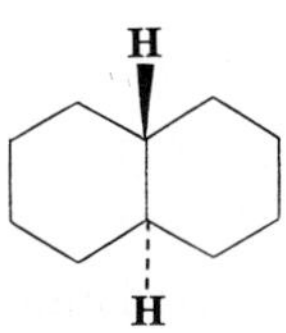

trans-decalin

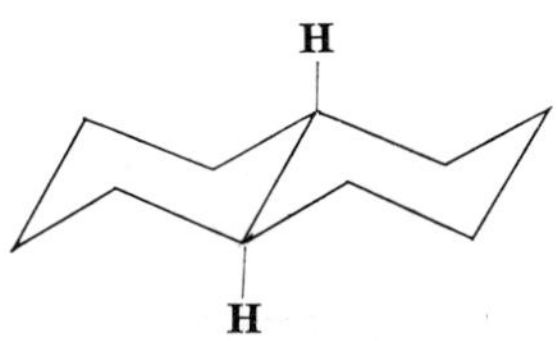

trans-decalin

The *cis* compound has a greater heat of combustion (it is *less* stable) than the *trans* compound. Suggest an explanation.

17. A 1:1 mixture of the (+)-rotating and (−)-rotating form of compound X (below) cannot be separated by ordinary physical means (crystallization, distillation, etc.). When this mixture of (+)-X and (−)-X is reacted with (+)-hydratropic acid (Y), *two* compounds corresponding to structure Z

are produced. These *can* be separated by crystallization. Each form of Z can then be converted back to (+)-Y and to one of the optically active forms of X. This is the way a dl mixture of a compound is separated or *resolved*.

a) Explain in terms of R and S configurations why the two isomers of Z are separable and the dl mixture of X is not.
b) Draw Fischer projections of the two isomers of Z.
c) What is the 1 : 1 mixture of (+)-X and (−)-X called?

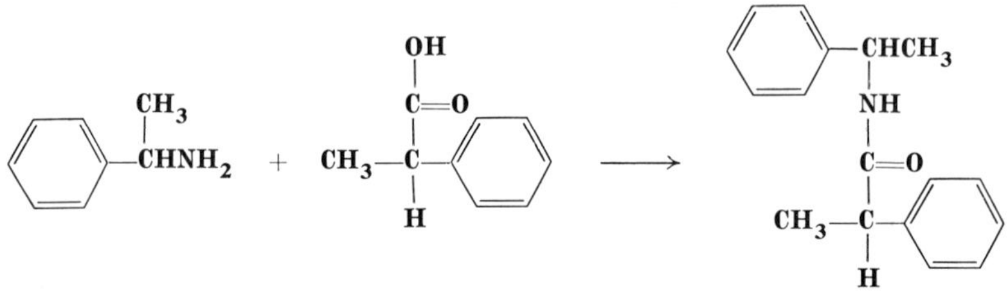

(±)-X (+)-Y Z (two compounds)

18. Draw the possible conformations of compound I in an increasing order of energy (decreasing order of stability). Include at least one boat form.

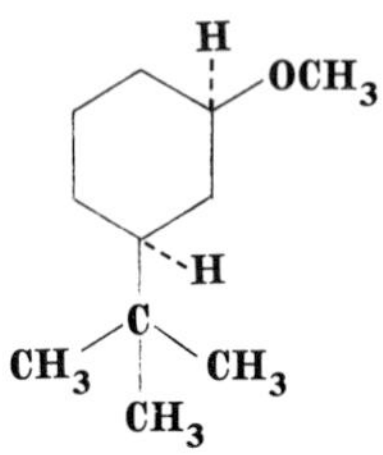

I

Conformational energy (axial vs. equatorial) of *t*-butyl = 6 kcal/mole.
Conformational energy (axial vs. equatorial) of OCH_3 = 0.74 kcal/mole.

19. An uninformed chemist attempted to prepare (+)-α-phenylethanol by reacting phenylmagnesium bromide with acetaldehyde, but he failed. The product he obtained was a racemic mixture, 50% (+), 50% (−), of α-phenylethanol. Explain.

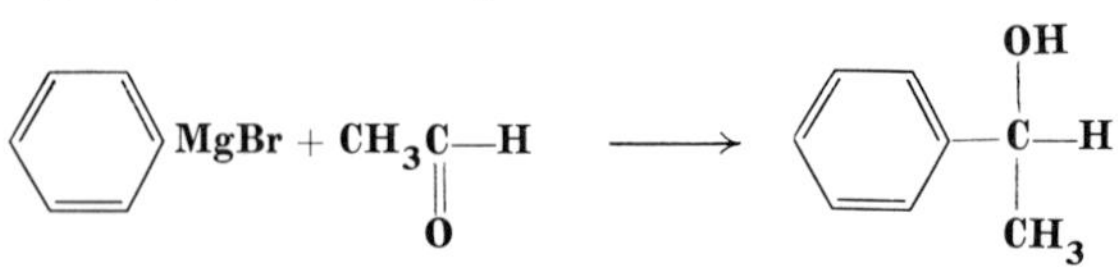

racemic

20. If one of the optical isomers of the compound below is treated with base, it loses its optical activity. Explain.

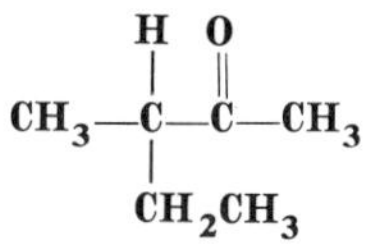

21. The *cis* isomer of 4-*t*-butylcyclohexanol is oxidized to the ketone by chromic acid (see Section 3.6) 2.97 times faster than the corresponding *trans* isomer at 25.0°. Cyclohexanol oxidizes 1.24 times faster than *trans*-4-*t*-butylcyclohexanol at 25.0°. What percentage of the cyclohexanol molecules have their hydroxyl groups in the *axial* position at 25.0°? Assume that the bulky *t*-butyl groups of both the *cis*- and *trans*-4-*t*-butylcyclohexanol are exclusively *equatorial*.

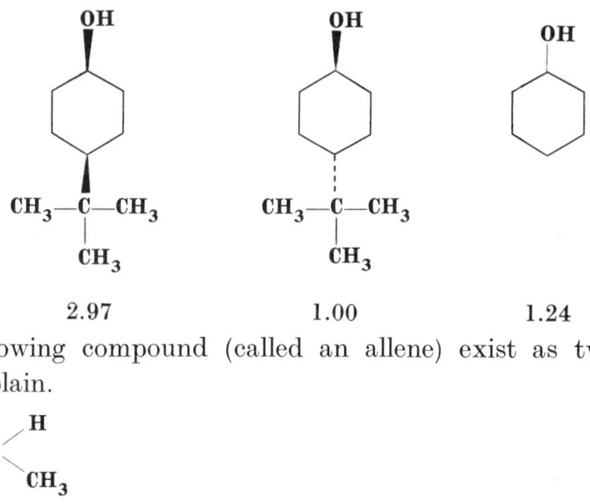

relative rate: 2.97 1.00 1.24

22. Can the following compound (called an allene) exist as two optical isomers? Explain.

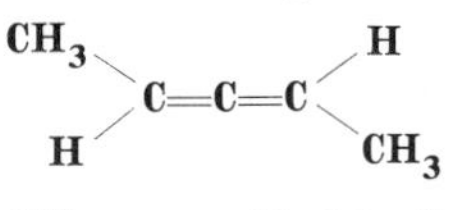

23. When an alkyl halide is treated with strong base, it is converted into an olefin according to the following reaction:

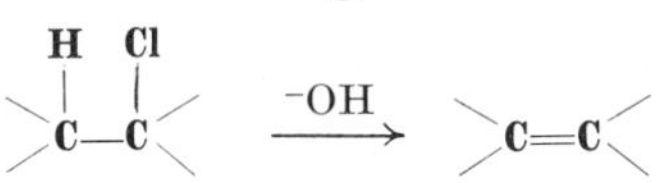

In order for this reaction to occur easily, the molecule must assume a conformation in which the departing hydrogen and chlorine have an "anti-parallel" arrangement:

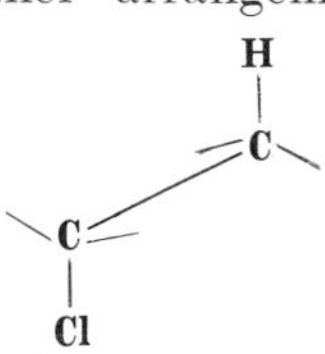

On this basis explain the following results:

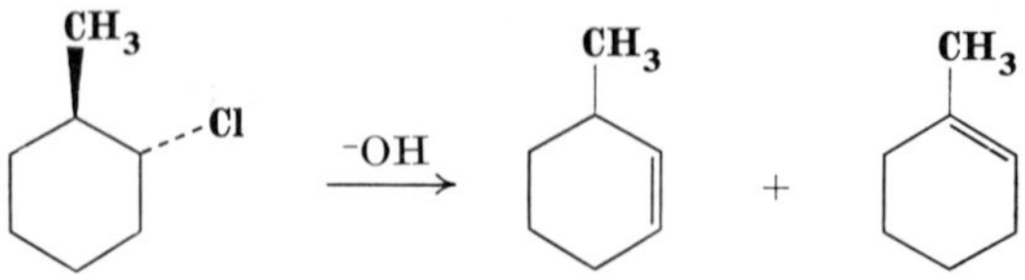

trans major product minor product

24. The following Grignard reaction yields two alcohols of identical formulas. Are they optically active? Are the two alcohols formed in identical amounts?

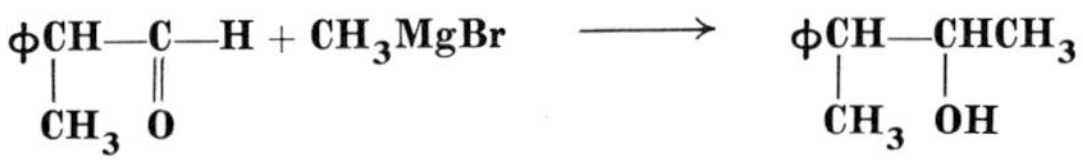

racemic

25. The following series of reactions was applied to the sugar threose in order to prepare a sugar alcohol of the general structure shown at the end of the sequence.

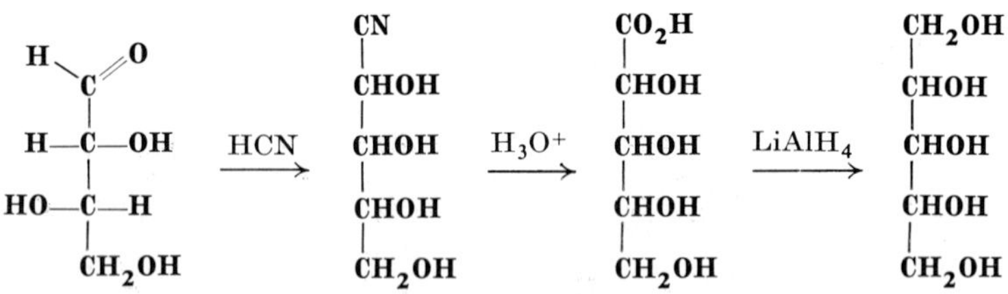

The final product actually obtained consisted of two compounds, one optically active and the other optically inactive. The latter could not be resolved into optically active forms. Draw Fischer projections of the two products. Explain why both are formed and why one of them is optically inactive.

26. Carbon chains of 50 or more atoms should be able to exist as knots:

Would you expect such a knot to be optically active, that is, nonsuperimposable on its mirror image?

6

STRUCTURE ELUCIDATION AND MOLECULAR SPECTROSCOPY

6.1 INTRODUCTION

In the preceding chapters we discussed various aspects of the structure and reactivity of organic molecules. We have been concerned with theories about the nature of chemical bonds, with the shapes of organic molecules, and with a few of the many reactions these substances undergo. However, we have not discussed how we know what the structure of a molecule is. When the organic chemist writes

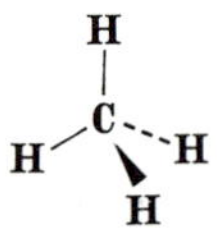

as the structure of a substance called methane, what empirical knowledge assures him that this picture has some real relationship to the actual constitution of methane? If we examine the physical properties of methane we find that it is a colorless, flammable gas. What other information can we acquire to buttress our faith in the structure we have accepted so readily to this point?

Our knowledge of the structures of most organic molecules has been gained through a combination of experimental data and deductive reasoning. Certain kinds of experimental data can answer quantitative questions about the structure of a molecule; questions such as, of what elements in what proportions does the molecule consist? what is the molecular weight of the compound? Other kinds of experimental data yield far less absolute answers. If we subject a compound of unknown structure to a variety of reaction conditions—acids, bases, oxidizing agents, reducing agents—can we propose a unique structure that will fit the results of all of these reactions? In chemical research we are constantly aware of the old saw that "the exception proves (tests) the rule." When a proposed structure fails the test of a reaction sequence whose outcome we know from myriad analogous compounds, we must modify our structural hypothesis accordingly.

A structural proposal must also fit the results of *molecular spectroscopy*. In molecular spectroscopy a molecule is allowed to interact with a given region of the electromagnetic spectrum. Depending upon its structure, the molecule absorbs specific frequencies of radiation within that region. In the second part of this chapter we will examine how information obtained from infrared, ultraviolet, and nuclear magnetic resonance spectroscopy may be used to determine the number and nature of functional groups within a molecule, and the interrelationship of these groups within a possible structure.

Two other kinds of instrumental analysis are also of great importance in determining organic structures. The first, *X-ray crystallography*, is often considered to be the ultimate method for determining structure. In X-ray crystallography a high-energy beam of X-rays is passed through a single crystal of a compound and projected onto a photographic plate. The X-rays are diffracted by the nuclei of the molecule's atoms, and analysis of the diffraction pattern obtained on the photographic plate reveals both the nature and the position of these nuclei. Although this method yields an "absolute" answer, its application is limited because it is necessary to use perfect crystals and the analysis is time-consuming (even with high-speed computers). In addition, molecules are occasionally susceptible to structural change upon X-ray bombardment and a false structure may be obtained.

The second method, *mass spectrometry*, involves bombarding a molecule in the gaseous state with a high-energy beam of electrons. The molecule becomes ionized by losing one or more of its own electrons. The resulting positive ion is then accelerated into a magnetic field where its path is deflected a certain amount, depending on the ion's mass-to-charge ratio. The strength of the magnetic field is set to deflect the ion along a predetermined path to a detector. Knowing the field strength and the accelerating voltage, we are then able to establish the actual mass-to-charge ratio. Since most molecules lose only a single electron in the mass spectrometer, we obtain the actual mass, or molecular weight, of the compound. We may also obtain a considerable amount of information about the structure of an organic compound from its mass spectrum. The energy imparted to the molecule by the ionizing electron beam is sufficiently great to cause the molecule to *fragment*; that is, to decompose into neutral particles and into a variety of positive ions whose mass is smaller than that of the original substance. Only the ions are detected and their masses can be measured from the value of the magnetic field strength. Fragmentation patterns for a variety of different molecules have been deduced from the mass spectra of many organic compounds, and the spectrum of an unknown compound can often be correlated with that of a known material.

6.2 COMBUSTION ANALYSIS

Let us examine the kinds of information we can gather to help formulate a structure for methane. First we need to know which atoms comprise the molecule, the proportions in which they are present, and the molecular weight of the substance. We can answer the first two questions by using one of the most common reactions of organic compounds, their susceptibility to oxidation. The only products of the combustion of methane in oxygen are carbon dioxide and water:

$$\text{Methane} + O_2 \longrightarrow CO_2 + H_2O$$

From this simple experiment we learn that methane contains carbon and hydrogen. We still do not know, however, whether oxygen was present in the original molecule. If we now carry out the same oxidation reaction in a quantitative manner, and calculate the percentages of hydrogen and carbon in the original molecule, we can demonstrate that methane contains only carbon and hydrogen, and that they are in a proportion of 1:4.

Quantitative determination of the percentages of carbon and hydrogen in an organic molecule is called *combustion analysis*. Combustion analysis involves, first, weighing a sample of an unknown material and then heating it at about 700°C in a stream of pure oxygen in a tube containing an oxidation catalyst such as copper oxide. The gaseous products of the combustion (water and carbon dioxide) are then passed through two absorption tubes, one containing a drying agent to absorb the water and the other containing a reagent (usually sodium hydroxide on asbestos) to absorb the carbon dioxide. From the increase in weight of these two tubes we obtain the weights of the water and carbon dioxide formed in the combustion reaction, and from these weights we calculate the percentages and proportions of carbon and hydrogen in our sample of methane. The following example shows how to obtain the percentage composition and the empirical formula of methane from a typical combustion analysis.

A sample of methane weighing 12.48 mg is burned in a stream of oxygen to yield 34.18 mg of carbon dioxide and 28.00 mg of water. The weight of carbon in the original methane sample is

$$34.18 \times \frac{12.01}{43.99} = 9.34 \text{ mg}$$

and the weight of hydrogen in the original sample is

$$28.00 \times \frac{2.02}{18.01} = 3.14 \text{ mg.}$$

The percentage composition of methane must then be:

$$\% \text{ carbon} = \frac{9.34}{12.48} \times 100 = 74.92$$

$$\% \text{ hydrogen} = \frac{3.14}{12.48} \times 100 = 25.08.$$

Since these percentages total 100%, methane cannot contain any elements other than carbon and hydrogen.

The percentages of the elements obtained in this combustion analysis also allow us to calculate an *empirical formula* for methane. The empirical formula tells us the relative number of carbon atoms to hydrogen atoms in methane. If we divide the percentage of each atom present by its gram atomic weight, we will obtain the relative number of atoms of that element in the compound. Thus:

$$\text{number of carbon atoms} = \frac{74.92}{12.01} = 6.23$$

$$\text{number of hydrogen atoms} = \frac{25.08}{1.008} = 24.88.$$

The empirical formula of methane is therefore $C_{6.23}H_{24.88}$.

Although this formula describes the correct ratio of carbon to hydrogen in methane, we know that molecules always contain whole numbers of atoms, and we therefore convert this formula into one containing the equivalent smallest whole numbers.

$$\text{number of carbon atoms} = \frac{6.23}{6.23} = 1$$

$$\text{number of hydrogen atoms} = \frac{24.88}{6.23} = 3.99 \approx 4.$$

The empirical formula of methane is therefore CH_4.

6.3 MOLECULAR WEIGHT AND MOLECULAR FORMULA

An empirical formula tells us the ratio of atoms in a molecule; it does not tell us the absolute number of each of these atoms. Thus the *molecular formula* of methane could be CH_4, C_2H_8, C_3H_{12}, or any other multiple of the empirical formula. In order to obtain the molecular formula of a compound, we must have both its empirical formula and its *molecular weight*. A number of methods are available for obtaining molecular weights. For example, in the Dumas method of molecular weight determination the volume of a known weight of a gas is measured under conditions of known temperature and pressure. Since we know that one gram molecular weight of a gas will occupy 22.4 liters at 0°C and 760 mm pressure, we can calculate what weight of our unknown would occupy this volume at STP. Another method for obtaining the molecular weight of our unknown is to measure its mass spectrum and to determine the mass of the molecular ion in that spectrum. In the case of methane either of these methods would yield a molecular weight of 16. (In actual practice the experimentally determined molecular weight would be measured with considerably more precision than that indicated by the whole number 16. Mass spectrometric determination can easily yield answers accurate to five or six decimal places.) On the basis of a molecular weight of 16, the only molecular formula for methane we can write is the same as its empirical formula, CH_4.

6.4 THE STRUCTURE OF METHANE

The molecular formula we have so far obtained for methane still tells us nothing about the *structure* of the molecule. In order to write a structure for this molecule we must apply some of the ideas we have gained from our experience with inorganic chemistry to some additional experimental information about methane. We know from the study of such compounds as water, ammonia, hydrogen chloride, and hydrogen sulfide that hydrogen is monovalent. Thus there are two possible kinds of structures we can write for methane, namely ionic or covalent:

ionic $(H^+)_4C^{-4}$ or $(H^-)_4C^{+4}$,

covalent

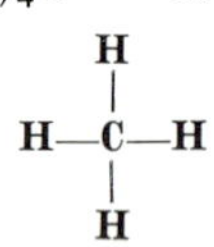

We also know the following physical and chemical characteristics of methane: it is a gas at normal temperatures, it is insoluble in water, and neither the liquid form of the substance nor its solutions will conduct electricity. Methane is, therefore, clearly nonionic, either as a pure substance or in solution. Our ionic structures must then be incorrect, and the covalent structure must be at least a first approximation of what the methane molecule looks like.

6.5 TETRAHEDRAL STRUCTURE

The covalent structure we wrote for methane is still only a partial structure, since it does not describe the geometry (bond angles) of the molecule. If we accept the idea that all four of the C—H bonds are identical (since no chemical experimentation with methane has ever indicated otherwise), we can describe a number of possible structures. These are shown in Fig. 6.1.

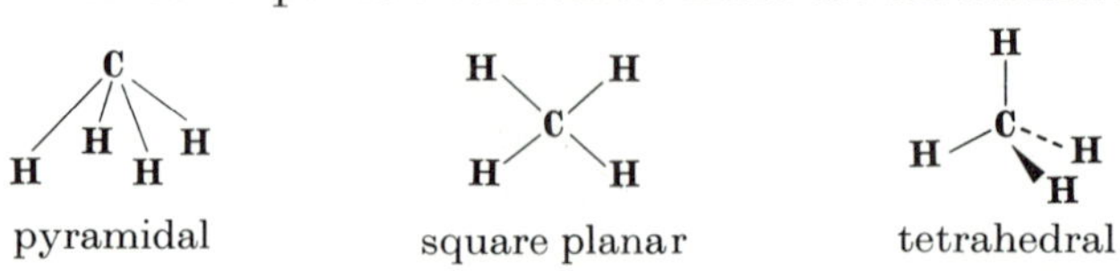

Fig. 6.1 Possible methane structures with four equivalent hydrogens

We can decide which of these structures best describes the three-dimensional nature of the methane molecule by examining the products obtained when methane undergoes light-catalyzed chlorination:

$$CH_4 + Cl_2 \longrightarrow CH_3Cl + CH_2Cl_2 + CHCl_3 + CCl_4$$

The empirical and molecular formulas of all of these products may be obtained the same way we obtained those of methane. Qualitative and quantitative analyses of the elements in each product are performed, and molecular weight measurements are made. We shall now focus our interest on just one of these chlorinated methanes, dichloromethane, CH_2Cl_2. Using the structures we suggested for methane we can write five possible structures for dichloromethane. These are shown in Fig. 6.2.

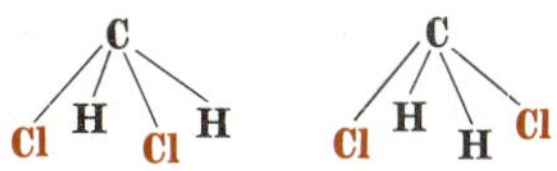

two pyramidal structures

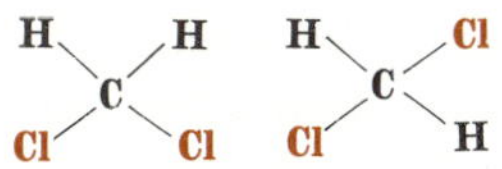

two square planar structures

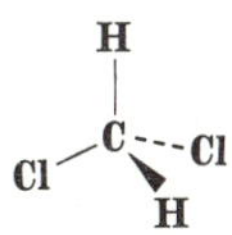

one tetrahedral structure

Fig. 6.2 Derived structures for dichloromethane

In contrast to methane, our structural possibilities for CH_2Cl_2 include two isomeric structures for both the pyramidal and square planar arrangements and only one for the tetrahedral configuration. It has been found that no matter how one chlorinates methane, only *one* substance of molecular formula CH_2Cl_2 is ever obtained. The tetrahedral structure must then be the correct "one". By analogy we can argue that methane itself should have a tetrahedral structure. Although this argument does not provide absolute proof of the tetrahedral structure of methane, it is a strongly presumptive case. In addition, all our other empirical knowledge of the structure of tetracoordinate carbon compounds agrees completely with this conclusion.

6.6 CITRONELLAL—SITES OF UNSATURATION

Let us now examine the proof for the structure of the compound called citronellal—a more complex organic molecule than methane—which is obtained from citronella oil (*Cymbopogon nardus* L.). In order to do this we must assume that we know something of the structural nature of organic compounds, and something about the way different kinds of compounds react under given conditions. Using combustion analysis and molecular weight determination, we obtain $C_{10}H_{18}O$ as the molecular formula of this compound. This formula alone tells us something of the nature of the substance. Any *saturated* carbon compound containing only the elements carbon and hydrogen (or carbon, hydrogen, and oxygen) must correspond to the general formula, C_nH_{2n+2}. (Consider CH_4 and CH_3CH_2OH, for example.) When the molecular formula of a substance falls short of this general formula by multiples of two hydrogens, each missing pair of hydrogens indicates the presence of either a double bond or a ring. For example, ethylene (C_2H_4) may be described by the formula $C_nH_{(2n+2)-2}$, and must therefore contain either one double bond or one ring. The molecular formula of cyclohexene (C_6H_{10}) is *four* hydrogens short of the number prescribed by the general formula. Hence it must contain one of the following: two double bonds, two rings, one double bond and one ring, or one triple bond (equivalent to two double bonds). For each pair of hydrogens by which the molecular formula of a compound differs from the formula of the corresponding *acyclic* hydrocarbon we say that the molecule has one *site of unsaturation*. In the case of citronellal, we must compare its formula ($C_{10}H_{18}O$) to the corresponding acyclic hydrocarbon ($C_{10}H_{22}$). Our formula for citronellal is four hydrogens shy of the hydrocarbon formula, and the molecule must therefore contain two sites of unsaturation.

6.7 SOME SIMPLE REACTIONS OF CITRONELLAL

The nature of the sites of unsaturation in the citronellal molecule may be gleaned from a few simple reactions this substance undergoes. As shown in Fig. 6.3, the substance reacts with bromine to yield a dibromo derivative, $C_{10}H_{18}OBr_2$.

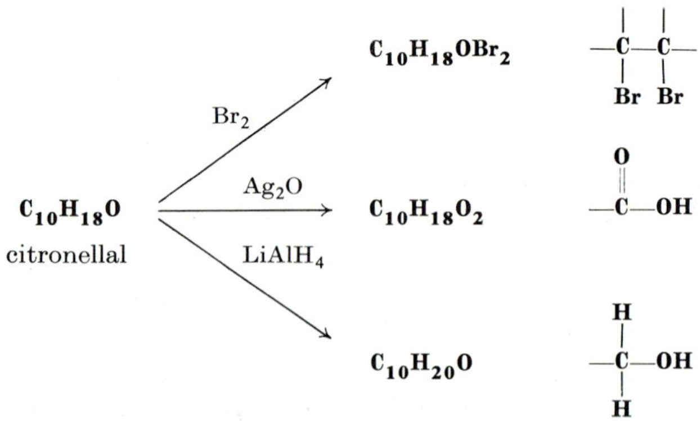

Fig. 6.3 Reactions of citronellal

This is a reaction common to compounds containing a double bond. In addition, we find that a mild oxidizing agent like silver oxide converts citronellal into a substance ($C_{10}H_{18}O_2$) which has the properties of an acid. In a third reaction, reduction by lithium aluminum hydride, citronellal is transformed into an alcohol, $C_{10}H_{20}O$.

On the basis of this evidence we may now say the citronellal molecule must be characterized by the groups,

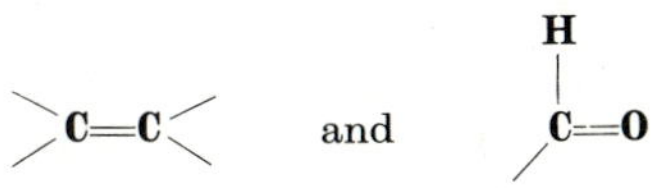

since our bromine reaction has shown the presence of a double bond, and our oxidation and reduction experiments are compatible with the presence of an aldehyde function.

6.8 OZONIZATION

When a compound containing a double bond is allowed to react with ozone (O_3), ozone is added to the double bond and cleavage occurs, leading to the formation of a substance called an ozonide:

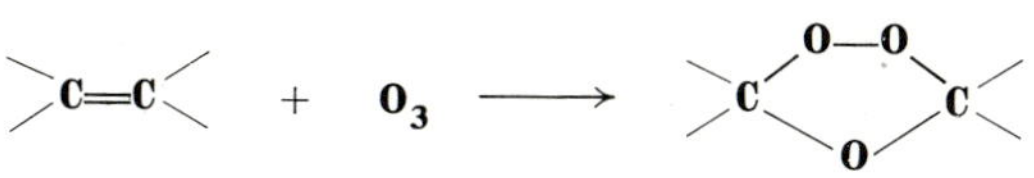

olefin ozone an ozonide

When the ozonide is then treated with water in the presence of zinc metal and acetic acid, two carbonyl groups are produced:

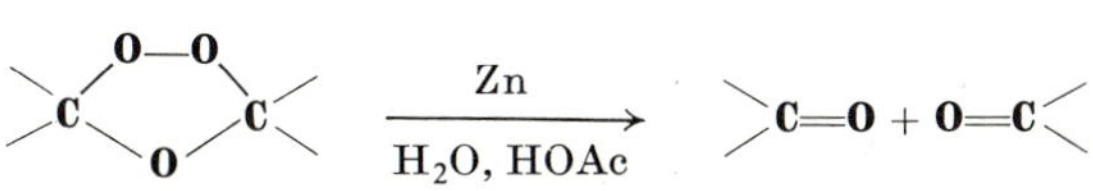

These carbonyl groups will appear either as aldehydes or as ketones, depending on the nature of the original substitution of the double bond, and they will be found either in two separate molecules or in the same molecule, depending on whether we start with a double bond in a chain or in a ring. Figure 6.4 illustrates the results of this process, called *ozonization* or *ozonolysis*, with a representative number of olefins.

If we examine the products of ozonolysis we realize that the carbonyl groups of these compounds occur at positions that were originally linked by a carbon-carbon double bond. Thus we can reconstruct the nature of an olefin

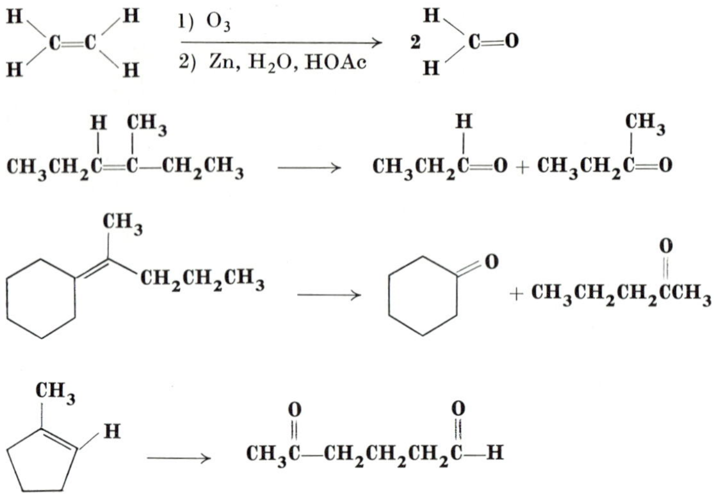

Fig. 6.4 Typical ozonization products of olefins

if we know what aldehydes and/or ketones it yields upon treatment with ozone. For example, if an unknown olefin ($C_{12}H_{24}$) yields

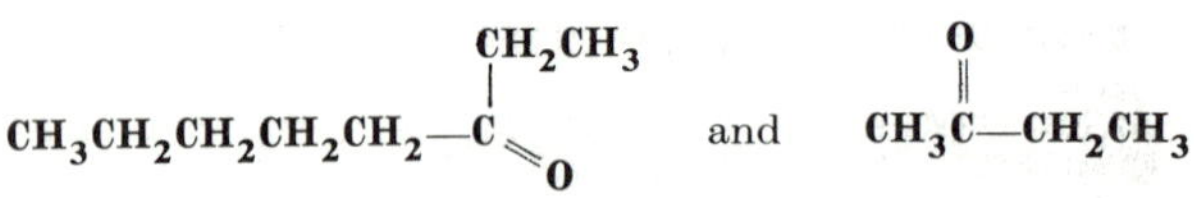

upon ozonization, then it must have had the structure

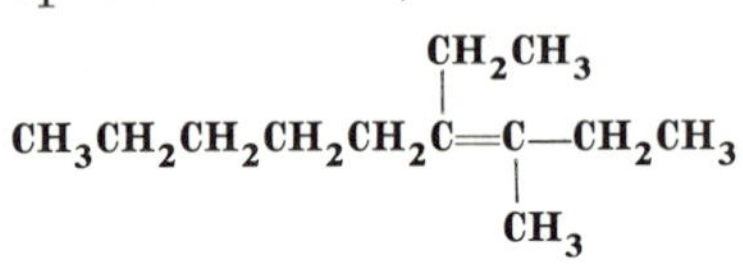

6.9 OZONIZATION OF CITRONELLAL

When citronellal reacts with ozone, two products are obtained:

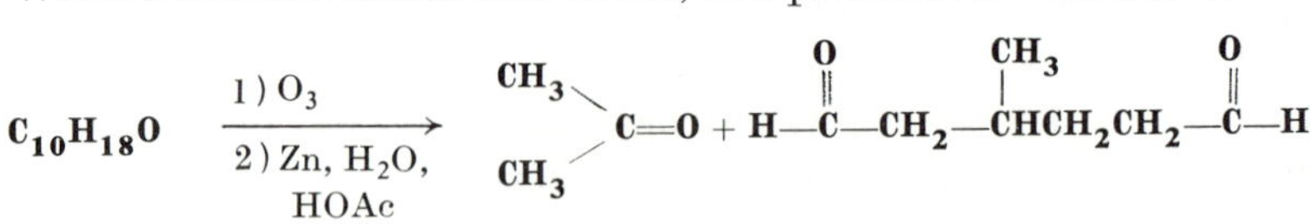

acetone

Unlike the examples in Fig. 6.4, however, this result does not allow us to write an unambiguous structure for our unknown. We can position one terminus of the double bond of citronellal at the carbon atom which appears as the carbonyl carbon of acetone. The position of the other terminus, however, is uncertain, since the second product of ozonolysis contains *two* aldehyde functional groups. One of these was of course present in the original molecule, but we do not know which one. Therefore at this stage we must write two possible structures for citronellal:

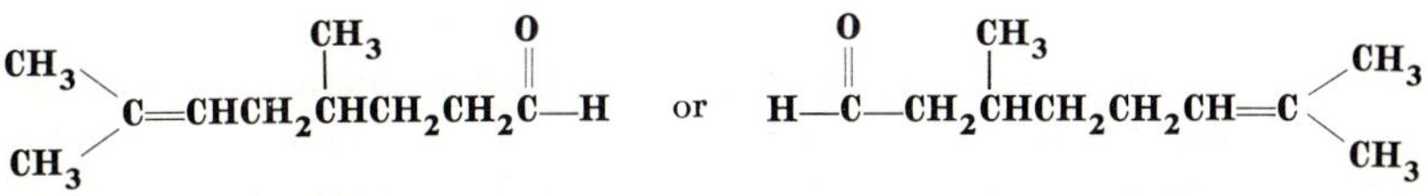

In order to solve our structural problem we must find some way to distinguish which of the aldehyde functions of the ozonolysis product was originally present in citronellal. One way to do this is to completely remove the aldehyde function of the starting material by subjecting citronellal to the following series of reactions. Letting R—CHO symbolize the structure of the starting material, where $R = C_9H_{17}$, we obtain $R—CH_3$:

$$\mathrm{RCHO} \xrightarrow{\mathrm{LiAlH_4}} \mathrm{RCH_2OH} \xrightarrow{\mathrm{PBr_3}} \mathrm{RCH_2Br} \xrightarrow{\mathrm{Mg}} \mathrm{RCH_2MgBr} \xrightarrow{\mathrm{H_3O^+}} \underset{C_{10}H_{20}}{\mathrm{RCH_3}}$$

Ozonization of the hydrocarbon $C_{10}H_{20}$ now leads to the two products

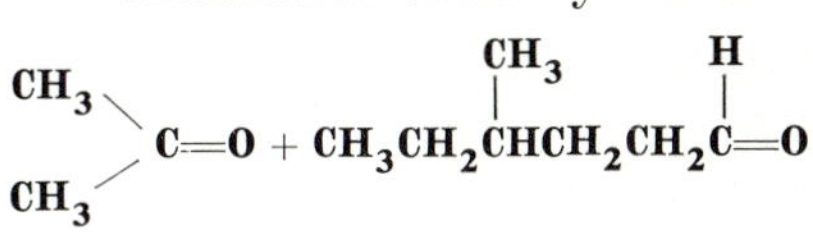

The structure of the hydrocarbon $C_{10}H_{20}$ must then be

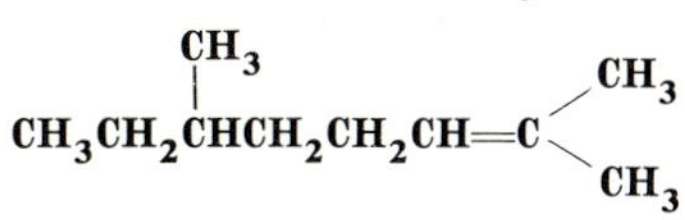

and citronellal can be unambiguously represented by

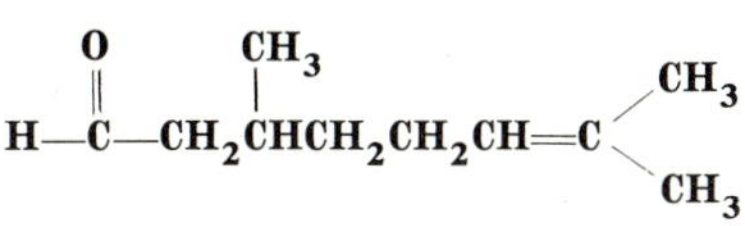

This example of a structure proof is admittedly an idealized one. We have had to assume that we can easily identify the structures of ozonolysis products. However, it is often necessary in many structure determinations to carry out a wide variety of degradative and transformation reactions in order to obtain substances of known structure from which one can logically deduce the composition of the unknown compound.

6.10 MOLECULAR SPECTROSCOPY

In molecular spectroscopy a beam of electromagnetic radiation is passed through a material. The absorption of energy over a range of frequencies is measured, and the specific frequencies at which absorption occurs are correlated with the structure of the sample compound.

The relationship between energy and frequency is given by

$$\Delta E = h\nu,$$

where ΔE is the difference in energy between the ground state and the excited state of the molecule, ν is the frequency of the radiation, and h is Planck's constant. We should recall also that frequency and wavelength (λ) are related by the expression

$$\nu = c(1/\lambda)$$

where c is the velocity of light.

Radiation in the ultraviolet region, 200 nm to 400 nm (1 *nanometer* = 10^{-9} meters), causes relatively high-energy transitions; infrared radiation, 1 μ to 25 μ (1 $\mu = 10^{-6}$ meters), effects lower-energy transitions. Figure 6.5 illustrates the various regions of the electromagnetic spectrum and the kind of molecular spectroscopy associated with them.

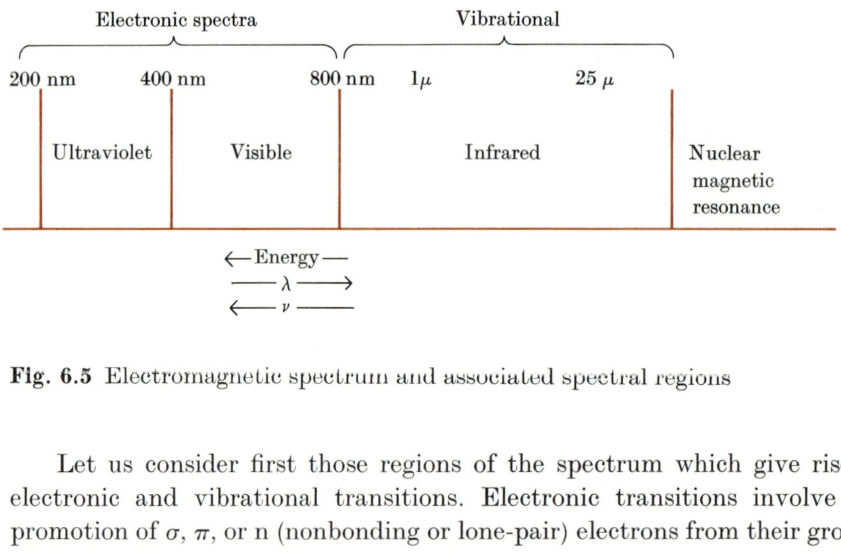

Fig. 6.5 Electromagnetic spectrum and associated spectral regions

Let us consider first those regions of the spectrum which give rise to electronic and vibrational transitions. Electronic transitions involve the promotion of σ, π, or n (nonbonding or lone-pair) electrons from their ground states to higher energy states. These processes require radiation in the relatively higher-energy ultraviolet and visible portions of the spectrum. In the infrared region vibrational transitions occur. The atoms of a molecule are in constant vibration about the bonds that join them. Infrared-induced transitions affect the amplitude of these vibrations. We should remember that all of these transitions are quantized: a given transition requires a specific amount of energy.

6.11 ULTRAVIOLET SPECTROSCOPY

In our discussion of bonding we pointed out that molecular orbitals may be constructed by combining atomic orbitals. It is a general rule that this combining process does not alter the number of orbitals. That is, for every two atomic orbitals that combine we produce two molecular orbitals. One of these molecular orbitals is called a bonding orbital and is associated with what we normally consider a "bond" between two atoms. The other is called an antibonding orbital and it is usually devoid of electrons. Thus, if two sp^3 orbitals (each with one electron) combine to form a σ-bonding orbital, they also produce an antibonding orbital designated as σ^*. Similarly, if two parallel p orbitals combine, they produce a π-bonding orbital and a higher energy π^*-antibonding orbital:

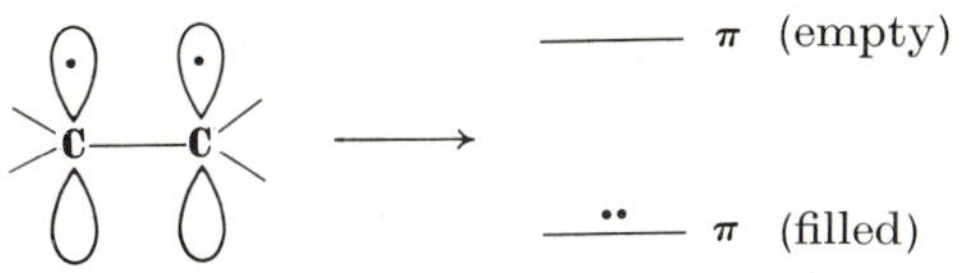

In organic chemistry we normally deal with the lower-energy bonding molecular orbitals since these are the ones occupied by the bonding electrons. However, energy absorption in the ultraviolet and visible regions results in the transition of electrons from the low-energy bonding orbitals to the unoccupied high-energy antibonding orbitals:

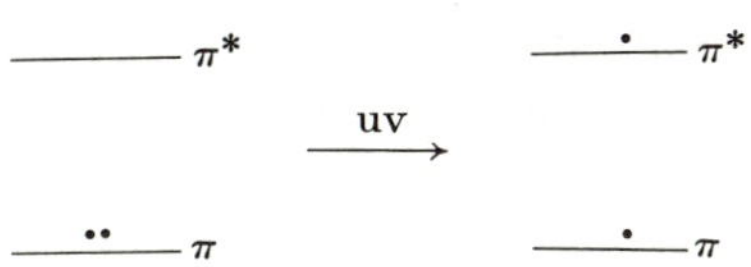

Even with intense ultraviolet light only a small fraction of the molecules are in the excited state at any given time.

Figure 6.6 illustrates the relative energy levels of the bonding, nonbonding, and antibonding orbitals (nonbonding orbitals contain unshared pairs of electrons). We can see that transitions from n and π orbitals to π^* orbitals (that is, $n \rightarrow \pi^*$ and $\pi \rightarrow \pi^*$ transitions) require less energy (and thus at higher wavelengths) than transitions from σ orbitals. For this reason the higher-energy σ-bond transitions are not obtained in most ultraviolet spectroscopy,

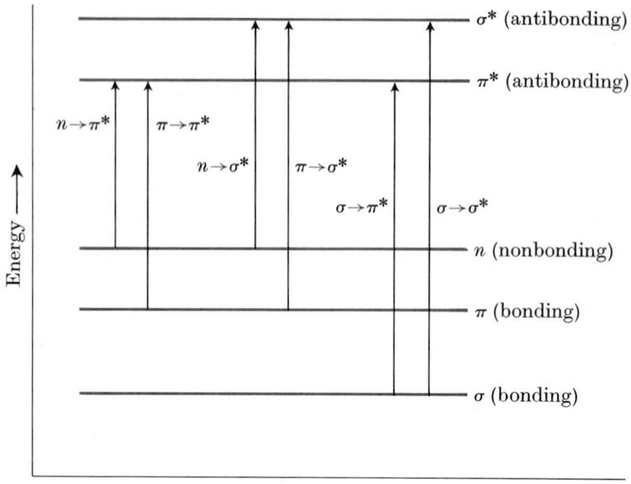

Fig. 6.6 Relative energy differences of electronic transitions

and this technique is most often used for obtaining structural information about unsaturated compounds.

Since the electronically excited states of organic molecules contain electrons in antibonding orbitals, we cannot represent these species by means of the usual structural formulas. We can, however, write approximate structures that aid us in the correlation of ultraviolet spectra if we take into account the apparent fact that the excited state has greater polarity (more charge separation) than the π-ground state. Thus the $\pi \rightarrow \pi^*$ absorption of acetone at 189 nm can be represented by a transition from the neutral acetone ground state to an excited state which more closely resembles a dipolar ion:

$$(CH_3)_2C{=}O \xrightarrow[\pi \rightarrow \pi^*\ 189\ \text{nm}]{h\nu} (CH_3)_2\overset{+}{C}{-}\overset{-}{O}$$

If we now place a double bond in conjugation with a ketone carbonyl group (i.e., C=C—C=O), resonance effects should stabilize the polar excited state more than the ground state, and therefore the difference in energy between the ground and excited states should be *less* than that for a simple ketone. This conclusion is borne out by the observation that the $\pi \rightarrow \pi^*$ transition of methyl vinyl ketone (217 nm)

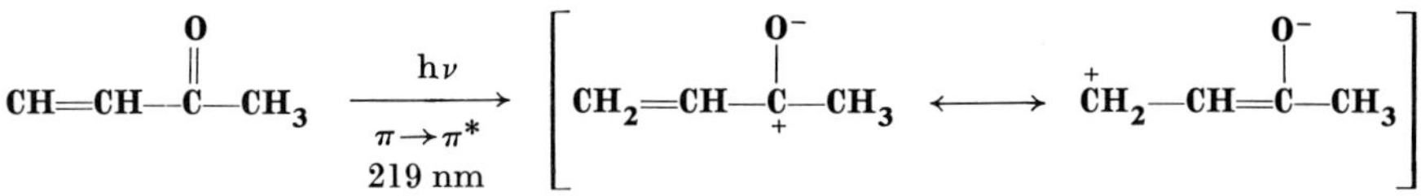

occurs at *lower* energy (longer wavelength) than does the transition of acetone. Thus one may correlate the position of the ultraviolet absorption with the degree of conjugation within a molecule (the greater the conjugation the longer the wavelength). Table 6.1 lists the wavelengths of absorption of some representative unsaturated compounds. A typical UV spectrum is shown in Fig. 6.7.

Table 6.1 Electronic transitions of some simple unsaturated organic compounds

Compound	Transition	λ_{max}*
$(CH_3)_2C{=}O$	$n \rightarrow \pi^*$	280
	$\pi \rightarrow \pi^*$	189
$CH_2{=}CH_2$	$\pi \rightarrow \pi^*$	162
$CH_2{=}CH{-}CH{=}CH_2$	$\pi \rightarrow \pi^*$	217
$CH_2{=}CH{-}C({=}O){-}CH_3$	$n \rightarrow \pi^*$	324
	$\pi \rightarrow \pi^*$	217

*λ_{max} = wavelength in nm units of maximum absorption of light.

The utility of ultraviolet spectroscopy in structure elucidation is nicely illustrated by a set of empirical rules which allow the prediction of λ_{max} for unsaturated ketones. According to these rules, an α,β-unsaturated ketone is given a base value of 215 nm; an alkyl group on the α carbon adds 10 nm; an alkyl group on the β carbon adds 12 nm; if there is a second conjugated double bond (in the γ,δ position), one adds 30 nm plus 18 nm for each γ or δ alkyl substituent. Thus, one calculates the following λ_{max} values for the two ketones below:

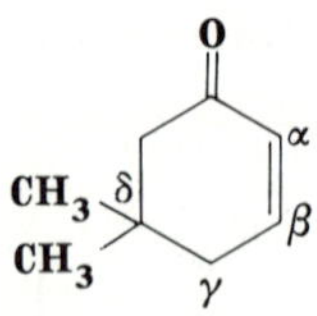

215 (base value)
12 (β alkyl group)
227 nm

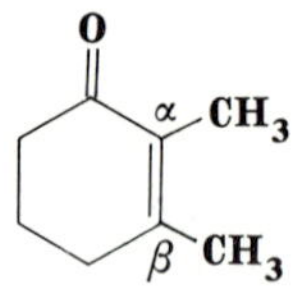

215 (base value)
10 (α alkyl group)
24 (2β alkyl groups)
249 nm

Note that the two δ methyl groups of the first compound do not add an increment to the base value of 215. This is because the methyl groups are not directly attached to the absorbing system. It is seen that the UV spectrum in Fig. 6.7 ($\lambda_{max} = 227$) is consistent with the first structure but not with the isomer absorbing at 249 nm.

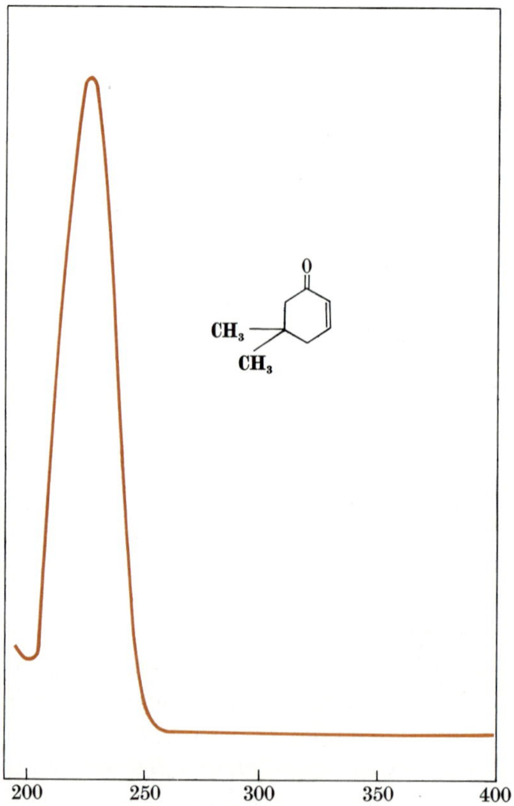

Fig. 6.7 Ultraviolet spectrum of 5,5-dimethylcyclohexenone. $\lambda_{max} = 227$ nm.

6.12 INFRARED SPECTROSCOPY

It was mentioned previously that the atoms of a molecule are in constant vibration. Consider a diatomic molecule where two atoms, A and B, are joined by a bond:

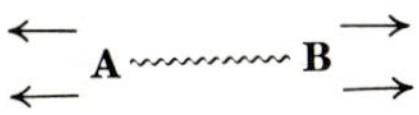

The two atoms vibrate much as two masses joined by a spring vibrate. The law that governs the motion of the spring is Hooke's Law, which states that the frequency of vibration, ν, is directly proportional to the spring constant, k (the tension of the spring), and inversely proportional to the mass m at the end of the spring:

$$\nu \propto \sqrt{\frac{k}{m}} \qquad \left(\nu \propto \frac{1}{\lambda}\right)$$

$$\therefore \lambda \propto \sqrt{\frac{m}{k}}$$

If we let k be a "force" constant for a bond (determined by the nature of the particular bond: the stronger the bond the greater the force constant, somewhat similar to the tension of a spring) and let m to be related to the masses of the bonding atoms then ν is the frequency of vibration of the two atoms. The frequency of vibration also corresponds to the frequency of the radiation needed to cause vibrational excitation of this bond. This spring-and-weight model of a molecule is a very useful one. For example, we can construct a model of a molecule by assembling a group of springs whose tensions are directly related to the force constants of the bonds that compose the molecule, and a group of weights whose masses are proportional to those of the atoms of the molecule. If we then set the model in vibration, we will find that every mode of vibration experienced by the model corresponds to a frequency of radiation absorbed by the compound in the infrared region. Therefore it is apparent that two compounds are not likely to have the same set of infrared absorptions. This procedure can be used to demonstrate the identity or non-identity of two materials.

The nature of the functional groups that give rise to absorption of radiation in the infrared is quite varied compared to those that absorb in the ultraviolet. Table 6.2 lists the wavelengths of exciting radiation for some functional groups. From the table we see, for example, that if a compound has a hydroxy group, there will be an absorption in the 3600 cm^{-1} region. Furthermore, we may rule out the presence of a hydroxy group in a material that does not exhibit this absorption.

Table 6.2 Some characteristic infrared absorption frequencies*

Bond	Type of compound	Frequency,† cm^{-1}	Intensity
—C—H (saturated C)	alkanes	2850–2960	strong
—C—D (saturated C)	alkanes	~2200	strong
=C—H	alkenes and arenes	3010–3100	medium
≡C—H	alkynes	3300	strong, sharp
O=C—H	aldehydes	2720	medium
—C—C— (saturated C)	alkanes	600–1500‡	weak
>C=C<	alkenes	1620–1680	variable
—C≡C—	alkynes	2100–2260	variable
—C≡N	nitriles	2000–2300	strong
—C—O (saturated C)	alcohols —C—OH, ethers —C—O—C—,	1000–1300	strong

>C=O	aldehydes —C(=O)—H	1720–1740	strong
>C=O	ketones —C—C(=O)—C—	1705–1725	strong
>C=O	acids —C(=O)—O—H, esters —C(=O)—O—C—	1700–1750	strong
—O—H	alcohols —C—O—H, phenols =C—O—H	3590–3650	variable, sharp
—O—H	hydrogen-bonded, alcohols and phenols —O—H···O<	3200–3400	strong, broad
—O—H	hydrogen-bonded, acids —O—H···O<	2500–3000	variable, broad
$—NH_2$	amines $—C—NH_2$	3300–3500 (double peak)	medium
>N—H	amines —C—N(H)—C—	3300–3500 (single peak)	medium

* Reproduced from J. D. Roberts and M. Caserio, *Basic Principles of Organic Chemistry*. W. A. Benjamin, Inc., New York, 1964.

† cm^{-1} is a frequency unit = $10{,}000/\lambda$, where λ is the wavelength in μ.

‡ In general, C—C single-bond stretching frequencies are not useful for identification.

Table 6.2 demonstrates that the relationships implied in Hooke's law do obtain. A carbon-carbon triple bond is stronger than a carbon-carbon double bond, which in turn is stronger than a carbon-carbon single bond. (The triple bond is a σ bond plus two π bonds; the double bond is a σ bond plus one π bond; and the single bond is just a σ bond). The triple bond will therefore have the greatest force constant and the single bond will have the smallest. We would therefore predict that the frequency of radiation needed for excitation of these groups will decrease (and the wavelength increase) as we go from a triple bond to a single bond. This prediction is borne out by the absorption frequencies of the three types of carbon-carbon bonds shown in Table 6.2.

The power of IR spectroscopy should be evident from Table 6.2 and the above discussion. IR spectra tell the chemist what functional groups are present in a molecule, an extremely useful piece of information. On the other hand, an IR spectrum gives little hint as to the arrangement of functional groups or the nature of the carbon skeleton, so that structures are seldom identified by IR alone. IR coupled with NMR (discussed below) is, however, frequently sufficient for complete structure proof.

The IR spectrum of diethyl ketone, $CH_3CH_2COCH_2CH_3$, is reproduced in Fig. 6.8. The strong absorption (an inverted "peak") at 1720 cm^{-1} indicates the presence of a carbonyl group (see Table 6.2). Although a ketone, an aldehyde, and an acid would all be consistent with an absorption at 1720 cm^{-1}, we can exclude the latter two possibilities due to the lack of certain spectral properties. The material cannot be an acid because there is no broad "O—H stretch" between 2500 and 3000 cm^{-1}, nor can it be an aldehyde because the spectrum lacks a medium-intensity peak near 2720 cm^{-1} (the fifth entry in

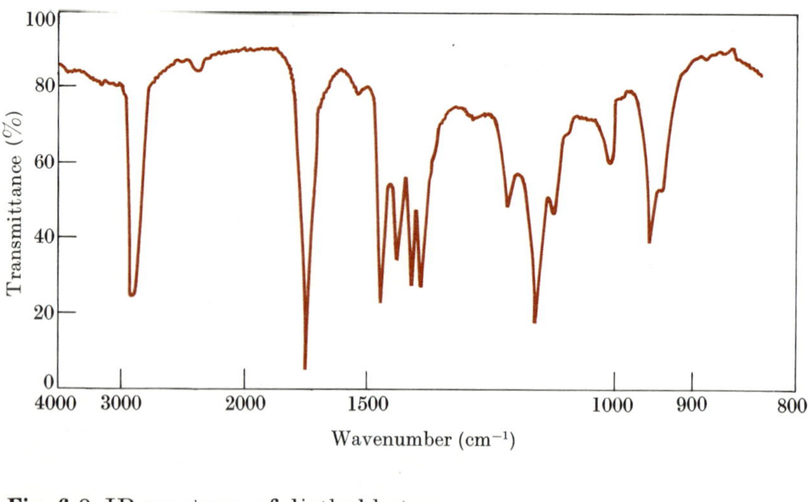

Fig. 6.8 IR spectrum of diethyl ketone

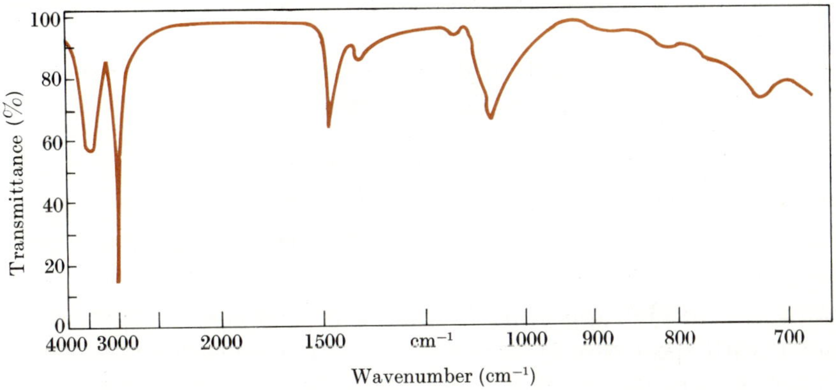

Fig. 6.9 IR spectrum of dodecanol

Table 6.2). The strong peak near 3000 cm^{-1} can be attributed to carbon-hydrogen stretch.

The IR spectrum of dodecanol, $CH_3(CH_2)_{11}OH$, is shown in Fig. 6.9. What does the spectrum tell us? The medium intensity peak near 3500 cm^{-1} is characteristic of both the O—H stretch of an alcohol and the N—H stretch of an amine (see Table 6.2). If the compound is an alcohol, then there should be a C—O stretch at 1000–1300 cm^{-1} (tenth entry in Table 6.2). Since there is such a peak at 1050 cm^{-1}, the IR evidence points strongly to an alcohol. The intense band just above 3000 cm^{-1} corresponds to C—H stretch and is found in most organic compounds. Almost as important as the peaks which are present are those which are *not* present. The absence of a peak at or near 1720 cm^{-1} proves that the compound in question is not a ketone.

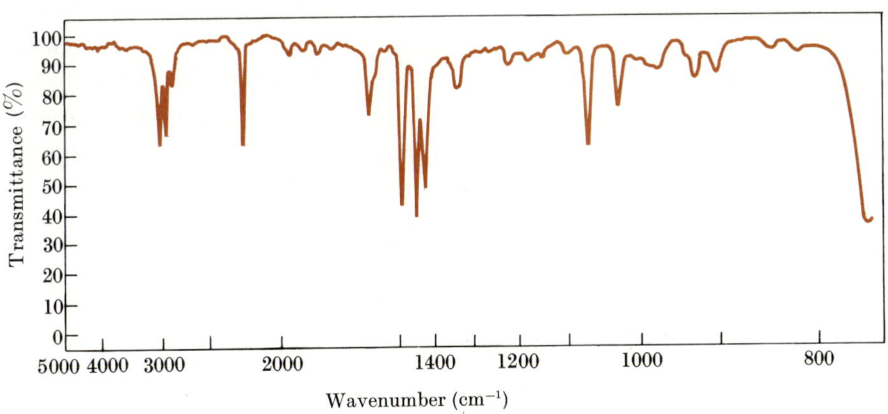

Fig. 6.10 IR spectrum of $C_6H_5CH_2CH_2CN$

The spectrum of β-phenylpropionitrile, $C_6H_5CH_2CH_2CN$, is given in Fig. 6.10. The peak at 2300 cm^{-1} corresponds to the nitrile group (—C≡N) and the peaks at 1480 cm^{-1} and 1600 cm^{-1} show that a phenyl group is present (the latter peak positions are not listed in Table 6.2).

6.13 NUCLEAR MAGNETIC RESONANCE SPECTROSCOPY

The phenomenon of nuclear magnetic resonance (NMR) was first demonstrated in 1946. Since then the technique has become so widely used that today at least 50% of the articles appearing in chemistry journals mention some application of NMR spectroscopy. It has allowed an understanding of structural organic chemistry that until now has been beyond the capability of the available physical approaches.

Nuclear magnetic resonance is possible because the nuclei of certain atoms like hydrogen behave as though they were charged spinning bodies. The spinning of this charge produces a magnetic moment. For example, the hydrogen nucleus appears to have two modes of spinning, each producing a magnetic moment. These moments are of equal strength but are oriented in opposite directions. The hydrogen nucleus therefore has two spin quantum numbers: $+\frac{1}{2}$ and $-\frac{1}{2}$. The associated magnetic moments may be considered to be the equivalents of the north and south poles of a bar magnet.

In the absence of a magnetic field, an average of half of the protons of an organic molecule will have magnetic moments (or spins) oriented in one direction and half in the opposite direction. When we place the molecule in a magnetic field, H, however, there will be a net excess of hydrogen nuclei aligned in the direction of that field. This orientation corresponds to a lower energy state than the one in which the proton "magnets" oppose the direction of the applied field. Energy may be absorbed, therefore, by protons oriented with the magnetic field (the lower energy state) to cause their excitation to an orientation opposing the magnetic field (the higher energy state). The energy required for this transition has been found to lie in the radiofrequency region of the electromagnetic spectrum. Thus, an average proton requires a frequency of 60 MHz (megahertz) to make transitions in a magnetic field of about 15,000 gauss. (The most common NMR instructions operate at either 60 or 100 MHz, with a corresponding difference in magnetic field strength.)

The nuclei composing a molecule are embedded in clouds of valence electrons. These electron clouds vary in density about the molecule because the nuclei have different electronegativities (desire for electrons). A highly electronegative atom will strongly induce electrons away from its neighbors. This inductive effect by a center of high electronegativity drops off rapidly with distance from that center. For example, in a molecule of acetaldehyde,

CH_3—CHO, the proton on the carbonyl carbon, being closer to the electronegative oxygen, will experience a greater electron removal (called *deshielding*) by the oxygen than the three methyl protons which are at a greater distance. The valence electrons act as shields against an applied magnetic field, and *the magnitude of the shielding effect depends upon the electron density at a particular site in the molecule*. Thus, the protons of a molecule of acetaldehyde that has been placed in a magnetic field of 15,000 gauss would not sense 15,000 gauss: the aldehyde proton (the proton on the carbonyl carbon) would sense a magnetic field of 15,000 gauss less some increment (due to its shielding electrons) and the methyl protons would sense a different and even lower magnetic field. To cause nuclear magnetic resonance of the protons in acetaldehyde, we would have to increase the strength of the applied magnetic field to compensate for the shielding effect about each of the different types of protons in the molecule. The net result of the shielding effect is that various sets of protons in a molecule will experience nuclear magnetic resonance at different applied magnetic field strengths: acetaldehyde, for example, will have the spectrum shown in Fig. 6.11. The abscissa in Fig. 6.11 represents changing field strength in the direction shown, and the ordinate is a detector signal indicating the occurrence of the resonance phenomenon. The magnitude of the detector signal is proportional to the number of protons undergoing resonance at the particular field strength. The low field peak is caused by the aldehyde proton and the high field peak by the methyl protons. This separation of proton resonances brought on by electron shielding effects is called the *chemical shift*.

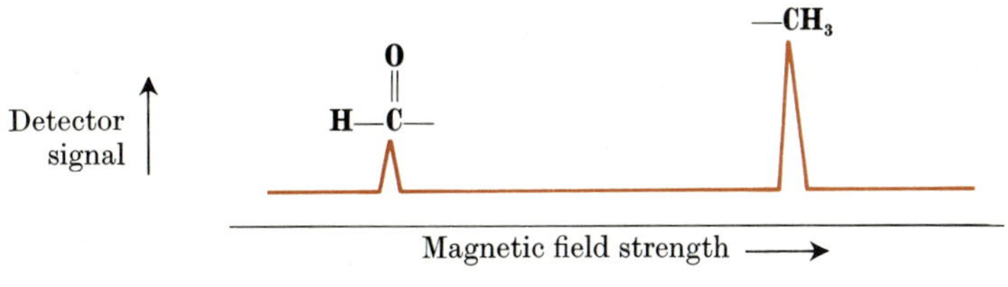

Fig. 6.11 NMR spectrum of acetaldehyde showing chemical shift

Nuclear magnetic resonance data have been standardized by reporting the chemical shifts observed in terms of the parts per million (ppm) variation in field strength between the resonance under observation and the resonance of a reference compound, usually tetramethylsilane (TMS)

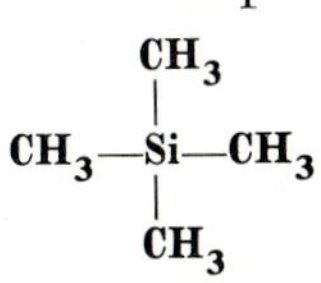

tetramethylsilane (TMS)

A small amount of TMS added to an acetaldehyde sample will produce a third peak in the spectrum (Fig. 6.12). We would report the positions of the aldehyde proton and the methyl protons as δ9.8 ppm and δ2.2 ppm lower field than TMS respectively. Table 6.3 shows typical chemical shifts of protons in a variety of structural environments.

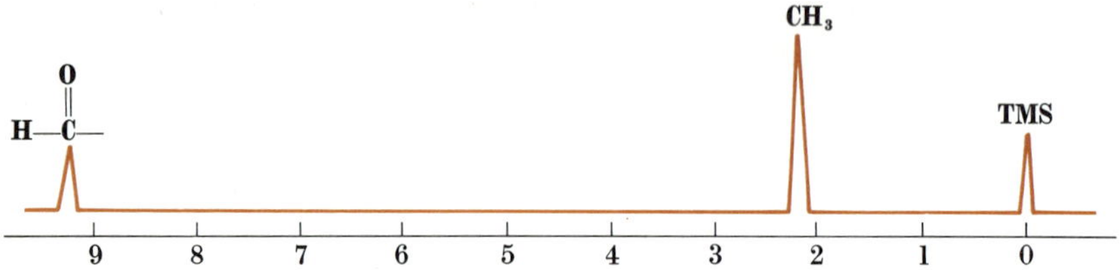

Fig. 6.12 Low resolution NMR spectrum of acetaldehyde showing chemical shift relative to tetramethylsilane

The low resolution spectrum of acetaldehyde (Fig. 6.11) is deceptively simple because NMR spectra are normally secured at high resolution where the peaks are seen to resolve into multiplets (Fig. 6.13). The low field proton splits into a quartet, and the high field resonance shows up as a doublet. This splitting of the peaks results from a phenomenon called *spin-spin coupling*. Spin-spin coupling arises because a proton senses not only the applied magnetic field, but the magnetic fields of neighboring protons as well. Consider, for example, the effect of the aldehyde proton on the methyl protons of acetaldehyde. The aldehyde proton generates a small magnetic field which can oppose or reinforce the applied magnetic field depending upon its orientation in the field. This orientation is a function of the spin quantum number of the

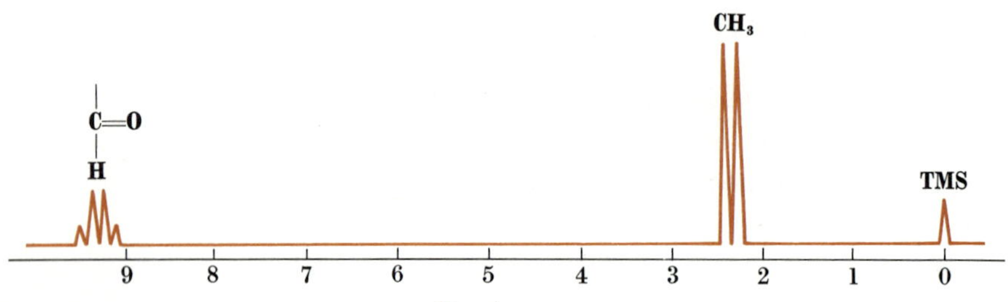

Fig. 6.13 High resolution spectrum of acetaldehyde showing the spin-spin coupling patterns

Table 6.3 Typical proton chemical shift values (dilute chloroform solutions)*

Type of proton[a]	Chemical shift, ppm[b,c]	Type of proton[a]	Chemical shift, ppm[b,c]
$R—\mathbf{CH_3}$	0.9	$O{=}C(R)—\mathbf{CH_3}$	2.3
$R—\mathbf{CH_2}—R$	1.3	$R—\mathbf{CH_2}—Cl$	3.7
$R_3C\mathbf{H}$	2.0	$R—\mathbf{CH_2}—Br$	3.5
$R_2C{=}\mathbf{CH_2}$	~5.0	$R—\mathbf{CH_2}—I$	3.2
$R_2C{=}\mathbf{CH}(R)$	~5.3	$RC\mathbf{H}(—Cl)_2$[d]	5.8
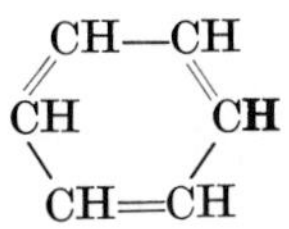	7.3	$R—O—\mathbf{CH_3}$	3.8
$R—C{\equiv}C—\mathbf{H}$	2.5	$(R—O—)_2\mathbf{CH_2}$[d]	5.3
$R_2C{=}C(R)—\mathbf{CH_3}$	~1.8	$R—C({=}O)—\mathbf{H}$	9.7
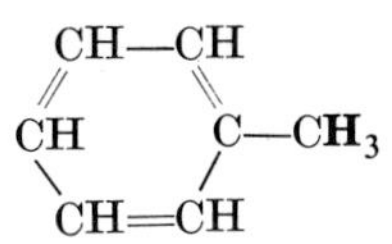	2.3	$R—O—\mathbf{H}$	~5[e]
		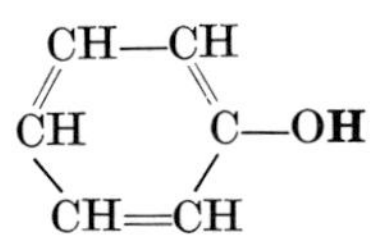	~7[e]
		$R—C({=}O)—\mathbf{OH}$	~11[e]

* Reproduced from J. D. Roberts and M. Caserio, *Basic Principles of Organic Chemistry*. W. A. Benjamin, Inc., New York, 1964.
[a] The proton undergoing resonance absorption is shown in heavy type. The group R denotes a saturated hydrocarbon chain.
[b] Relative to tetramethylsilane as 0.00 ppm.
[c] Spectrometer frequency, 60 MHz.
[d] Note how the shift produced by two chlorines or two RO— groups is greater than, but by no means double, that produced by one chlorine or RO— group.
[e] Sensitive to solvent, concentration, and temperature.

nucleus ($+\frac{1}{2}$ or $-\frac{1}{2}$ for a single proton). Approximately half the molecules will have aldehyde protons with spins that reinforce the applied magnetic field and half the molecules will have aldehyde protons with spins that oppose the applied magnetic field. Thus we see two signals for the acetaldehyde methyl group, one for a methyl in each type of acetaldehyde molecule described above.

The methyl protons will have a similar effect on the aldehyde proton. There are four net spins possible for the methyl protons: $+\frac{3}{2}$, $+\frac{1}{2}$, $-\frac{1}{2}$, and $-\frac{3}{2}$. This is seen by summing the spins of three protons in the various spin orientation combinations possible (Fig. 6.14).

↑↑↑	↑↑↓ ↑↓↑ ↓↑↑	↑↓↓ ↓↑↓ ↓↓↑	↓↓↓
$+\frac{3}{2}$	$+\frac{1}{2}$	$-\frac{1}{2}$	$-\frac{3}{2}$

Fig. 6.14 Possible spin orientation combinations for a methyl group (↑ = spin $+\frac{1}{2}$ and ↓ = spin $-\frac{1}{2}$)

Thus, four lines are exhibited for the aldehyde proton, the position for a proton in a specific molecule being determined by which of the above combination of spins of its neighboring methyl protons it experiences. The ratio of the intensity of the four resonances will be 1:3:3:1 since there are three times as many ways of having methyls with net spins of $+\frac{1}{2}$ and $-\frac{1}{2}$ as there are of having methyls with net spins of $+\frac{3}{2}$ and $-\frac{3}{2}$. The magnitude of the splitting is called a spin-spin coupling constant.

Spin-spin coupling can take place only between magnetically nonequivalent nuclei and in general only between protons on adjacent carbon atoms. When two nuclei or sets of nuclei do couple, the coupling constants for each multiplet will be the same. Coupling constants are independent of the strength of the applied magnetic field.

With molecules more complicated than acetaldehyde the multiplet patterns become more complex. Consider the hypothetical compound, 1, where the substituents at C-1 are electronegative groups that may cause a chemical shift of the three sets of protons, so that the proton at C-1 is at lowest field

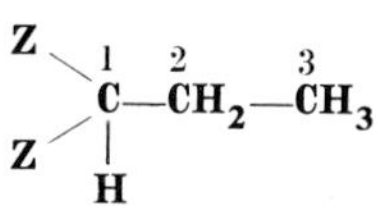

1

and the proton at C-3 is at highest field. The molecule would have the spectrum shown in Fig. 6.15. The pattern can be analyzed as follows. By spin-spin coupling with the central methylene group the methyl protons and the C-1 proton signals are split into triplets. The methylene group signal in turn is split twice, first by the methyl group into a quartet and then each peak of the quartet into a doublet by the C-1 proton.

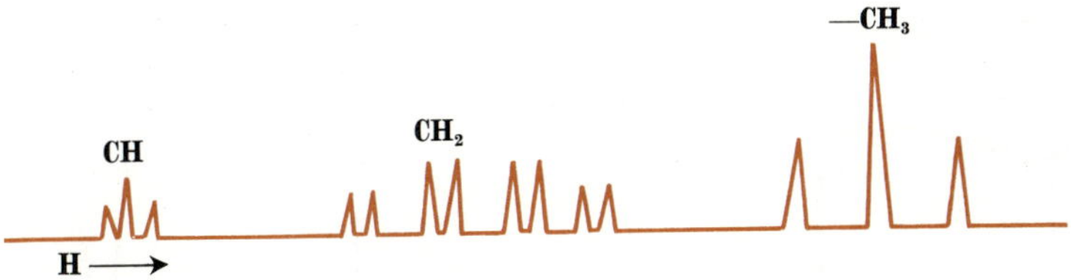

Fig. 6.15 NMR spectrum for three adjacent nonequivalent sets of protons

In summary, the following important structural information is secured from an NMR spectrum.

1) The *number* of signals indicates how many different kinds of protons there are in a molecule.
2) The *area* of the signals indicates the relative number of each type of proton.
3) The *chemical shift* of the signals tells something about the electronic environment of the protons (*e.g.*, whether a proton is aromatic, etc.).
4) The *splitting* of the signals provides information as to the number of nonequivalent nearby protons of a given proton.

6.14 APPLICATION OF SPECTROSCOPY TO STRUCTURE DETERMINATION

Let us now return to the structure of citronellal and apply the results of spectroscopy to its solution. If we examine the infrared spectrum of this compound (Fig. 6.16), we find the two characteristic peaks of an aldehyde group. The absorption at 2720 cm^{-1} in the carbon-hydrogen stretching region is indicative of the aldehyde C—H bond, and the band at 1730 cm^{-1} suggests a carbonyl group. The position of this latter peak also reveals that the carbonyl group is not conjugated to other unsaturation which may be in the molecule. The presence of another site of unsaturation is indicated, however, by the very weak band at 1640 cm^{-1}. This band lies in the region of carbon-carbon double bond stretching absorption. The infrared spectrum of citronellal also indicates that the molecule contains the grouping

$$>C—CH_3$$

since such groups reveal a C—H bending band in the 1380 cm^{-1} region.

The NMR spectrum of citronellal (Fig. 6.17) yields even more structural information than the infrared spectrum. The presence of an aldehyde group is apparent from the appearance of a signal at 9.7 ppm relative to tetramethylsilane (TMS). In addition, this peak is split into a triplet. The aldehyde

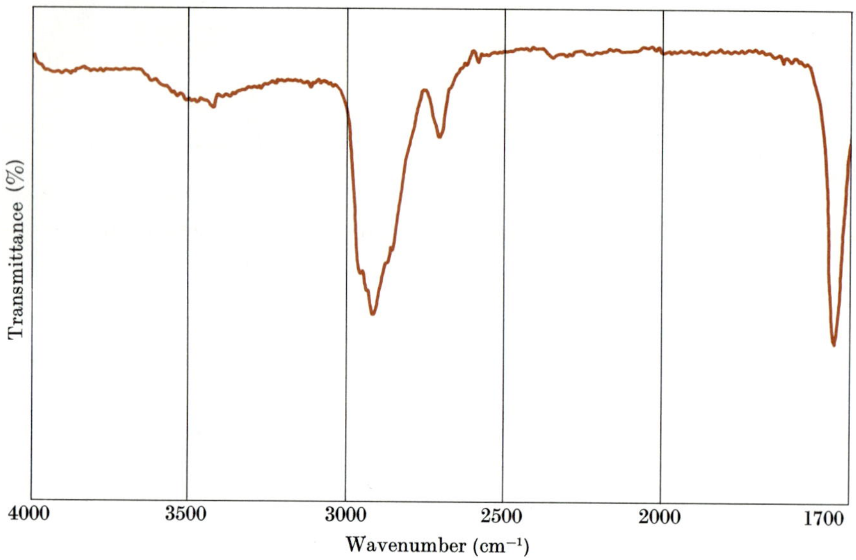

Fig. 6.16 IR spectrum of citronellal

group, therefore, must be joined to a carbon which bears two hydrogens, and we can write the partial structure,

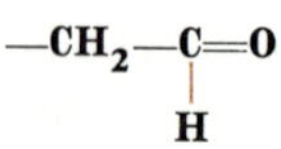

The following structural features are also readily apparent from the NMR spectrum. The molecule contains a methyl group attached to a carbon bearing a hydrogen,

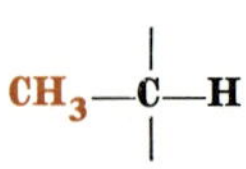

since we find a doublet centered at 0.93 ppm. It also contains at least one methyl group attached to a double bond, as indicated by the two peaks at 1.57 and 1.66 ppm. If we now integrate the areas under the peaks of this spectrum, we find that the peaks in the olefinic methyl group region are equal to six hydrogens relative to the aldehyde peak. Therefore, there must be two methyl groups attached to double bonds in citronellal, and since our knowledge of the molecular formula of this compound reveals that only one carbon-carbon double bond can be present, we can write two possible partial structures,

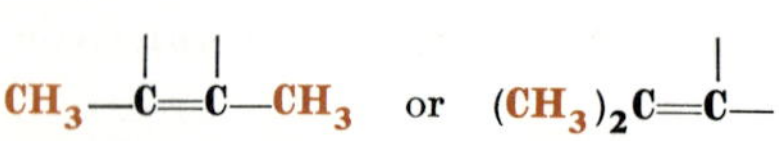

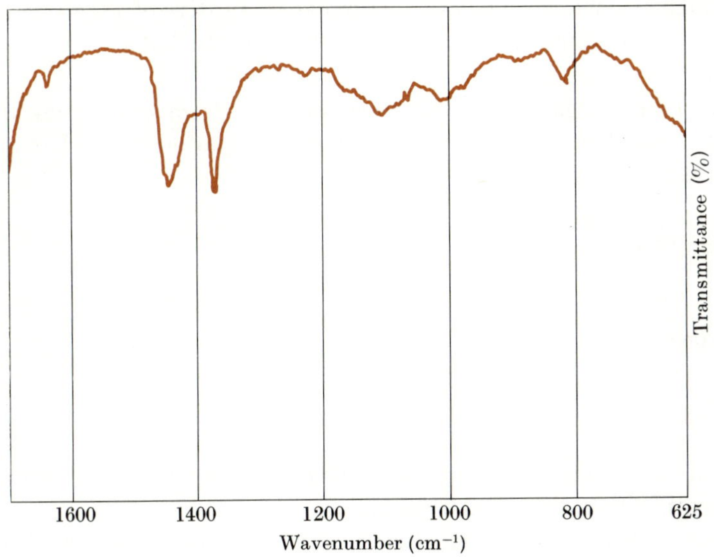

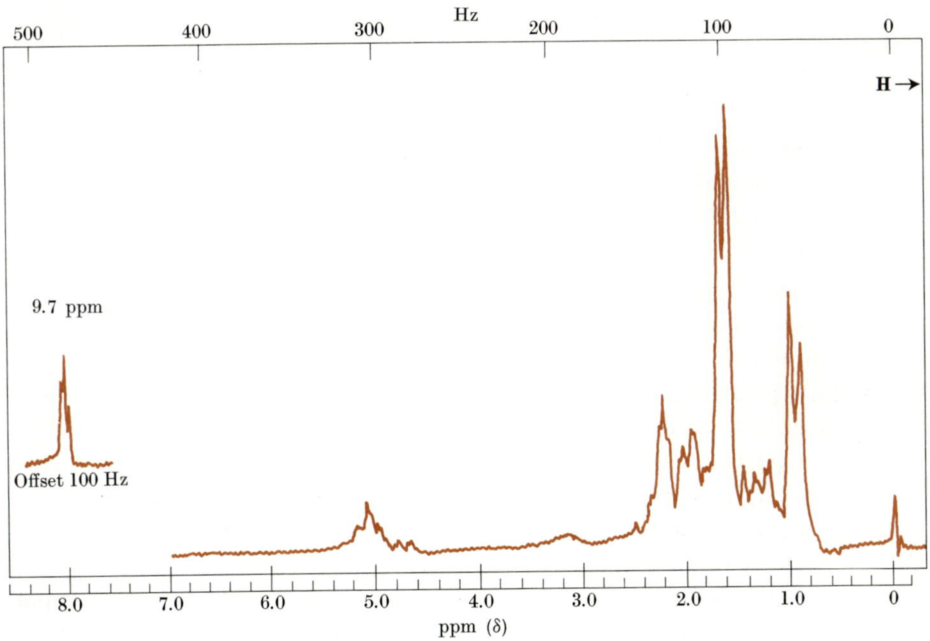

Fig. 6.17 NMR spectrum of citronellal

Finally, the resonance lines centered at 5.07 ppm are indicative of a hydrogen on a double bond, and integration of this multiplet reveals that it is due to just one proton.

If we now compare the information we obtained from our chemical examination of citronellal with our spectroscopic information we find that both are compatible with the following structure:

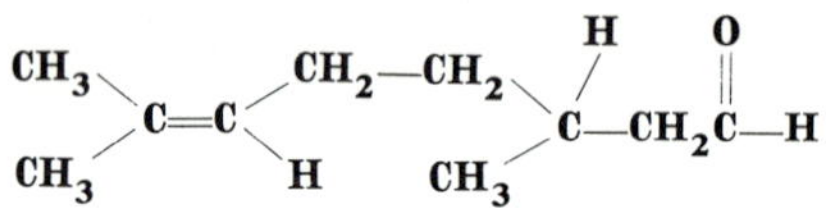

citronellal

ADDITIONAL READING

N. L. Allinger and J. Allinger, *Structures of Organic Molecules*, Englewood Cliffs, N.J.: Prentice-Hall, 1965. Paperback.
A. Ault, *Problems in Organic Structure Determination*, New York: McGraw-Hill, 1967. Paperback.
J. R. Dyer, *Applications of Absorption Spectroscopy of Organic Compounds*, Englewood Cliffs, N.J.: Prentice-Hall, 1965. Paperback.

PROBLEMS

1. A compound has a medium intensity IR peak at 2722 cm^{-1} and a strong IR peak at 1730 cm^{-1}. What type of compound is it?
2. A compound with three carbons has a λ_{max} in the UV at 280 nm. There is a strong IR band at 1710 cm^{-1} but none at 2720 cm^{-1}. What is the compound?
3. A compound has an IR peak at 1705 cm^{-1}. Its NMR spectrum shows a 5-proton signal at 7.4 ppm and a 3-proton signal at 2.3 ppm. What is the compound?
4. Identify the following compounds from their formulas and NMR spectra.
   a) C_2H_4O—doublet at 2.2 ppm and quartet at 9.8 ppm.
   b) C_3H_6O—quintet at 2.7 ppm and triplet at 4.7 ppm.
   c) C_4H_8O—triplet at 1.1 ppm, singlet at 2.1 ppm, and quartet at 2.5 ppm.
   d) $C_4H_{11}N$—9-proton signal at 1.1 ppm and 2-proton signal at 1.2 ppm. Both are singlets.
   e) $C_6H_{12}O_2$—9-proton singlet at 1.5 ppm and 3-proton singlet at 2.0 ppm.
   f) C_7H_8—3-proton singlet at 2.3 ppm and 5-proton signal at 7.2 ppm.
   g) C_8H_6—proton singlet at 3.0 ppm and a 5-proton multiplet at 7.4 ppm.
   h) $C_{12}H_{10}$—A complex multiplet at 7.5 ppm.

i) $C_3H_6Cl_2$—2-proton quintet at 2.2 ppm and a 4-proton triplet at 3.7 ppm.

5. A compound with an empirical formula C_4H_7N has a strong IR peak at 2100 cm^{-1}. The NMR consists of a large doublet and a small septet. What is the compound?
6. A compound with weakly acidic properties ($pK_a = 10$) displays an NMR with a 4-proton multiplet at 7.4 ppm and a 3-proton singlet at 2.3 ppm. What is the compound?
7. Suggest possible structures for the hydrocarbons which would yield the following ozonolysis products. Assume that one mole of each product is formed for each mole of compound ozonized.

a) 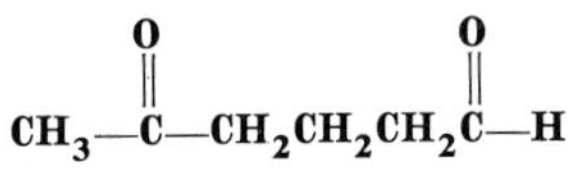

b) 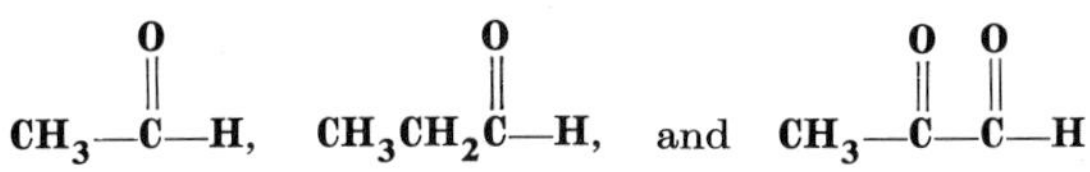

c) 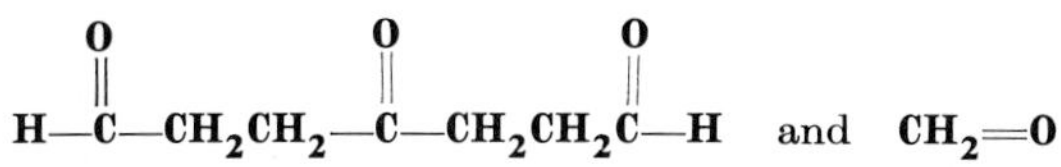

d) 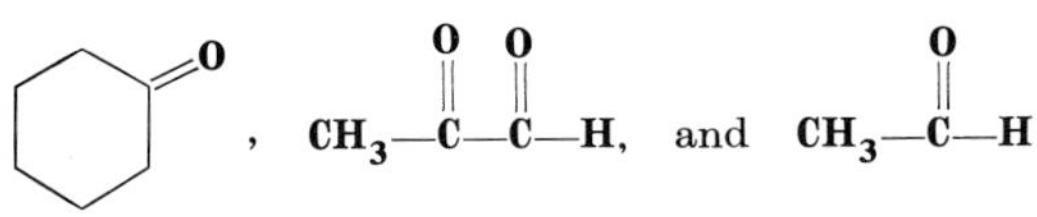

8. What feature of the infrared spectrum could one use to distinguish the two members of each of the following pairs of compounds? Give roughly the position of absorption in cm^{-1}.

a) CH_3CHO and $CH_3CH(OCH_3)_2$

b) $(CH_3)_3N$ and $(CH_3)_2CHNH_2$

c) $CH_3CH_2CH_2CH_3$ and $CH_3CH_2CH{=}CH_2$

d) $CH_3C{\equiv}CCH_3$ and $CH_3CH_2C{\equiv}CH$

e) CH_3CH_2OH and CH_3CO_2H

f) CH_3CH_2CHO and CH_3COCH_3

9. Draw five possible structures with normal valences for a compound of formula C_3H_3Cl. A compound of this formula has a strong, sharp band at 3300 cm^{-1}; what is its structure?

10. How many magnetically different types of protons are there in each of the following compounds?

a) CH_3CH_3 b) $(CH_3)_2CHCH_2CH_3$ c) $ClCH_2CH_2CH_2Cl$

d) 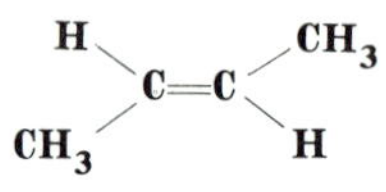

e) 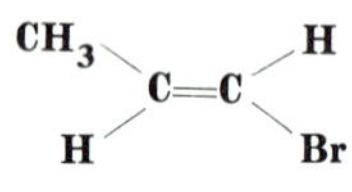

f) 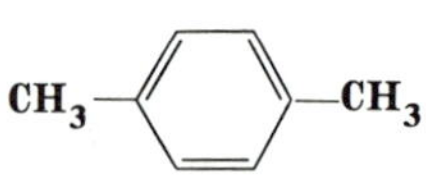

g) 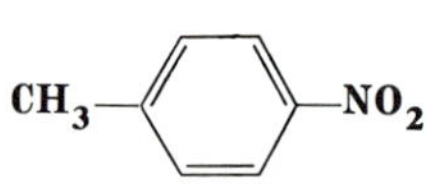

h) 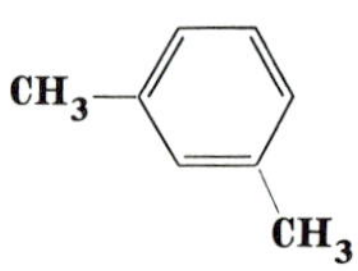

i) 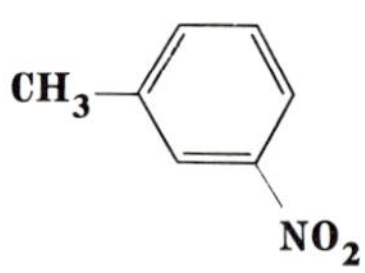

11. Given below are the combustion analyses and NMR spectra of three compounds. Deduce the structures from these data. The numbers appearing over the NMR peaks are the relative integrated areas for each type of proton signal in the spectrum.

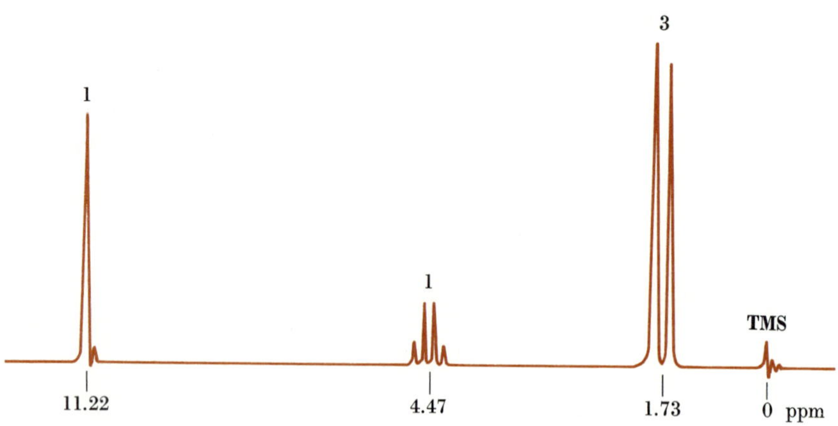

a) *Compound A:* 33.03% carbon, 4.59% hydrogen, 33.03% chlorine. The molecular weight is 109.

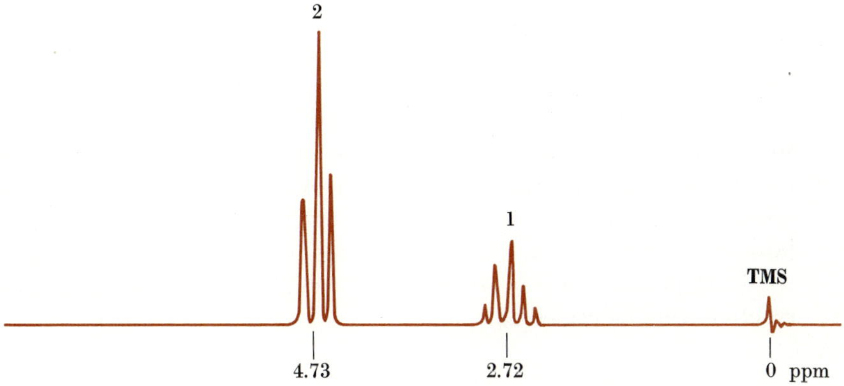

b) *Compound B:* 62.07% carbon, 10.34% hydrogen. The molecular weight is 58.

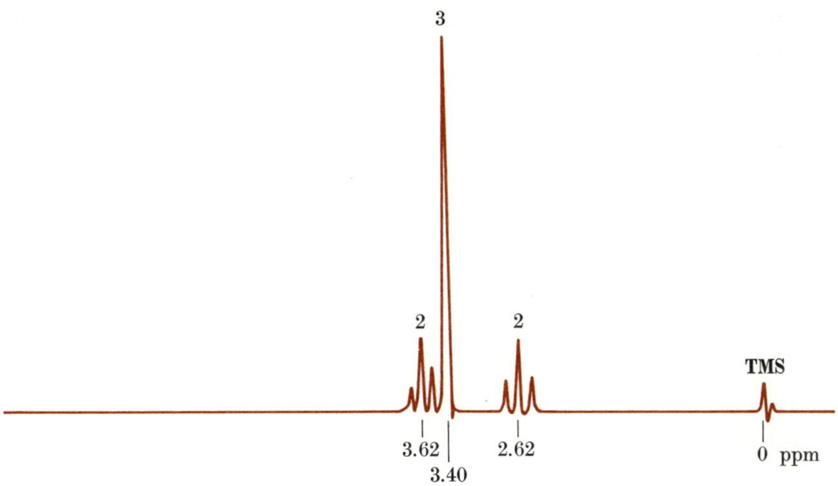

c) *Compound C:* 56.50% carbon, 8.24% hydrogen, 16.49% nitrogen. The molecular weight is 85.

12. Review the reactions in Chapter 3 and solve the following problem. Compound A of unknown structure was subjected to combustion analysis. The percentages of carbon and hydrogen were found to be 82.35% and

9.80%, respectively. The molecular weight was determined to be 204. Compound A reacted readily with one mole of bromine in carbon tetrachloride. When A was treated with chromic acid (CrO_3) in acetone, a new material B was produced. The infrared spectrum of B showed an intense band at 1720 cm^{-1}. This band was absent in the spectrum of A. In addition, the spectrum of A contained a peak at 3600 cm^{-1} which was not present in that of B. When compound B was reacted with phenylmagnesium bromide (C_6H_5MgBr), compound C was formed. The infrared spectrum of C displayed a band at 3600 cm^{-1} but no longer contained the peak at 1720 cm^{-1}. Compound C was treated with sulfuric acid, and the resulting product D was subjected to ozonolysis. The products of this reaction were identified as

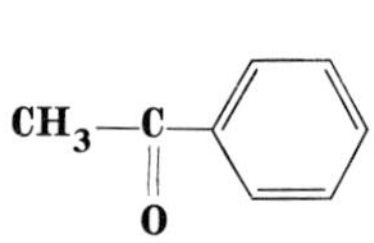

and

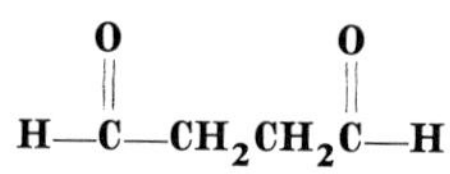

Suggest plausible structures for A, B, C, and D.

13. Assign one of the infrared spectra (Figs. 6.18 through 6.22) to each of the structures shown below. Identify as many bands in each spectrum as you can.

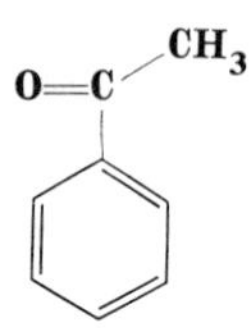

I

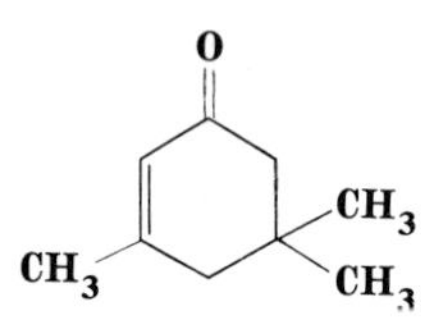

II

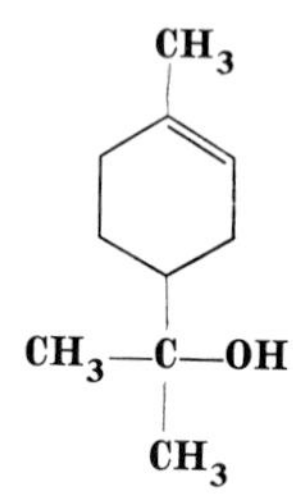

III

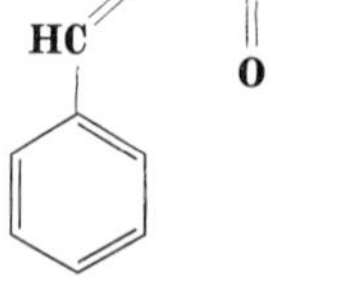

IV

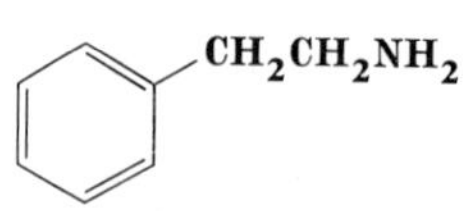

V

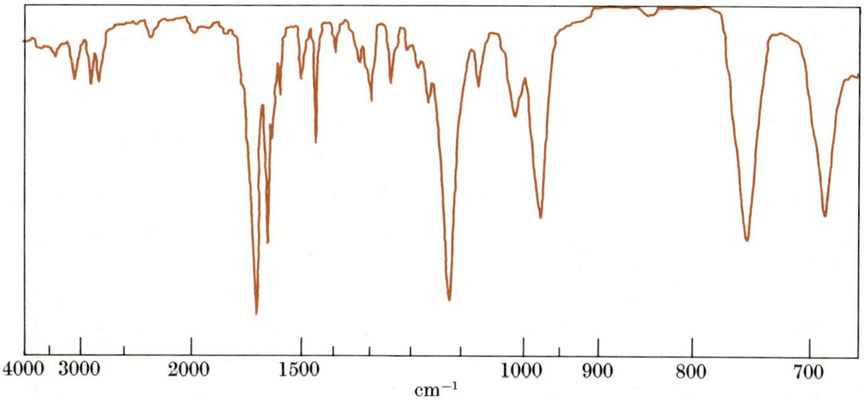

Figure 6.18

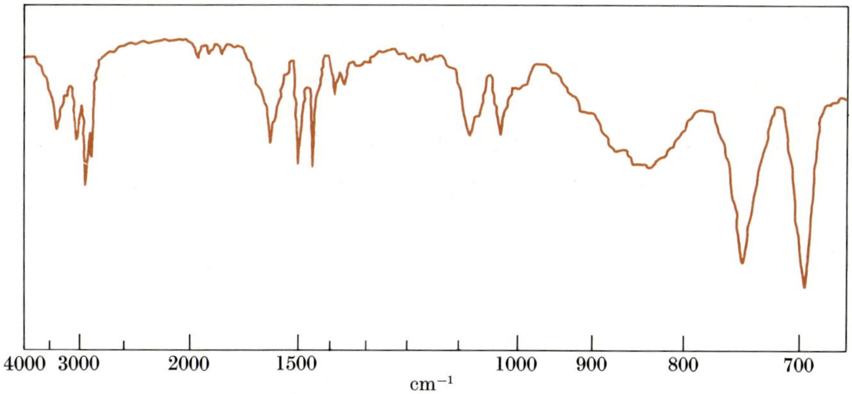

Figure 6.19

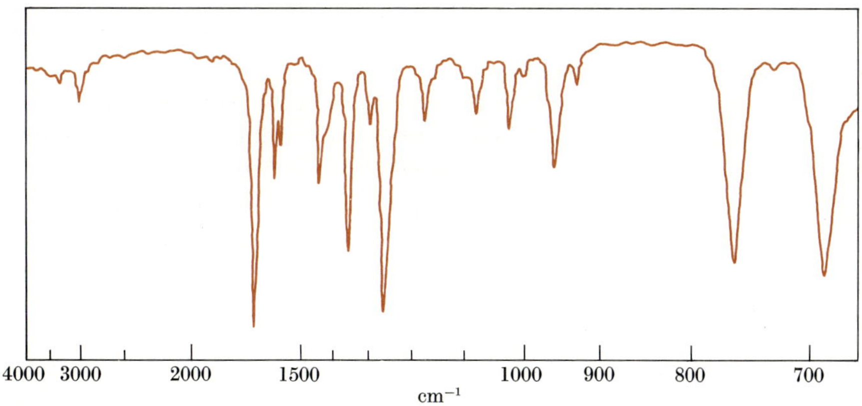

Figure 6.20

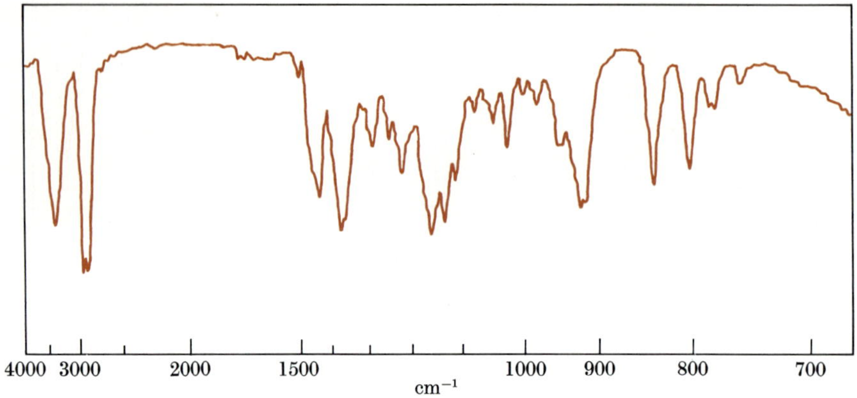

Figure 6.21

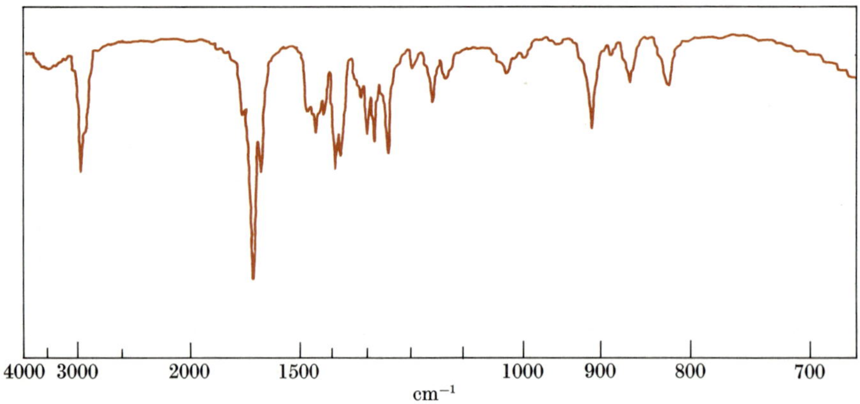

Figure 6.22

14. Explain the following facts.
    a) The infrared absorption stretching frequency for the carbonyl group of an amide ($RCONR_2$) is about 1680 cm^{-1}, while the corresponding band for a ketone carbonyl (R_2CO) is about 1710 cm^{-1}.
    b) The chemical shift for the methyl protons of $ROCH_3$ is 3.8 ppm downfield from tetramethylsilane, whereas the corresponding value for the methyl protons of R—CH_3 is 0.9 ppm (see Table 6.3).

7
SUBSTITUTION REACTIONS

7.1 CLASSES OF ORGANIC REACTIONS

Heterolytic cleavage has been defined as bond breakage in which one of the fragments retains both electrons. All heterolytic organic reactions fall into one of four reaction categories: *substitution, elimination, addition,* and *rearrangement.* As we shall see, the principles of the substitution reaction are exceedingly general and lead directly to an understanding of the remaining three types of reactions. We will therefore start our discussion of the reactions of organic chemistry by considering the substitution (or displacement) reaction.

7.2 SUBSTITUTION REACTIONS

Many reactions can be classified as substitution processes. Some of these are listed below:

1) 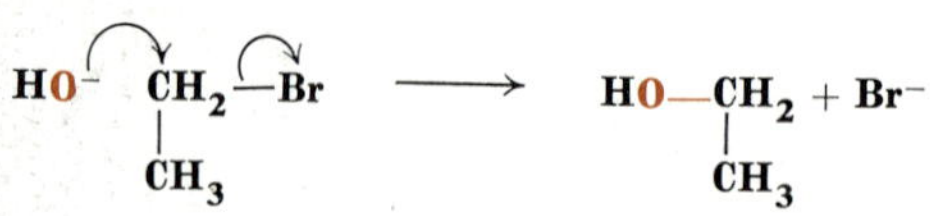

2) 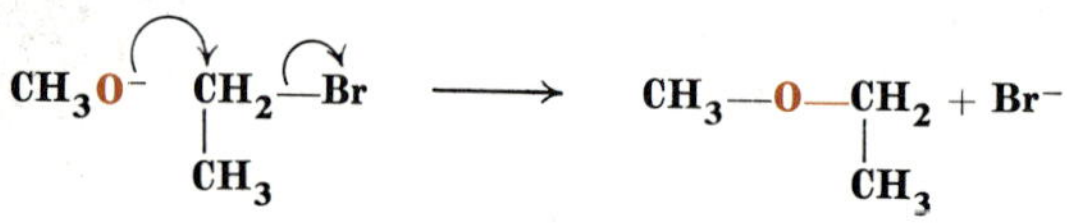

3) 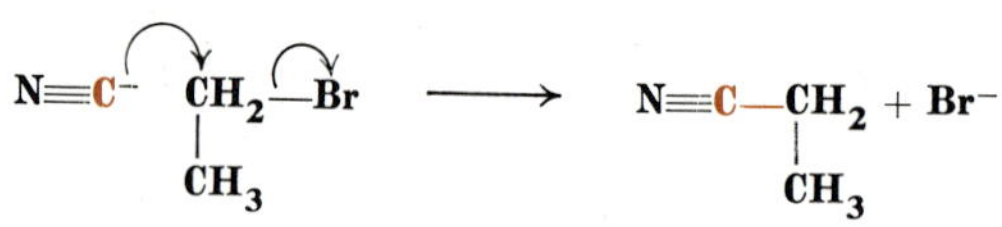

4) 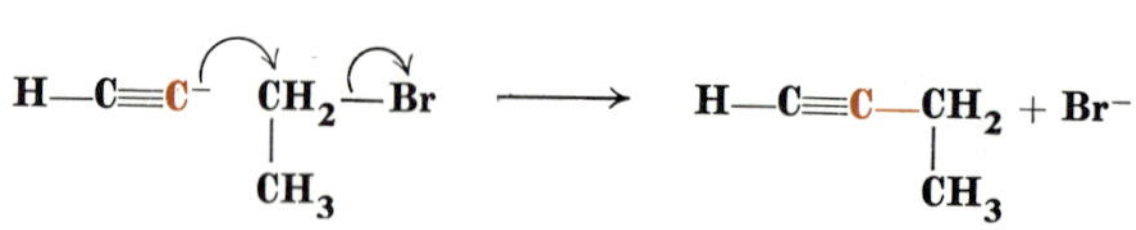

5) 

6) 

7) 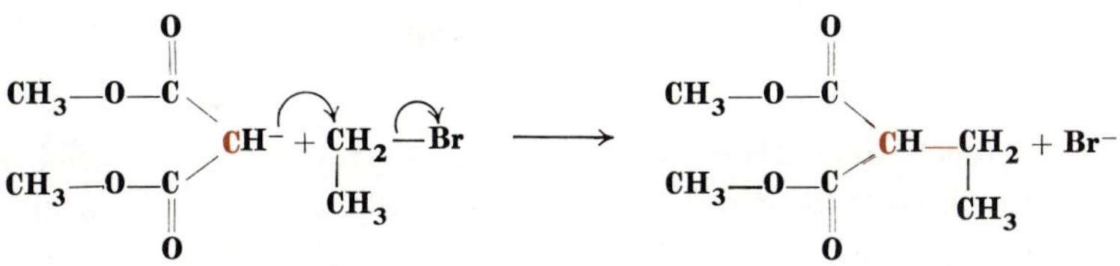

8) 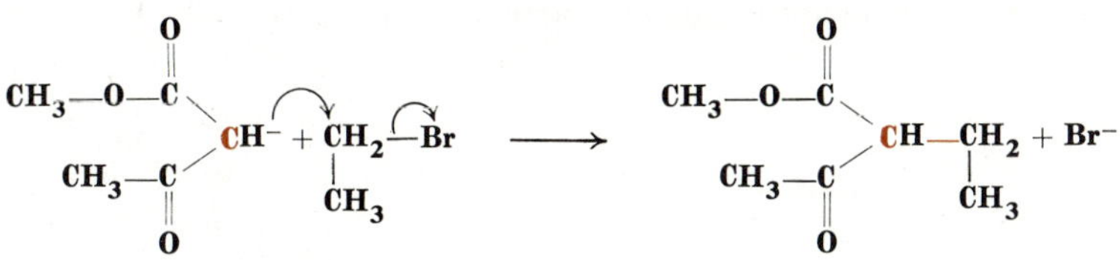

9) 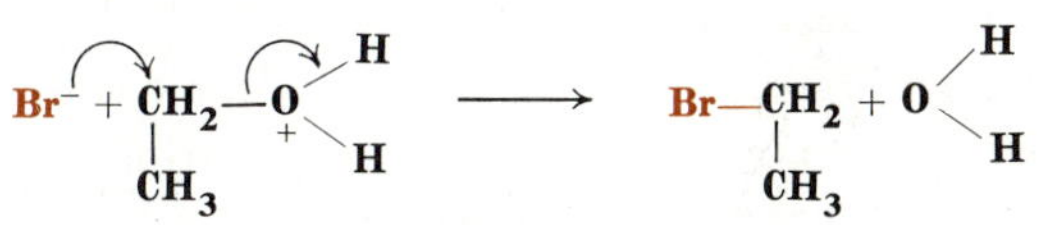

10) 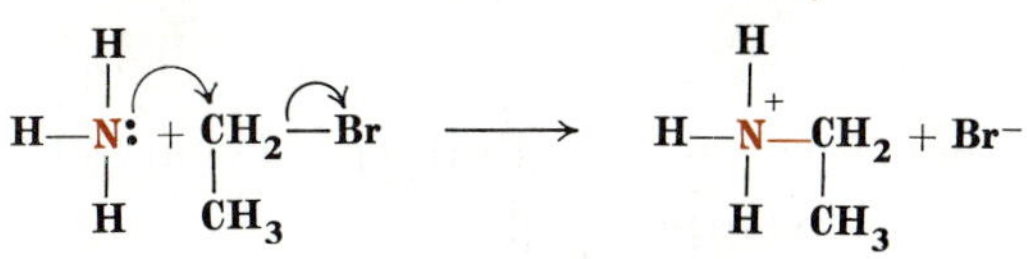

The usefulness of these reactions for synthetic purposes will be discussed later. For the time being, we shall concern ourselves with the mechanism of substitution reactions.

In each of the above reactions, a species with an unshared pair of electrons (a nucleophile) becomes attached to an alkyl group via the electron pair. Furthermore, a fragment (a *leaving group*) is displaced from the alkyl group, taking with it the electrons that once bound it to the carbon. This process is indicated by the curved arrows in the ten reactions above. We may generalize the displacement reaction in the following way:

$$Y^- + R—X \longrightarrow Y—R + X^-$$

We might think of three ways this process could occur:

1. Initial formation of the Y—C bond, followed by C—X bond breakage.
2. Simultaneous formation of the Y—C bond and breakage of the C—X bond.
3. Initial heterolytic breakage of the C—X bond, followed by formation of the Y—C bond.

These are illustrated below:

1) $Y^- + R{-}X \longrightarrow [Y{-}R{-}X]^- \longrightarrow Y{-}R + X^-$

2) $Y^- + R{-}X \longrightarrow [\overset{-\delta}{Y}\cdots R\cdots \overset{-\delta}{X}] \longrightarrow Y{-}R + X^-$

3) $RX \longrightarrow X^- + [R^+] \xrightarrow{Y^-} RY$

Note that only mechanism 2 does not involve an intermediate (a species, often of high energy, which is formed during the course of a reaction but which has no partial bonds).

We may rule out mechanism 1 because it requires an intermediate species with ten electrons in the outer shell of carbon:

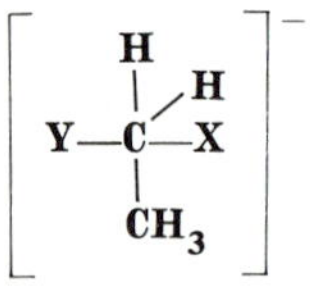

This is an impossible situation because first-row elements cannot incorporate more than eight electrons in their outer shells.

Both the second and third mechanisms do, in fact, occur. In mechanism 2 the rate of the reaction must be proportional to the concentration of both Y^- and RX, since they are both involved in the transition state:

rate of formation of RY $= k_2[Y^-][RX]$.

This process is called an *S_N2 reaction*; *S* for substitution, $_N$ for nucleophilic, and 2 for the bimolecular nature of the reaction.

Mechanism 3 is a two-step process involving formation of a carbonium ion intermediate. If we assume that the first step is slow and rate-determining, then the rate of the reaction depends only on the rate of ionization of RX:

rate of formation of RY $= k_1[RX]$.

The rate is independent of the concentration of Y^- (see Chapter 4 and the detailed discussion of the ketone bromination which is independent of Br_2). Reactions that proceed via mechanism 3 are called *S_N1 reactions*; S_N for nucleophilic substitution and 1 for the unimolecular kinetic behavior.

A consideration of the details of the S_N1 and S_N2 processes is most informative. We shall first examine the S_N2 reaction.

7.3 S_N2 MECHANISM

As we pointed out above, the essence of an S_N2 reaction is the displacement of a leaving group by a nucleophile. One can imagine two geometries for this process, front-side attack and back-side attack (Fig. 7.1).

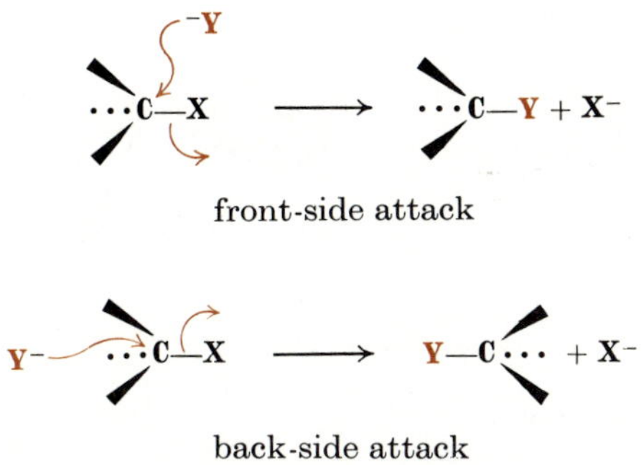

Fig. 7.1 Two modes of attack in a displacement reaction

There is overwhelming evidence that S_N2 reactions proceed exclusively by back-side attack. Consider, for example, the displacement of a chloride ion by a hydroxide ion in *cis*-1-chloro-3-methylcyclopentane (Fig. 7.2):

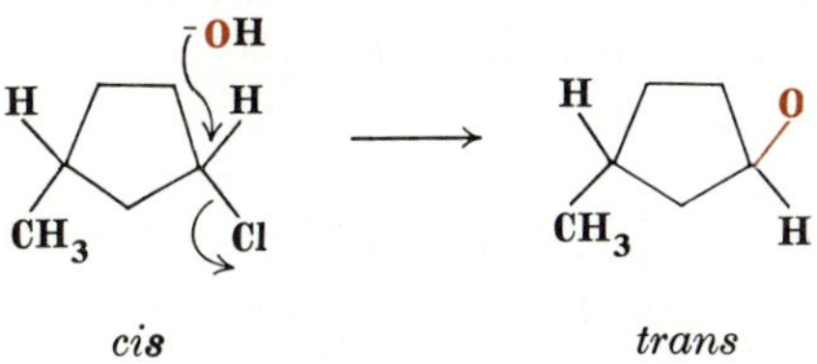

Fig. 7.2 Stereochemistry of a displacement reaction

The sole product is the *trans*-alcohol. If there had been any front-side attack, *cis*-alcohol would have been formed. Another example is the reaction between an optically active alkyl iodide and radioactive iodide ion. Front-side attack would lead to incorporation of radioactivity with no change in optical activity. It is found, however, that the radioactive alkyl iodide product is of inverted configuration, as expected for a back-side attack:

$$(+)\mathbf{R{-}I} + {}^{*}\mathbf{I}^- \longrightarrow (-)\mathbf{R{-}I}^{*} + \mathbf{I}^-$$

* = radioactive

Back-side attack is preferred because the nucleophile can be bonded via the back-side lobe of the relevant molecular orbital at the moment leaving group departs (Fig. 7.3). The resulting transition state resembles an sp^2 hybridized carbon with the p-orbital perpendicular to a plane defined by the central carbon and its three substituents. X and Y are partially bonded to the lobes of the p-orbital. The central carbon undergoes a type of flipping, called *Walden inversion*, which demands that if we start with an optically active material, we will end with an optically active product of opposite configuration.

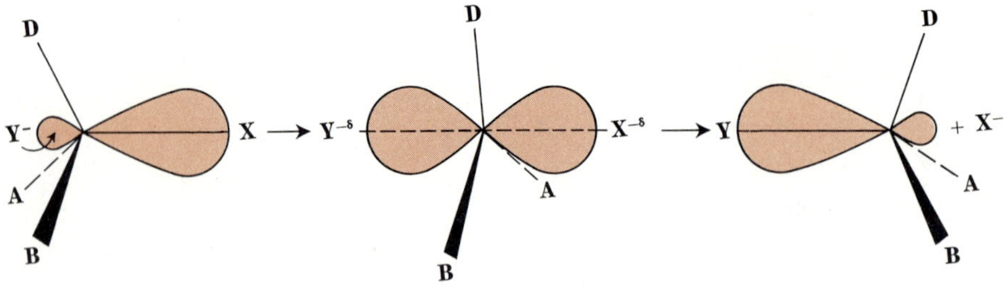

Fig. 7.3 Walden inversion in an S_N2 displacement

The rate of an S_N2 reaction depends on the number and type of substituents on the central carbon, as in Table 7.1, which lists the reaction rates of various alkyl bromides (RBr) with excess chloride ion in acetone. The most obvious explanation of these data is that the alkyl groups sterically impede the approach of the nucleophile to the back-side of the reactive carbon. This steric effect gives rise to the following order of S_N2 reactivity for alkyl halides: methyl > primary > secondary > tertiary (corresponding to zero, one, two, and three alkyl substituents). Of course, the magnitude of the relative rates, as given above, depends on the exact reaction under study. In particular, the larger the nucleophile, the greater the sensitivity of the rates to the number of substituents. No matter what the reaction conditions or nucleophile size, methyl and primary alkyl halides almost always substitute by an S_N2 mechanism, whereas tertiary halides react exclusively by the alternate pathway, the S_N1 mechanism.

Table 7.1

R	Relative rate of S_N2
CH_3—	30
CH_3CH_2—	1
$(CH_3)_2CH$—	0.03
$(CH_3)_3CCH_2$—	10^{-5}
$(CH_3)_3C$—	Too slow to measure

The reader might find it helpful if we conclude this discussion of the S_N2 reaction with an actual synthetic procedure:

$$CH_3(CH_2)_3Br + CH_3O^- \longrightarrow CH_3(CH_2)_3OCH_3 + Br^-$$

n-Butyl bromide (0.1 mole) and sodium methoxide (0.2 mole) are dissolved in 50 ml methanol and boiled under reflux for three hours. The solution is cooled, poured into 300 ml water, and the resulting organic layer is separated and purified by distillation. Note that if we want to determine how fast the reaction is proceeding, we would merely have to titrate the methoxide with standardized acid. Whereas we start with 0.2 mole of methoxide, at the completion of the reaction we are left with only 0.1 mole. We could obtain relative rates of reactions for different alkyl bromides by determining the rate at which the methoxide is consumed.

7.4 S_N1 MECHANISM

An S_N1 mechanism, in its simplest form, involves slow ionization of an alkyl halide to a carbonium ion, followed by bonding of the carbonium ion to a species with an unshared pair of electrons (Fig. 7.4).

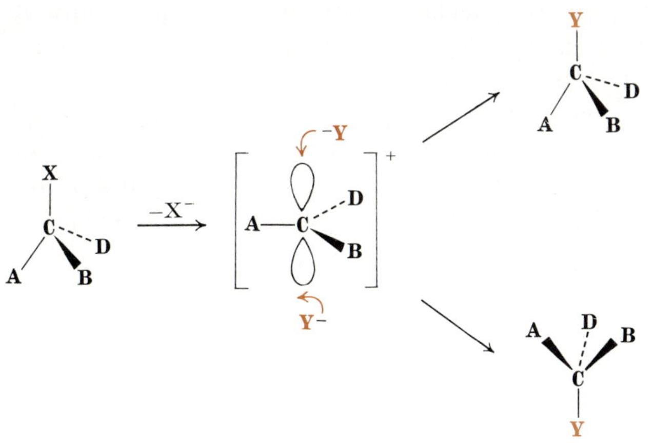

Fig. 7.4 Stereochemistry of an S_N1 reaction

In the ideal case, the sp^2 hybridized carbonium ion has a plane of symmetry and both sides are equally likely to be attacked by the nucleophile. This means that if we effect an S_N1 reaction with an optically active halide, the product will be racemic (50% l and 50% d). In practice, one usually realizes only partial racemization of an optically active halide via an S_N1 pathway. This may be because the leaving group did not completely dissociate from the carbonium ion when the nucleophile attacked, causing a slight preference for back-side bonding. The close relationship between S_N1 and S_N2 mechanisms should be apparent from this last statement.

The order for S_N1 reactivity of an alkyl halide is: tertiary > secondary > primary. We can illustrate this order in the reaction of alkyl halides with acetic acid (see also Table 7.2):

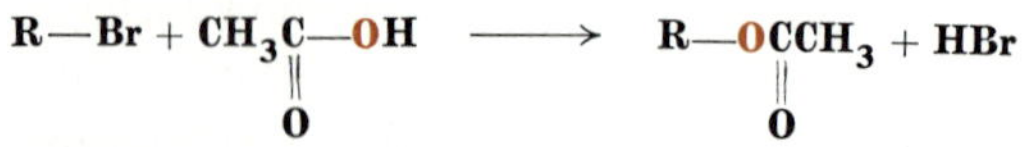

Table 7.2

R	Relative rate of S_N1
CH_3CH_2Br	Too slow to measure
$(CH_3)_2CHBr$	1
$(CH_3)_3CBr$	10^5

The reaction is carried out by heating a solution of the alkyl halide in acetic acid. Since the solvent itself acts as the nucleophilic reagent, this substitution is called a *solvolysis reaction.* This is a useful system for comparing relative rates of S_N1 reactions, because acetic acid is a very poor nucleophile (compared, for example, to hydroxide, methoxide, or iodide) and therefore the chances of a competing S_N2 reaction are minimal.

As the above data demonstrate, unlike the S_N2 mechanism, the S_N1 mechanism shows a striking rate increase with tertiary halides. The reason for this is twofold. In the first place, the alkyl groups are electron donors and therefore inductively stabilize the transition state leading to the carbonium ion intermediate:

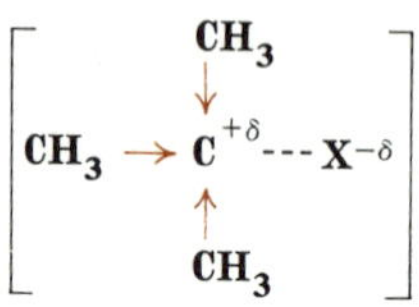

The fast S_N1 reactions of tertiary substrates may also be ascribed to *steric acceleration.* The three alkyl groups of a tertiary halide are separated by 109° (the tetrahedral angle). As the leaving group departs, the bond angles expand until they reach 120° in the sp^2 hybridized intermediate. Since the groups are farther apart in the transition state leading to the intermediate than they are in the ground state, strain is relieved when the substrate reacts. This tends to lower the activation energy for highly substituted alkyl halides compared to unsubstituted ones.

Interestingly, solvolyses (substitution reactions in which the only nucleophile is the solvent) have a peculiar property. It is impossible to tell whether this type of reaction is S_N1 or S_N2 simply by determining kinetic order. The rate expression for an S_N1 reaction is

$$\text{rate} = k_1\,[\text{RX}].$$

The rate expression for an S_N2 solvolysis reaction is

rate = k_2[solvent][RX].

But the solvent concentration is virtually constant and the latter expression can be written as

rate = k_2'[RX],

where the constant k_2' contains the solvent concentration. Clearly, both S_N1 and S_N2 solvolyses appear to be first-order reactions. The mechanism of a solvolysis reaction must be determined from stereochemical data (inversion for S_N2 and racemization for S_N1) and from the dependence of reaction rates on solvent polarity and nucleophilicity (S_N2 solvolyses are much more sensitive to solvent nucleophilicity than S_N1 solvolyses).

A typical S_N1 reaction would be carried out as follows: 25 ml *t*-Butyl alcohol is stirred vigorously with 100 ml concentrated HBr.

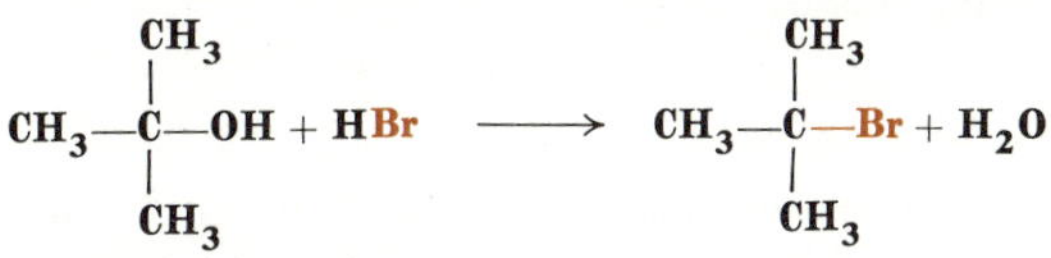

After 20 minutes the aqueous layer is discarded, and the remaining crude product is purified by distillation. It is important to realize that the leaving group in this reaction is water, not hydroxide ion. It is not possible to displace a hydroxide ion or alkoxide ion in either an S_N1 or an S_N2 reaction. Substitution of alcohol and ether groups must always be carried out in acidic media where the leaving groups are protonated.

7.5 MISCELLANEOUS COMMENTS

Elimination reactions are generally found to compete with substitution reactions. In an S_N1 substitution, the intermediate carbonium ion can lose a proton as well as pick up a nucleophile:

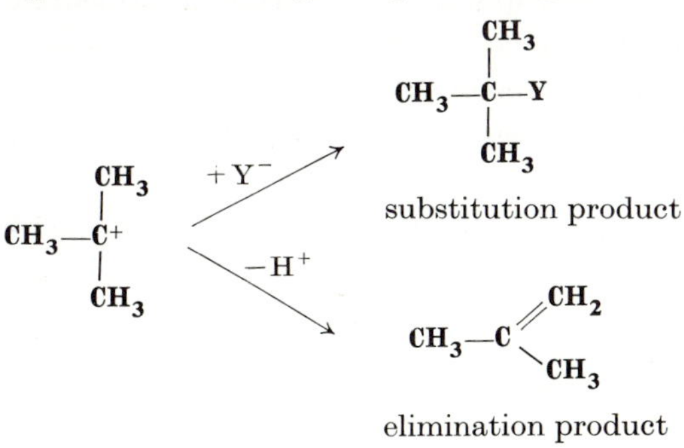

Elimination products (olefins) are also found in S_N2 substitutions, sometimes in appreciable quantities.

Another point worth mentioning is that aryl and vinyl halides are inert to substitution by both S_N1 and S_N2 mechanisms.

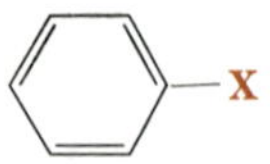

$CH_2{=}CH{-}X$

aryl halide vinyl halide

The S_N2 mechanism is impeded by electrostatic repulsion between the nucleophile and the π cloud in the halide. The S_N1 reaction is unfavorable because heterolytic cleavage of the carbon-halogen bond is resisted by the relatively electronegative sp^2 hybridized carbon. On the other hand, allyl and benzyl halides are

$CH_2{=}CH{-}CH_2{-}X$

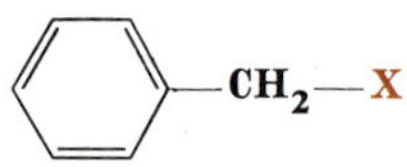

allyl halide benzyl halide

particularly reactive to S_N1 substitution because the carbonium ions (and hence the transition states leading to them) are resonance stabilized:

$$CH_2{=}CH{-}\overset{+}{C}H_2 \longleftrightarrow \overset{+}{C}H_2{-}CH{=}CH_2$$

Among the halogens, iodine is the best leaving group and fluorine the poorest ($I > Br > Cl > F$). The bonding orbitals in iodine are in the 5-shell and far away from the positive nucleus. Consequently, the orbitals can be easily distorted to accommodate the stretched carbon-halogen bond found in the transition states of the S_N1 and S_N2 substitutions. In other words, iodine is a much better leaving group (and nucleophile) than fluorine because the former is more *polarizable*.

Finally, we must mention the Law of Continuity. This law states that there are no sharp breaks in nature. For example, while materials are classified as solids, liquids, and gases, there are substances which have intermediate properties such as so-called "liquid crystals." The point is that definitions and classifications are artificial conveniences, and the S_N1 and S_N2 categories are no exception. Substitution reactions are undoubtedly comprised of a whole spectrum of mechanisms with varying degrees of S_N1 and S_N2 character. Thus the transition state below might be considered that of an S_N1 reaction with a small degree of S_N2 character.

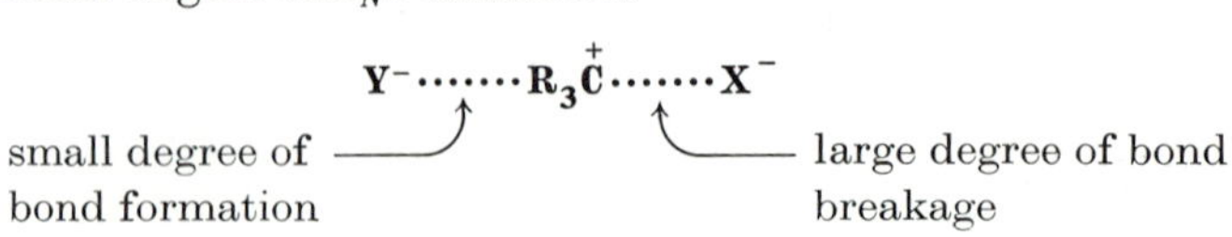

Substitution reactions are of great synthetic value, and we shall now illustrate how they may be used in the preparation of various functional groups.

APPLICATIONS OF SUBSTITUTION REACTIONS

7.6 ALCOHOLS

The formation of an alcohol by a substitution reaction is typified by the reaction of a primary or secondary alkyl halide with dilute aqueous base (a tertiary halide gives too much olefin to be useful):

$$\mathbf{HO^- + RX \longrightarrow HOR + X^-,\quad (X = Cl, Br, I).}$$

In practice, this reaction is not often employed in syntheses; it is much more common to convert alcohols into alkyl halides:

$$\mathbf{ROH + HX \longrightarrow RX + H_2O}$$

Recall that displacement of an alcohol group requires prior protonation of the oxygen:

$$\mathbf{CH_3OH + HBr \rightleftharpoons Br^- + CH_3{-}\overset{+}{O}H_2 \longrightarrow Br{-}CH_3 + H_2O}$$

In order for heterolytic cleavage of the carbon-oxygen bond to be feasible, the electronegativity of the oxygen must be enhanced by conversion into a positive species.

Although this method of preparing alkyl halides is general for bromides and iodides, it is too slow to be convenient for synthesizing alkyl chlorides (chloride is too weak a nucleophile). Other methods have been developed for this purpose, involving the use of phosphorus trichloride (PCl_3) or thionyl chloride ($SOCl_2$) in pyridine as a solvent:

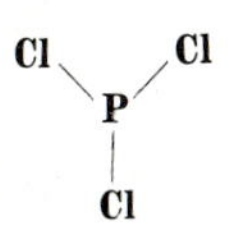

phosphorus trichloride

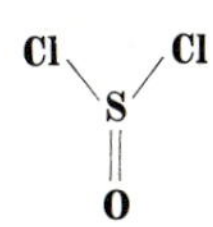

thionyl chloride

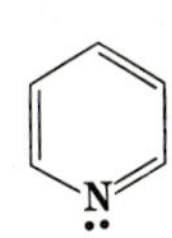

pyridine

$$\mathbf{R{-}OH + PCl_3 \longrightarrow R{-}Cl}$$

$$\mathbf{R{-}OH + SOCl_2 \longrightarrow R{-}Cl + HCl + SO_2}$$

Thionyl chloride is a particularly convenient reagent to use since the only by-products are HCl and SO_2. The HCl is neutralized by the pyridine, and the SO_2 is expelled as a gas. The mechanism is a two-stage process:

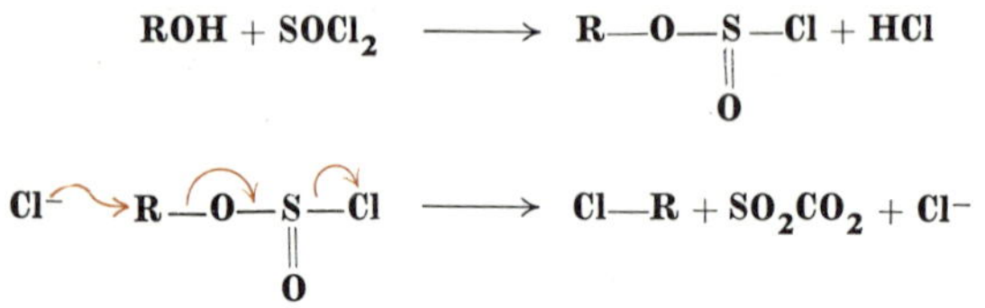

In the first step, the hydroxyl group is converted into a good leaving group. The second step is a substitution reaction with chloride as the nucleophile. The following chloride has been prepared in 79% yield using this procedure:

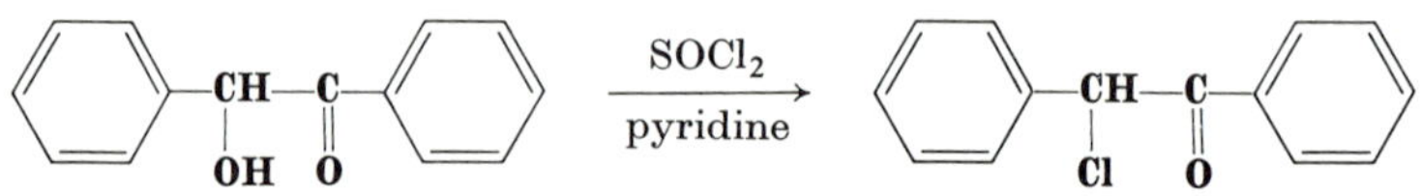

79%

7.7 ETHERS

There are two types of ethers: symmetrical and unsymmetrical:

R—O—R	R—O—R′
symmetrical	unsymmetrical

The method employed to synthesize a particular ether depends on the type of ether being prepared. Both synthetic routes utilize the substitution reaction, one under acidic conditions and the other under basic conditions.

We can illustrate the preparation of symmetrical ethers by the synthesis of diethyl ether from ethanol in the presence of acid:

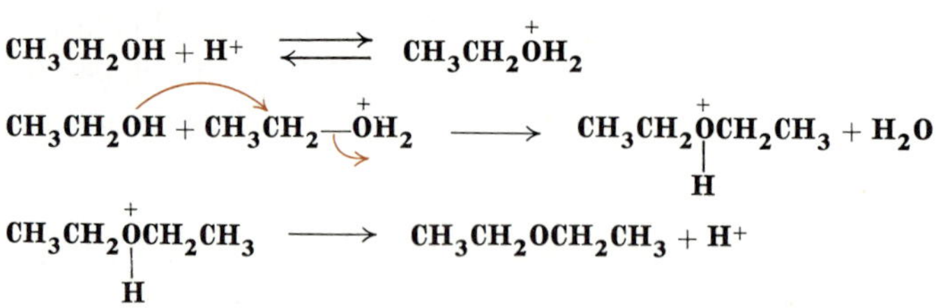

The mechanism involves protonation of an alcohol oxygen, displacement of water by a second alcohol molecule, and proton loss from the resulting intermediate to give the ether. The net effect is the loss of one mole of water from two moles of ethanol:

$$2CH_3CH_2OH \longrightarrow CH_3CH_2OCH_2CH_3 + H_2O$$

The acid of choice for this reaction is sulfuric acid, not HCl or HBr, which would give alkyl halides via nucleophilic attack by halide ion on the protonated alcohol.

The procedure used for the synthesis of unsymmetrical ethers is known as the *Williamson synthesis*; it is illustrated below in the preparation of ethyl *t*-butyl ether.

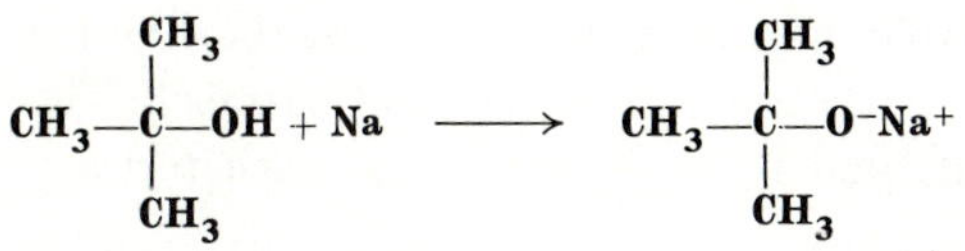

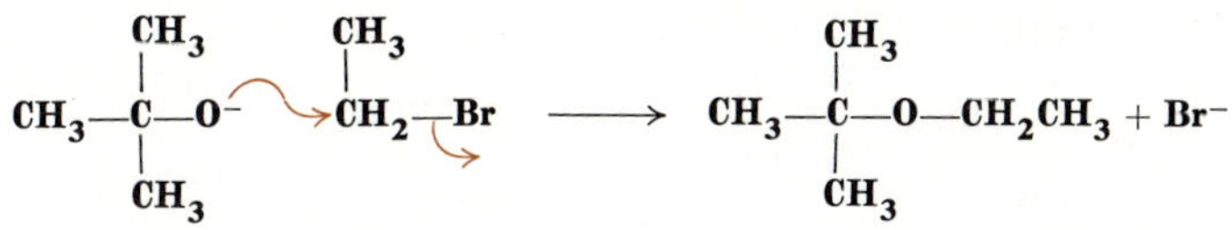

In the Williamson synthesis it is desirable to prepare the sodium salt of the more substituted alcohol and to use the less substituted halide. This is because substitution and elimination reactions compete, and (as will be explained in Chapter 8) the alternative route gives major amounts of olefin:

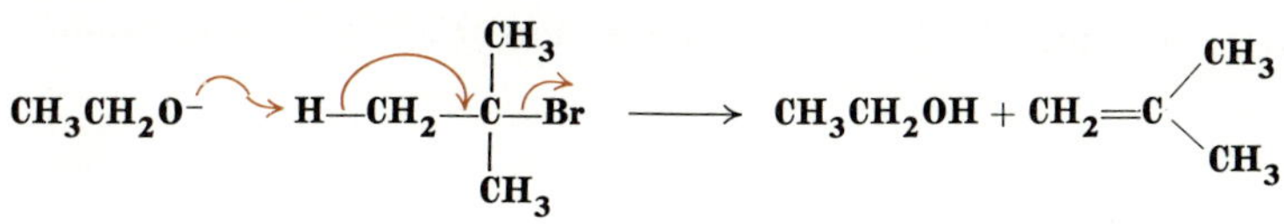

7.8 CARBOXYLIC ACIDS

Carboxylic acids cannot be prepared directly by nucleophilic substitution. However, *nitriles* can be prepared by an S_N2 reaction between a primary or secondary halide and sodium cyanide:

$$CH_3{-}X + {}^{-}C{\equiv}N \longrightarrow \underset{\text{a nitrile}}{CH_3{-}C{\equiv}N} + X^-$$

The nitrile group can then be hydrolyzed to a carboxylic acid in either strong acid or strong base:

$$CH_3C{\equiv}N \xrightarrow[HO^- \text{ or } H_3O^+]{H_2O} CH_3COOH$$

The mechanism of the hydrolysis of the nitrile involves several intermediates and will not be discussed here.

The cyanide displacement reaction fails with tertiary halides because, as in the Williamson synthesis, elimination is the predominant reaction:

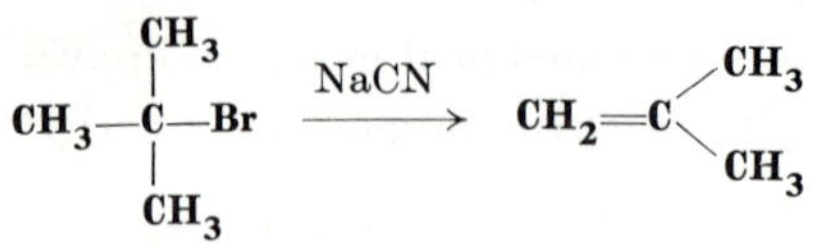

An alternative, and often preferable, means of converting an alkyl halide into a carboxylic acid with one more carbon was discussed in Chapter 3. The method employs a Grignard reagent, and it works well even with tertiary halides:

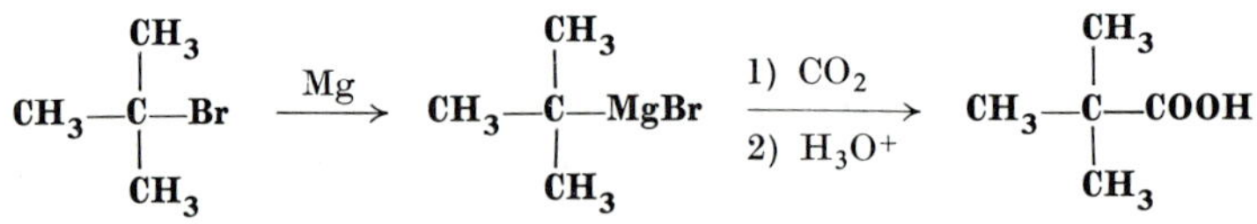

7.9 ALKYNES, ALKENES, AND ALKANES

Alkynes can be prepared via a substitution reaction known as *alkylation*. In this procedure an alkyl group is added to acetylene to produce a substituted acetylene. This is accomplished with sodium amide ($NaNH_2$), an extremely strong base.

$$\mathbf{Na + NH_3 \xrightarrow{Fe^{+3}} \overset{+}{Na}\overset{-}{N}H_2 + H_2}$$

Sodium amide is sufficiently strong a base to remove a proton from acetylene (aliphatic and olefinic C—H bonds are normally inert to sodium amide):

$$\mathbf{H{-}C{\equiv}C{-}H + \overset{+}{Na}\overset{-}{N}H_2 \longrightarrow H{-}C{\equiv}C^{-}Na^{+} + NH_3}$$

The sodium acetylide can then be used as a nucleophile to displace a halogen from a primary alkyl halide:

$$\mathbf{H{-}C{\equiv}C^{-}Na^{+} + CH_3{-}Br \longrightarrow H{-}C{\equiv}C{-}CH_3 + NaBr}$$

Secondary and tertiary halides give too much elimination product for this procedure to be generally useful. The reaction can be repeated with another primary halide to effect a second alkylation. For example, methylacetylene can be converted into ethylmethylacetylene:

$$\mathbf{CH_3{-}C{\equiv}C^{-}Na^{+} + CH_3CH_2{-}Br \longrightarrow CH_3{-}C{\equiv}C{-}CH_2CH_3}$$

The preparation of alkynes also provides a route to alkenes and alkanes. For example, it is possible to convert a disubstituted alkyne into the *cis* olefin by reduction with hydrogen and a palladium catalyst:

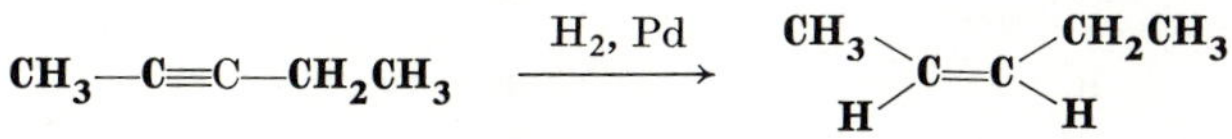

cis

The *trans* olefin can be obtained by reacting the alkyne with sodium dissolved in liquid ammonia (ammonia liquefies at –33°):

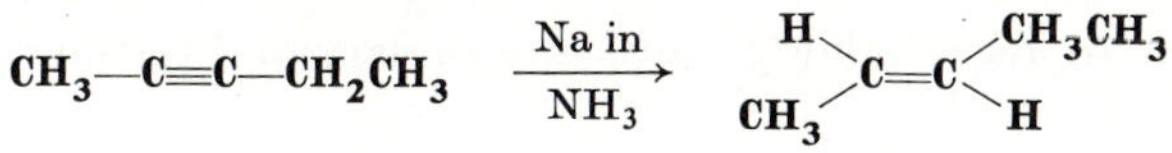

trans

Further reduction of the hydrocarbons with hydrogen and platinum gives the alkane:

$$\mathrm{CH_3CH{=}CHCH_2CH_3 \xrightarrow{H_2,\ Pt} CH_3CH_2CH_2CH_2CH_3}$$

cis or *trans*

The details and mechanisms of these important reactions will be discussed later. For the moment it is necessary only to realize that alkynes provide an entry to the alkene and alkane series.

7.10 AMINES

Ammonia is a potent neutral nucleophile which readily participates in substitution reactions. If an alkyl halide is treated with ammonia, an alkylammonium halide results:

methylammonium bromide

The free amine can be released by neutralization of the salt with sodium hydroxide:

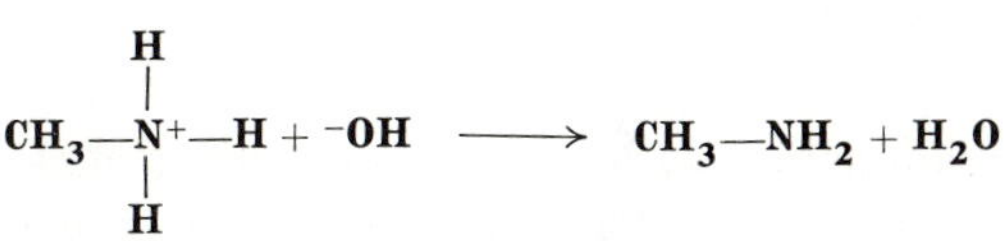

This latter reaction is the same as that between ammonium bromide and base:

$$\mathrm{NH_4Br + NaOH \longrightarrow NH_3 + H_2O + NaBr}$$

The substitution reaction can also be used to prepare secondary and tertiary amines:

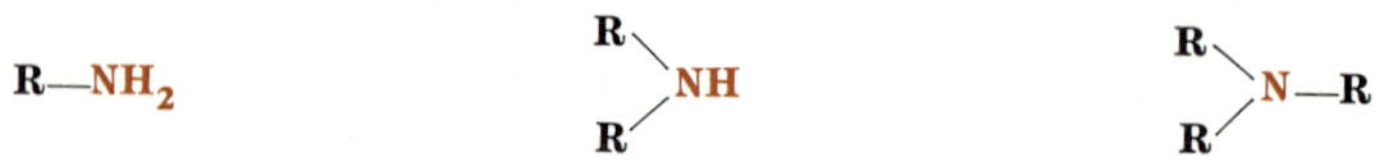

primary amine — secondary amine — tertiary amine

(Primary, secondary, and tertiary amines differ in the number of groups on the nitrogen; primary, secondary, and tertiary alcohols differ in the number of groups on the carbon bonded to the hydroxyl). Thus, a primary and secondary amine react with an alkyl halide to give a secondary and tertiary amine, respectively.

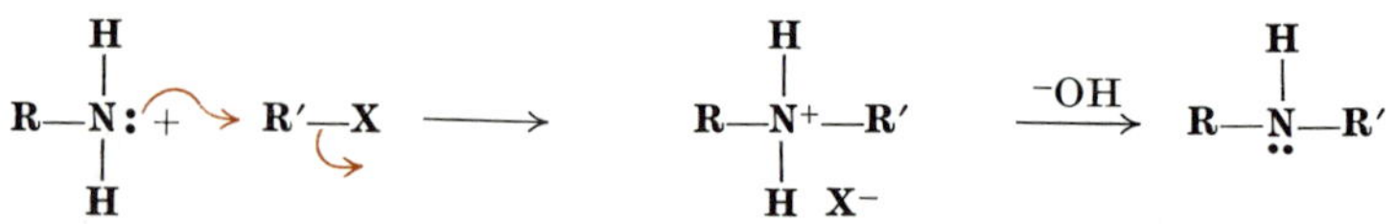

primary amine — dialkylammonium halide — secondary amine

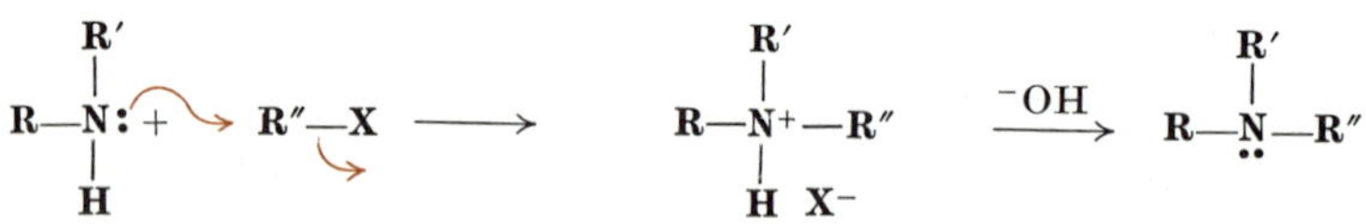

secondary amine — trialkylammonium halide — tertiary amine

Tertiary amines, in turn, will undergo further alkylation to afford tetraalkylammonium halides (also called *quaternary ammonium salts*):

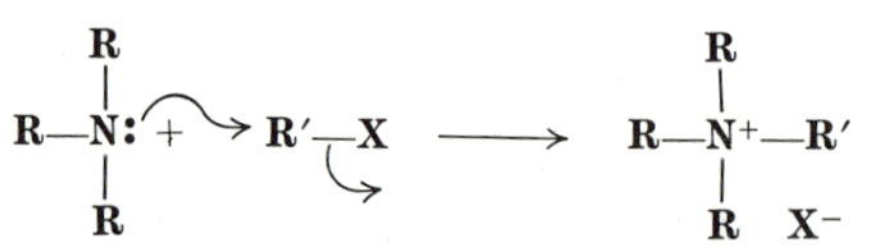

tertiary amine — quaternary ammonium salt

Thus, all the hydrogens of the ammonium ion can be replaced by alkyl groups.

Simple alkylation as a means of securing amines has serious limitations. For example, it is difficult to control the alkylation of a primary amine so that it will yield the secondary amine exclusively. Tertiary amine and even tetraalkylammonium halide may be formed. The secondary ammonium halide that forms in the initial substitution reaction may lose its proton to a primary amine, freeing the secondary amine for further alkylation:

$$CH_3NH_2 + CH_3I \longrightarrow (CH_3)_2\overset{+}{N}H_2 + I^-$$

$$(CH_3)_2\overset{+}{N}H_2 + CH_3NH_2 \rightleftharpoons (CH_3)_2NH + CH_3\overset{+}{N}H_3$$

$$(CH_3)_2NH + CH_3I \longrightarrow (CH_3)_3\overset{+}{N}H + I^-$$

To avoid having to separate mixtures, synthetic routes have been devised that can give rise to primary, secondary or tertiary amines exclusively. The *Gabriel synthesis* is a general method for producing primary amines to the exclusion of higher alkylation products. The starting material for this sequence is the readily available substance, phthalimide:

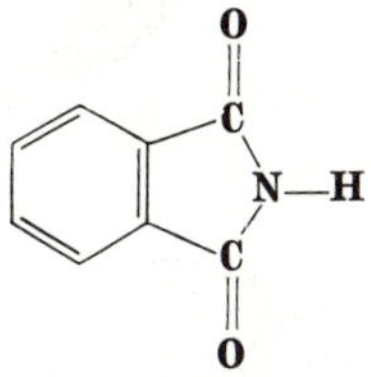

phthalimide

Phthalimide may be viewed as a derivative of ammonia in which two of the hydrogens have been replaced by carbonyl groups. The nitrogen is no longer basic because the lone pair electrons on the nitrogen are now involved in amide resonance and therefore are not available for attack by a proton:

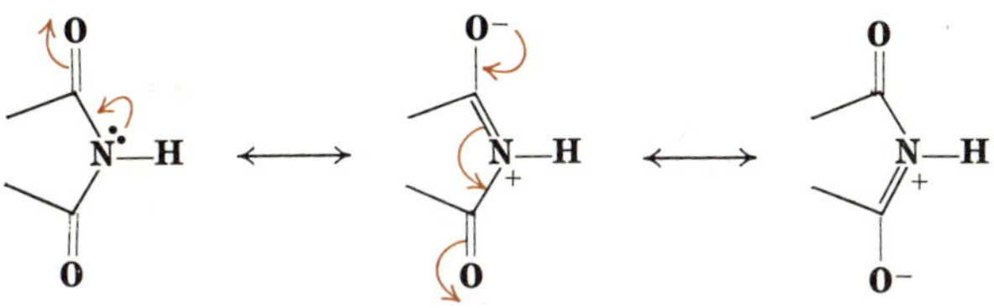

Consequently, the nucleophilic character of the nitrogen is also diminished relative to ammonia, for it was the availability of the lone pair electrons on the ammonia nitrogen that gave the nitrogen its nucleophilic character.

The imide group is not only a very weak base, but it is an *acid* of $pK_a = 10$ (equivalent to the acidity of phenol). The acidity of the proton on the imide nitrogen is the result of a highly resonance-stabilized anion:

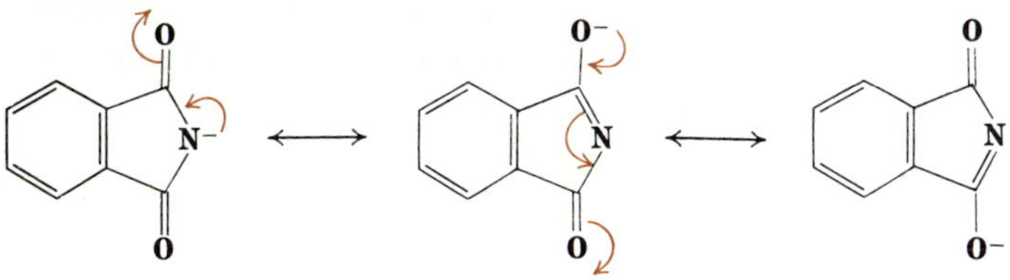

This anion can be alkylated by an alkyl halide in a typical S_N2 reaction:

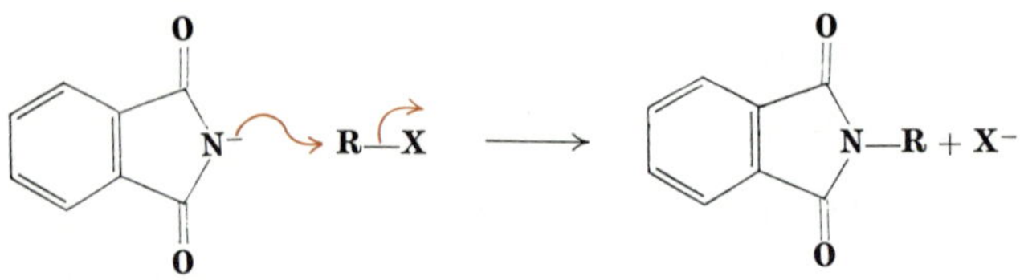

The product is a substituted imide which can be hydrolyzed to provide the primary amine uncontaminated with higher alkylation products:

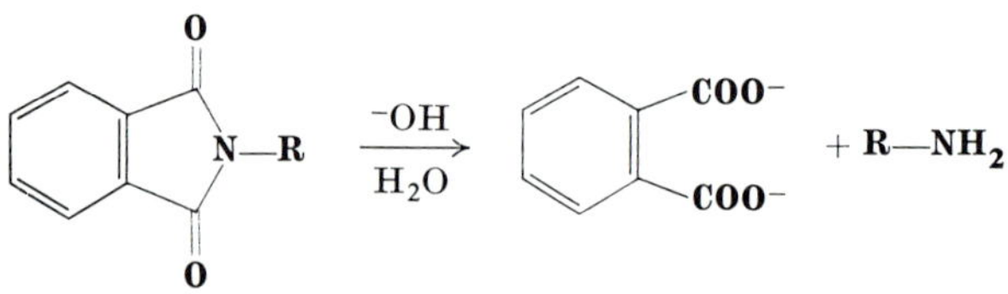

No details of the hydrolysis mechanism will be given. The important point is that only one alkyl group can bond to the imide anion. Therefore there is no danger of contamination by secondary or tertiary amines. In the synthesis of primary amines the Gabriel synthesis is preferred to the simple alkylation of ammonia. A specific example is shown below:

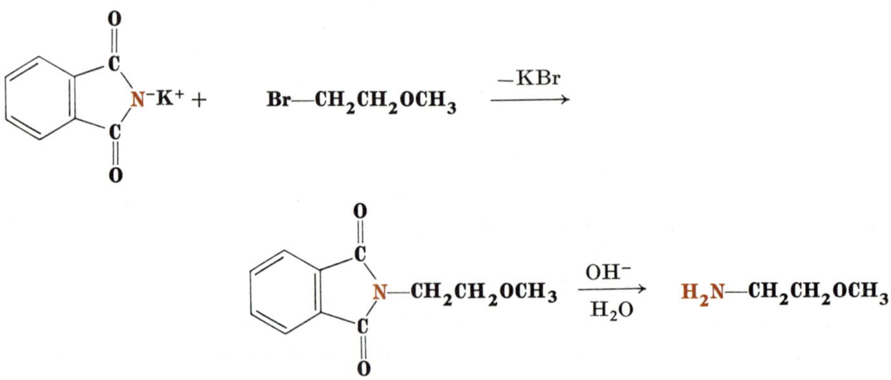

Other selective methods for synthesizing primary, secondary, and tertiary amines have been developed. For example, amides can be reduced with lithium aluminum hydride to produce amines:

$$R-\overset{\overset{O}{\|}}{C}-NH_2 \xrightarrow{LiAlH_4} R-CH_2-NH_2$$

$$R{-}\overset{O}{\overset{\|}{C}}{-}NHR \xrightarrow{LiAlH_4} R{-}CH_2{-}NHR$$

$$R{-}\overset{O}{\overset{\|}{C}}{-}NR_2 \xrightarrow{LiAlH_4} R{-}CH_2{-}NR_2$$

Lithium aluminum hydride will also reduce nitriles to primary amines:

$$R{-}C{\equiv}N \xrightarrow{LiAlH_4} R{-}CH_2{-}NH_2$$

Examples of $LiAlH_4$ reductions are given below:

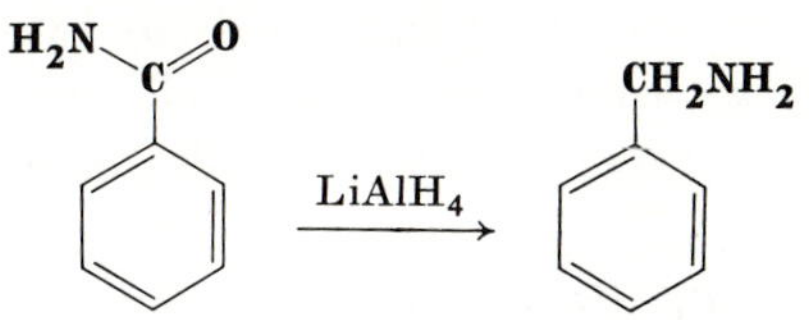

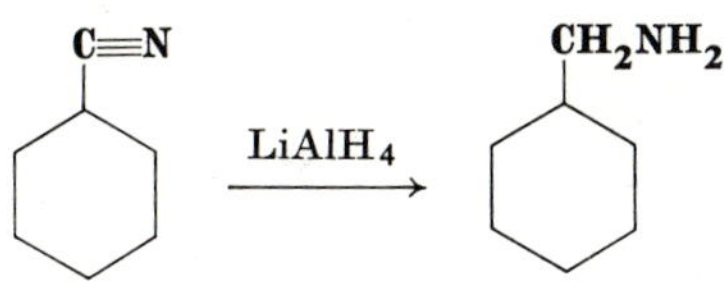

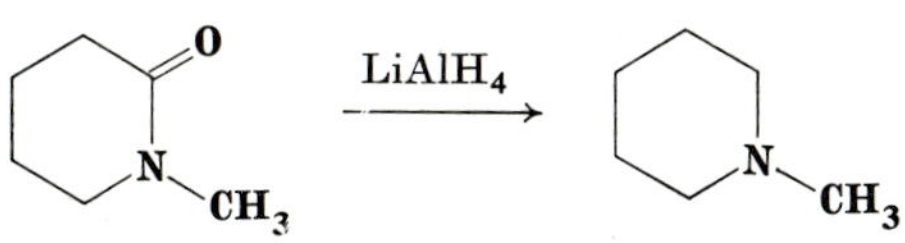

7.11 EPOXIDE RING OPENING

An *epoxide* is a three-membered cyclic ether that can be prepared by treating an olefin with an oxidizing agent called a peracid:

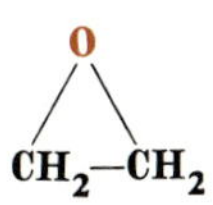

an epoxide
(ethylene oxide)

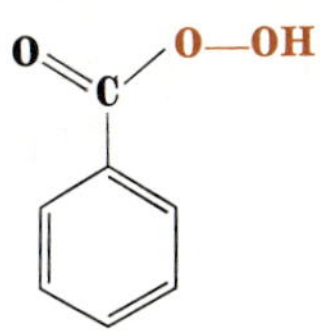

a peracid
(perbenzoic acid)

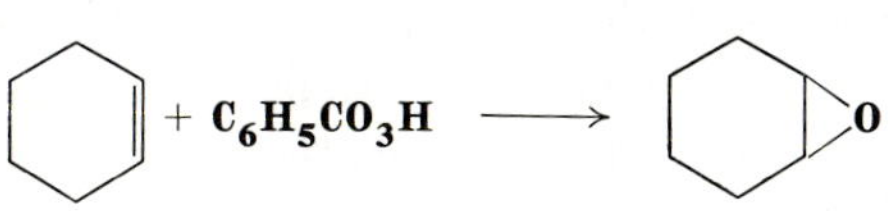

Epoxides open up readily in both acidic and basic media because of the strain associated with the three-membered ring. For example, a Grignard reagent will react with ethylene oxide to give an alcohol (after work-up of the reaction product in acid):

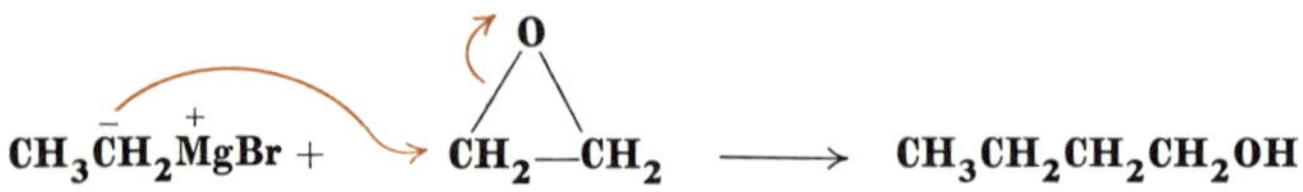

Therefore the organic chemist has an excellent way to extend a chain by two carbons: alcohol, to alkyl halide, to Grignard, to alcohol of two additional carbons. This type of reaction is, however, limited to ethylene oxide; more complicated epoxides give side reactions.

If the epoxide is derived from a cyclic olefin, the reaction with a nucleophile yields a *trans*-disubstituted ring:

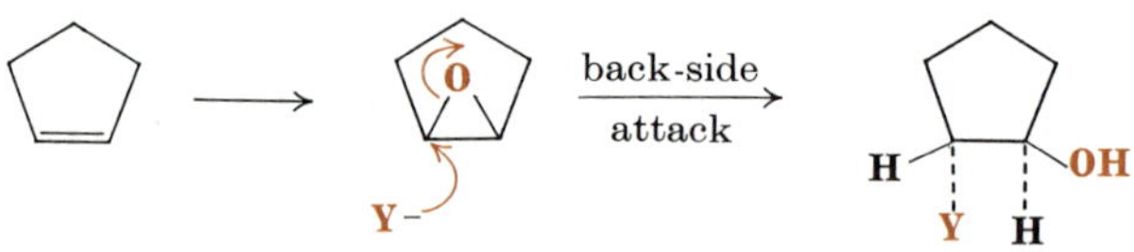

trans

Examples:

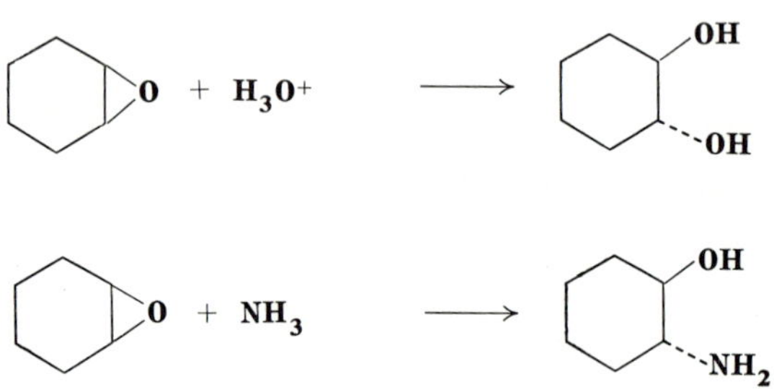

7.12 MALONIC ESTER AND ACETOACETIC ESTER SYNTHESES

These two related syntheses, the key steps of which are substitution reactions, are widely used in the construction of organic molecules. In our discussion of the alkylation of acetylide anion we mentioned that aliphatic C—H linkages are not usually acidic enough to allow anion formation. This is not true for the methylene (—CH_2—) group of malonic ester and acetoacetic ester:

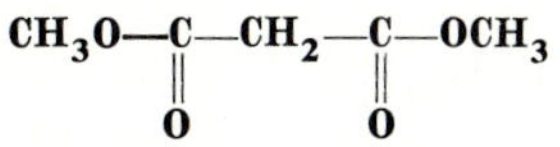

malonic ester

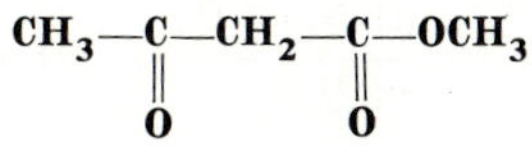

acetoacetic ester

The methylene groups are flanked by *two* carbonyl groups. Reaction of the esters with an alkoxide generates a highly resonance-stabilized carbanion in which the bulk of the negative charge resides on the electronegative oxygens:

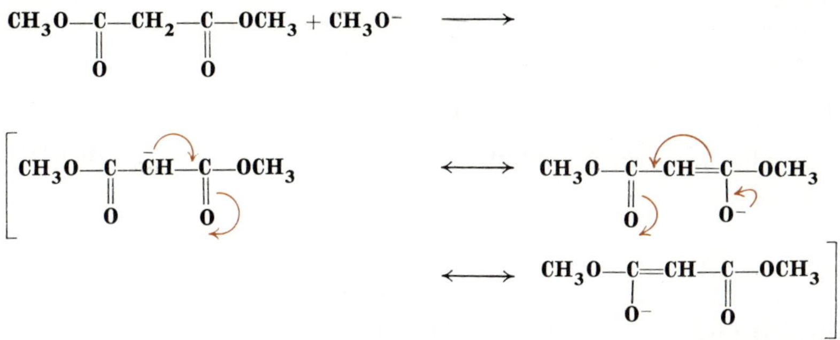

The carbanions derived from malonic ester and acetoacetic ester are good nucleophiles and may be used in substitution reactions. For example, the anions may be alkylated at the carbon when treated with an alkyl halide:

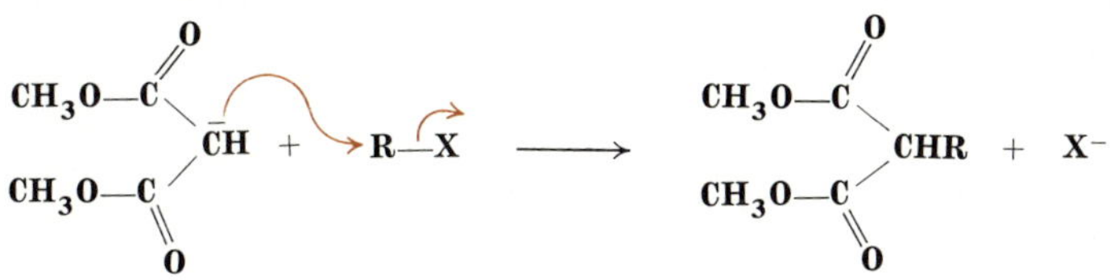

malonic ester anion

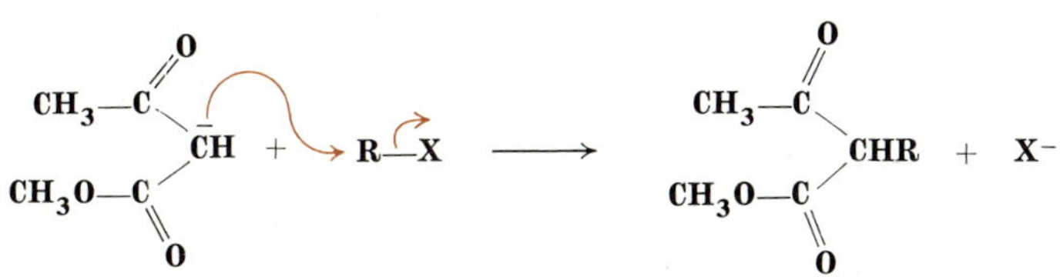

acetoacetic ester anion

The alkylation procedure can be repeated to afford dialkylated malonic esters and acetoacetic esters:

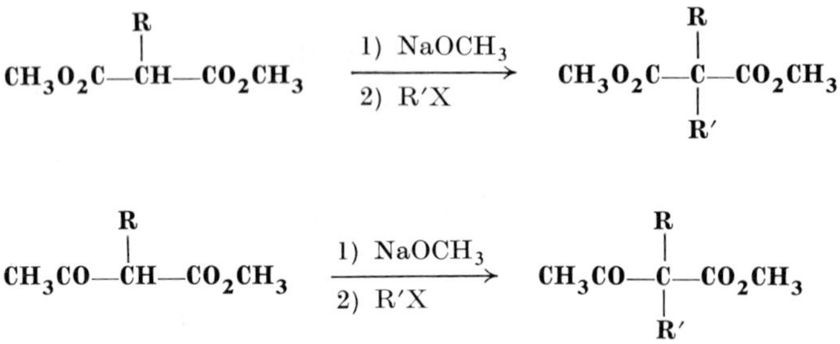

These alkylations are useful because the substituted malonic ester can be converted into a carboxylic acid, and the substituted acetoacetic acid can be converted into a ketone:

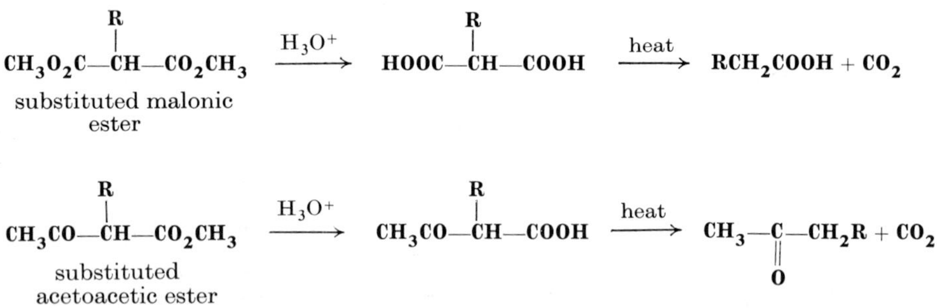

The first step is an acid-catalyzed *hydrolysis* of an ester to a carboxylic acid. The second step is a *decarboxylation* (loss of carbon dioxide). Carboxylic acids are normally inert to decarboxylation except when there is a carbonyl group in the β position:

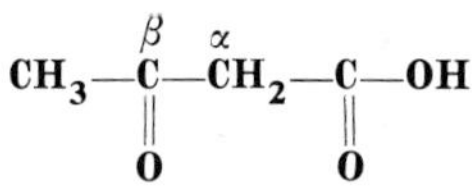

Such ketoacids decarboxylate on gentle warming. While a detailed discussion of ester hydrolysis and decarboxylation is premature, the student should recognize the value of these reactions in carrying out the following transformations:

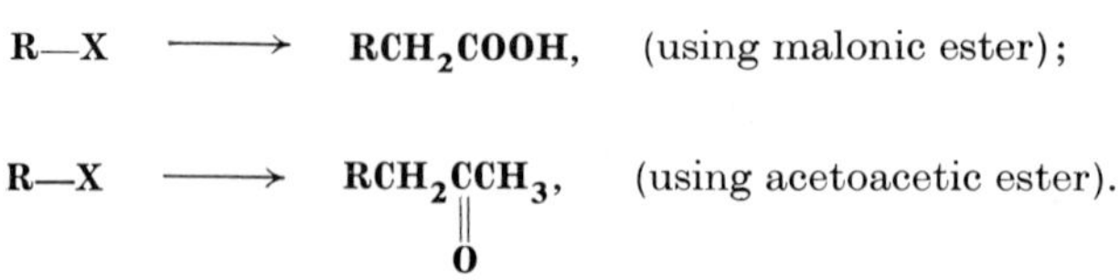

Two examples are shown below:

1) 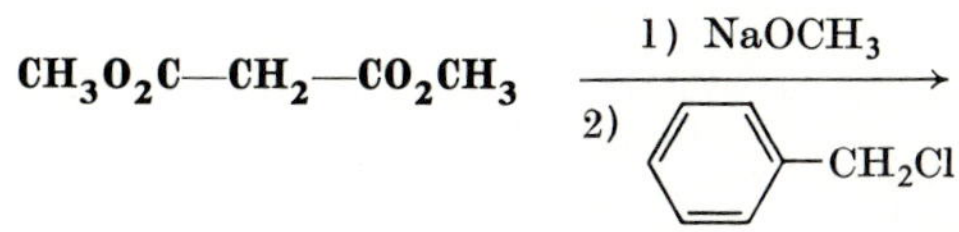

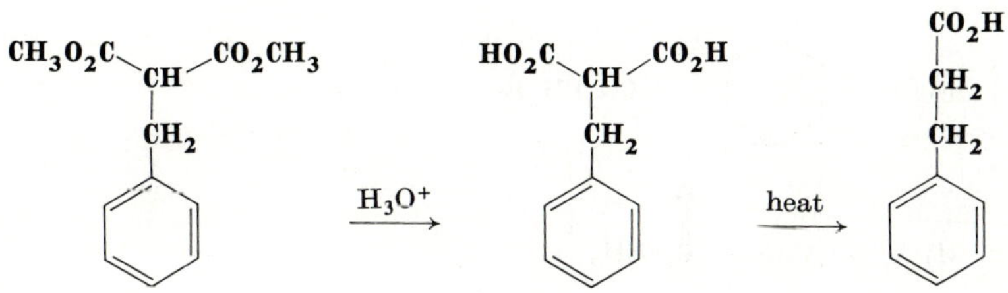

Net reaction:

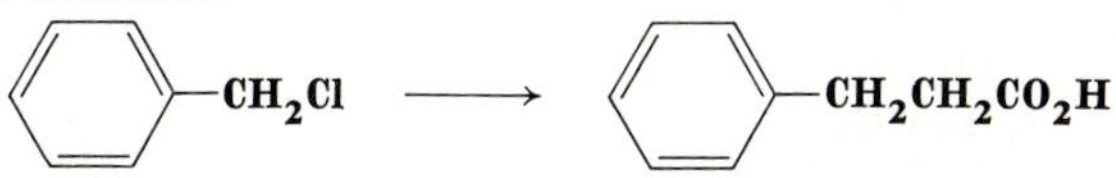

2) 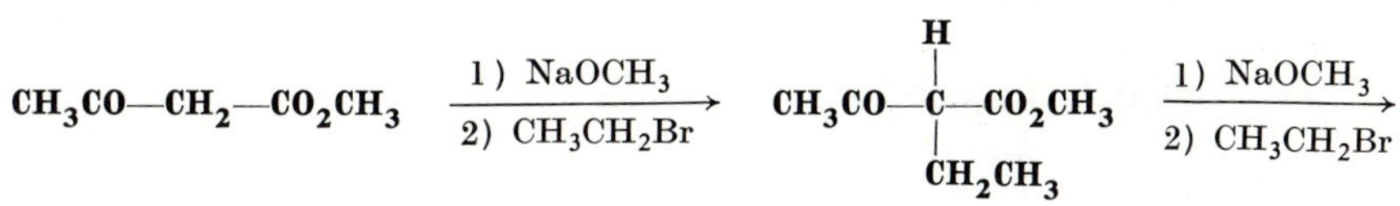

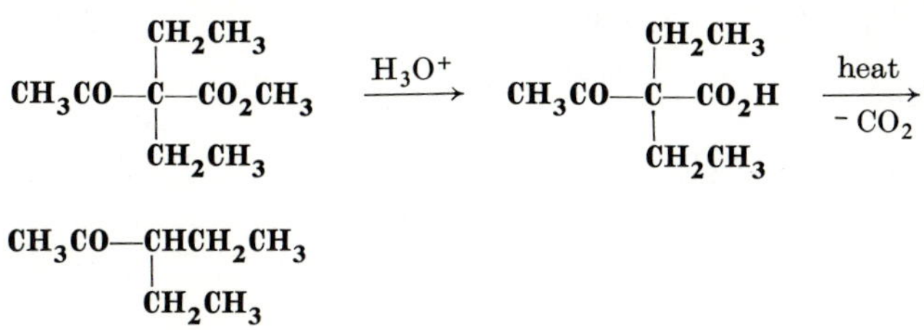

ADDITIONAL READING

R. W. Alder, R. Baker, and J. M. Brown, *Mechanism in Organic Chemistry*, New York: Wiley, 1971. Chapters 2 and 3.

E. R. Thornton, *Solvolytic Mechanisms*, New York: Ronald Press, 1964.

PROBLEMS

1. Supply the correct reaction products.

a) 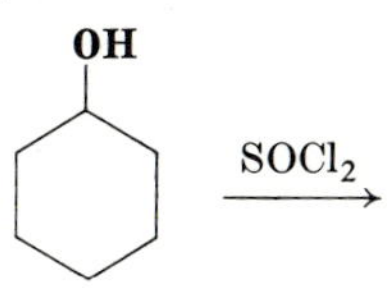

b) 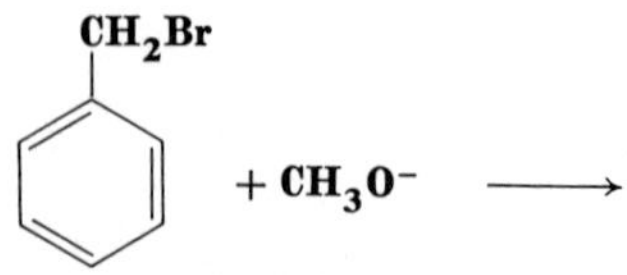

c) 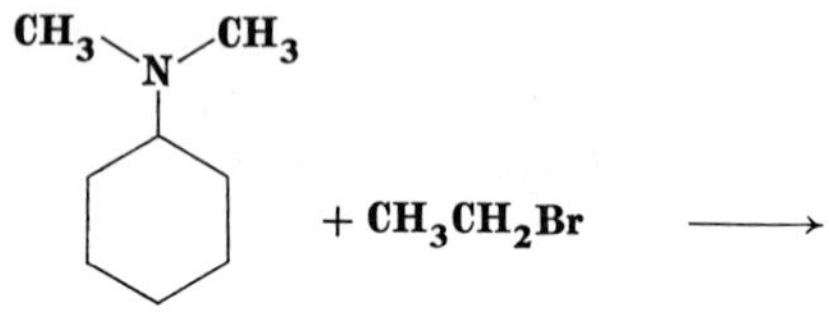

d) 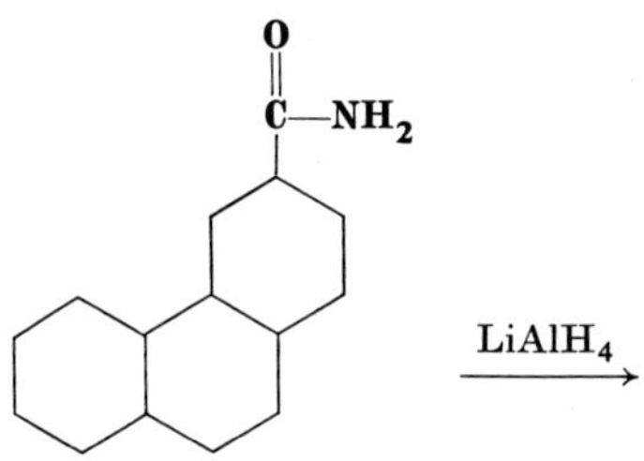

e) 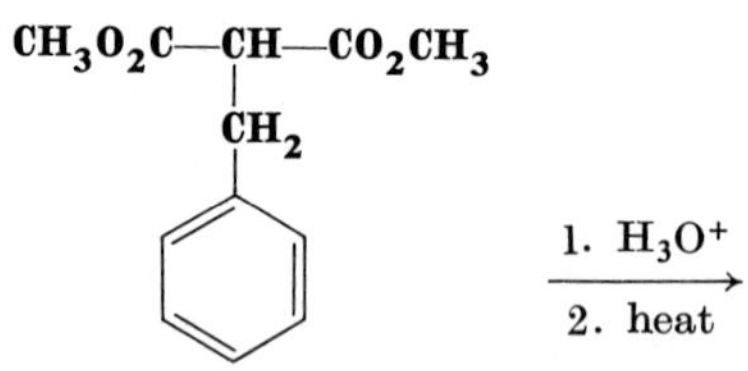

f) 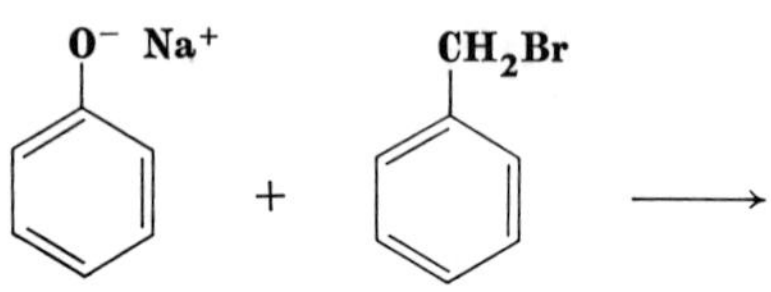

g) 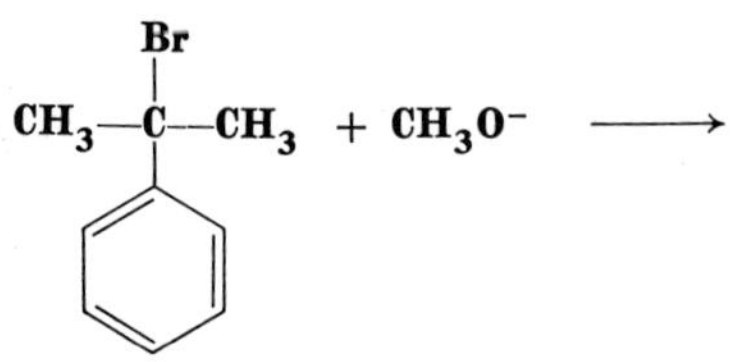

h) 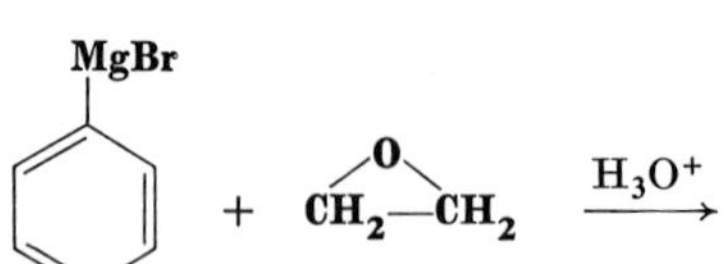

i) 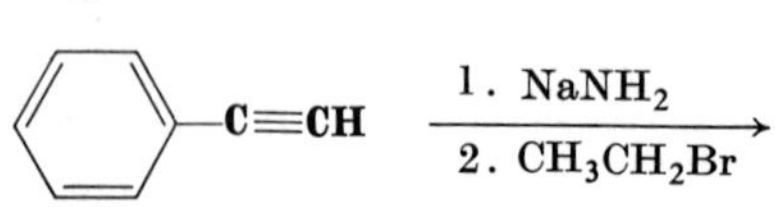

2. Fill in the missing blanks.

   a) Primary halides are ——— reactive than tertiary halides in S_N2 substitutions.
   b) The S_N1 reaction involves a ——— ion intermediate.
   c) ——— reactions compete with substitution reactions.
   d) A ——— amine reacts with one mole of methyl iodide to form a quarternary ammonium salt.
   e) S_N2 reactions occur exclusively by ——— attack of the nucleophile.
   f) An alkyl halide can be converted into an acid containing two more carbons by means of the ——— ester synthesis.
   g) A three-membered cyclic ether is called an ———.
   h) The ——— synthesis is a general method form making primary amines.
   i) Sodium amide is made by adding sodium metal to ———.
   j) The number of transition states involved in an S_N1 process is ———.
   k) Iodide is a much better leaving group than fluoride because iodide is more ———.
   l) Thionyl chloride has the formula ———.
   m) ——— is a bromide which is inert to substitution by S_N1 and S_N2 mechanisms.

3. Complete the following reactions by supplying one of the reactants, the major organic product, or the required catalyst.

   a) $(CH_3)_2CHCH_2CH_2OH \xrightarrow{?} (CH_3)_2CHCH_2CH_2OCH_2CH_2CH(CH_3)_2$

   b) $(CH_3)_2CHCH_2Br + ? \longrightarrow (CH_3)_2CHCH_2OCH_2CH_3$

   c) $(CH_3CH_2)_3C{-}Br + {}^{-}CN \longrightarrow ?$

   d) $HC{\equiv}C^- \; Na^+ + CH_2{=}CHCH_2CH_2Br \longrightarrow ?$

   e) 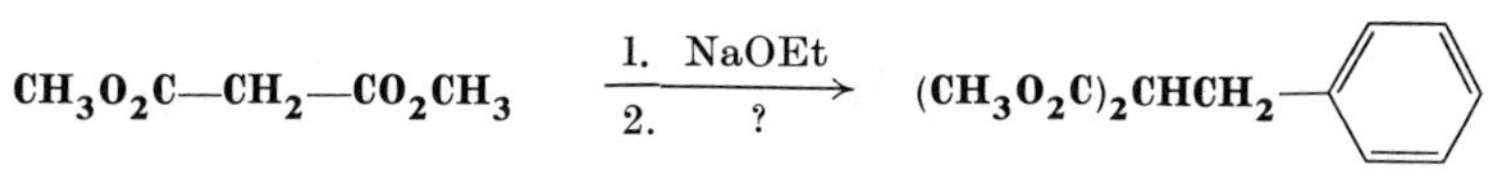

   f) 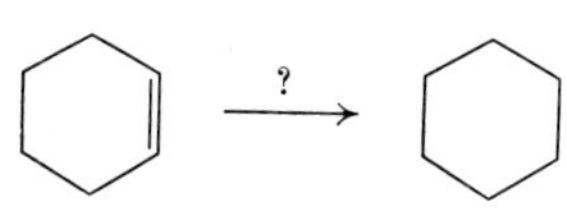

   g) 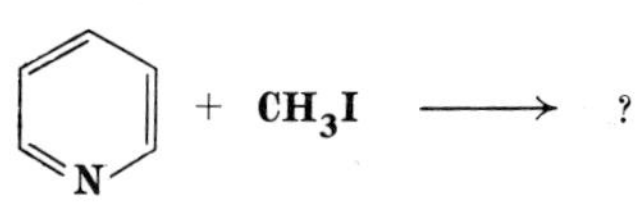

h) 

4. Predict the relative reactivity of the following under S_N1 conditions.

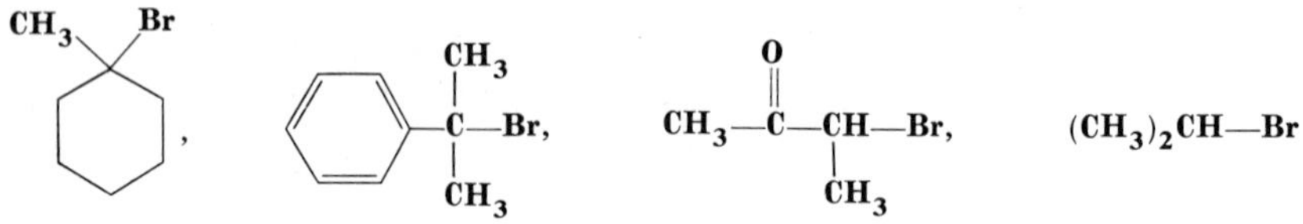

5. Solvents can be classified according to a physical property called the *dielectric constant*. The dielectric constant is a measure of how well the solvent insulates two charges. Solvents such as water with large dielectric constants are good insulators and can dissolve ionic materials. Nonpolar solvents such as heptane have low dielectric constants and are unable to dissolve salts. The dielectric constants of a few common solvents are given below.

Solvent	Dielectric constant
water	78
methanol	31
ethanol	25
benzene	2.3
hexane	1.9

Would you expect an S_N1 reaction to proceed faster in methanol or benzene? Explain.

6. The bicyclic alkyl bromide drawn below resists substitution by both S_N1 and S_N2 mechanisms. Explain.

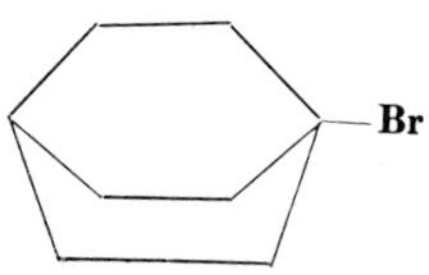

7. For a given atom, the stronger the base, the better the nucleophile. Arrange the following nucleophiles in decreasing order of reactivity toward methyl iodide.

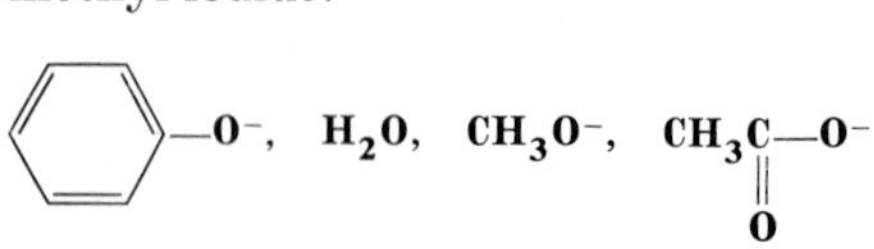

8. Cells contain enzymes (see Problem 26, Chapter 4) which catalyze the substitution of a group Y for a group X in the molecule RX. In certain cases, it is found that an optically active RX gives an optically active RY of the *same* configuration (i.e., the biological substitution goes with *retention* of configuration). This result contrasts with simple S_N1 and S_N2 reactions which give *racemization* and *inversion*, respectively. Explain how retention of configuration might occur.

9. Convert:

a) $CH_3CH_2CH_2C(CH_3)_2{-}Br$ into $CH_3CH_2CH_2C(CH_3)_2{-}COOH$

b) 

c) 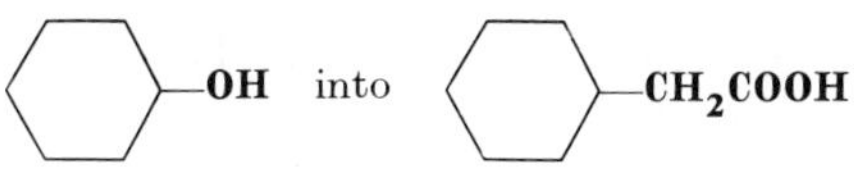

d) 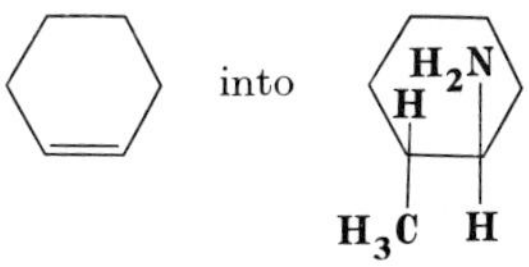

10. Nitriles can be hydrolyzed in strong base to carboxylic acids (see Section 7.8). Write a mechanism for this reaction using the following guidelines:

$$CH_3C{\equiv}N \xrightarrow{^{-}OH} CH_3COO^- + NH_3$$

Hydroxide ion attacks the partially positive nitrile carbon, while the nitrogen picks up a proton from a water molecule. The resulting intermediate tautomerizes (see Section 4.12) to an amide. Hydroxide ion then adds to the amide carbonyl group. A proton is transferred from the hydroxyl to the nitrogen, and the unstable intermediate then collapses to the carboxylate anion and ammonia.

11. Nitrogen mustard is a biologically active alkylating agent which has been used in cancer chemotherapy. The compound

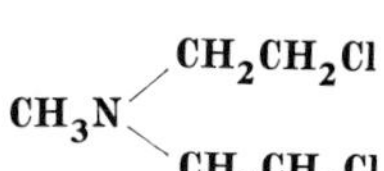

inhibits mitosis in tumor cells. It is believed that nitrogen mustard alkylates DNA (the material of which genes are composed) and prevents the DNA synthesis needed for cell division. Nitrogen mustard is a much more powerful alkylating agent than ethyl chloride, although both are primary alkyl halides. Suggest an explanation for this fact.

12. Explain the following:

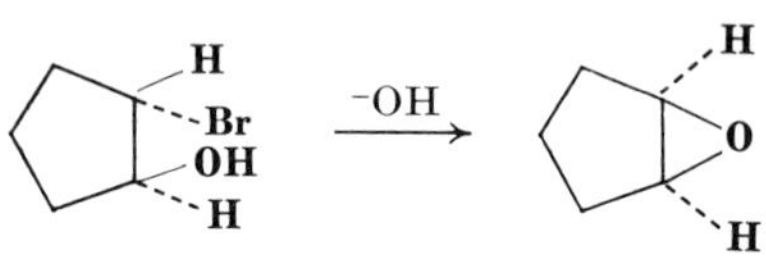

while

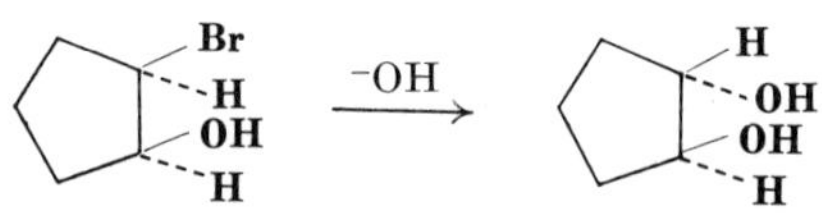

13. The rate of solvolysis of benzyl bromide in aqueous acetone is inhibited by lithium bromide but not by lithium perchlorate. This inhibition is called a *common ion effect*. The simple

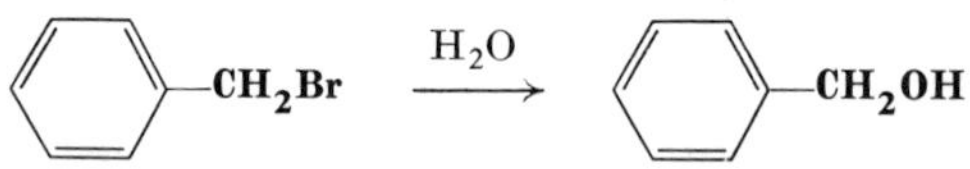

mechanism for an S_N1 reaction does not explain the common ion effect because the rate is proportional only to the amount of benzyl bromide:

$$\mathbf{RBr} \xrightarrow[\text{slow}]{k_1} [\mathbf{R}^+] + \mathbf{Br}^-$$

$$[\mathbf{R}^+] \xrightarrow[\text{fast}]{\mathrm{H_2O}} \mathbf{ROH} + \mathbf{H}^+$$

$$\mathbf{rate} = k_1[\mathbf{RBr}]$$

On the other hand, the experimental results *are* consistent with an S_N1 mechanism in which the first step is reversible:

$$\mathbf{RBr} \underset{k_2}{\overset{k_1}{\rightleftharpoons}} [\mathbf{R}^+] + \mathbf{Br}^-$$

$$[\mathbf{R}^+] \xrightarrow{k_3} \mathbf{ROH} + \mathbf{H}^+$$

$$\mathbf{rate} = \frac{k_1k_3[\mathbf{RX}]}{k_3 + k_2[\mathbf{Br}^-]}$$

Derive the rate expression for this mechanism using the steady state assumption (see Section 4.20) for the carbonium ion. How would you determine experimentally the rate of benzyl bromide solvolysis?

14. An optically pure secondary iodide (rotation = +20°) is converted into the corresponding acetate in 0.1*M* sodium acetate in acetic acid:

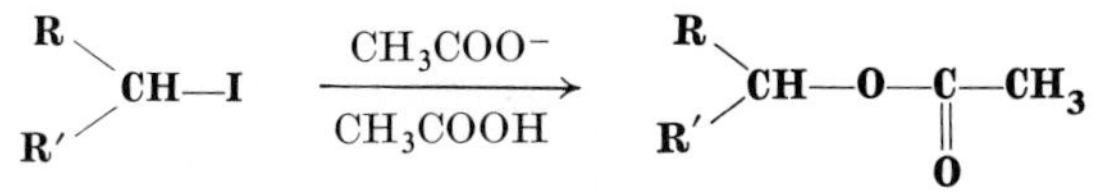

The acetate is isolated and found to have an optical rotation of –12°. Optically pure acetate is known to have a rotation of –16°. When the secondary iodide is reacted in a 1.0*M* sodium acetate solution, the isolated product is found to have a rotation of –16°. Explain the difference in optical rotations between the products formed in 0.1*M* and 1.0*M* sodium acetate in acetic acid.

15. The stereochemistry of the reaction between an alcohol and thionyl chloride to give an alkyl chloride depends on the reaction conditions. When pyridine is used as the solvent, the substitution occurs with inversion. In the absence of pyridine, retention predominates. Was pyridine used in the second reaction of the first sequence shown below? Assume that all other displacement reactions proceed via S_N2 mechanisms.

$$1)\quad \underset{(+)}{\mathbf{RCl}} \xrightarrow{^{-}OH} \mathbf{ROH} \xrightarrow{SOCl_2} \mathbf{RCl} \xrightarrow{CH_3O^-} \underset{(+)}{\mathbf{ROCH_3}}$$

$$2)\quad \underset{(-)}{\mathbf{RCl}} \xrightarrow{^{-}OH} \mathbf{ROH} \xrightarrow{Na} \mathbf{RO^-} \xrightarrow{CH_3Br} \underset{(-)}{\mathbf{ROCH_3}}$$

16. Starting from organic compounds of three carbons or less, malonic ester or acetoacetic ester, and any inorganic reagent, synthesize the following compound.

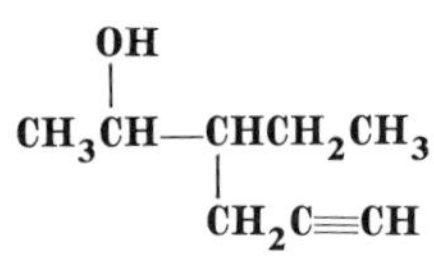

Before attempting this problem, review the reactions in Chapter 3.

17. Virtually all antihistaminic drugs contain the structural unit R_2N—C—C—X where X is nitrogen or oxygen. One of the earliest of these drugs (discovered while searching for antimalarials) is shown below along with its synthesis. The compound protects animals from bronchial

spasm caused by histamine. Write a mechanism for the first synthetic step.

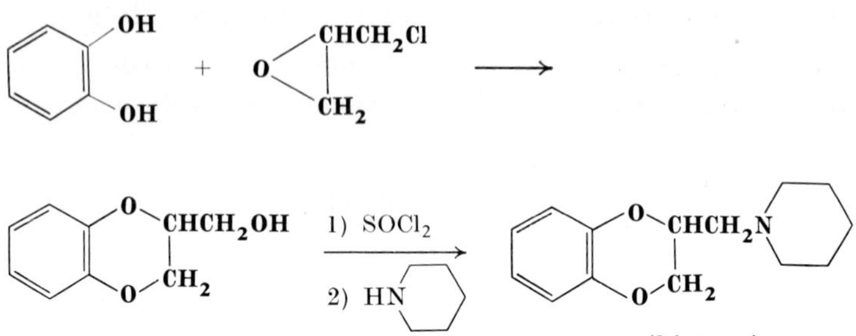

an antihistamine

8
ELIMINATION REACTIONS

8.1 ELIMINATION REACTIONS

The elimination reaction is the second of the fundamental heterolytic organic processes. In an elimination reaction, a small fragment is ejected from a molecule, and unsaturation (multiple bonding) is introduced. Some elimination reactions are illustrated below:

1) 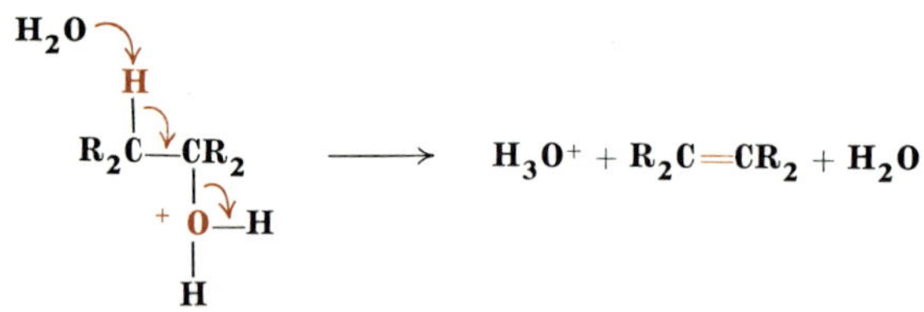

2) 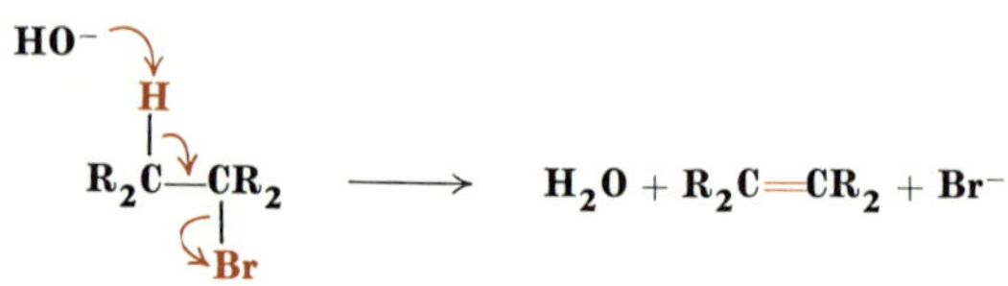

3) 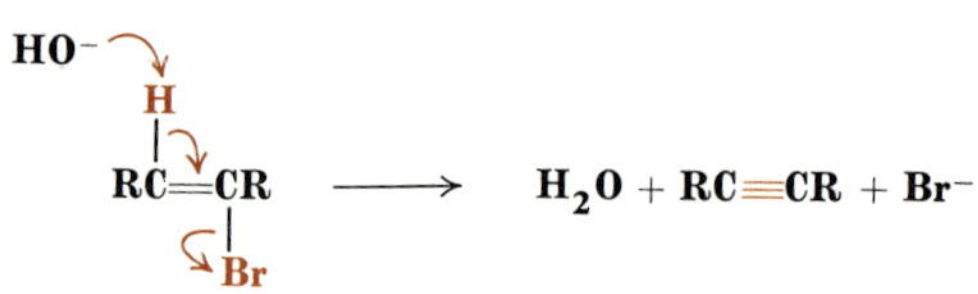

4) 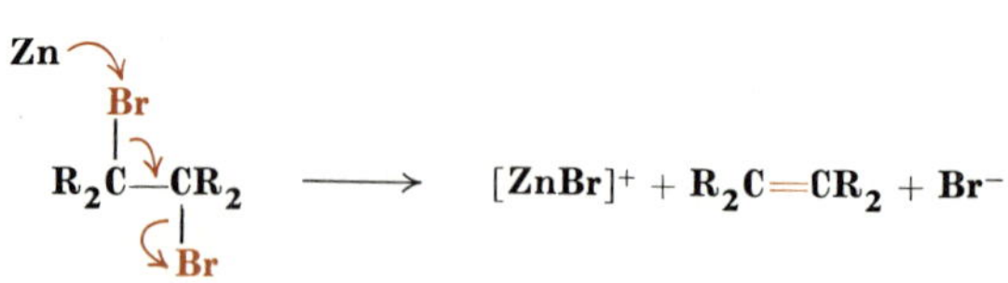

5) 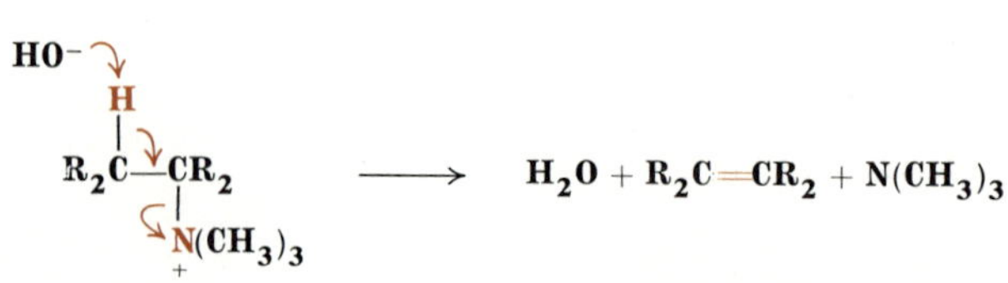

6) 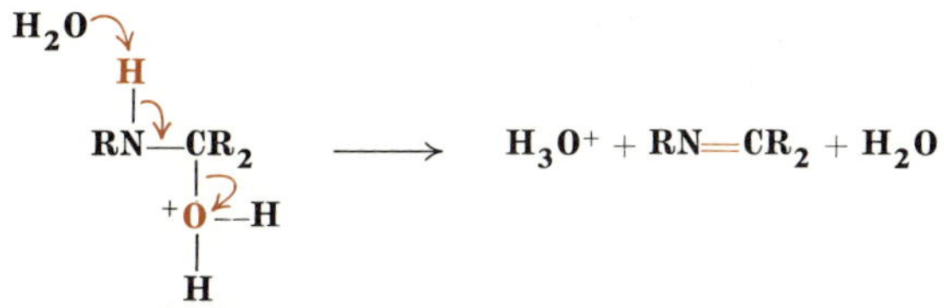

7) 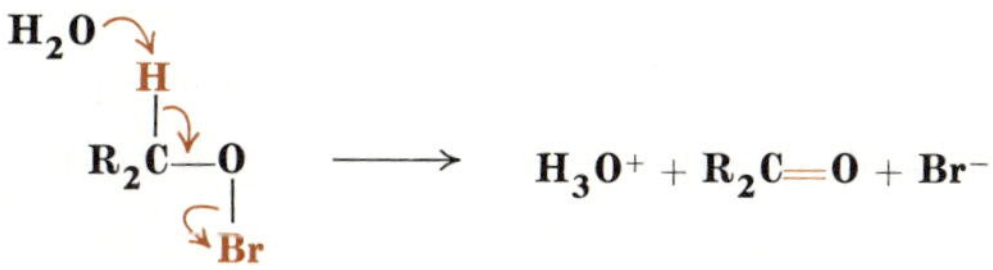

Note that not only are carbon-carbon multiple bonds generated in elimination reactions but carbon-oxygen and carbon-nitrogen multiple bonds as well. We will consider specific applications of these reactions later; now we will examine the mechanism of this process.

8.2 MECHANISM

There are three convenient categories of eliminations (E2, E1cb, and E1). The first of these is depicted in Fig. 8.1.

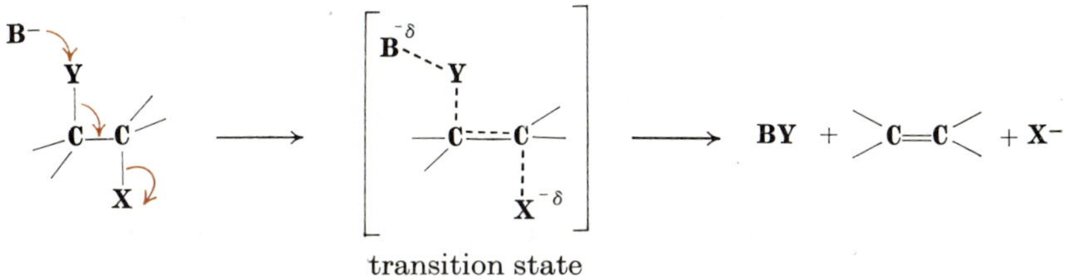

Fig. 8.1 E2 mechanism

A base, B^-, attacks Y and removes it as a positive fragment. This releases a pair of electrons (those that bound Y to carbon) which may then establish the carbon-carbon multiple bond by ejecting X as an anion. The transition state for the process is shown in brackets; B—Y and carbon-carbon bond formation, as well as Y-carbon and X-carbon bond cleavage, all occur simultaneously.

The above mechanism entails no discrete intermediate and is called an *E2 elimination*. The 2 refers to the two reactants involved in the transition state. *Dehydrohalogenation* (loss of HX) in the following reaction undoubtedly takes place via an E2 mechanism:

$$ClCH_2CH_2CH(OCH_3)_2 + KOH \longrightarrow CH_2{=}CHCH(OCH_3)_2 + KCl + H_2O$$

The reaction is carried out by heating the compound over powdered KOH at 220°. Another example of an E2 elimination is the loss of two moles of HX from styrene dibromide (in the presence of sodium amide in liquid ammonia) to form phenyl acetylene:

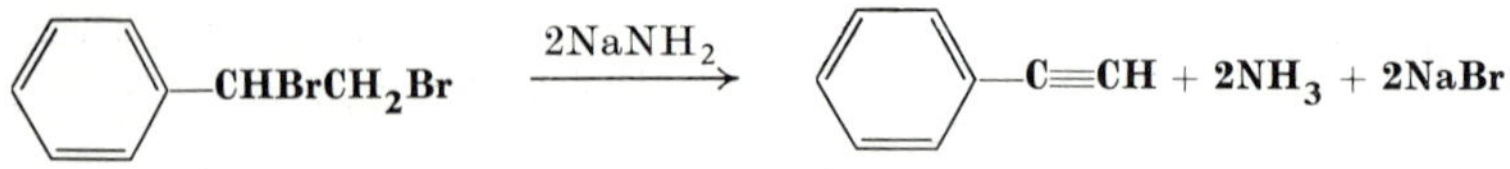

The E2 process involves simultaneous bond cleavage and bond formation. Alternatively, one might imagine a step-wise mechanism such as that shown in Fig. 8.2.

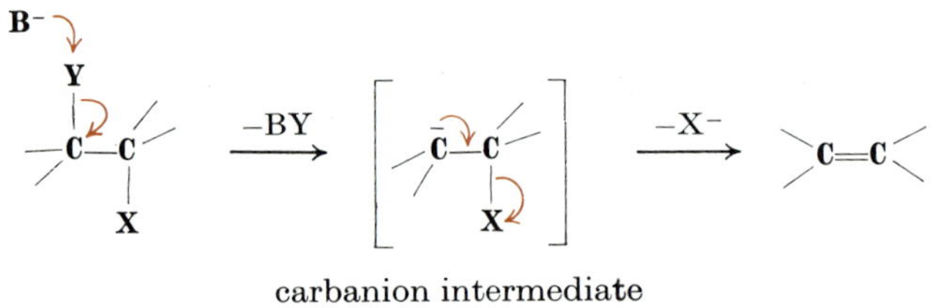

Fig. 8.2 E1cb mechanism

The Y-carbon bond cleaves heterolytically, while the X-carbon bond remains intact. The carbanion intermediate subsequently decomposes to olefin. This mechanism is called an *E1cb elimination* (referring to a first-order elimination from a carbanion). The E1cb is the least common of the three mechanisms. It occurs mainly in compounds in which the carbanion is stabilized by resonance or inductive effects:

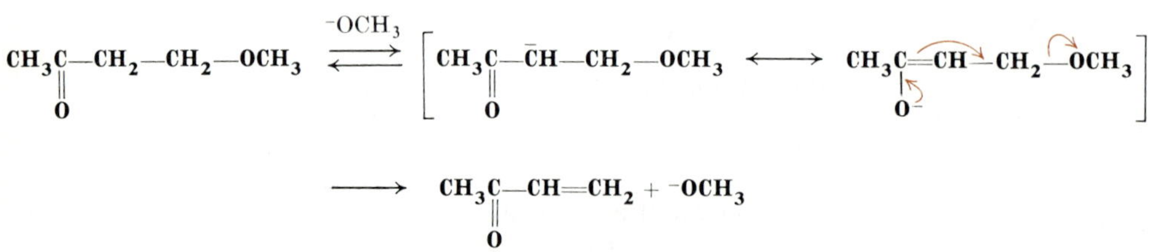

The third possible mechanism, called an *E1 elimination*, involves initial loss of X^- with no cleavage of the Y-carbon bond (Fig. 8.3).

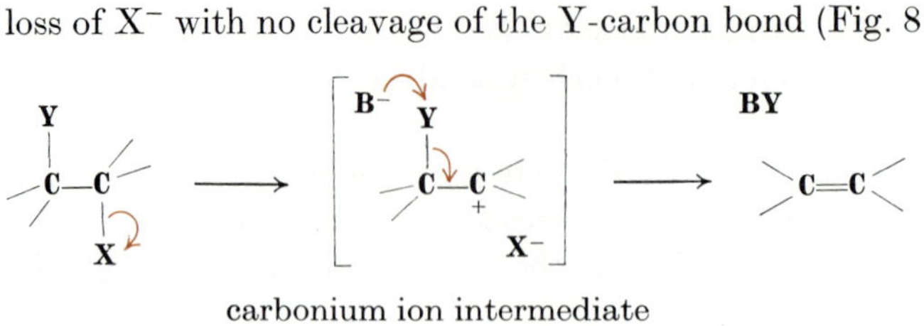

Fig. 8.3 E1 mechanism

The resulting carbonium ion intermediate subsequently loses Y to form an olefin. If carbonium ion formation is the slow rate-determining step in the elimination, then the rate will be independent of the concentration of B^-. This mechanism is the counterpart of the S_N1 substitution, which also entails carbonium ion formation. Indeed, the S_N1 and E1 reactions compete with each other. E1 eliminations (and S_N1 substitutions) are favored by any factor which stabilizes the carbonium ion, such as alkyl substituents and resonance with aromatic rings. A typical example of E1 elimination is the alcohol *dehydration* (loss of water) performed in strongly acidic solutions:

$$(C_6H_5)_2C(OH)CH_2(C_6H_5) \xrightarrow{20\%\ H_2SO_4} (C_6H_5)_2C{=}CH(C_6H_5) + H_2O$$

It is unlikely that this dehydration proceeds via an E2 mechanism because of the absence of a moderately strong base to participate in such a process. An E1cb mechanism is absolutely out of the question; carbanions do not form in acidic media.

We must emphasize that the three categories of elimination reactions are somewhat contrived. This becomes clear when one considers an E2 transition state (Fig. 8.1) in which there is a great deal of cleavage of the X-carbon bond, and very little cleavage of the Y-carbon bond. A reaction with such a transition state might be considered an E2 elimination with considerable E1 character. Thus a spectrum of mechanisms is possible; there are no sharp breaks in nature.

How do we tell whether a particular elimination reaction resembles one of the three mechanistic extremes? Both kinetic and stereochemical evidence have been brought to bear on this question.

8.3 KINETIC EVIDENCE

As the name implies, the rate of an E2 elimination is proportional to the concentration of both base and substrate ("substrate" being any reactive compound):

$$\text{rate}_{E2} = k_2[\text{base}][\text{substrate}].$$

In the E1cb reaction, the rate is also directly dependent on the concentration of base and substrate. This is because the higher the concentration of base, the greater the concentration of carbanion, and the greater the rate of the reaction. Therefore, the rate expression is the same as that for an E2 reaction. The rate of an E1 reaction, on the other hand, is proportional only to the substrate, since the first step in the mechanism (formation of the carbonium ion) is rate determining:

$$\text{rate}_{E1} = k_1[\text{substrate}].$$

If the *half-life* of a substrate (the time required for half the material to react) in 1*N* base is twice what it is in 2*N* base, then the rate is proportional to the base, and an E2 (or E1cb), rather than an E1 mechanism, is operative.

The carbonium ion formed in an E1 mechanism can either eliminate a proton to give an olefin or bond to a nucleophile to give an S_N1 substitution product. The ratio of elimination to substitution product should be the same (as long as carbonium ion formation is rate determining) no matter what the leaving group happens to be. Both of the substrates shown in Fig. 8.4 give the same ratio of elimination to substitution products (in 80% aqueous ethanol at 65°) despite the fact that the sulfur compound reacts much faster (it forms a carbonium ion more readily). Such a result constitutes evidence for an E1 mechanism.

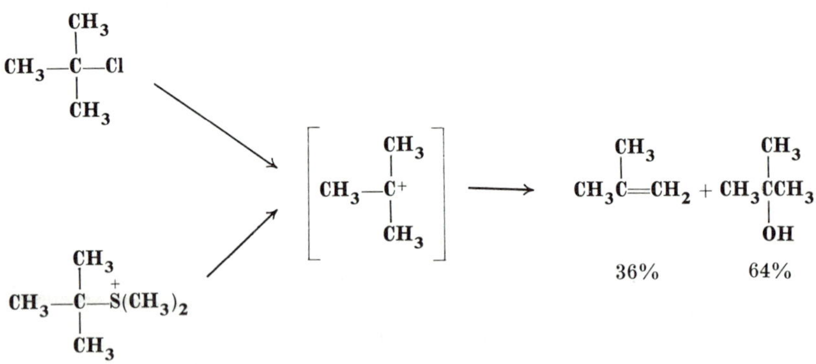

Fig. 8.4 E1 elimination of two compounds

The determination of the presence or absence of a *kinetic isotope effect* is another way to establish the mode of elimination. This method is based on the fact that it is easier to break a carbon-hydrogen bond than it is to break a carbon-deuterium bond (deuterium is a nonradioactive isotope of hydrogen with an atomic weight of two). Thus, in an E2 elimination, deuterated material reacts two to seven times slower than nondeuterated compounds because a carbon-hydrogen bond is broken in the rate-determining step:

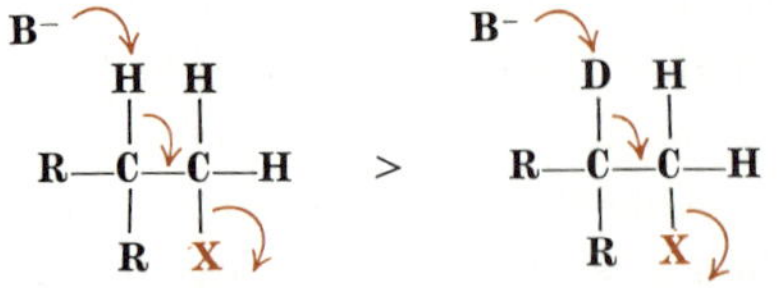

A substrate that reacts via an E1 mechanism displays only a small kinetic isotope effect (a change of only a few percent) since there is no carbon-hydrogen cleavage in the rate-determining step.

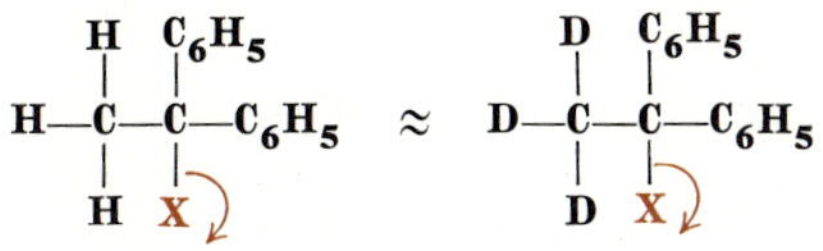

8.4 STEREOCHEMICAL EVIDENCE

The E2 elimination (Fig. 8.1) may be viewed as a substitution reaction in which the nucleophile happens to be on a carbon atom adjacent to the carbon on which the displacement is occurring. As in the substitution reaction, a pair of electrons (released from the carbon-Y bond) displaces a pair of electrons (those of the carbon-X bond). Since an S_N2 reaction requires back-side attack, it is not surprising that an E2 elimination requires a *trans-coplanar* relationship (in which X, Y, and the two central carbons are in the same plane, and X and Y are on opposite sides of the carbon-carbon bond):

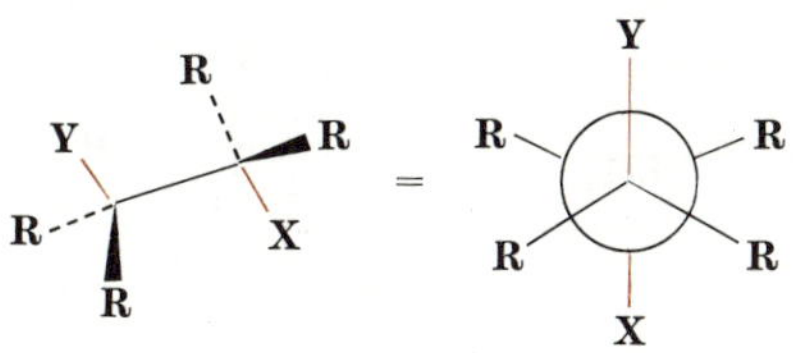

The back-side nature of the elimination may, perhaps, be seen more clearly in terms of an E1cb mechanism.

E1cb S_N2

The stereochemical consequences of the preferred *trans*-coplanar arrangement of the eliminating groups in an E2 reaction can be seen by examining closely the iodide-catalyzed elimination of bromine from 2,3-dibromobutane. There are three possible optical isomers for 2,3-dibromobutane; the *meso* compound and its diastereoisomers, l- and d-2,3-dibromobutane (Fig. 8.5). See page 184. For the *meso* compound to undergo elimination via the lowest energy transition state, the two carbon-bromine bonds should orient themselves in a *trans*-coplanar relationship to each other. This is the situation in conformation (1) of Fig. 8.6. Note that as a consequence of eliminating the two bromines from this conformation, the two methyl groups are oriented in a *trans* relationship to each other. This means that the 2-butene formed from this elimination must be *trans*-2-butene (Fig. 8.7).

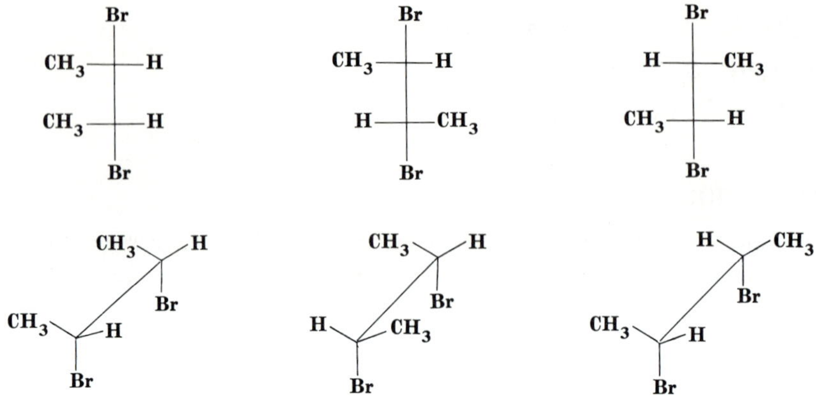

meso d-l 2,3-dibromobutane

Fig. 8.5 Isomers of 2,3-dibromobutane

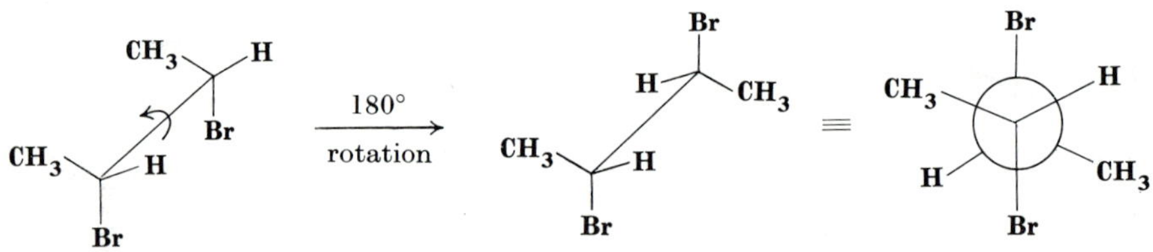

(1)

Fig. 8.6 Conformation of *meso*-2,3-dibromobutane favorable to elimination of bromine

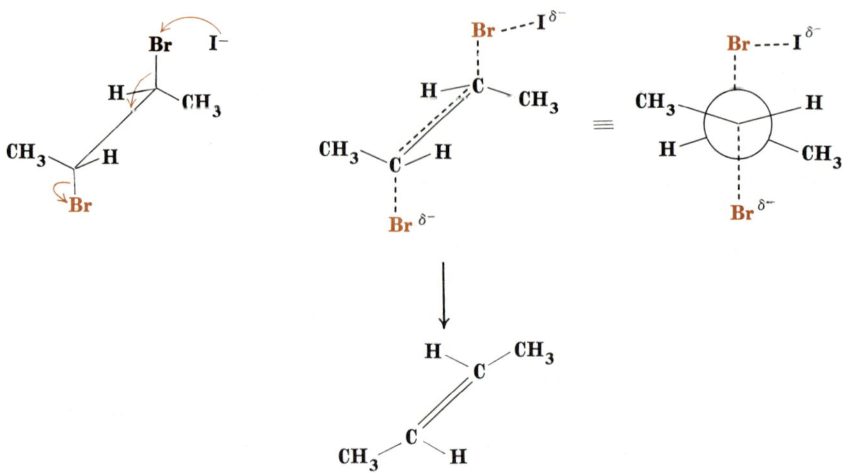

Fig. 8.7 Elimination of *meso*-2,3-dibromobutane

From similar considerations one would expect l- and d-2,3-dibromobutane to yield *cis*-2-butene. These stereochemical expectations are realized experimentally.

The E1 reaction, on the other hand, need not display stereospecificity. The rate-determining step in this mechanism is loss of X^- to form a carbonium ion (Fig. 8.8).

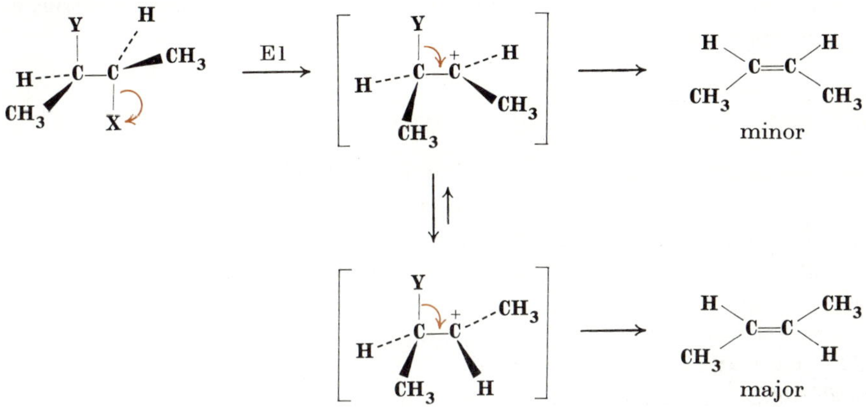

Fig. 8.8 Stereochemistry of E1 elimination

If we assume that rotation about the central carbon-carbon single bond of the carbonium ion is extremely fast, then there is no best disposition of Y relative to X. After the carbonium ion forms, it equilibrates rapidly to a variety of conformations so that all "memory" of X is lost. The different conformations may lead to either *cis* or *trans* material, but the major product will be the more stable *trans*-olefin. (The reaction leading to the most stable product in a series of closely related reactions usually has the lowest activation energy and therefore the most stable product is formed the fastest.) The *trans* isomer is the more stable because steric interference between the two methyl groups destabilizes the *cis* molecule. Thus, under E2 conditions, l- and d-2,3-dibromobutane yield *cis*-2-butene (Fig. 8.7), whereas under E1 conditions one would expect mainly *trans*-olefin (Fig. 8.8). Clearly, stereochemistry can be a powerful tool in the elucidation of reaction mechanisms.

8.5 E1 VERSUS E2

The mechanistic course of a given elimination depends on a number of factors. For example, the rate law for an E2 mechanism includes a base term, whereas the rate law for an E1 mechanism does not. Consequently, the absence of a strong base (such as OH^- or NH_2^-) favors an E1 mechanism. Clearly, a substrate that can form a good carbonium ion, because of stabilization by alkyl or aromatic groups, is a likely candidate for an E1 elimination. This is especially true if X is a good leaving group (such as water, iodide, or acetate). Carbonium ion formation is also favored by a polar solvent (such as water, methanol, or acetic acid) which can solvate the carbonium ion intermediate and the transition state leading to it. All these factors must be assessed in guessing the most likely mechanism for a given system. Performing an experiment, however, is the best means of establishing a mechanism.

8.6 ELIMINATION VERSUS SUBSTITUTION

It has already been pointed out that elimination and substitution are competitive reactions. (In fact, E1 and S_N1 reactions have a common carbonium ion intermediate.) For second-order reactions, a strong base favors elimination over substitution (as well as E2 over E1). This is why sodium amide ($NaNH_2$), a very strong base, is often used to promote eliminations. For first-order reactions, it has been found that branching of the substrate favors elimination over substitution (E1 over S_N1):

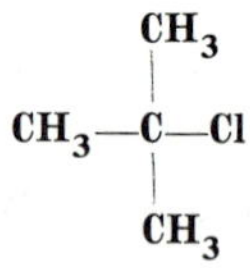

16% elimination
(in aqueous ethanol)

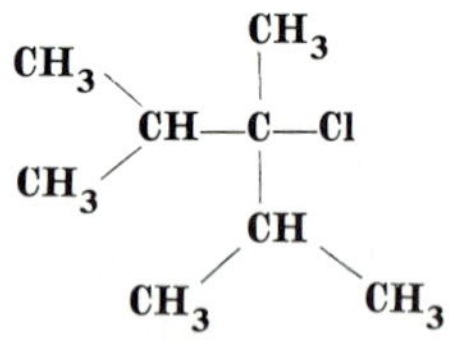

78% elimination
(in aqueous ethanol)

8.7 SAYTZEFF AND HOFMANN RULES

Consider the acid-catalyzed dehydration of 2-butanol:

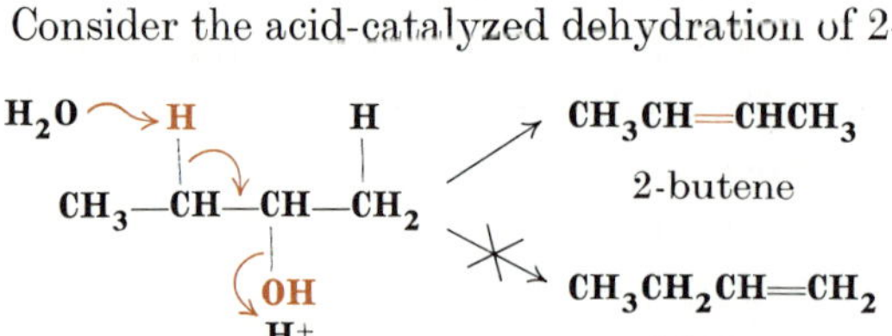

Although there are two possible products, 2-butene and 1-butene, the major product is 2-butene. This mode of elimination (to form the most highly-substituted olefin) is called *Saytzeff elimination*, and reflects the finding that alkyl groups tend to stabilize a double bond (unless, of course, bulky alkyl groups in the *cis* configuration sterically interfere with one another). Thus, of the possible transition states for a given elimination reaction, that which has the largest number of alkyl groups on the developing double bond will have the lowest energy, and will be the one through which the reaction will proceed. The majority of eliminations, with one notable exception, give the most substituted olefin (*Saytzeff's Rule*).

The main exception to Saytzeff's rule is the *Hofmann elimination* in which an amine is eliminated from a quaternary ammonium hydroxide salt upon heating at 100–200°:

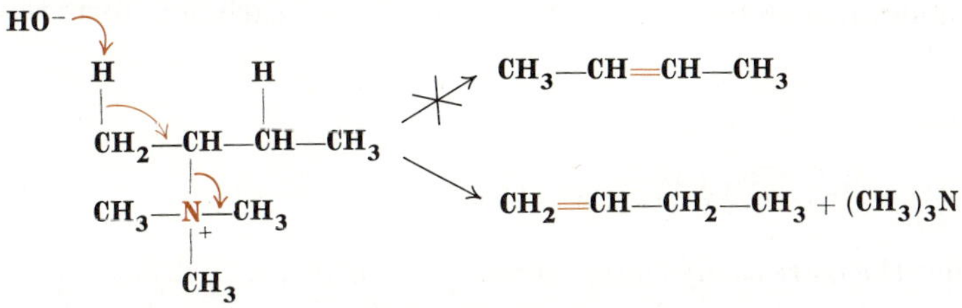

The production of the *least* substituted olefin may be related to differences in access of the various hydrogens to the attacking base, although this point is not firmly established.

This terminates our discussion of the mechanism of the elimination reaction. We have placed considerable emphasis on this reaction, not so much because of its synthetic importance, but because it serves to illustrate several important principles of physical organic chemistry, the study of how organic systems behave. Synthetic applications of the elimination reaction will be considered next.

APPLICATIONS OF ELIMINATION REACTIONS

8.8 DEHYDRATION OF ALCOHOLS

As indicated earlier, dehydration of alcohols (especially tertiary alcohols) is useful in the synthesis of alkenes. A specific example was given in Section 8.2, and another is shown below:

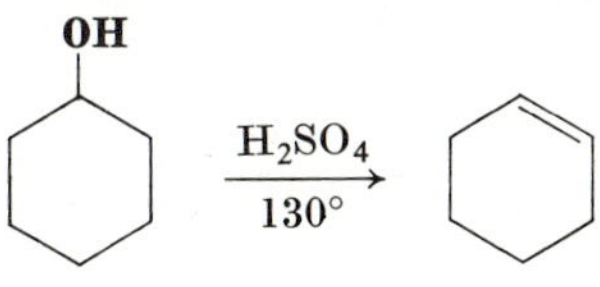

The first step in the process is a reversible protonation of the hydroxyl oxygen to yield an *oxonium ion* (a species bearing a trivalent positive oxygen):

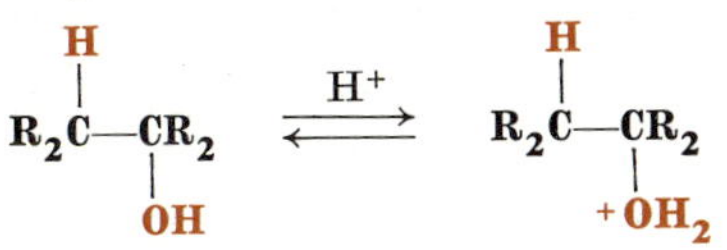

The oxygen-carbon bond then cleaves heterolytically to form a carbonium ion in typical E1 fashion:

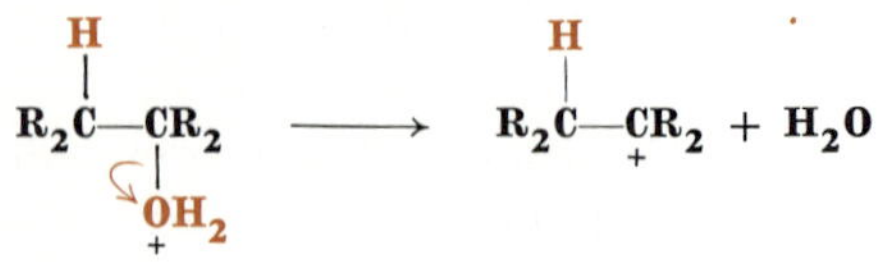

Finally, the carbonium ion loses a proton from an adjacent carbon to form the olefin:

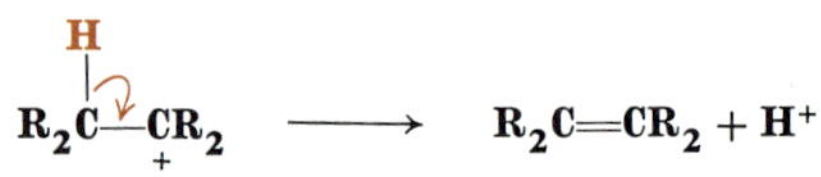

It is apparent that the more easily the carbon atom bearing the hydroxyl group can stabilize the positive charge, the more reactive the particular alcohol will be toward dehydration. Since the order of stability of carbonium ions is tertiary > secondary > primary, one would expect tertiary alcohols to be considerably more subject to dehydration than primary alcohols. This is indeed the case. In fact, some tertiary alcohols dehydrate so readily that only a mild acid catalyst need be employed at room temperature.

In practice, alcohols (especially the slower reacting primary and secondary ones) are dehydrated in acid only when their molecular weights are small so that the olefin can be distilled from the reaction medium as it is being formed. Otherwise, the olefin would remain too long in the acidic medium where it tends to polymerize. For this reason higher-molecular-weight secondary and tertiary alcohols are frequently dehydrated by an alternate procedure using phosphorus oxychloride ($POCl_3$) or thionyl chloride ($SOCl_2$) in pyridine as the solvent:

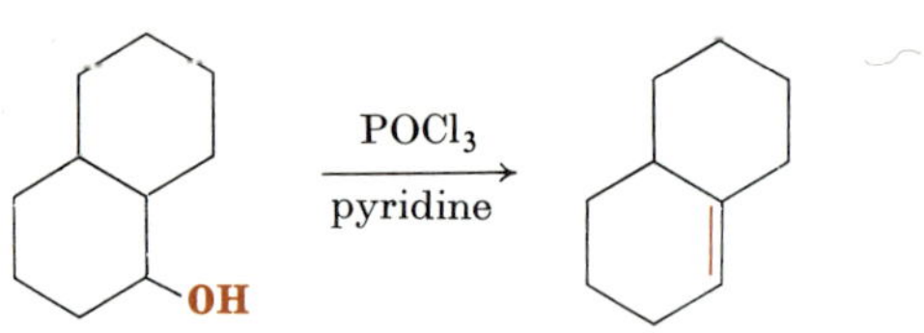

The first step in this conversion involves derivatization of the hydroxyl group into a good leaving group by the reagent:

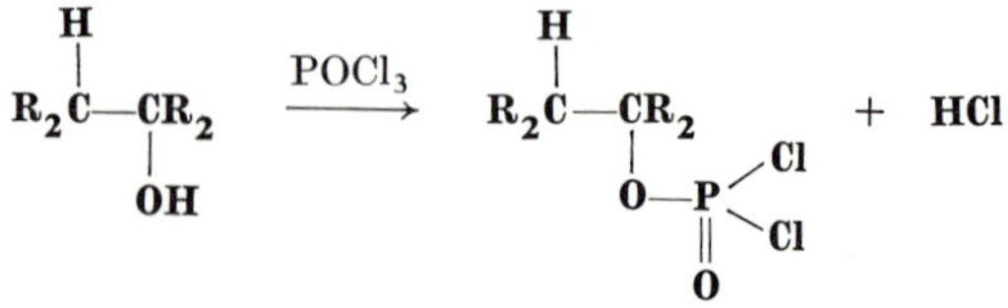

Then pyridine (a weak base whose conjugate acid has a pK_a near 5) removes a proton in an E2 elimination. Since the reaction is carried out under basic conditions, polymerization of the product is less of a problem.

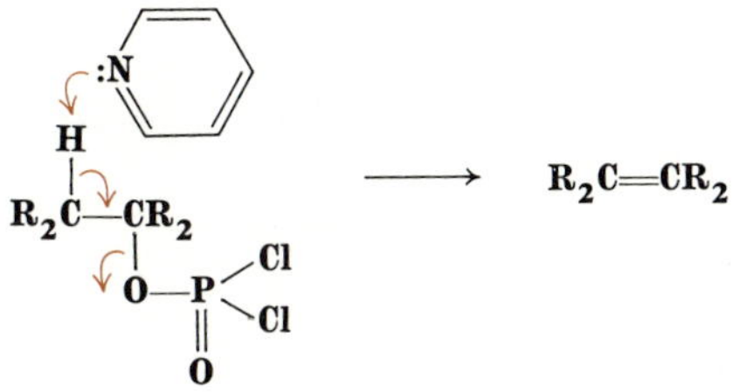

8.9 DEHYDROHALOGENATION OF ALKYL HALIDES

This reaction

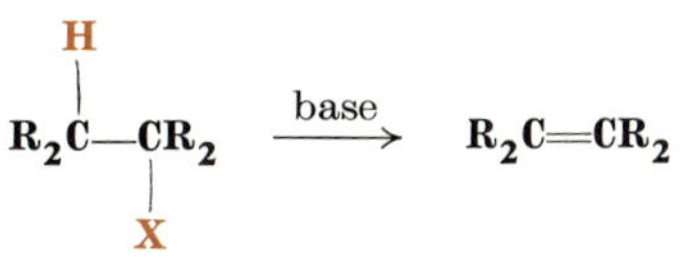

is as useful for the synthesis of alkenes as the dehydration of alcohols. It complements the dehydration in that this is a base-catalyzed process whereas dehydration is performed in acid. See Section 8.2 for two examples; another one is shown below:

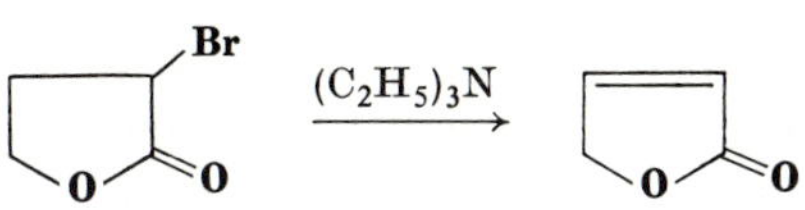

8.10 ELIMINATION OF VICINAL DIHALIDES

This method is not widely used for the synthesis of alkenes, because in most cases vicinal dihalides (halogens on adjacent carbon atoms) are derived from alkenes rather than vice versa:

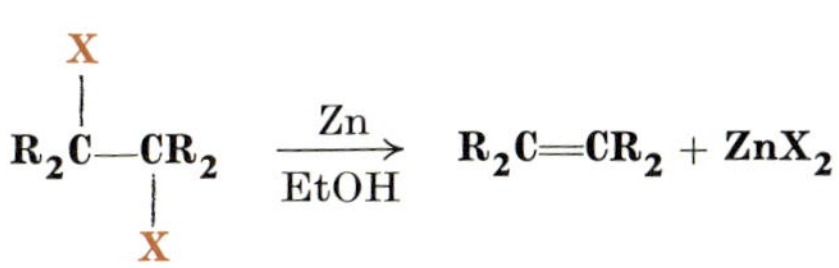

This reaction has been used to purify alkenes. For example, cholesterol and dihydrocholesterol occur together in nature, and are quite difficult to separate (Fig. 8.9).

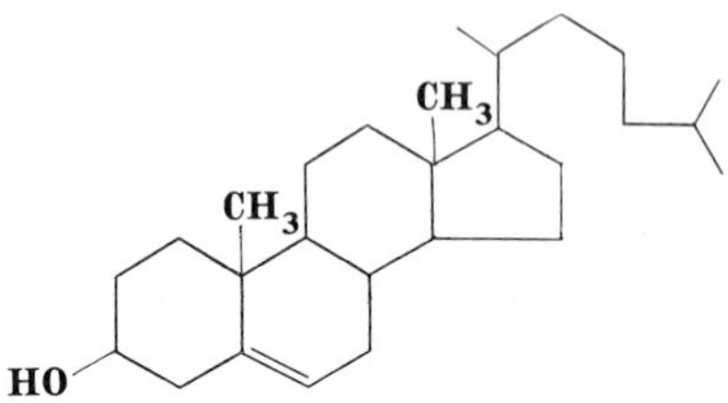

cholesterol

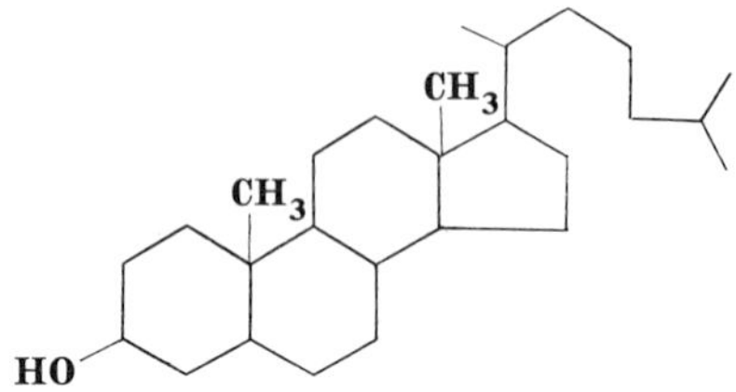

dihydrocholesterol

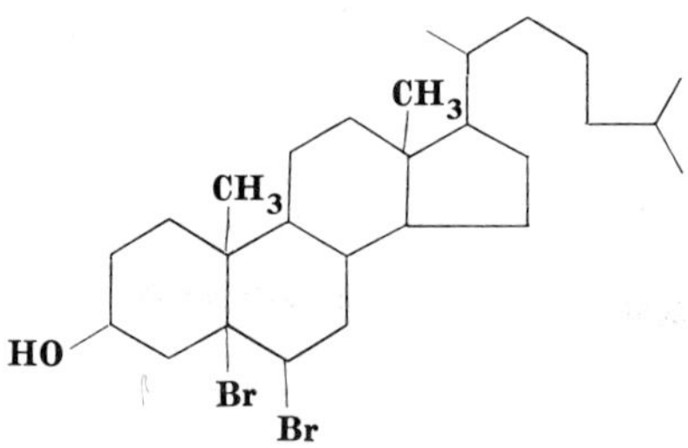

dibromocholesterol

Figure 8.9

However, it is possible to convert cholesterol into dibromocholesterol (an addition reaction that will be discussed in Chapter 9) by treating a mixture of the two with bromine. Dihydrocholesterol, not having a double bond, remains unchanged. It is relatively easy to separate dibromocholesterol from dihydrocholesterol since they have very different solubility properties. Once separation is achieved, the dibromocholesterol can be reconverted into pure cholesterol by treating it with powdered zinc in warm ethanol (Fig. 8.10).

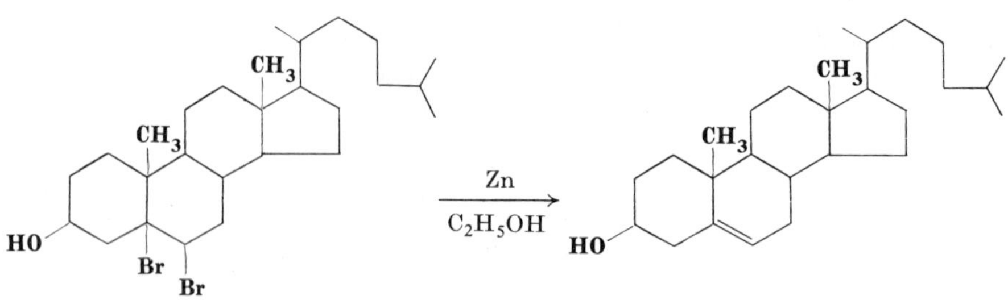

Fig. 8.10 Conversion of 5,6-dibromocholesterol to cholesterol

Another example of a dehalogenation is the formation of allene from 2,3-dichloropropene:

$$CH_2{=}C(Cl){-}CH_2{-}Cl \xrightarrow{Zn} CH_2{=}C{=}CH_2$$

8.11 HOFMANN ELIMINATION

The Hofmann elimination has been used to prepare alkenes and to degrade organic compounds in order to determine their structure. The reaction is useful because it provides an alternative to the Saytzeff routes, namely the dehydration of alcohols and the dehydrohalogenation of alkyl chlorides. Thus one may convert secondary butyl bromide ($CH_3CHBrCH_2CH_3$) into either 2-butene or 1-butene depending upon the method chosen for elimination. If the bromide is treated with alcoholic potassium hydroxide, the major product is 2-butene:

$$CH_3{-}CH(Br){-}CH_2{-}CH_3 \xrightarrow[\text{alc.}]{\text{KOH}} CH_3{-}CH{=}CH{-}CH_3$$

Alternatively, the bromide can be treated first with trimethylamine to yield *s*-butyltrimethylammonium bromide via a substitution reaction. The bromide can then be converted into the quaternary ammonium hydroxide by reaction with silver oxide and water. Heating the quaternary hydroxide salt results in the formation of 1-butene:

$$CH_3CH(Br)CH_2CH_3 + (CH_3)_3N \longrightarrow CH_3CH(\overset{+}{N}(CH_3)_3)CH_2CH_3\ Br^- \xrightarrow[H_2O]{Ag_2O} CH_3CH(\overset{+}{N}(CH_3)_3)CH_2CH_3\ {}^-OH$$

$$\xrightarrow{\text{heat}} CH_2{=}CH{-}CH_2{-}CH_3 + (CH_3)_3N$$

The Hofmann elimination has been widely used in the degradation of naturally occurring amines into simpler compounds that can be more easily analyzed:

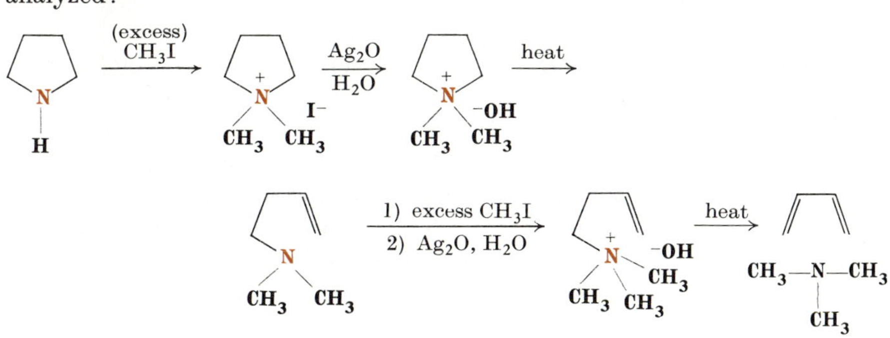

Pyrrolidine is treated with excess methyl iodide until the fully alkylated quaternary salt is formed. The iodide salt is then converted into the hydroxide salt and heated, forming an olefin. The process is repeated on the amine-containing olefin, yielding a nitrogen-free material with two double bonds (butadiene).

8.12 ACETATE PYROLYSIS

A chemical transformation produced by applying heat to a single compound is called a *pyrolysis reaction*. A number of compounds undergo pyrolytic eliminations to form olefins, and two of these deserve particular mention. The first is the pyrolysis of acetates. Acetates are esters of acetic acid which can be prepared, among other ways, by treating alcohols with acetic anhydride:

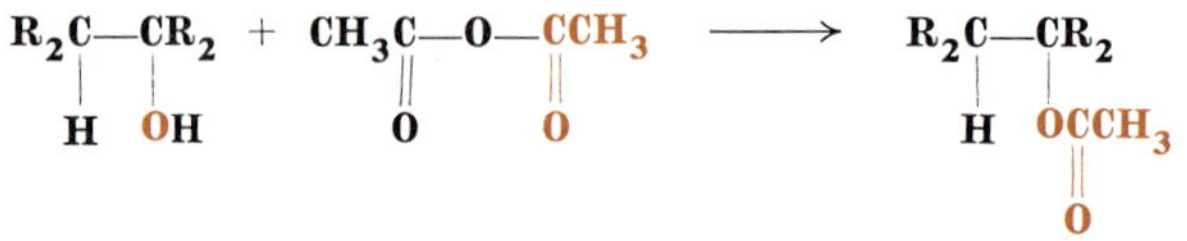

alcohol — acetic anhydride — acetate

Heating the acetate to high temperatures results in the elimination of acetic acid:

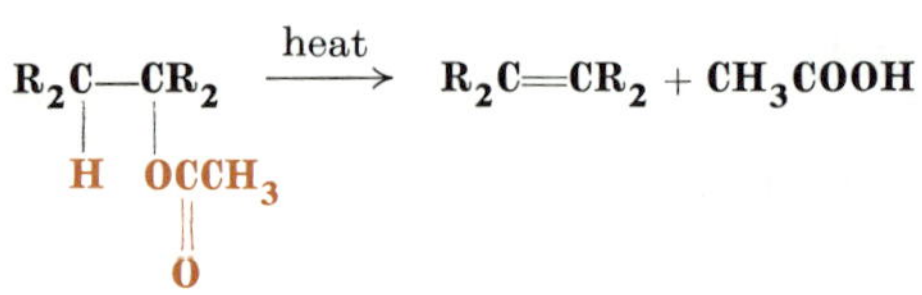

For example, 1,4-pentadiene has been prepared by the thermal decomposition of a diacetate at 575°:

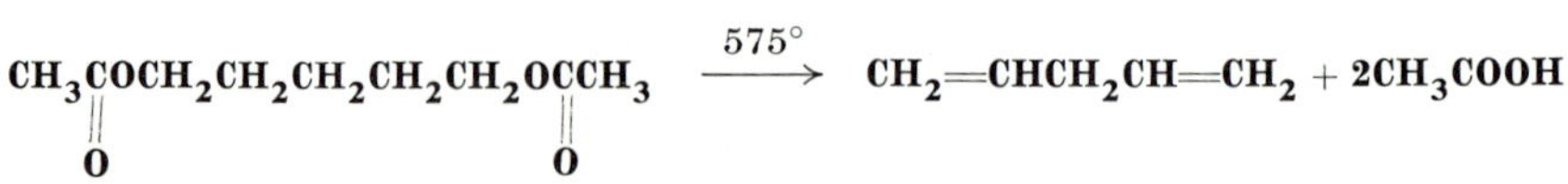

The reaction is carried out by adding the diacetate drop by drop to the top of a vertical column packed with glass beads and encased in an electric furnace which heats the beads to the required temperature. The compound pyrolyzes on its way down the column, and the product is collected in an ice-cooled trap attached to the bottom of the reaction tube.

The most likely mechanism for acetate pyrolysis is a *cyclic concerted process* ("concerted" meaning that bond formation and breakage occur in unison):

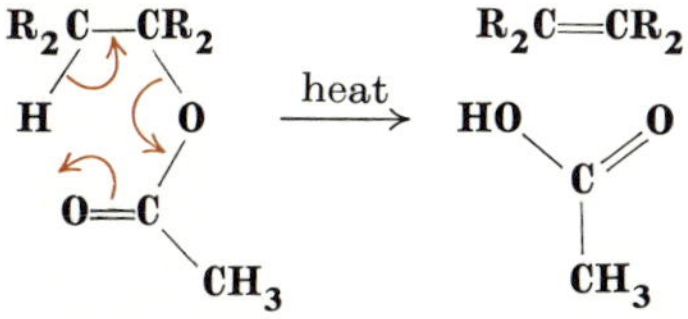

This mechanism, unlike the E2 elimination, predicts that the proton and the leaving group (acetate) must be in a *cis* relationship, and this is what is actually observed. Thus the acetate below pyrolyzes to give a Hofmann-type product (the least substituted olefin), because the Saytzeff mode of elimination is precluded by the absence of a *cis* hydrogen on the substituted side of the molecule.

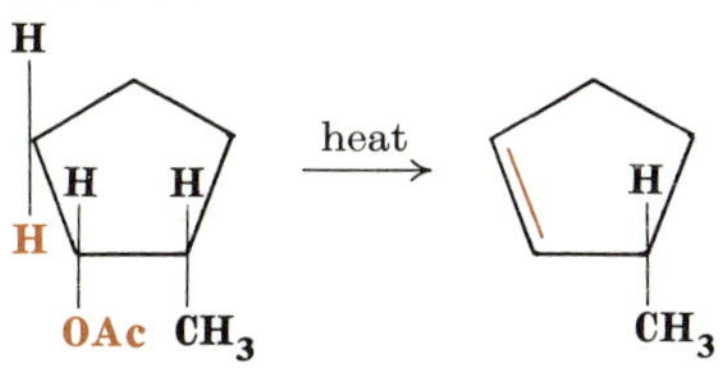

Acetate pyrolyses are carried out in the absence of a solvent, which is an important difference between this type of elimination and the E1 and E2 reactions. Since there is no solvent to solvate and stabilize charge, the production of charged intermediates is inhibited. The cyclic concerted mechanism involves only a single transition state with very little charge creation.

8.13 COPE REACTION

When a tertiary amine is treated with 30% hydrogen peroxide, an *amine oxide* is formed.

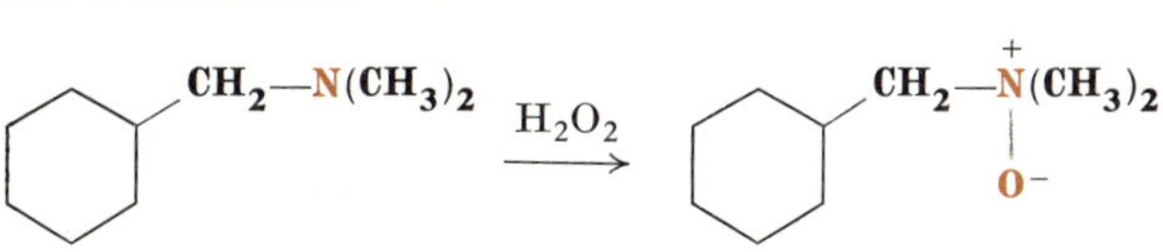

On heating to 160°, the amine oxide decomposes within about two hours:

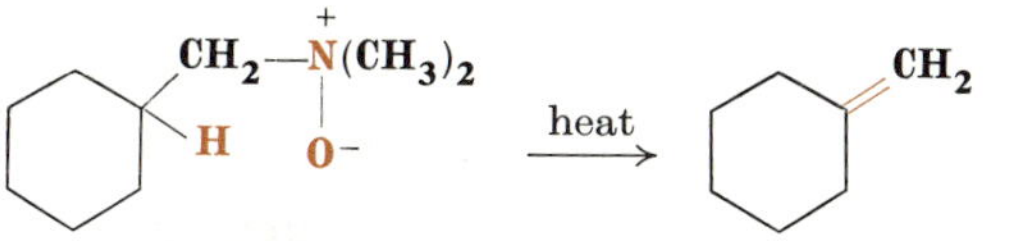

This elimination is known as the *Cope reaction*, and it provides an excellent means of converting an amine into an olefin. In general, the yields are better than those of the Hofmann elimination which performs the same function. As with the acetate pyrolysis, the pyrolysis of amine oxides is characterized by *cis* elimination:

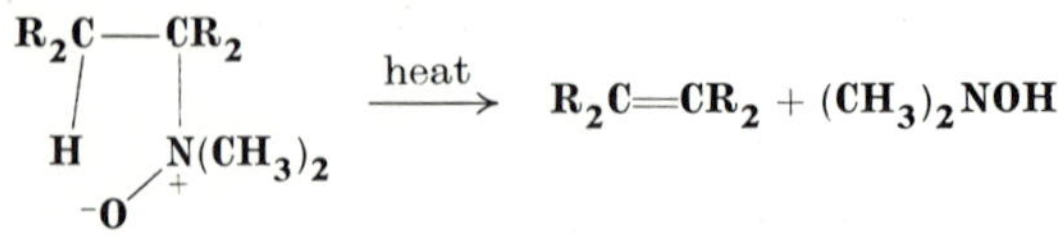

ADDITIONAL READING

W. H. Saunders and A. F. Cockerill, *Mechanisms of Elimination Reactions*, New York: Wiley, 1973.

M. S. Newman, Ed., *Steric Effects in Organic Chemistry*, New York: Wiley, 1956. Chapter 6.

PROBLEMS

1. Provide the necessary reagents for the following reaction "network."

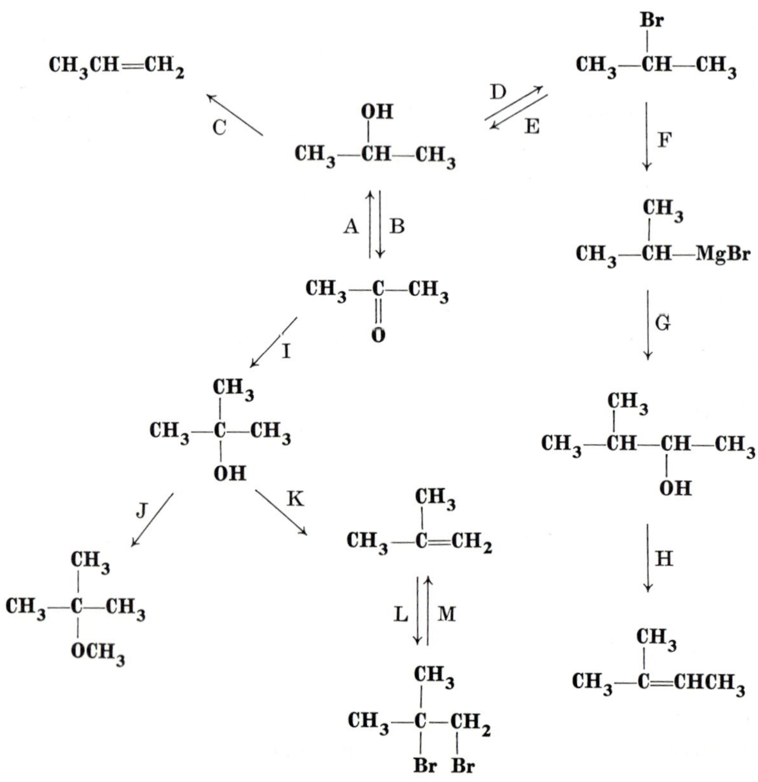

2. Fill in the blanks.
   a) A monomolecular reaction carried out at elevated temperatures is called a ———.
   b) Elimination to give the most highly substituted olefin is called ——— elimination.
   c) Elimination of water from an alcohol is called a ———.
   d) E2 eliminations prefer a *trans* ——— arrangement of X, Y, and the two relevant carbons.
   e) Tertiary amine plus H_2O_2 yields an ———.
   f) An ——— elimination involves a carbonium ion intermediate.
   g) The ——— isomer of 2-butene is the more stable.
   h) Zinc plus ——— dihalide gives olefin.
   i) Substitution and elimination reactions of alcohols are carried out in ——— media because water is a much better leaving group than hydroxide.
   j) The slow ——— step in an E1 reaction is cleavage of the C—X bond.
   k) E2 elimination of bromine from *meso*-2,3-dibromobutane gives ——— 2-butene.
   l) Elimination from a tertiary halide would be expected to proceed by an ——— mechanism.
3. How would you carry out the following transformations using elimination reactions in addition to other reactions you have learned in previous chapters? Several steps may be required.

a)

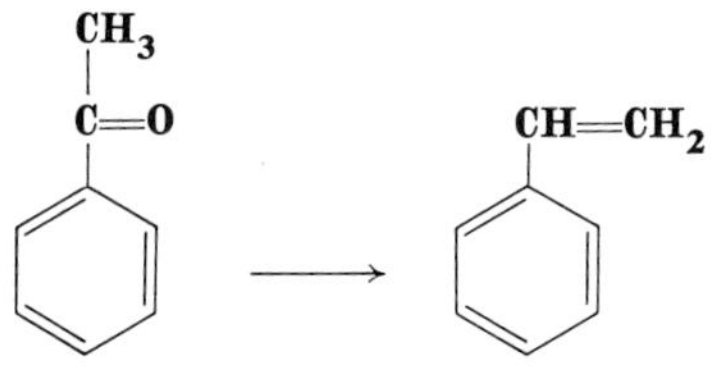

b)

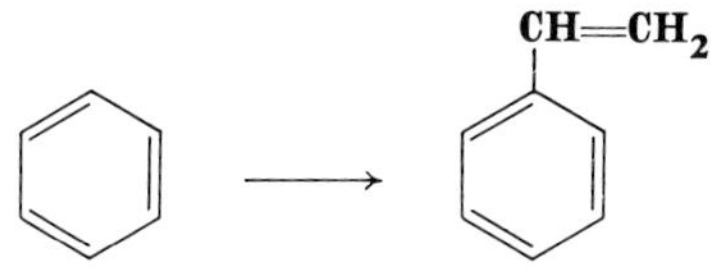

c)

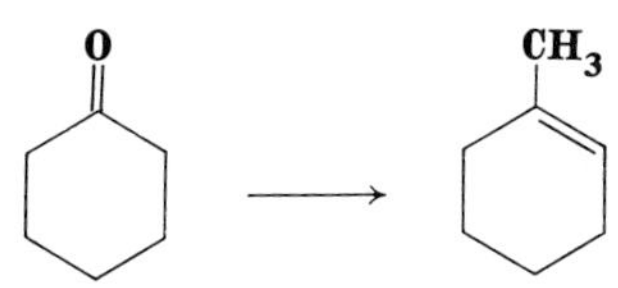

d) 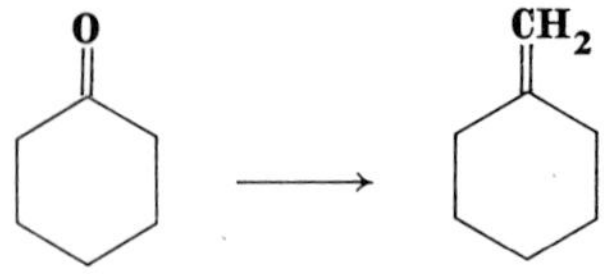

e) 

f) 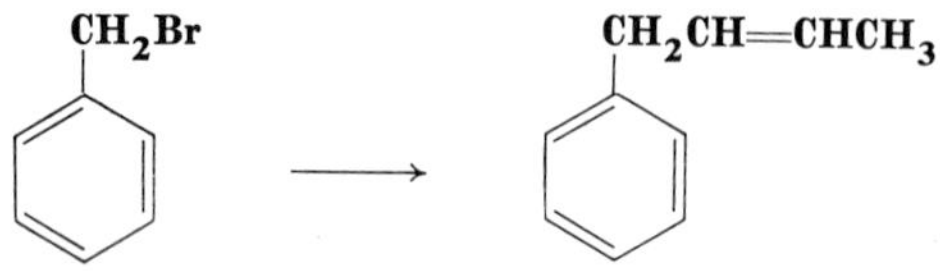

g) 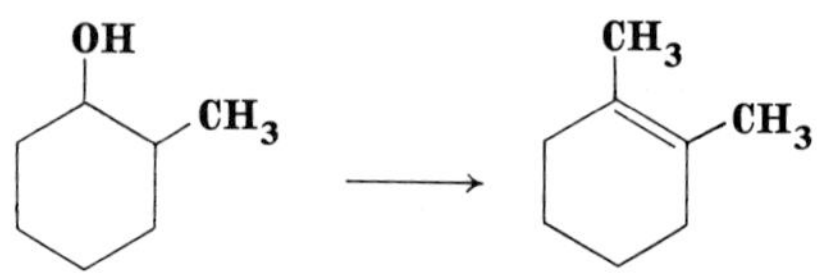

h) 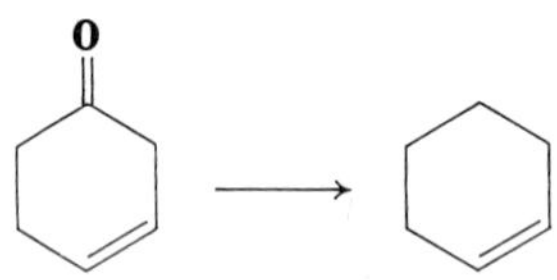

4. Complete the following reactions by supplying the structure of the principal organic product.

a) 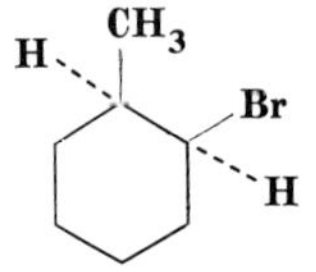

$\xrightarrow[C_2H_5OH,\ heat]{KOH}$

Ag_2O

b) $CH_3CH_2CH(CH_3)—CH(\overset{+}{N}(CH_3)_3)—CH_3$ I^- $\xrightarrow[2)\ heat]{1)\ Ag_2O,\ H_2O}$

c) 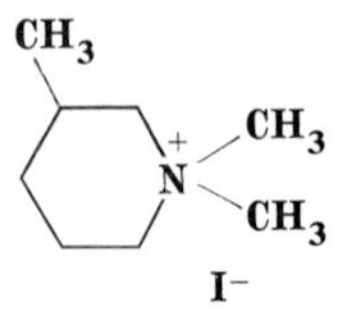

$\xrightarrow[2)\ heat]{1)\ Ag_2O,\ H_2O}$

d) 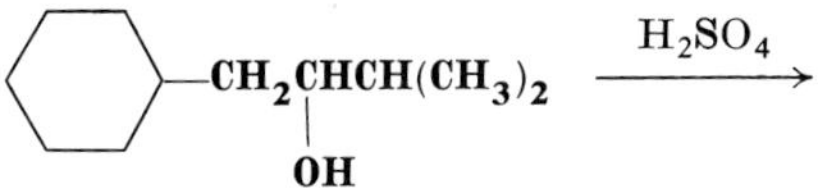

e) 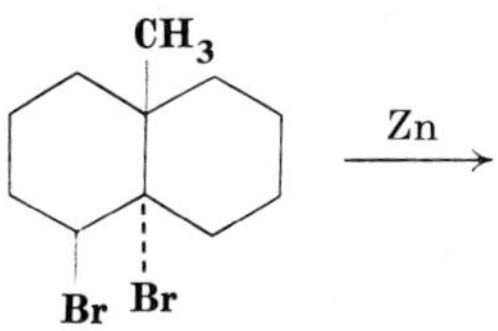

5. Menthyl chloride, A, when treated with potassium hydroxide in hot alcohol yields olefin, B. In contrast, neomenthyl chloride, C, under the same conditions affords D. Consider the three-dimensional structures of A and C and suggest an explanation of these results.

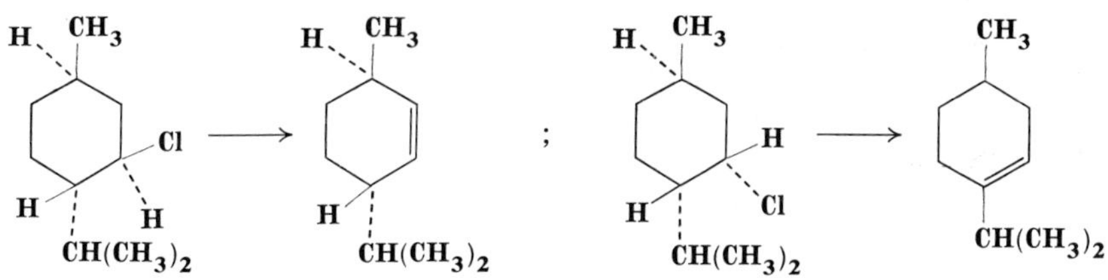

A B C D

6. Starting with any organic compound of three or less carbons, and with any inorganic reagents, synthesize the following substances. Before attempting this problem, review briefly the reactions in Chapters 3, 7, and 8.

a) $CH_3CH_2CH_2CH(CH_3)CH_3$

b) $CH_3CH_2CH_2CH_2CH{=}CH_2$

c) $(CH_3)_2CHCH_2CH_2CH{=}CHCH_2CH_3$

d) $CH_2(OH)—CH(OH)—CH_2—CH(OCH_3)—CH_3$

7. Propose a mechanism for the decomposition of trimethylamine-N-oxide in an aqueous solution of sulfur dioxide:

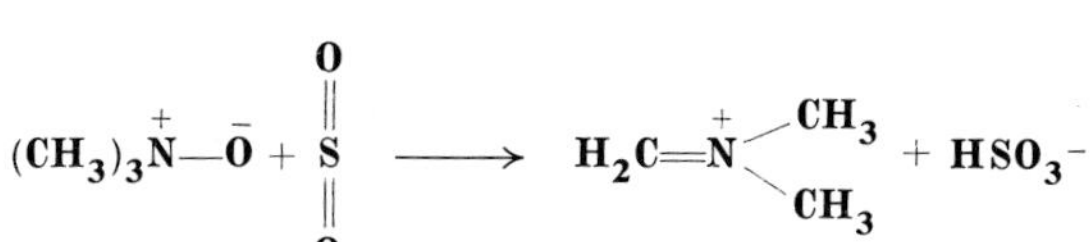

The cationic product is unstable in water and hydrolyzes to dimethylamine and formaldehyde. Write a mechanism for this reaction.

8. Predict the major product of the following reactions. If geometric isomerism is possible, be sure to indicate whether the product is *cis* or *trans*.

a) 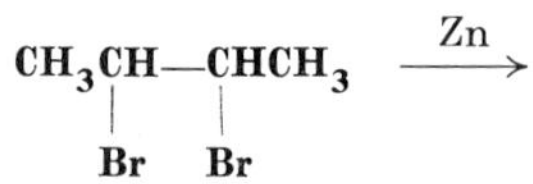

meso

b) 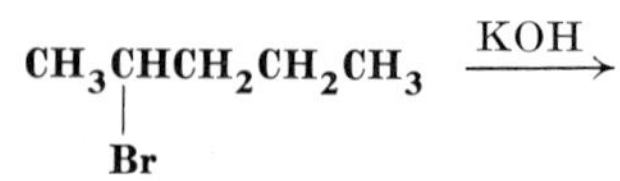

c) 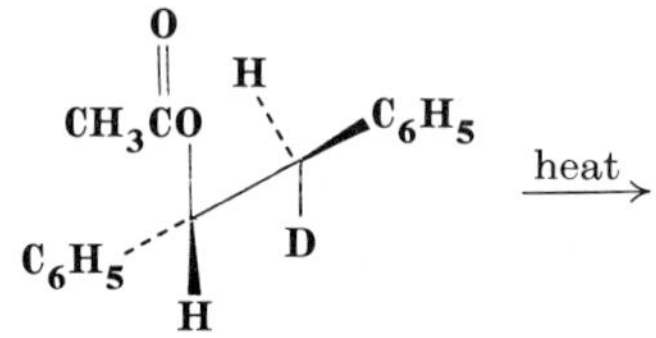

erythro

d) 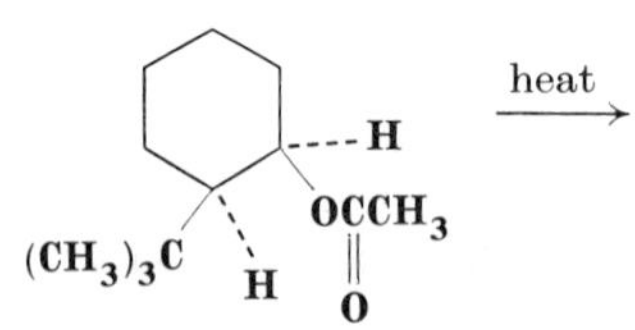

e) 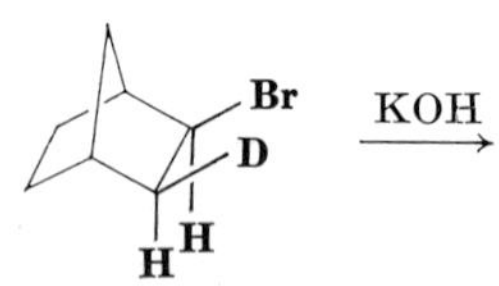

9. When 2-phenylethyl trimethylammonium bromide is heated in sodium ethoxide/ethanol, styrene, and trimethylamine are produced:

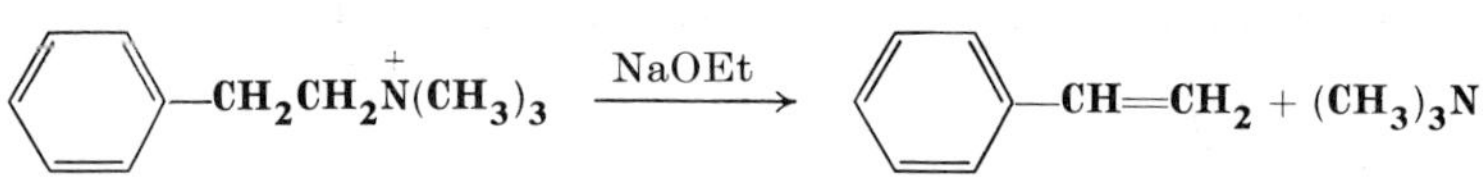

The Hofmann elimination is then performed on deuterated material:

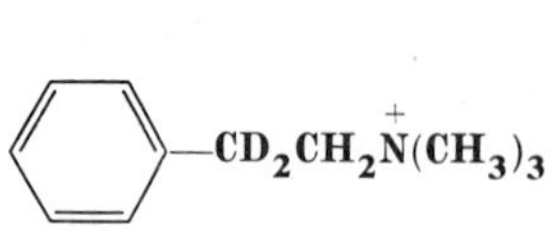

The reaction is allowed to go to 50% completion, at which point it is quenched with acid, and the two products and unreacted starting material are isolated. If we analyze the quaternary salt, styrene, and trimethylamine for deuterium, we will obtain the following results:

Compound	Atoms of deuterium per molecule
$C_6H_5CD_2CH_2\overset{+}{N}(CH_3)_3$	
a) original	2
b) recovered	2
styrene	1
$(CH_3)_3N$	0

Show how the data rule out the mechanisms shown below for the Hofmann elimination and support the usual E2 mechanism.
Note: ϕ represents the phenyl group.

a) 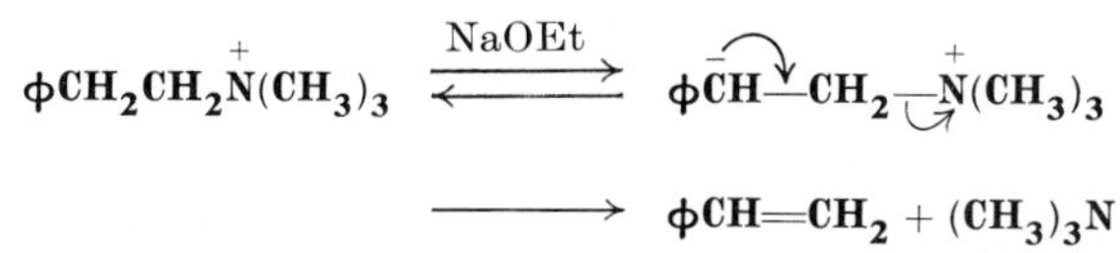

b) 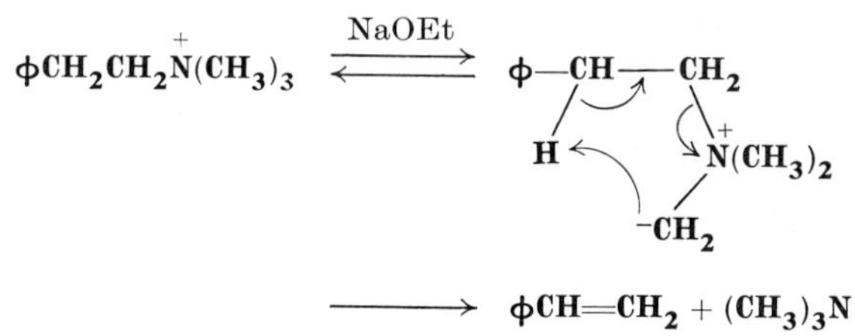

10. A compound $C_9H_{17}N$ is found to take up no hydrogen in the presence of platinum. The material is subjected to degradation with excess methyl iodide, silver oxide, and heat. The sequence is repeated twice until a nitrogen-free compound is produced with an empirical formula of C_9H_{14}. This compound is found to possess no methyl groups, and it takes up three moles of hydrogen when catalytically reduced with hydrogen and platinum. NMR spectroscopy shows that there are 8 vinyl hydrogens, and ultraviolet spectroscopy shows that there is no double bond conjugation. What is the structure of the unknown amine?

11. The Elcb mechanism involves a rapidly attained equilibrium followed by a rate-determining step:

$$\mathbf{EtO^- + R_2CH{-}CXR_2 \overset{K}{\rightleftharpoons} R_2\bar{C}{-}CXR_2 + EtOH}$$

$$\mathbf{R_2\bar{C}{-}CXR_2 \xrightarrow{k} R_2C{=}CR_2 + X^-}$$

Derive the following rate expression which shows that the overall E1cb process is a second-order reaction:

$$\text{rate} = kK[\text{EtO}^-][\text{R}_2\text{CH—CXR}_2].$$

12. One of the main disadvantages of alcohol dehydrations in strong acid is that isomeric olefins are sometimes obtained. Thus, in the dehydration of 1-butanol, it is possible to isolate some 2-butene. Explain how 2-butene might arise.

$$CH_3CH_2CH_2CH_2OH \xrightarrow{H_2SO_4} CH_3CH_2CH{=}CH_2 + CH_3CH{=}CHCH_3$$

Why is this side reaction of no concern in the dehydration of the two cyclohexanols drawn below?

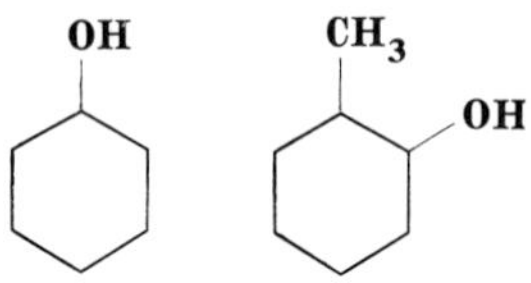

13. Xanthate ester pyrolysis, a means of obtaining an olefin from an alcohol, is often used in preference to acetate pyrolysis. Heating a xanthate ester to 100–200° is usually sufficient to secure elimination, whereas acetate pyrolysis requires much higher temperatures. The sequence employed in the xanthate ester method is given below. What are the structures of the missing species (A, B, and C)?

$$RCH_2CH_2OH + Na \longrightarrow A + H_2$$

$$A + CS_2 \longrightarrow B$$

$$B + CH_3I \longrightarrow C + NaI$$

$$C \xrightarrow[\text{(cyclic concerted mechanism)}]{\text{heat}} RCH{=}CH_2 + \left[HS{-}\overset{\overset{\displaystyle O}{\|}}{C}{-}SCH_3\right]$$

$$\left[HS{-}C({=}O){-}SCH_3\right] \longrightarrow COS + CH_3SH$$

14. Chlorobenzene treated with sodium amide in liquid ammonia produces aniline:

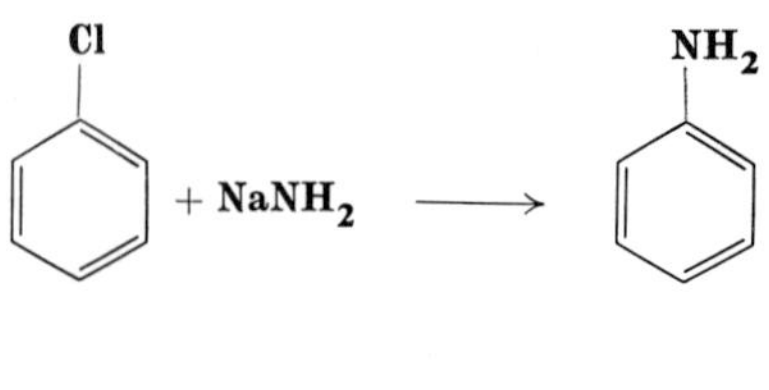

This reaction is known *not* to involve an S_N1 or S_N2 substitution. Instead, an elimination occurs to form a highly unstable *benzyne* intermediate:

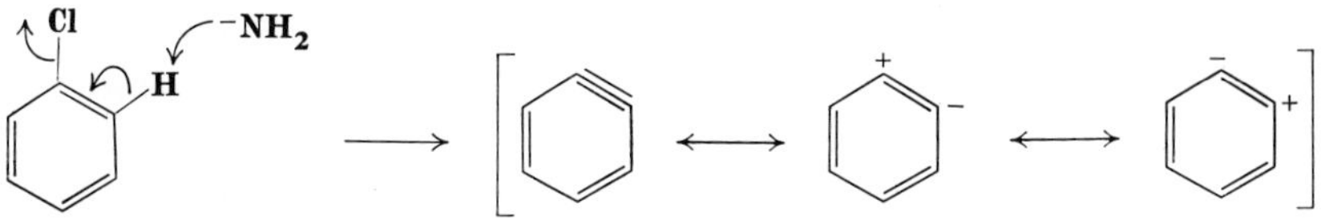

benzyne

The benzyne subsequently reacts with ammonia:

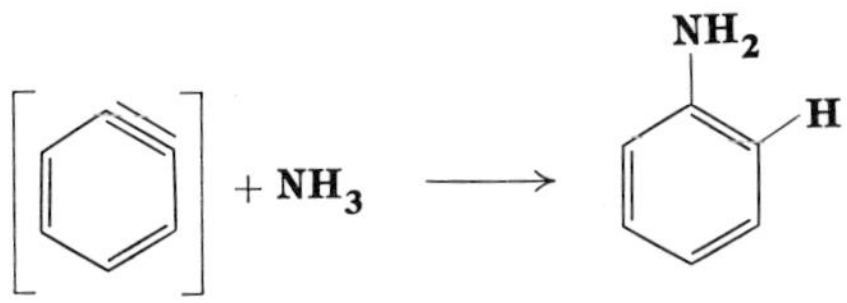

Suggest an experiment which will show that the reaction of an aromatic halide with sodium amide to give an aromatic amine does not involve an S_N1 or S_N2 mechanism.

15. In 1930 a steroid, estriol, was isolated from the urine of pregnant females and converted into the hormone estrone. Suggest a mechanism for this reaction.

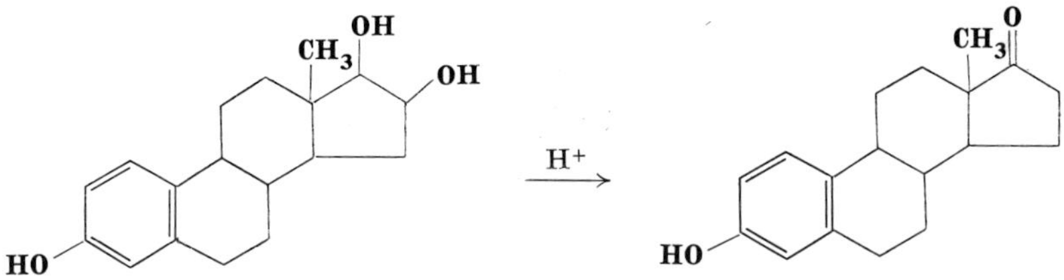

estriol estrone

16. Commercial vitamin A, almost all of which is synthetic, can be prepared, among other ways, by the route given below. The reader should be able to recognize each of the reactions. Review the reactions in Chapters 3, 7, and 8, study the vitamin A synthesis carefully, and answer the following questions.

a) What is a reasonable mechanism for converting 1 into 2?
b) What is a reasonable mechanism for converting 4 into 5?
c) Is the dehydration step, in which vitamin A acetate is formed from 6, an E1 or an E2 elimination?

d) Why does only one of the alcohol groups of 5 react with acetic anhydride to give a monoacetate?

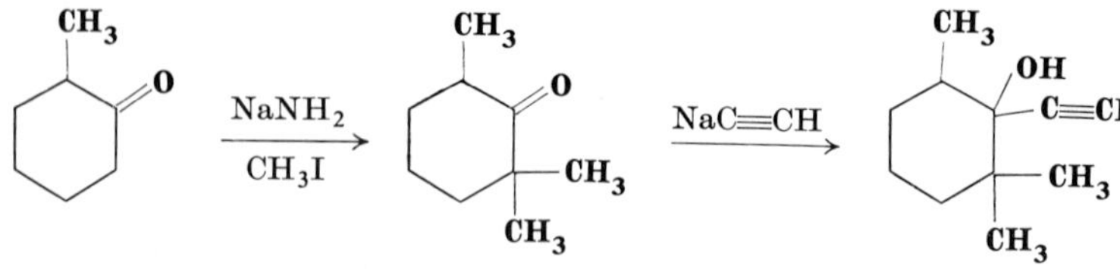

1 2 3

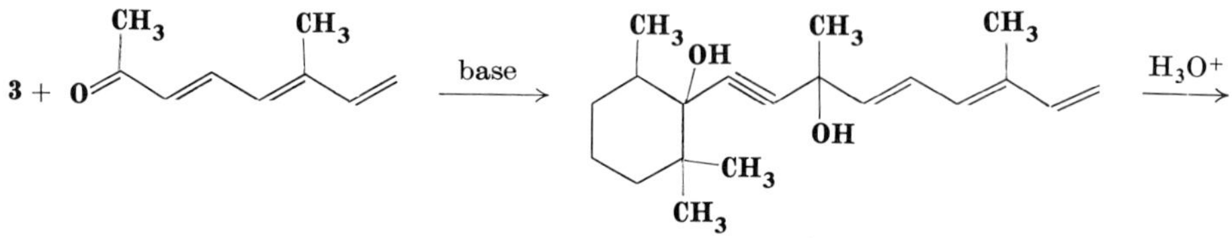

4

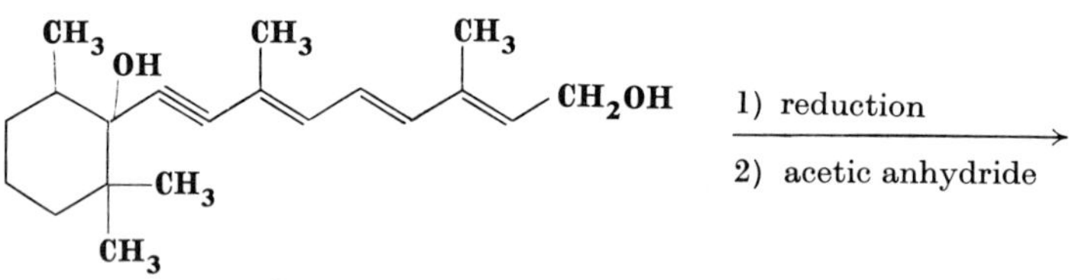

5

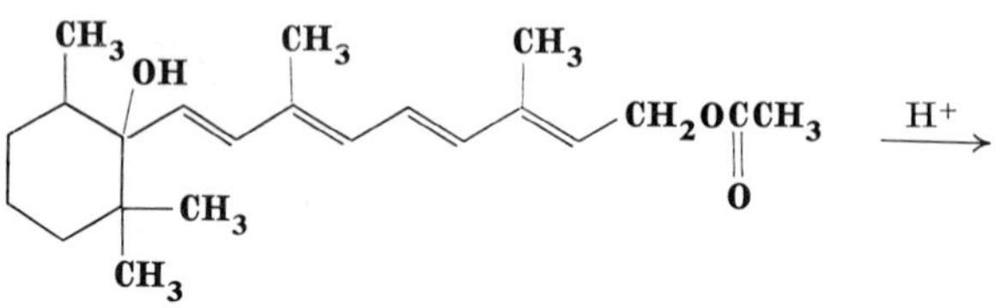

6

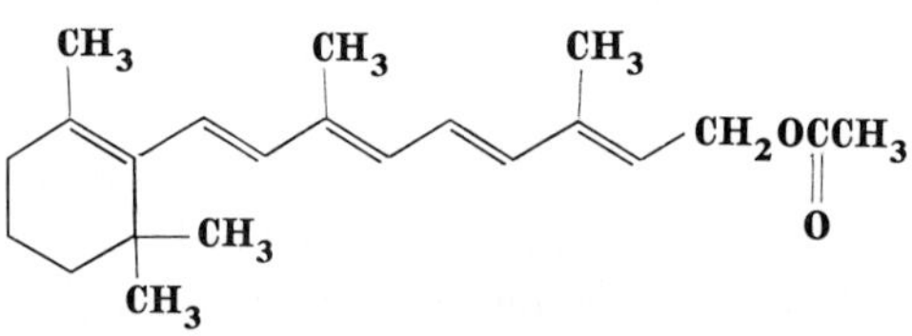

vitamin A acetate

17. The common dehydrohalogenation reaction is called a β *elimination* because the labile hydrogen is on the β carbon of the alkyl halide:

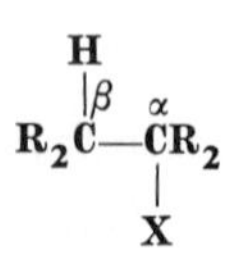

It is possible with chloroform ($CHCl_3$) to effect an α *elimination* in which the hydrogen on the carbon bonded to the halogen is removed:

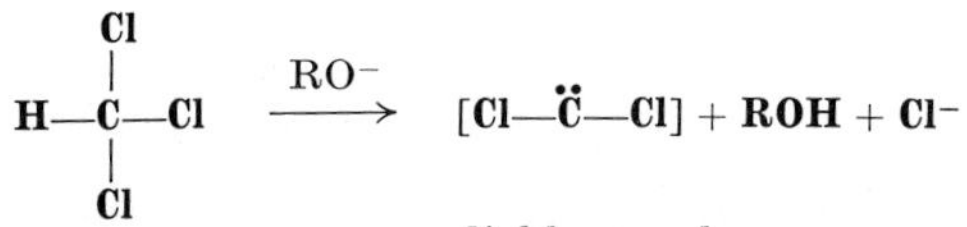

dichlorocarbene

The reaction produces an unstable intermediate, called a carbene, which has a divalent neutral carbon with only six electrons in the bonding shell. Write an E2 and E1cb type mechanism for the α elimination.

9

ADDITION REACTIONS

9.1 ADDITION REACTIONS

The addition reaction is the third fundamental organic process. Some addition reactions are illustrated below:

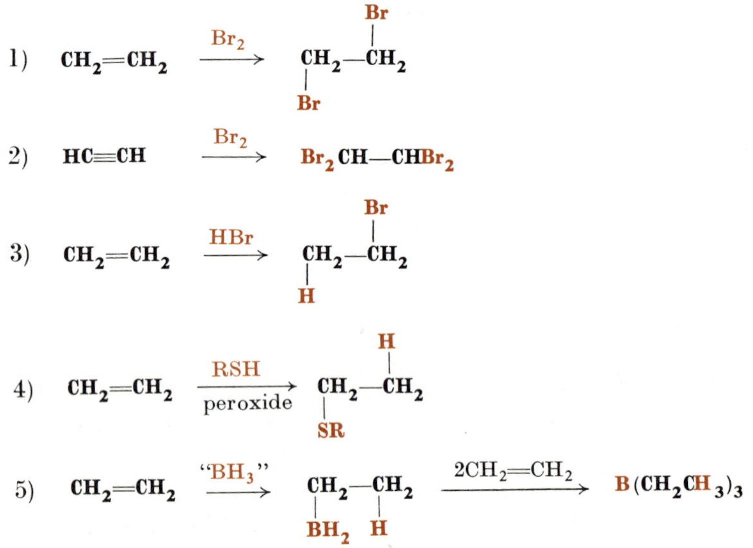

We will examine these reactions more closely later in this chapter. First, we will discuss the mechanism of addition to simple olefins.

9.2 MECHANISM

The detailed mechanism for the addition of X—Y to an olefin is shown in Fig. 9.1. At first glance this mechanism may appear complex. However, now that the reader has a feeling for the mechanistic features of two important reactions, substitution and elimination, he is in a position to grasp the more sophisticated presentation of Fig. 9.1 (which has the virtue of being consistent with most known kinetic and stereochemical data regarding ionic addition reactions).

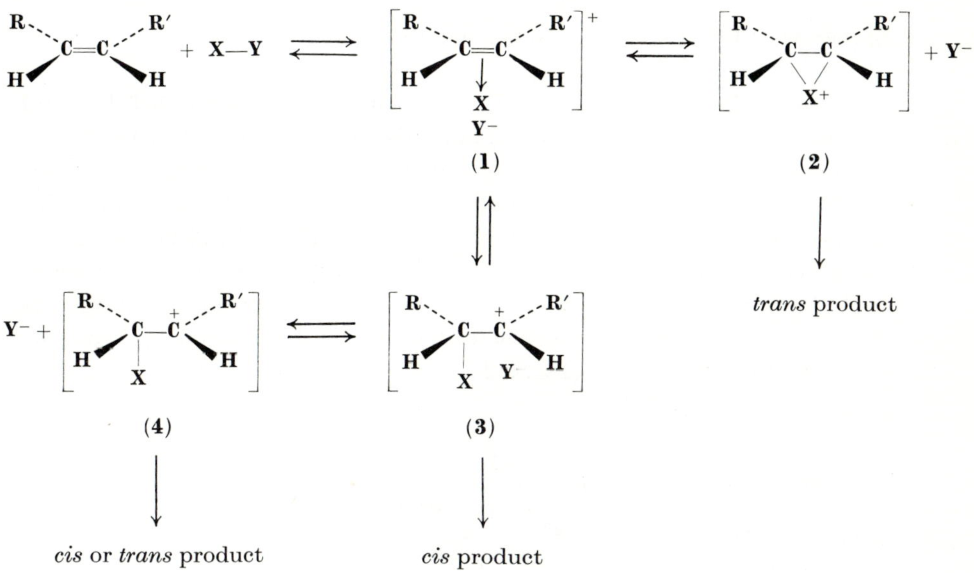

Fig. 9.1 Mechanism of X—Y addition to an olefin

In our discussion of the covalent bond we found that a unique feature of multiple bonds is the π cloud, a far more reactive electron system than a σ system. Thus, we expected the electrons of multiple bonds to behave as "bases" toward Lewis acids. The addition reaction confirms this expectation.

Consider the reaction between an alkene and a reagent X—Y, where Y is more electronegative than X (that is, $X^{+\delta}—Y^{-\delta}$). The first step of the addition mechanism occurs when the π cloud of the alkene attacks the partially positive portion of X—Y, namely X. Actually, this first step is a variant of the substitution reaction; the nucleophile is the π cloud and the leaving group is Y [compare (a) and (b) below]:

a) 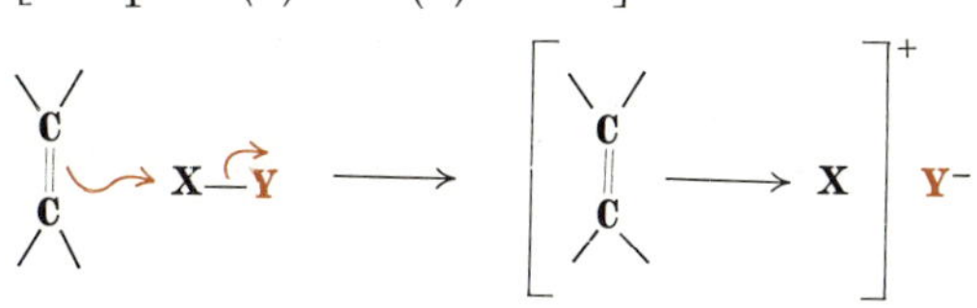

(1)

b) 

Intermediate 1, whose positive fragment X is embedded in the π cloud, is called a π *complex*.

The anion Y^- is closely associated with the positively charged π complex. When two ions are in close proximity (with no solvent between them), the species is called an *ion pair*. Intermediate 1 is viewed as an ion pair between Y^- and the π complex.

Intermediate 1 is in equilibrium with intermediates 2 and 3 (Fig. 9.1). The relative importance of 2 and 3 depends on structural factors (such as the nature of X, and the substituents on the olefin). Intermediate 2 is a three-membered ring with covalent bonds joining X and the two central carbons (rather than the loose complexation of intermediate 1). Intermediate 2 is not an ion pair; Y^- and the intermediate are separate entities (much like sodium and chloride ions which are separately solvated in aqueous solutions). Reaction product arises from *back-side attack* (S_N2) of Y^- on one of the central carbons of intermediate 2 (Fig. 9.2):

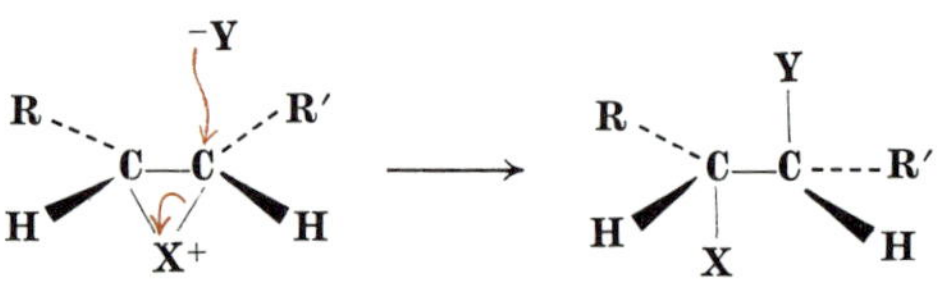

Fig. 9.2 Back-side attack in opening of "onium" ion intermediate

If R and R′ (the olefin substituents) are identical, the two carbon atoms are equally liable to attack by Y^-; if they are different, one carbon may be preferred over the other.

Most addition reactions form products from an intermediate such as that illustrated in Fig. 9.2. Let us explore the stereochemical consequences of this mechanism using a specific example, the addition of bromine to *cis*-2-butene. The initial π complex forms a cyclic intermediate known as a *bromonium ion* (Fig. 9.3):

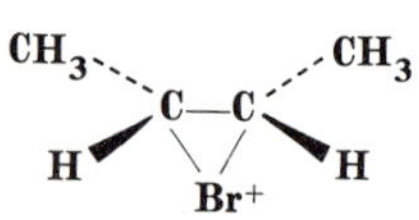

Fig. 9.3 Bromonium ion

Bromide ion may now attack one of the two equivalent carbons of the bromonium ion shown in Fig. 9.4 and produce one of the two optically active forms of 2,3-dibromobutane. If the bromide ion attacks the other carbon atom, the other enantiomer is produced. Since substitution at the two carbon atoms is equally likely, a racemic mixture (50% d, 50% l) is formed.

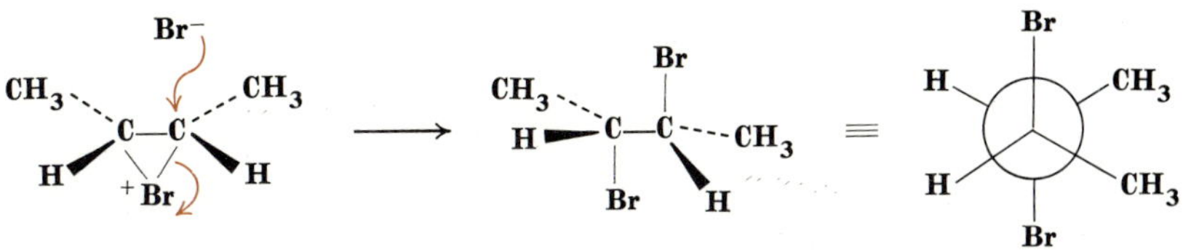

Fig. 9.4 Opening of bromonium ion by bromide anion

Similarly, we expect *trans*-2-butene to yield *meso*-2,3-dibromobutane, and experiment proves this to be the case.

The stereochemical consequences of the bromonium ion intermediate is even more evident in the case of cyclic olefins. The addition of bromine to cyclopentene yields *trans*-1,2-dibromocyclopentane, not the *cis* isomer:

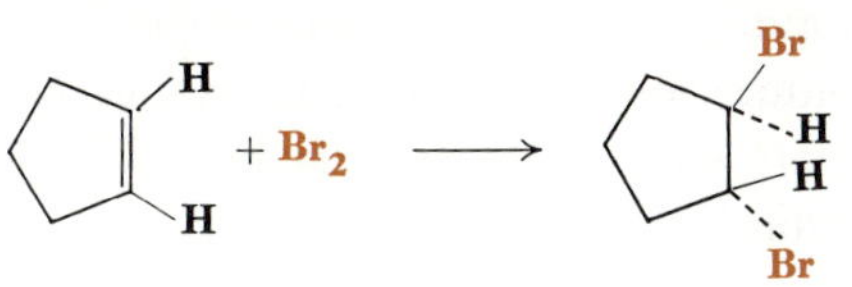

trans

Although *trans* addition of X—Y to an olefin is by far the most common mode of behavior, *cis* addition is also known to occur. For example, addition of deuterium bromide (DBr) to acenaphthene gives predominantly the *cis* addition product:

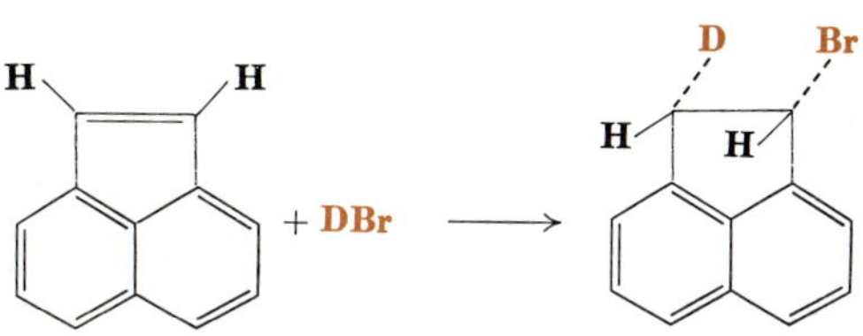

cis

Intermediate 3 (Fig. 9.1) has been invoked to explain the *cis* stereochemistry. This intermediate involves ion pairing between Y^- and a carbonium ion formed by covalent bonding of X to a central carbon:

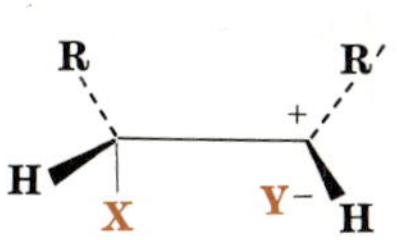

(3)

In terms of the acenaphthene example, intermediate 3 has the structure shown in Fig. 9.5.

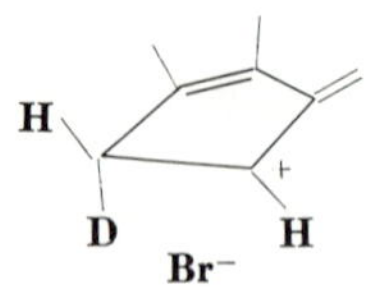

Fig. 9.5 "Ion pair" intermediate from acenaphthene

The important point is that D and Br^- are on the *same* side of the molecule. Then, before Br^- has a chance to migrate to the other side of the acenaphthene molecule, it bonds to the carbonium ion to form a *cis* product.

Most addition reactions yield little *cis* product and intermediate 2 predominates over intermediate 3 (Fig. 9.1). As one might expect, *cis* stereochemistry is found primarily in systems having a relatively stable carbonium ion. In the acenaphthene case, the carbonium ion is stabilized by resonance with the aromatic ring. In contrast, *trans* stereochemistry is favored by the ability of X to form a cyclic "onium" ion (Br > Cl > F > H).

Interestingly, if the carbonium ion in the ion pair (intermediate 3) is sufficiently stabilized by resonance, then the ion pair can dissociate into free carbonium ion and free Y^- (intermediate 4):

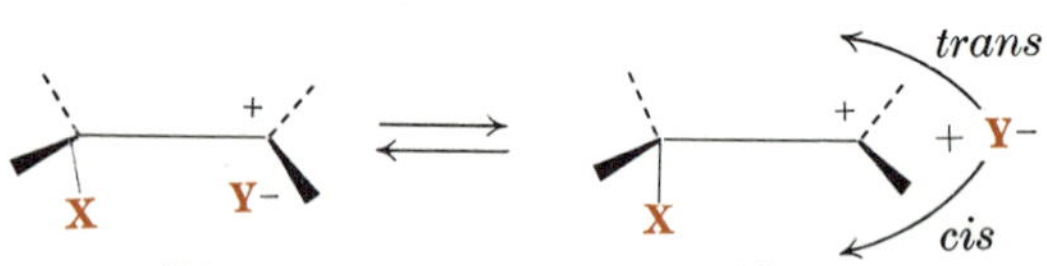

(3) (4)

When this happens, Y^- can attack on either side of the carbonium ion, and the exclusive *cis* stereochemistry is lost. In such a case, the products will form according to the relative stability of the isomers.

9.3 REACTIVITY

We can deduce the effect of substituents on the relative reactivity of alkenes from the mechanism in Fig. 9.1. Since the addition reaction begins when a positive reactant attacks the π electrons of the double bond, anything that increases the density of the electrons on the double bond will enhance its reactivity toward this attack. Electron-donating substituents on the double bond, for example, will have this effect, while electron-withdrawing substituents will have the opposite effect. If we increase the alkyl groups (which are electron releasing) on a double bond, reactivity toward bromination increases:

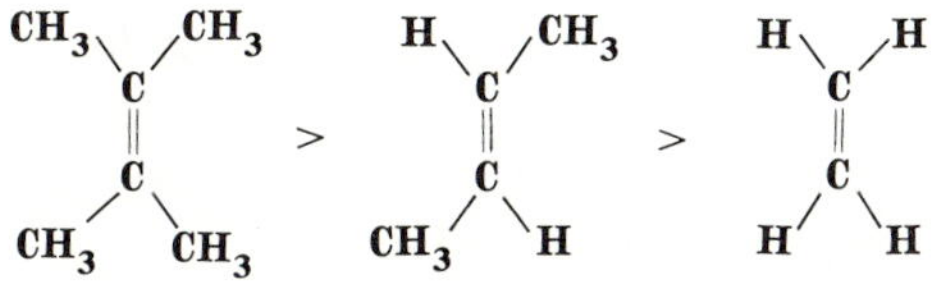

9.4 MARKOVNIKOV RULE

An interesting problem arises in the addition of unsymmetrical reagents (such as HBr) to unsymmetrically substituted alkenes (such as propene). Two products (1-bromopropane and 2-bromopropane) are possible, but only one is formed. The mode of addition that normally obtains is governed by the *Markovnikov rule* which states that adding an unsymmetrical reagent to a simple olefin will bond the negative portion of the reagent to the more substituted carbon.

Thus, in the addition of HBr to propene, the product is 2-bromopropane:

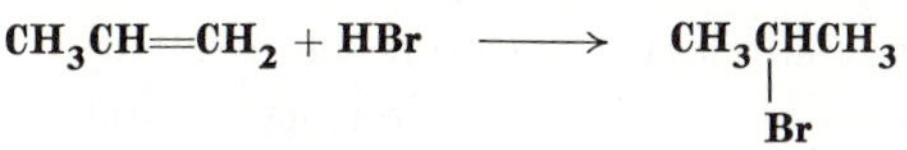

This result can be rationalized in terms of the mechanism in Fig. 9.1. The π complex between HBr and the propene is in equilibrium with the cyclic intermediate:

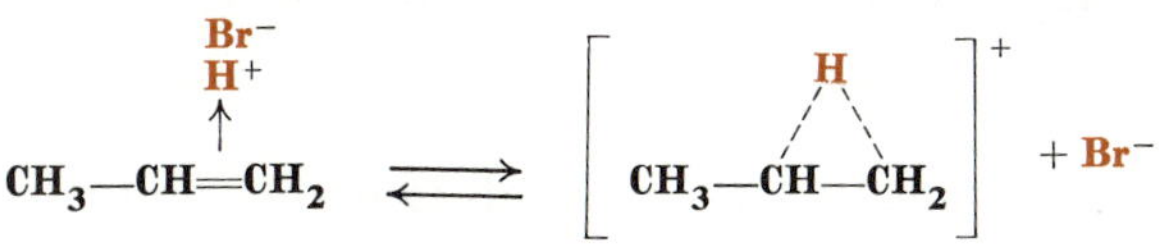

The cyclic intermediate may be viewed as a resonance hybrid of two contributors:

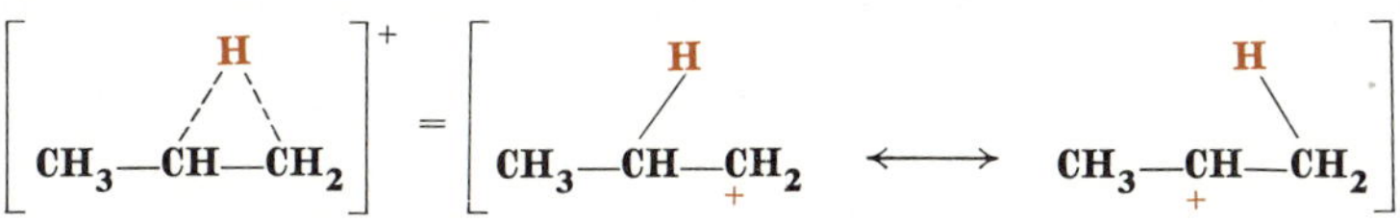

(1) (2)

The two contributors are not identical and they do not contribute equally to the hybrid. Contributor 2 is more important than contributor 1 because a secondary carbonium ion is more stable than a primary carbonium ion. (The reader should recall that the order of carbonium ion stability is tertiary > secondary > primary.) To claim that resonance contributor 2 is more important than 1 is tantamount to saying that bond formation in the cyclic intermediate is greater between hydrogen and the primary carbon than between hydrogen and the secondary carbon. Hence, the bromide preferentially attacks the secondary carbon site (in a back-side attack, effecting *trans* addition). From this mechanism we may restate the Markovnikov rule: an unsymmetrical reagent will add to a double bond with the negative fragment bonding to the carbon better able to support positive character.

9.5 MICHAEL ADDITION

When HX adds to a double bond conjugated with a carbonyl group, the X ends up on the β carbon:

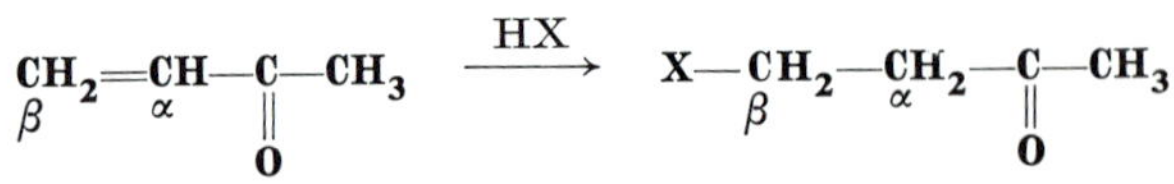

methyl vinyl ketone

Conjugation of a double bond with a carbonyl group markedly changes the nature of the double bond. There is little similarity between the mechanism of addition to this type of olefin and the addition to a simple isolated double bond. Under acidic conditions the first step is protonation of the carbonyl group which enhances the positive character of the β carbon:

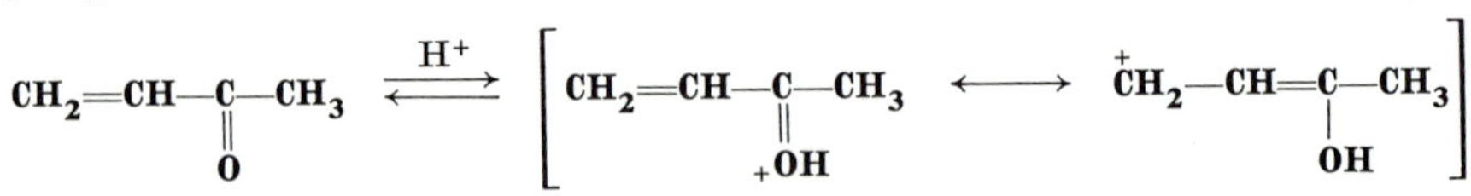

Attack by X^- on the β carbon gives the enol which rapidly tautomerizes to the more stable keto form:

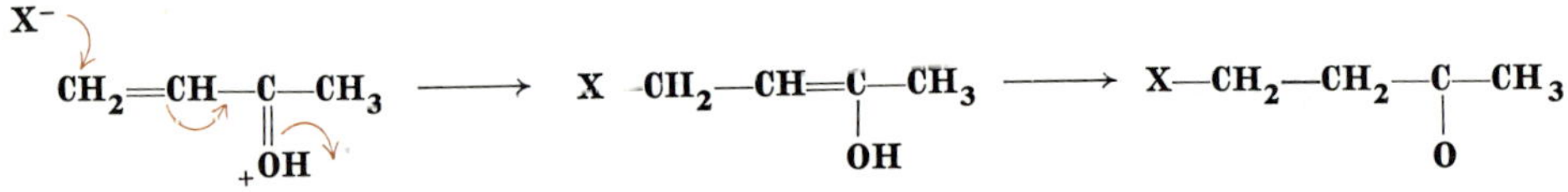

enol ketone

Addition to an α,β-unsaturated ketone under basic conditions, called the *Michael reaction*, is very useful in synthesis. In particular, the anions of malonic ester and acetoacetic ester (review Section 7.12) can add to a compound such as methyl vinyl ketone (subsequent protonation of the resulting enolate gives the final product):

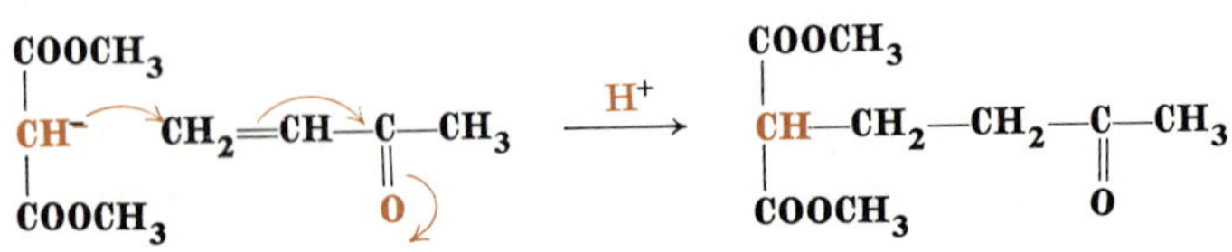

The Michael addition product can be hydrolyzed and decarboxylated to give a ketoacid:

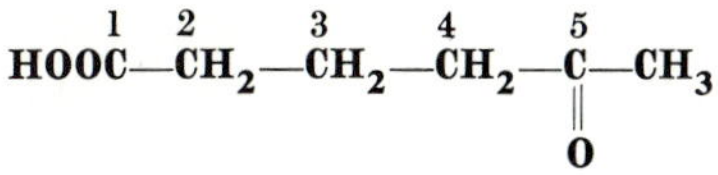

This is a standard way to prepare a compound which has two carbonyl groups in a 1,5 relationship.

We must emphasize that neither malonic ester anion, chloride, hydroxide, nor any other anion will attack an ordinary isolated double bond. An unconjugated double bond is electron rich and susceptible to attack only by electron-deficient species.

9.6 RADICAL ADDITION

The addition of hydrogen bromide to double bonds in the presence of peroxide (RO—OR) is another reaction that appears to violate the Markovnikov rule:

$$CH_3CH{=}CH_2 \xrightarrow[\text{peroxide}]{\text{HBr}} CH_3CH_2CH_2Br$$

In the absence of peroxide, ordinary Markovnikov addition takes place to give the isomeric 2-bromopropane. The reaction with peroxide does not really represent a reversal of the normal mode of addition: the two addition reactions are completely unrelated. This new reaction involves a *free radical* mechanism which is faster than the ionic process shown in Fig. 9.1.

The first step in free radical addition is homolytic cleavage of the peroxide molecule to give two oxygen radicals:

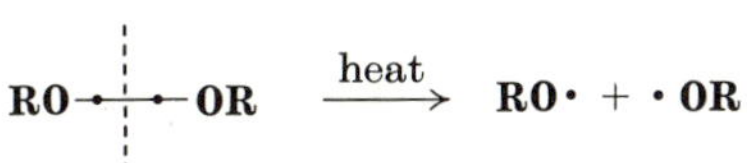

The oxygen-oxygen bond of the peroxide is labile to cleavage, because the bonding of two electronegative atoms is an unstable situation. The oxygen radical then attacks hydrogen bromide to give a bromine atom:

The bromine atom, which is one electron short of a complete octet, attacks the π system of a double bond:

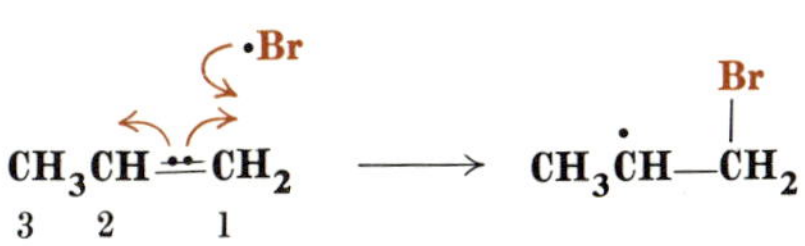

Although there are two carbons which could conceivably be attacked by the bromine atom, only the 1-carbon bonds to the bromine because this produces a *secondary* radical. Radical stability follows the order tertiary > secondary > primary. Thus, if a bromine atom attacked the 2-carbon, the less stable *primary* carbon radical would result. The final step is the reaction between the carbon radical and hydrogen bromide to give the "anti-Markovnikov" addition product plus a bromine atom:

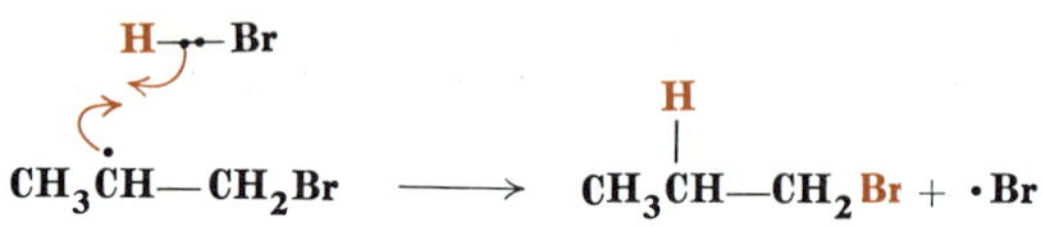

The bromine atom can then attack another olefin molecule. This type of self-perpetuating reaction is called a *chain reaction.*

It should be noted that hydrogen chloride and hydrogen iodide do not undergo free radical addition: the second step in the reaction sequence (cleavage of the hydrogen-chlorine bond) is too unfavorable because the hydrogen chloride bond is a stronger bond than the hydrogen-bromine bond. With hydrogen iodide, the second step of the sequence takes place readily because the hydrogen-iodine bond is weak. However, coupling of the iodine radical to form iodine (I_2) is preferred over formation of a comparatively unstable carbon-iodine bond:

$$\mathrm{H{-}I} \xrightarrow{\text{peroxide}} \mathrm{I\cdot} \xrightarrow{\mathrm{I\cdot}} \mathrm{I_2}$$

Some reagents will add to double bonds only by this free radical mechanism and not by the ionic mechanism. Thus mercaptans (R—S—H) are not acidic enough to yield ionic addition. With peroxide catalysis, however, addition can take place:

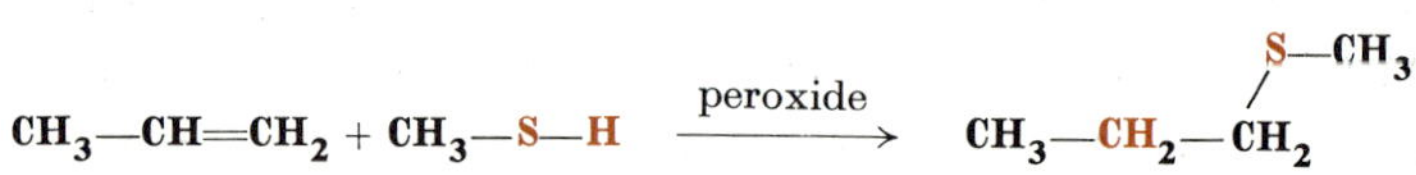

APPLICATIONS OF ADDITION REACTIONS

9.7 ADDITION OF HALOGEN

Chlorine and bromine add to double bonds to form vicinal dihalides. Iodine does not add in a similar manner, as the reverse reaction (elimination of iodine) proceeds too rapidly:

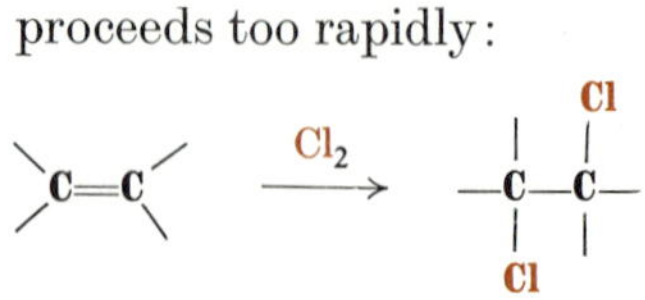

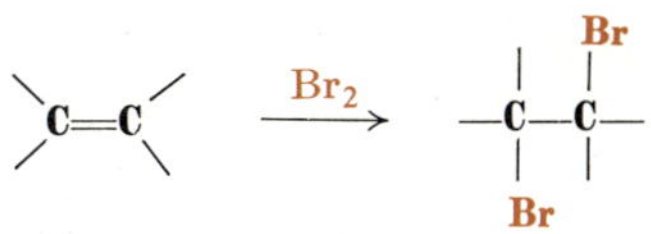

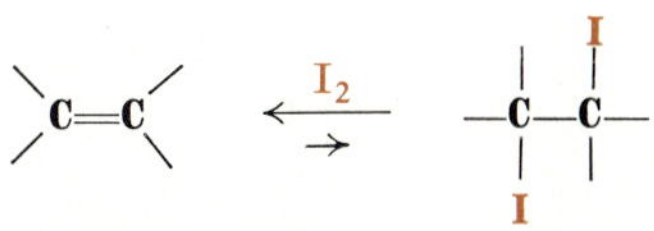

The preparation of tetrabromostearic acid provides a specific example of a bromine addition:

$$CH_3(CH_2)_4CH{=}CHCH_2CH{=}CH(CH_2)_7COOH$$

$$\xrightarrow{2Br_2} CH_3(CH_2)_4CHBrCHBrCH_2CHBrCHBr(CH_2)_7COOH$$

Bromine is added drop by drop to a heptane solution of the unsaturated acid at 10–15°, and the product precipitates out in good yield. The consumption of bromine is obvious because the reddish-brown color of the reagent disappears.

If brominations are carried out at higher temperatures, a considerable amount of substitution product is formed. One way of effecting such a reaction preparatively is by using a commercially available reagent called N-bromosuccinimide (NBS).

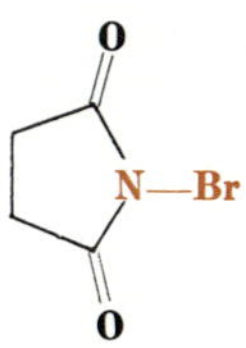

NBS

This reagent brominates an olefin at a carbon *adjacent* to the double bond. Thus an olefin such as propene is mixed with NBS in carbon tetrachloride (CCl_4) and warmed. The allylic bromide forms under these mild conditions:

$$NBS + CH_2{=}CH{-}CH_3 \longrightarrow CH_2{=}CH{-}CH_2Br$$

The reaction is known to be a free radical process.

9.8 ADDITION OF HYDROGEN HALIDE

Hydrogen chloride, hydrogen bromide, and hydrogen iodide readily add across multiple bonds. Thus, 2-chloro-2-phenylpropane is generated when hydrogen chloride gas is passed through α-methylstyrene at 0°:

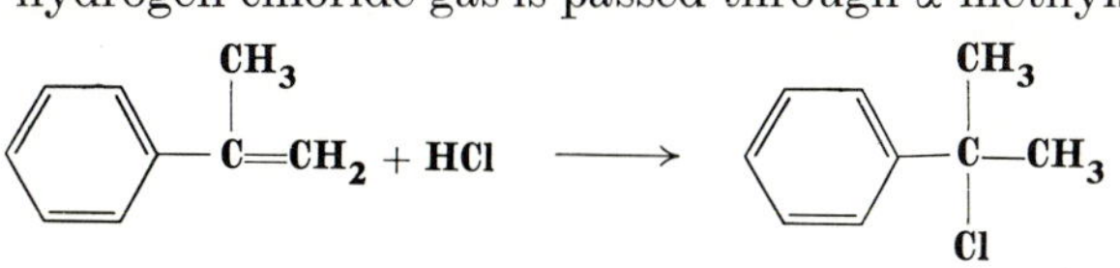

Recall that addition usually occurs with *trans* stereochemistry:

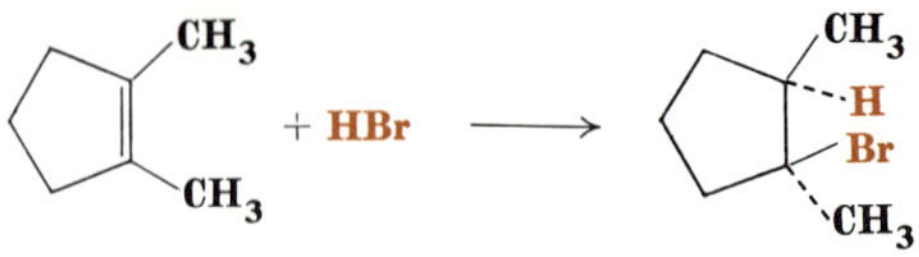

9.9 ADDITION OF HYPOHALOUS ACIDS

The oxygen of hypohalous acids (HOX) bears a partial negative charge ($HO^{-\delta}$—$X^{+\delta}$) because oxygen is more electronegative than the halogens except for fluorine. Therefore, Markovnikov addition of HOX results in the bonding of the oxygen to the more substituted carbon:

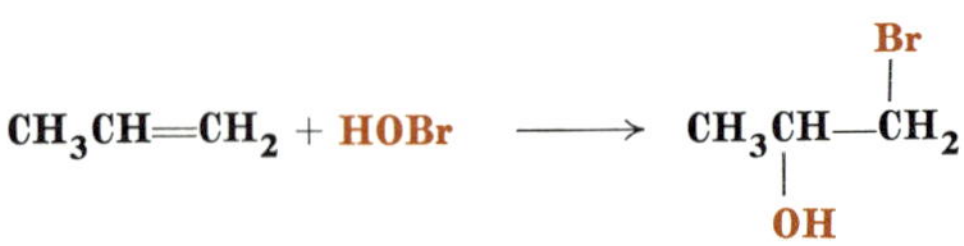

The products of these additions are called halohydrins; the above addition product, for example, is a bromohydrin. Halohydrins are useful in the synthesis of epoxides. If the bromohydrin of cyclopentene is treated with sodium hydroxide, the epoxide forms via an intramolecular displacement reaction. Back-side attack in this S_N2 reaction is possible only by virtue of the fact that HOX undergoes *trans* addition:

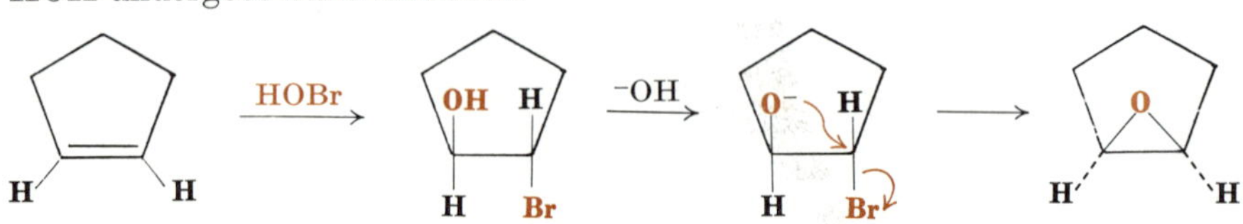

The student may recall an alternative method for preparing epoxides (Section 7.11):

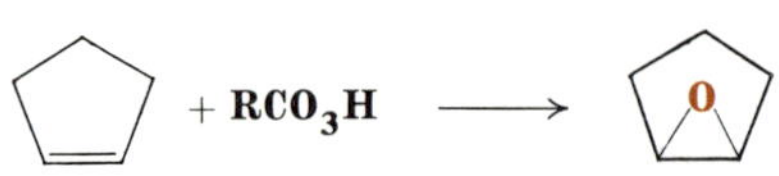

9.10 HYDRATION OF ALKENES

It is possible to hydrate alkenes in the presence of strong acid to form alcohols:

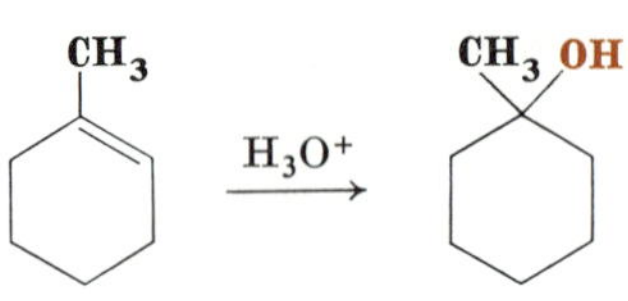

This reaction is the reverse of the acid-catalyzed dehydration of alcohols to form alkenes. Many hydrations are simply carried out by dissolving the olefin in cold concentrated H_2SO_4 and then quenching the mixture with water. The reaction conditions contrast with those of acid-catalyzed dehydrations of alcohols in which the amount of water is kept to a minimum. Note from the above example that water adds in a Markovnikov fashion. In the next section we will describe a method which hydrates a double bond in the opposite sense.

9.11 HYDROBORATION

Hydroboration was developed only recently and it has already become one of the most valuable reactions in the synthetic organic chemist's arsenal. Diborane (B_2H_6) adds to alkenes in the following manner (for the sake of simplicity, diborane is written "BH_3"):

$$3CH_3CH{=}CH_2 \xrightarrow{BH_3} (CH_3\overset{H}{\overset{|}{C}}H{-}CH_2)_3B$$

Note that the boron atom bonds to the *least* substituted carbon. The carbon-boron bond can then be converted into the alcohol by treatment with hydrogen peroxide and sodium hydroxide:

$$(CH_3CH_2CH_2)_3B \xrightarrow[NaOH]{H_2O_2} 3CH_3CH_2CH_2OH$$

The net effect of this reaction sequence is the apparent "anti-Markovnikov" hydration of the double bond. The conversion of 1-methylcyclopentene into *trans*-2-methylcyclopentanol may be carried out in an ether solvent at 0° in 85% yield:

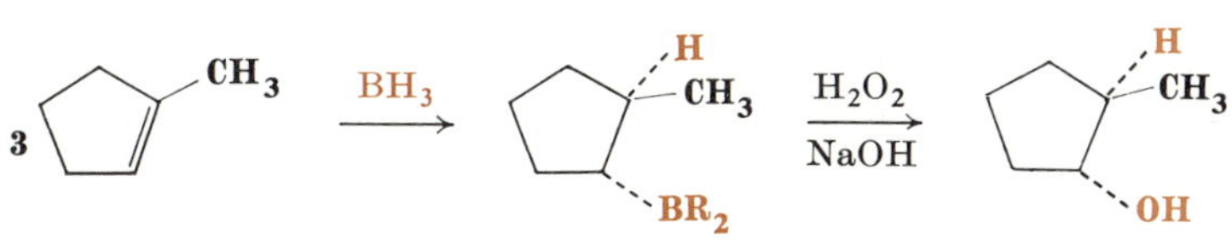

Both steps can be performed in one pot; the trialkylborane need not be isolated.

The student should commit this reaction to memory. To facilitate this task we will explain the mechanism. Boron (atomic number 5) has three electrons in its bonding shell. When it forms three covalent bonds with three atoms, to form BR_3, there are still only six electrons in the outer 2-shell.

Since a full octet of electrons is desirable, trivalent boron readily accepts a pair of electrons from an electron donor:

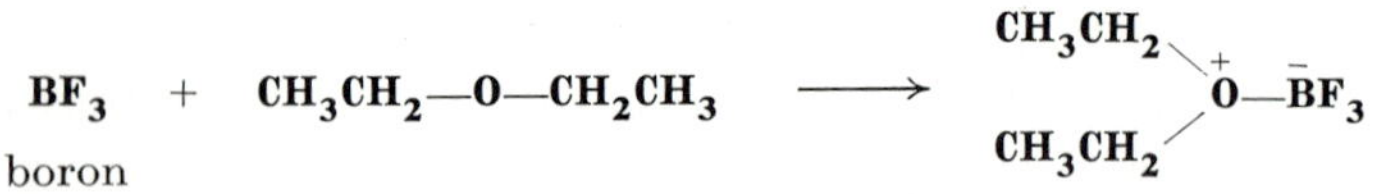

boron trifluoride — boron trifluoride etherate

In a similar manner, BH_3 accepts a pair of electrons from a double bond:

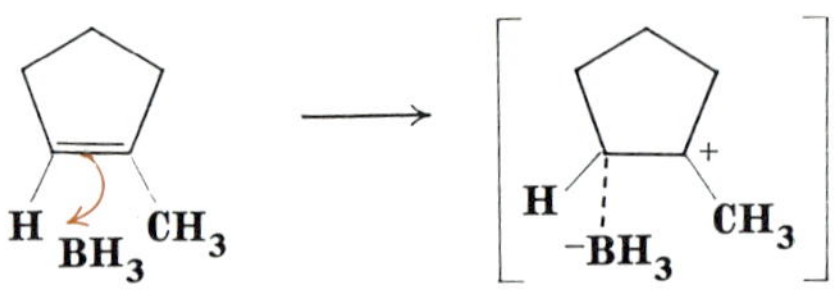

The boron bonds to the less substituted carbon in order to form the more substituted and more stable carbonium ion. Transfer of a hydride ion (H^-) then produces the alkyl borane:

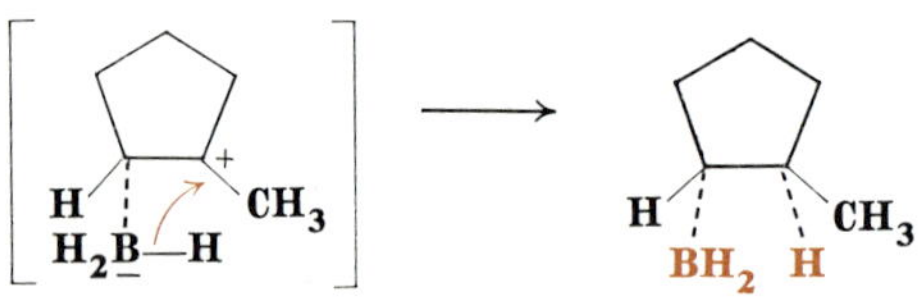

The mechanism can also be written as a one-step cyclic concerted process:

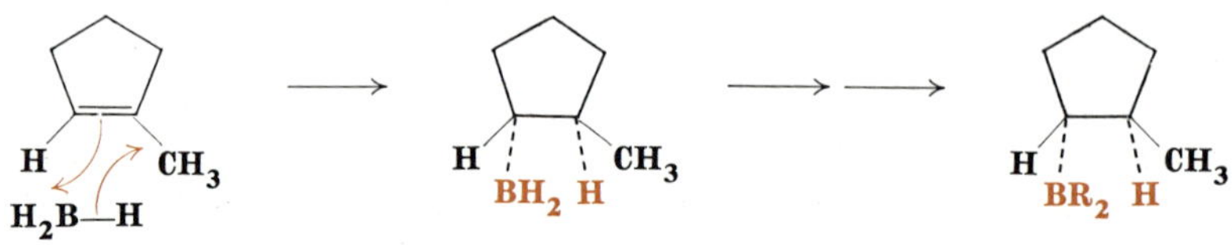

Note that the boron and the hydrogen are delivered by *cis* addition in both mechanisms.

When hydrogen peroxide and sodium hydroxide are added to the reaction mixture, peroxide anion bonds to the electron-deficient boron:

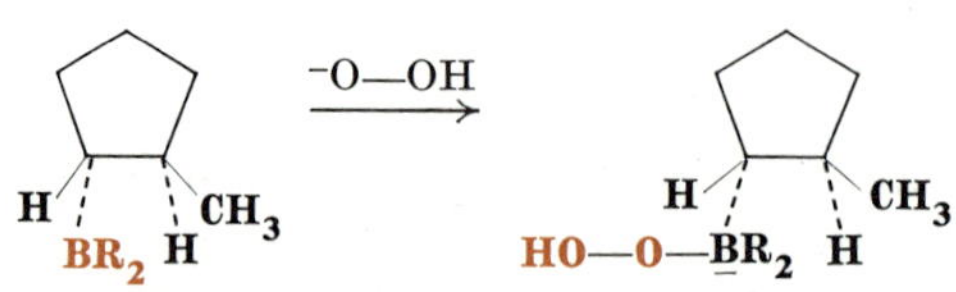

The final step is an *alkyl migration* in which a carbon moves from the boron to an oxygen with the ejection of a hydroxide ion. (This reaction will be better understood after studying Chapter 12, Rearrangement Reactions.) For the moment, it is necessary only to realize that the resulting boron-oxygen complex is unstable and readily forms the final reaction product, an alcohol, when the reaction mixture is exposed to acid during the isolation procedure.

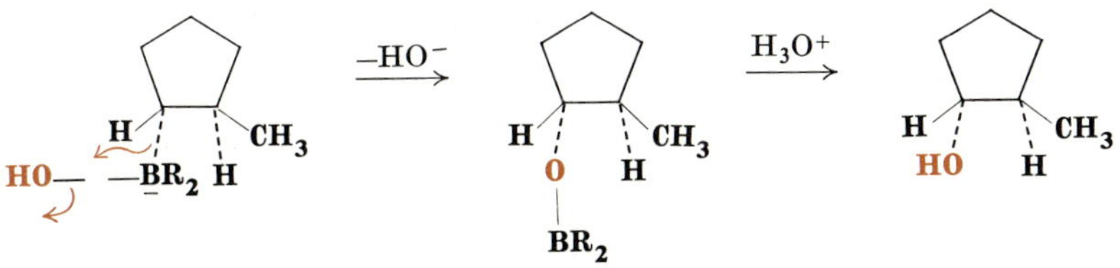

Note that the migration occurs with retention of configuration. That is to say, the oxygen of the alcohol ends up exactly where the boron used to be (*trans* to the methyl group). Thus, the overall stereochemical course of the hydroboration sequence is an anti-Markovnikov *cis* addition of water to a double bond.

Treating the "BH_3" addition product with acid instead of with basic hydrogen peroxide produces the alkane:

$$(CH_3CH_2CH_2)_3B \xrightarrow{H_3O^+} CH_3CH_2CH_3$$

9.12 DIELS-ALDER ADDITION

Two reactions have been considered important enough to warrant awarding the Nobel Prize to the chemists who discovered them. These are the Grignard reaction, with which we are already familiar, and the Diels-Alder reaction. The Diels-Alder reaction is exceedingly useful for synthesizing 6-membered cyclic systems. The reaction requires a conjugated *diene* and an alkene (usually substituted with an electron-withdrawing substituent) called a *dienophile*.

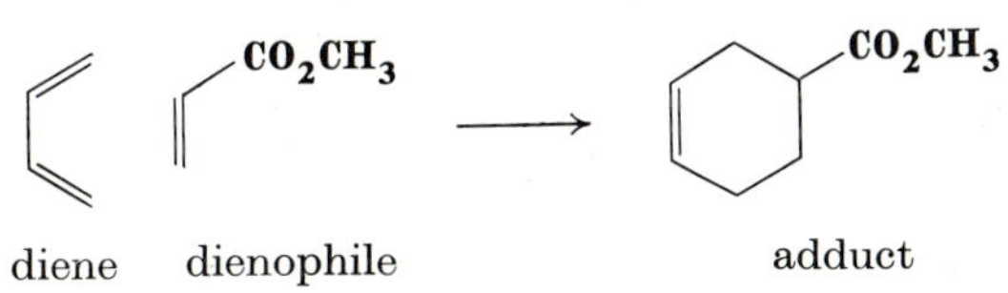

The mechanism of this reaction, which has been studied intensively for a long time and which is just beginning to be understood, is beyond the scope of this text. For the sake of simplicity, the mechanism is illustrated here as a one-step cyclic concerted process. This may not be too far from the truth since the reaction rate is often insensitive to the polarity of the solvent, suggesting the absence of charged transition states or intermediates.

Two more examples of the Diels-Alder reaction are given below. Note that bicyclic products can be produced from cyclic dienes.

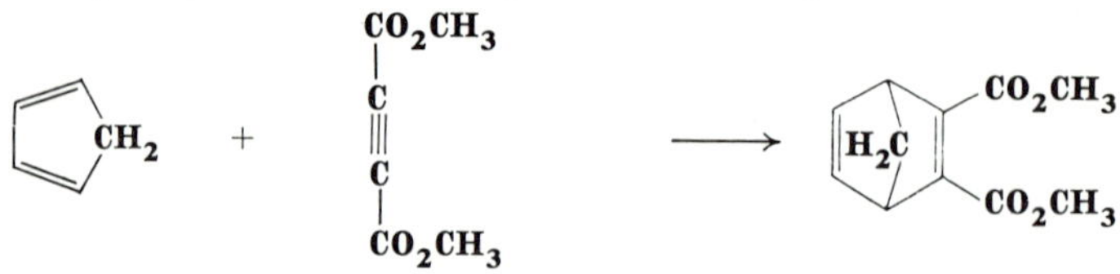

cyclopentadiene dimethyl acetylene-dicarboxylate

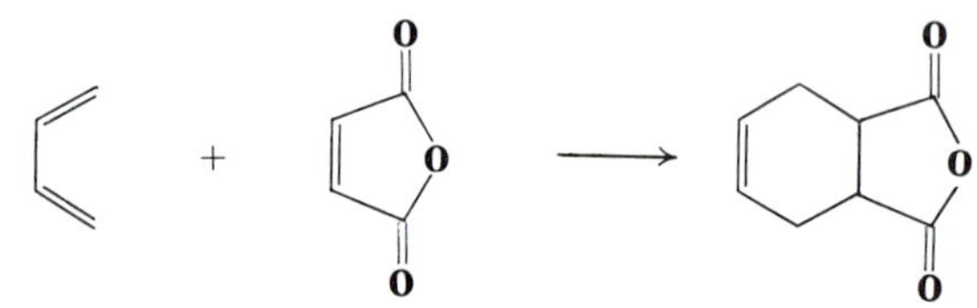

butadiene maleic anhydride

The Diels-Alder reaction is usually easy to carry out. Thus, the adduct of butadiene and maleic anhydride is obtained quantitatively when butadiene gas is passed through melted maleic anhydride (53°). Higher temperatures must be used with less reactive compounds.

While we will not dwell on the mechanism of the Diels-Alder reaction, we will discuss briefly its stereochemistry. For this purpose, we will first have to understand the difference between kinetic and thermodynamic control of a reaction.

9.13 KINETIC VERSUS THERMODYNAMIC CONTROL

Consider two reactions, A→B and A→C, with the energy diagrams shown in Fig. 9.6. Product C will form faster because it requires less activation energy

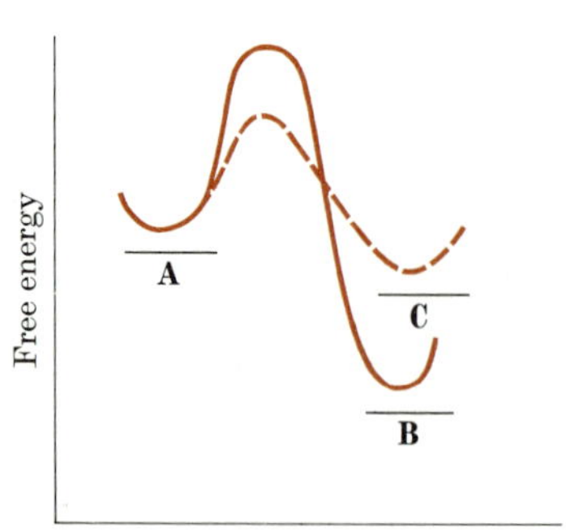

Fig. 9.6 Two reaction coordinates for A going to B and A going to C

than product B. Hence C is called the *kinetic product*. The energy relationship in Fig. 9.6 is interesting because, although C is formed first, B is more stable. If the reaction is allowed to proceed long enough, C will revert back to A, allowing slow formation of B. This circumstance will allow the reagents to come into equilibrium, and the relative amounts of B and C will then depend on the stabilities of B and C. Since B is more stable than C, it will be the predominant *thermodynamic product*. In other words, at mild temperatures and a short reaction time product C will form, while at higher temperatures and a longer reaction time product B will result. Of course, in many instances the more stable product has the smaller activation energy (the kinetic and thermodynamic products are the same). However, if this is not the case, then the nature of the reaction product depends on the reaction conditions. As we shall see, the Diels-Alder reaction is a good example of this situation.

9.14 STEREOCHEMISTRY OF THE DIELS-ALDER REACTION

Cyclopentadiene reacts with maleic anhydride to form a bicyclic adduct:

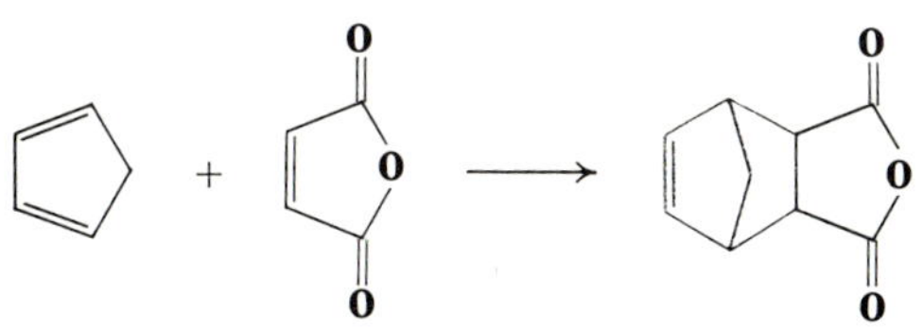

The adduct can exist as either an *exo* or an *endo* isomer (Fig. 9.7).

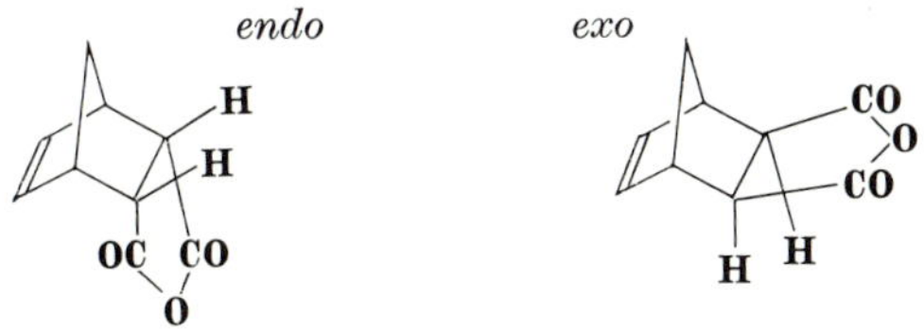

Fig. 9.7 *Endo* and *exo* cyclopentadiene–maleic anhydride Diels-Alder products

The *exo* isomer is more stable than the *endo* isomer (probably because there is less steric interaction in the *exo* adduct). When the *endo* adduct is heated, it isomerizes to the *exo* form:

$$endo \underset{\leftarrow}{\xrightarrow{\text{heat}}} exo$$

Nevertheless, at low temperatures the Diels-Alder reaction of cyclopentadiene and maleic anhydride produces exclusively the *endo* isomer. If the reaction is carried out at high temperatures, considerable *exo* adduct is formed. Clearly, the *endo* adduct is the kinetic product, and the *exo* adduct is the thermodynamic product.

The usual explanation for the more rapid formation of *endo* adduct is that this form arises from an arrangement of diene and dienophile in which there is more extensive orbital overlap than there is in the *exo* isomer (Fig. 9.8). This tends to stabilize the *endo* transition state and reduce its activation energy.

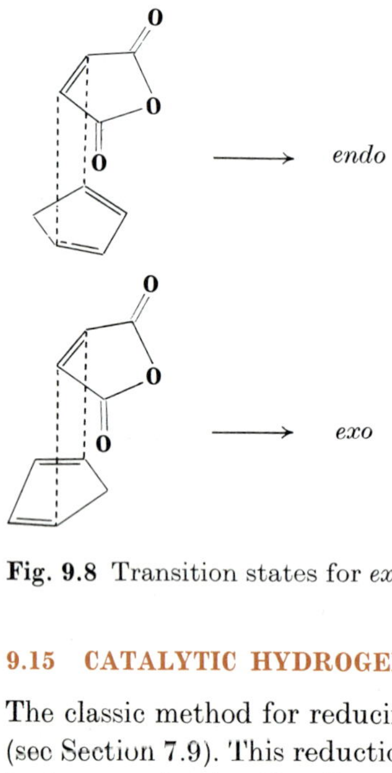

Fig. 9.8 Transition states for *exo* or *endo* Diels-Alder addition products

9.15 CATALYTIC HYDROGENATION

The classic method for reducing alkenes to alkanes is *catalytic hydrogenation* (see Section 7.9). This reduction will be discussed in detail in Chapter 13, but in the meantime it will suffice to point out that hydrogen adds to carbon-carbon double and triple bonds in the presence of a catalyst such as platinum or palladium:

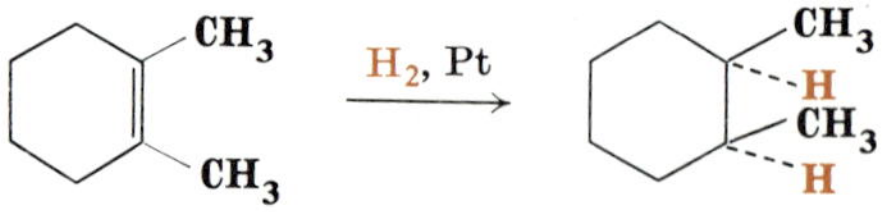

cis

The reaction is usually run by stirring an alcoholic solution of the olefin over a finely divided metal catalyst in a hydrogen atmosphere. The catalyst adsorbs the hydrogen gas, and in a *cis* addition delivers two hydrogen atoms to the olefin. This is an example of a *heterogeneous reaction*, a reaction involving more than one phase.

Triple bonds are easier to reduce than double bonds, so that a triple bond can be reduced to a double bond without carrying the reaction all the way to the alkane. This can be done with hydrogen and *Lindlar catalyst* (consisting of palladium partially deactivated with lead acetate):

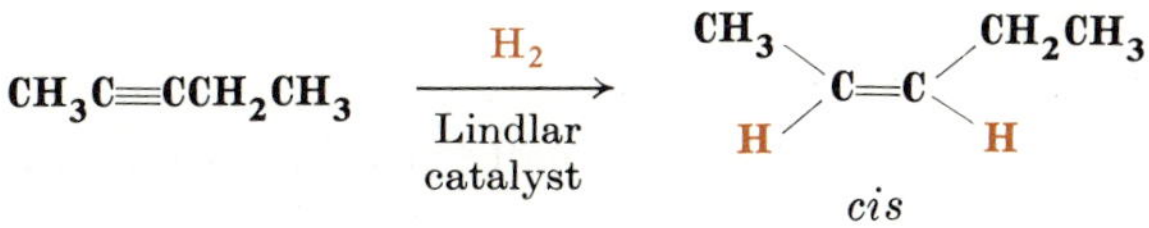

The less stable *cis* isomer always results from such a reduction. Obviously, catalytic reduction is kinetically controlled.

9.16 OZONIZATION

Ozone (O_3), which is made by passing oxygen through an electric discharge, reacts rapidly with alkenes to add across the double bond and form a so-called molozonide:

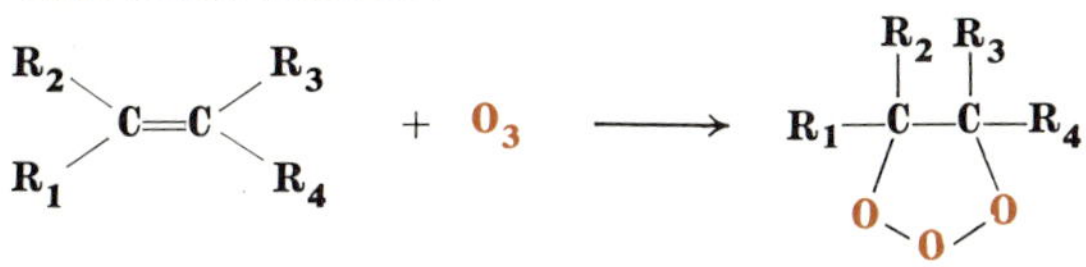

The molozonide is unstable and spontaneously isomerizes to an *ozonide*:

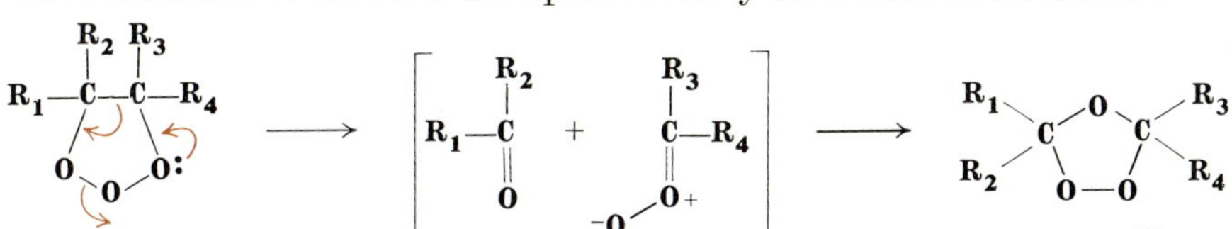

an ozonide

Note that this reaction breaks both the π and the σ bonds of the double bond. Ozonization is useful because ozonides can be cleaved with zinc and water to form aldehydes and/or ketones (depending on the substitution of the olefin):

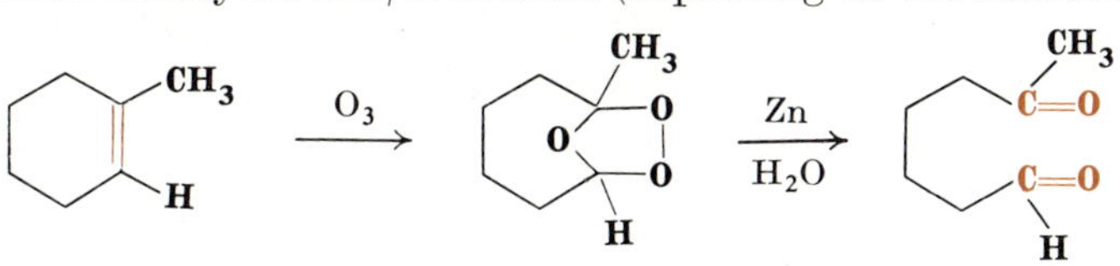

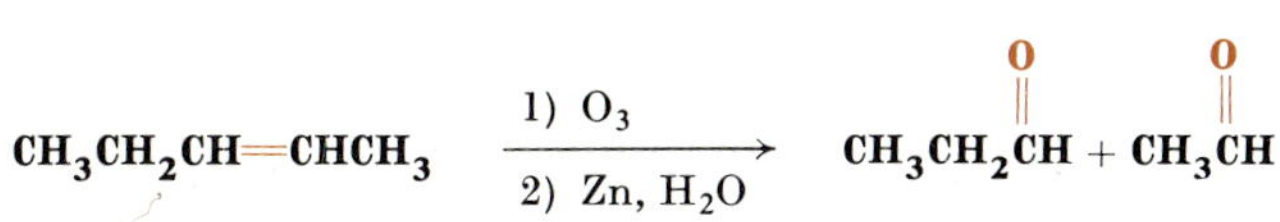

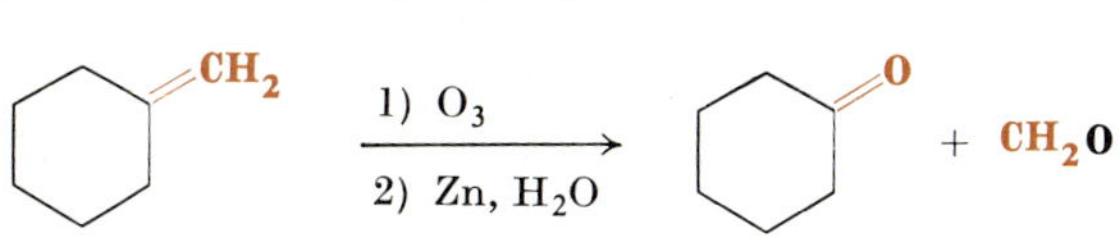

9.17 PERACID OXIDATION

The reader already knows that peracids react with olefins to yield epoxides:

The mechanism probably resembles that shown in Fig. 9.1 for the addition of XY to a double bond:

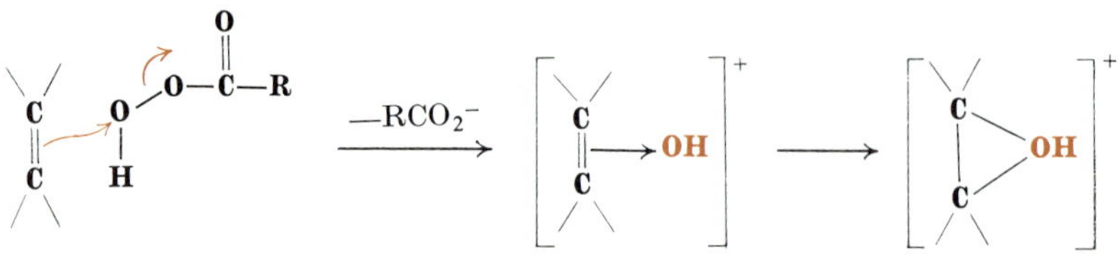

π complex

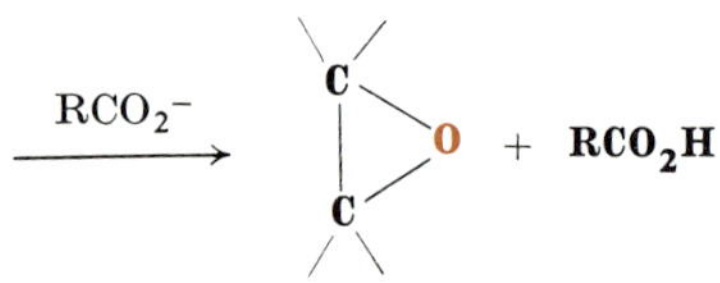

Recall that epoxides are attacked by nucleophiles to give the *trans*-1,2-disubstituted compounds:

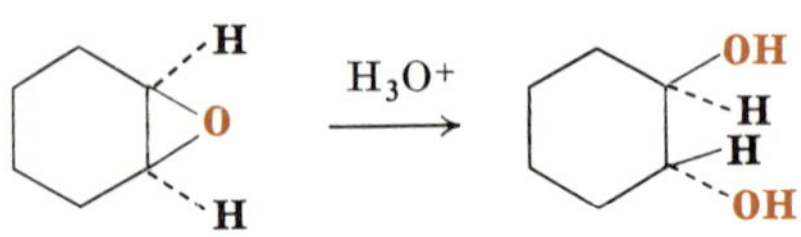

trans

9.18 GLYCOL FORMATION

Potassium permanganate or the more expensive osmium tetroxide can directly convert an olefin into a 1,2-glycol (a 1,2-dialcohol):

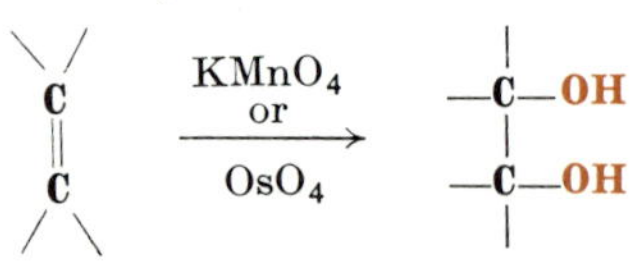

The glycol produced from a cyclic olefin is exclusively the *cis* form. This stereochemical course is explained by the following mechanism:

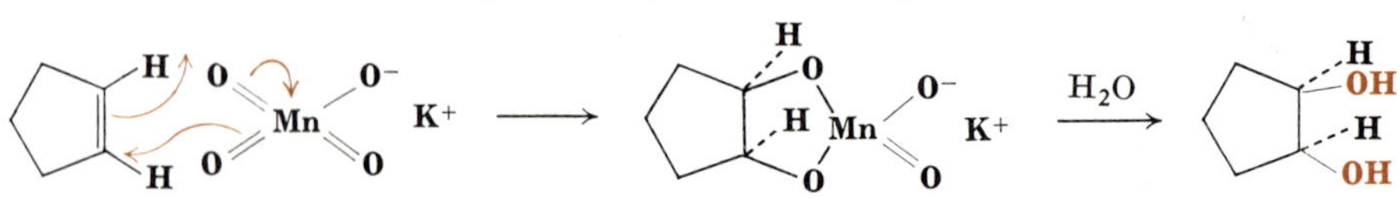

Thus, to make a *cis*-glycol, one uses permanganate oxidation; to make a *trans*-glycol, one uses the peracid method.

A specific example of a permanganate oxidation is shown below:

$$CH_2{=}CHCH(OC_2H_5)_2 \xrightarrow{KMnO_4} \underset{OH}{CH_2}{-}\underset{OH}{C}HCH(OC_2H_5)_2$$

The reaction is carried out by adding an aqueous solution of potassium permanganate (neutral pH) to a suspension of the olefin in water at 5°. The product can be obtained in 67% yield. More vigorous conditions (higher temperatures and acidic pH values) result in the cleavage of the glycol.

This terminates our discussion of addition reactions. Obviously, we have been exposed in this chapter to an abundance of new reactions and concepts. These are the tools with which an organic chemist works and, alas, they must be memorized just as anyone wishing to speak a new language must learn the vocabulary by rote. The student of organic chemistry has an advantage over the language student, however. Organic chemists can explain their discipline in terms of a relatively small number of mechanistic principles, and these are of considerable aid in learning the reactions and in understanding the behavior of organic systems.

9.19 SYNTHESIS

The preceding three chapters have dealt with a variety of reactions that are useful for synthesis. These include new ways to make carbon-to-carbon bonds, and a number of functional group preparations. Let us now examine how some of these may be used in the solution of synthetic problems.

9.20 GYPSY MOTH PHEROMONE

Chemical agents used by insects for communication are called pheromones. One of these, the gypsy-moth sex attractant, has the following structure:

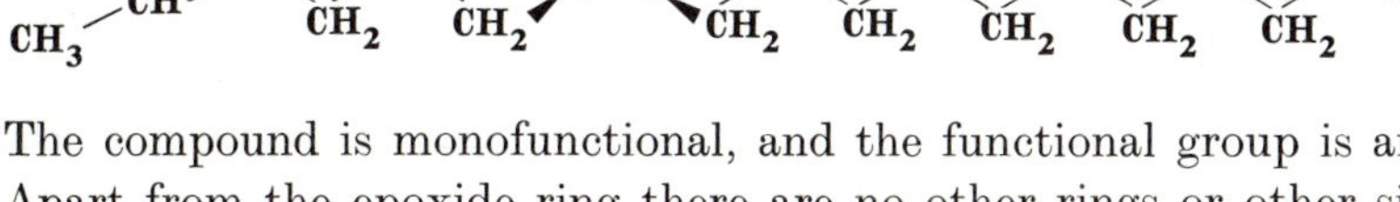

The compound is monofunctional, and the functional group is an epoxide. Apart from the epoxide ring there are no other rings or other sites of unsaturation in the molecule. There is one branch, a methyl group, on one of the saturated alkyl groups attached to the epoxide ring. In addition, the two alkyl chains appended to that ring have a *cis*-relationship.

Two methods are available to us for preparing an epoxide group. These are, peracid epoxidation of a double bond (Section 9.17):

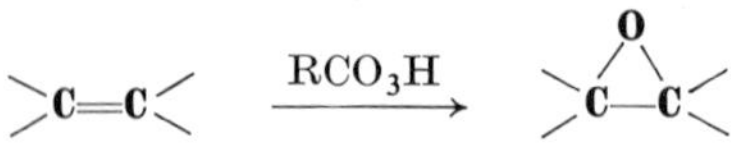

and treatment of an olefin with hypobromous acid followed by reaction of the resulting bromohydrin with base (Section 9.9.):

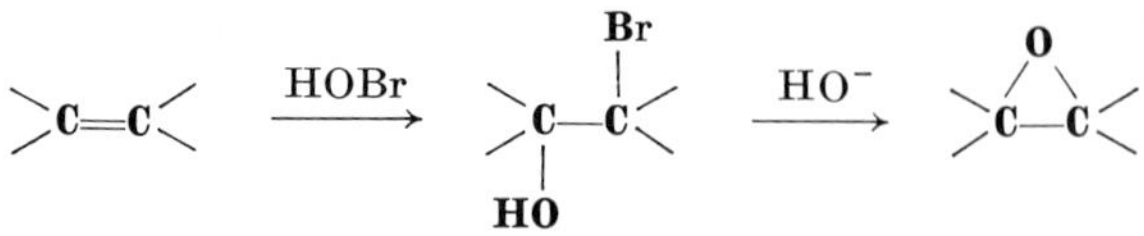

Both of these methods are addition reactions of olefins. Moreover, in each case, the geometry of the epoxide product is the same as the starting olefin. Thus a *trans*-substituted olefin yields a *trans*-substituted epoxide, and a *cis*-substituted olefin yields a *cis*-substituted epoxide. As a consequence, the first step in building a synthetic plan for the gypsy-moth pheromone is to consider the preparation of an olefinic "functional-group equivalent" of the target epoxide. The structure and the *stereochemistry* of the precursor olefin must be:

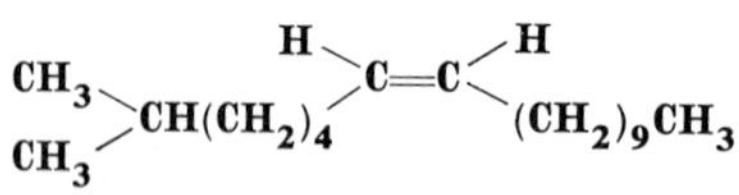

Several methods may be considered for the preparation of this olefin. Construction of the carbon skeleton by the Grignard synthesis, for example, could lead to the following alcohol:

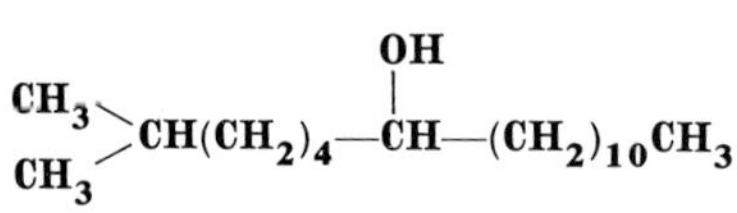

The latter upon treatment with acid should dehydrate to yield an olefin. We realize, however, that dehydration of such an alcohol can give two isomeric olefins. There does not appear to be any clear reason why one direction of dehydration should be preferred over the other. Hence, a synthesis based on this approach would not be specific for the desired olefin. (The reader might also consider the following and suggest an explanation. Dehydration of the above alcohol would not only yield two different olefins, but each of these would most likely have *trans*-stereochemistry, not *cis*.)

Since our first approach to constructing a synthetic plan has reached a "dead end," we must go back at least to our first precursor molecule, the olefin, and consider additional methods for its construction. When we rejected a dehydration reaction as a way of introducing the double bond we also implicitly rejected the Grignard synthesis as a way of constructing the entire alkyl chain. We must therefore find another way of preparing both the functional group and the carbon skeleton.

In looking for ways to prepare our present target, the olefin, we note that the double bond is an *internal* one; it is located within the carbon chain rather than at either of the termini. One of the newer methods for carbon-carbon bond formation that we have acquired is the alkylation of acetylene and substituted acetylenes (Section 7.9):

$$HC{\equiv}C^-Na^+ + R{-}X \longrightarrow HC{\equiv}C{-}R + NaX$$

$$R{-}C{\equiv}C^-Na^+ + R'{-}X \longrightarrow R{-}C{\equiv}C{-}R' + NaX$$

This method can be used to prepare alkyl chains with either terminal or internal unsaturation.

An acetylene group is also a *functional equivalent* of an olefin. Catalytic hydrogenation of an internal triple bond (Section 9.15) yields a 1,2-disubstituted double bond. Furthermore, the double bond is produced with *cis*-geometry:

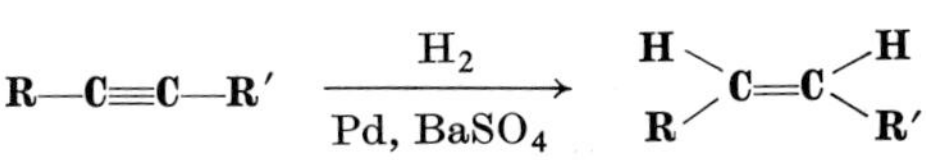

With this knowledge we can construct a partial synthetic "tree" for the gypsy moth pheromone (Fig. 9.9).

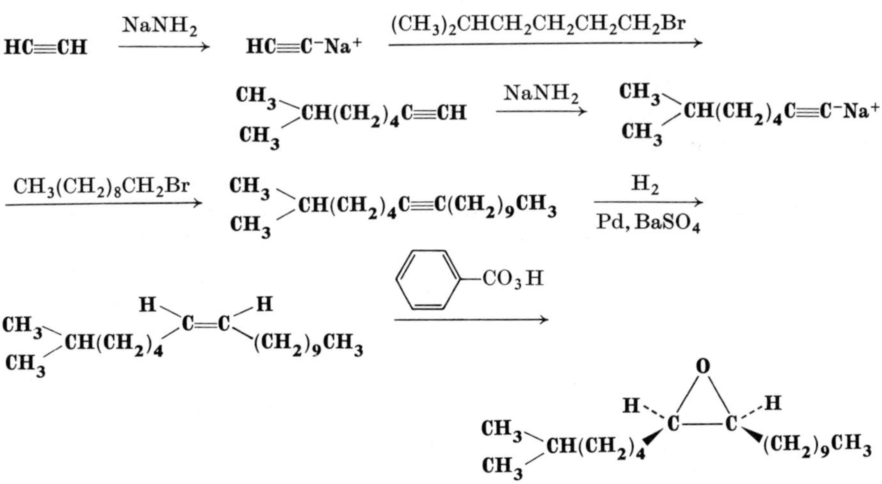

Fig. 9.9 Conversion of acetylene to a *cis*-substituted epoxide

The branches of this tree that still require construction are those calling for preparation of the two appropriate alkyl bromides, (Fig. 9.9). Both of these bromides may be prepared by a conventional Grignard synthesis. Figure 9.10 illustrates the synthesis of the branched alkyl group.

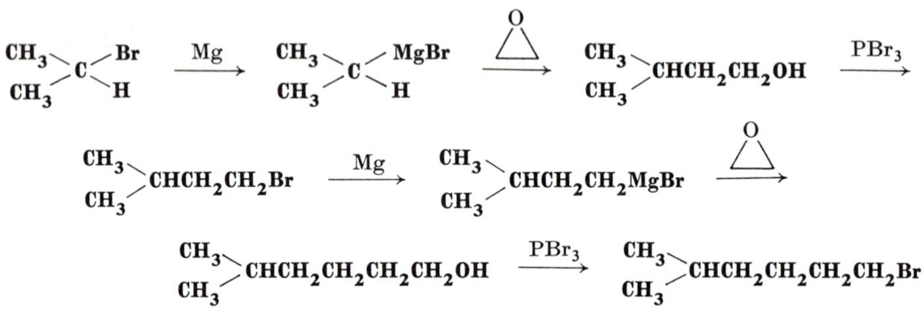

Fig. 9.10 Construction of the isoheptyl chain of the gypsy-moth pheromone

9.21 A BIFUNCTIONAL COMPOUND

Propose a synthesis for 2-methyl-3-ethylhexane-2,6-diol:

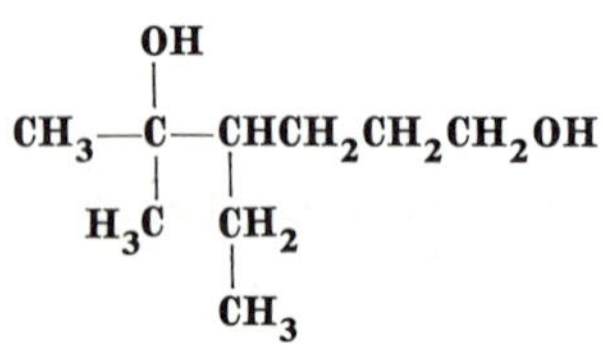

2-methyl-3-ethylhexane-2,6-diol

This problem differs from the others we have considered since the target molecule has *two* functional groups. In dealing with multifunctional compounds we cannot arbitrarily pick one reactive site on which to perform some transformation and simply ignore what might occur at other sites. A rational synthesis of a multifunctional compound must use one or more of the following synthetic principles: symmetry, selectivity, and protection.

9.22 THREE SYNTHETIC PRINCIPLES

If a bifunctional molecule is symmetrical it can often be synthesized by carrying out identical transformations at both reactive sites of a symmetrical bifunctional precursor. For example, the dicarboxylic acid, succinic acid, can be prepared from sodium cyanide and dibromoethane as shown in Fig. 9.11.

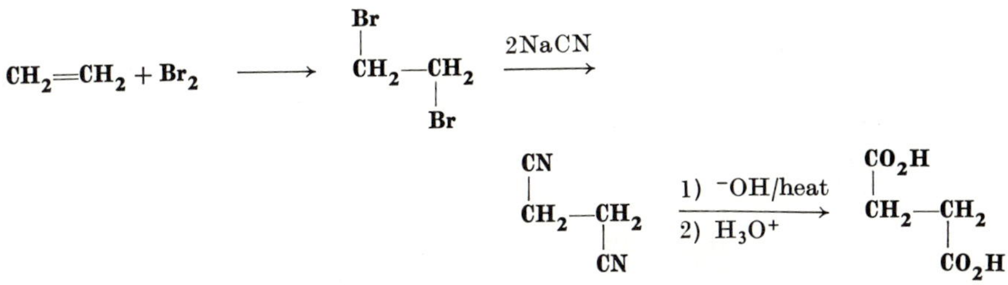

Fig. 9.11 A synthesis of succinic acid

Selectivity in synthesis means that one functional group reacts under a given set of reaction conditions in preference to another. This preference may be the result of a difference in the rate of reaction between two similar functional groups. For example, when the two double bonds of a diene differ in their degree of substitution, the more highly substituted one will react more rapidly with a peracid. When 2,6-dimethyl-2,6-heptadiene is treated with one equivalent of perbenzoic acid the *mono*epoxide shown below is obtained:

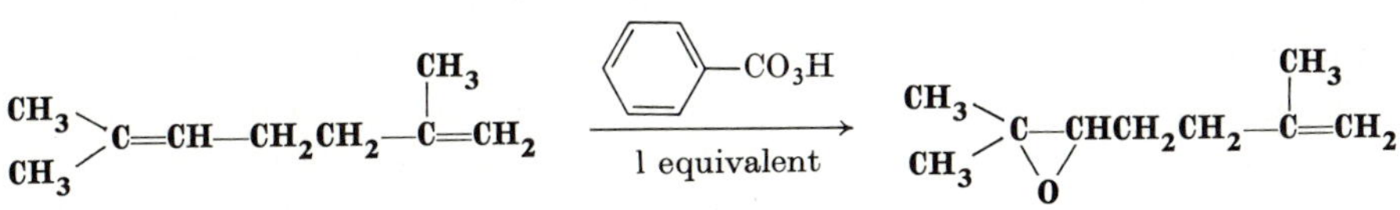

Selectivity also occurs as a result of differences in functional group type. A Grignard reagent, for example, will react with a carbonyl group but not with a carbon-carbon double bond. This type of selectivity will be employed in our synthesis of a bifunctional molecule.

Protection implies that by a selective reaction one functional group is converted into an equivalent function. The latter will be unreactive under the conditions used to perform another necessary transformation elsewhere in the molecule. Examples of this technique will be given in Sections 11.14 and 15.18.

9.23 A SYNTHETIC PLAN FOR A BIFUNCTIONAL MOLECULE

The two functional groups of our bifunctional target molecule, 2-methyl-3-ethylhexane-2,6-diol, are both hydroxyl groups. One is tertiary and the other primary. The carbon skeleton is substituted by two branches, a methyl group, and an ethyl group. Since the molecule contains little symmetry it is unlikely that both functional groups can be introduced simultaneously. As a starting point let us consider how we might try to introduce one functional group in

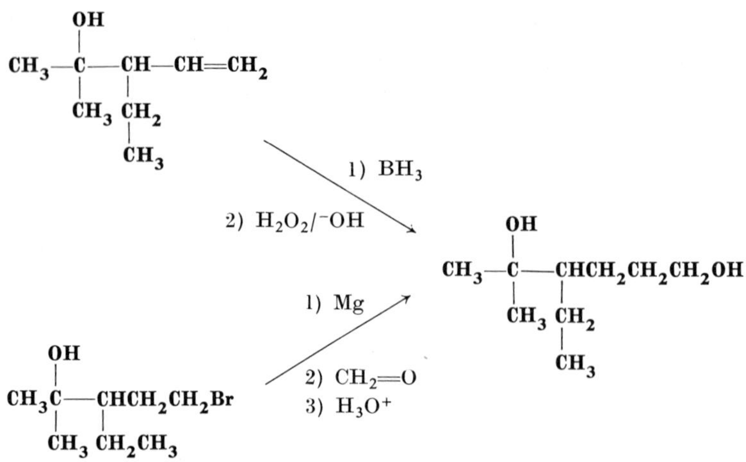

Fig. 9.12. Alternative methods for the preparation of a primary alcohol

the presence of the other. Arbitrarily let us try to introduce a primary alcohol group into a precursor molecule which contains the tertiary group. Figure 9.12 illustrates two possibilities; hydroboration-oxidation of a terminal olefin, and Grignard addition to formaldehyde. The second pathway appears attractive at first glance since it involves both functional group and carbon-carbon bond formation in a single step. It suffers, however, from an error often made in proposed solutions for synthetic problems. A Grignard reagent prepared from a bromo-alcohol would "bite its own tail." Acidic functions like hydroxyl groups react with Grignard reagents:

$$R{-}OH + R'{-}MgBr \longrightarrow R'{-}H + RO^{-}\,\overset{+}{M}gBr$$

Thus the only product we could expect from the first step of the second pathway is:

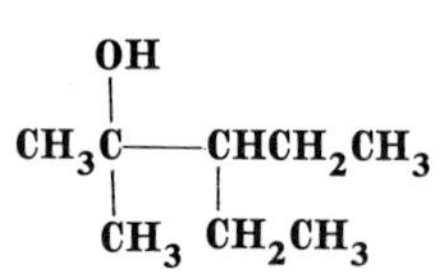

The hydroboration-oxidation reaction of the first pathway (Fig. 9.12) is a selective process. It does not affect hydroxyl groups already existing in the molecule. Now we must consider how to make the olefinic tertiary alcohol required for the hydroboration reaction. Since this alcohol also contains two functional groups we must again consider how we can introduce one functional group in the presence of another. Tertiary alcohols are commonly made by the addition of a Grignard reagent to a ketone. As we have noted earlier this is a selective reaction with regard to carbon-carbon double bonds. Thus the alcohol may be prepared from an unsaturated ketone precursor:

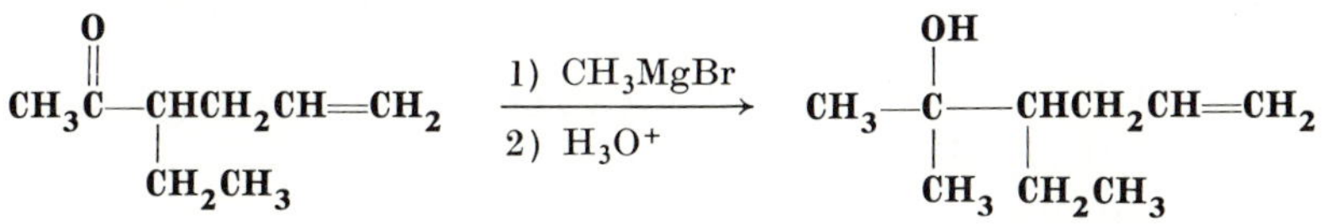

At this stage we have still done little in constructing the carbon skeleton. The olefinic ketone shown above could presumably be made by a Grignard sequence. However, if we realize that this ketone is structurally related to acetoacetic ester, a much more economical approach can be used. The typical structure obtainable from acetoacetic ester is either a methyl substituted ketone or a transformation product of a methyl ketone. Alkylation of acetoacetic ester, therefore, first with ethyl bromide, and then with allyl bromide, followed by hydrolysis and decarboxylation yields our desired intermediate. The preparation of the bifunctional intermediate, allyl bromide is shown in Section 9.7. Figure 9.13 illustrates the complete synthesis.

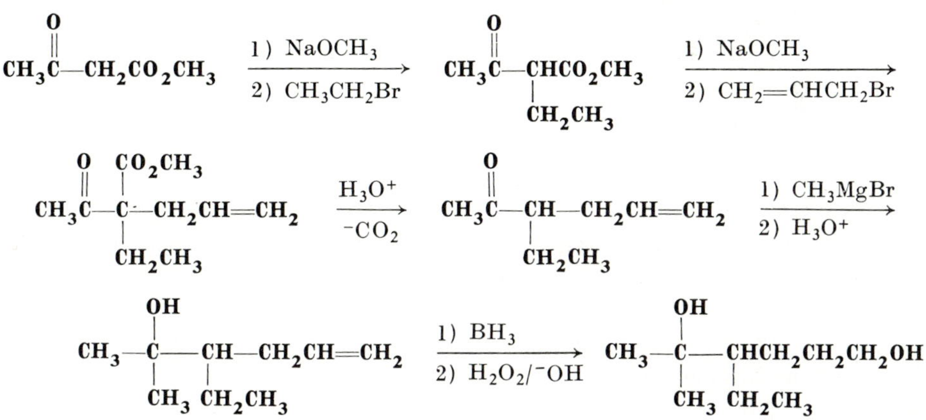

Fig. 9.13 The complete synthesis of a bifunctional molecule

ADDITIONAL READING

W. H. Saunders, Jr., *Ionic Aliphatic Reactions*, Englewood Cliffs, N.J.: Prentice-Hall, 1965. Paperback.

J. March, *Advanced Organic Chemistry: Reactions, Mechanisms, and Structure*, New York: McGraw-Hill, 1968. Chapter 15.

PROBLEMS

1. Fill in the blanks.
   a) An alkene plus bromine gives a vicinal ———.
   b) Treatment of an alkene with BH_3 followed by basic H_2O_2 gives an ———.
   c) The Diels-Alder reaction is a good way to prepare ——— systems.
   d) Addition to the double bond of an α,β-unsaturated ketone is called ——— addition.
   e) Radical addition of hydrogen bromide to a double bond does not occur in a ——— manner.
   f) An alkene plus $KMnO_4$ gives a ———.
   g) The reaction of tetramethylethylene with ozone and zinc gives two moles of ———.
   h) An alkyne may be converted into the corresponding *cis*-alkene using H_2 and ——— catalyst.
   i) Allylic bromination of an alkene is best accomplished with the reagent called ———.
   j) Treatment of an epoxide with aqueous acid gives a glycol with a ——— stereochemistry.
   k) When a cation and an anion are closely associated in solution, the species is called an ———.
   l) In the Diels Alder reaction, a diene reacts with a ——— to form an adduct.
   m) The ——— product refers to the product that is formed the fastest even though it may not be as stable as the ——— product.
   n) Alkene plus ——— gives epoxide.
2. Write the products of the reaction of cyclohexene with the following reagents. In some instances cyclohexene does not react with the reagent, and this should be indicated by "NR."
   a) bromine
   b) hydrogen and platinum
   c) hydroxide ion
   d) N-bromosuccinimide
   e) ozone followed by zinc and water
   f) Lindlar catalyst and hydrogen
   g) diborane followed by basic hydrogen peroxide
   h) osmium tetroxide
   i) peracetic acid
   k) ammonia
   l) sodium bromide

3. A compound C_4H_6O absorbs in the UV at 219 nm. The NMR consists of three peaks of 1, 2, and 3 protons each. Only the 3-proton peak is unsplit. The IR has an intense peak at 1695 cm^{-1}. The compound reacts with dimethylamine to form a new compound which has a strong IR peak at 1710 cm^{-1}. What are the two compounds?

4. Draw an example of:
   a) a dienophile
   b) an ozonide
   c) a glycol
   d) a peracid
   e) a bromonium ion intermediate
   f) a peroxide
   g) an α,β-unsaturated ketone
   h) a hypohalous acid
   i) a brominating agent
   j) an epoxide

5. Complete the following reactions by supplying the structure of the starting material, the principal organic product, or the necessary catalyst.

a) $(CH_3)_2C{=}CHCH_2CH_3 + HBr \longrightarrow$?

b) 

cis

c) 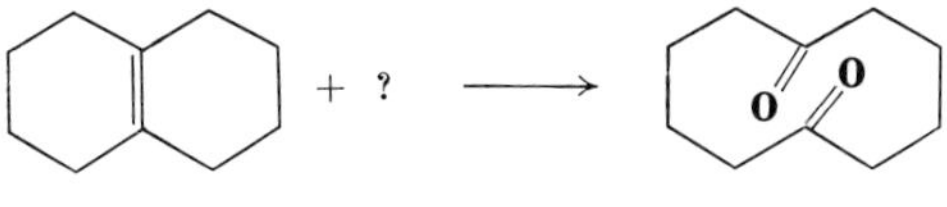

d) $CH_3CH{=}CH\overset{O}{\overset{\|}{C}}CH_3 + CH_3NH_2 \longrightarrow$?

e) 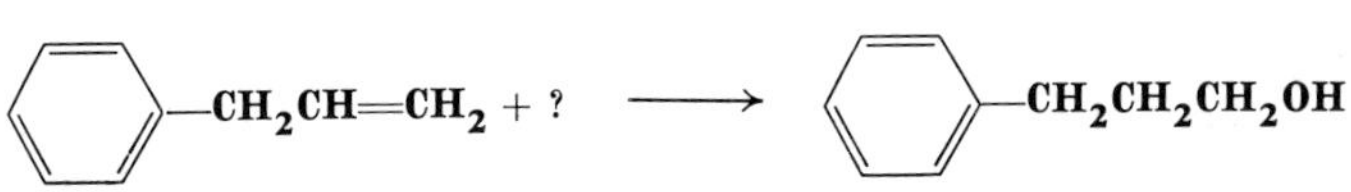

f) 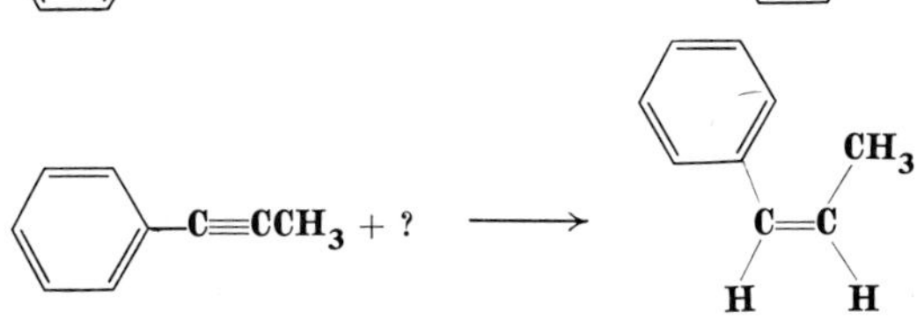

g) 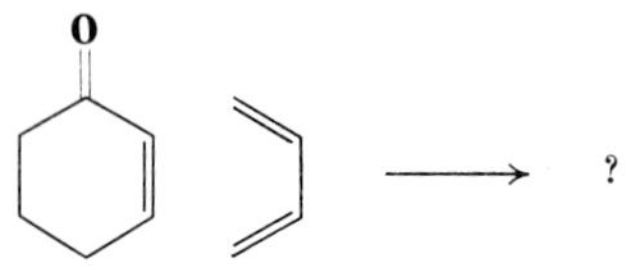

h) $CH_2{=}CH{-}CH_2Br + ? \longrightarrow BrCH_2CH_2CH_2Br$

i) 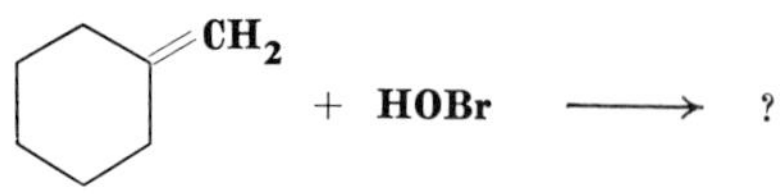

j) $CH_3CH{=}CHCOOCH_3 + ? \longrightarrow BrCH_2CH{=}CHCOOCH_3$

6. Hydration of an alkyne, which is more difficult than hydration of an alkene, requires the presence of a mercuric sulfate catalyst:

$$R{-}C{\equiv}C{-}R \xrightarrow[H_3O^+]{HgSO_4} R{-}\underset{\underset{O}{\|}}{C}{-}CH_2{-}R$$

The hydration product of an alkyne is a ketone. Explain.

7. A trialkyl borane treated with NH_2Cl forms a primary amine:

$$(CH_3CH_2)_3B + 3NH_2Cl \longrightarrow 3CH_3CH_2NH_2$$

Write a mechanism for this reaction which, incidentally, is a good way to convert an alkene into an amine. Ammonia, which is a nucleophile and not an electrophile, does not add directly to a double bond:

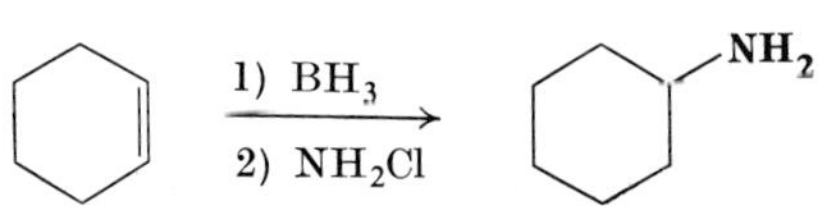

8. Glance over the reactions you have learned in the previous chapters. Then show how you would carry out the following transformations:

a) 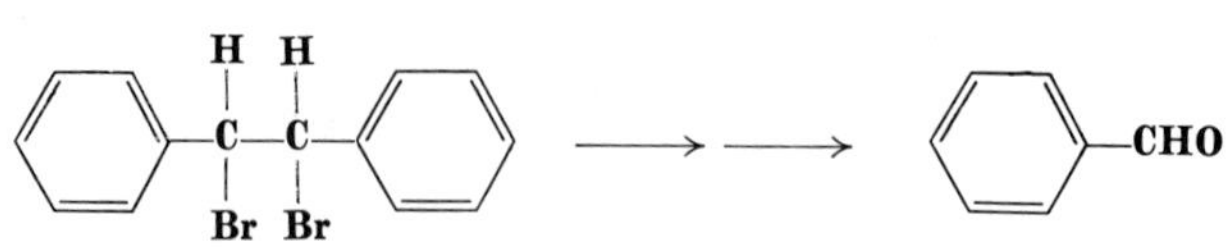

b) 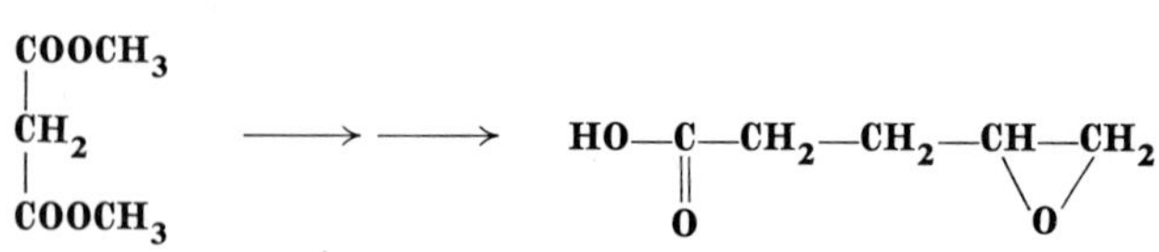

c) 

d) 

e) 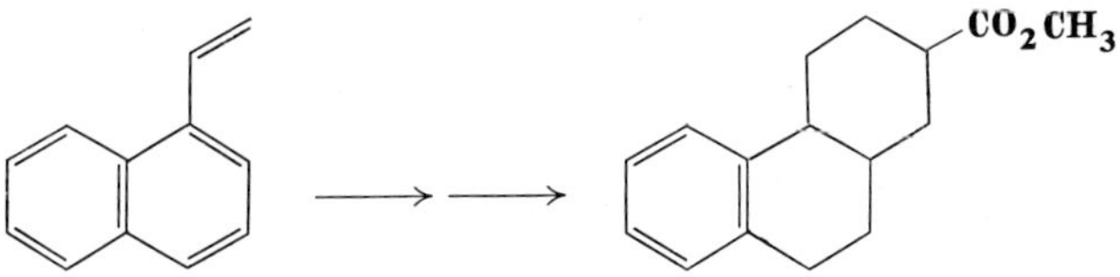

f) 

g) 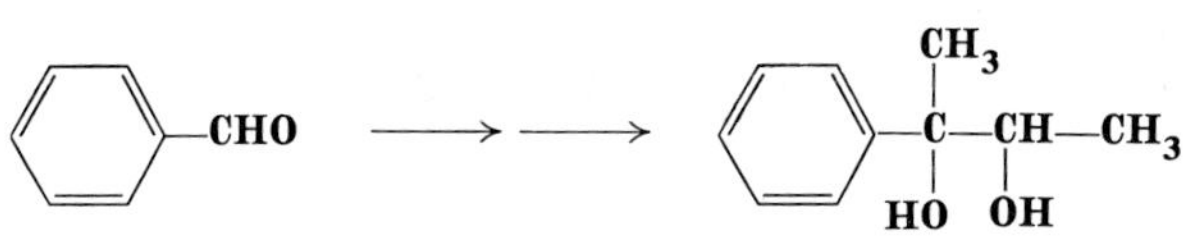

9. Hydrogen bromide adds to conjugated dienes to give both *1,2 addition* and *1,4 addition*:

$$\overset{4}{CH_2}{=}\overset{3}{CH}{-}\overset{2}{CH}{=}\overset{1}{CH_2} + HBr \longrightarrow \underset{\text{1,2 addition}}{CH_2{=}CH{-}\overset{\overset{\displaystyle Br}{|}}{CH}{-}CH_3} + \underset{\text{1,4 addition}}{\overset{\overset{\displaystyle Br}{|}}{CH_2}CH{=}CHCH_3}$$

Assume that the additions pass through a cyclic onium ion involving a proton and one of the double bonds (see intermediate 2 in Fig. 9.1). Show how both products could arise from such an intermediate. The 1,2 addition product forms predominantly at low temperatures, whereas at higher temperatures the 1,4 addition product is the major isomer. Which product is more stable? Why does the 1,2 product form faster than the 1,4 product?

10. From the information given below suggest plausible structures for compounds A through H shown on p. 262.

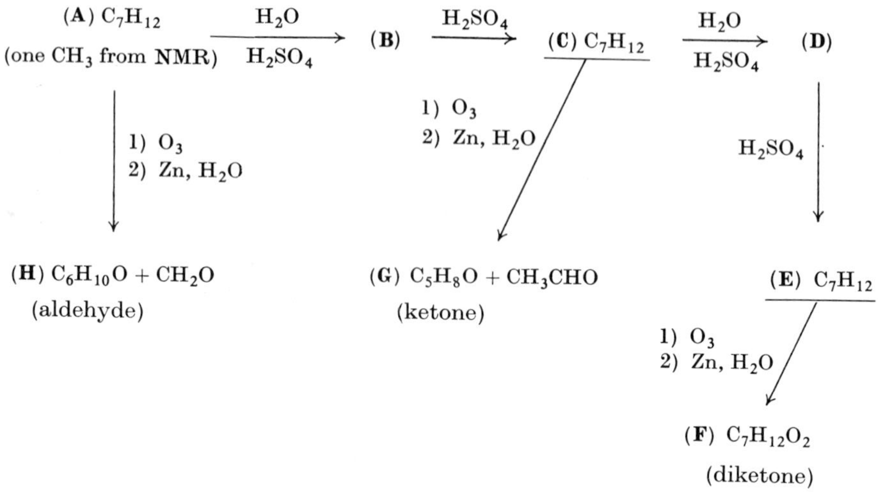

11. Give the major product of the following reactions. Indicate the correct stereochemistry when more than one isomer is possible.

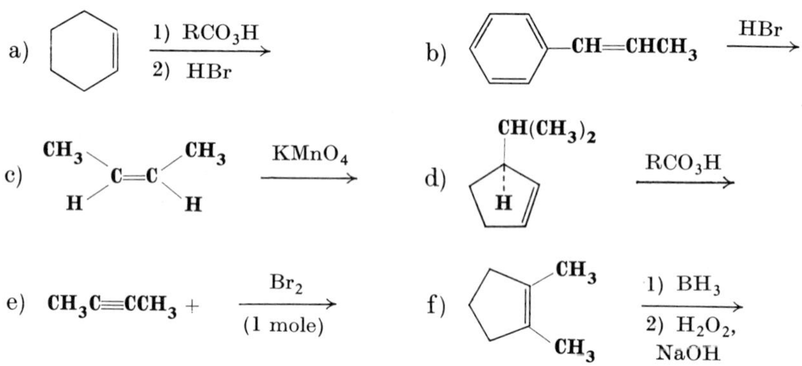

12. Reaction of Cl_2 with *cis*-1-phenylpropene gives 25% *trans* and 75% *cis* addition:

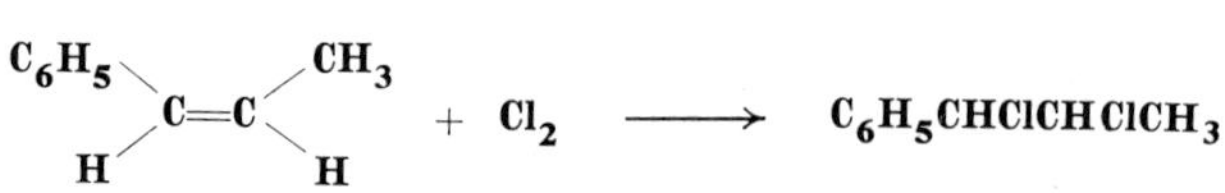

Reaction of Br_2 with the same olefin gives 83% *trans* and 17% *cis* addition. In terms of the mechanism in Fig. 9.1, why does bromine yield so much more *trans* addition than chlorine? Remember that chlorine is more electronegative than bromine.

13. One possible mechanism for the acid-catalyzed hydration of an olefin is a fast reversible carbonium ion formation followed by a slow rate-determining addition of a water molecule:

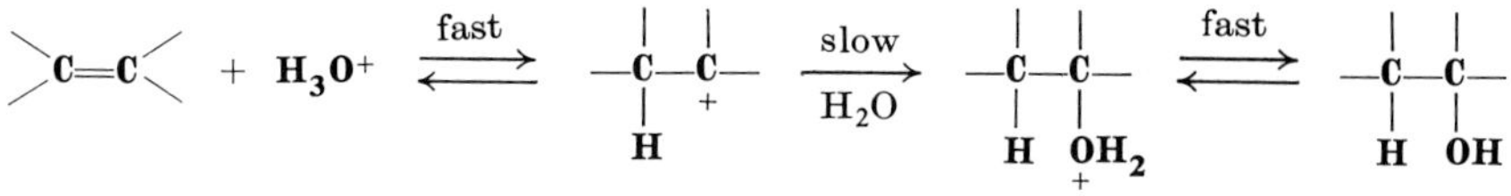

When 2-methyl-2-butene is hydrated in D_2O and the olefin recovered after 10–30% of the reaction has been completed, no deuterium is found in the olefin. Is this result consistent with the above mechanism? Explain.

14. *Polymerization* refers to a chemical reaction in which many small molecules join to form a substance of high molecular weight, called a *polymer*. Olefins are commonly polymerized by means of a free radical chain reaction. For example, vinyl chloride plus a small amount of peroxide forms polyvinyl chloride, a plastic that is used for phonograph records and many other items:

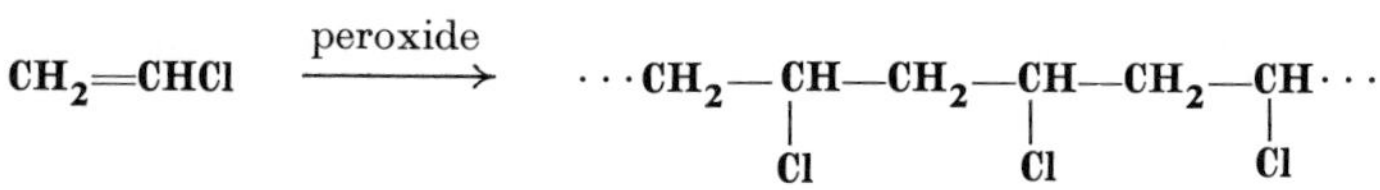

vinyl chloride polyvinyl chloride

Write a mechanism for this polymerization starting with a hydroxy radical.

Extra credit: It is found that polymers often have more desirable properties (higher melting points, etc.) if the chains are joined or *cross-linked*. Suggest a way to prepare polystyrene so that the chains are occasionally cross-linked to form a two-dimensional molecular network.

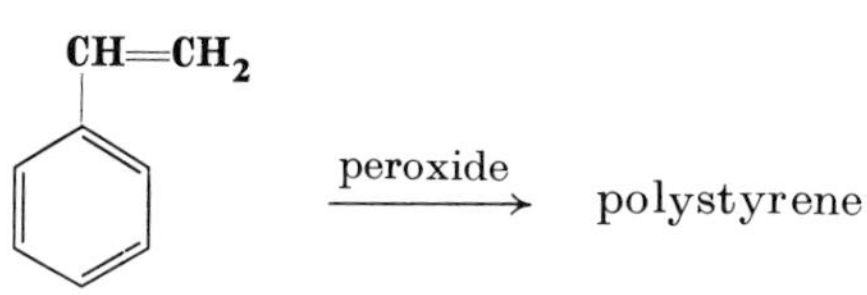

styrene

15. There is an important enzyme (biological catalyst) which promotes the hydration of fumaric acid to malic acid:

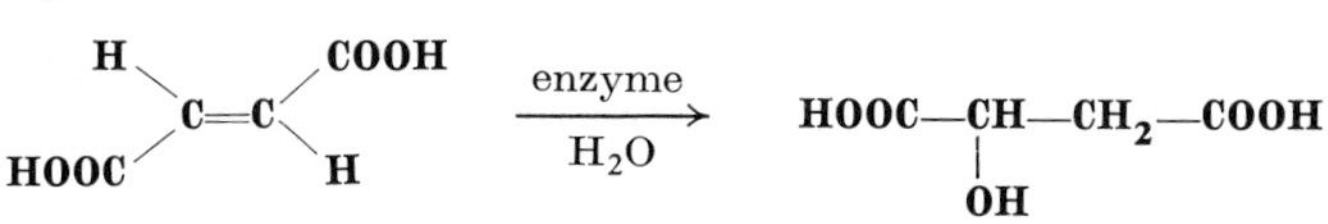

fumaric acid malic acid

When the "fumarase reaction" was run in deuterium oxide (D_2O), a deuterated malic acid was isolated:

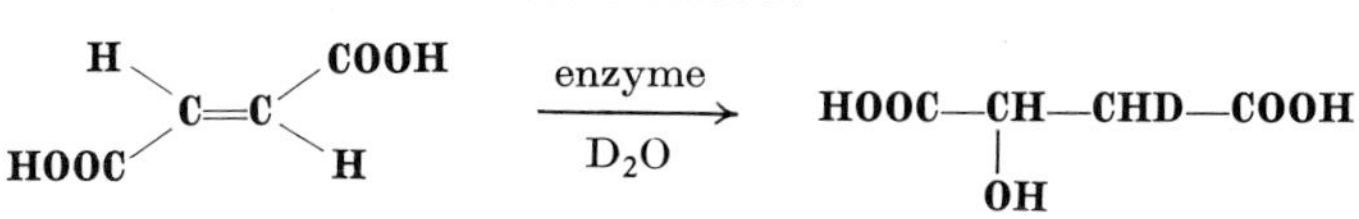

$J = 7.3$ Hz

The NMR coupling constant, J, between the two central protons was found to be 7.3 Hz. Monodeuterated malic acid was also prepared non-enzymatically by the following route:

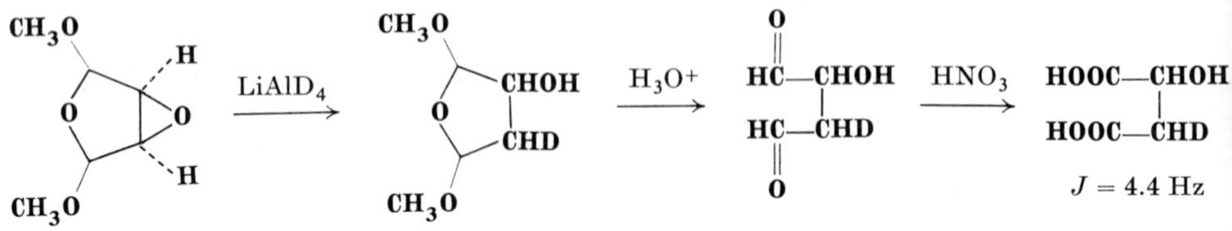

The product had a coupling constant of 4.4 Hz, distinctly different from that of the enzymatically produced malic acid. Does the enzyme-catalyzed hydration of fumaric acid exhibit a *cis* or a *trans* stereochemistry? Explain.

16. The butenolide structure occurs in a number of antibiotics obtained from molds:

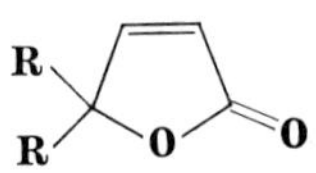

This system can be synthesized by the following route:

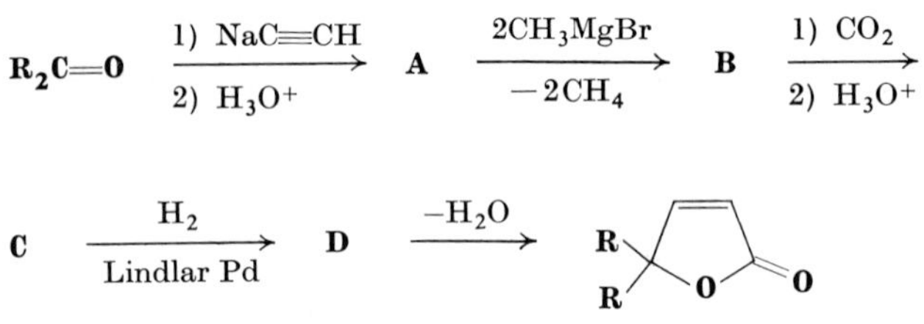

What are the structures of A, B, C, and D?

17. Cantharidin (Spanish fly) has the following structure:

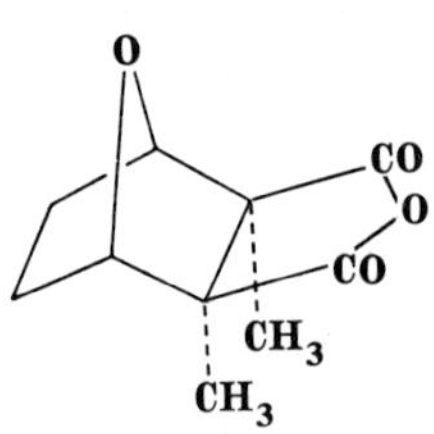

Propose a synthesis for this compound.

Note: Do not try to carry out in the laboratory the synthesis you will undoubtedly propose, because it will not work. F. von Bruchhausen and H. W. Bersch in Münster were disappointed to find this out in 1928. A perfectly good "paper synthesis" frequently does not work out in practice.

It will be a little while yet before you are in a position to understand the 10-step synthesis that was eventually required to make cantharidin.

18. Isobutylene undergoes an acid catalyzed self addition:

$$(CH_3)_2C{=}CH_2 \xrightarrow{85\%\ H_2SO_4} (CH_3)_3CCH_2\underset{\displaystyle CH_3}{\underset{|}{C}}{=}CH_2 + (CH_3)_3CCH{=}C(CH_3)_2$$

Write a mechanism for this reaction.

19. Homopentaprismane has been synthesized by the following route:

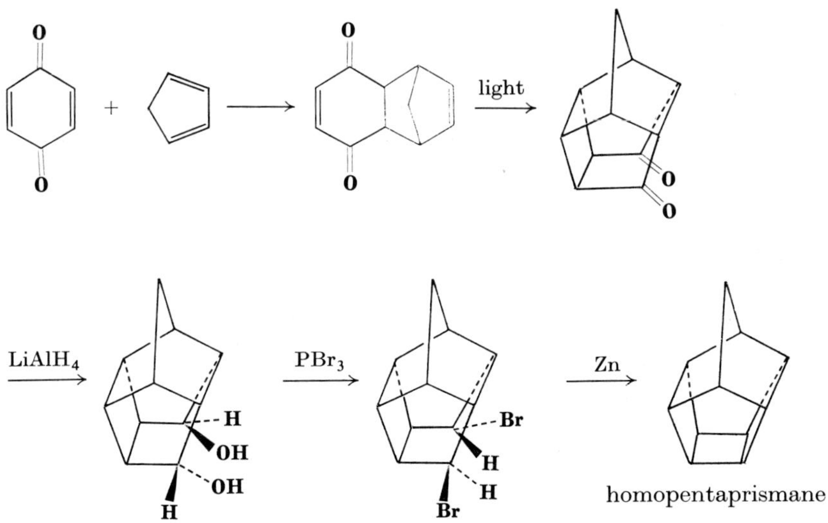

a) Construct the Diels-Alder adduct with molecular models if they are available. Satisfy yourself that the product of the photochemical reaction is an isomer of the adduct.

b) In the $LiAlH_4$ reaction, the only glycol formed is that shown (in which the two hydroxyl groups are directed toward rather than away from each other). Explain.

c) Does the PBr_3 reaction go with inversion?

d) Write a mechanism for the last step of the sequence. Note the similarity to the elimination reaction described in Section 8.10.

10

AROMATIC SUBSTITUTION REACTIONS

10.1 INTRODUCTION

This chapter discusses the synthetic and mechanistic chemistry of aromatic systems. The subject is more than an intellectual exercise; the ability to manipulate aromatic rings has relieved hunger and disease and substantially upgraded our standard of living. A few familiar commercial aromatic compounds are illustrated in Fig. 10.1.

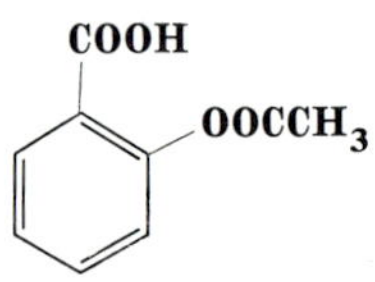

aspirin

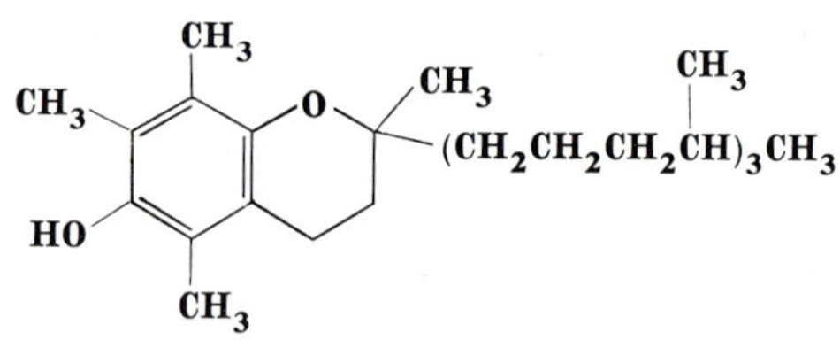

vitamin E

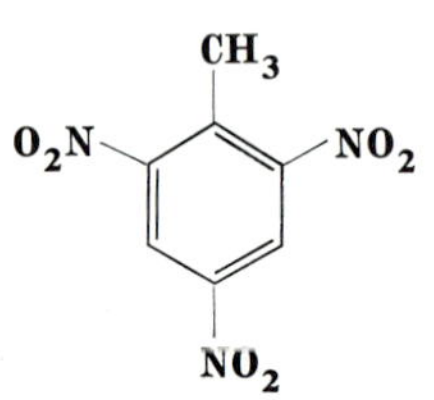

TNT

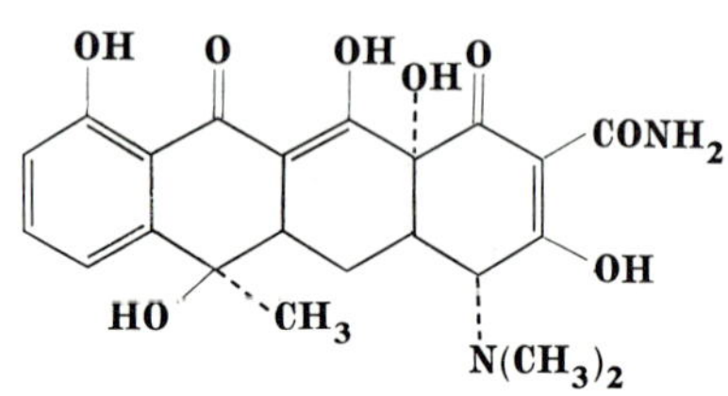

tetracycline
(an antibiotic)

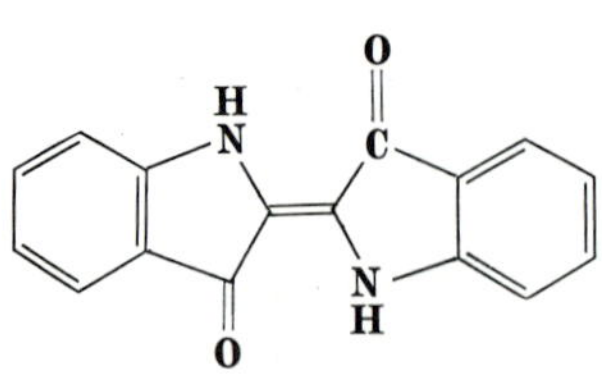

indigo

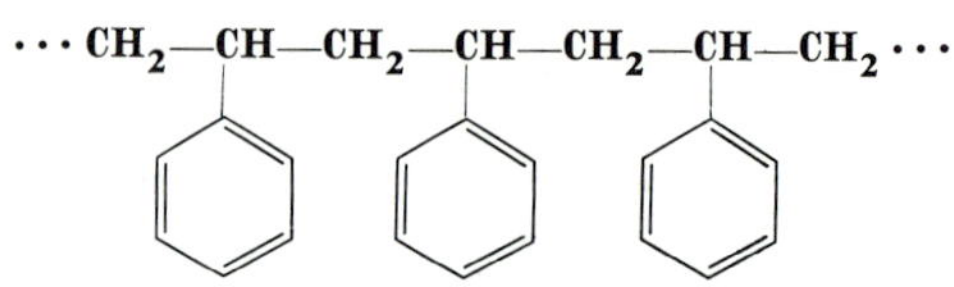

polystyrene

Fig. 10.1 Some commercial aromatic compounds

Unfortunately, not all of these compounds have been used exclusively for humanitarian purposes. So far, no one has found a way to limit the application of scientific discoveries to the improvement of this planet.

10.2 AROMATICITY

Benzene and related compounds possess special properties which are ascribed to its *aromaticity* or *aromatic character* (see Section 2.3). Three of these properties are cited below:

1) Benzene displays unusual thermodynamic stability. For example, the heat of hydrogenation of benzene (*i.e.*, the heat evolved when benzene is hydrogenated to cyclohexane in the presence of a metal catalyst) is 49.8 kcal/mole:

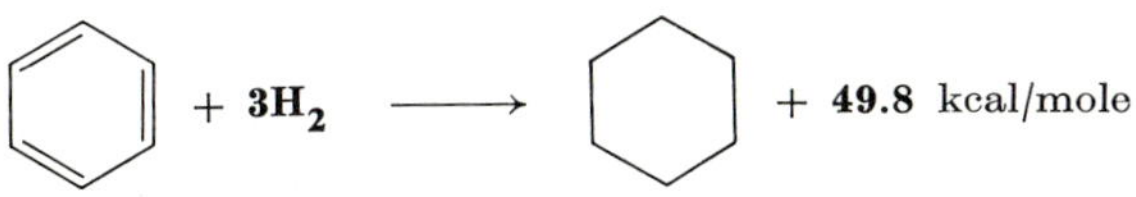

This value is less than would be expected in the absence of resonance, considering that the heat of hydrogenation for cyclohexene is 28.6 kcal/mole:

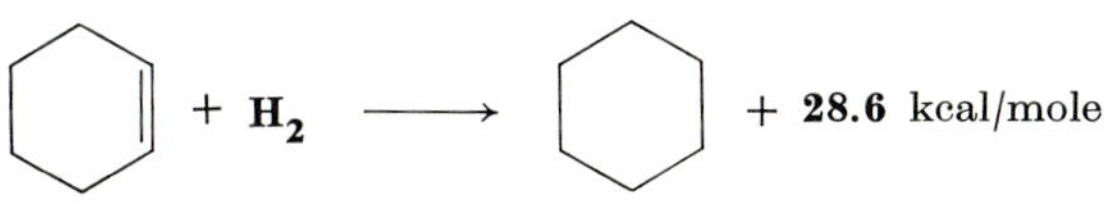

If benzene were merely a six-membered ring with three separate noninteracting double bonds, we would have expected benzene to have a heat of hydrogenation roughly three times that of cyclohexene, namely 85.8 kcal/mole. The difference between the hypothetical heat of hydrogenation of a "nonresonating" benzene and that actually found for benzene (85.8 and 49.8 kcal/mole, respectively) equals 36 kcal/mole. This latter value is called the *resonance energy* of benzene. Benzene is stabilized by resonance to the extent of 36 kcal/mole.

2) The NMR signal of the benzene proton appears at 7.3 ppm (see Table 6.3) which is considerably downfield from that of olefinic protons (about 5 ppm). The π electrons of benzene "circulate" in the applied magnetic field thereby creating a small local field which adds to the applied field at the site of the protons. Therefore a smaller applied field is required to induce resonance of the benzene protons. This unusual downfield NMR shift is frequently used as a criterion for aromaticity.

3) Benzene is generally less reactive than simple alkenes. For example, benzene is inert to bromine under conditions in which alkenes add bromine rapidly. As we shall see in the next section, benzene reacts with bromine only in the presence of an iron catalyst, and even so the process is a *substitution* and not an addition.

Hückel predicted that aromatic character in monocyclic systems would be found whenever there are $(4n + 2)$ π resonating electrons ($n = 0, 1, 2, 3 \ldots$). Thus, benzene with its six π electrons has a Hückel number of electrons ($n = 1$) and is, of course, aromatic. Tropilium cation (six π electrons) is stable, whereas the corresponding anion (eight π electrons) is not:

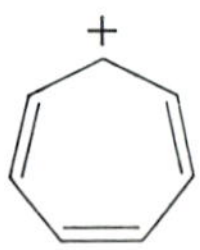

aromatic

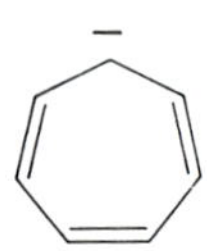

non-aromatic

Note that the stabilities of the cation and the anion could not be predicted merely by drawing resonance contributors (each has seven identical contributors). Cyclobutadiene (four π electrons) is extremely unstable, whereas the so-called [18]-annulene (eighteen π electrons) appears to be aromatic from its NMR spectrum:

Cyclobutadiene
non-aromatic

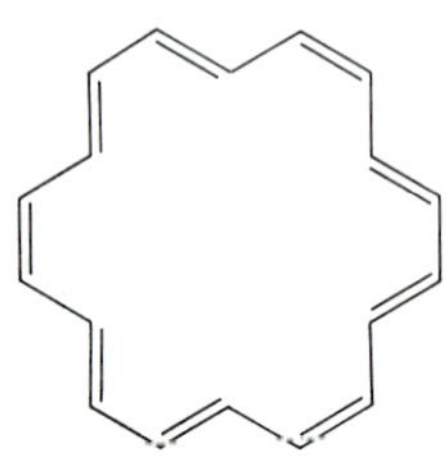

[18]-Annulene
aromatic

10.3 AROMATIC SUBSTITUTION

By far the most common aromatic reaction is *electrophilic aromatic substitution*. In this reaction, examples of which are illustrated below, a hydrogen of an aromatic ring is replaced by an atom or group of atoms:

1) 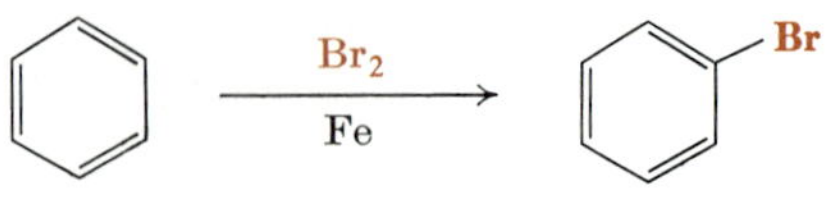

2) 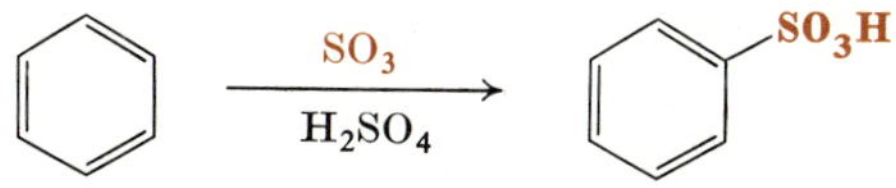

3) 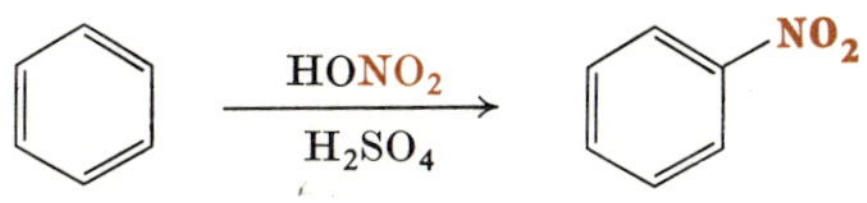

4) 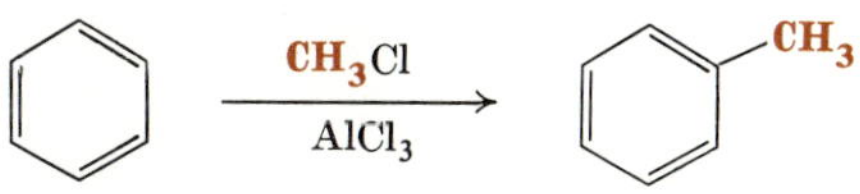

5) 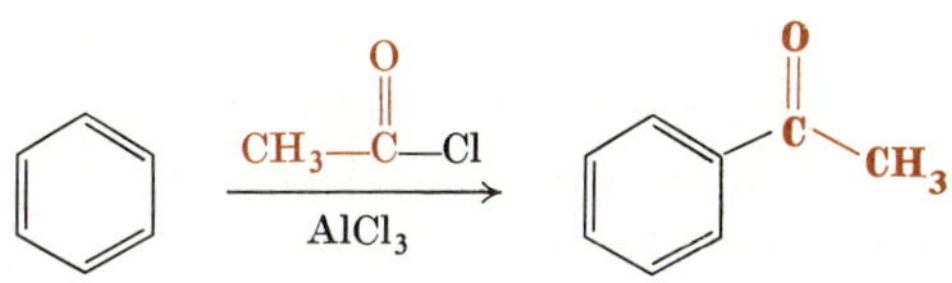

These substitution reactions are called *electrophilic* because, as we shall presently see, the species attacking the aromatic rings are electron deficient. It is not surprising that aromatic rings, with their electron-rich π clouds, are subject to attack by electrophiles (H^+, Br^+, NO_2^+) rather than by nucleophiles (NH_3, OH^-, Br^-). A nucleophile will react with an aromatic system only under unusual circumstances (when the aromatic ring is highly substituted with electron-withdrawing groups). Electrophilic substitution, on the other hand, occurs readily and is a standard way to prepare substituted aromatic compounds.

10.4 MECHANISM

We shall first discuss the mechanism of *nitration* of aromatic rings, as illustrated by the conversion of benzene into nitrobenzene:

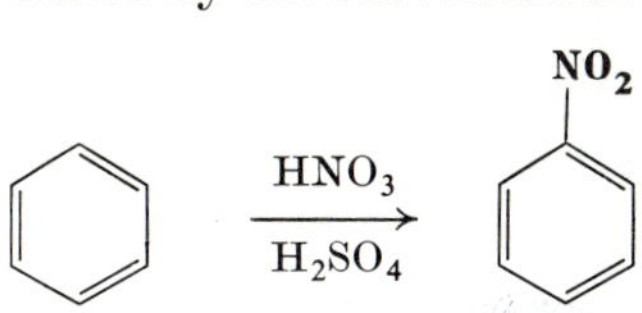

nitrobenzene

This reaction can be carried out with good yield by adding benzene to a mixture of fuming nitric acid and concentrated sulfuric acid at 10–15°. The nitrating agent in such a mixture is believed to be nitronium ion (NO_2^+)

which is formed in the following manner:

$$HNO_3 + 2H_2SO_4 \longrightarrow NO_2^+ + H_3O^+ + 2HSO_4^-$$

The mechanism of this reaction involves protonation of nitric acid by sulfuric acid followed by loss of water to form a nitronium ion. The water is subsequently protonated by the sulfuric acid:

$$HO{-}NO_2 + H_2SO_4 \xrightarrow{-HSO_4} H{-}\overset{+}{\underset{H}{O}}{-}NO_2 \xrightarrow{-H_2O} NO_2^+$$

$$H_2O + H_2SO_4 \longrightarrow H_3O^+ + HSO_4^-$$

We know that nitronium fluoroborate ($NO_2^+\ BF_4^-$), an isolable salt, will also nitrate benzene and this supports our hypothesis that nitronium ion is the attacking electrophile in aromatic nitrations. We can imagine two simple mechanisms by which nitronium ion could effect substitution:

a) Direct displacement:

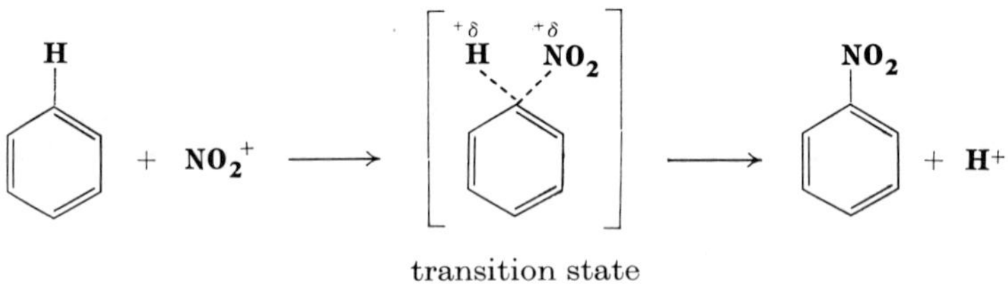

b) Addition-elimination:

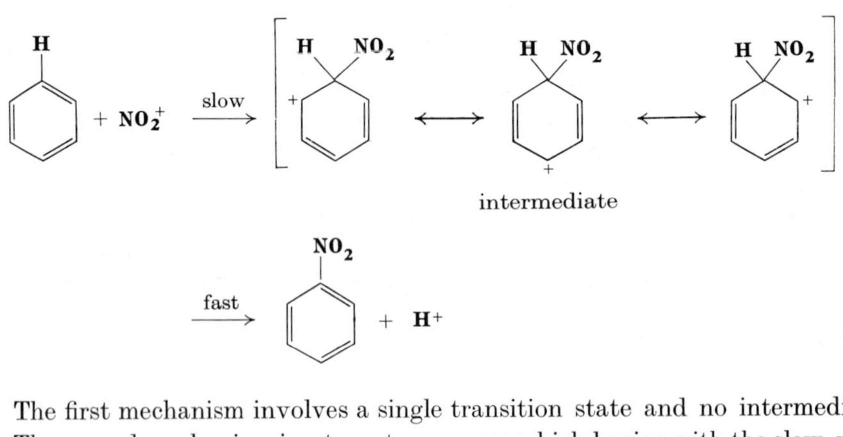

The first mechanism involves a single transition state and no intermediate. The second mechanism is a two-step process which begins with the slow, rate-determining addition of nitronium ion to the benzene ring, producing a resonance-stabilized intermediate. This is followed by the fast ejection of a proton, yielding the product (nitrobenzene). Since deuterated benzene (C_6D_6) reacts at the same rate as benzene we know that the second mechanism is in fact correct. If the direct displacement mechanism were correct, the deuterated benzene would react more slowly because it is more difficult to break a carbon-deuterium bond than a carbon-hydrogen bond. The absence of an isotope effect, on the other hand, is consistent with a rate-determining addition of nitronium ion in which there is no carbon-hydrogen cleavage.

Rate studies never really prove a mechanism; they may, however, provide conclusive evidence that a mechanism is *incorrect*. Thus, the absence of an isotope effect noted above disproves the direct displacement mechanism. If it is valid to assume that there are two *a priori* reasonable mechanisms, then we can accept the alternative mechanism.

Most electrophilic aromatic substitutions entail an addition intermediate called a σ *complex* (a name derived from the fact that a new σ bond forms):

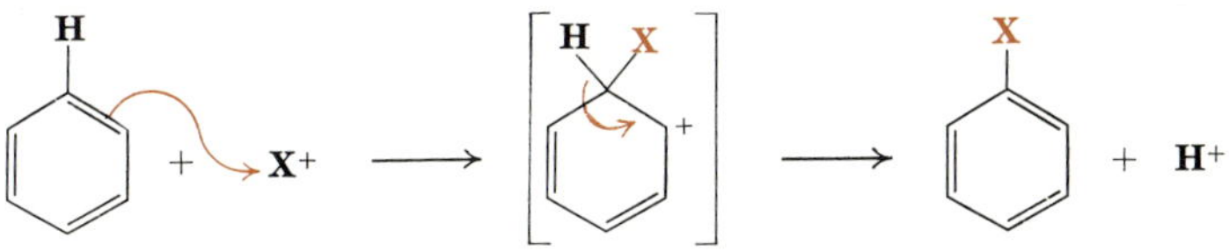

σ complex

In certain cases the σ complex can be directly detected. For example, when benzene is added to a mixture of HF and BF_3 at low temperatures,

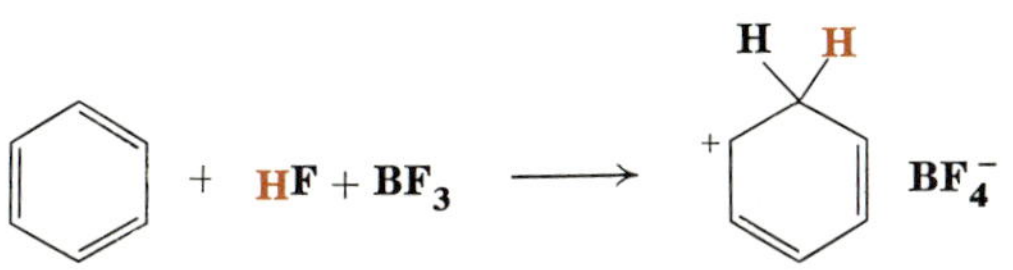

the solution becomes highly colored and electrically conducting, as would be expected if a σ complex were forming. While this does not mean that σ complexes are necessarily involved in nitrations and other aromatic substitutions, it is comforting to know that the postulated σ complex intermediate is a chemically observable species.

Bromination of aromatic systems occurs in the presence of bromine and an iron catalyst. For example, nitrobenzene may be brominated to *m*-bromonitrobenzene by adding iron powder and dry bromine to nitrobenzene at 135°.

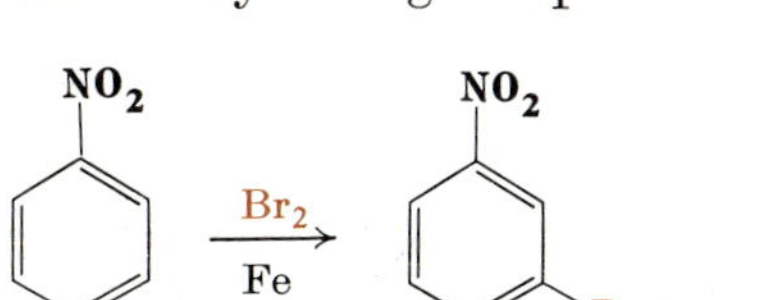

m-bromonitrobenzene

(The reason the bromine bonds to the *meta* position—see Section 4.8—will be explained later.) Iron must be present because bromine is too poor an electrophile to react with nitrobenzene by itself. Under the reaction conditions, iron is converted into ferric bromide, a Lewis acid which complexes with bromine. Complexation enhances the electrophilic nature of the bromine:

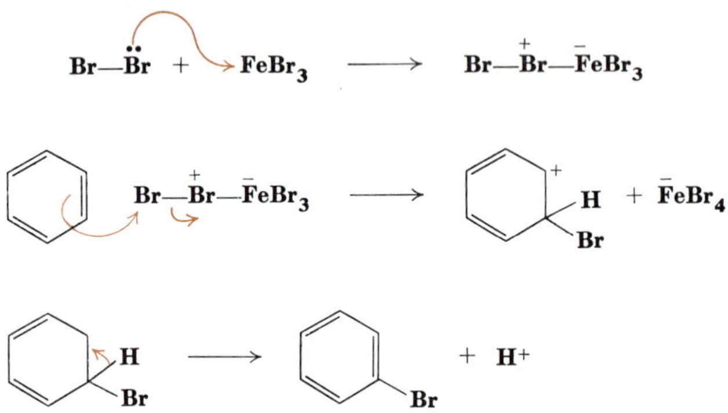

The second step may be viewed as a "push-pull" process: the push by the benzene is complemented by the pull of the ferric bromide. Certain aromatic systems, such as naphthalene, are so reactive that no iron catalyst is necessary:

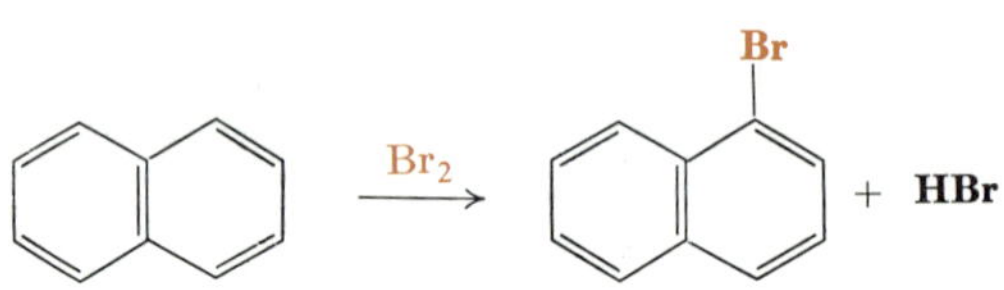

Aromatic systems can be both alkylated and acylated by means of the important *Friedel-Crafts reaction.* For example, benzene is converted into toluene by methyl bromide and a Lewis acid catalyst such as aluminum bromide:

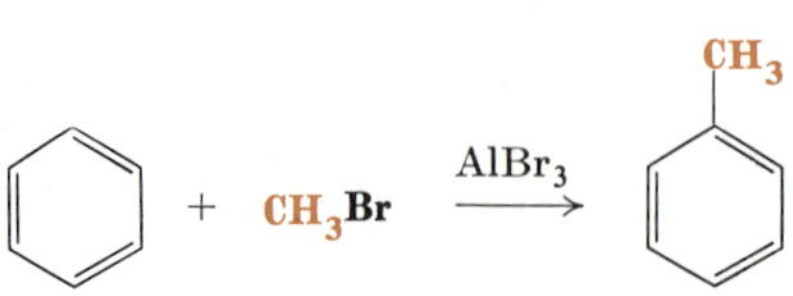

toluene

There are two possible mechanisms for this reaction which differ only in the nature of the attacking electrophile.

1) $CH_3Br + AlBr_3 \rightleftharpoons CH_3^+ + AlBr_4^-$

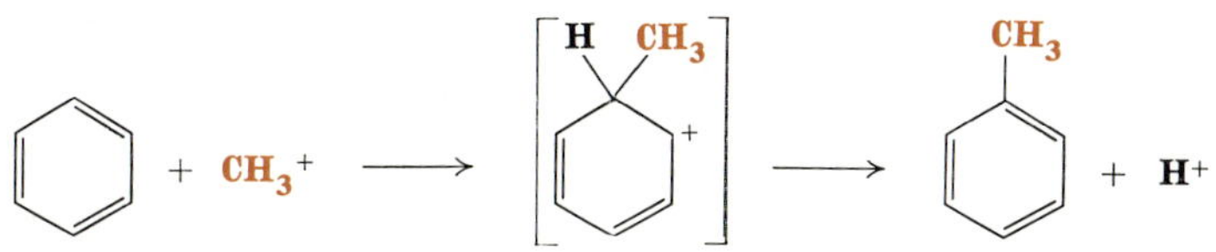

2) $CH_3Br + AlBr_3 \rightleftharpoons CH_3\text{—}\overset{+}{Br}\text{—}\overset{-}{Al}Br_3$

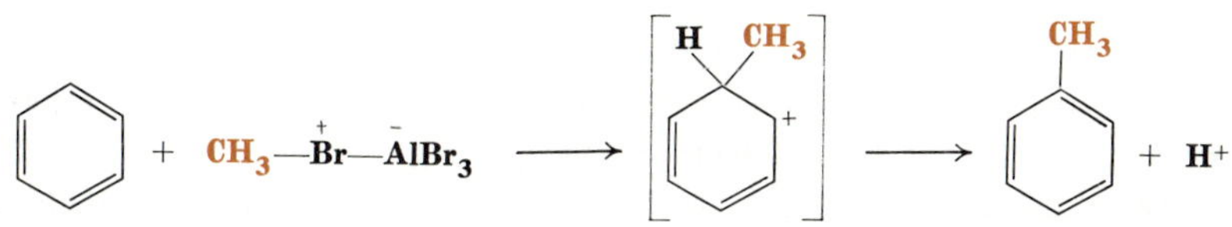

In the first mechanism a carbonium ion attacks the benzene ring. In the second mechanism, the electrophile is a complex between methyl bromide and aluminum bromide. The following experiment verifies the validity of the second mechanism. Toluene methylated with methyl iodide and methyl bromide yields a mixture of *ortho*, *meta*, and *para* isomers:

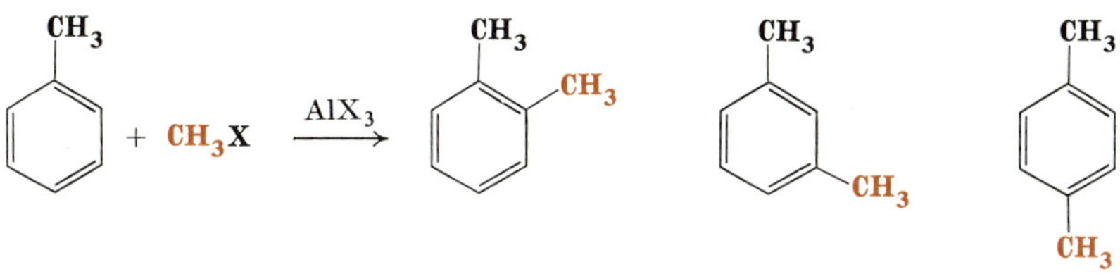

ortho *meta* *para*

Methyl iodide produces 49% *ortho*, 11% *meta*, and 40% *para* isomers; methyl bromide yields 54% *ortho*, 17% *meta*, and 29% *para* isomers. We will explain the ratio of products in the next section, but for the moment we need emphasize only that methyl bromide and methyl iodide do *not* give the same product mixture. If the methyl carbonium ion mechanism were correct, the two systems would give identical amounts of each product because both reactions involve the *same* methyl carbonium ion intermediate. Since methyl carbonium ion is very unstable compared to the more highly-alkylated carbonium ions, the experimental results are not very surprising. A discrete carbonium ion is, however, believed to be important in the alkylation of aromatic compounds by tertiary halides such as *t*-butyl chloride. A carbonium ion is probably also formed in Friedel-Crafts acylations in which aromatic rings are substituted with acyl groups (RCO—):

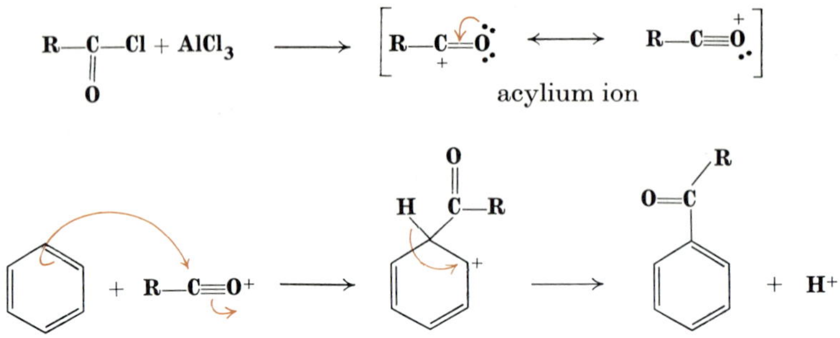

The reaction between an acid chloride (RCOCl) and aluminum chloride produces a special type of resonance-stabilized carbonium ion, called an *acylium ion*, which acts as the electrophilic species.

For the sake of completeness, it is worth mentioning that certain electrophilic substitutions may entail an additional step, namely association of the electrophile with the aromatic ring prior to the actual addition:

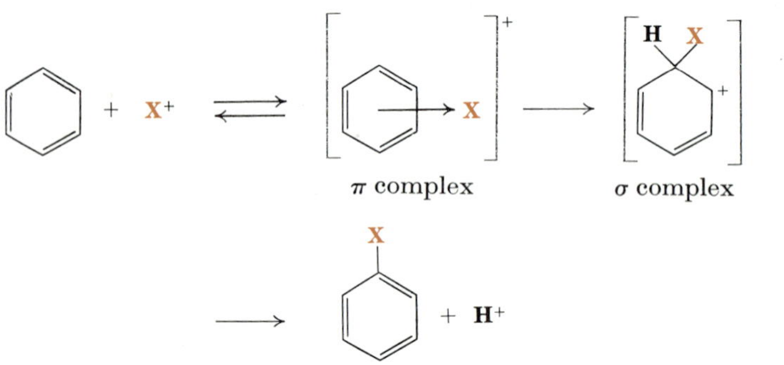

The initial noncovalent complex, called a π *complex*, is reminiscent of the complex formed in electrophilic additions to simple double bonds (Section 9.2).

10.5 DIRECTING EFFECTS

A problem arises when substitution occurs on a benzene ring that already bears a substituent. In such a compound there are three nonequivalent positions on the ring (*ortho*, *meta*, and *para*; see Section 4.8). It is therefore important to be able to predict the predominant substitution site(s) so that one can design rational syntheses of polysubstituted aromatic compounds. Fortunately, this is usually not a difficult task.

There are three classes of substituents:

1) *Meta* directing and deactivating. Examples:

nitro	$—NO_2$
ammonium	$—\overset{+}{N}R_3$
nitrile	—CN
aldehyde	—CHO

2) *Ortho-para* directing and activating. Examples:

hydroxyl	—OH
alkoxyl	—OR
alkyl	—R
phenyl	$—C_6H_5$
amino	$—NH_2$

3) *Ortho-para* directing and deactivating. Examples:

halogens	—Cl, —Br, —I

Electrophilic substitution on an aromatic compound which is monosubstituted with a *meta*-directing substituent yields a *meta*-disubstituted product. An *ortho-para*-directing substituent gives a mixture of *ortho-* and *para*-disubstituted compounds. Thus, bromination of nitrobenzene produces mainly *m*-bromonitrobenzene. Bromination of toluene gives a mixture of *o*-bromotoluene and *p*-bromotoluene:

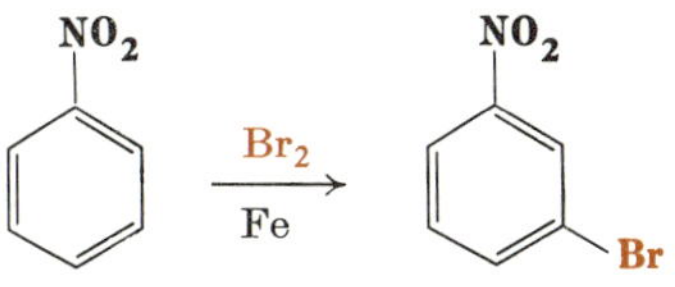

meta

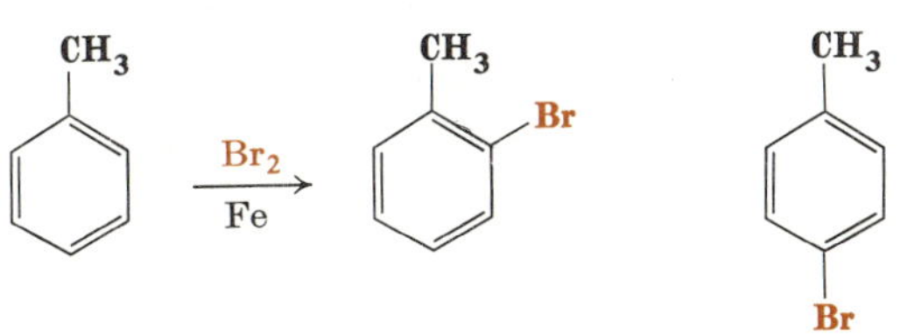

ortho *para*

A nitro group is not only *meta* directing but also deactivating. This means that the substituted benzene reacts more slowly than benzene itself. Similarly, an activating group enhances the reaction rate of the substituted benzene ring. Thus, toluene brominates much faster than nitrobenzene.

The *meta*-directing and deactivating effects of the nitro, carbonyl, and nitrile groups have a straightforward explanation. These groups are all electron withdrawing. For example:

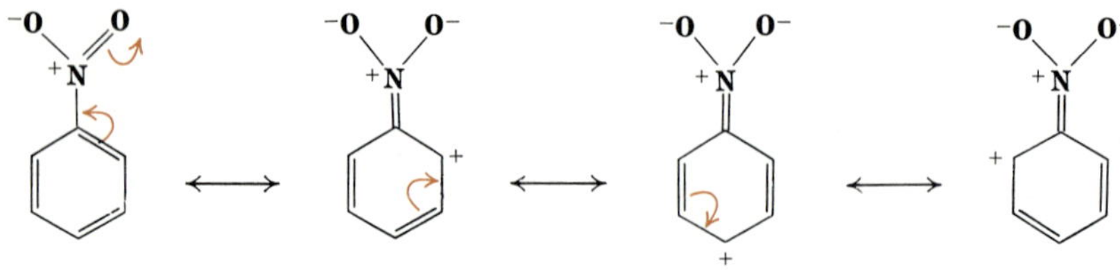

A nitro group would, therefore, be expected to *destabilize* a σ complex intermediate as well as the transition state leading to the complex (which are both positively charged and electron deficient):

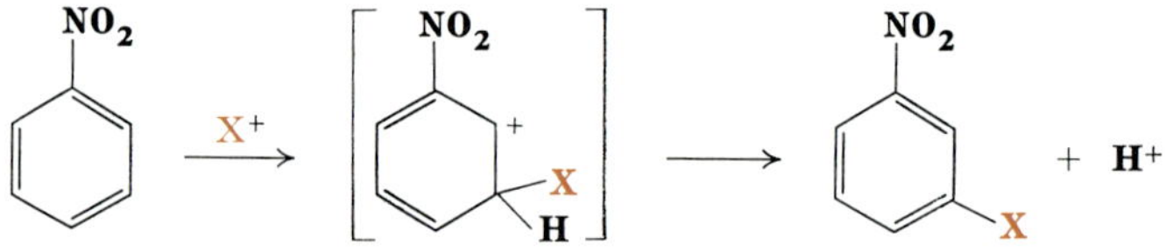

The energy diagrams for the reactions of benzene and nitrobenzene are shown in Fig. 10.2. The energies of benzene and nitrobenzene are arbitrarily placed at the same level. Clearly, the activation energy for required substitution on nitrobenzene is much greater than that for benzene. In other words, the nitro group is deactivating.

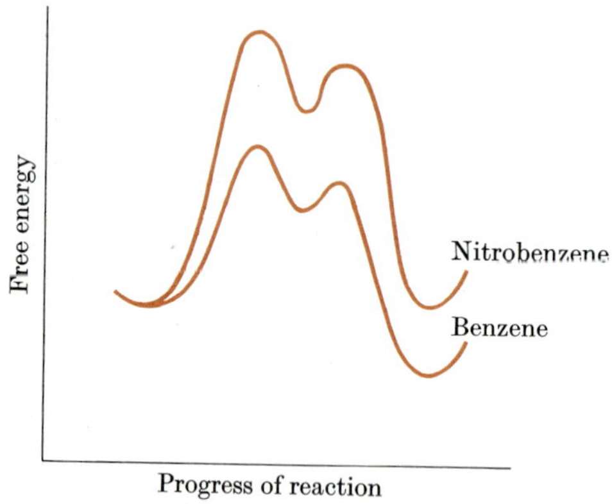

Fig. 10.2 Energy diagrams for reaction of benzene and nitrobenzene

All positions of nitrobenzene are deactivated relative to benzene. Electrophiles add to the *meta* position because the *meta* position is deactivated less than the *ortho* and *para* positions. We can see this by comparing the resonance contributors of the intermediates formed by *meta* and *para* addition (Fig. 10.3).

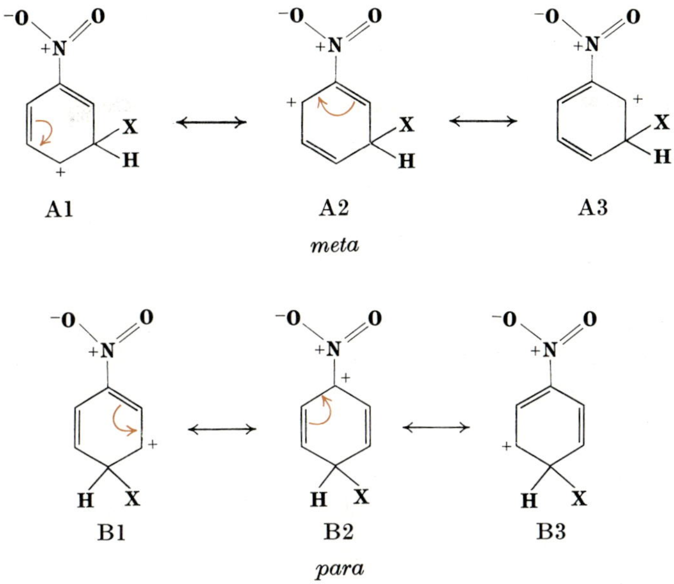

Fig. 10.3 Resonance contributions to the intermediate formed during substitution at *meta* and *para* positions of nitrobenzene

Three resonance extremes contribute to each of the intermediates. Resonance contributor B2 is of relatively high energy compared to the other structures, due to the repulsive interaction of the positive charges on two adjacent atoms. Consequently, the *meta*-addition intermediate, with three resonance contributors of comparable energy, is more stable than the *para*-addition intermediate with two forms of comparable energy and one of much greater energy. If we make the reasonable assumption that the transition state leading to a σ complex resembles the σ complex, then we must conclude that *meta* substitution has the lower transition state energy and therefore predominates. Note that the ground state energies for *meta* and *para* addition are identical and need not be considered in the above comparison of activation energies.

The hydroxyl, amino, and methyl groups are examples of *ortho-para*-directing and activating substituents. Figure 10.4 shows the resonance contributors for the complexes formed by *para* and *meta* addition to phenol. (Whatever resonance interactions occur for the *para* complex are also possible for the *ortho*; we have arbitrarily chosen to illustrate the *para* system). The major difference between the *meta* and *para* complexes is that the positive charge can be transferred to the oxygen (contributor A3) in the *para* complex but not in the *meta* complex. Contributor A3 is of lower energy than any of the other resonance structures because all the atoms contain a full outer shell

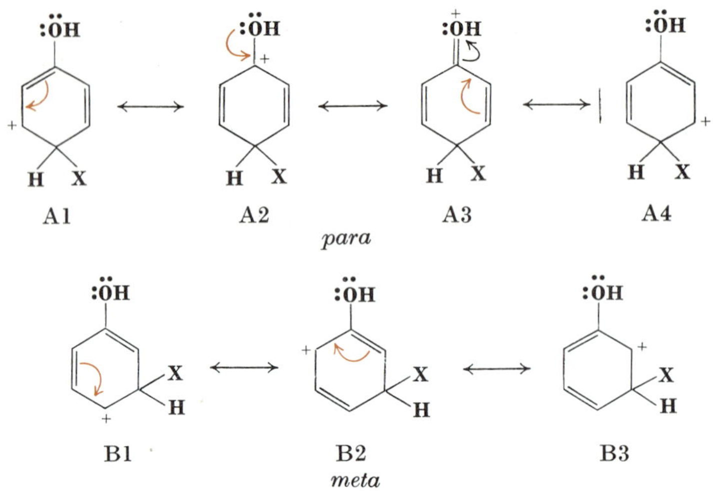

Fig. 10.4 Resonance contributions to the intermediate formed during substitution at *meta* and *para* positions of phenol

of electrons. Thus, the *para*- (and *ortho*-) addition complex is more stable than the *meta* complex. The same is true of the transition states leading to the complexes, and therefore *ortho* and *para* addition predominate. (The ground states for *ortho*, *meta*, and *para* addition are, of course, identical since they all involve phenol and an electrophile.)

The enhanced reactivity of phenol compared to benzene can be attributed to resonance stabilization of the positive charge on the transition state by the electron-donating hydroxyl group. Hydroxyl, amino, alkyl, and other activating groups are all electron donors and can therefore stabilize carbonium ions.

The halogens are exceptions to the rule that electron-donating groups are *ortho-para* directing and activating, while electron-withdrawing groups are *meta* directing and deactivating. Halogens are *ortho-para* directing, but deactivating. We can rationalize this anomaly by invoking both resonance and inductive effects. If an unshared pair of electrons on the halogen is available for resonance stabilization (similar to that in phenol), preferred *ortho-para* addition results. Yet halogens are less basic than oxygen, and less prone to donate an electron pair for carbonium ion stabilization. The resonance effect of halogens is apparently not sufficient to compensate for their electron-withdrawing inductive effect which, of course, deactivates the ring.

One final point must be made about the relative amount of *ortho* and *para* material formed by reaction of an aromatic ring substituted with an *ortho-para*-directing group. Usually both isomers are formed. If either the *ortho-para*-directing substituent or the electrophile is bulky, then *para* substitution is favored for steric reasons. Thus, Friedel-Crafts reactions tend to go primarily *para* in *ortho-para*-directed substitutions. Nitration of *t*-butylbenzene gives considerably more *para*-substituted material than nitration of toluene under identical conditions:

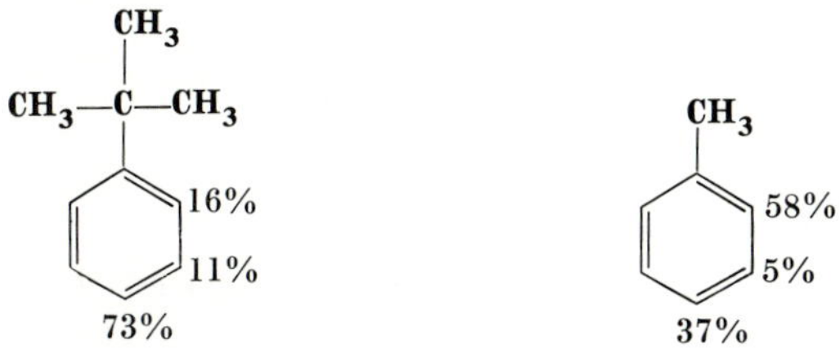

APPLICATIONS OF ELECTROPHILIC SUBSTITUTIONS

10.6 HALOGENATION

Chlorobenzene and bromobenzene can be easily prepared by treating benzene with chlorine and bromine, respectively, in the presence of a Lewis acid catalyst:

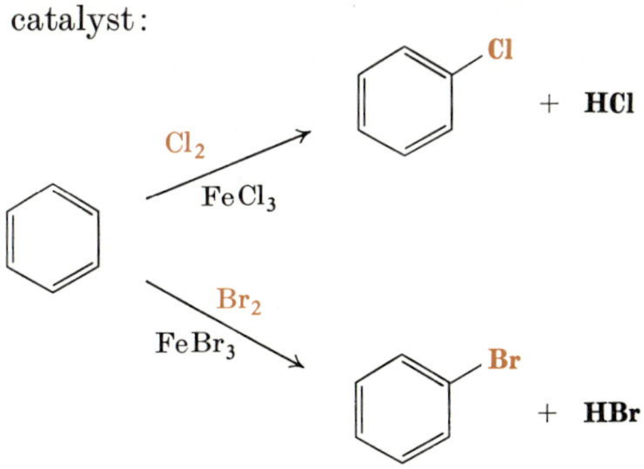

Preparation of fluorobenzene and iodobenzene requires special procedures which will be discussed later.

Bromobenzene is useful for preparing aromatic derivatives via the Grignard reaction:

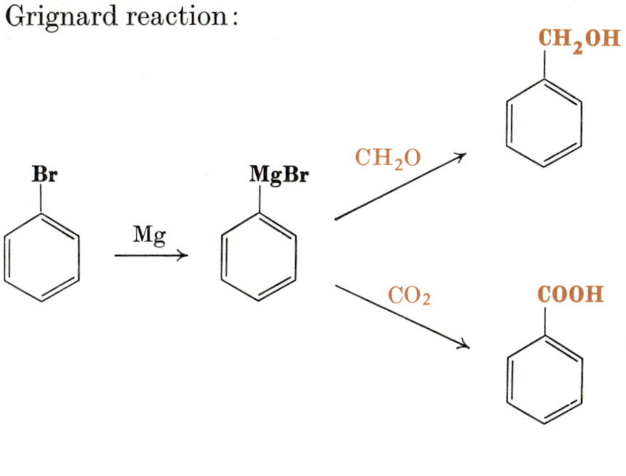

Aryl amines and phenols are so reactive toward electrophilic aromatic substitution that multiple halogenation often occurs:

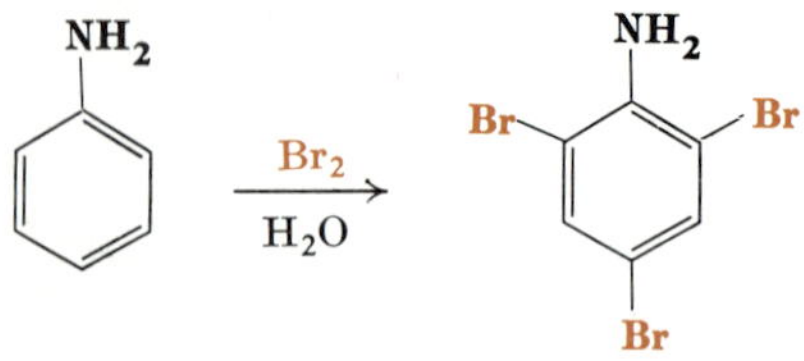

aniline 2,4,6-tribromoaniline

In order to prepare *p*-bromoaniline, aniline must first be acylated with acetyl chloride:

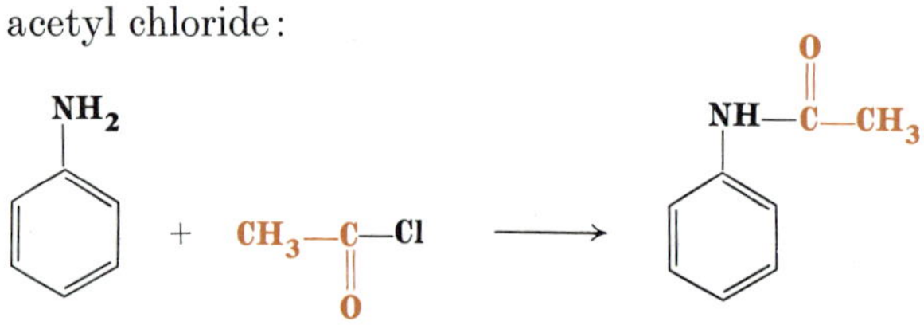

aniline acetyl chloride acetanilide

The resulting acetanilide is much less activated because the unshared pair of electrons on the nitrogen interacts with the carbonyl of the amide group and is thus less available for resonance interaction with the benzene ring.

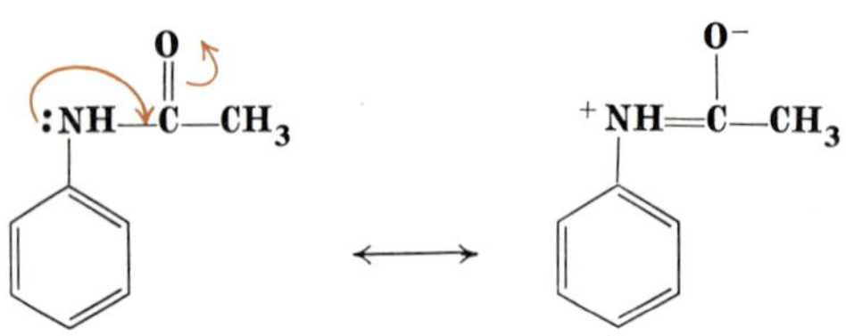

Nevertheless, the amide group is still *ortho-para* directing and activating:

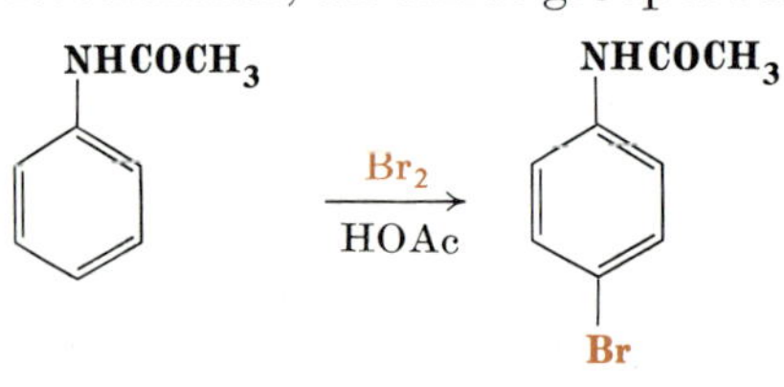

The amino group is reformed by hydrolysis of the amide:

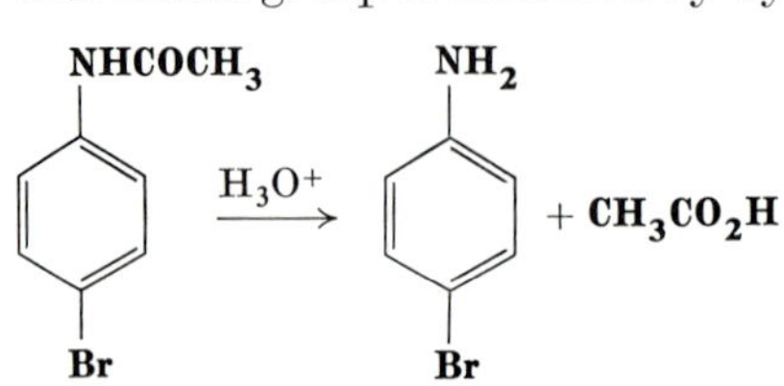

p-bromoaniline

The same procedure is used for the bromination of phenols.

When the above three-step sequence is carried out with *p*-toluidine, the product is 2-bromo-4-methylaniline:

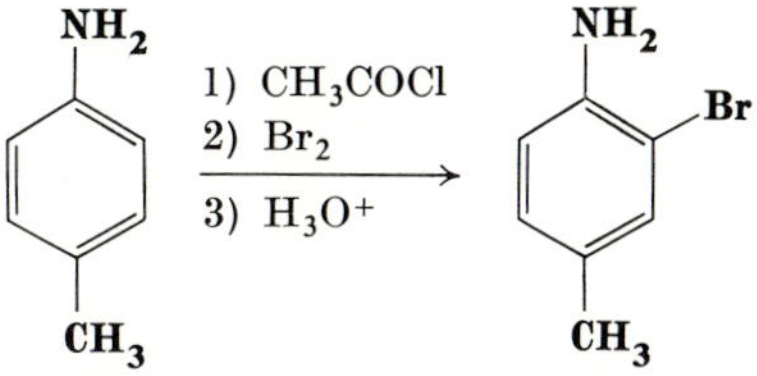

The amide group is obviously a more powerful *ortho-para* director than the methyl group.

Bromination of naphthalene gives mainly 1-bromonaphthalene:

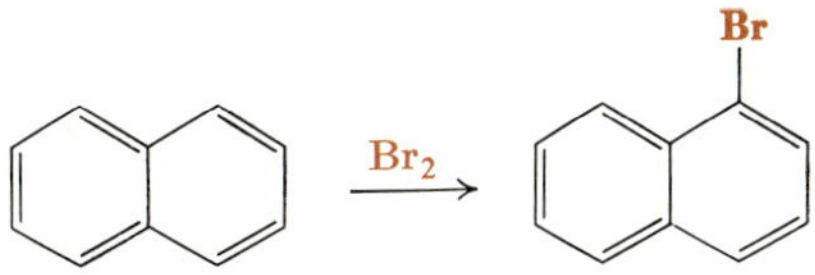

naphthalene 1-bromonaphthalene

The reason 1-substitution is favored over 2-substitution may be that only the 1-addition intermediate can be resonance stabilized without disturbing the benzenoid character of the second ring:

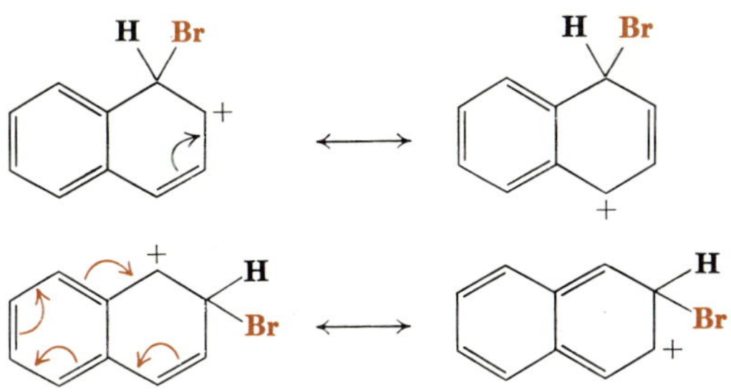

10.7 SULFONATION

Sulfonation, the substitution of a sulfonic acid or —SO_3H group, is carried out by heating an aromatic compound in fuming sulfuric acid (concentrated sulfuric acid to which sulfur trioxide has been added). The sulfonating agent is probably sulfur trioxide:

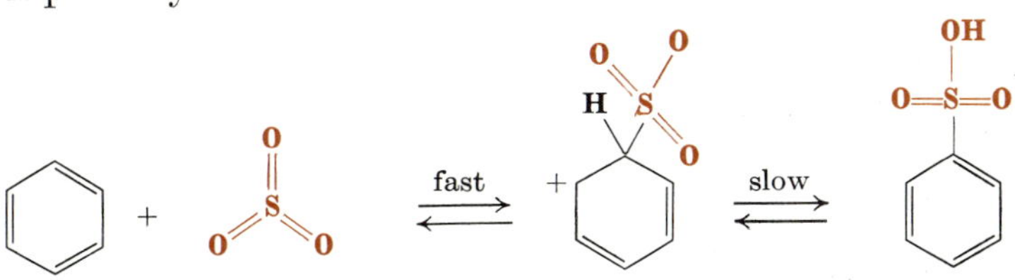

sulfur trioxide benzenesulfonic acid

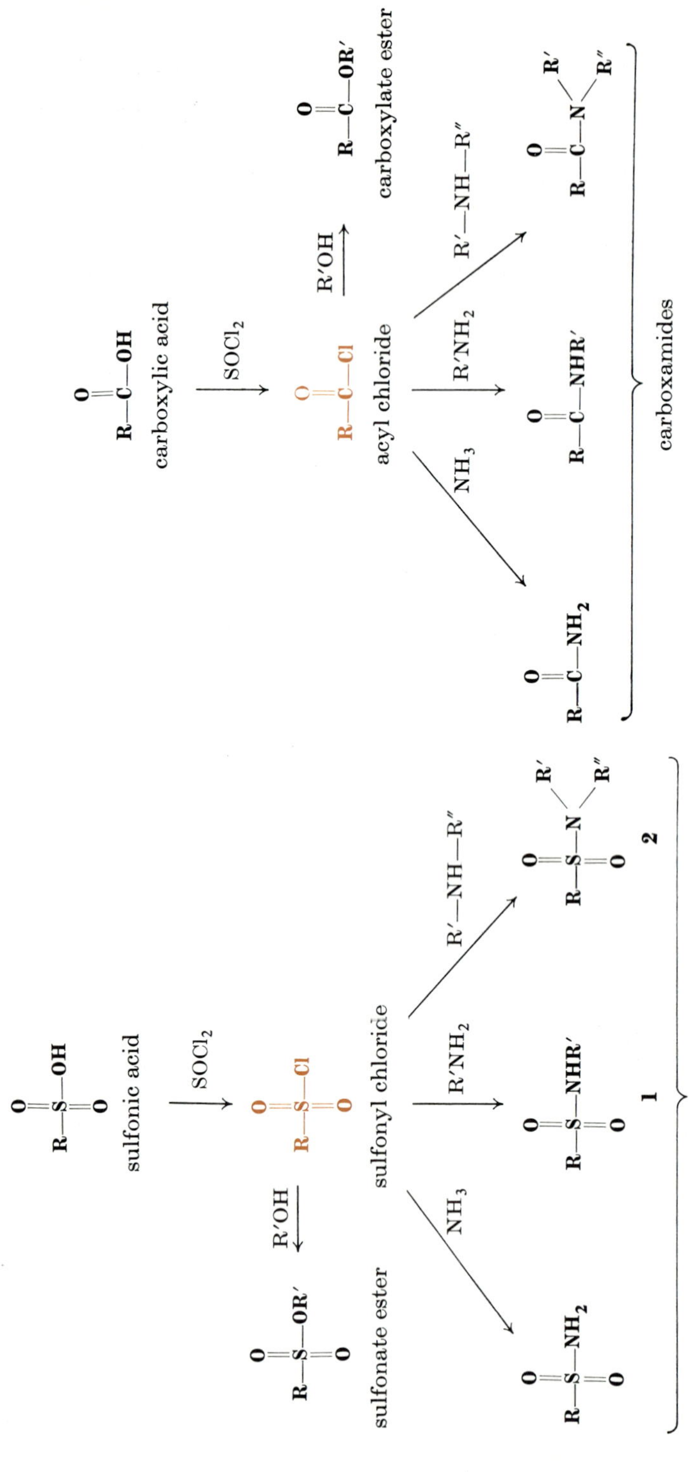

Fig. 10.5 Reactions of sulfonic acids and carboxylic acids

The product, a sulfonic acid, is strongly acidic (pK_a ≈ 1). Interestingly, there is evidence that formation of the σ complex is fast and reversible (the rate-determining step being the collapse of the intermediate). This conclusion is based, in part, on the observation that benzene reacts faster than deuterobenzene (C_6D_6).

Sulfonic acids can be converted into sulfonyl chlorides with thionyl chloride (Fig. 10.5). Sulfonyl chlorides can in turn be converted into sulfonamides and sulfonate esters. These reactions all have their counterpart in the carboxylic acid series (Fig. 10.5 and Chapter 11).

The sulfonation of naphthalene is interesting from a theoretical point of view. When naphthalene is sulfonated at low temperatures (80°), the product is 1-naphthalenesulfonic acid. At higher reaction temperatures (160°), the major product is 2-naphthalenesulfonic acid.

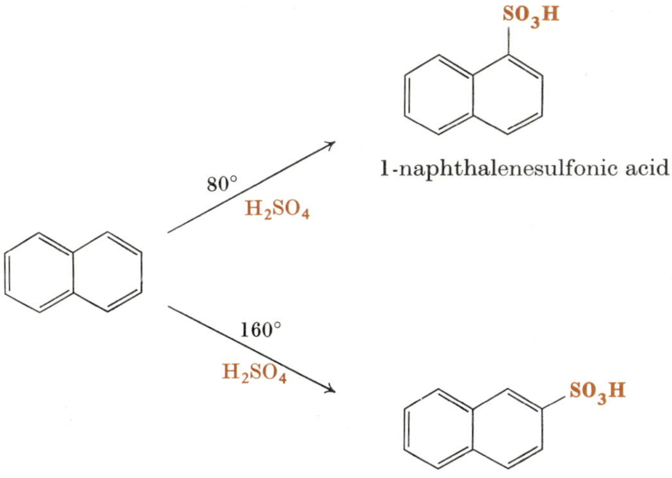

The formation of 1-naphthalenesulfonic acid proceeds faster than the formation of the other isomer for reasons explained earlier in connection with the bromination of naphthalene. The 1-isomer is therefore the *kinetic product* (see Section 9.13). Even though it is formed faster, 1-naphthalenesulfonic acid is less stable than 2-naphthalenesulfonic acid (the *thermodynamic product*), because of the 1–8 steric interaction:

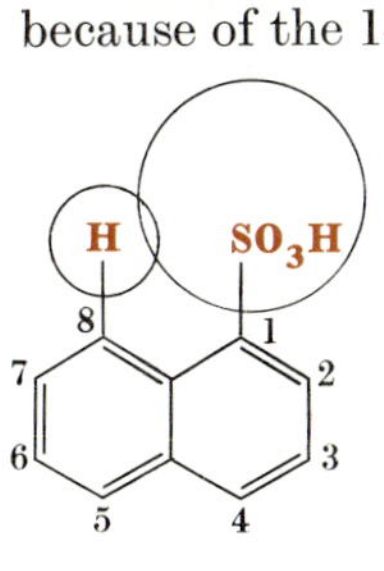

Thus, when the sulfonation is carried out at higher temperatures, where equilibrium may be established, the more stable 2-naphthalenesulfonic acid predominates. This is confirmed by heating the 1-isomer to 160° in sulfuric acid, whereupon it equilibrates to produce mainly the 2-isomer.

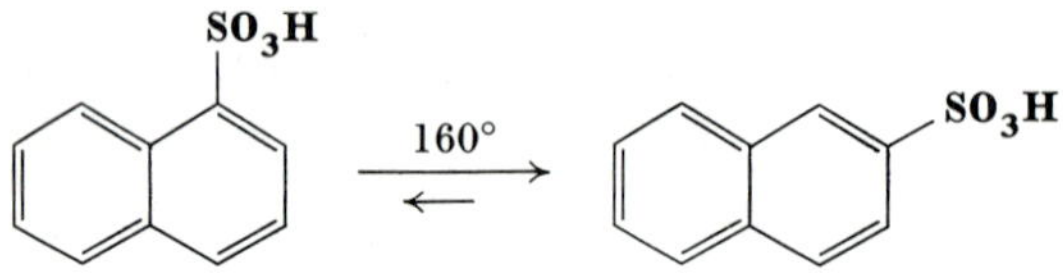

Sulfonic acids are not in themselves very important. However it is usually easy to convert sulfonic acids into phenolic compounds which do have great synthetic, commercial, and biological importance. This reaction is carried out by melting or "fusing" the sulfonic acid in the presence of NaOH and/or KOH. Although the conditions seem drastic, the yields are often quite good:

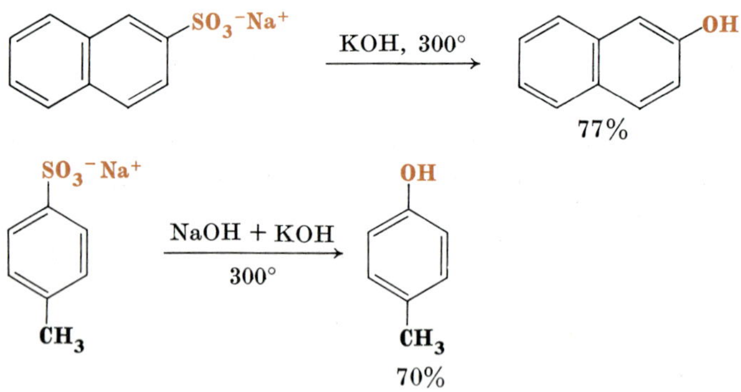

Note that in the above reaction the sulfonic acids are used in the form of their sodium salts.

10.8 NITRATION

Aromatic rings are usually nitrated by warming the aromatic compound with a mixture of concentrated nitric acid and sulfuric acid. Since the nitro group is deactivating, there is little danger of substituting the aromatic ring with more than one nitro group (unless forcing conditions, such as elevated temperatures and long reaction times, are employed). Two specific examples of nitrations are given below:

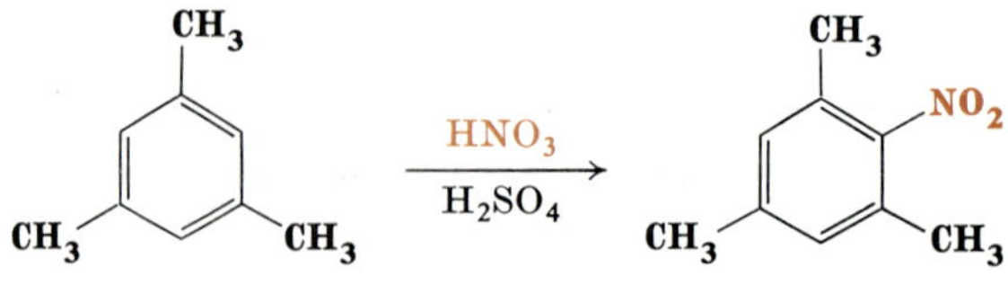

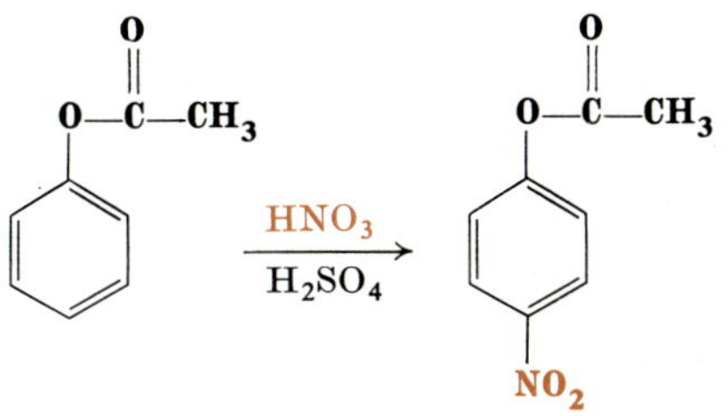

The reduction of aromatic nitro compounds affords a prime route for the synthesis of aromatic amines:

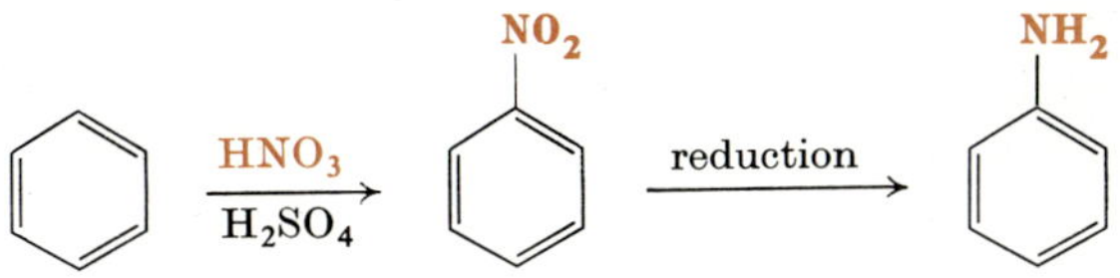

The nitro group is exceedingly easy to reduce; zinc plus hydrochloric acid and catalytic hydrogenation (H_2 and Pt) are often used to prepare amines from nitro compounds. As we shall see when we discuss the Sandmeyer reaction, the aromatic amine group can be converted into a halide, nitrile, hydroxyl, or hydrogen. Thus nitro compounds, which are precursors of aromatic amines, are important intermediates in organic syntheses.

Ammonium hydrosulfide is a useful reducing agent, because it can selectively reduce one nitro group in the presence of a second nitro group:

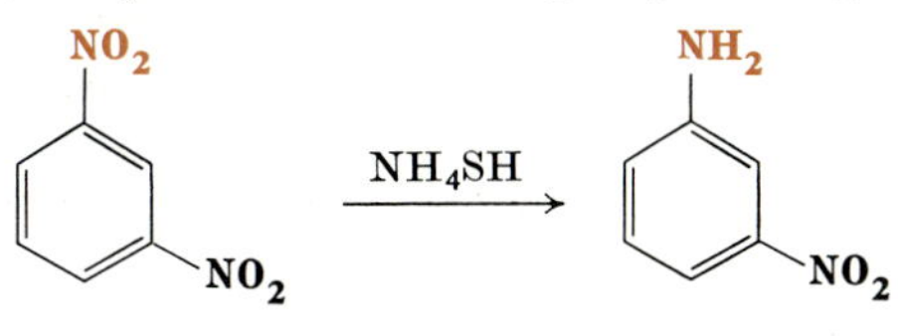

10.9 FRIEDEL-CRAFTS ALKYLATION

The Friedel-Crafts reaction provides a direct path for the preparation of alkylated aromatic compounds. We can illustrate this process with the reaction of benzene and ethyl bromide in the presence of aluminum trichloride:

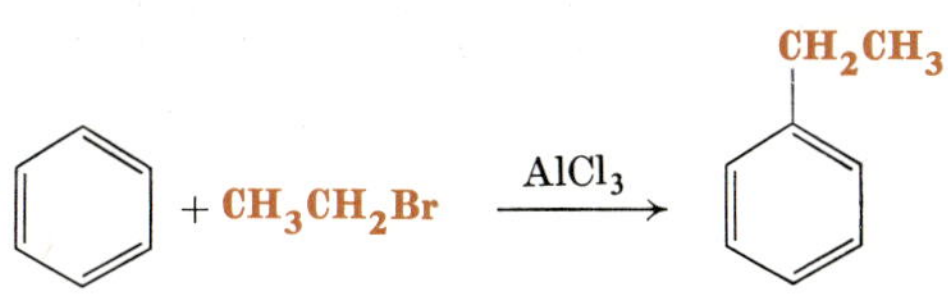

Unfortunately, there are complications which severely limit the utility of Friedel-Crafts alkylations in syntheses. The reaction cannot be used for the preparation of long-chain aromatic derivatives since isomerization may occur when the alkylating agent contains more than two carbon atoms:

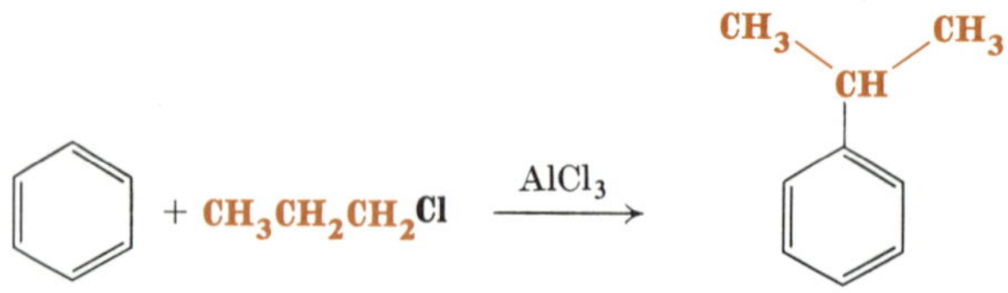

In general, only alkyl halides whose halide is located at the most substituted carbon of the alkyl group can be employed without fear of rearrangement:

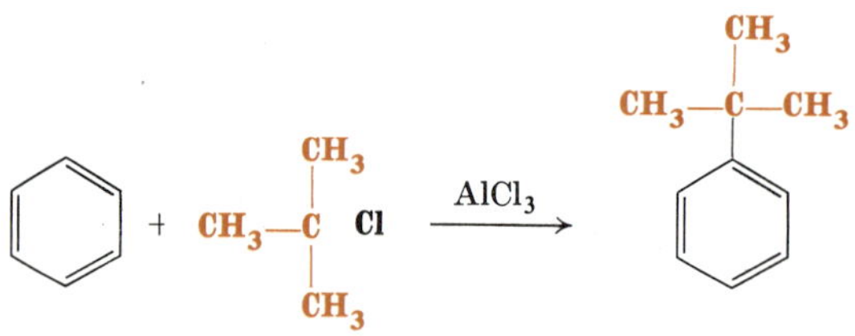

Friedel-Crafts alkylation will usually not occur on a benzene ring substituted with a deactivating group. In fact, nitrobenzene is a good solvent for carrying out the reaction.

Since the product of a Friedel-Crafts alkylation is more reactive than the original aromatic compound (alkyl groups are activating), multiple alkylation is a distinct possibility. This is especially true when excess alkyl halide is used. For example, alkylation of benzene with excess ethyl bromide produces 1,3,5-triethylbenzene:

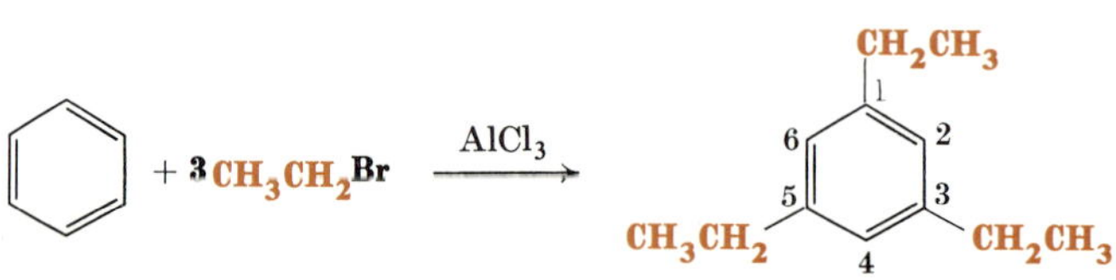

The *meta* relationships of the alkyl groups suggest that 1,3,5-triethylbenzene is a thermodynamic product. The material we would expect to form most rapidly—1,2,4-triethylbenzene—apparently rearranges to the more stable isomer having less steric interference between substituents.

10.10 FRIEDEL-CRAFTS ACYLATION

Friedel-Crafts acylation provides an important and convenient route for preparing aromatic ketones, that is, ketones whose carbonyl is attached directly to the aromatic ring. The acylation of anthracene is typical of this reaction:

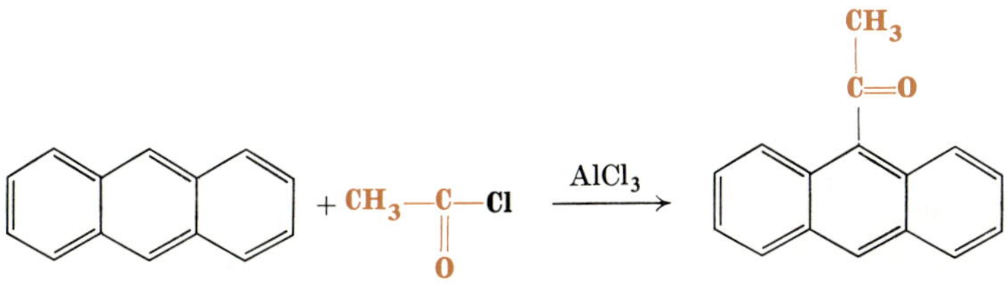

The acylation is carried out by adding anhydrous aluminum chloride slowly to a suspension of anthracene in a benzene-acetyl chloride mixture (0°).

Many acid chlorides, such as acetyl chloride, can be obtained commercially; when the desired acid chloride is not available, it can be prepared from the corresponding acid and thionyl chloride:

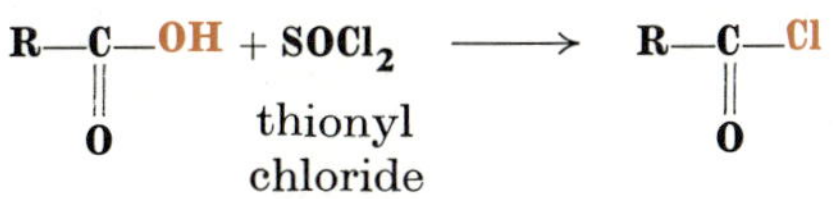

This reaction is similar to that between an alcohol and thionyl chloride to produce an alkyl chloride (Section 7.6).

Acid anhydrides can also be used to acylate aromatic rings:

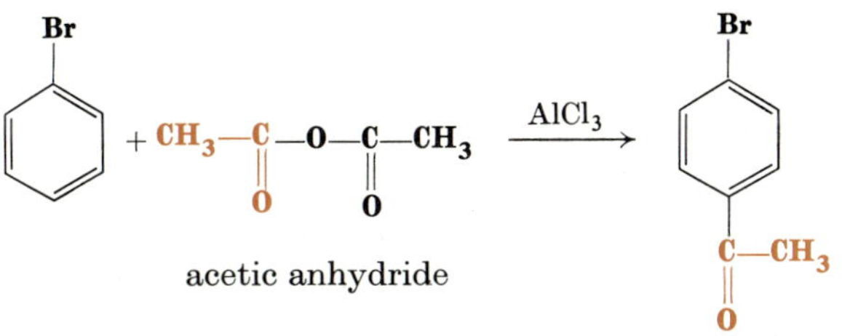

Acylation by acetic anhydride and acetyl chloride both involve electrophilic attack by the acylium ion, $CH_3\overset{+}{C}{=}O$.

Intramolecular Friedel-Crafts acylations produce cyclic ketones:

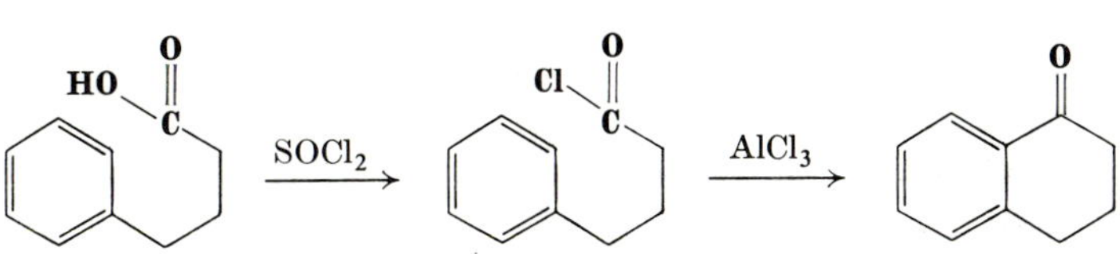

The carbonyl group of an aromatic ketone can be converted directly into a methylene group by reduction with zinc amalgam and hydrochloric acid (*Clemmensen Reduction*):

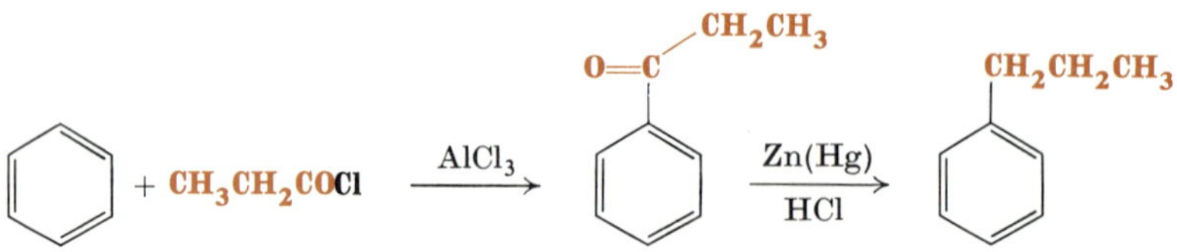

This two-step sequence can be used to synthesize straight-chain alkylated aromatics which are difficult to prepare by Friedel-Crafts alkylations because of isomerization and polysubstitution problems.

A synthesis of 1-propyl-3,4-dimethoxybenzene involving a Friedel-Crafts acylation is shown below:

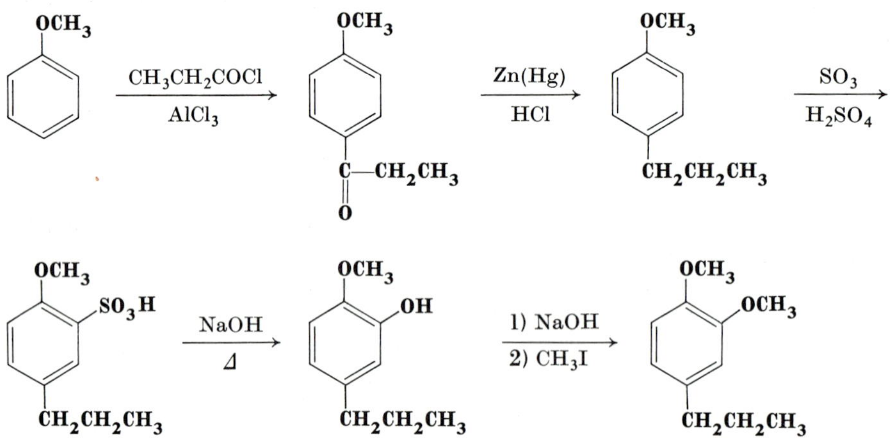

The third step in the sequence, the sulfonation, brings up an interesting point. Will the sulfonic acid group go *ortho* to the propyl group or *ortho* to the methoxy group? The answer is the latter, because the methoxy group is a stronger *ortho-para* activator than the propyl group. The propyl group donates electrons by means of a rather small inductive effect, whereas the methoxy group donates electrons to the *ortho* and *para* positions by resonance. Of course, the methoxy group cannot in this case direct the sulfonic acid group to the *para* position because that position is already occupied.

We have considered in detail the mechanistic and synthetic aspects of electrophilic aromatic substitutions. We now conclude this chapter with two reactions which do not fall under this category but which are very useful for preparing aromatic compounds.

10.11 DIAZOTIZATION OF AMINES

The reaction of aromatic primary amines with nitrous acid affords unstable products called diazonium salts:

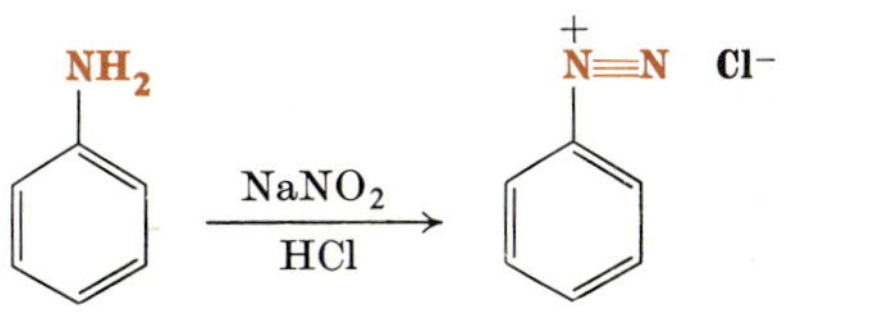

aniline benzenediazonium chloride

Diazotization is usually carried out by mixing the amine and sodium nitrite in aqueous hydrochloric acid at 0°. One possible mechanism for this reaction is illustrated in Fig. 10.6. A similar reaction can be effected with aliphatic primary amines; the resulting aliphatic diazonium salts, however, are so unstable that they spontaneously decompose to carbonium ions and nitrogen:

$$R{-}NH_2 \xrightarrow{HNO_2} \left[R{-}\overset{+}{N}{\equiv}N\right] \longrightarrow R^+ + N_2 \xrightarrow{H_2O} ROH + H^+$$

Aromatic diazonium salts can be converted into a variety of compounds, and therein lies their utility:

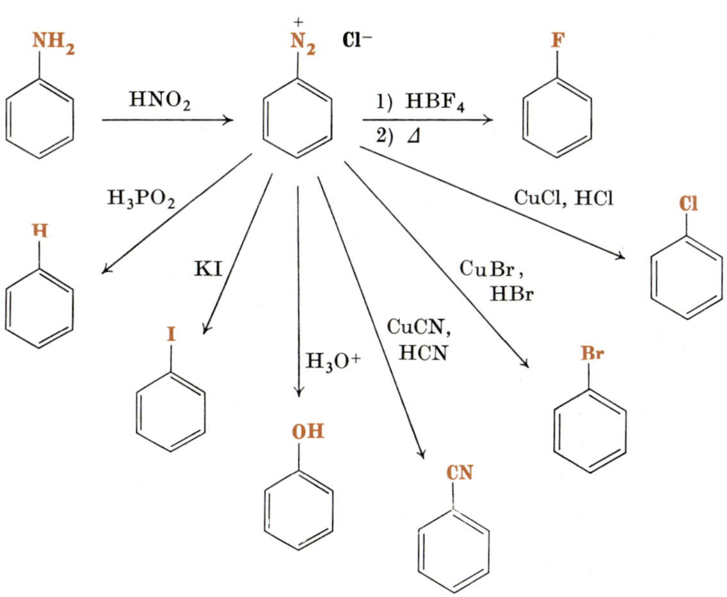

$$NaNO_2 + HCl \longrightarrow HO{-}N{=}O + NaCl$$

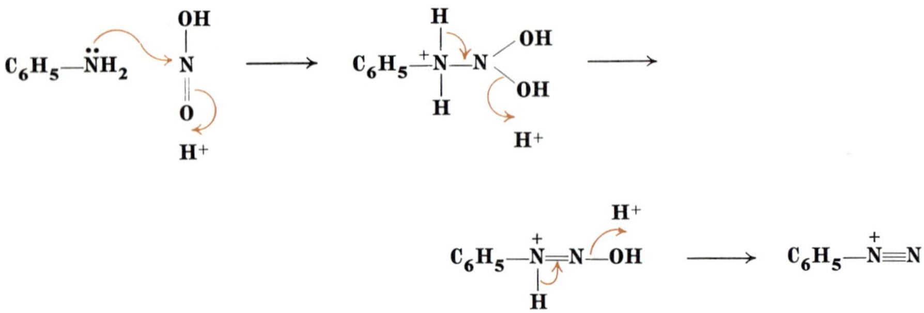

Fig. 10.6 Mechanism of amine diazotization

The reactions catalyzed by cuprous ion are known as *Sandmeyer reactions.* The mechanism of the Sandmeyer reaction is complicated; it appears that an aryl radical is formed by the transfer of an electron from cuprous ion to the diazonium salt:

$$C_6H_5{-}\overset{+}{N}{\equiv}N + Cu^{+} \longrightarrow C_6H_5\cdot + N_2 + Cu^{+2}$$

$$C_6H_5\cdot + Cu^{+2} + Cl^{-} \longrightarrow C_6H_5Cl + Cu^{+}$$

Diazonium salts have been used to prepare compounds which would otherwise have been difficult to obtain:

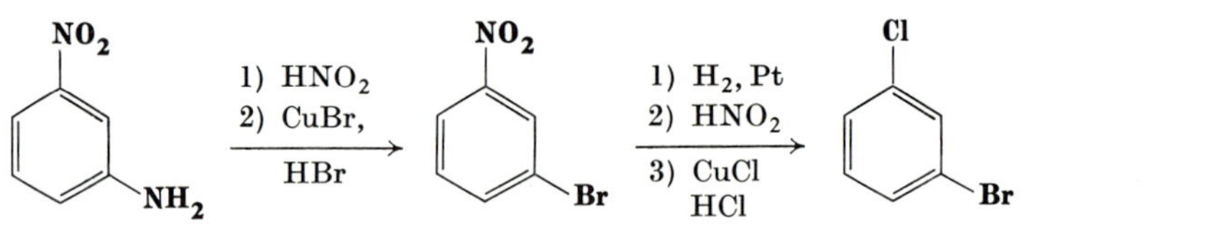

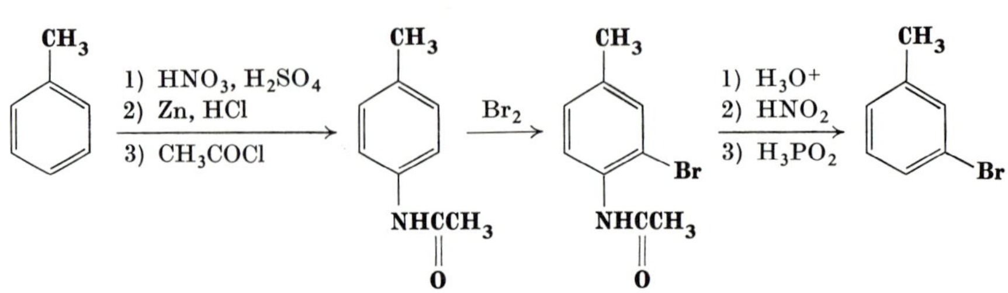

10.12 REACTIONS OF AROMATIC SIDE CHAINS

Aromatic compounds can be prepared either by introducing a new substituent or by modifying a substituent that is already present. Side-chain oxidation of alkylbenzenes with chromic acid is an example of the latter:

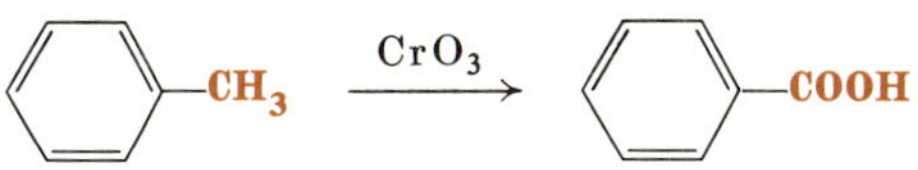

Higher alkyl groups are likewise degraded by chromic acid to produce the aromatic carboxylic acid:

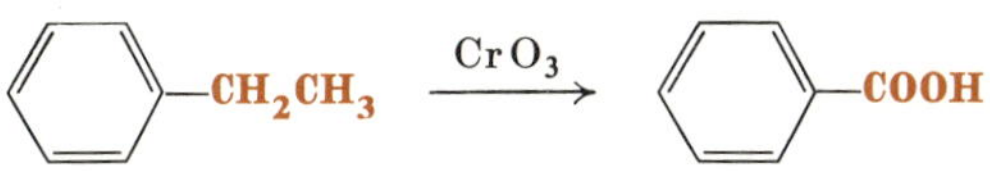

Carbon atoms adjacent to an aromatic ring are subject to free radical chlorination (see Section 4.20).

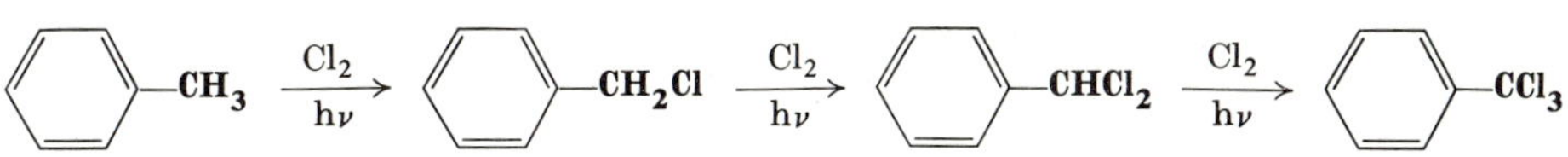

The ease of chlorination reflects the stability of the benzyl radical ($C_6H_5CH_2\cdot$). Benzyl halides are reactive toward nucleophilic substitution (as explained in Chapter 7) and may be used to prepare a variety of other compounds:

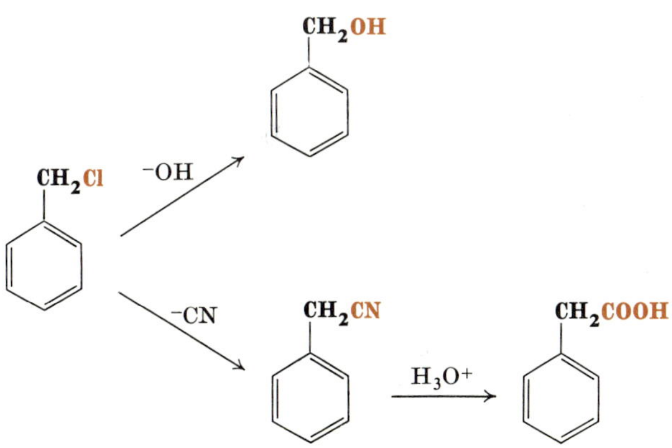

ADDITIONAL READING

L. M. Stock, *Aromatic Substitution Reactions*, Englewood Cliffs, N.J.: Prentice-Hall, 1968. Paperback.

M. Tomlinson, *An Introduction to the Chemistry of Benzenoid Compounds*, Elmsford, N.Y.: Pergamon Press, 1971.

PROBLEMS

1. Alkenes react readily with bromine to form dibromides. On the other hand, benzene does not add bromine. In the presence of iron catalyst, benzene reacts with bromine in a *substitution* reaction to form bromobenzene. Explain the difference between benzene and alkenes.
2. Write the correct reaction product or products.

a) 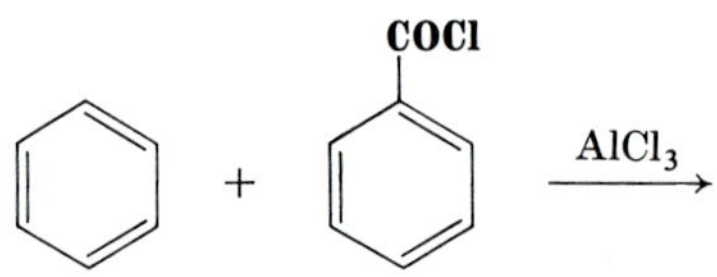

b) 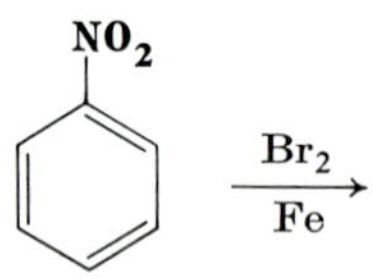

c) 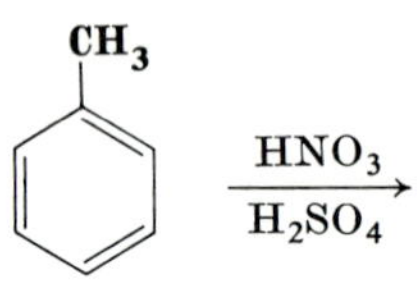

d) 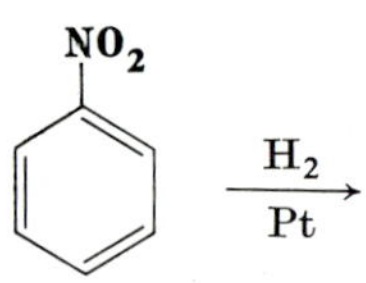

e) 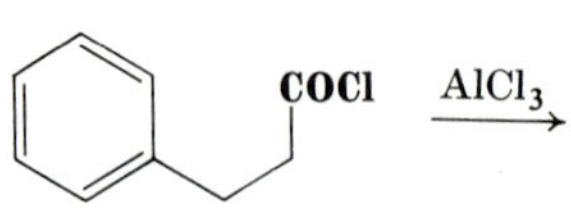

f) 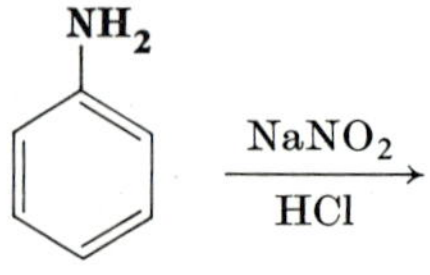

g) 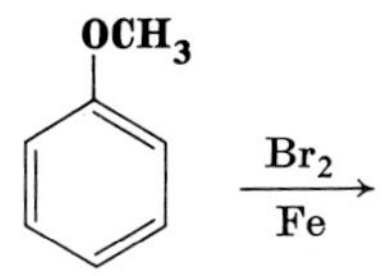

h) 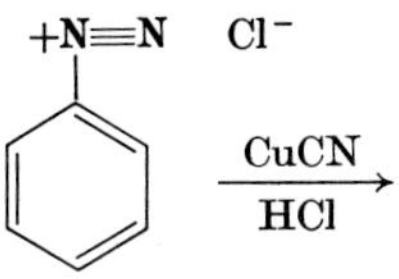

3. Fill in the blanks with the names of the correct compounds.

a) Nitration of toluene gives ________.
b) Aniline plus acetyl chloride gives ________.
c) Bromobenzene plus magnesium gives ________.
d) Benzenesulfonic acid plus thionyl chloride gives ________.
e) Aniline plus sodium nitrite and HCl gives ________.
f) Nitration of nitrobenzene gives ________.
g) Bromination of naphthalene gives ________.
h) Photochemical chlorination of toluene gives ________.

4. Give an example of:

a) a *meta* directing group
b) a σ complex
c) a Lewis acid catalyst
d) a sulfonamide
e) an acid anhydride
f) a benzyl halide
g) an activating substituent
h) a 1,3,5-trialkylbenzene (see Nomenclature section in Appendix)
i) a 2-substituted naphthalene
j) an acylium ion intermediate
k) a substituted phenol which is more acidic than phenol itself.
l) a substituted aniline which is more basic than aniline itself.

5. A compound $C_{10}H_{10}O$ has an intense IR peak at 1725 cm^{-1}. The NMR shows a 4-proton multiplet in the region of 7.0 ppm and a 6-proton singlet at 1.1 ppm. What is the compound?

6. Orcinol, a constituent of many substances found in lichens, is readily prepared from toluidine. Supply the necessary reagents (A through F).

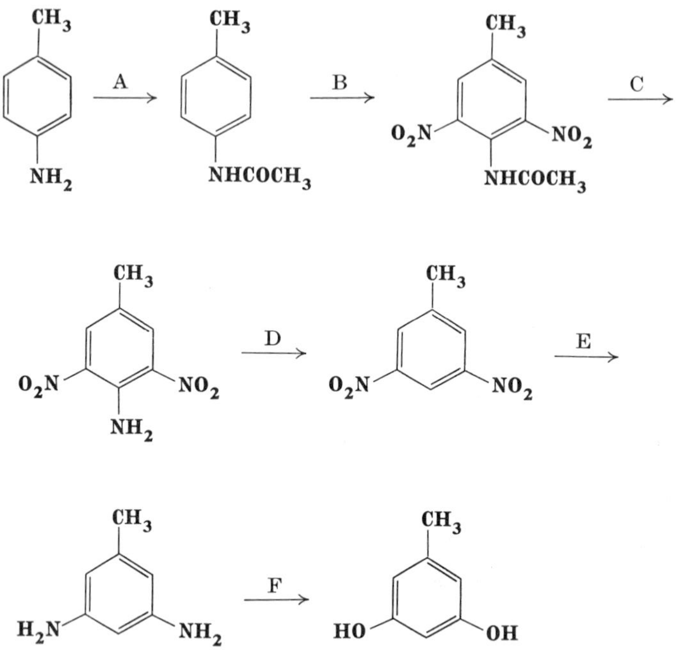

7. Predict the relative bromination rates of the following aromatic compounds:

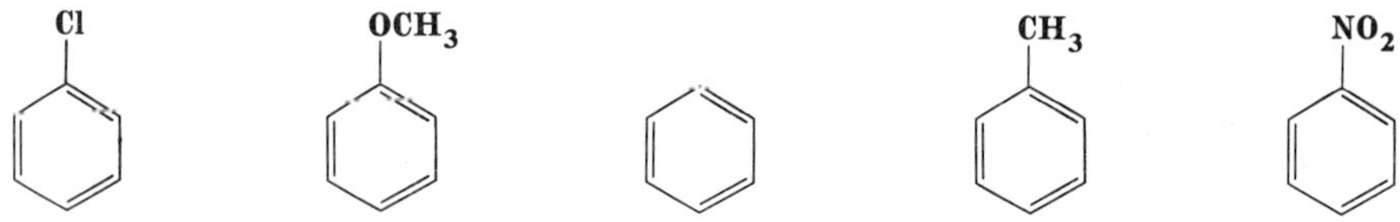

8. Show how you would carry out the following conversions with as many steps as required. Indicate all reagents and intermediate products.

a) 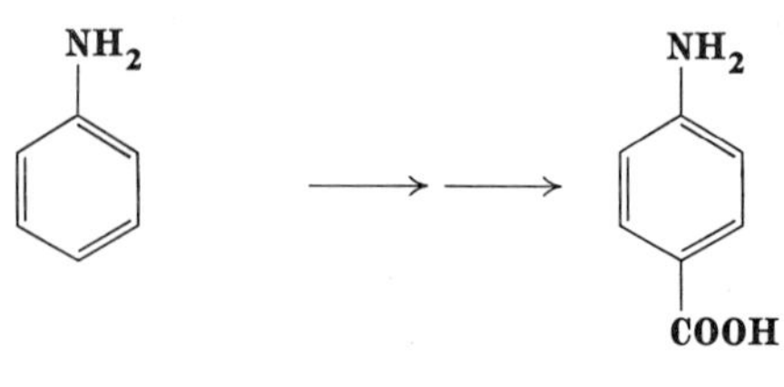

b) 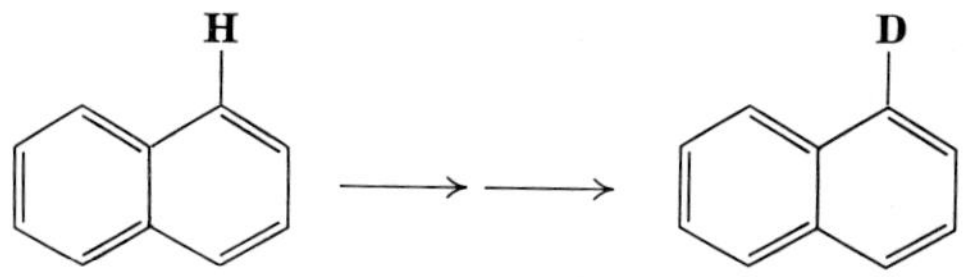

c) 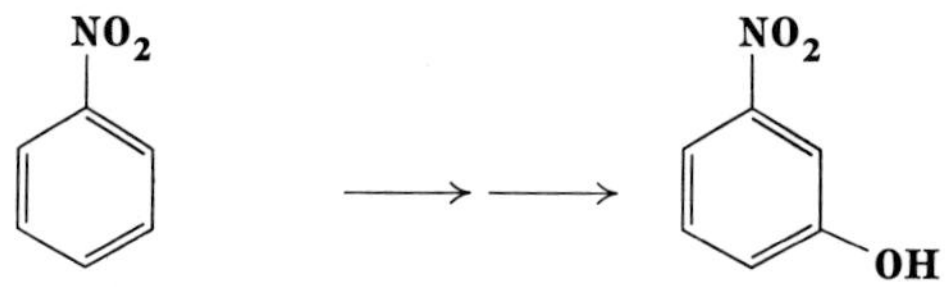

d) 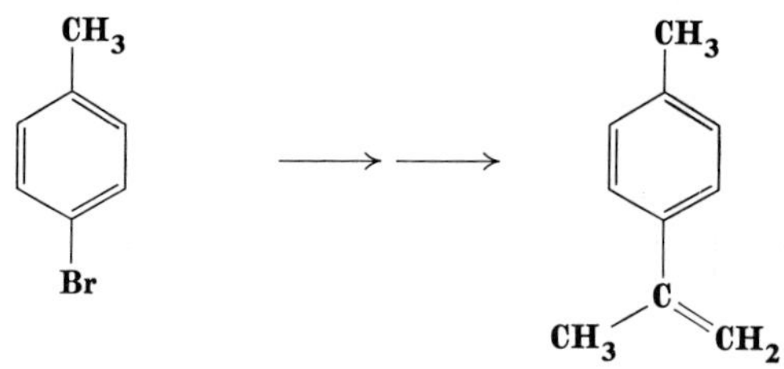

e) 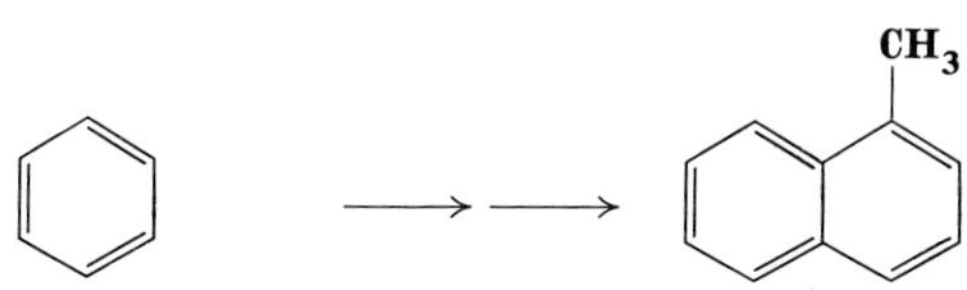

f) 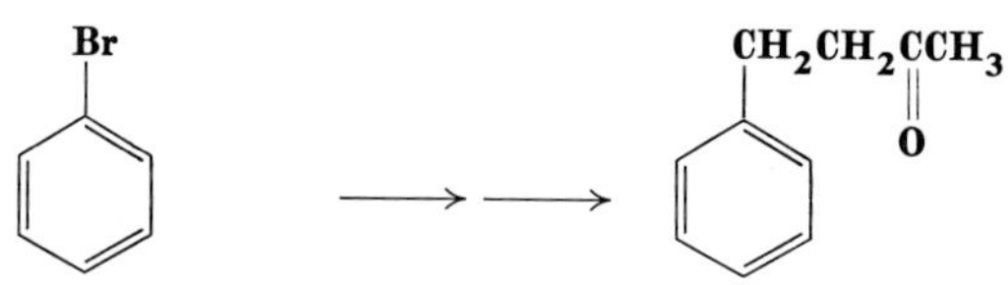

9. Predict the products of the following reactions.

Note: These reactions were not specifically covered in the preceding chapter. Nevertheless, with a firm grasp of reaction mechanisms, you should be able to predict a reasonable course for all of these "new" reactions. This is one reason reaction mechanisms are emphasized in this book.

a) 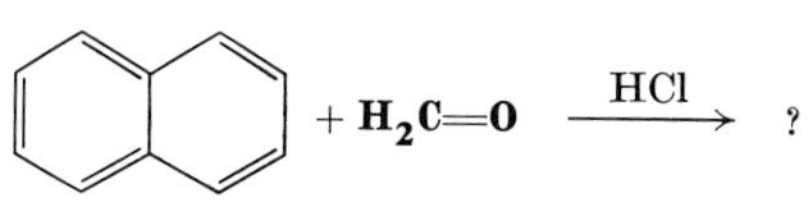

b) 

c) 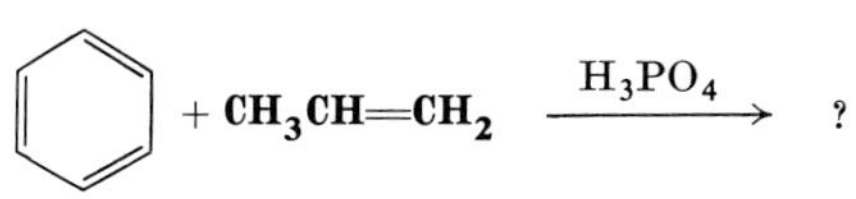

d) 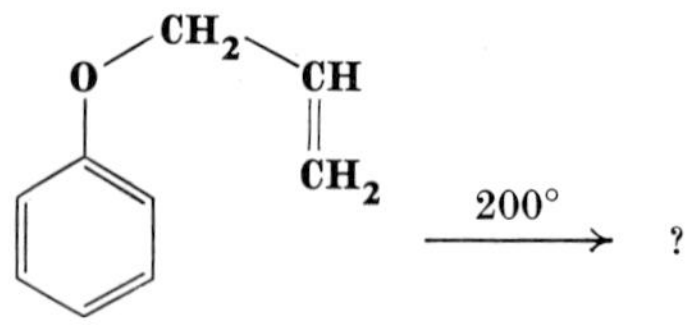

Hint: The product is a phenol. (See Section 8.12.)

10. Compounds in which the two *para* positions of a benzene ring are joined by a long chain are called paracyclophanes. One of the early routes to these interesting substances involved an intramolecular Friedel-Crafts acylation:

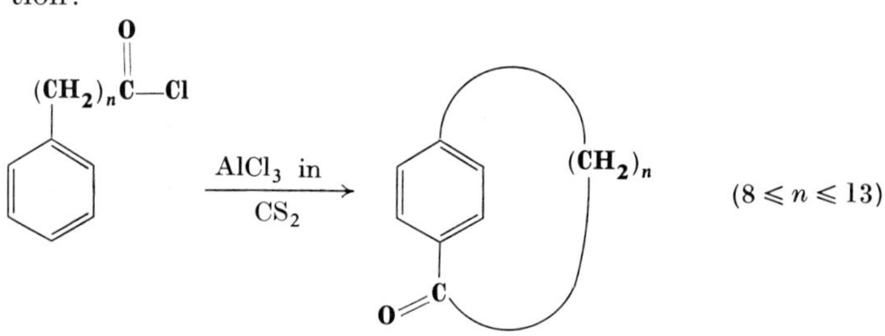

This reaction proceeds in good yield only if the acid chloride is in very high dilution. Explain.

11. Phenols and aromatic amines couple with diazonium salts to form azo compounds:

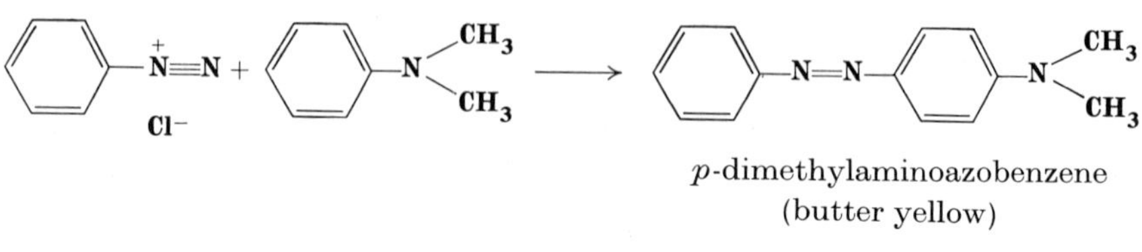

Azo compounds are colored and important to the dye industry. The above azo dye is known as butter yellow because it was once used to color fats (until it was discovered to be carcinogenic). Write a mechanism for the formation of butter yellow.

12. Equilenin was the first steroidal hormone to be totally synthesized.

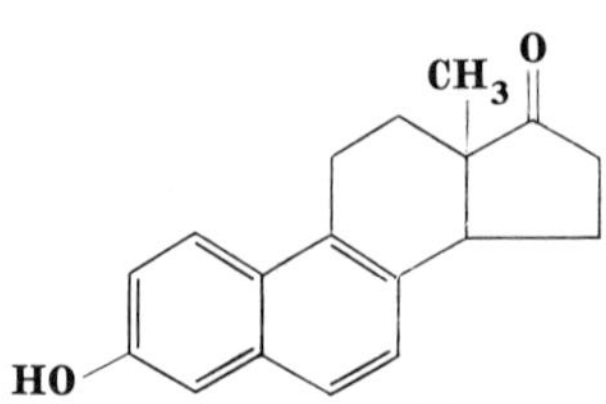

equilenin

This synthesis is shown below with several missing reagents (A through I). What are these reagents?

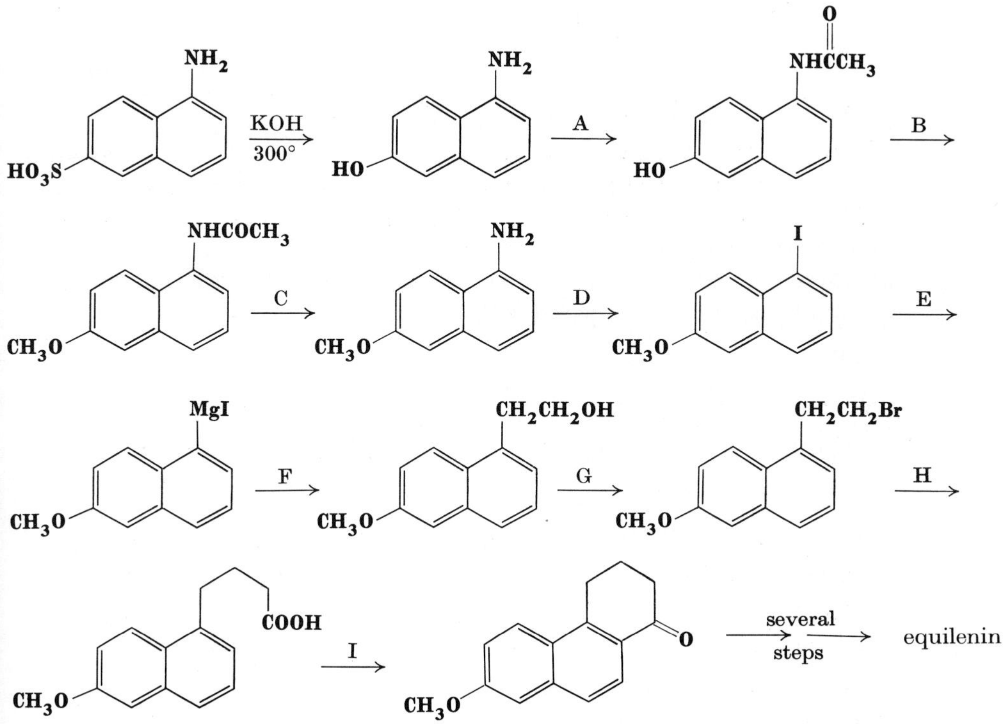

The formation of the final five-membered ring is not given in detail because it requires reactions that are not covered until the next chapter.

13. *Nucleophilic* aromatic substitution occurs only when the aromatic system is substituted with highly electron-withdrawing groups. An example of such a reaction is the conversion of 2,4-dinitrochlorobenzene into 2,4-dinitrophenol:

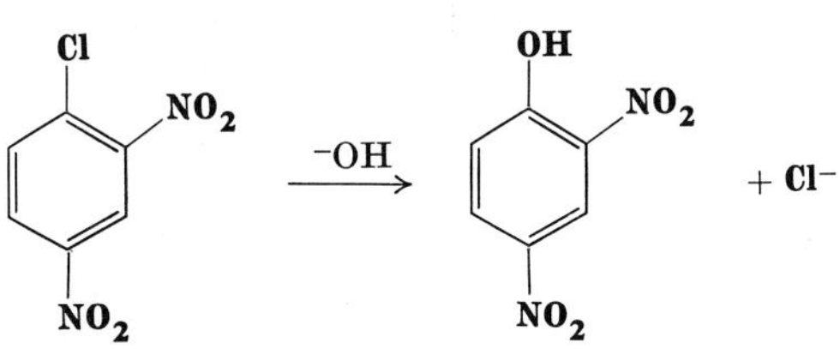

Nucleophilic aromatic substitutions have been shown to proceed via an addition intermediate. Write a mechanism for the above reaction.

14. Synthesize the following compounds starting with either toluene or benzene. It will be helpful, particularly for (c), to review the reactions in previous chapters.

a) 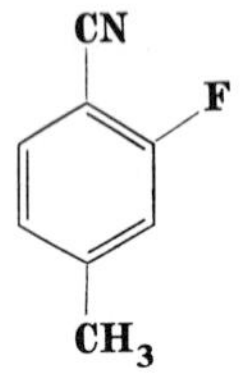

b) 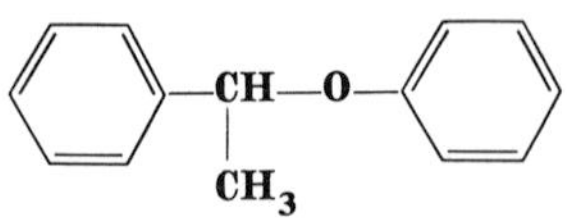

c) 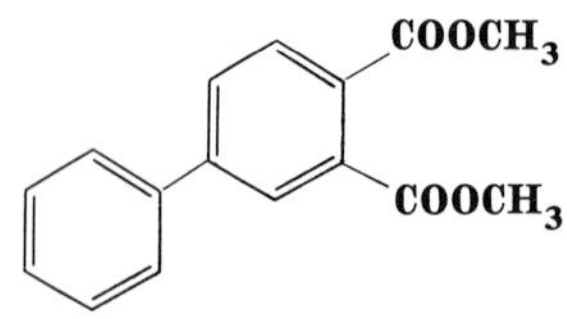

15. Explain the following observations.
   a) There is a marked change in the *ortho* to *para* ratio for the sulfonation of toluene at 100° compared with that at 0°.

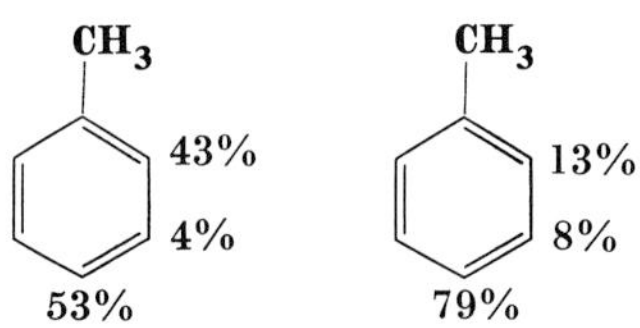

0° 100°

   b) Nitration of chlorobenzene gives 30% *ortho* substitution, whereas nitration of fluorobenzene gives only 12%.
   c) Nitration of *p*-nitrobenzoic acid gives 2,4-dinitrobenzoic acid rather than 3,4-dinitrobenzoic acid.

16. The preparation and isolation of benzocyclobutadiene,

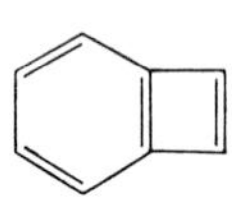

first attempted in 1907, has still not been achieved. The compound is apparently too strained and unstable to "bottle." The reaction below, however, indicates that benzocyclobutadiene can probably have a fleeting existence:

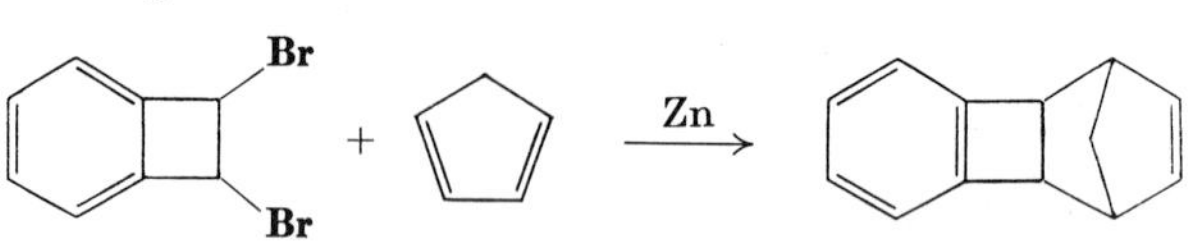

Write a mechanism for the reaction.

17. The compound, known commercially as 2,4-D, was one of the most widely used herbicides. It is absorbed by the plant, accumulates in the vacuoles of the growing plant tissues, and induces tumerous growths which ultimately kill the plant. Propose a synthesis of 2,4-D from benzene.

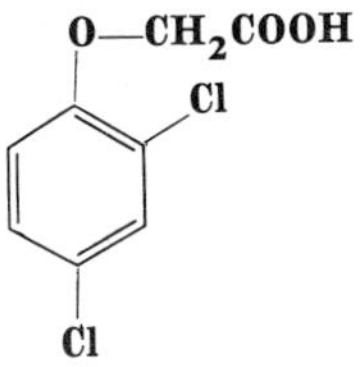

2,4-D

18. The synthesis of isotopically labeled compounds is very important in chemical, biochemical, and medical research. Syntheses of *p*-amino-benzoic-C^{14} acid and 2-methylnaphthalene-8-C^{14} are shown below. The starred (*) atoms refer to C^{14}, a radioactive isotope of carbon (half-life 5600 years). Provide the missing reagents in these two schemes.

a)

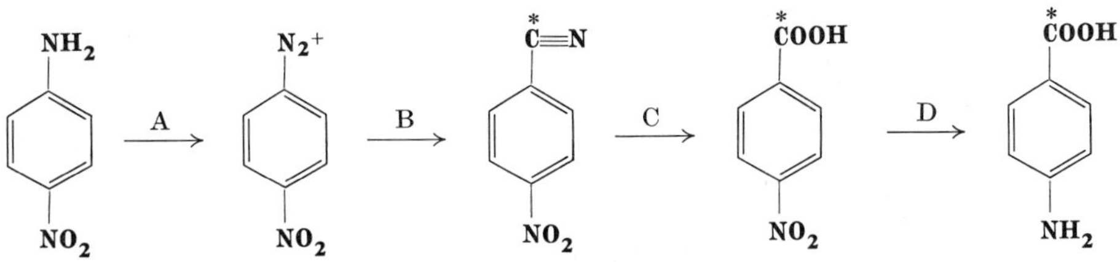

b)

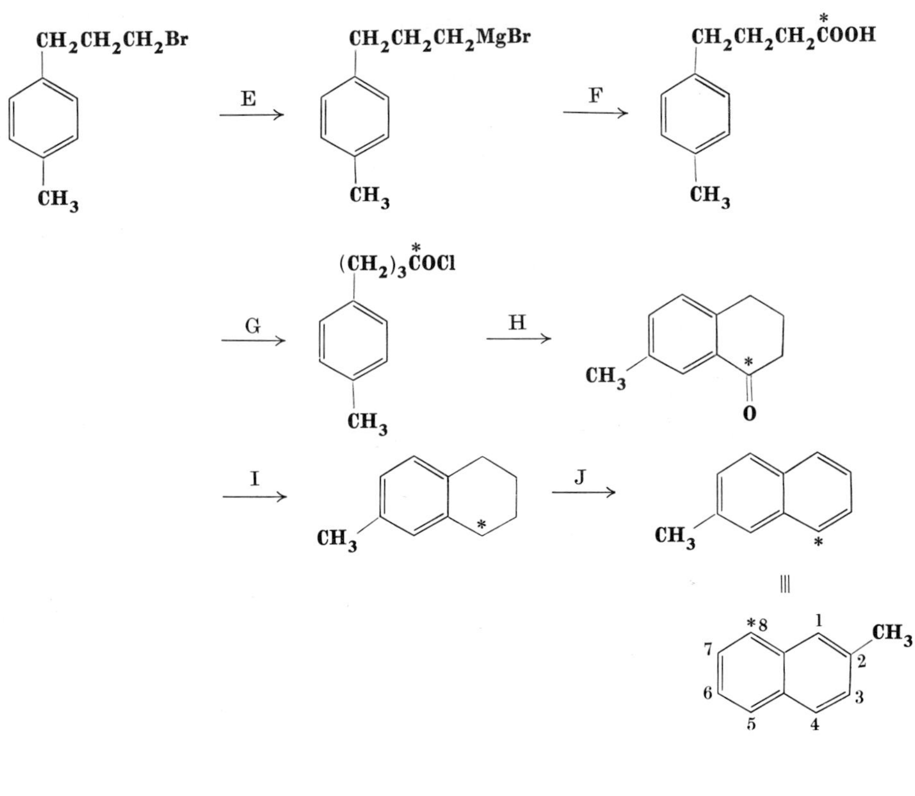

19. Nialamide belongs to a group of drugs called "psychic energizers" which are used to combat depression.

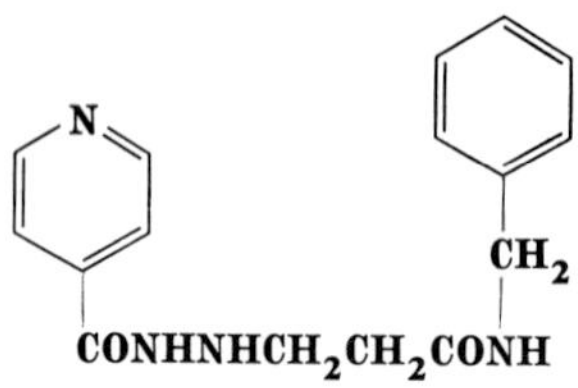

nialamide

Nialamide is believed to function by inhibiting an enzyme (monoamine oxidase) which catalyzes the destruction of certain substances in the brain that elevate the mood. Propose a synthesis of the drug using the following compounds:

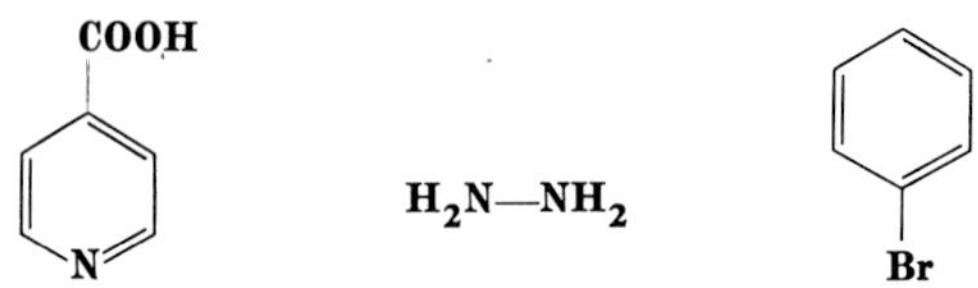

4-picolinic acid | hydrazine | bromobenzene

20. Write a mechanism for the following anomalous Grignard reaction:

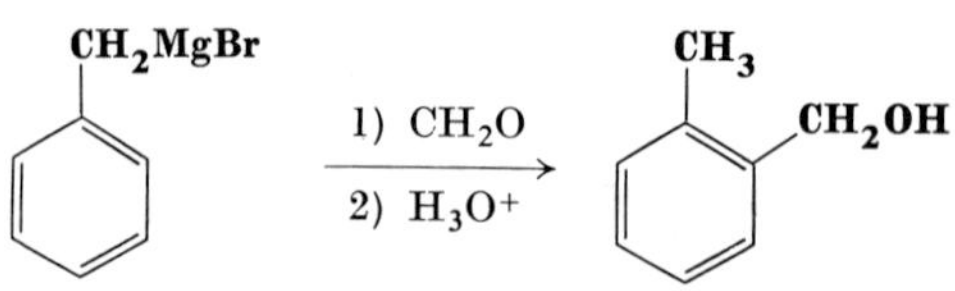

21. Synthesize the following compound from benzene.

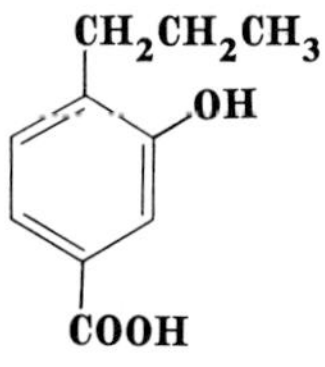

11

CARBONYL REACTIONS

11.1 CARBONYL COMPOUNDS

Aldehydes, ketones, carboxylic acids, and acid derivatives comprise one of the most important classes of organic compounds. These compounds all contain a carbonyl group (>C=O) and are of enormous utility in the synthesis of a wide variety of structures that have theoretical, biological, and commercial interest (Fig. 11.1). Although the chemistry of carbonyl compounds encompasses a large number of reactions, almost all of these can be understood in terms of three basic reaction types: *addition*, *addition–elimination*, and *enolization–ketonization*. We will now discuss each of these in turn.

11.2 CARBONYL ADDITION

Nucleophiles can add to carbonyl groups under either acidic or basic conditions. In a basic medium, a nucleophile such as hydroxide ion attacks the partially positive carbon atom of the carbonyl group in the following way:

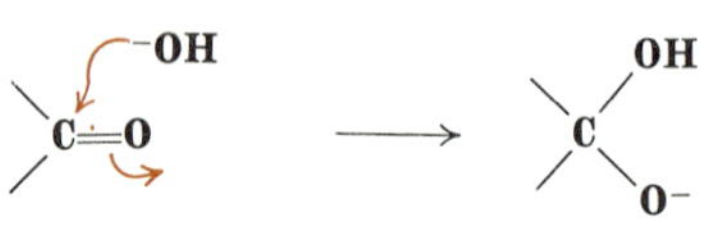

The final neutral product is formed when the solvent or acid added in the work-up donates a proton:

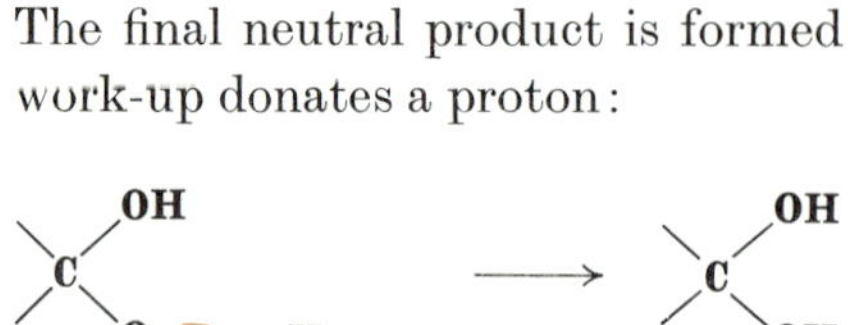

Certain carbonyl additions, such as the one above, are reversible; that is, the addition product is unstable and can easily revert back into the starting material:

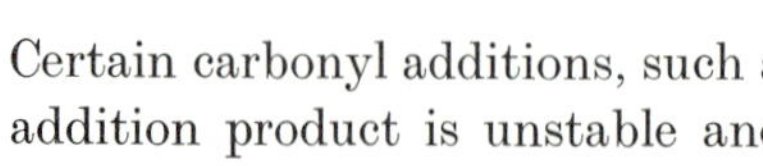

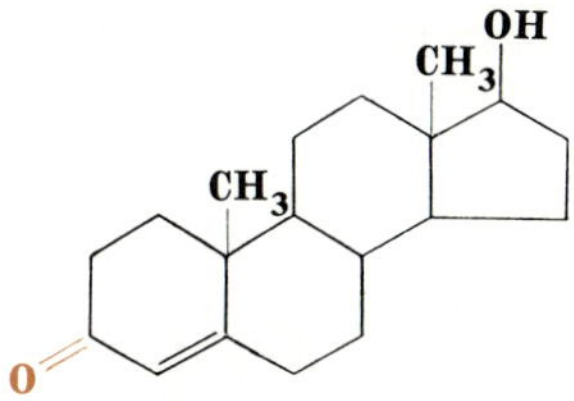

testosterone
(ketone, hormone)

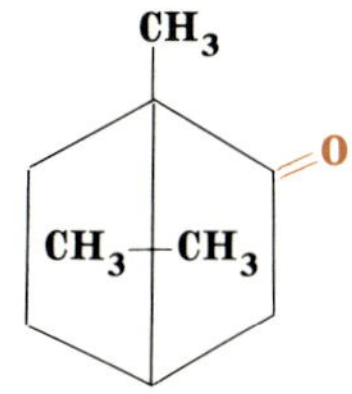

camphor
(ketone, antiseptic)

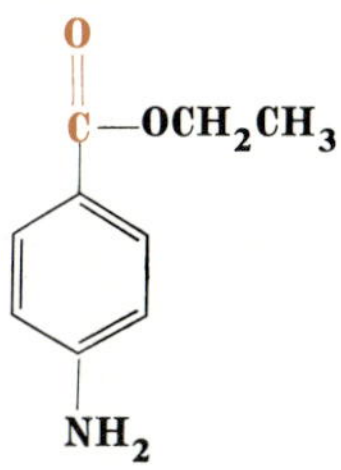

benzocaine
(ester, local anesthetic)

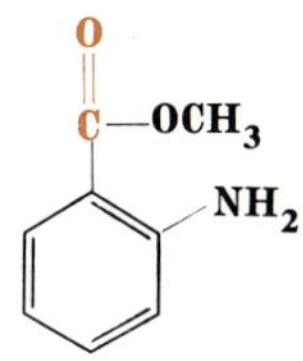

methyl anthranilate
(ester, grape flavor)

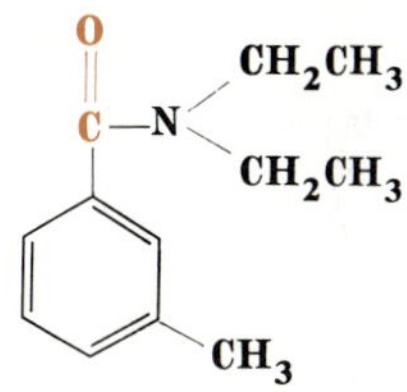

N,N-diethyl-*m*-toluamide
(amide, insect repellent)

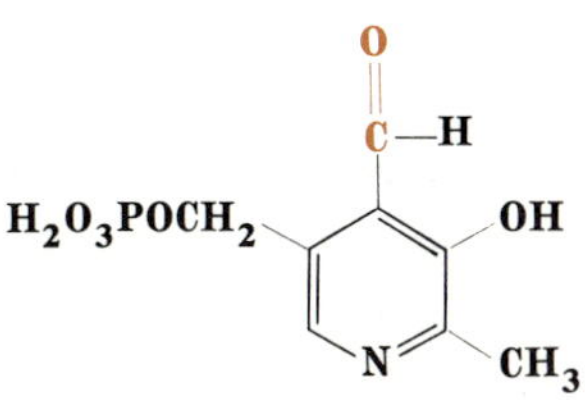

pyridoxal phosphate
(aldehyde, vitamin B_6)

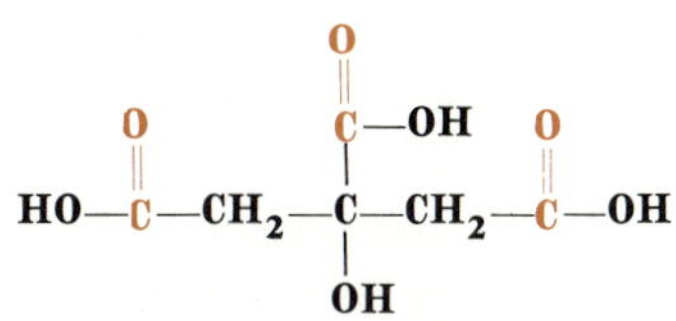

citric acid
(acid, food additive)

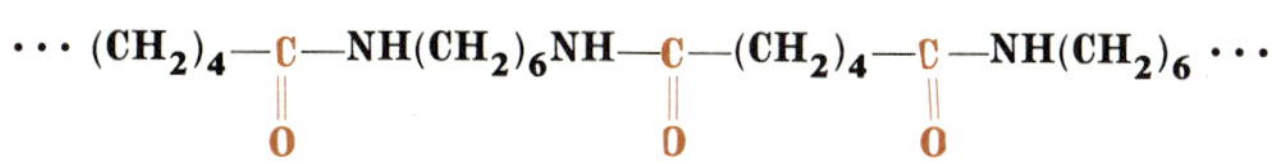

nylon
(amide, polymer)

Fig. 11.1 Some important carbonyl containing compounds

In many instances, however, the addition product is perfectly stable. The addition of a Grignard reagent and a ketone illustrates this situation:

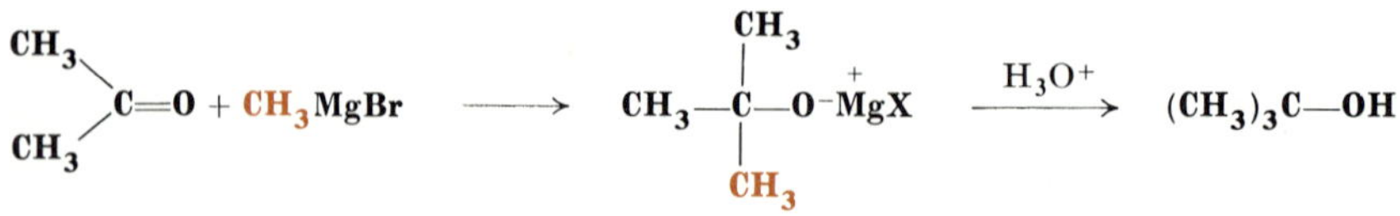

In acid solution, the reaction begins with protonation of the carbonyl oxygen and proceeds by the attack of a nucleophile:

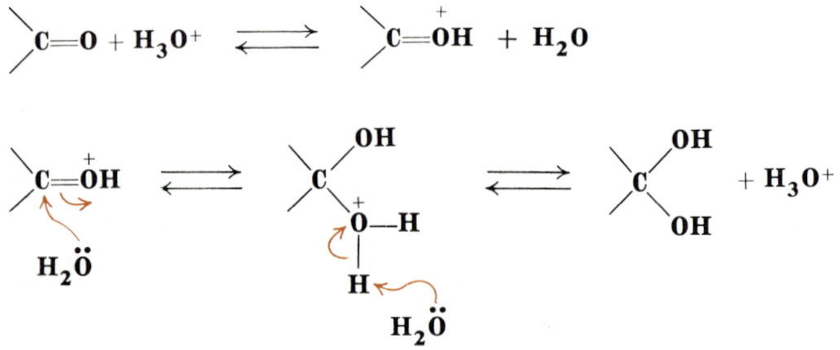

Note that there is no net consumption of a proton in this reaction: the proton that is consumed in the first step is regenerated in the final step. Such a reaction is said to be *acid catalyzed*.

Carbonyl groups in ketones, aldehydes, and esters, are weakly basic, so that only a small percentage of the molecules are protonated even in strong aqueous acid. A protonated carbonyl group, however, is particularly susceptible to nucleophilic attack because protonation enhances the partial positive charge on the carbonyl carbon:

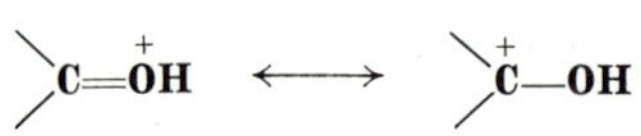

Carbonyl compounds differ widely in their reactivity toward nucleophilic attack:

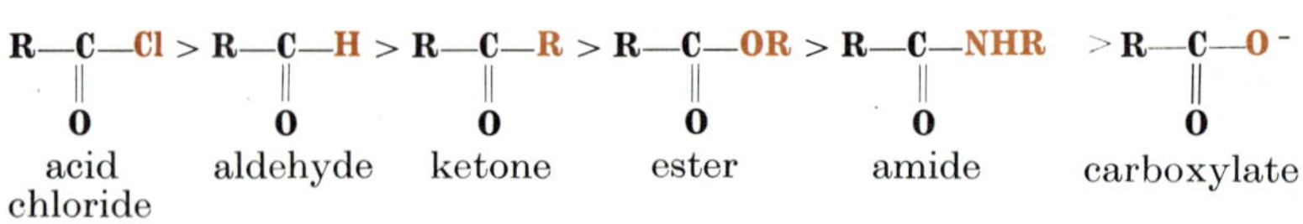

The above sequence is easily explained in terms of resonance, inductive, and steric effects. For example, an amide is a relatively unreactive group because of the high degree of resonance stabilization resulting from the donation of a pair of nitrogen electrons:

The chlorine inductively withdraws electrons from the carbonyl carbon of acid chlorides, rendering them extremely reactive.

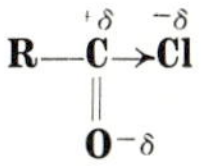

Ketones are somewhat less reactive than aldehydes because alkyl groups donate electrons and provide some degree of steric hindrance to nucleophilic addition.

11.3 ADDITION–ELIMINATION

Addition to a carbonyl group is often reversible, as we have already seen. However, an added nucleophile may be *eliminated*, and the carbonyl group re-formed, using a pair of oxygen electrons:

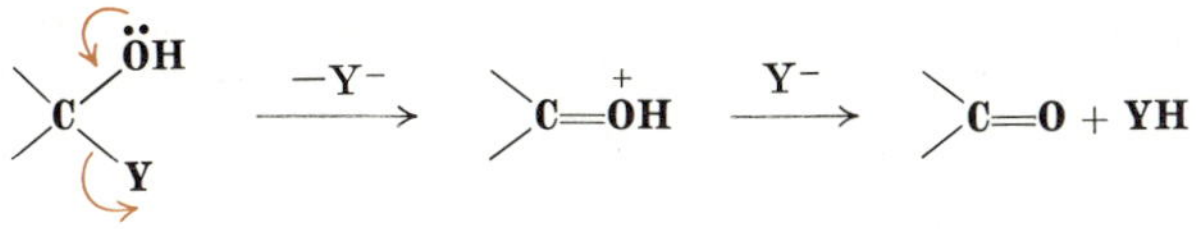

Conversely, if the added nucleophile contains any unshared electrons, it may donate them to form a carbon-nucleophile double bond, thereby eliminating the original carbonyl oxygen atom. A specific example of this type of behavior is the addition–elimination reaction of a ketone with a primary amine:

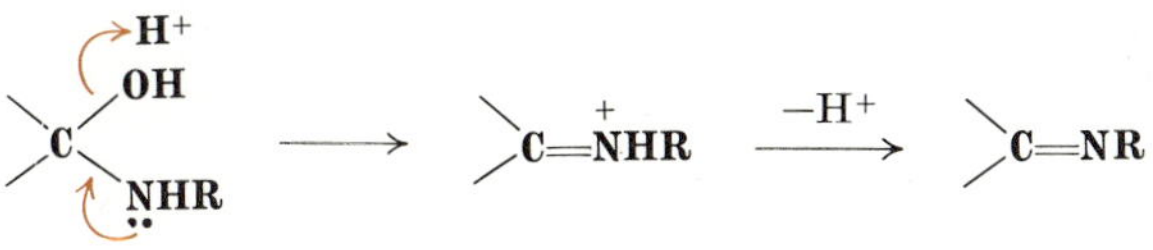

Groups other than the carbonyl oxygen or the added nucleophile can also be eliminated. For example, if X in the addition product shown below can bear a negative charge, it too may be expelled:

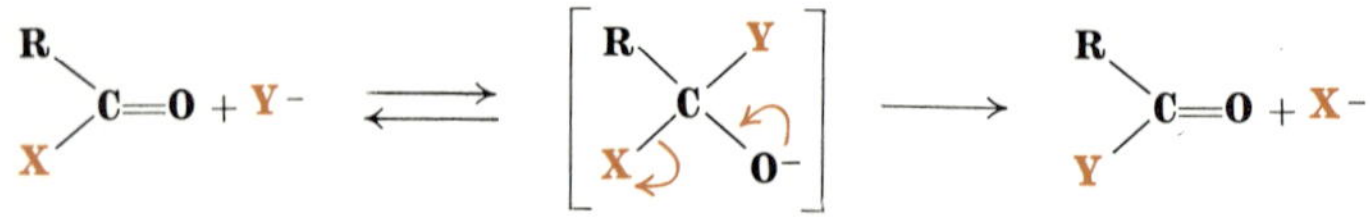

Although the overall result appears to be the displacement of group X by the nucleophile Y^-, the course of the reaction is an *addition* to the carbonyl group (to form a so-called *tetrahedral intermediate*) followed by *elimination* to form a new carbonyl compound. The hydrolysis of an ester to an acid in aqueous base illustrates this type of process:

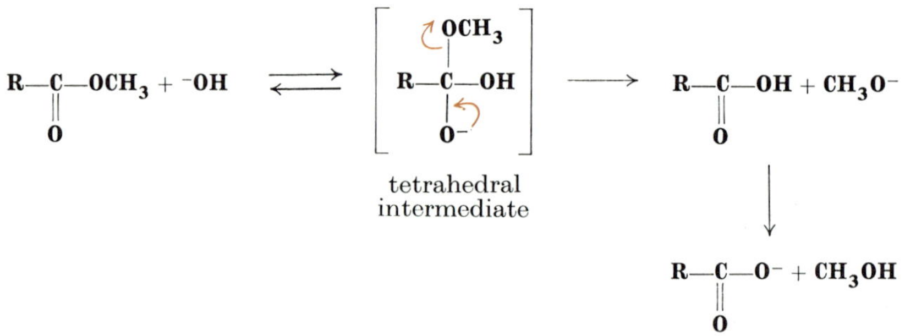

11.4 ENOLIZATION-KETONIZATION

In Sections 3.6 and 4.11 we examined the conversions of aldehydes and ketones into enols or enolates. A ketone such as cyclohexanone treated with base may lose an α hydrogen to form a resonance-stabilized enolate:

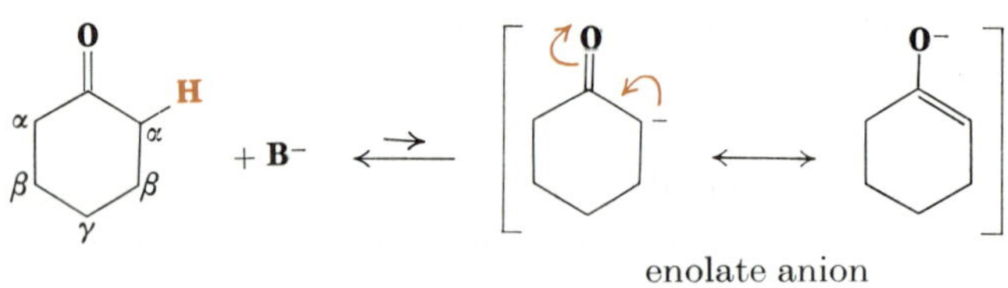

Since ketones are not very acidic, the above equilibrium lies far to the left, even in strongly basic aqueous solutions. The enolate may pick up a proton to re-form the ketone (*ketonization*), or it may react with a species AB:

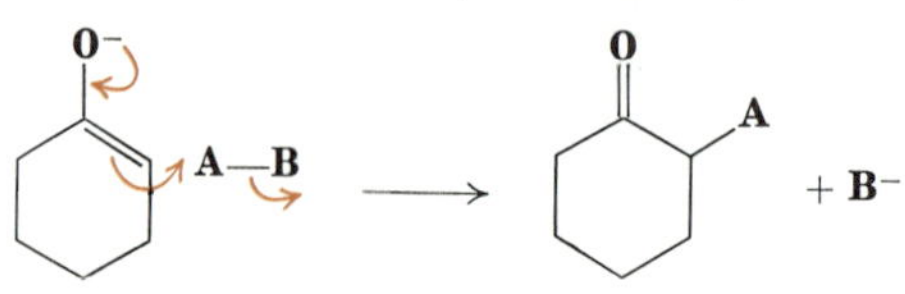

The same type of reaction may occur with an enol; this is illustrated by the reaction of cyclohexanone enol with bromine:

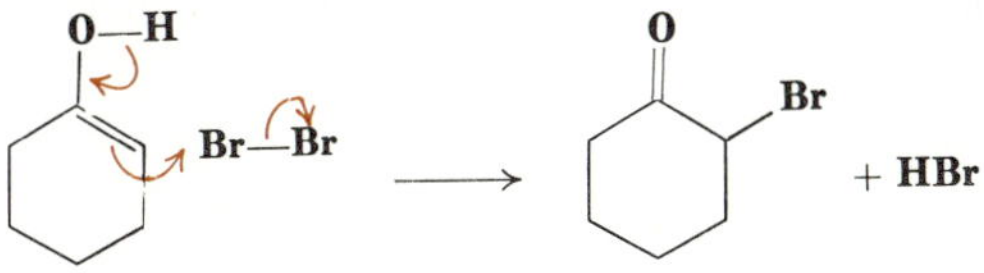

For the sake of convenience, the bromination was given in a single step, but the reader should understand that this is equivalent to a more detailed description:

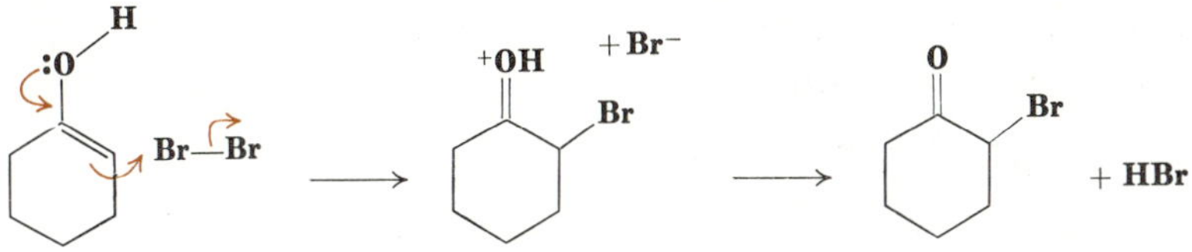

Now that you have been introduced to the three major types of carbonyl reactions, we shall look at specific examples of these reactions, emphasizing their mechanisms and utility in syntheses.

ADDITION REACTIONS

11.5 HYDRATE FORMATION

The simplest carbonyl addition is the addition of the elements of water to an aldehyde or ketone:

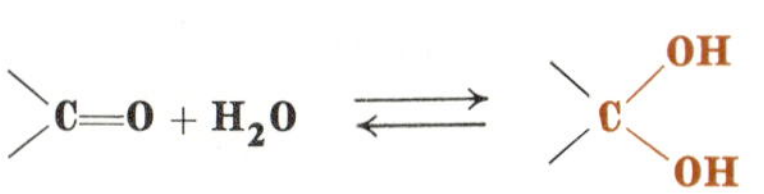

The reaction may be either acid or base catalyzed (Section 11.2).

Aldehyde and ketone hydrates are seldom isolated; they are in equilibrium with their respective carbonyl compounds in water solution, with the carbonyl form generally predominating. Some carbonyl compounds, however, do form stable hydrates. The most familiar example is chloral hydrate, the classic Mickey Finn:

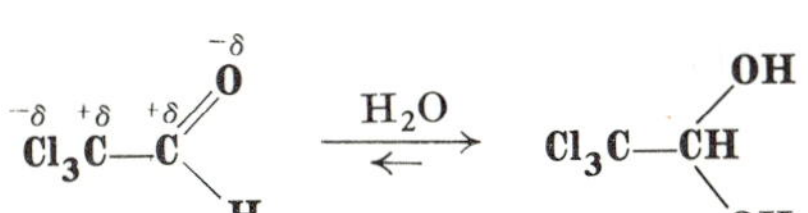

chloral chloral hydrate

Chloral hydrate is stable compared to the aldehyde because the aldehyde contains two adjacent carbon atoms, both bearing large partial positive charges. The electronic repulsion between these atoms is diminished in the hydrate form.

11.6 HEMIACETALS AND HEMIKETALS

Aldehydes and ketones react with alcohols to form adducts called *hemiacetals* and *hemiketals*, respectively:

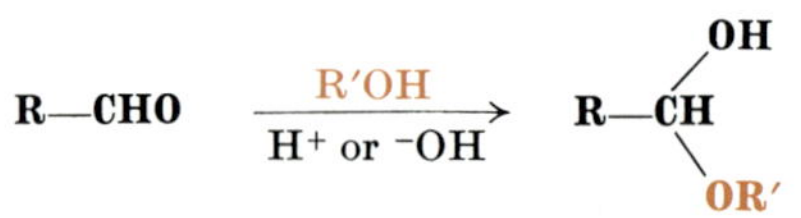

hemiacetal

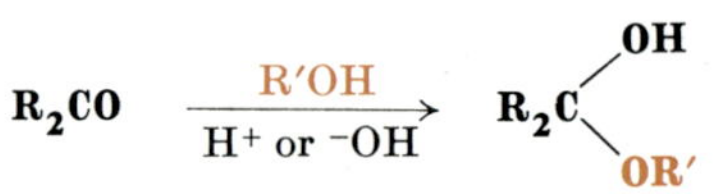

hemiketal

Like hydrates, hemiacetals and hemiketals are unstable species, except in special cases (such as when the alcohol and carbonyl groups are in the same molecule and react to form five- or six-membered rings). A good example of this is the intramolecular hemiacetal formation in the simple sugar glucose, whose equilibrium favors the cyclic hemiacetal form (Fig. 11.2).

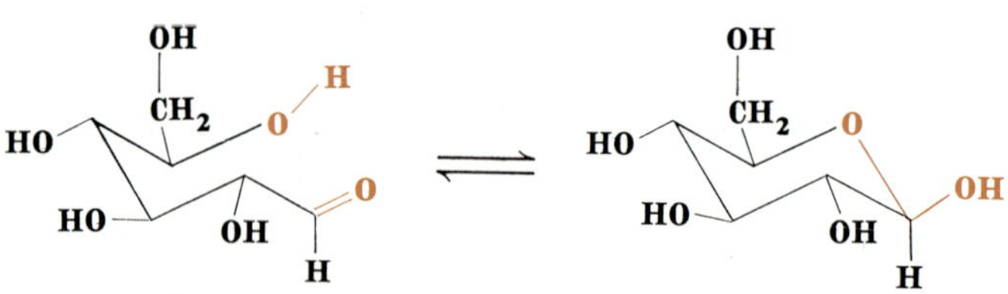

glucose aldehyde form — glucose hemiacetal form

Fig. 11.2 Hydroxy aldehyde-hemiacetal equilibrium

The mechanism for acid-catalyzed hemiacetal formation is illustrated in Fig. 11.3.

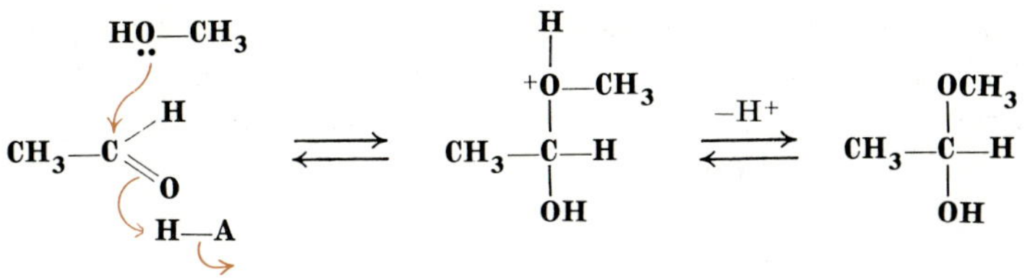

Fig. 11.3 Acid-catalyzed hemiacetal formation

11.7 CYANOHYDRINS

If hydrocyanic acid (HCN) is added to an aldehyde or ketone, a *cyanohydrin* results. The reaction rate increases with the addition of base or cyanide ion, showing that cyanide anion is the actual nucleophilic species:

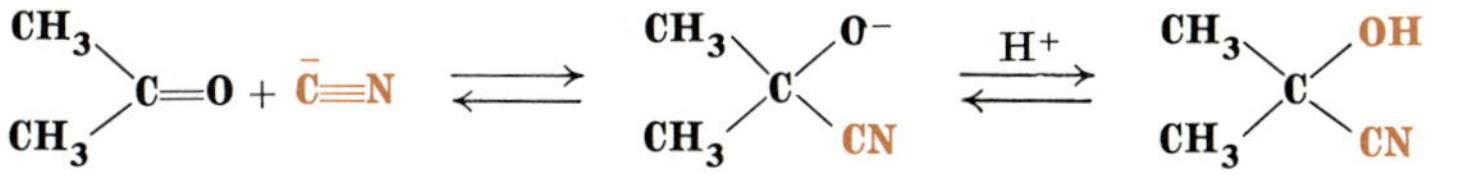

a cyanohydrin

Cyanohydrin is formed by an equilibrium reaction which lies to the right for aldehydes and aliphatic ketones.

Benzaldehyde cyanohydrin can be converted into mandelic acid by hydrolysis of the cyano group (also called a nitrile group):

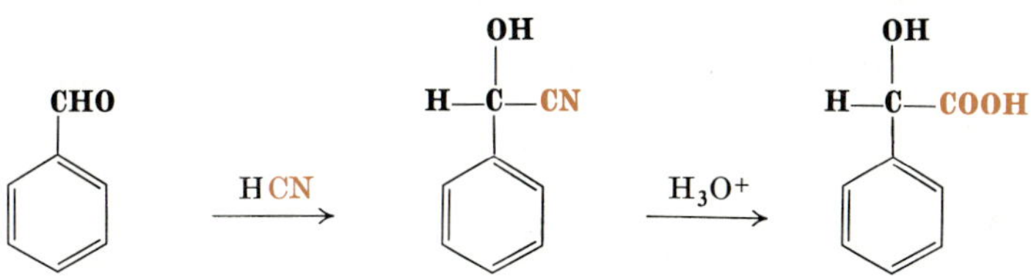

benzaldehyde — benzaldehyde cyanohydrin — mandelic acid

Aliphatic cyanohydrins, in addition to providing hydroxy acids, can also be converted into unsaturated nitriles by eliminating water:

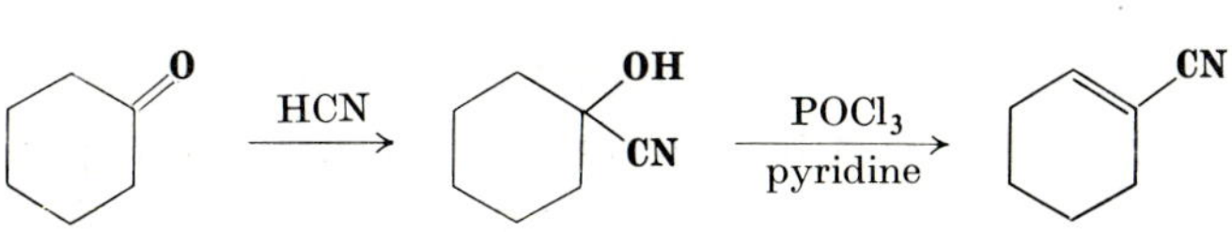

These compounds can, in turn, be converted into both unsaturated and saturated carboxylic acids:

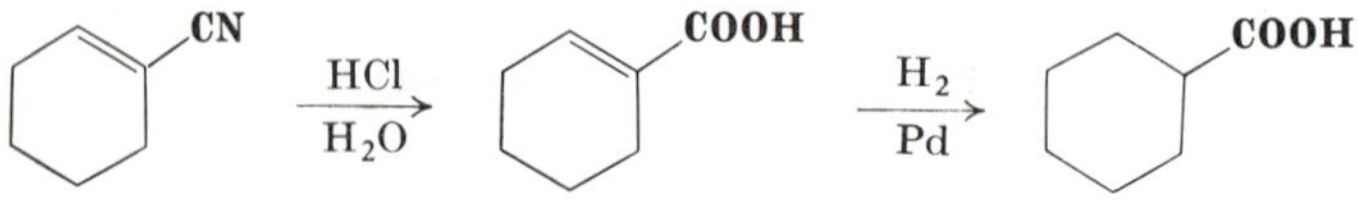

11.8 CARBINOLAMINES

Aldehydes and ketones react with amines to form compounds called *carbinolamines*:

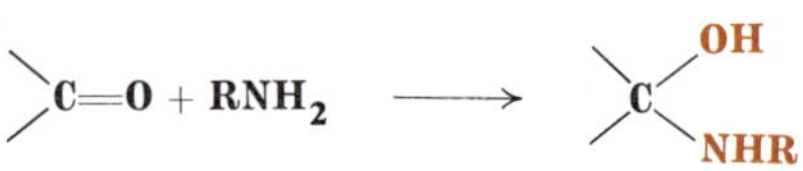

Carbinolamines are usually unstable with respect to starting materials, but can be converted into stable products by the elimination of water. We shall return to these reactions later in this chapter.

11.9 ADDITION OF GRIGNARD REAGENTS

Grignard reagents (R—MgX) add to aldehydes and ketones, yielding (after work-up with aqueous acid) alcohols in which a new carbon-carbon bond has been formed (see Section 3.7). This important synthetic reaction can be represented generally as

(after work-up)

A simplified picture of the mechanism of Grignard addition involves coordination of the carbonyl oxygen with magnesium, followed by transfer of the organic portion of the Grignard reagent to the positive carbon of the aldehyde or ketone. We can therefore consider the R-group of the Grignard reagent as a nucleophile; for practical purposes, the Grignard reagent reacts as if it were $R^- \, MgX^+$:

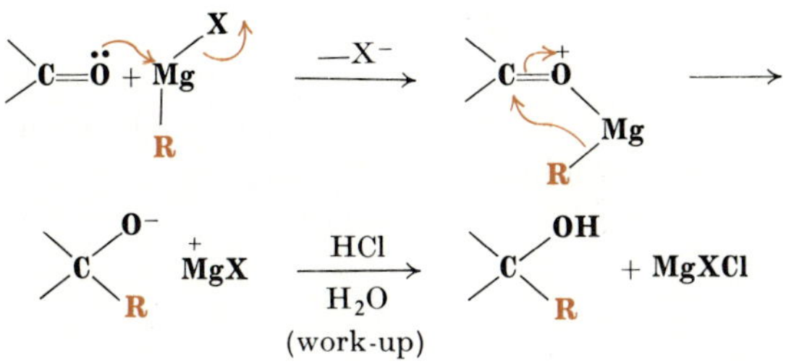

Figure 11.4 shows how a Grignard reagent, methyl magnesium bromide, reacts with carbonyl compounds. The reaction with ethylene oxide is included in the figure because it is a useful way to make primary alcohols and to lengthen a carbon chain two atoms at a time (see Section 7.11).

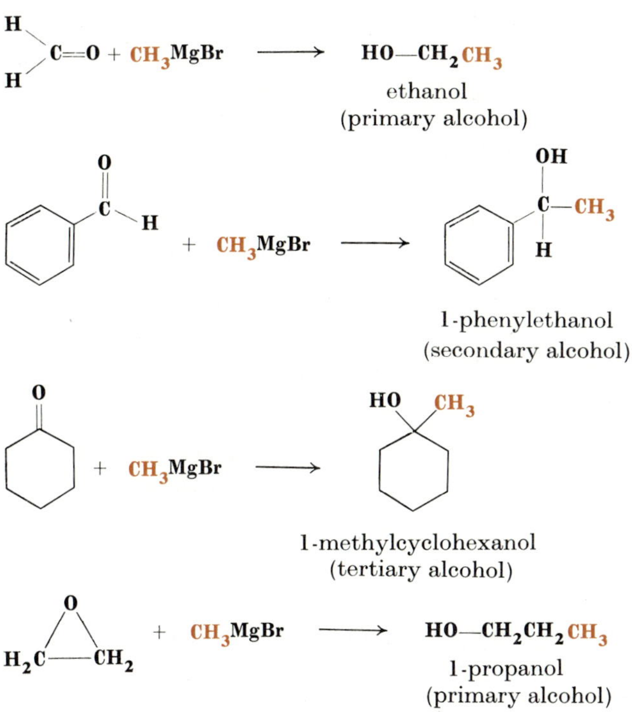

Fig. 11.4 Grignard reactions

The reactions in Fig. 11.4 can be summarized by the following rules. A Grignard reagent added to formaldehyde or ethylene oxide yields a primary alcohol; added to any other aliphatic or aromatic aldehyde it yields a secondary alcohol; added to a ketone, it yields a tertiary alcohol.

There are certain limitations to the usefulness of Grignard addition which must be kept in mind. Since Grignard reagents are salts of strongly basic carbanions, they react with functions containing acidic hydrogen:

$$R^{-}\ \overset{+}{Mg}X + HA \longrightarrow RH + MgXA$$

Thus, any carbonyl compound that contains an even slightly acidic function (OH, NH, SH, or CH in a β-ketoester) will react first with its acidic group to form a magnesium salt and an alkane:

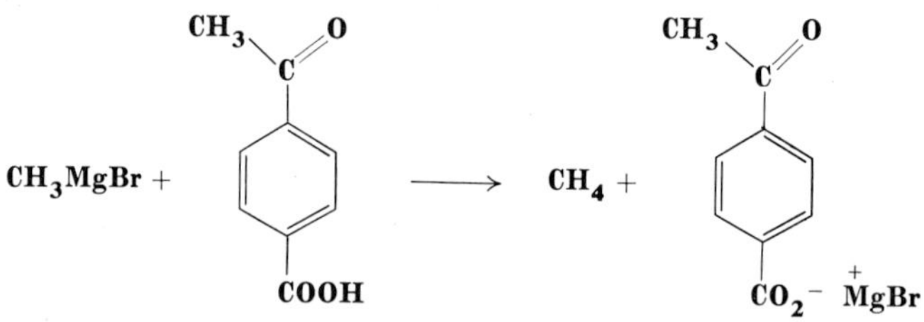

The salts are often precipitated in the solvents used for the Grignard reaction (ether or tetrahydrofuran) preventing the desired addition even when excess Grignard reagent is used. As a corollary, it is clear that the halogen compound precursor of the Grignard reagent cannot contain groups which react with an organomagnesium halide.

Grignard addition with aldehydes and ketones provides a straightforward way to build carbon chains. When this reaction is coupled with the oxidation of alcohols to carbonyl compounds, a wide variety of organic molecules can be constructed. Some typical examples are illustrated in Fig. 11.5.

Esters also react with Grignard reagents to form tertiary alcohols. For example, phenyl magnesium bromide and methyl benzoate produce triphenylcarbinol in good yield:

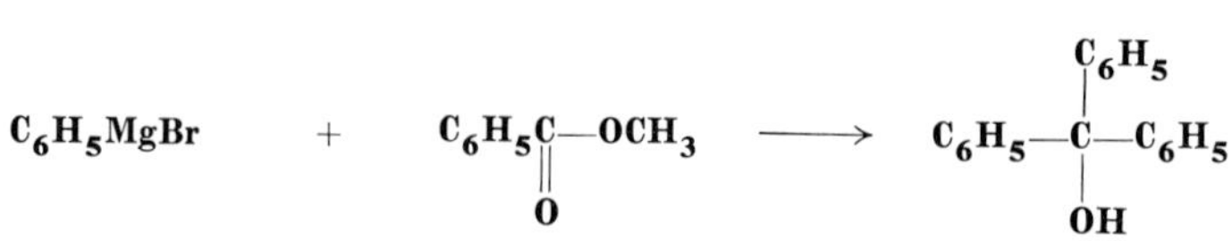

phenyl magnesium bromide — methyl benzoate — triphenylcarbinol

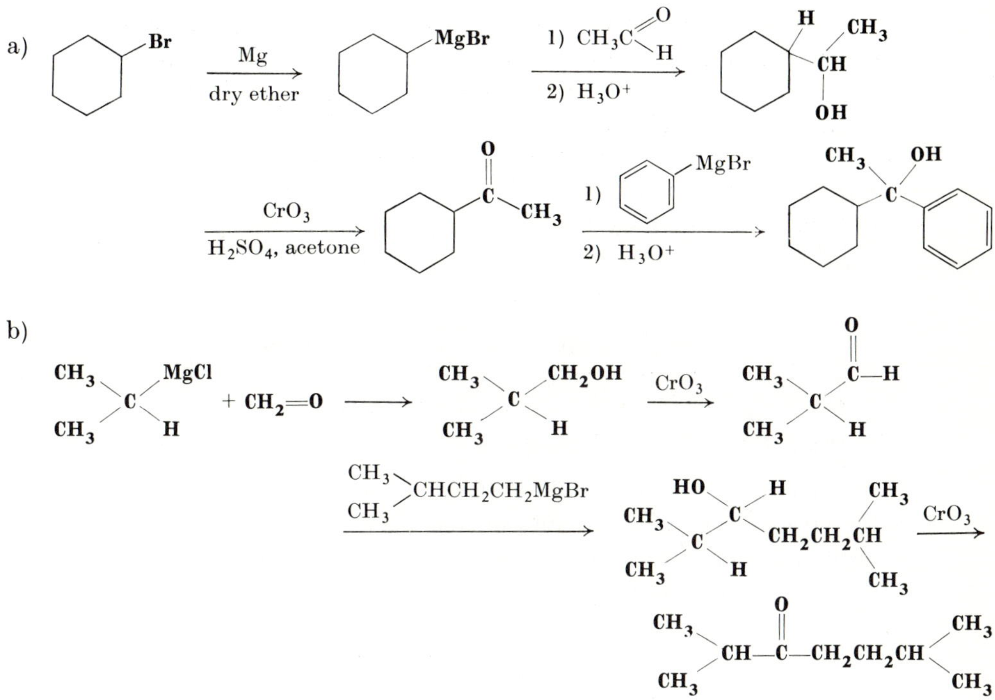

Fig. 11.5 Grignard syntheses

We may understand this reaction in terms of an ester and a Grignard reagent reacting initially to form a ketone via an addition elimination process:

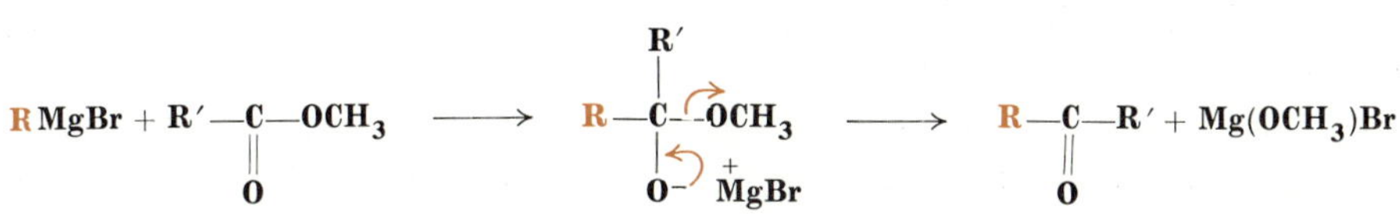

Since ketones are more reactive than esters (Section 11.2), the ketone thus formed reacts with another mole of Grignard reagent to give the tertiary alcohol:

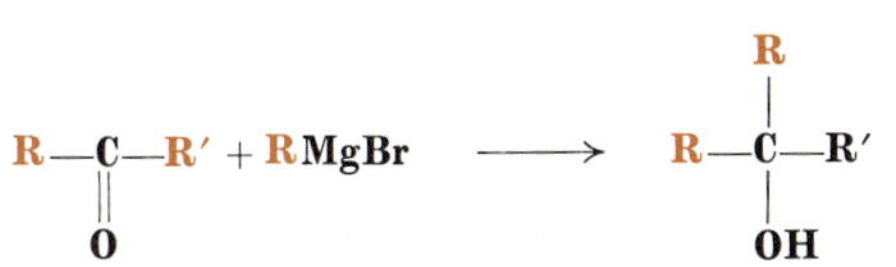

It is virtually impossible to isolate a ketone from the addition of a Grignard reagent and an ester.

Grignard addition with carbon dioxide is very useful in the synthesis of acids. The reaction is easy to carry out; the ether solution of the organomagnesium compound is poured onto solid carbon dioxide (Dry Ice).

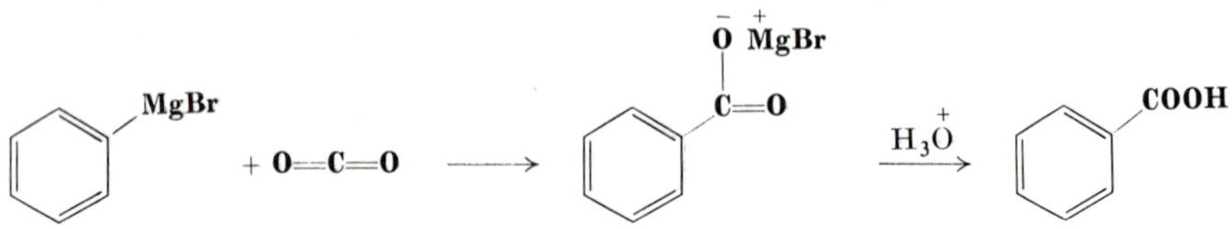

$$(CH_3)_3C\text{—}MgCl \xrightarrow[\text{2) } H_3O^+]{\text{1) } CO_2} (CH_3)_3CCOOH$$

11.10 ADDITION OF HYDROGEN

Aldehydes and ketones can be reduced to alcohols by molecular hydrogen in the presence of metal catalysts such as platinum, palladium, or nickel. The reaction is similar to the catalytic reduction of alkenes to alkanes with which the reader is already familiar.

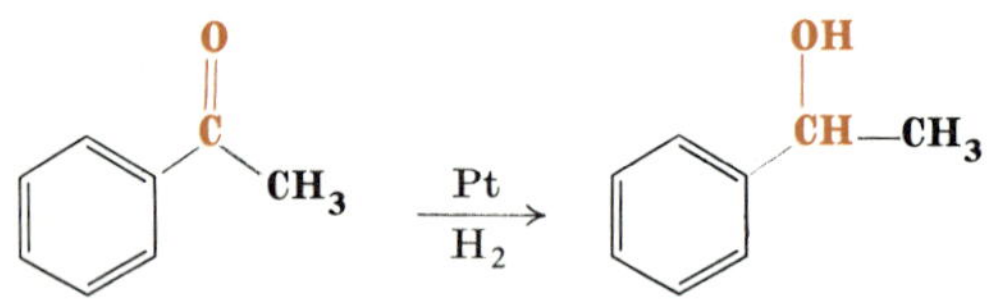

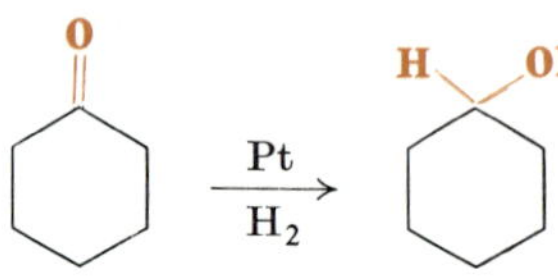

Carbonyl compounds can also be reduced by hydride-donating reagents (H^-), as we shall see in the next section.

11.11 LITHIUM ALUMINUM HYDRIDE AND SODIUM BOROHYDRIDE

Among the many hydride reagents, lithium aluminum hydride and sodium borohydride are the most widely used:

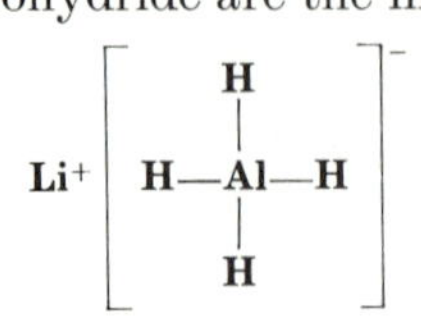

lithium aluminum hydride

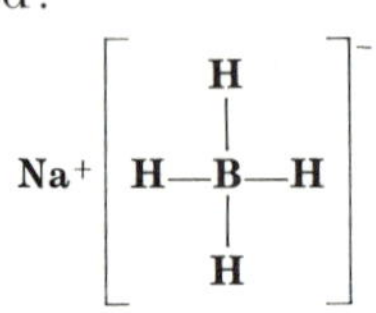

sodium borohydride

Lithium aluminum hydride reduces aldehydes and ketones to alcohols:

$$R_2C{=}O + LiAlH_4 \longrightarrow [(R_2CHO)_4Al]^- \quad Li^+$$

The initial product is the aluminum salt of the alcohol (aluminum alkoxide). In the isolation procedure (i.e., the work-up), the salt is treated with aqueous acid to yield the free alcohol:

$$[(R_2CHO)_4Al]^- \quad Li^+ \xrightarrow{H_3O^+} R_2CHOH$$

Note that one mole of lithium aluminum hydride will reduce four moles of carbonyl compound. The overall reaction may be written

$$R_2C{=}O \xrightarrow[\text{2) } H_3O^+]{\text{1) } LiAlH_4} R_2C(H){-}OH$$

The following is a typical example of lithium aluminum hydride reduction:

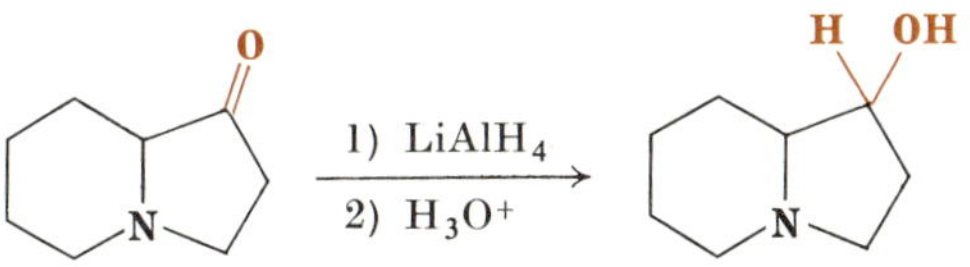

The mechanism of this reduction involves adding hydride ion (H^-) to the carbonyl carbon atom in a reaction analogous to the Grignard addition (Fig. 11.6).

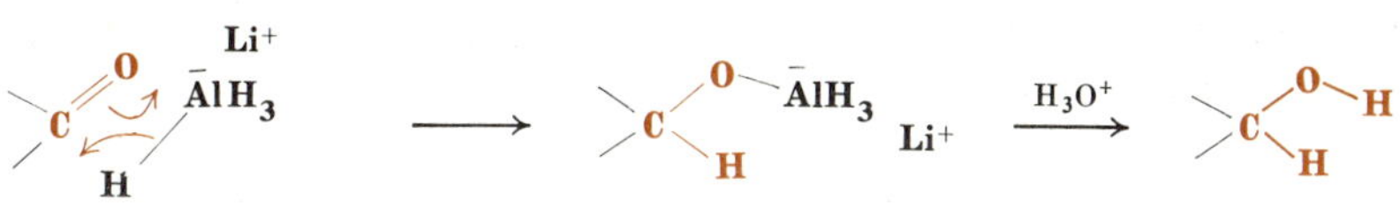

Fig. 11.6 Mechanism of $LiAlH_4$ reduction

There is a useful rule in organic chemistry which says that the *milder* a reagent, the more *selective* it is. Comparison of lithium aluminum hydride and sodium borohydride illustrates this generalization. Sodium borohydride is a much milder and less reactive reagent than lithium aluminum hydride. For example, in cold water, sodium borohydride is stable, whereas lithium aluminum hydride decomposes vigorously. Lithium aluminum hydride reduces not only aldehydes and ketones, but other types of carbonyl functions as well (esters and amides, for example, see Section 11.20). Sodium borohydride, being a milder reagent, is much more selective. It will react only with ketones and aldehydes, but not with less reactive carbonyl groups (see Section 11.2).

Thus, a compound such as the keto ester below may be *selectively* reduced to a hydroxyester using borohydride:

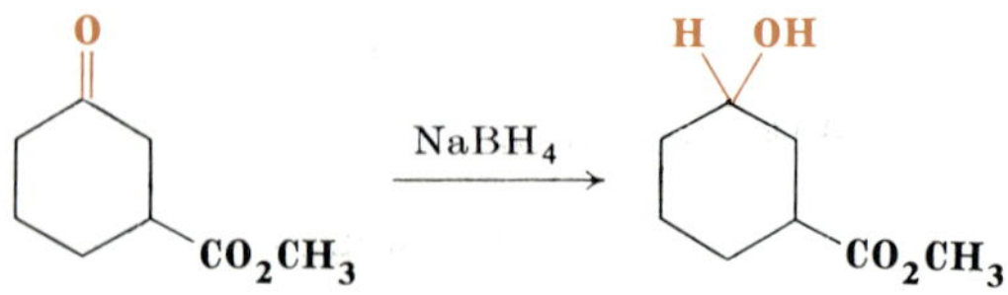

Most interesting organic compounds contain more than one functional group. Obviously, it is important to know which functional groups will and which will *not* react with a particular reagent. For example, since hydride-donating reagents reduce ketones *but not isolated carbon-carbon double bonds*, we could reasonably expect to carry out the following transformation in good yield:

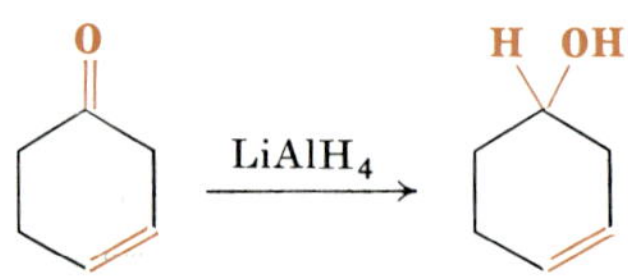

ADDITION–ELIMINATION

11.12 IMINES AND RELATED COMPOUNDS

As we noted earlier in our general discussion of carbonyl reactions many products of addition reactions undergo a further step, eliminating a group in order to re-form a trigonal carbon atom. Primary amines react with aldehydes and ketones via such an addition–elimination pathway:

$$R{-}NH_2 + \gt C{=}O \xrightarrow{H^+} \gt C{=}NR + H_2O$$

In the above equation R—NH_2 may be any substance with a primary amino group. Figure 11.7 lists some of the specific types of compounds that can be made by this reaction. Oximes, semicarbazones, and dinitrophenylhydrazones are most often used to characterize (identify) aldehydes and ketones since these nitrogen compounds are, in general, crystalline solids whose characteristic melting points can be compared with those of known substances.

When ketones and aldehydes are added to aqueous solutions of hydroxylamine (NH_2OH) at pH = 7, the IR carbonyl stretching band and the UV carbonyl absorption (see Chapter 6) rapidly disappear. Subsequently, IR and UV absorption corresponding to that of the oxime slowly appear. This latter absorption increase can be accelerated by the addition of acids. These results are consistent with the mechanism illustrated in Fig. 11.8. The carbonyl

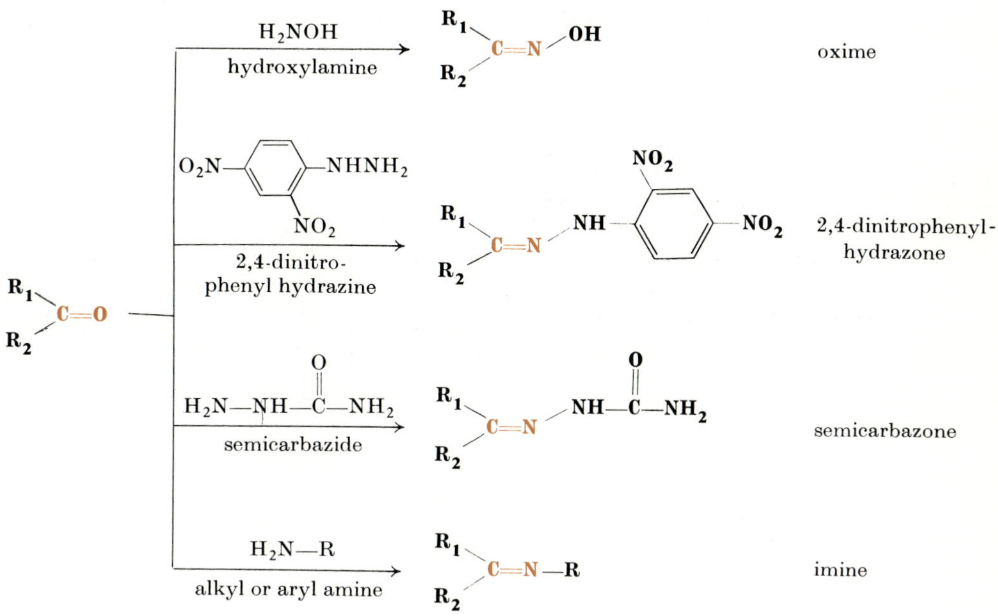

Fig. 11.7 Imine derivatives of the carbonyl group

compound and hydroxylamine are in *fast* equilibrium with the addition product. The slow reaction is the acid-catalyzed elimination of water from the neutral adduct.

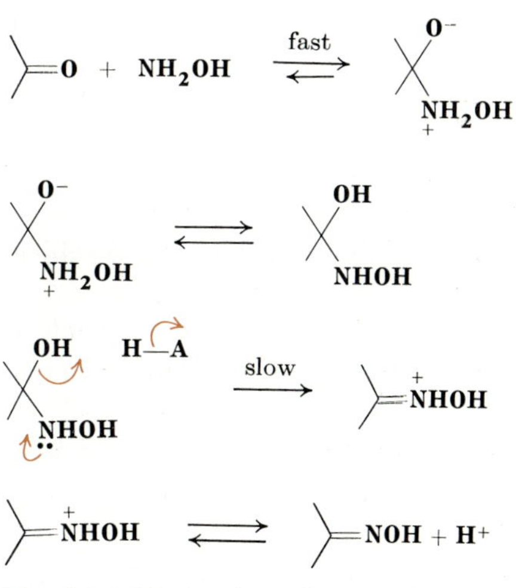

Fig. 11.8 Mechanism of oxime formation

11.13 WITTIG REACTION

The reactions of alkyl halides with trivalent phosphorus compounds are analogous to their reactions with tertiary amines to form quaternary salts (see Section 7.10). For example, methyl bromide and triphenylphosphine react to form a phosphonium bromide:

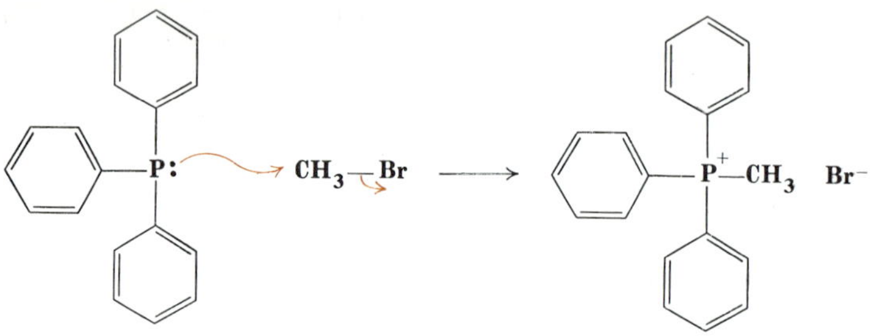

triphenylphosphine triphenylphosphonium bromide

The phosphonium salt is a weak acid. If the salt is treated with a powerful base such as butyllithium, a dipolar molecule called an *ylide* is formed:

$$(C_6H_5)_3\overset{+}{P}—CH_3\ Br^- + BuLi \longrightarrow (C_6H_5)_3\overset{+}{P}—\overset{-}{C}H_2 + BuH + LiBr$$

The methyl group of the phosphonium salt is weakly acidic because the negative charge on the conjugate base is stabilized by the inductive effect of the neighboring positive atom and also because the ylide is resonance stabilized:

$$[(C_6H_5)_3\overset{+}{P}—\overset{-}{C}H_2 \longleftrightarrow (C_6H_5)_3P{=}CH_2]$$

triphenylphosphonium methylide

Note that phosphorus, with its bonding electrons in the 3-shell, is not subject to the octet rule.

The phosphorus ylide can be added to a carbonyl compound such as cyclohexanone to yield a new salt which may subsequently undergo elimination.

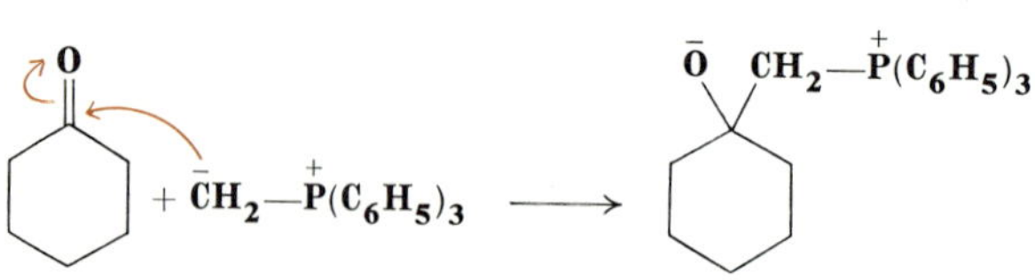

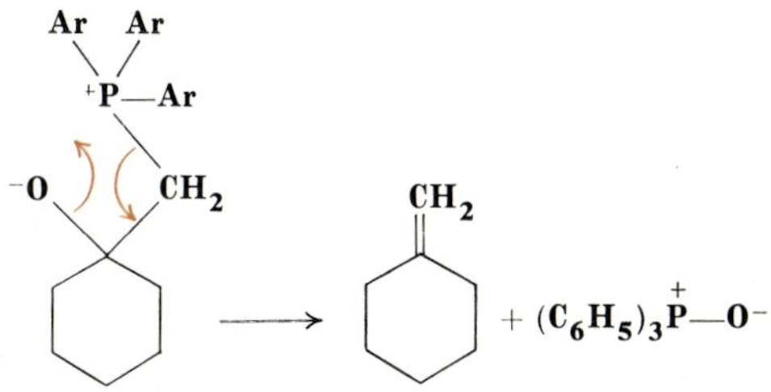

The final product is an olefin in which the carbon-oxygen double bond of the carbonyl compound has been replaced by a carbon-carbon double bond. This type of olefin-forming reaction is called the *Wittig reaction*; it is particularly useful in syntheses because it allows the construction of a double bond at a specific location in a molecule. For example, 1,2-distyrylbenzene can be prepared using a dibromide, triphenylphosphine, and benzaldehyde:

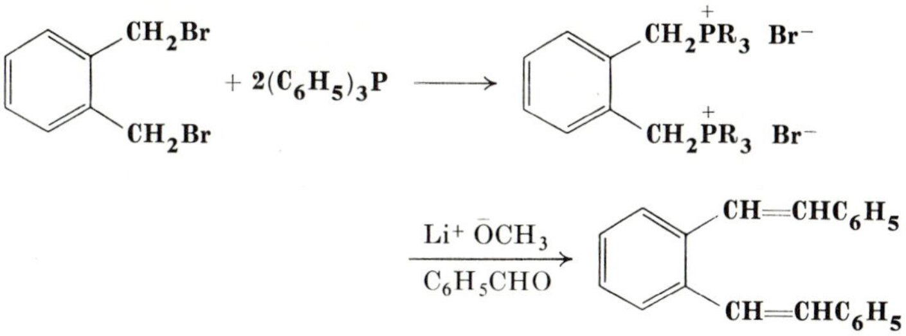

11.14 ACETAL AND KETAL FORMATION

We have already examined the formation of hemiacetals and hemiketals. These compounds can, in turn, be converted into dialkoxy compounds called *acetals* and *ketals* using a pathway that requires an elimination step.

Hemiacetals can undergo acid-catalyzed elimination of water:

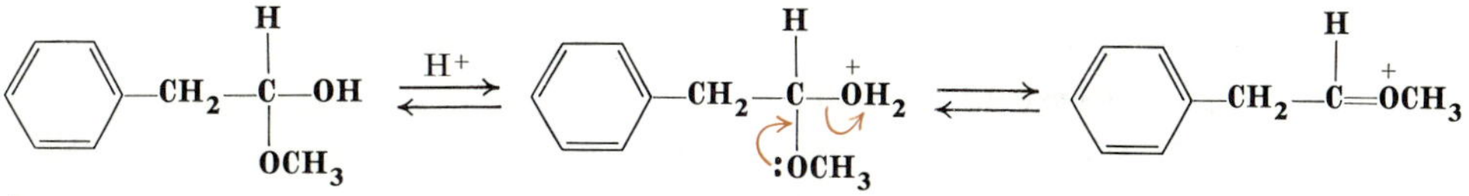

In the presence of methanol, the resulting oxonium ion intermediate adds alcohol to form an acetal:

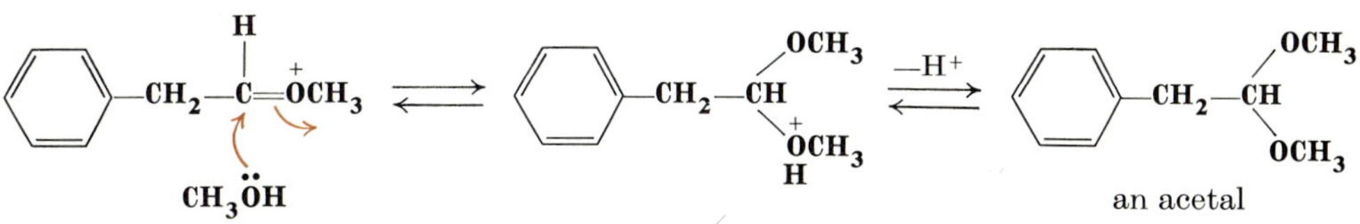

One of the most useful types of ketals is the cyclic ketal formed from a ketone and ethylene glycol:

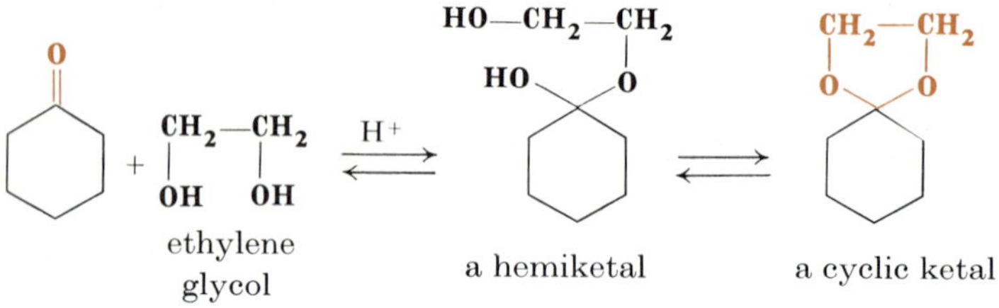

A ketone is usually converted into its cyclic ketal to *protect* a ketone function in a basic medium. Ketals are stable to base, but they will hydrolyze back to the ketone in aqueous acid. If a reaction is to be carried out with a nonacidic reagent on one part of a molecule, while another part of the molecule carries a labile ketone function which would interfere with the reaction, the ketone could be protected by being converted into its cyclic ketal. After the reaction the ketone could be regenerated by acid hydrolysis. A good example of this procedure is given below. If the ketone had not been protected, the Grignard reagent would have added to the ketone.

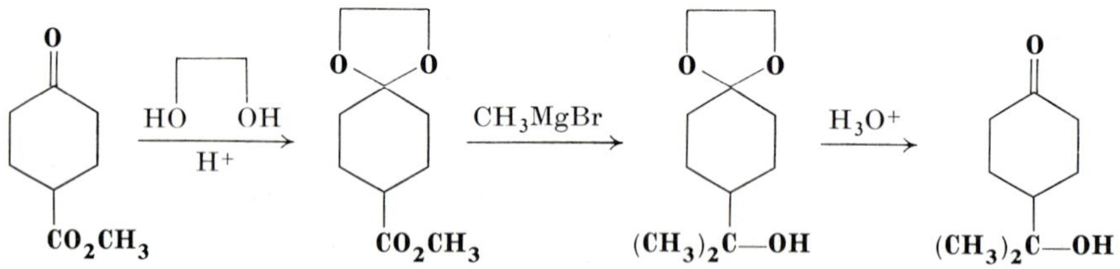

11.15 ADDITION–ELIMINATION REACTIONS OF ACIDS AND THEIR DERIVATIVES

Carboxylic acids and their derivatives undergo a variety of reactions which can be classified as addition–elimination sequences. The mechanism for most of these involves the addition of a nucleophile—either neutral or ionic—to the carbonyl carbon, followed by loss of one of the original groups on this carbon. In general form, this mechanism is

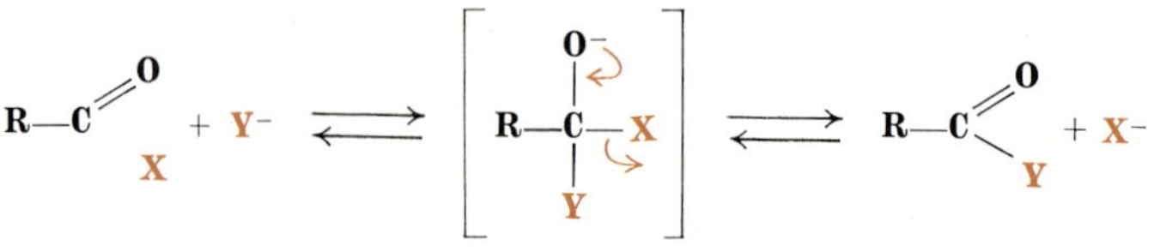

These reactions are potentially reversible since the eliminated group (X^-) is itself a nucleophile, and in many cases where the initial reactant and

the product do not differ greatly in energy, the equilibrium may not be overly rich in product. Before citing important specific examples of this reaction, we will outline the mechanism of one particular reaction, the basic hydrolysis of an ester to an acid:

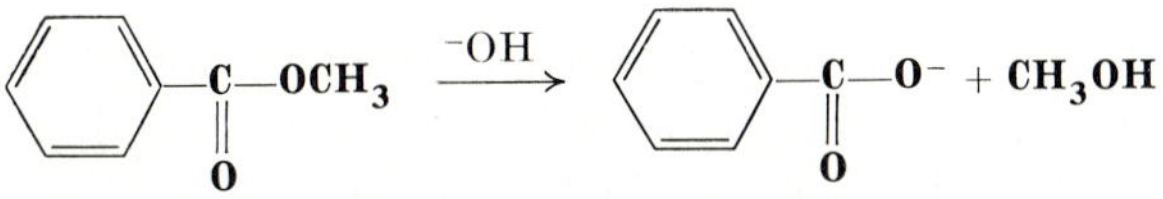

methyl benzoate benzoate anion

First we should point out that the basic hydrolysis of an ester happens to be irreversible. The reaction product (benzoic acid, in the above example) is obviously in an anionic state in the strongly basic solution. Because the negatively charged group is resonance stabilized, carboxylate anions are unreactive toward nucleophilic attack, precluding the possibility of a reverse reaction.

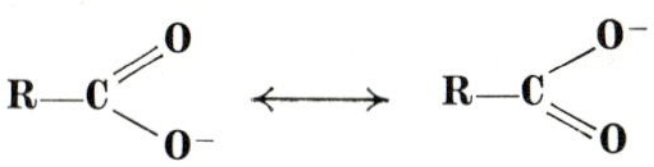

Basic ester hydrolysis is found to be second order; that is,

$$\text{rate} = k_2[\text{ester}][^{-}\text{OH}]$$

One might imagine two mechanisms for a hydrolysis involving an ester molecule and a hydroxide ion in the rate-determining step:

1) methyl-oxygen cleavage

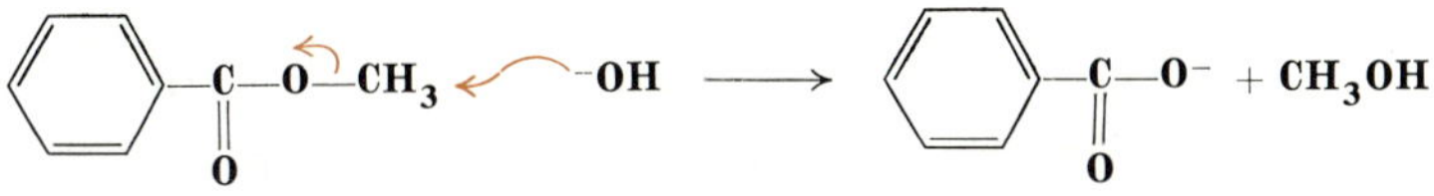

2) carbonyl carbon-oxygen cleavage

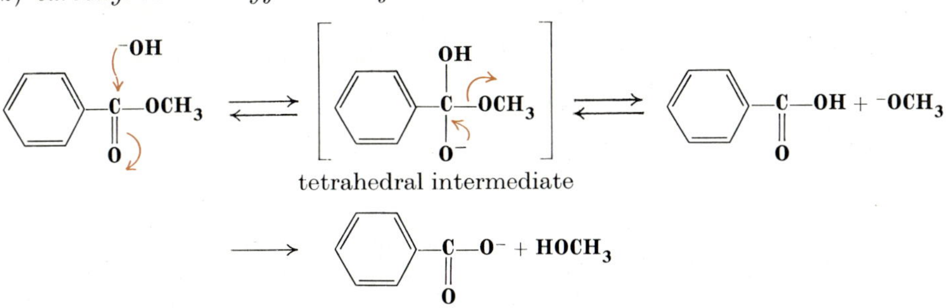

tetrahedral intermediate

We can demonstrate that the second mechanism is correct by hydrolyzing methyl benzoate labeled with ^{18}O.

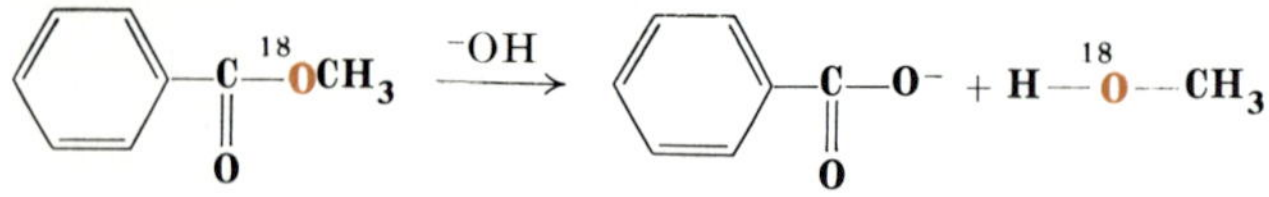

Of the two products, methanol and benzoate, only the methanol was found to contain ^{18}O. If we examine the two alternatives we will find that this result is consistent only with the second mechanism, in which the hydroxide adds to the carbonyl to form a *tetrahedral intermediate* which subsequently collapses to products.

The ^{18}O results are also consistent with a direct S_N2 displacement at the carbonyl carbon:

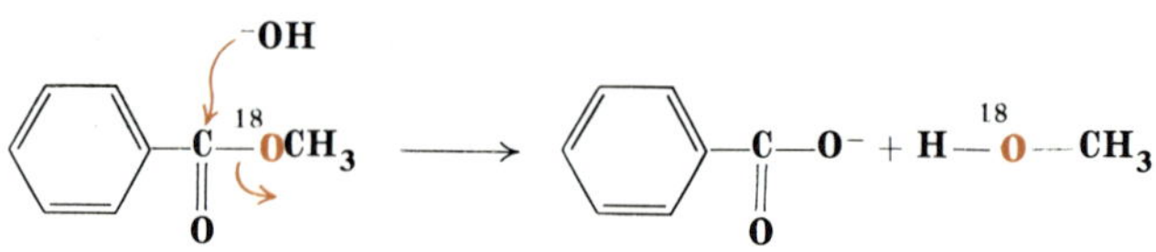

There is good evidence, however, that this mechanism is incorrect.

11.16 ESTER HYDROLYSIS AND FORMATION IN ACIDIC MEDIA

Esters can be hydrolyzed in aqueous acid as well as in aqueous base:

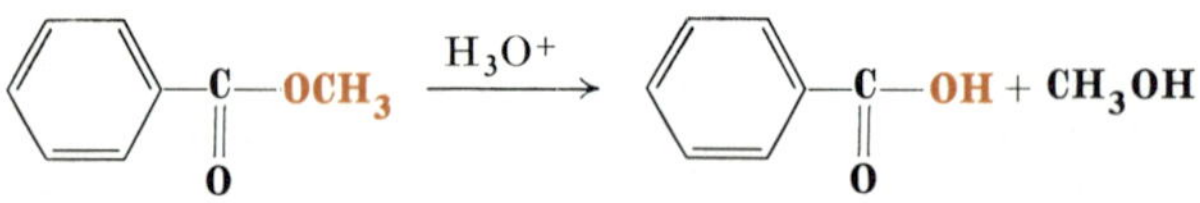

The mechanism for this acid-catalyzed ester hydrolysis is given in Fig. 11.9. Note that every step is reversible.

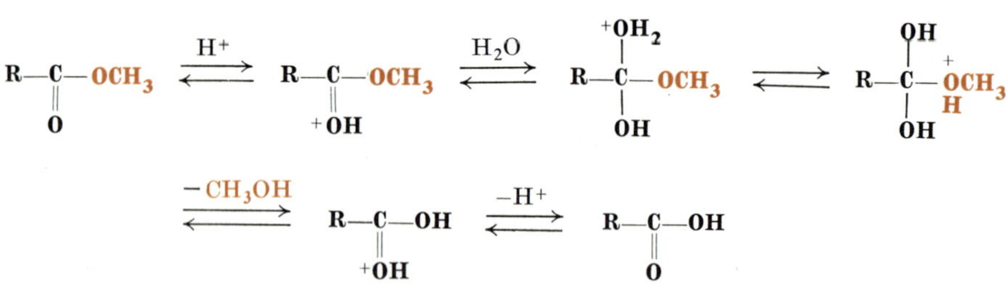

Fig. 11.9 Mechanism of acid-catalyzed ester hydrolysis

In order to complete the reaction and obtain a good yield of product, it is necessary to use a large excess of water. Esters can also be produced from acids using an excess of alcohol. The mechanism for preparing an ester from an acid is the exact reverse of hydrolysis. For example, methyl benzoate can be prepared, by adding benzoic acid to methanol saturated with dry HCl gas. The amount of water formed in the esterification is too small, compared to the amount of methanol, to cause any serious back reaction.

In the mechanism for acid-catalyzed ester hydrolysis (Fig. 11.9), the ester carbonyl group is initially protonated thereby increasing the partial positive charge on the carbonyl carbon and accelerating the nucleophilic addition by water. The tetrahedral intermediate has three weakly basic oxygens, any one of which can be protonated and bear a positive charge. Thus in acidic solutions there are steady-state concentrations of three tetrahedral intermediates which are in equilibrium with each other and differ only in which oxygen happens to be protonated. Collapse of a tetrahedral intermediate to the acid product undoubtedly involves the one possessing a protonated methoxy group. With this intermediate, methanol rather than methoxide (a poor leaving group) is lost.

11.17 ACID CHLORIDES

Acid chlorides, which are very useful for preparing carboxylic acid derivatives, are generally prepared by treating carboxylic acids with thionyl chloride ($SOCl_2$) or phosphorus pentachloride (PCl_5):

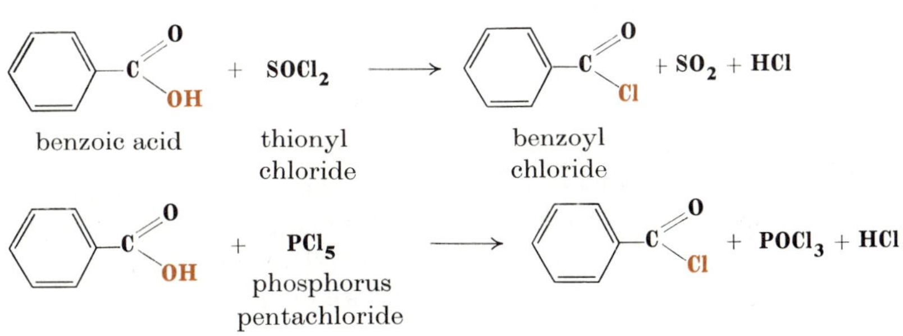

The thionyl chloride method is particularly convenient since the inorganic products are gases, which are easily removed.

Acid chlorides can be converted into amides, esters, and acid anhydrides by treating them with primary or secondary amines, alcohols, or salts of carboxylic acids, respectively (Fig. 11.10).

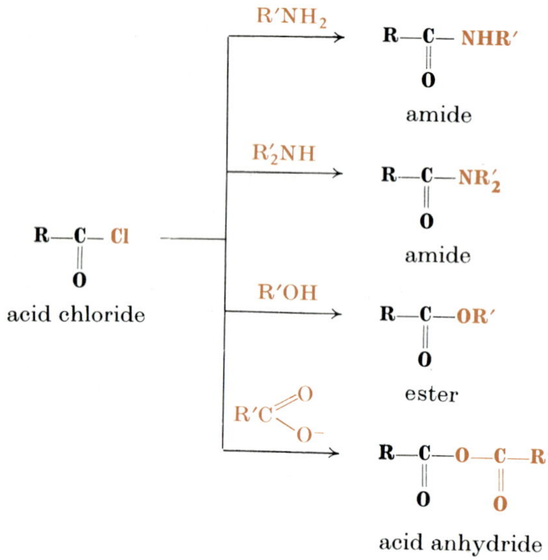

Fig. 11.10 Reactions of acid chlorides

Note that it is not possible to prepare an amide by simply mixing a carboxylic acid with an amine, because the product is a stable salt in which the amine is protonated and therefore no longer a nucleophile, and the acid is anionic and therefore no longer subject to nucleophilic attack:

$$RCOOH + R'NH_2 \longrightarrow RCOO^- \; R'\overset{+}{N}H_3$$

Amines will react with esters to form amides.

$$RCOOCH_3 + R'NH_2 \longrightarrow RCONHR' + CH_3OH$$

However, the reaction is slow and an excess of amine must be used. If the amine is valuable, this is prohibitive. Preparation of acid derivatives via acid chlorides is often the method of choice.

Using acid chlorides to prepare acid derivatives has two main advantages: the reactions are fast and they are irreversible. The reactivity of acid chlorides is related to the highly electron-deficient carbonyl carbon resulting from polarization of the electrons in the direction of both oxygen and chlorine:

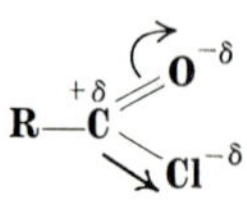

The reaction of, for example, an acid chloride with an alcohol to form an ester is irreversible because the chloride ion is a weak nucleophile and cannot add to the ester carbonyl group in order to displace the alcohol.

Esters of tertiary alcohols are made by employing an alkoxide ion rather than the alcohol itself:

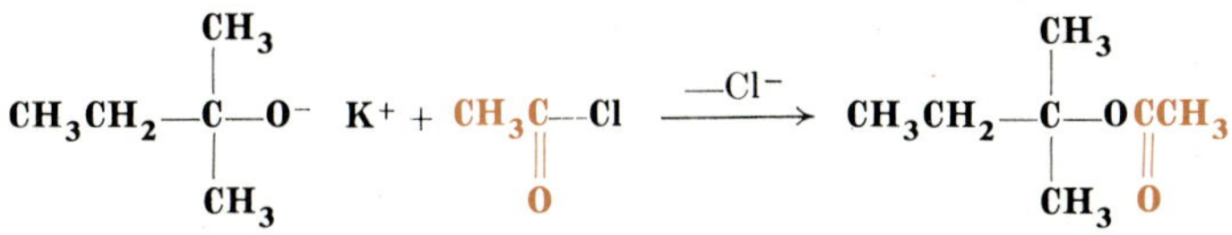

This way, the by-product is potassium chloride rather than hydrogen chloride. Esters of tertiary alcohols are prone to acid-catalyzed eliminations, and the alkoxide method precludes danger from this reaction.

Specific examples of reactions involving acid chlorides are shown below:

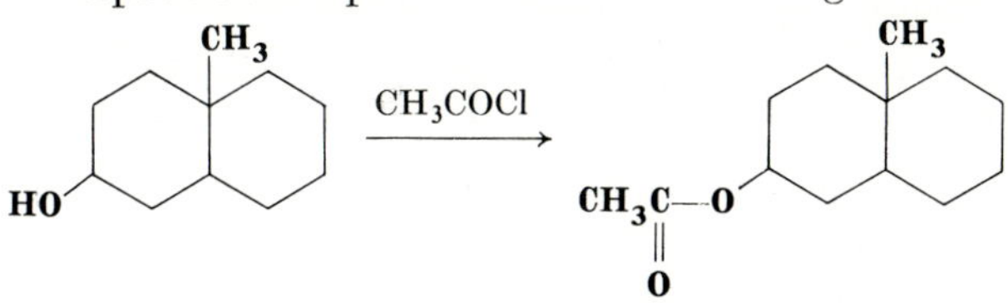

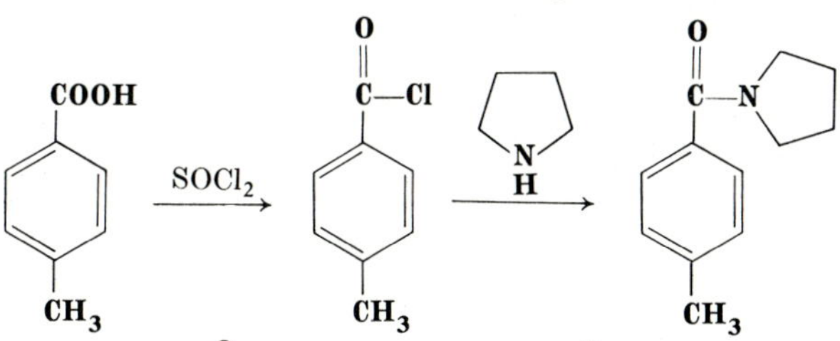

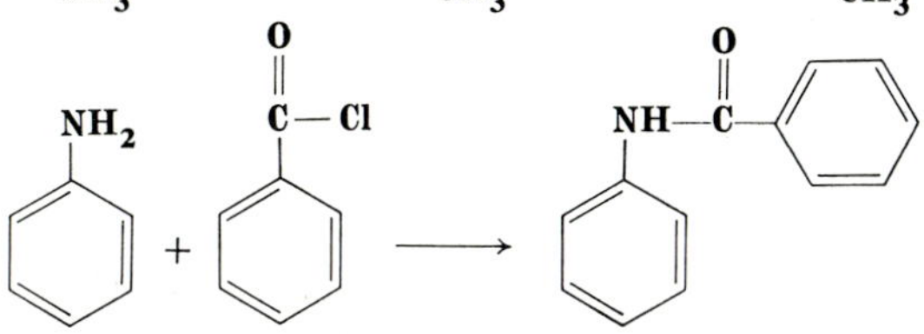

Two other acid chloride reactions should be mentioned. The first is the reduction of acid chlorides to aldehydes using a modified hydride reagent, lithium tri-*t*-butoxyaluminum hydride:

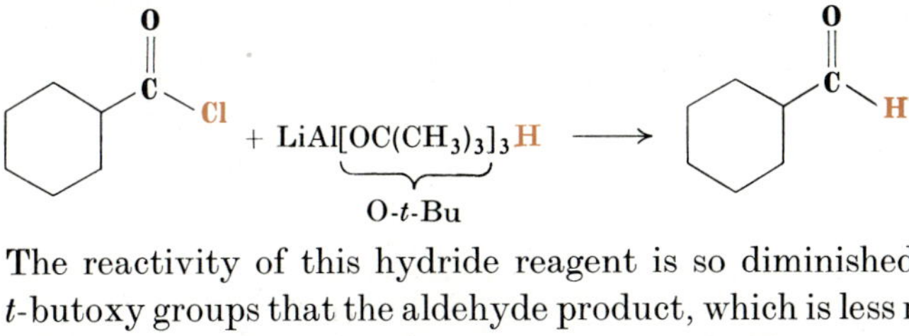

The reactivity of this hydride reagent is so diminished by the three bulky *t*-butoxy groups that the aldehyde product, which is less reactive than the acid chloride (see Section 11.2), is not reduced. Here is another example of a relatively unreactive reagent displaying selectivity.

The second reaction is the conversion of acid chlorides into ketones by treating them with organocadmium compounds:

$$RMgCl + CdCl_2 \longrightarrow RCdCl + MgCl_2$$

$$R'COCl + RCdCl \longrightarrow R'-\underset{\underset{\displaystyle O}{\|}}{C}-R$$

Organocadmium compounds are less reactive than the corresponding magnesium compounds, and they add very slowly to ketones compared to acid chlorides.

11.18 ACID ANHYDRIDES

Acid anhydrides, prepared from acid chlorides and acid anions (Fig. 11.10), are efficient acylating agents. For example, a good way to prepare an amide of acetic acid is to treat an amine with commercially available acetic anhydride:

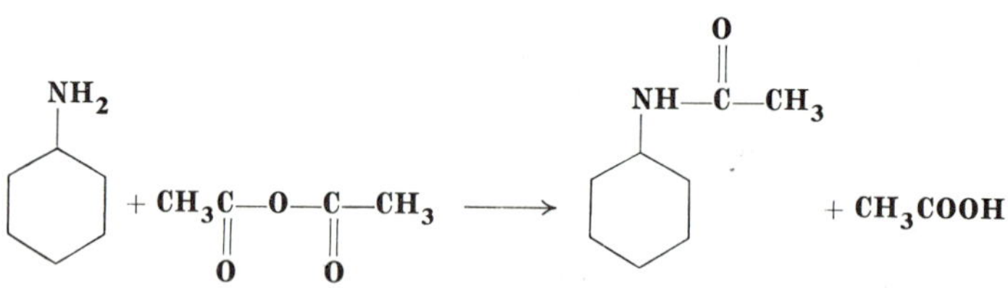

acetic anhydride
(Ac_2O)

11.19 AMIDE HYDROLYSIS

Acid chlorides, esters, and amides can be hydrolyzed back to carboxylic acids (see Section 11.16), the process being fastest for chlorides and quite slow for amides. Thus, acid chlorides decompose when added to water at room temperature. In contrast, the hydrolysis of amides requires hot aqueous acid or base. Nevertheless, the amide hydrolysis proceeds with good yield so that acylation is useful for protecting an amino group:

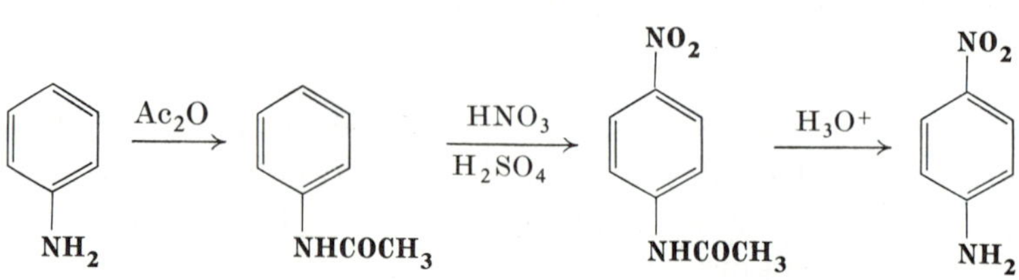

The amide group is extremely important biologically because proteins are composed of amino acids linked by amide bonds. We will discuss the amide group in biological systems in Chapter 15.

11.20 REDUCTION OF ACID DERIVATIVES

Esters are reduced by lithium aluminum hydride to primary alcohols via an addition–elimination mechanism:

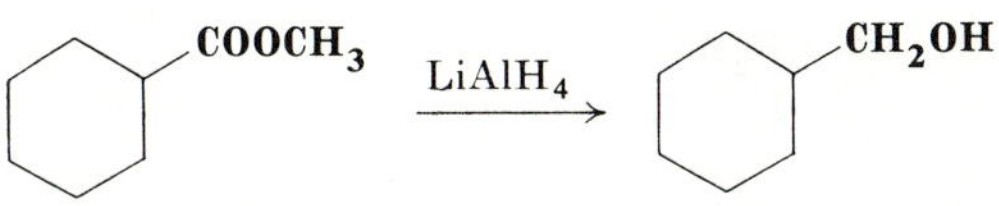

The mechanism of this reduction is shown below (with the reducing agent represented as hydride ion for the sake of simplicity):

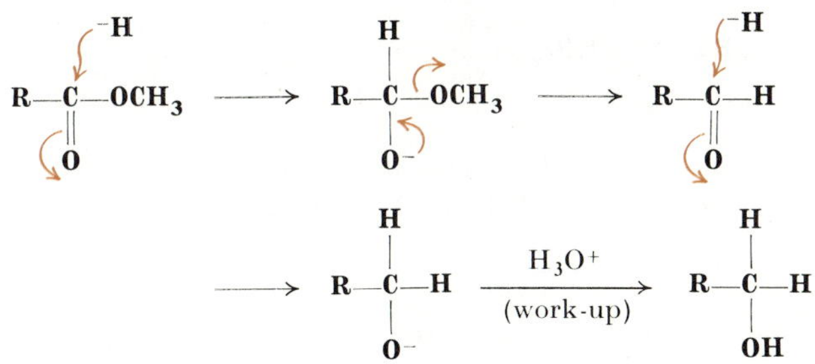

Amides are converted into amines with lithium aluminum hydride:

ENOLIZATION–KETONIZATION REACTIONS OF CARBONYL COMPOUNDS

11.21 HALOGENATION

When acetophenone is treated with bromine in a sodium hydroxide solution, two products are formed, sodium benzoate and bromoform:

$$\underset{\text{acetophenone}}{C_6H_5—\overset{\displaystyle O}{\overset{\|}{C}}—CH_3} + Br_2 \xrightarrow{\text{NaOH}} \underset{\text{sodium benzoate}}{C_6H_5—\overset{\displaystyle O}{\overset{\|}{C}}—\bar{O}Na^+} + \underset{\text{bromoform}}{CHBr_3}$$

This is an example of the so-called *haloform reaction*. It also occurs with chlorine and iodine, and is limited to methyl ketones ($RCOCH_3$).

The course of the haloform reaction depends on the formation of the enolate ion of the methyl ketone, in this case, acetophenone:

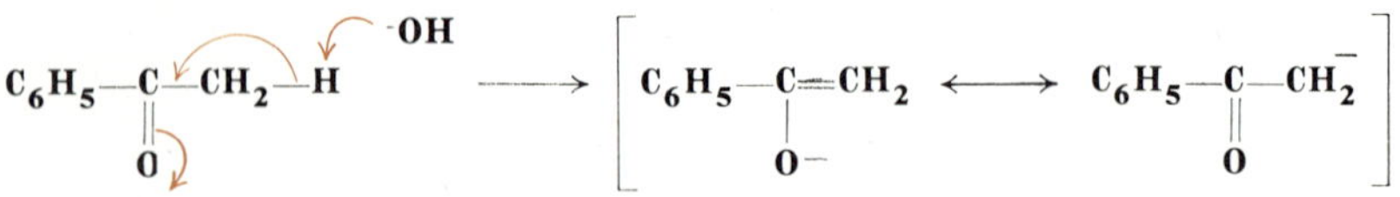

an enolate anion

The enolate reacts with bromine in the following manner:

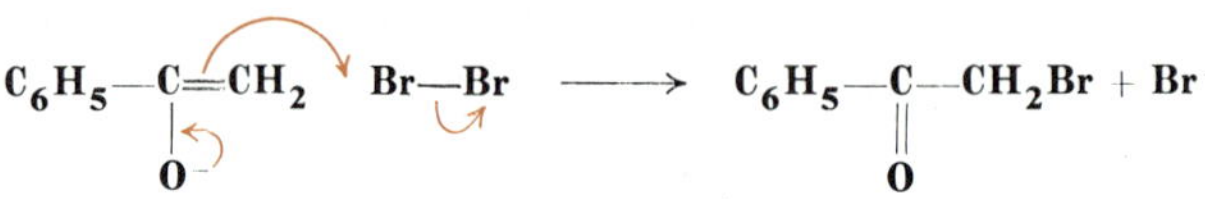

The initial product of the haloform reaction, monobromoacetophenone, is more acidic than acetophenone itself (why?), and thus the sequence repeats, finally yielding tribromoacetophenone:

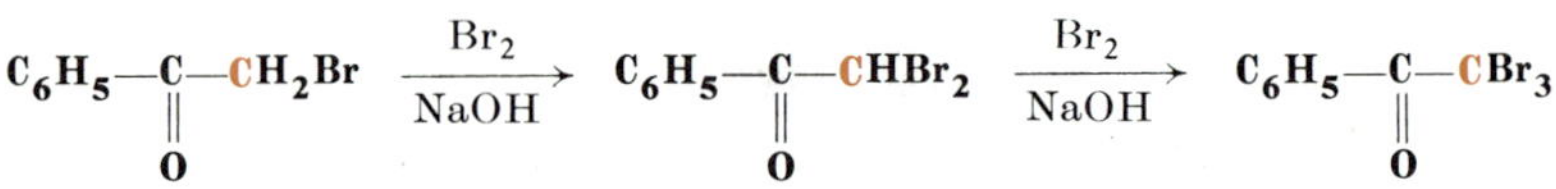

Tribromoacetophenone then undergoes cleavage via an addition–elimination pathway, giving bromoform and the salt of the carboxylic acid.

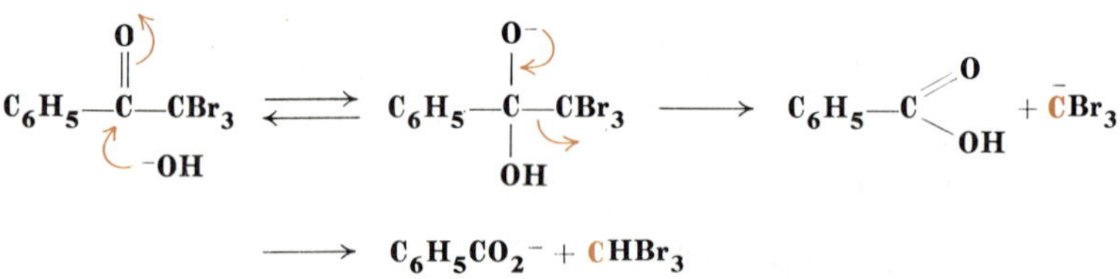

Carbon-carbon cleavage, a key step in the haloform mechanism, is made possible by the presence of three electronegative halogen atoms which inductively stabilize the developing carbanion. In other words, the halogens reduce the negative charge on the nonelectronegative carbon by bearing some of it themselves via inductive withdrawal:

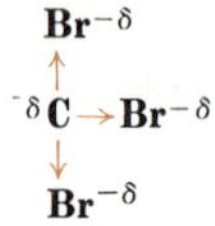

Resonance stabilization may also play a role in reducing the energy of the carbanion and facilitating its formation:

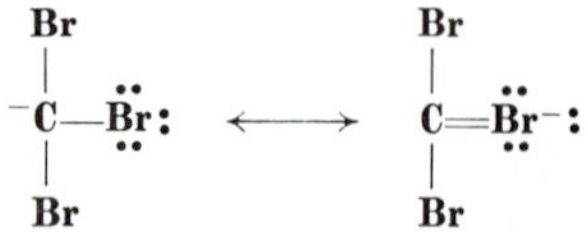

Note that the bromine in the second contributor is surrounded by 10 electrons. This is possible only with large atoms whose bonding electrons are in the 3-shell (which can hold a maximum of 18 electrons, as opposed to 8 in the 2-shell).

A third factor which may account for the carbon-carbon cleavage is *steric acceleration*. Both of the carbons engaged in the cleavage are fully substituted with bulky groups. When the carbon-carbon bond is cleaved, the steric interactions between the groups are relieved.

Thus, the production of bromoform is nicely rationalized in terms of inductive, resonance, and steric effects, the three most important factors chemists use to explain relative reactivities of organic compounds.

The haloform reaction proceeds in good yield and therefore can be used for synthetic purposes. For example, 5-methylcaproic acid can be prepared from the corresponding methyl ketone by treating the ketone with bromine and sodium hydroxide and then acidifying the product:

$$(CH_3)_2CHCH_2CH_2CH_2\overset{\overset{O}{\|}}{C}-CH_3 \xrightarrow[2)\ H_3O^+]{1)\ Br_2,\ NaOH} (CH_3)_2CHCH_2CH_2CH_2COOH$$

Halogenation of ketones in base is usually not useful for other than methyl ketones. A ketone such as cyclohexanone, with four α hydrogens, yields a mixture of halogenated products under haloform conditions. In contrast, *acid*-catalyzed bromination of ketones is a controllable process, and a single hydrogen can often be substituted:

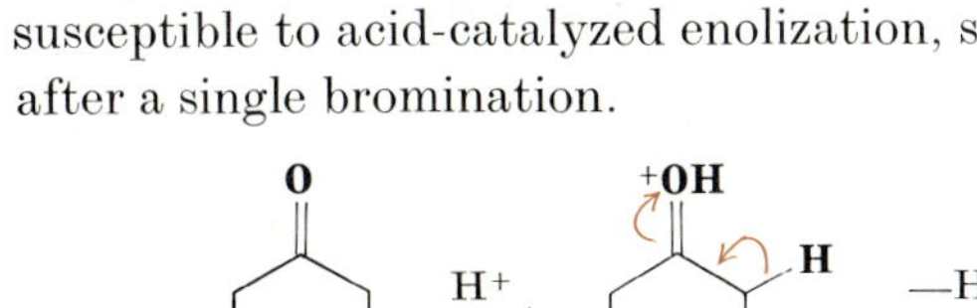

In this reaction the enol, rather than the enolate, undergoes the bromination. The product is a *weaker* base than the starting material, and therefore less susceptible to acid-catalyzed enolization, so that the reaction can be stopped after a single bromination.

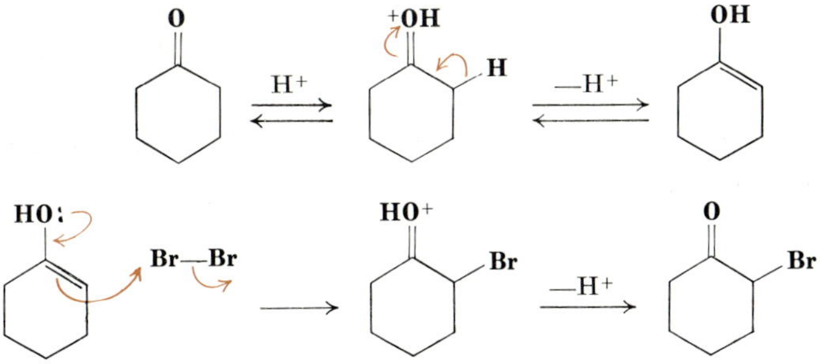

11.22 ALKYLATION

Enolate ions of ketones can be alkylated by alkyl halides in a substitution reaction.

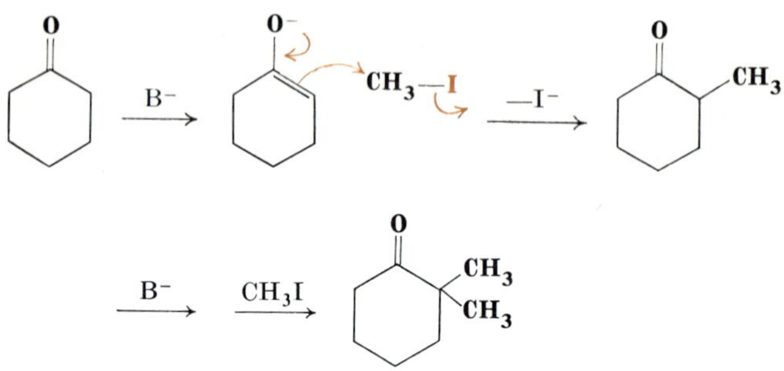

A strong base, such as sodium ethoxide ($\overset{+}{Na}\overset{-}{O}CH_2CH_3$), is commonly employed. However, the problem with this reaction is that it is difficult to stop at the desired monoalkyl stage and usually continues until it yields dialkylation product.

11.23 ALDOL CONDENSATION

Enolate ions (and often enols) of aldehydes, ketones, esters, and acid anhydrides undergo *condensation reactions*. Condensations can be viewed as a combination of enolization and addition pathways. The simplest condensation reaction is called the *aldol condensation* because the product is an *ald*ehyde-alcoh*ol* (Fig. 11.11).

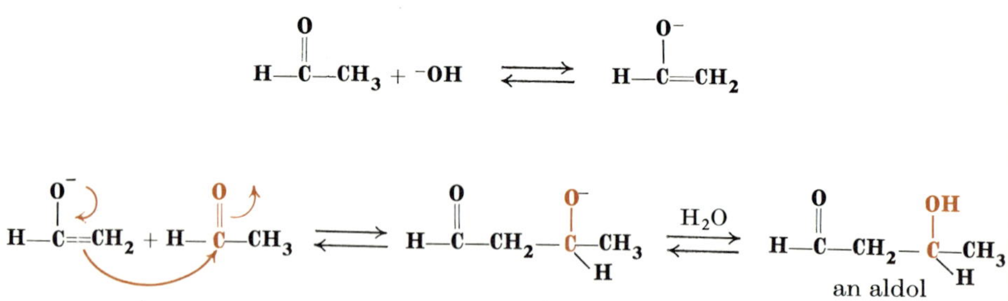

Fig. 11.11 Aldol condensation

The aldol condensation results in the formation of a new carbon-carbon bond and is therefore very useful in the synthesis of carbon chains and rings.

Frequently, particularly at elevated temperatures, the aldol dehydrates under the basic conditions of the condensation:

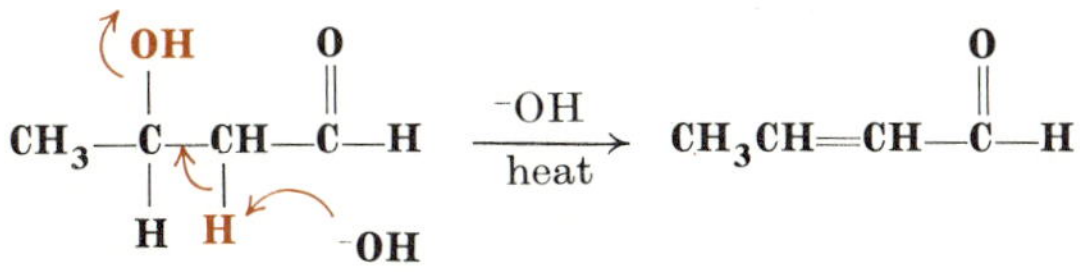

In fact, dehydration is often necessary, because the aldol condensation is a reversible reaction, and the dehydration may be needed to drive the process to completion.

Aldol condensations are usually practical for synthesis only when the ketone or aldehyde is allowed to condense with itself. If the aldol condensation is carried out with two different carbonyl compounds, four products can be formed. Intramolecular mixed aldol condensations are an exception. A prime example is the formation of a bicyclic six-membered ring system illustrated below:

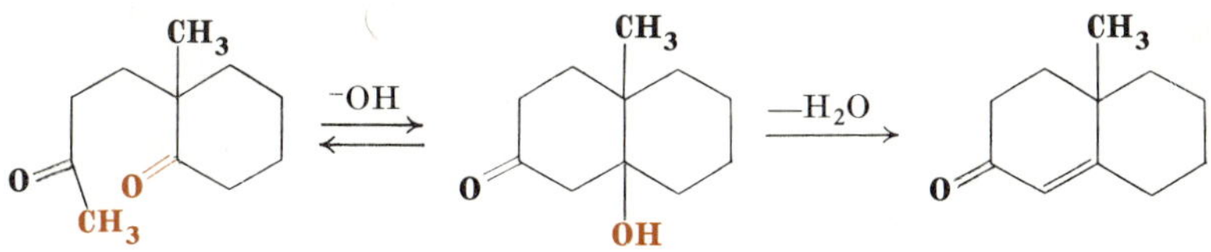

This reaction (combined with the Michael condensation which we will discuss soon) has been widely applied to the synthesis of biologically important compounds.

11.24 CLAISEN-SCHMIDT CONDENSATION

Claisen-Schmidt reaction is a name given to a special type of mixed aldol condensation in which the "acceptor" is benzaldehyde. The product is usually the dehydrated α,β-unsaturated ketone:

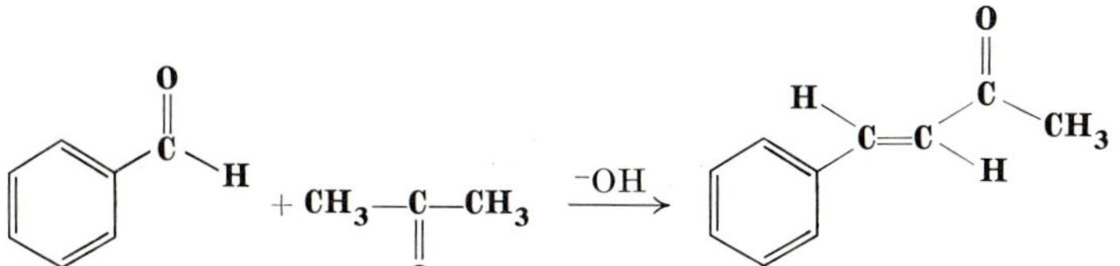

This reaction proceeds well because the benzaldehyde, with no α hydrogens, cannot undergo self-condensation.

11.25 MANNICH CONDENSATION

We have already discussed the reaction of amines with ketones and aldehydes to form nitrogen analogues of carbonyl compounds:

$$\rangle C{=}O + RNH_2 \longrightarrow \rangle C{=}NR$$

The product of this reaction, an imine, will undergo condensation with an aldehyde or ketone in a manner similar to the aldol condensation. The re-

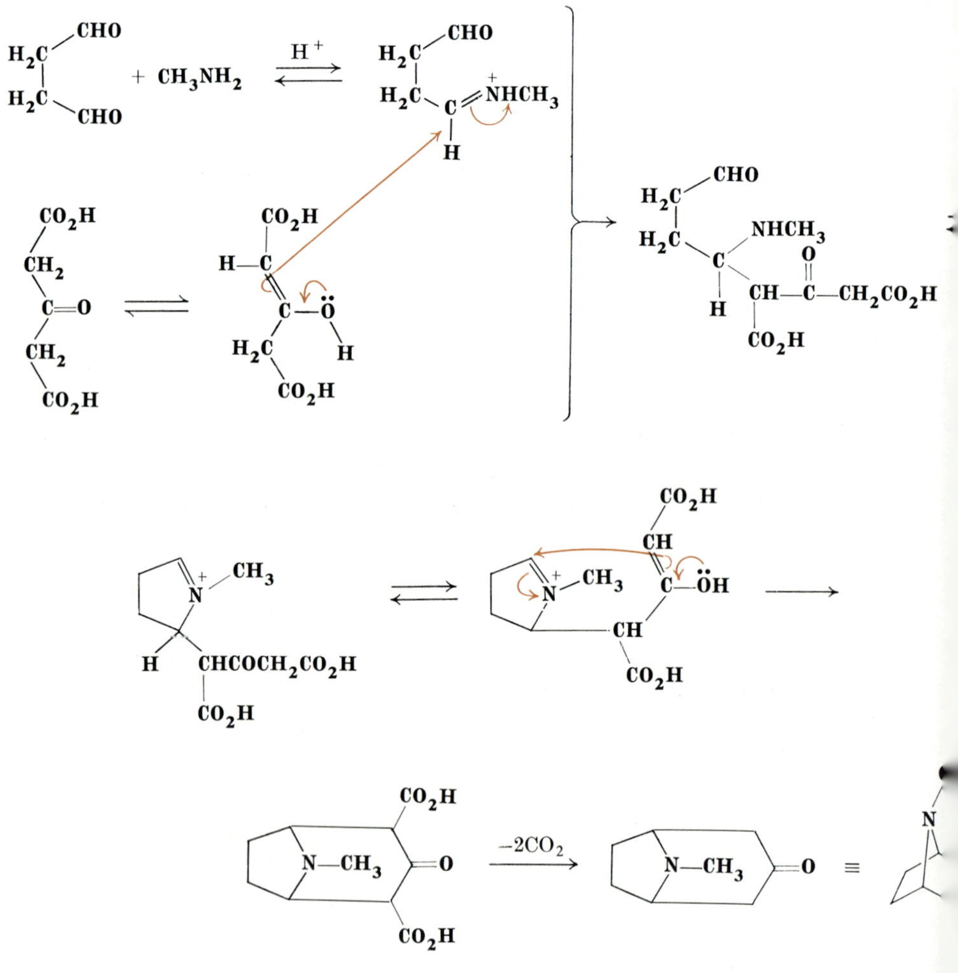

Fig. 11.12 Application of Mannich condensation to synthesis

action, which is general-acid catalyzed, is called the *Mannich condensation*:

$$CH_3{-}\overset{\overset{\displaystyle O}{\|}}{C}{-}CH_3 + CH_2{=}O + NH_2CH_3 \xrightarrow{H^+} CH_3{-}\overset{\overset{\displaystyle O}{\|}}{C}{-}CH_2CH_2NHCH_3$$

The first step is the formation of the imine, which in acid solution will exist as its *immonium* salt:

$$H_2C{=}O + CH_3NH_2 \xrightarrow{H^+} H_2C{=}\overset{+}{N}HCH_3 + H_2O$$

This salt behaves like a carbonyl group (consider the electronegativity of the protonated nitrogen), and will condense with the *enol* of acetone:

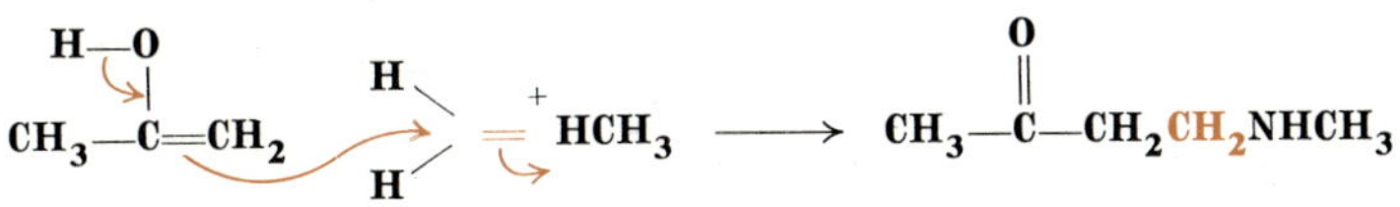

The Mannich condensation is particularly interesting because it appears to be similar to routes by which many compounds are synthesized in living systems. For example, the ring system of cocaine and related compounds,

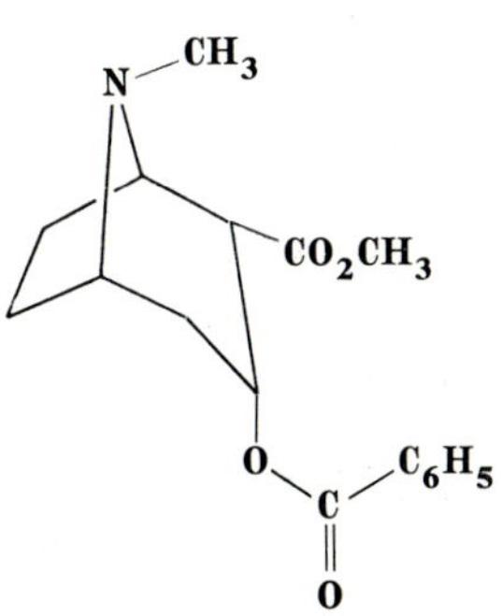

can be prepared from a dialdehyde, methylamine, and acetone dicarboxylic acid under "physiological conditions" (pH values not too far from neutrality, at room temperature).

11.26 PERKIN CONDENSATION

Acid anhydrides undergo aldol-like condensations with aromatic aldehydes called *Perkin condensations*. Acetic anhydride can be converted into its enolate in low concentration with sodium acetate (sodium hydroxide cannot be used because it would destroy the anhydride):

$$CH_3\underset{O}{\underset{\|}{C}}-O-\underset{O}{\underset{\|}{C}}-CH_3 + NaOAc \rightleftharpoons CH_2{=}\underset{O^-}{\underset{|}{C}}-O-\underset{O}{\underset{\|}{C}}-CH_3 + HOAc$$

The enolate then adds to a reactive aldehyde, such as benzaldehyde, and the initial product dehydrates to form an unsaturated mixed anhydride. The latter can then be hydrolyzed to cinnamic acid (Fig. 11.13).

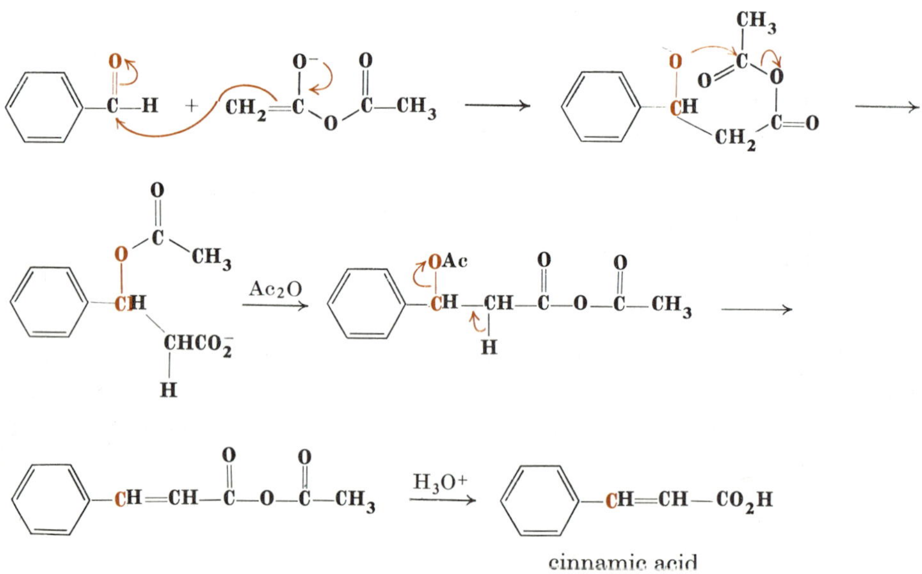

Fig. 11.13 Perkin condensation

11.27 CLAISEN CONDENSATION

Esters with at least two α hydrogens undergo a reaction similar to the aldol condensation. For example, when ethyl acetate is treated with a strong base, usually sodium ethoxide, ethyl acetoacetate (acetoacetic ester) is produced:

$$\underset{\text{ethyl acetate}}{2CH_3\overset{O}{\overset{\|}{C}}-OEt} \xrightarrow{^-OEt} \underset{\text{acetoacetic ester}}{CH_3\overset{O}{\overset{\|}{C}}-CH_2-\overset{O}{\overset{\|}{C}}-OEt}$$

The pathway for this reaction, the *Claisen condensation*, again requires the formation of an enolate anion. The enolate reacts with another molecule of ester by an addition–elimination reaction (Fig. 11.14).

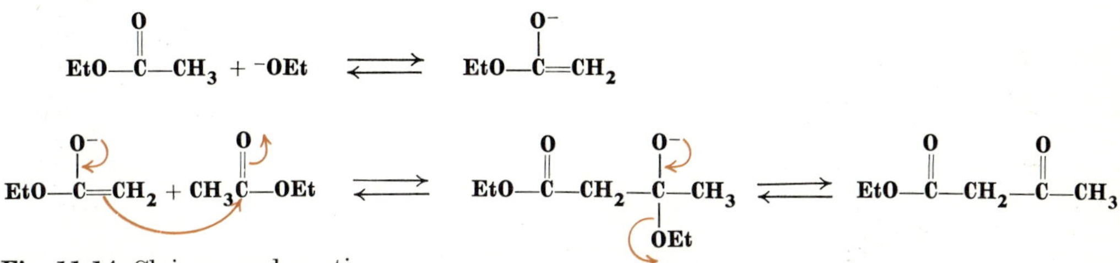

Fig. 11.14 Claisen condensation

The Claisen condensation, like the other condensations we have discussed, is an equilibrium process. The reaction is driven to completion only by converting the product into its conjugate base. As we noted earlier (Section 4.11), β-dicarbonyl compounds are reasonably acidic ($pK_a \approx 10$), and the equilibrium between acetoacetic ester and ethoxide with their respective conjugate base and acid forms lies completely in the direction of acetoacetate anion and ethanol:

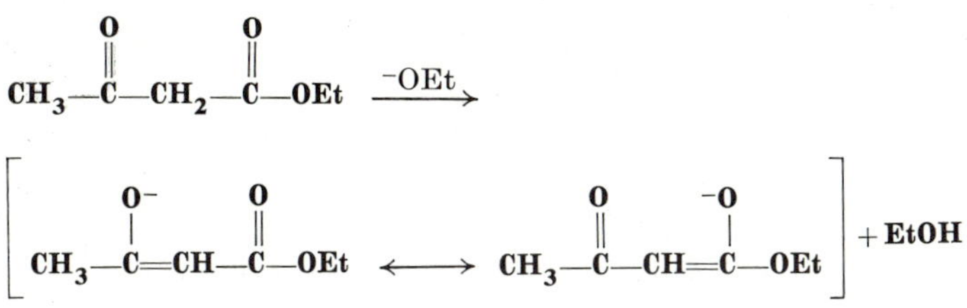

Since this last step is required for the success of the Claisen condensation, it is clear why esters which have only one α hydrogen do not undergo the reaction. The resulting β ketoester would have no hydrogen-bearing carbon adjacent to two carbonyl groups:

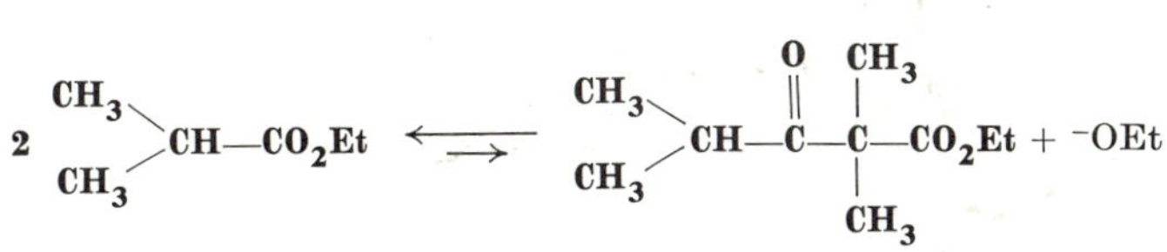

A variation of the Claisen condensation, known as the *Dieckmann cyclization*, is useful in the synthesis of five- and six-membered rings. When both ester functions are contained in the same molecule, intramolecular condensation leads to a cyclic ketoester:

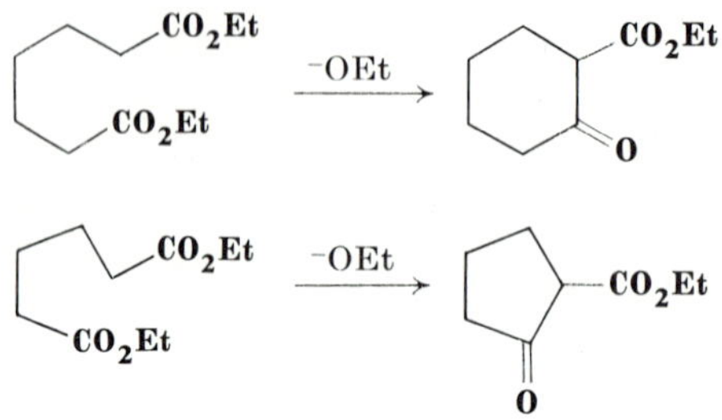

β-Ketoesters are very useful in organic syntheses. As we saw in Section 7.12, in the discussion of displacement reactions, β-ketoesters (and related β-dicarbonyl systems such as malonic ester) are readily alkylated:

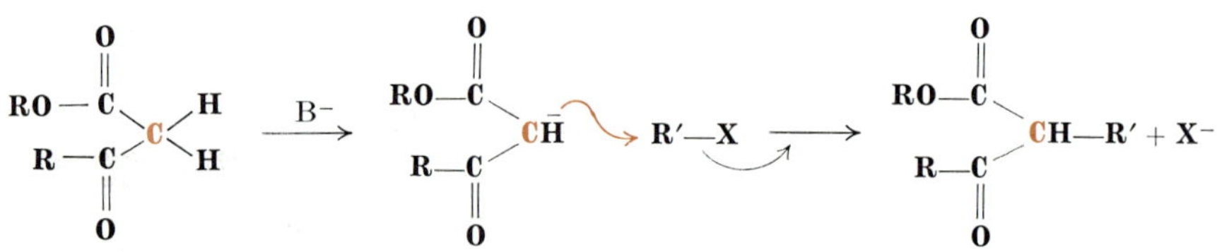

Acid-catalyzed hydrolysis of an alkylated β-ketoester yields a β-ketoacid:

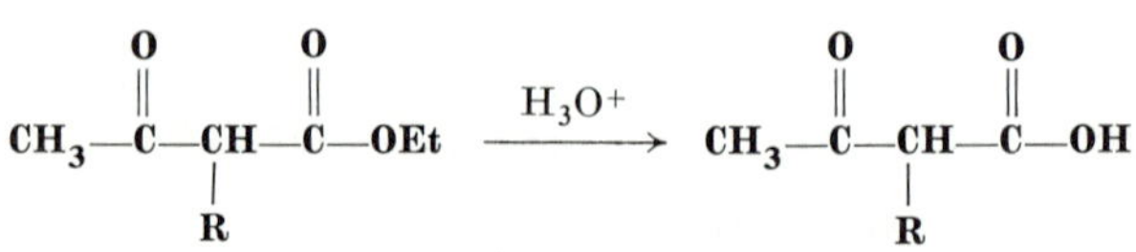

β-Ketoacids in turn lose carbon dioxide on warming, thereby forming a ketone:

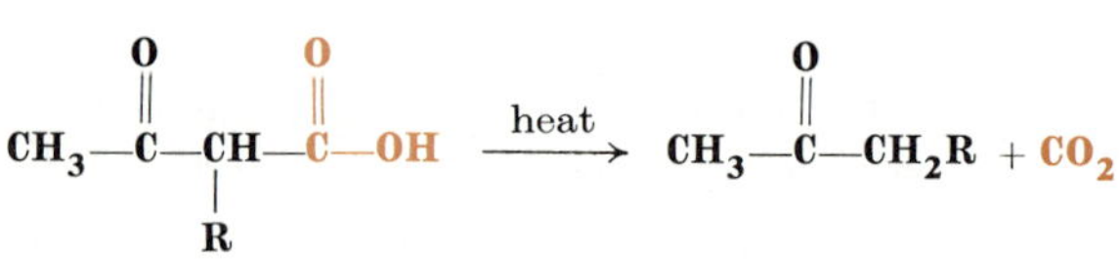

The decarboxylation mechanism has been shown to be that given in Fig. 11.15.

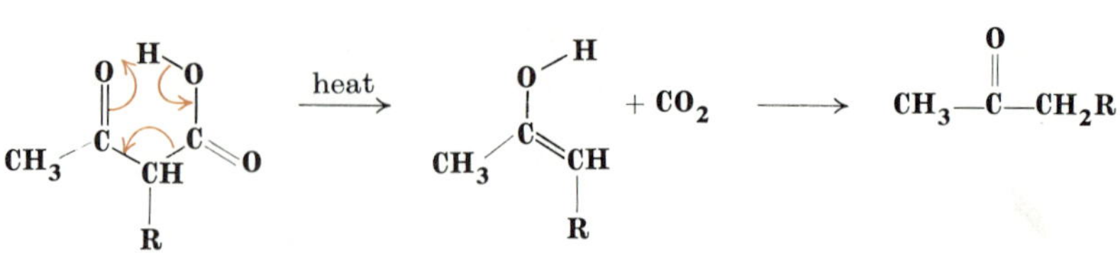

Fig. 11.15 Mechanism of decarboxylation of a β-ketoacid

This mechanism is consistent with the fact that decarboxylation of a β-ketoacid in the presence of bromine leads to a monobrominated ketone, confirming the presence of an enol intermediate. Moreover, the decarboxylation rates of β-ketoacids are insensitive to solvent polarity, as would be expected from a cyclic mechanism in which no charge is created or destroyed.

The synthesis of 2-benzylcyclopentanone in Fig. 11.16 illustrates the synthetic utility of the condensation-alkylation-hydrolysis-decarboxylation sequence.

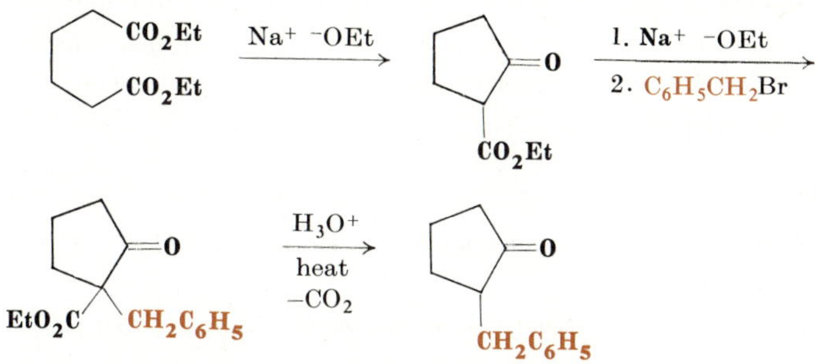

Fig. 11.16 Synthesis of 2-benzylcyclopentanone

β-Dicarbonyl systems can be alkylated not only by alkyl halides, but also by α,β-unsaturated carbonyl compounds via a Michael addition (see Section 9.5). The product is always a 1,5-dicarbonyl compound:

$$C_6H_5CH{=}CHCHO + CH_2(CO_2Et)_2 \xrightarrow{^-OEt} C_6H_5{-}\underset{\displaystyle CH(CO_2Et)_2}{CH}{-}CH_2CHO$$

$$\xrightarrow[-CO_2]{H_3O^+} C_6H_5{-}\underset{\displaystyle CH_2CO_2H}{CH}{-}CH_2CHO$$

If a strong base such as potassium *t*-butoxide is used, even a simple ketone can be alkylated by Michael addition to an unsaturated carbonyl compound (Fig. 11.17).

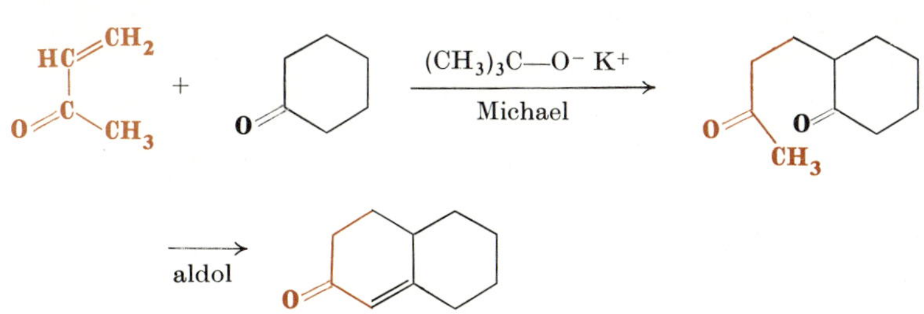

Fig. 11.17 Use of Michael addition and aldol reaction for ring synthesis

11.28 CONCLUDING REMARKS

In this chapter we have discussed about three dozen reactions. The task of learning these reactions is considerably simplified by the fact that most of them are variations on a few basic reaction types.

Learning organic chemistry is much like learning to write. In writing one has to begin by memorizing words and learning the rules of grammar. Once this is done, it becomes possible to write meaningful sentences. Creative writers can produce poetry. And so it is with the chemist. Once the reactions are learned and the mechanisms understood, it is possible to formulate simple organic syntheses. Creative chemists can devise fantastic molecules never before known on the face of this earth.

ADDITIONAL READING

C. D. Gutsche, *The Chemistry of Carbonyl Compounds*, Englewood Cliffs, N.J.: Prentice-Hall, 1967. Paperback.

R. W. Alder, R. Baker, and J. M. Brown, *Mechanism in Organic Chemistry*, New York: Wiley, 1971. Chapter 5.

PROBLEMS

1. To help you remember the important reactions in Chapter 11, we have compiled the following list. Give a specific example of each item. Refer to the text to ensure that your examples are accurate. Then use the list as a study guide.

Ketones and Aldehydes

a) Cyanohydrin formation
b) Grignard reaction
c) Carbonyl reduction with $LiAlH_4$
d) Oxime formation
e) Wittig reaction
f) Acetal or ketal formation
g) Halogenation
h) Aldol condensation
i) Mannich condensation
j) Michael addition

Esters

a) Grignard addition
b) Acid and base catalyzed hydrolysis
c) Synthesis from acid chlorides
d) Reduction by $LiAlH_4$
e) Claisen and Dieckmann condensations

Amides

a) Formation from acid chlorides
b) Formation from acid anhydrides
c) Hydrolysis
d) Reduction by $LiAlH_4$

Acid Chlorides

a) Formation from acids
b) Reactions with amines and alcohols

2. State the specific compound or class of compounds (alcohol etc.) which acts as the nucleophile in the following reactions of aldehydes and ketones.
   a) Cyanohydrin formation
   b) Oxime formation
   c) Hemiacetal formation
   d) Wittig reaction
   e) Imine formation
   f) Perkin condensation

3. Give the correct name for each of the following "name" reactions.

a) 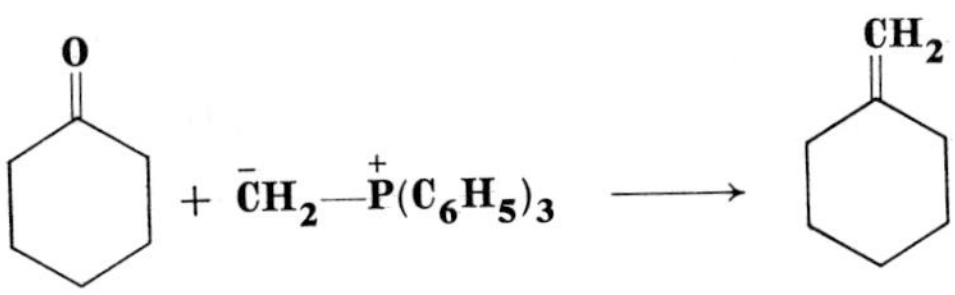

b) 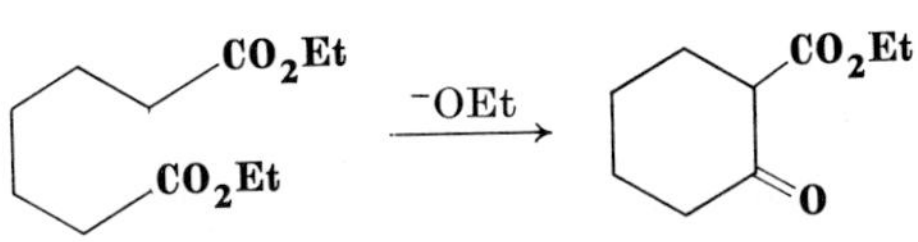

c) $CH_3{-}\overset{\displaystyle O}{\overset{\|}{C}}{-}CH_3 + CH_2{=}O + NH_2CH_3 \xrightarrow{H^+} CH_3{-}\overset{\displaystyle O}{\overset{\|}{C}}{-}CH_2CH_2NHCH_3$

d) 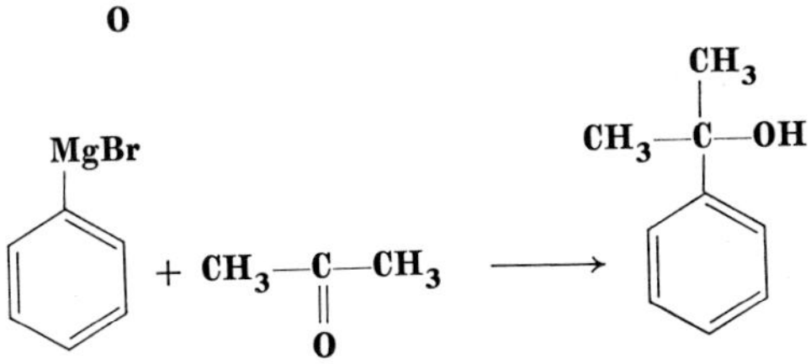

e) $2CH_3{-}\overset{\displaystyle O}{\overset{\|}{C}}{-}OEt \xrightarrow{^-OEt} CH_3\overset{\displaystyle O}{\overset{\|}{C}}{-}CH_2{-}\overset{\displaystyle O}{\overset{\|}{C}}{-}OEt$

4. Provide the correct reagents for the following reaction "network."

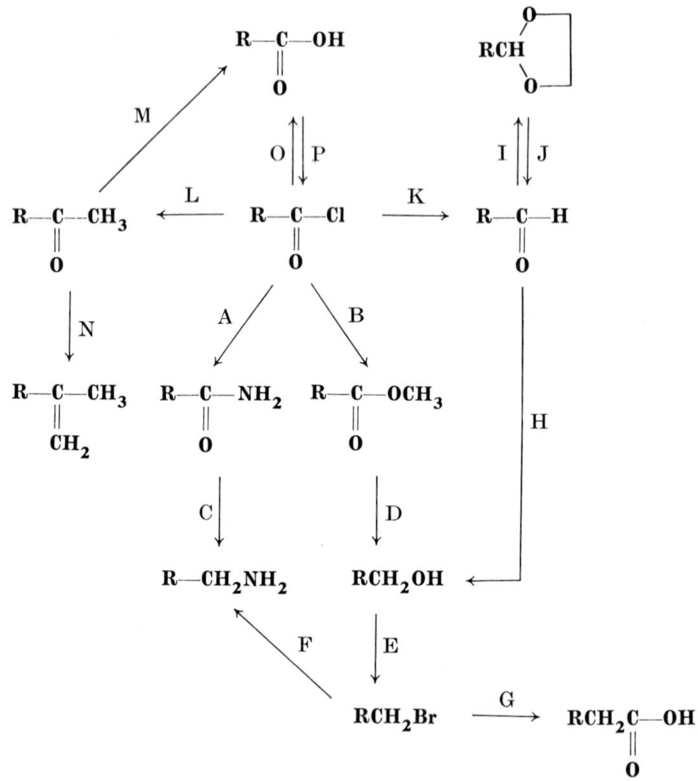

5. Supply the structure of the major organic product of each of the following reactions.

a) $CH_3CH_2CH_2CHO + HCN \longrightarrow$

b) 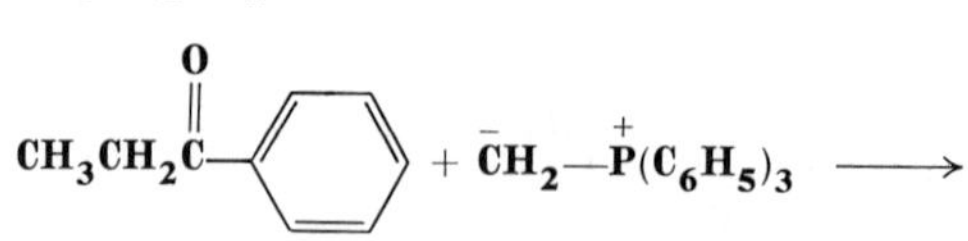

c) 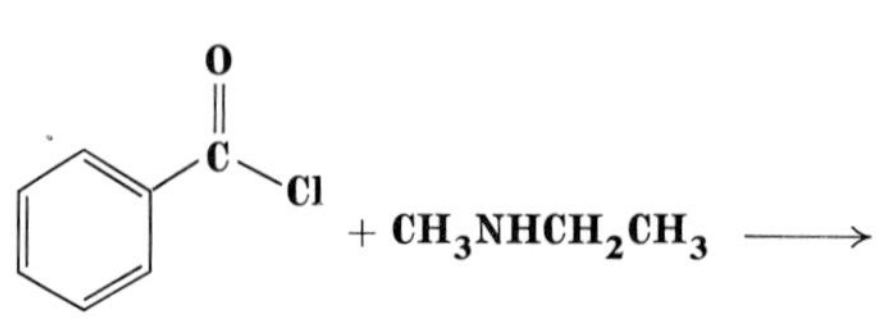

d) 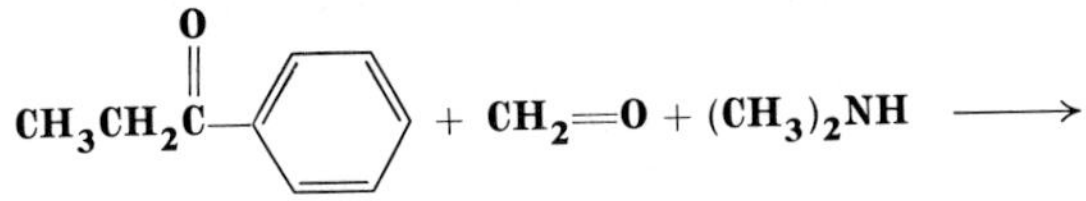

e) 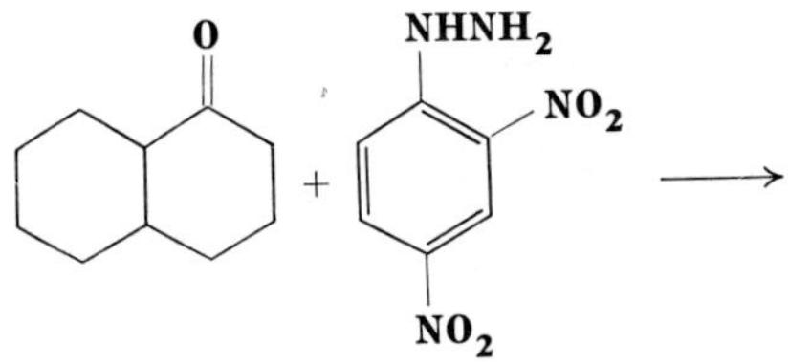

f) 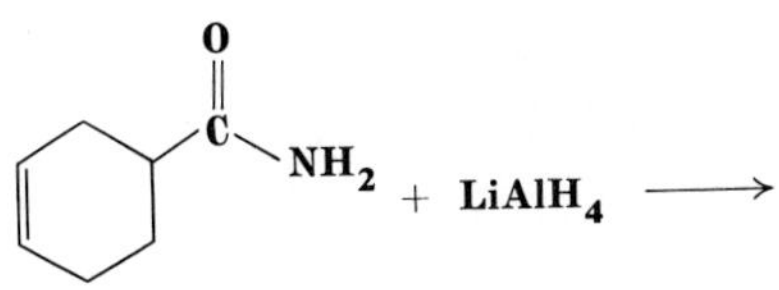

g) 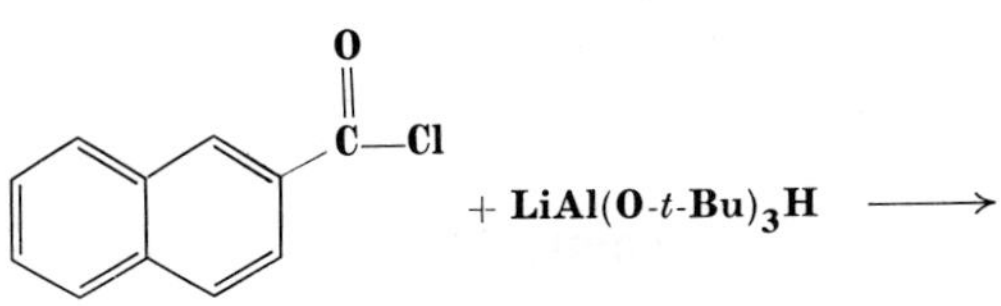

h) 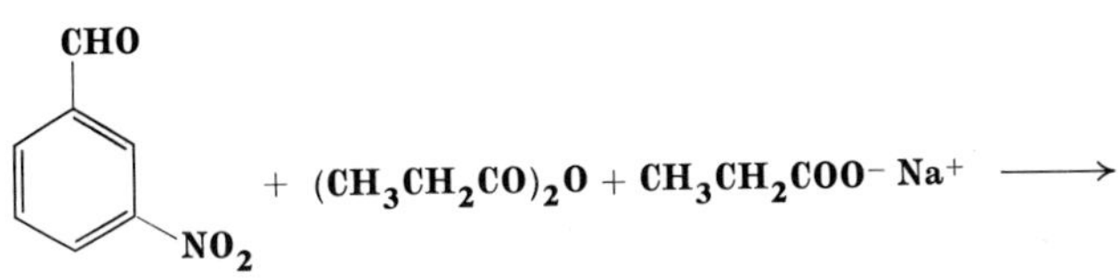

i) 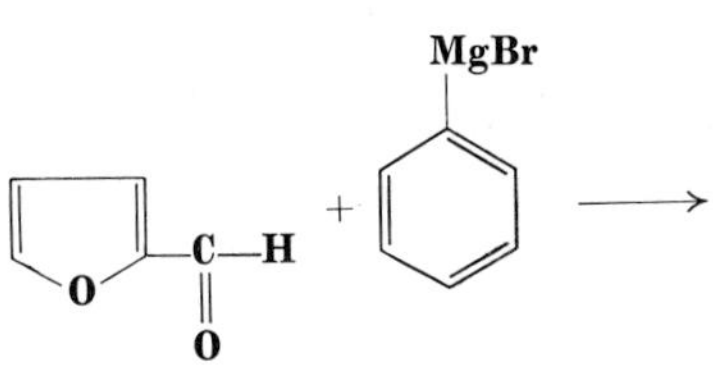

j) 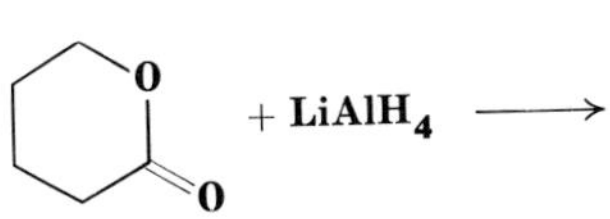

6. Supply the missing reagents in the following reactions. *Note*: A letter may represent more than one reagent.

a)

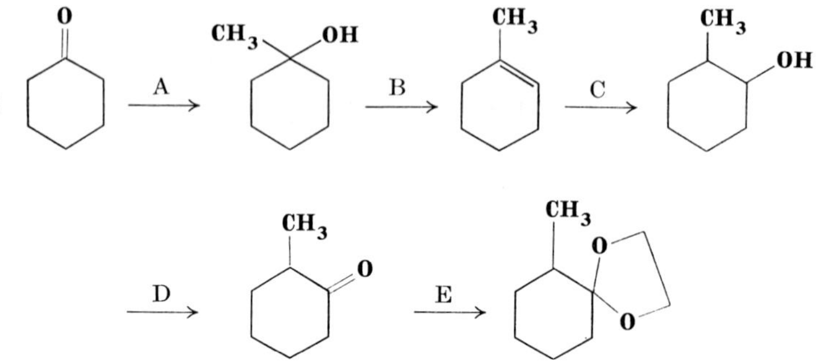

b)

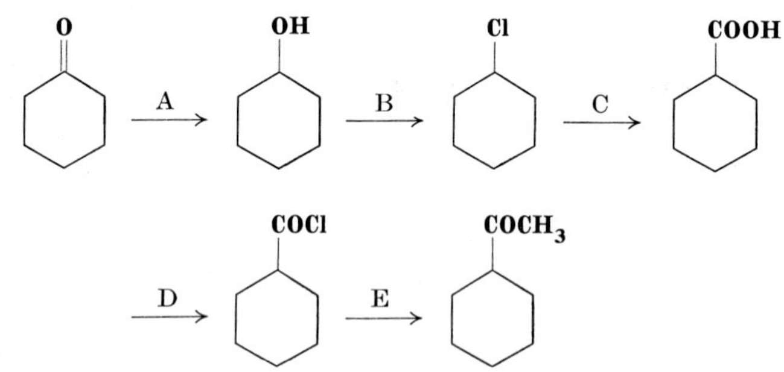

c)

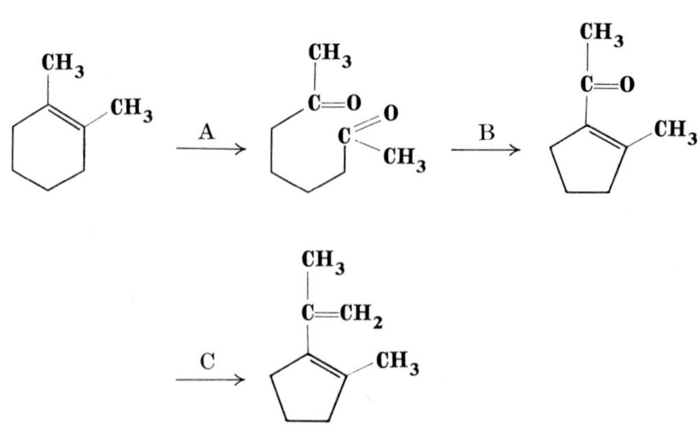

d) 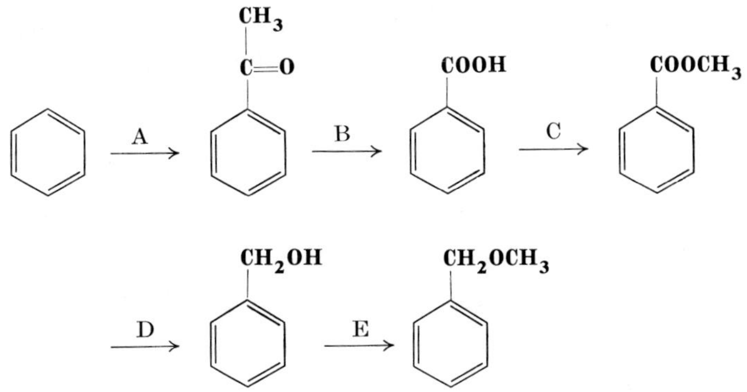

e) 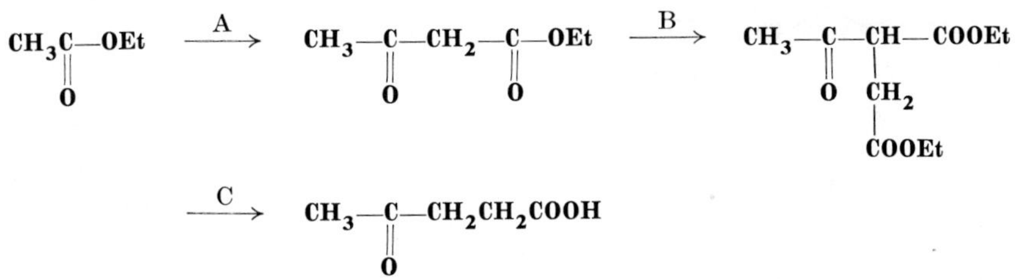

f) 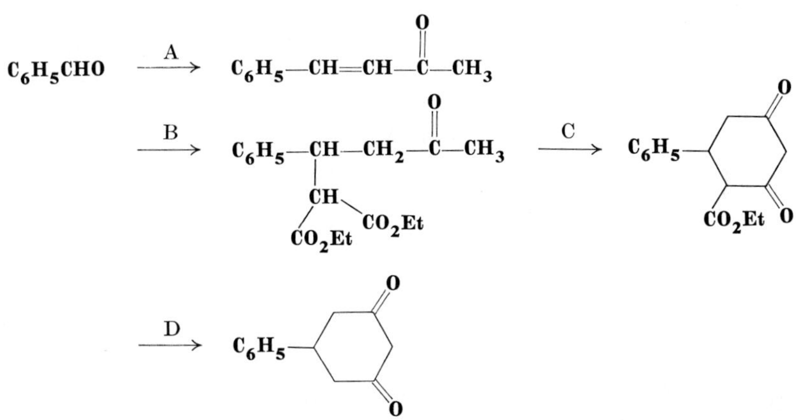

7. Zearalenone, a substance with pronounced uterotrophic activity, was isolated in 1962 from a mold growing on corn.

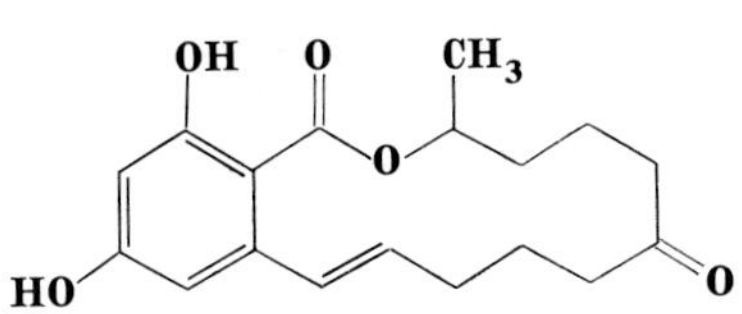

zearalenone

A group of chemists at Merck and Co., Research Laboratories synthesized the compound in 1968 by the route shown below. Supply the missing reagents (A through M).

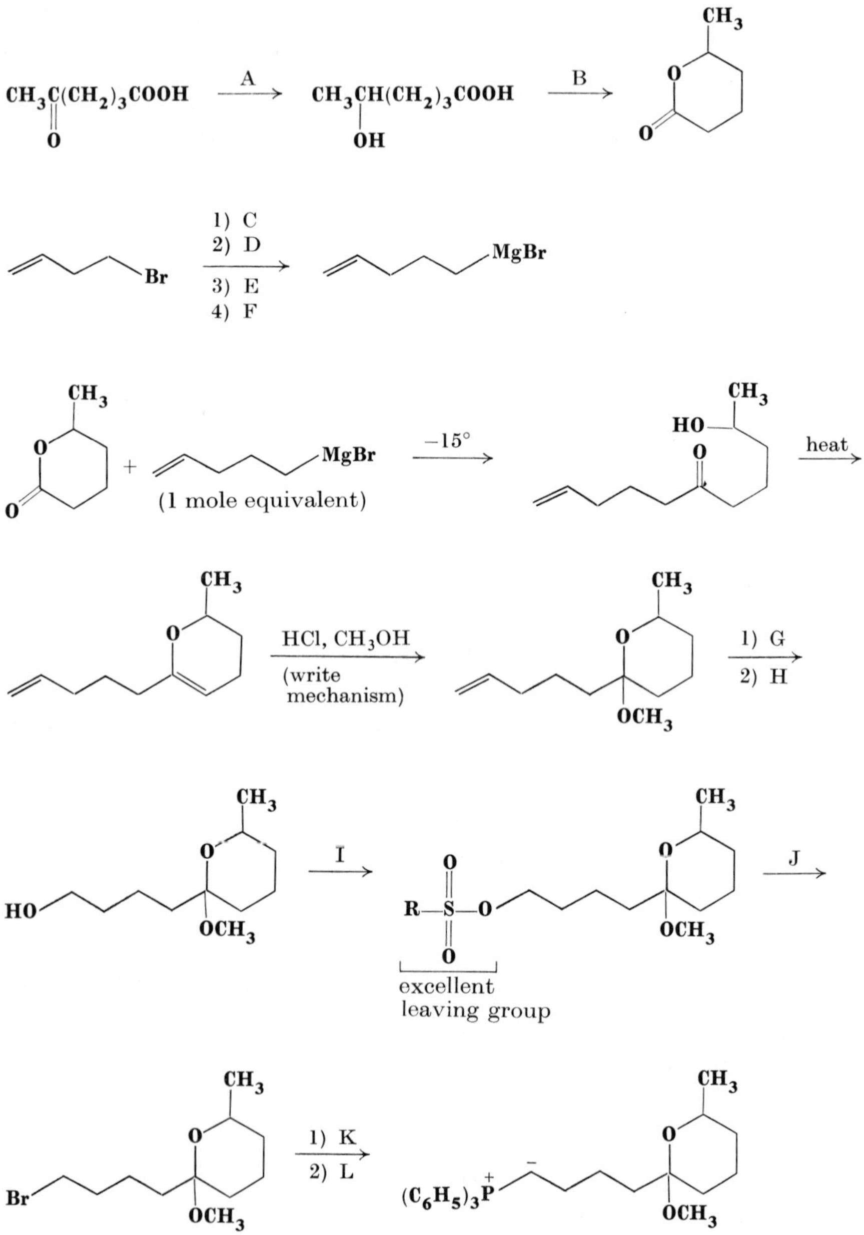

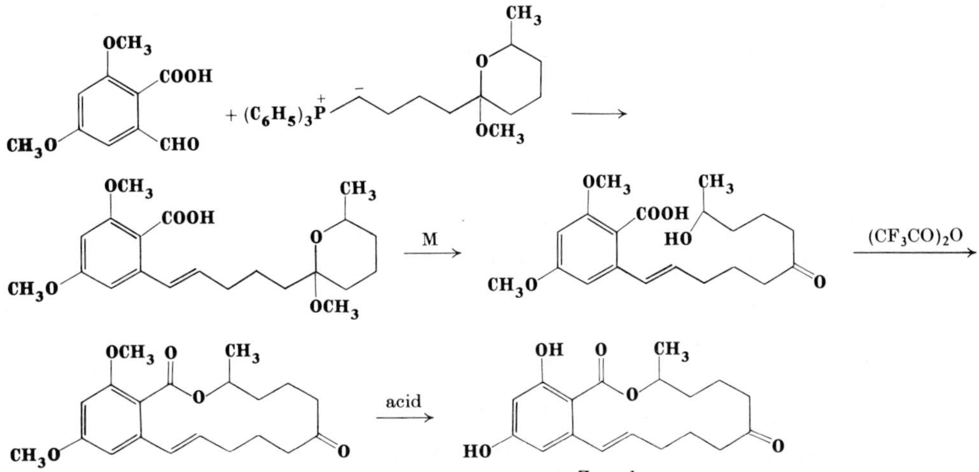

Zearalenone

8. The *Reformatsky reaction* involves the addition of an organozinc derivative of an α-bromoester to a carbonyl group:

$$BrCH_2CO_2Et + Zn \longrightarrow BrZnCH_2CO_2Et$$

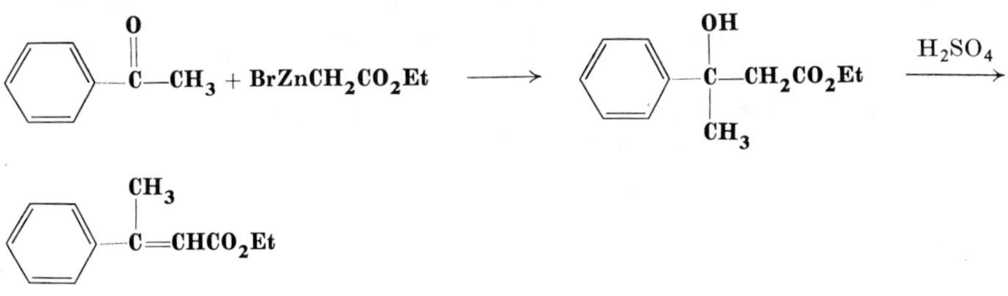

Why is zinc used in preference to magnesium?

9. The reaction of a ketone with a primary amine produces an *imine* (Fig. 11.7). The reaction of a ketone with a secondary amine produces a compound known as an *enamine*:

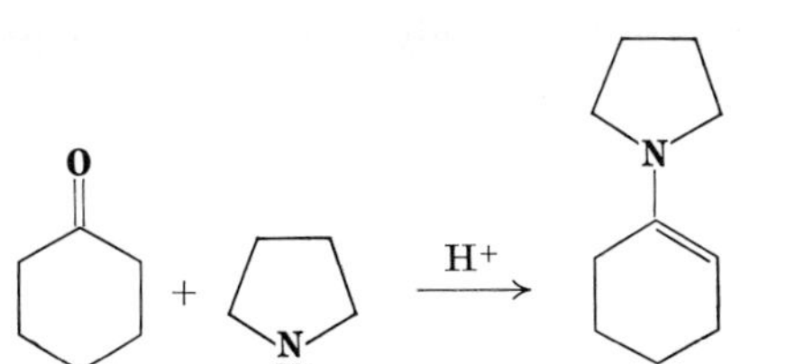

an enamine

Write a mechanism for this reaction.

10. Suggest a plausible mechanism for the following reaction:

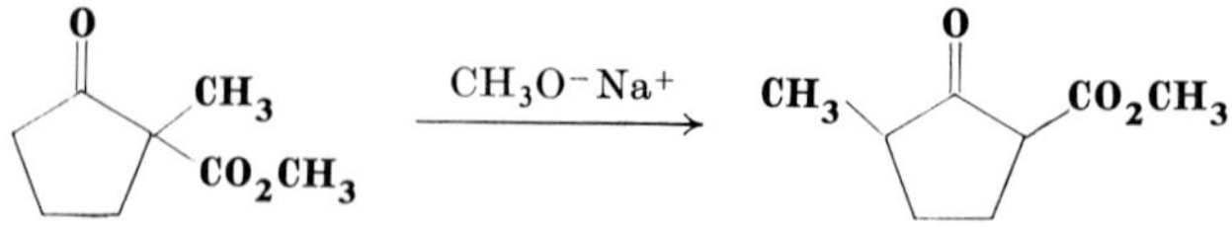

11. "Queen substance," a ketoacid secreted by the mandibular glands of the queen bee, plays a remarkable role in the life of a hive.

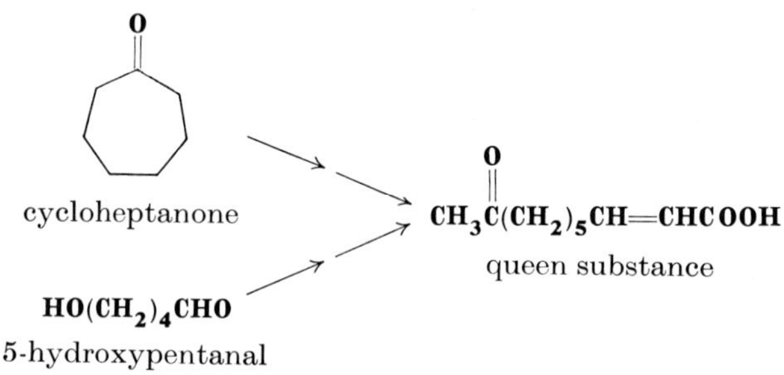

Queen substance covers the body of the queen bee and spreads to the worker bees by body contact during social feeding. An extremely small amount of queen substance inhibits the development of the ovaries of the worker bees and prevents the workers from rearing new queen bees. Propose two syntheses of queen substance, one starting from cycloheptanone and the other starting with 5-hydroxypentanal.

12. "Twistane" is an interesting hydrocarbon composed entirely of twisted cyclohexane rings:

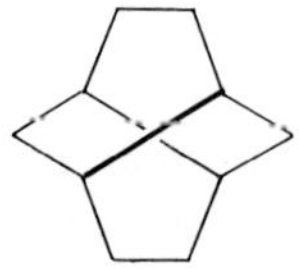

twistane

The synthesis of twistane, given below, involves only two reactions which have not yet been discussed:

$$R\text{—}I \xrightarrow{Pt,\ H_2} R\text{—}H$$

$$R_2C{=}O \xrightarrow[KOH,\ 150^\circ]{H_2NNH_2} R_2CH_2$$

We will deal with these reactions in more detail in Chapter 13 when we discuss oxidations and reductions. After studying the synthesis of twistane, provide the missing reagents and answer the questions which follow.

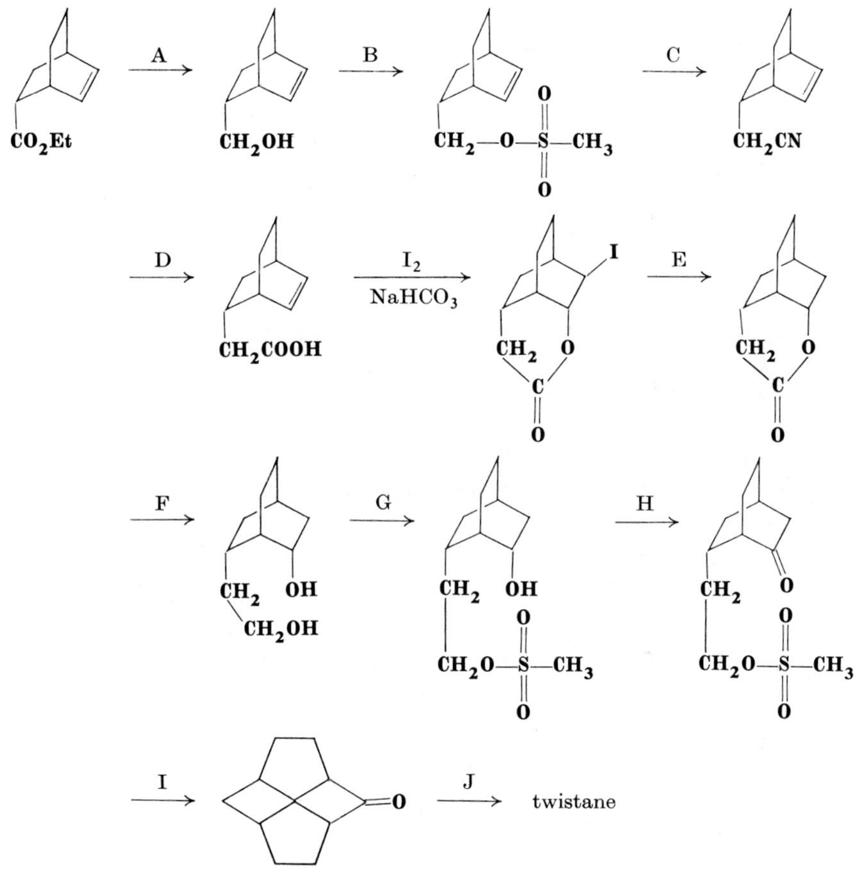

a) In the preparation of the nitrile (R—CN), why is the alcohol converted into the sulfonate ester prior to the displacement reaction?
b) Write a mechanism for the reaction involving I_2.
c) Reaction G involves only one of the two alcohol groups. Why is this possible?
d) Make a molecular model of twistane.

13. Write a mechanism for the *Cannizzaro reaction* (which is known to involve a *hydride transfer*):

$$2CH_2{=}O + NaOH \xrightarrow[H_2O]{\text{heat}} CH_3OH + H{-}C{\overset{O}{\lesssim}}O^-Na^+$$

14. Write the major organic product of each of the following reactions:

a) 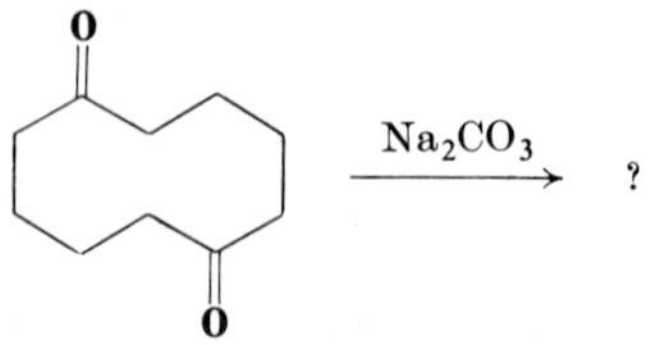

b) $C_6H_5COCH_3 + CO_2 \xrightarrow{Mg(OCH_3)_2}$?

c) 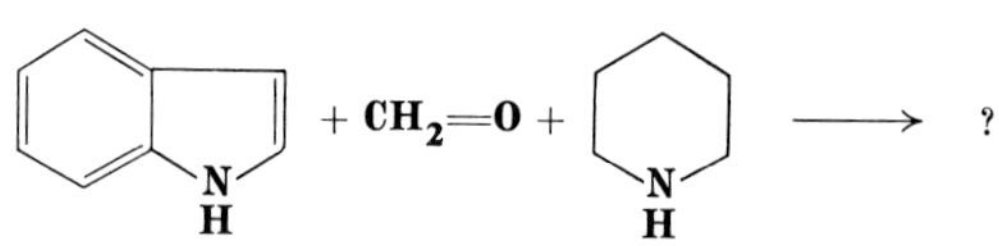

d) 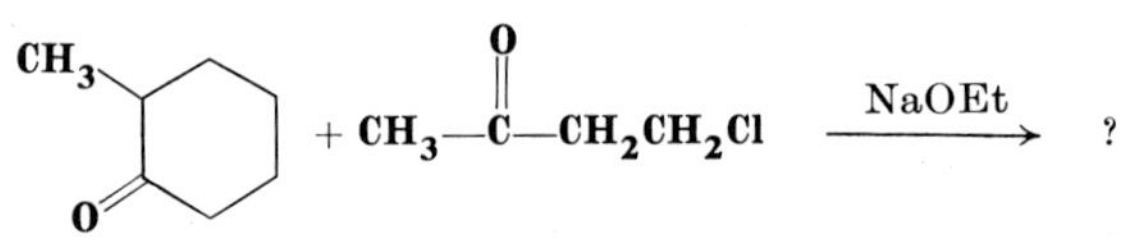

e) 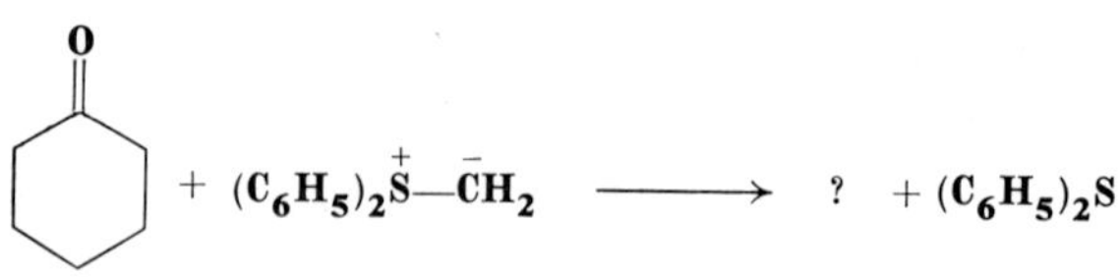

15. Compound I ($C_9H_{18}O$) decolorizes bromine in CCl_4 and reacts with hydrogen in the presence of a platinum catalyst to give compound II ($C_9H_{20}O$). Compound II does not react with bromine. When compound I is treated with concentrated HCl, compound III ($C_9H_{17}Cl$) is obtained. The latter reacts with KOH in hot ethanol to yield compound IV (C_9H_{16}). Compound IV, reacted with ozone and then hydrolyzed in the presence of zinc, yields compounds V and VI:

V: 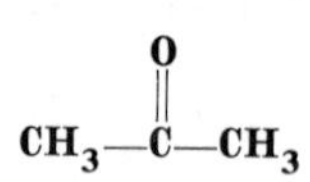

VI: OHC—CH$_2$—CHO

Draw the structures of compounds I through IV.

16. Starting with cyclohexanone (organic reagents of no more than six carbon atoms) and any inorganic reagents, synthesize the following compound:

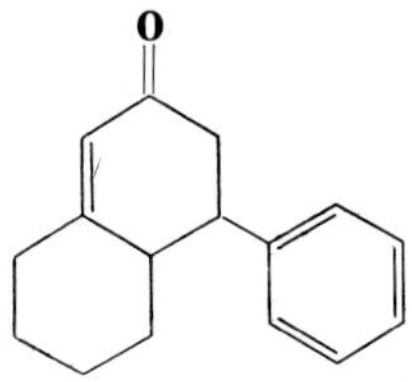

17. Write a mechanism for the following reaction:

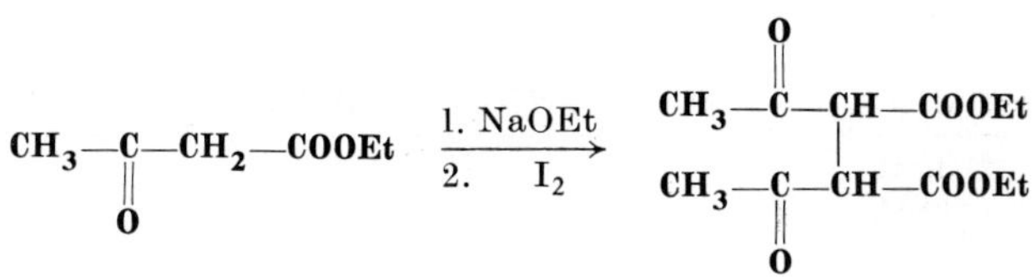

18. Show by means of specific compounds how the following conversions might be accomplished:

a) An alkyl bromide to an acid with one additional carbon.
b) An alkyl bromide to an acid with two additional carbons.
c) An acid to an aldehyde with the same number of carbons.
d) A methyl ketone to an acid with one less carbon.
e) A diester to a ketone.
f) An olefin to an alcohol with the same number of carbons.
g) An acid to a ketone with one additional carbon.
h) An acid to an amine with the same number of carbons.
i) An alkyl bromide to an alcohol with two additional carbons.
j) A ketone to an olefin with one additional carbon.
k) An olefin to an unsaturated alkyl bromide with the same number of carbons.
l) An olefin to a glycol with the same number of carbons.
m) A noncyclic diene to a six-membered ring.

19. Biotin (vitamin H) is a growth factor which participates in biological carboxylation.

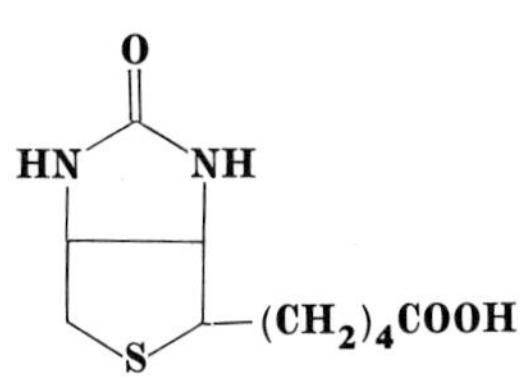

One milligram of the substance was first isolated in 1935 from 500 pounds of egg yolk. Biotin has been synthesized by a series of reactions shown below. Fill in the missing structures (A through J).

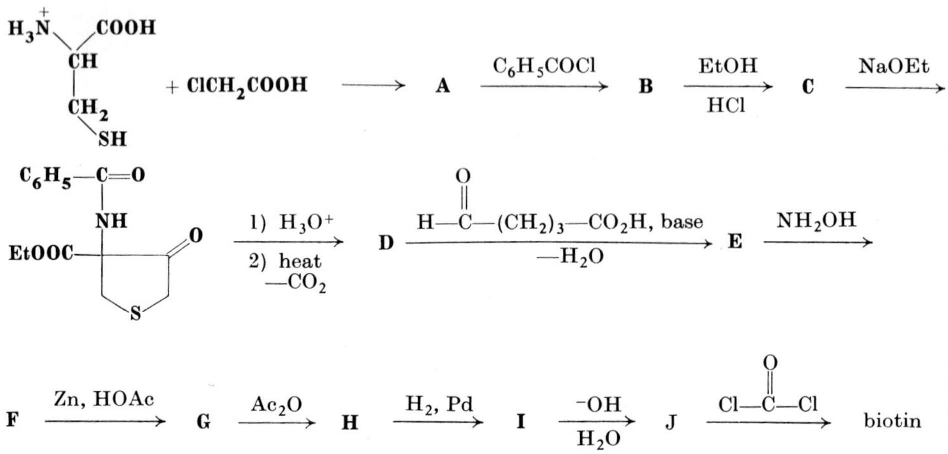

12
REARRANGEMENT REACTIONS

12.1 REARRANGEMENT REACTIONS

Almost all the organic reactions we discussed in the previous chapters fall into two categories: those which involve the interchange of functional groups, and those which involve the formation of carbon-carbon bonds. The oxidation of alcohols to ketones, the reduction of amides to amines, and the conversion of acid chlorides to esters are examples of the first category. The Grignard synthesis, the Diels-Alder reaction, and the aldol condensation are examples of the second category. In this chapter, we shall discuss a third category, the *rearrangement* reaction, in which the atoms within a molecule are rearranged through the combined making and breaking of bonds.

12.2 REARRANGEMENT TO AN ELECTRON-DEFICIENT CENTER

When neopentyl alcohol is treated with strong acid, 2-methyl-2-butene is produced:

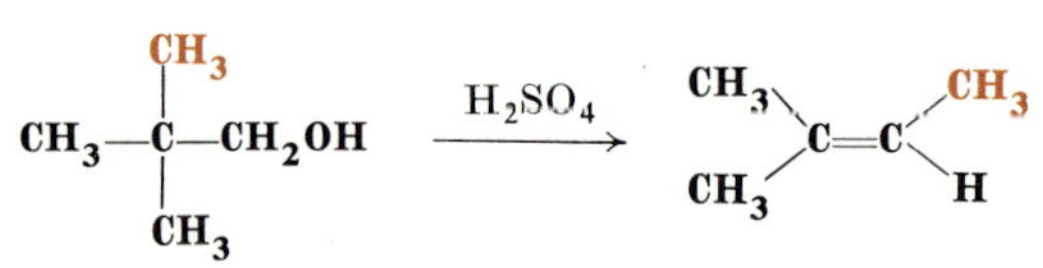

neopentyl alcohol   2-methyl-2-butene

We can rationalize this result in the following way. Protonation of neopentyl alcohol yields an oxonium ion:

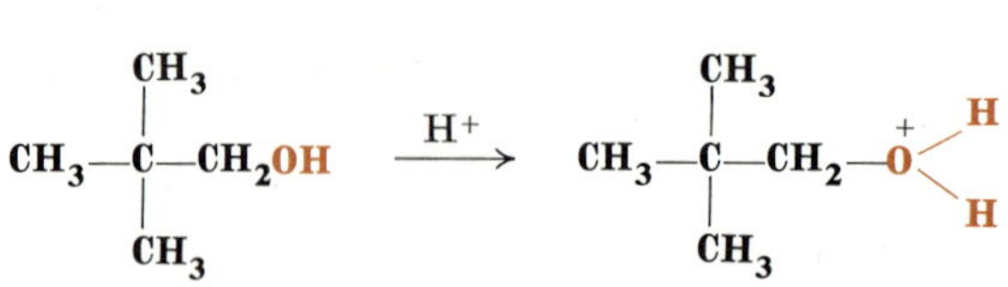

Recall that the corresponding oxonium ion formed from isobutyl alcohol can undergo elimination to form an olefin:

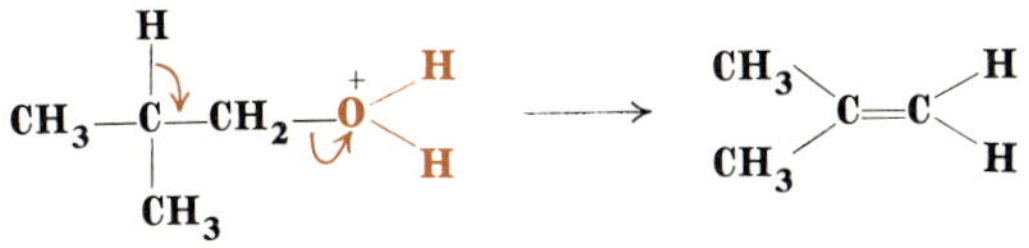

In the neopentyl system, however, no proton is available for elimination. Instead, a methyl group and its bonding electrons migrate to the electron-deficient primary carbon as the water molecule departs:

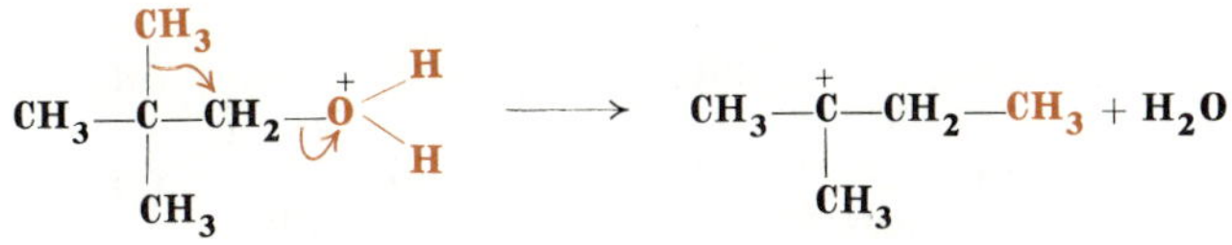

This results in a tertiary carbonium ion with a *rearranged* skeleton. The final step is loss of a proton to yield the 2-methyl-2-butene:

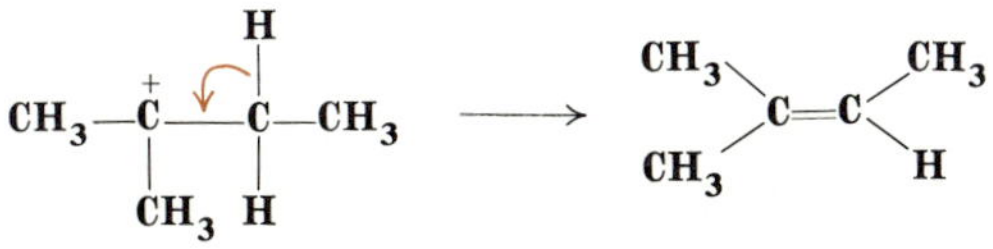

This type of reaction, in which an alkyl or aryl group migrates to an electron-deficient carbon, is called a *Wagner-Meerwein rearrangement*.

The rearrangement of neopentyl alcohol might involve water loss prior to methyl migration:

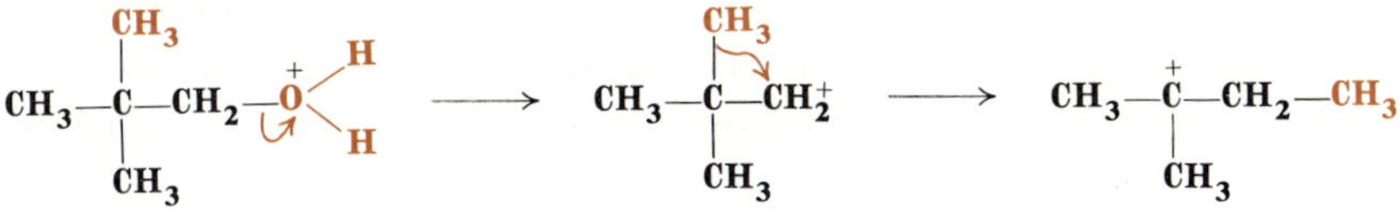

However, since a primary carbonium ion is very unstable, it is more likely that migration and water loss are concerted (occur simultaneously).

The migration of the methyl group from one atom to another is not surprising if one considers two factors. First, the carbonium ion resulting from migration is tertiary and therefore relatively stable. Second, steric repulsion between the three methyl groups of the starting material is reduced when one of the methyl groups moves to the adjacent carbon.

12.3 PINACOL-PINACOLONE REARRANGEMENT

Pinacol (1,2-dimethyl-1,2-dihydroxybutane) rearranges when treated with sulfuric acid:

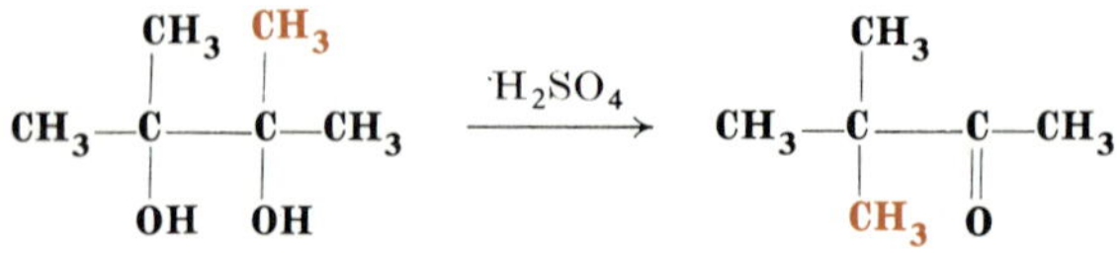

pinacol pinacolone

The mechanism of this Wagner-Meerwein rearrangement is shown in Fig. 12.1.

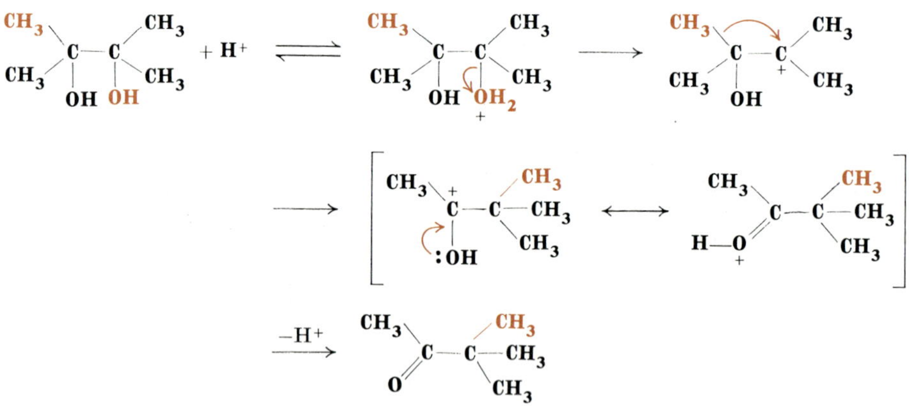

Fig. 12.1 Mechanism of pinacol-pinacolone rearrangement

Note that although both the initial and the rearranged carbonium ions are tertiary, the rearranged cation is a resonance-stabilized oxonium ion (a protonated carbonyl). The resonance stabilization is undoubtedly an important driving force for the rearrangement. Loss of a proton from the rearranged ion produces pinacolone.

In the pinacol rearrangement the migrating group must be capable of supplying electrons to the electron-deficient site, in order to stabilize the positive charge in the transition state. Thus, in 1,2-diphenyl-1,2-di-*p*-anisyl-1,2-dihydroxy-ethane (Fig. 12.2), the *p*-anisyl group migrates, since this aryl group can delocalize the positive charge more easily than the phenyl group.

Other interesting examples of Wagner-Meerwein rearrangements are illustrated in Fig. 12.3. Note that here again the reactions proceed from less stable to more stable carbonium ions.

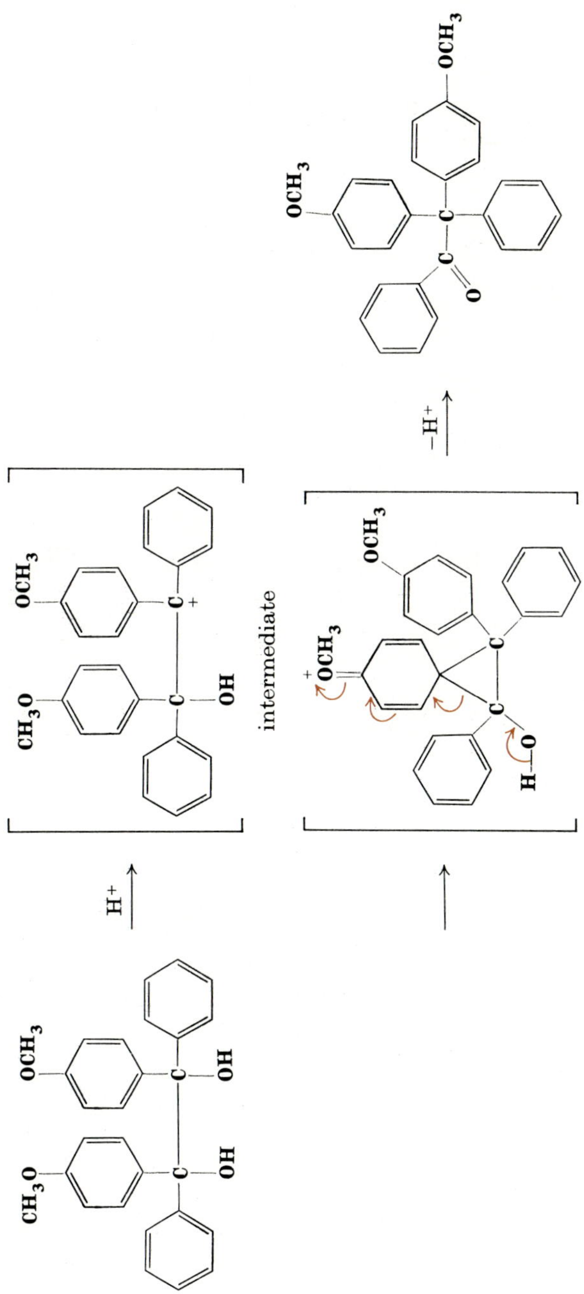

Fig. 12.2 Mechanism of pinacol-type rearrangement with an aromatic migrating group

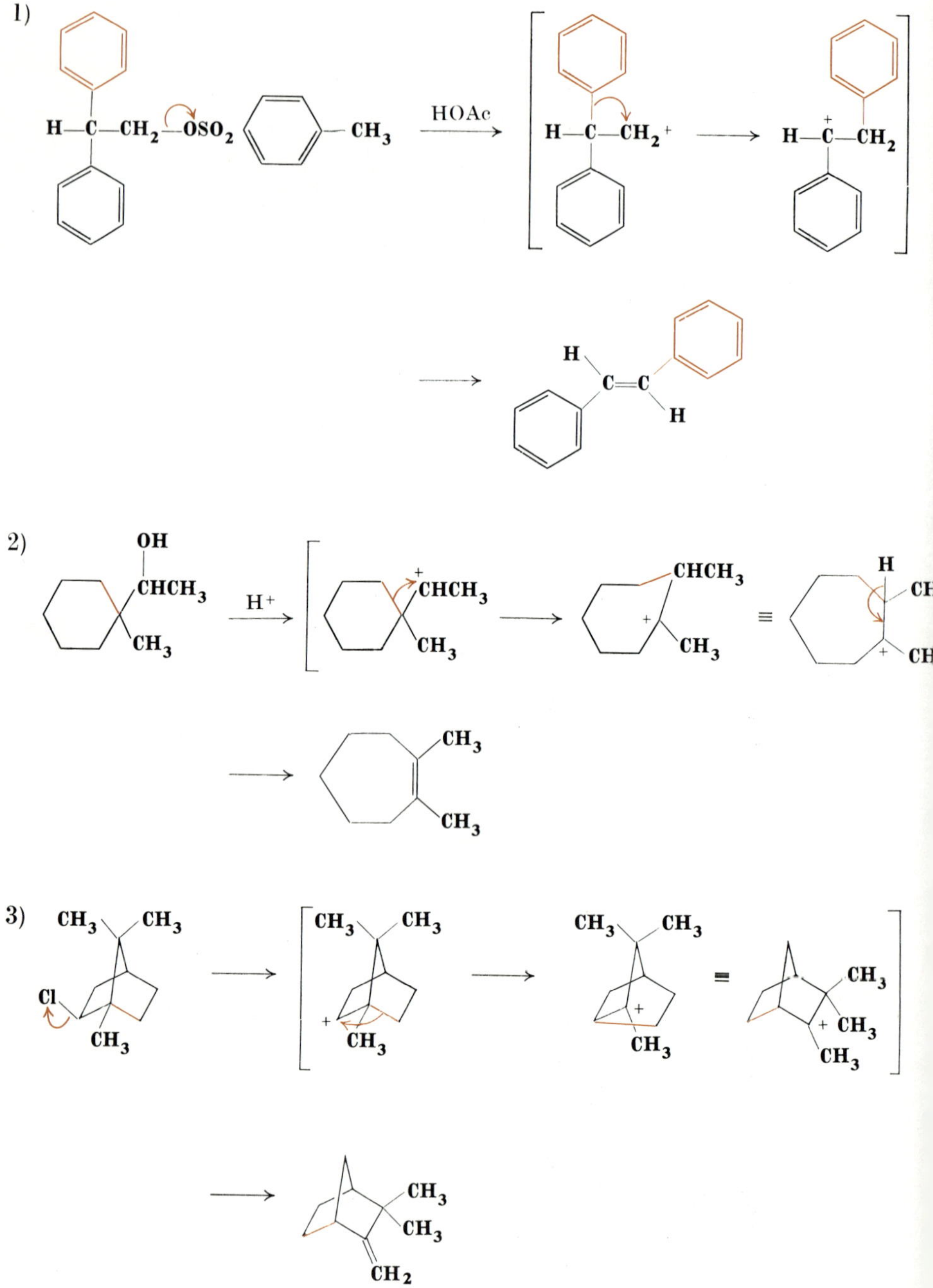

Fig. 12.3 Some Wagner-Meerwein rearrangements

12.4 REARRANGEMENT TO ELECTRON-DEFICIENT OXYGEN: THE BAEYER-VILLIGER OXIDATION

When ketones are treated with peroxyacids, they are rearranged into esters. This reaction, known as the *Baeyer-Villiger oxidation*, is illustrated by the conversion of cyclohexyl methyl ketone into cyclohexyl acetate:

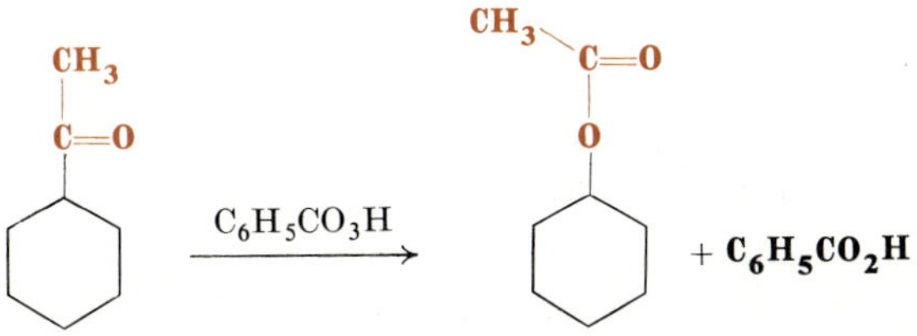

The pathway for this process, illustrated in Fig. 12.4, involves addition of the peroxy acid to the ketone carbonyl group, followed by migration of an alkyl group to an electron-deficient peroxide oxygen atom.

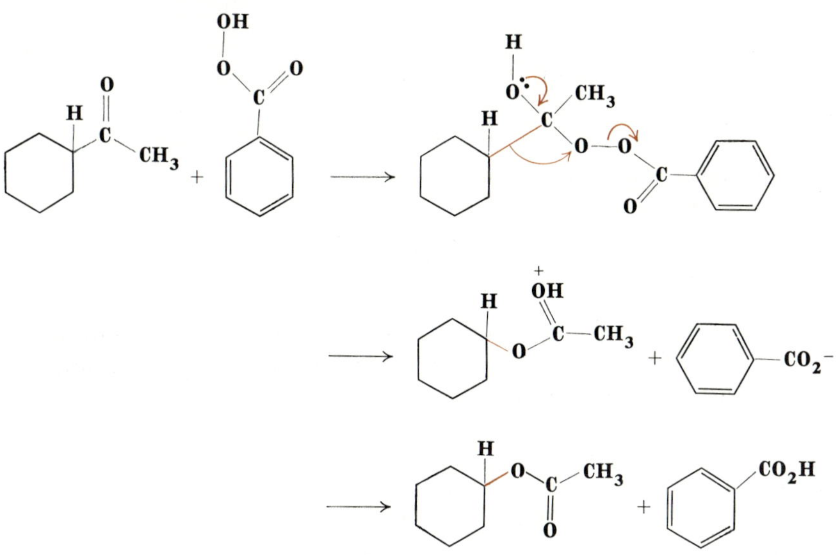

Fig. 12.4 Baeyer-Villiger oxidation

This electron deficiency results from the facile heterolytic cleavage of two bonded highly electronegative oxygen atoms. We should note also that the migrating group is usually the one most able to support the positive charge in the transition state; thus, the secondary cyclohexyl group migrates in preference to the primary methyl group.

Further examples of the Baeyer-Villiger reaction are illustrated in Fig. 12.5.

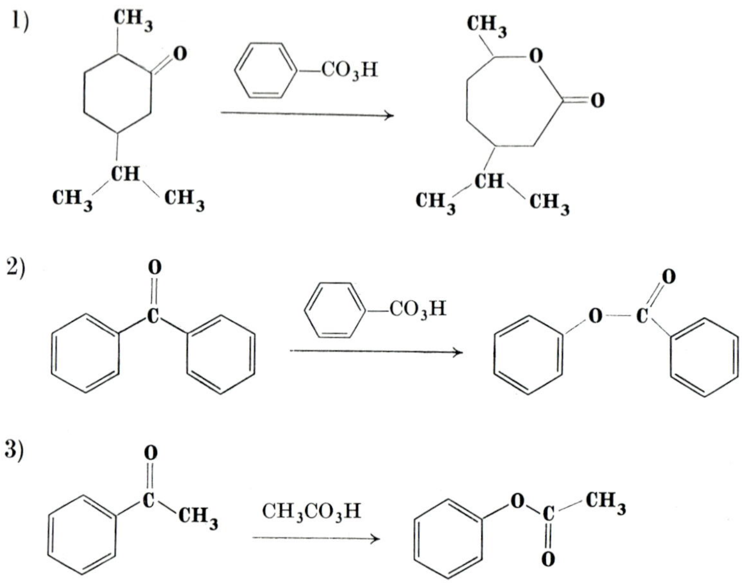

Fig. 12.5 Examples of the Baeyer-Villiger reaction

12.5 REARRANGEMENT TO ELECTRON-DEFICIENT NITROGEN: BECKMANN REARRANGEMENT

When an oxime is treated with strong concentrated acid, such as sulfuric or polyphosphoric acid, it is converted into an amide. Acetone oxime, for example, rearranges to N-methylacetamide:

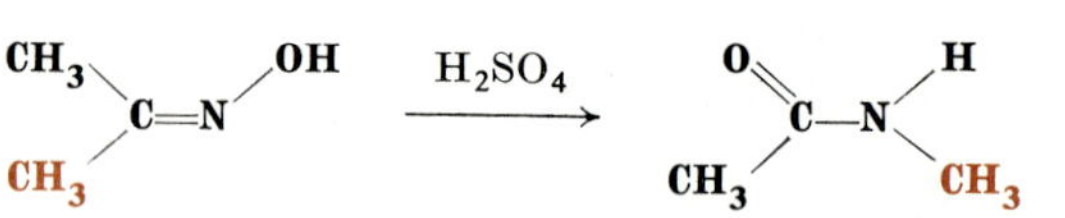

acetone oxime N-methylacetamide

The first step in this *Beckmann rearrangement* (Fig. 12.6) is protonation of the oxime oxygen.

Since water is a much better leaving group than hydroxide ion, protonation facilitates cleavage of the nitrogen-oxygen bond. Moreover, the positive charge on the oxygen inductively removes electrons from the nitrogen, causing methyl migration to an electron-deficient atom, as in the Wagner-Meerwein rearrangement. Rearrangement and water loss are concurrent, producing a

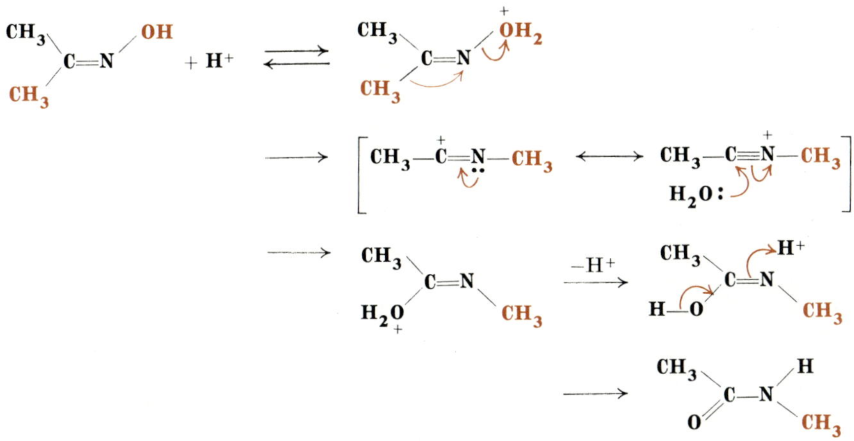

Fig. 12.6 Beckmann rearrangement

resonance-stabilized cation. Addition of water at the carbon site is followed by deprotonation and tautomerization, producing the amide product.

Amines are generally more basic than the corresponding alcohols, and one would expect the oxime nitrogen to be more basic than the oxygen. However, protonation of the nitrogen is reversible and does not result in any further reaction.

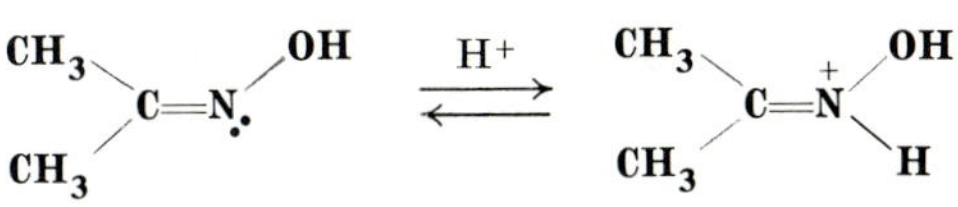

The Beckmann product results entirely from oxime molecules protonated at the oxygen.

The *stereochemistry* of the starting oxime determines which alkyl group of the oxime will migrate in the Beckmann rearrangement:

1)
2)

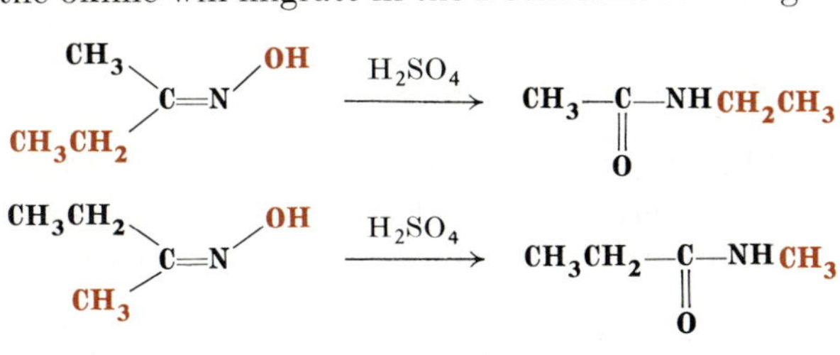

The group which is in a *trans* (or anti) relationship to the hydroxyl group is the one that migrates. This finding supports the mechanism in Fig. 12.6 in which migration and water loss occur simultaneously. If the alkyl group migrated to a positive nitrogen only *after* water loss, the reaction would not be stereospecific. Then the group with the greater "migratory aptitude" would be the one to migrate. This is clearly not the case.

The reaction of a ketone and hydroxylamine forms an oxime whose hydroxyl group is in a predominantly *trans* relationship to the larger of the two groups on the ketone.

Figure 12.7 presents some other examples of the Beckmann rearrangement.

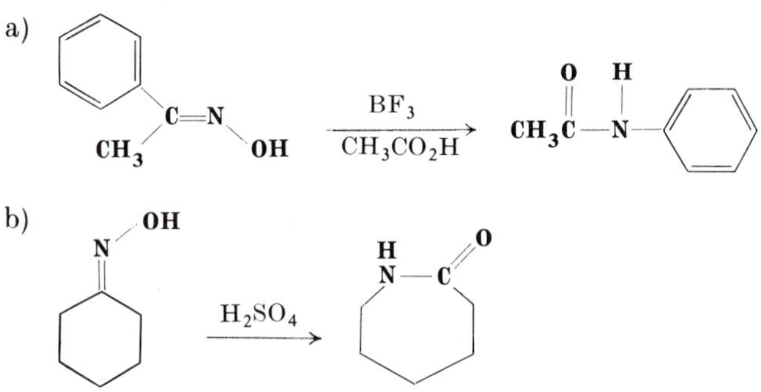

Fig. 12.7 Examples of the Beckmann rearrangement

12.6 HOFMANN REARRANGEMENT

All of the rearrangements we have discussed so far are carried out in acid media and involve cationic intermediates. The *Hofmann rearrangement*, however, occurs in basic solutions and involves a *neutral* electron-deficient nitrogen species known as a *nitrene*.

When an unsubstituted amide such as butyramide is treated with bromine in a strongly basic solution, it rearranges to a primary amine:

$$\underset{\text{butyramide}}{CH_3CH_2CH_2C(=O)NH_2} \xrightarrow[Br_2]{NaOH} \underset{n\text{-propylamine}}{CH_3CH_2CH_2{-}NH_2}$$

Note that the amine product has one carbon atom less than the starting amide. The Hofmann rearrangement, therefore, provides a good way to prepare amines from acids of one additional carbon atom, since the latter may be readily converted into amides:

$$CH_3CH_2CH_2CO_2H \xrightarrow{SOCl_2} CH_3CH_2CH_2COCl \xrightarrow{NH_3} CH_3CH_2CH_2CONH_2$$

Figure 12.8 shows the mechanism of the Hofmann rearrangement. The steps are similar to reactions with which the reader is already familiar.

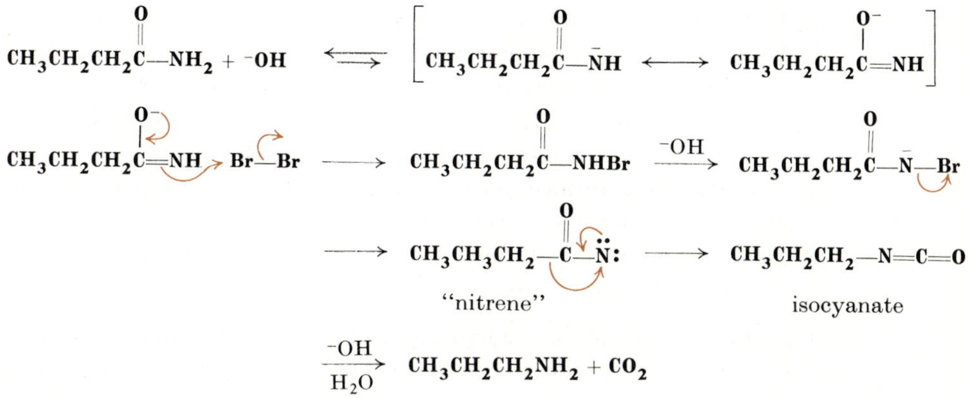

Fig. 12.8 Hofmann rearrangement

The reaction of an amide with base and bromine is similar to the bromination of a ketone in base (see Section 11.21). Like a bromoketone, an N-bromoamide is more acidic than the starting carbonyl compound, and it readily reacts with strong base to yield a new anion. Loss of bromide ion from this intermediate, via a process called *α-elimination*, produces the unstable species known as a nitrene. The nitrogen atom of the nitrene is neutral and electron deficient, having only six electrons in its outer shell. Rearrangement of the nitrene leads to an isocyanate which is unstable in the reaction conditions and therefore spontaneously hydrolyzes to the amine and carbon dioxide.

ADDITIONAL READING

M. S. Newman, Ed., *Steric Effects in Organic Chemistry*, New York: Wiley, 1956. Chapter 5.

J. March, *Advanced Organic Chemistry: Reactions, Mechanisms, and Structure*, New York: McGraw-Hill, 1968. Chapter 18.

PROBLEMS

1. State whether the following rearrangements are to an electron-deficient carbon, oxygen, or nitrogen.

   a) Hofmann b) Wagner-Meerwein c) Baeyer-Villiger d) Beckmann

2. Answer the following questions.

   a) What are two rearrangements that are catalyzed by strong acid?
   b) In the Beckmann rearrangement of oximes, what determines which of the two possible alkyl groups migrates?

c) Why is hydroxide ion necessary in the Hofmann rearrangement?
d) Why can the peroxide oxygen atom that receives the migrating group in the Baeyer–Villiger oxidation be considered electron deficient?
e) Would you expect a Wagner–Meerwein rearrangement from a tertiary to a primary carbonium ion to be a facile process?
f) Does the amine formed in the Hofmann rearrangement have the same number of carbons as the original amide?
g) What by-product is formed from the perbenzoic acid used in the Baeyer–Villiger reaction?

3. Suggest a mechanism for the hydrolysis of an isocyanate in aqueous base (Fig. 12.8). What product would you expect from the reaction of an isocyanate in methanolic sodium methoxide?

4. Suggest a mechanism for the following reaction:

Note that the expected Baeyer-Villiger product does not form here.

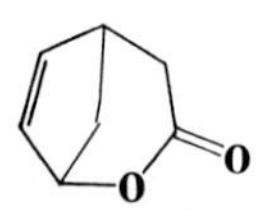

expected
Baeyer-Villiger
product

5. Show how you would carry out the following conversions. Employ a rearrangement reaction in each case.

a)
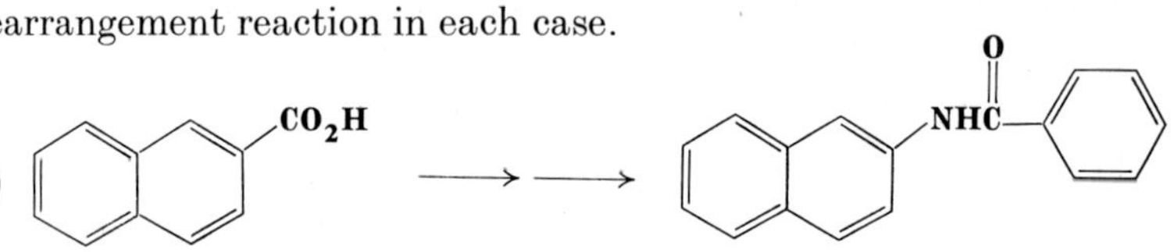

b)
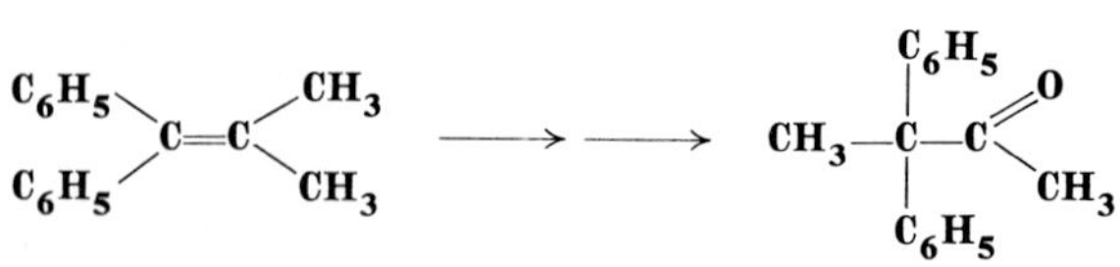

c)
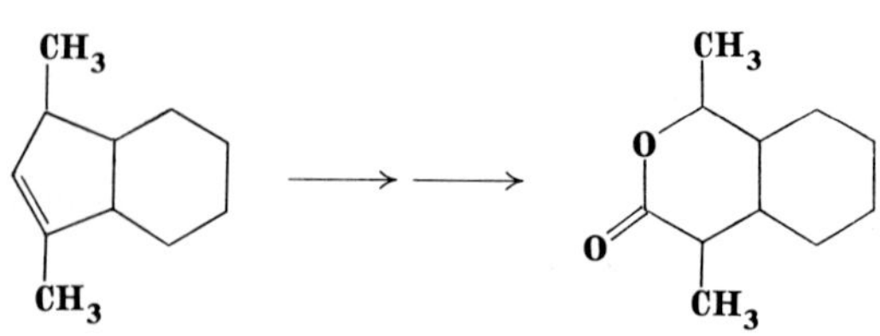

d) 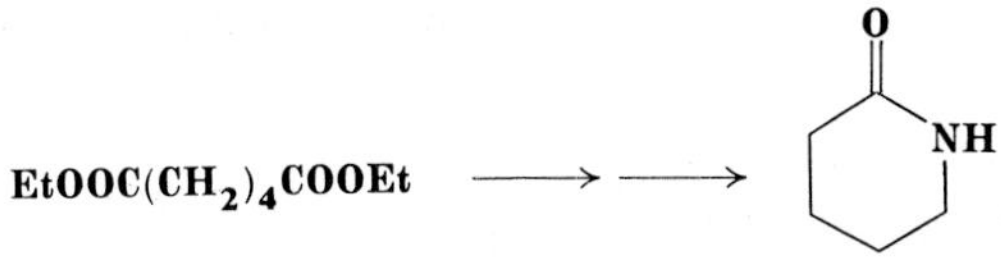

6. Suggest a mechanism for the following reactions:

a) 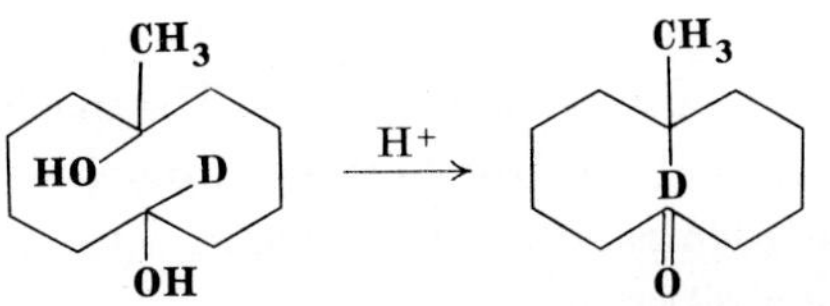

(a transannular or "across the ring" rearrangement)

b) 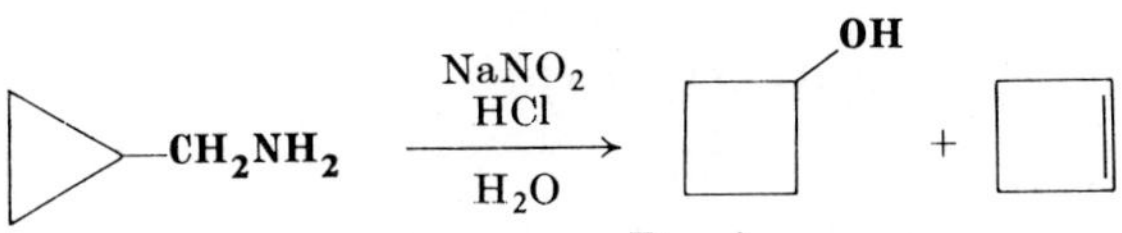

Demjanov rearrangement

Hint: See Section 10.11.

c) 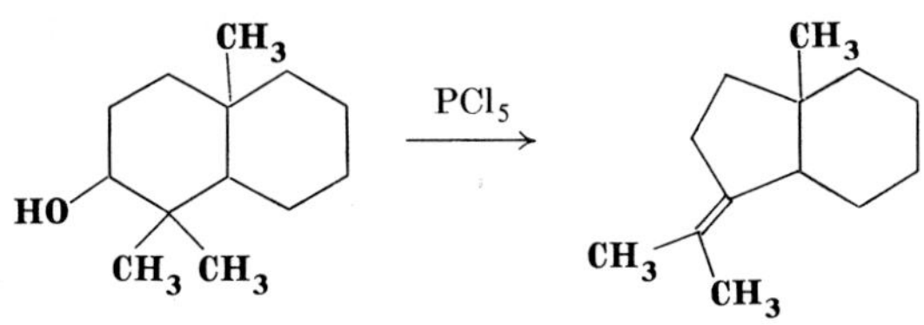

d) 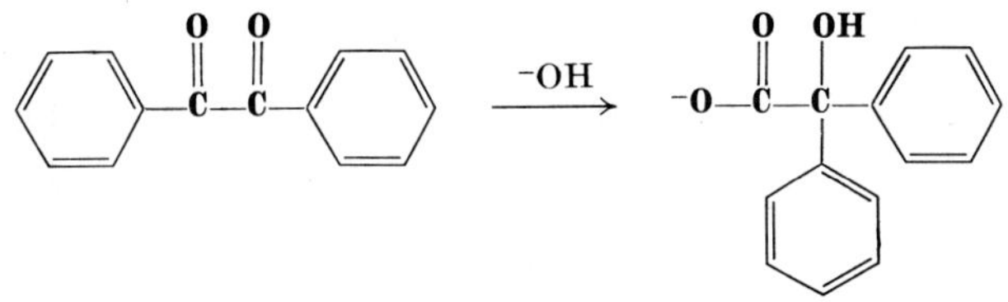

(benzilic acid rearrangement)

e) 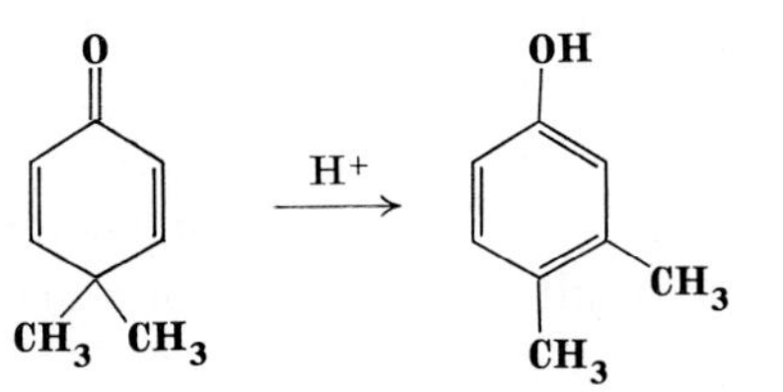

(dienone-phenol rearrangement)

7. The Arndt-Eistert synthesis converts an acid into a new acid with one additional carbon:

$$\mathrm{RCOOH} \xrightarrow{SOCl_2} \mathrm{RCOCl} \xrightarrow{CH_2N_2} \mathrm{R{-}\overset{O}{\overset{\|}{C}}{-}CHN_2} \xrightarrow[Ag_2O]{H_2O} \mathrm{RCH_2COOH}$$

Draw two major resonance contributors for the key reagent, diazomethane (CH_2N_2). Write the mechanism for the final step in the Arndt-Eistert synthesis, ignoring the role of the silver oxide catalyst. The mechanism is thought to be very similar to that of the Hofmann rearrangement except that a carbene intermediate (neutral divalent carbon with six outer electrons) is formed instead of a nitrene.

8. Two mechanisms have been proposed for the rearrangement of phenyl glyoxal to phenyl glycollic acid:

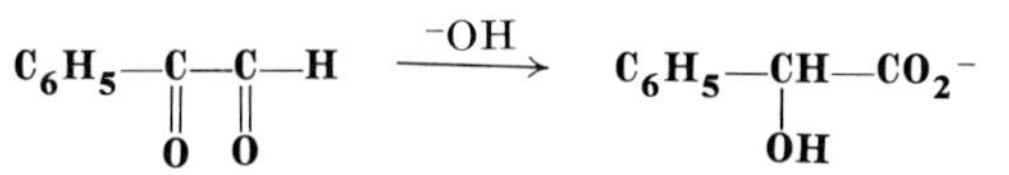

phenyl glyoxal phenyl glycollic acid

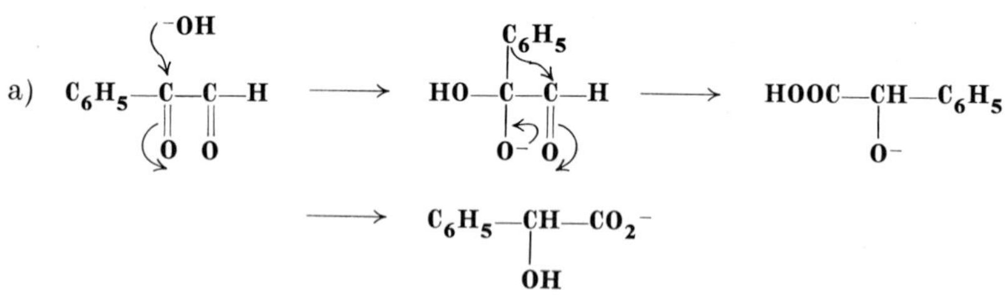

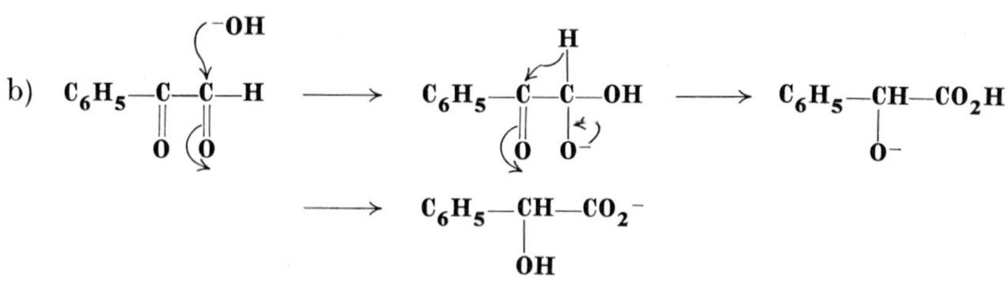

Design an experiment which will distinguish the two mechanisms. The second mechanism was shown to be the correct one.

9. Deduce from the reactions of the following optically active compounds whether the alkyl group in the Hofmann rearrangement migrates with retention of configuration. Are the results consistent with a mechanism in which the migrating alkyl group becomes free from the rest of the molecule before bonding to the nitrogen?

a) 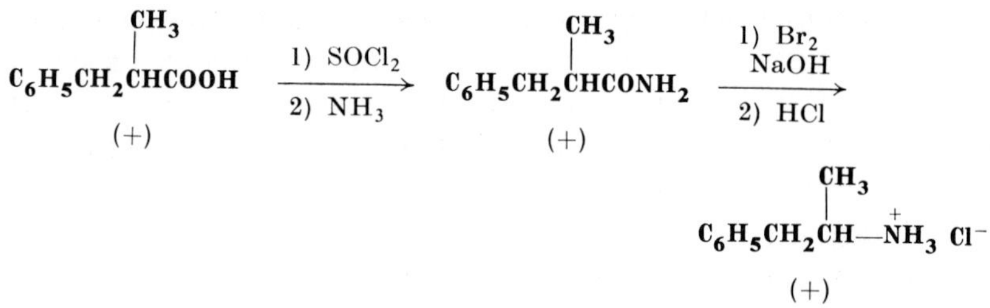

b) 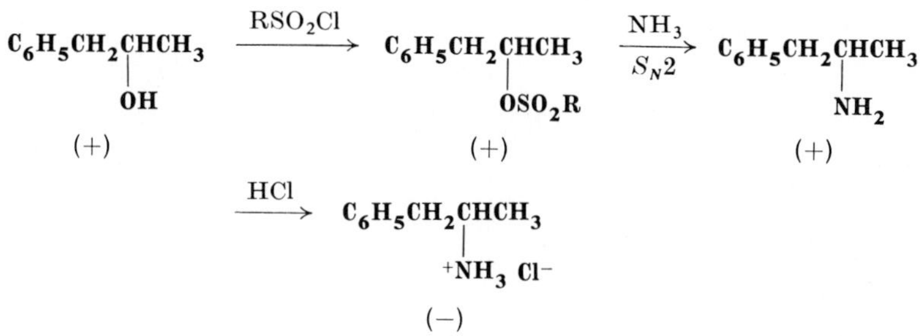

c) 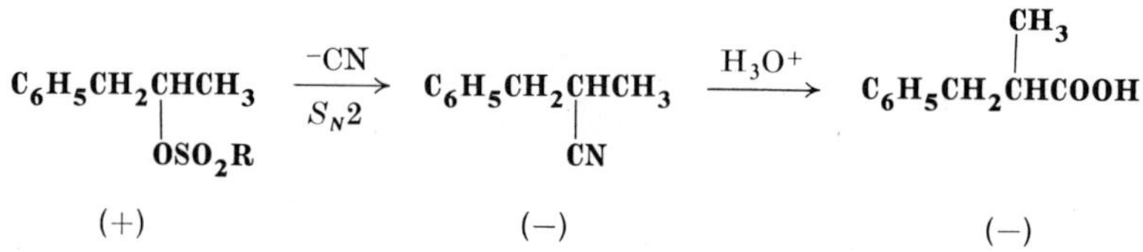

10. One type of rearrangement reaction we have not previously mentioned is sometimes called a "no mechanism" reaction. An important biological example of this reaction is the first step in the photochemical conversion of ergosterol into vitamin D_2:

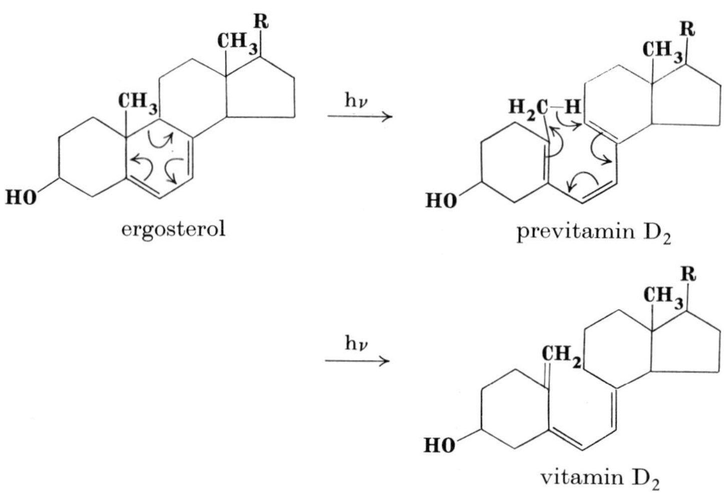

Vitamin D deficiency, leading to rickets, can be prevented in children by exposing them to sunshine (which promotes the above reaction in the skin) or by feeding them cod liver oil which is rich in vitamin D.

Propose a mechanism for the following reaction which also involves a light-induced rearrangement.

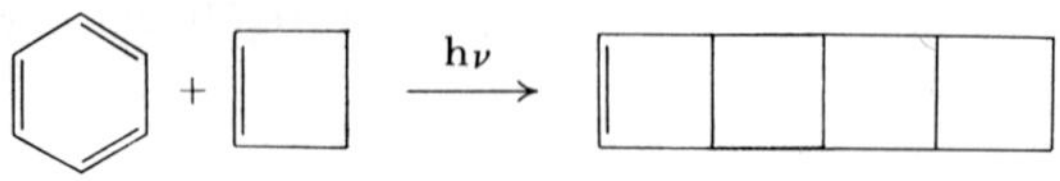

13
OXIDATION–REDUCTION REACTIONS

13.1 OXIDATION–REDUCTION

Many of the reactions we have discussed can be classified as either oxidations or reductions. There are a number of criteria which identify these oxidation–reduction reactions. For example, oxidation of almost all organic compounds involves either gaining oxygen atoms or losing hydrogen atoms. Thus, the following may all be considered oxidation reactions.

1) 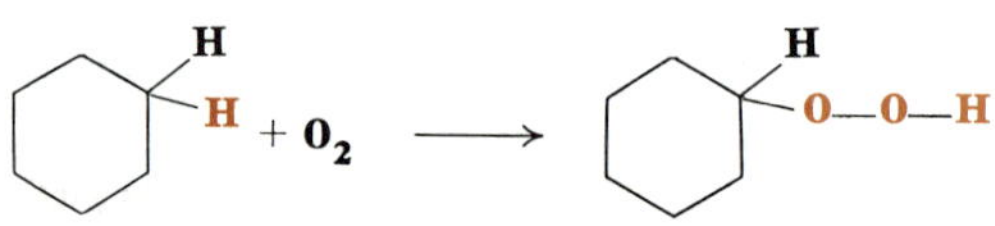

2) 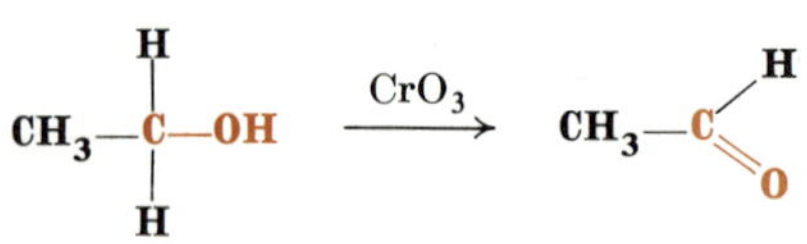

3) 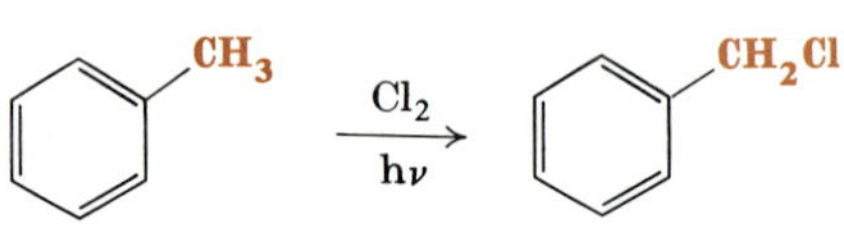

4) 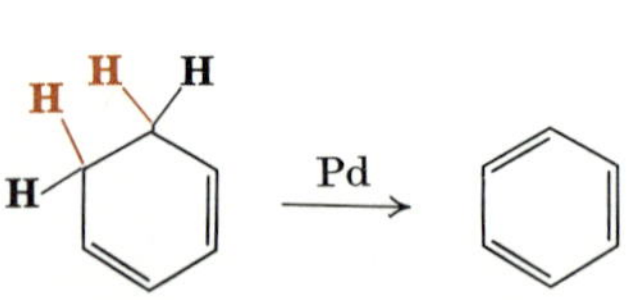

Reduction reactions usually involve gaining hydrogen and, in many cases, losing a heteroatom (such as oxygen, nitrogen, halogen). Some typical reductions are illustrated below:

1) $CH_3CH{=}CHCH_2CH_3 \xrightarrow[Pt]{H_2} CH_3CH_2CH_2CH_2CH_3$

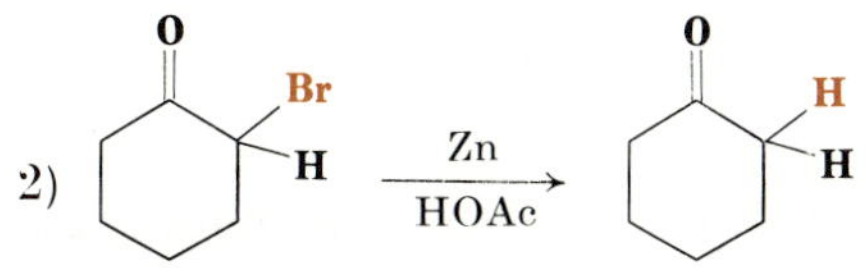

$$\text{3) } CH_3\underset{\underset{O}{\|}}{C}{-}OCH_3 \xrightarrow{LiAlH_4} CH_3CH_2OH + CH_3OH$$

There is a simple way to judge the degree of oxidation of a molecule. First, imagine adding a water molecule to all unsaturated units in the compound. If the compound has a ring, imagine that the ring is broken with the addition of water. Next, count the heteroatoms (oxygen, nitrogen, etc.) after the addition of the water molecules. The larger the number of heteroatoms, the more highly oxidized the molecule.

It should be clear from the above that the dehydration of *n*-propanol to propylene is *not* an oxidation.

$$\underset{n\text{-propanol}}{CH_3CH_2CH_2OH} \xrightarrow{H^+} \underset{\text{propylene}}{CH_3CH{=}CH_2} + H_2O$$

The oxidation count of both compounds is one. The oxidation counts of a number of compounds are given in Fig. 13.1.

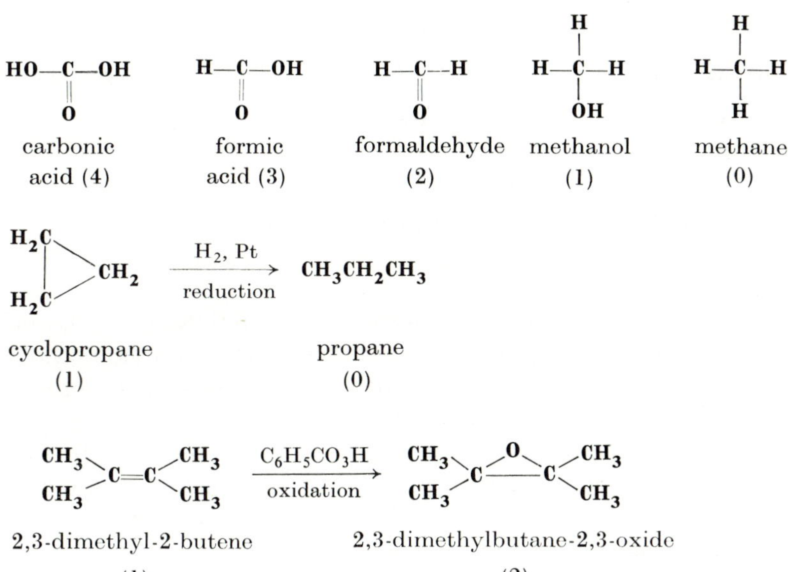

Fig. 13.1 Oxidation states for carbon in some organic compounds

OXIDATION REACTIONS

13.2 ALCOHOLS

The oxidation of alcohols by CrO_3 was one of the first reactions we discussed (Section 3.6). The reaction is usually carried out in acid solution, but a basic solvent, pyridine, can also be employed:

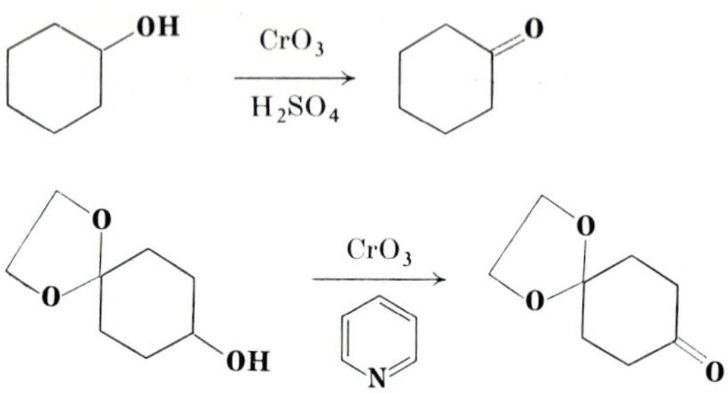

The mechanism of the chromate oxidation is illustrated in Fig. 13.2.

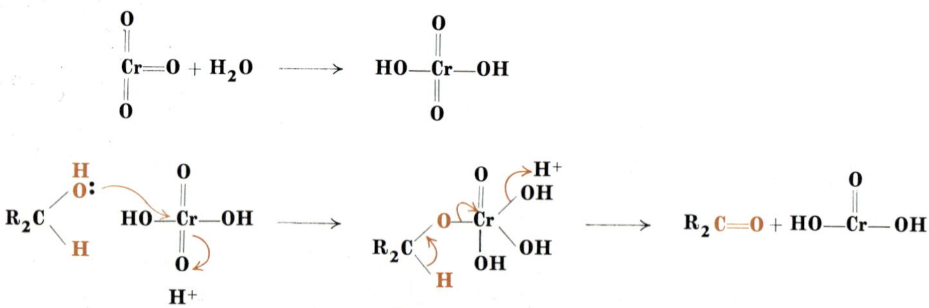

Fig. 13.2 Oxidation of an alcohol to an aldehyde by chromate

The actual oxidizing agent is H_2CrO_4, which is in equilibrium with CrO_3 in the strong aqueous acid. The H_2CrO_4 forms a chromate ester with the alcohol, in a reaction similar to the addition of an alcohol to an aldehyde to form a hemiacetal (see Section 11.6). Decomposition of the chromate ester (an elimination reaction mechanism) produces the carbonyl compound. In the overall reaction, the alcohol is oxidized to a carbonyl group, and the Cr^{+6} is reduced to Cr^{+4}.

CrO_3 oxidizes primary alcohols to aldehydes:

$$RCH_2OH \xrightarrow[H_2SO_4]{CrO_3} RCHO$$

It is usually difficult to isolate the aldehydes from such reactions, because aldehydes readily oxidize further, to acids:

$$\mathbf{RCHO} \xrightarrow[H_2SO_4]{CrO_3} \mathbf{RCOOH}$$

The mechanism for the oxidation of an aldehyde to an acid is similar to that in Fig. 13.2 except that the species which undergoes oxidation is the hydrate of the aldehyde [$RCH(OH)_2$] rather than an alcohol.

13.3 ALDEHYDES

In addition to CrO_3, two other substances are particularly useful in the oxidation of aldehydes. Many aldehydes are readily converted into acids by molecular oxygen. For example, the aldehyde shown below is quantitatively converted into an acid by bubbling O_2 through a solution of the aldehyde:

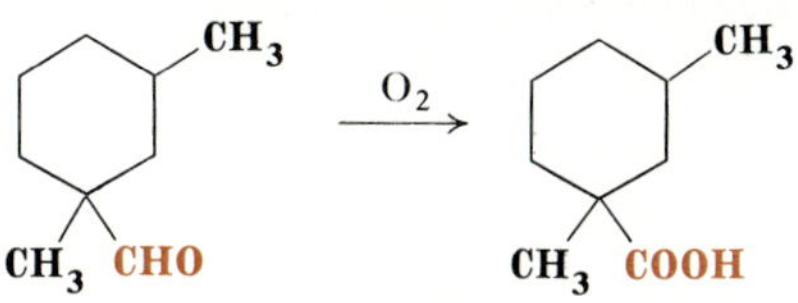

Silver oxide is more commonly used for oxidizing aldehydes:

$$\mathbf{R{-}C({=}O){-}H + Ag_2O \longrightarrow R{-}C({=}O){-}OH + 2Ag^0}$$

This reaction also serves as a qualitative test for the aldehyde function. In the so-called *Tollens' test*, the suspected aldehyde is shaken with silver oxide dissolved in aqueous ammonia. If a silver mirror is deposited on the walls of the vessel, oxidation has occurred:

$$\mathbf{RCHO + Ag(NH_3)_2OH \longrightarrow \underset{\text{(silver mirror)}}{Ag^0\downarrow}}$$

A similar test for aldehydes is the *Fehling's test*. Here a basic copper solution is used, and if red cuprous oxide (Cu_2O) precipitates out, the presence of an aldehyde is confirmed.

$$\mathbf{RCHO + Cu^{+2} + OH^- \longrightarrow \underset{\text{(red precipitate)}}{Cu_2O\downarrow}}$$

Cupric ion is too weak an oxidizing agent to oxidize a simple alcohol. However, a positive Fehling's test is obtained with α-hydroxy carbonyl compounds.

Thus, sugars which contain a free carbonyl function, either an aldehyde or a ketone, can be characterized by their reaction with Fehling's reagent:

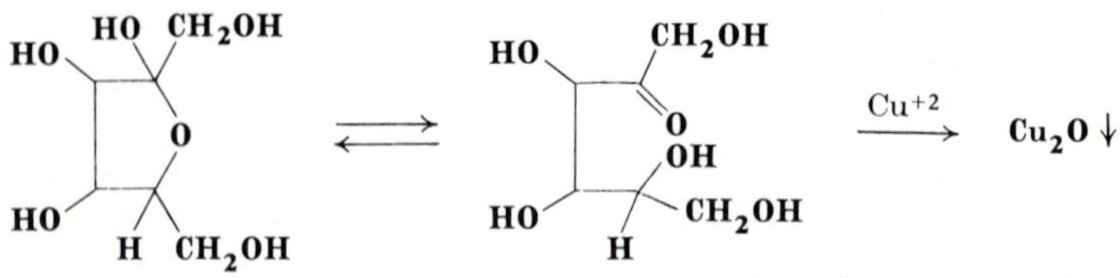

13.4 DOUBLE BONDS

The π system of a carbon-carbon double bond is readily accessible to a variety of oxidizing agents. We have already discussed some of these (ozone, peracid, potassium permanganate, and osmium tetroxide) in Chapter 9. The reactions in Fig. 13.3 will serve as a review of these oxidation reactions.

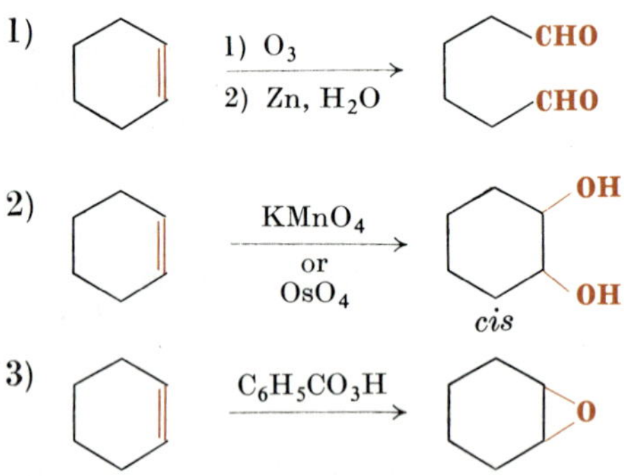

Fig. 13.3 Some oxidations of olefins

The products of permanganate and osmium tetroxide oxidations are 1,2-diols or glycols. These glycols can be further oxidized by either periodic acid (HIO_4) or lead tetraacetate [$Pb(OAc)_4$]:

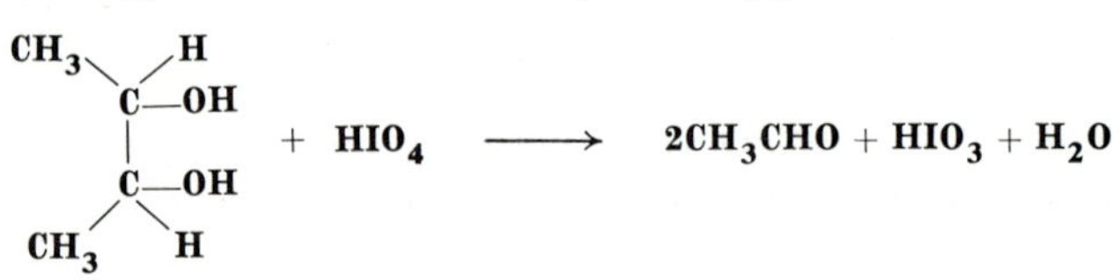

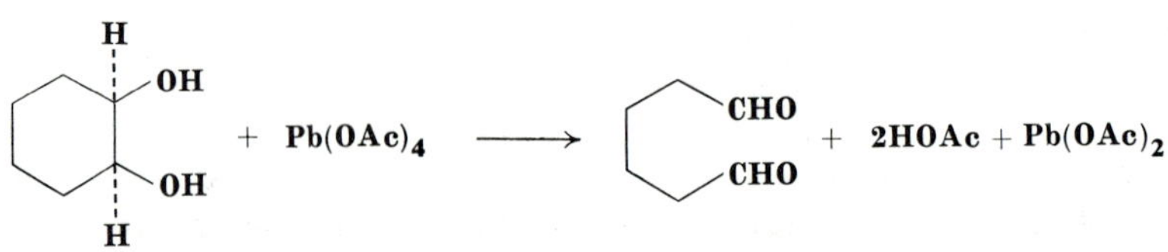

These two reactions proceed by similar mechanisms, in that they both depend on the formation of a cyclic intermediate. Figure 13.4 illustrates the mechanism of $Pb(OAc)_4$ oxidation; the periodate mechanism is left as an exercise for the reader.

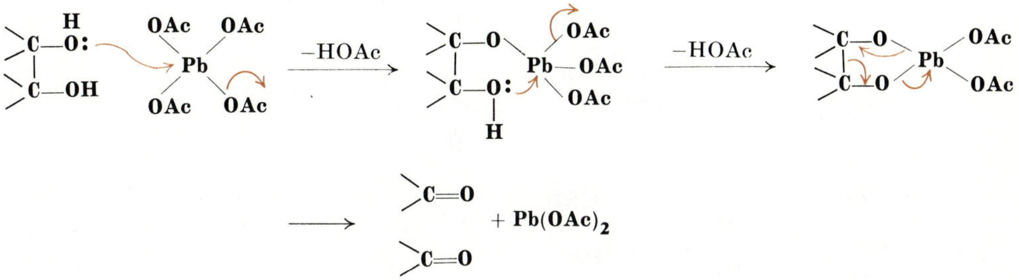

Fig. 13.4 Oxidation of a 1,2-glycol by $Pb(OAc)_4$

When small-ring cyclic glycols are oxidized by these reagents, the *cis* isomer is transformed at a greater rate than the *trans* isomer (why?). Thus *cis*-cyclohexane-1,2-diol can be oxidized to the dialdehyde while the *trans* compound remains unchanged:

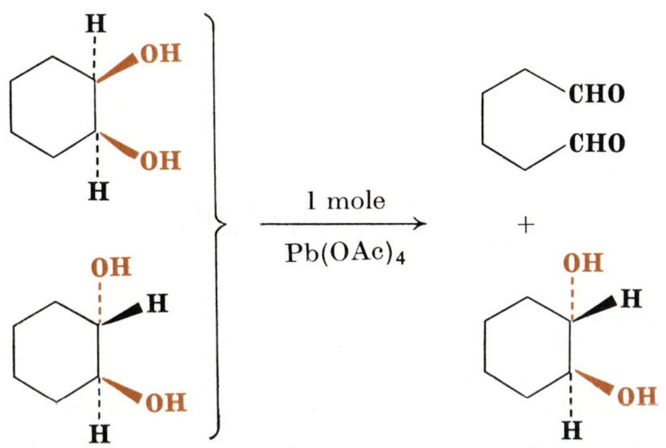

REDUCTIONS

13.5 CATALYTIC HYDROGENATION

The reader is already familiar with hydrogenation (addition of H_2 to an unsaturated grouping). A number of metals are used as catalysts in the hydrogenation of unsaturated compounds, the most common being palladium, platinum, nickel, and rhodium. Platinum and nickel catalysts are generally used as finely divided pure metals, while rhodium and palladium are employed as metal deposits on inert substances such as barium sulfate. Depending upon the choice of catalyst, solvent, temperature, and pressure, one may catalytically hydrogenate double bonds, carbonyl groups, nitriles, nitro groups, and aromatic rings. Some typical examples are illustrated in Fig. 13.5.

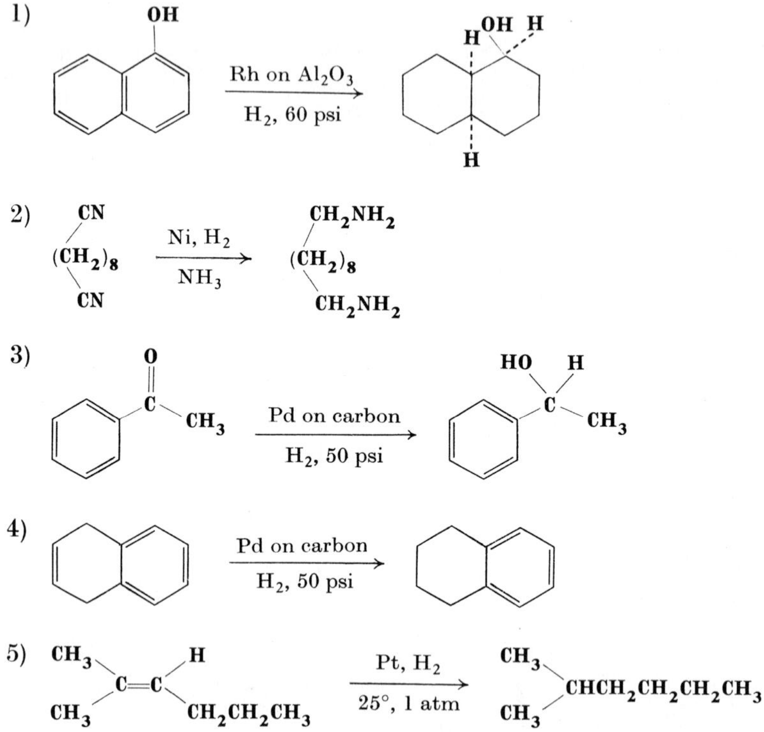

Fig. 13.5 Some catalytic hydrogenations

It is instructive to give a few experimental details of a typical catalytic reduction (involving Raney nickel, a special type of porous nickel with a large surface area):

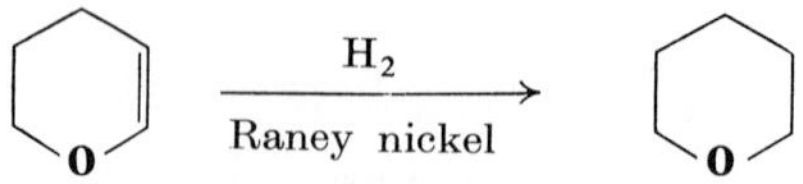

dihydropyran tetrahydropyran

Dihydropyran (50 g, 0.6 mole) and Raney nickel catalyst (8 g) are mixed in a bottle which has a rubber stopper with a tube leading to a hydrogen gas source. Hydrogen gas is admitted into the system until the pressure reaches 40 lb. The bottle is then shaken until the pressure drops to that corresponding to the consumption of 0.6 mole of hydrogen (about 20 minutes). A nearly quantitative yield of product can then be decanted from the catalyst.

Although there are exceptions, hydrogenation usually occurs with *cis* addition of hydrogen. Thus 1,2-dimethylcyclohexene yields *cis*-1,2-dimethylcyclohexane:

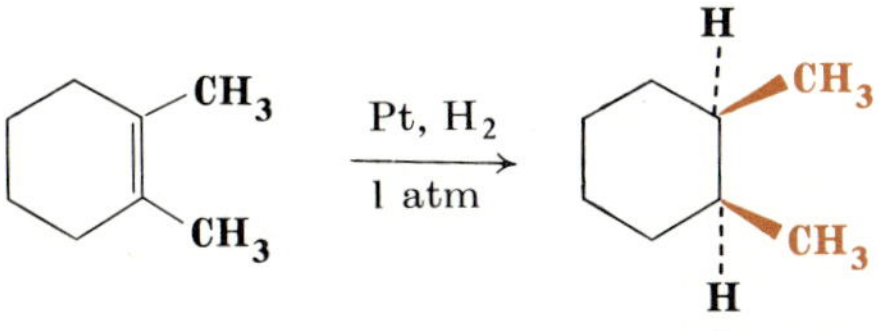

The accepted mechanism for this reaction involves the adsorption of hydrogen on the surface of the metal, followed by adsorption of the unsaturated compound (Fig. 13.6). The binding of the organic substrate to the metal involves the donation of π electrons to orbitals of the metal atom. As might be expected, the greater the concentration of π electrons, the more readily the compound is adsorbed and reduced. Thus acetylenes are reduced more rapidly than olefins (see Section 9.15).

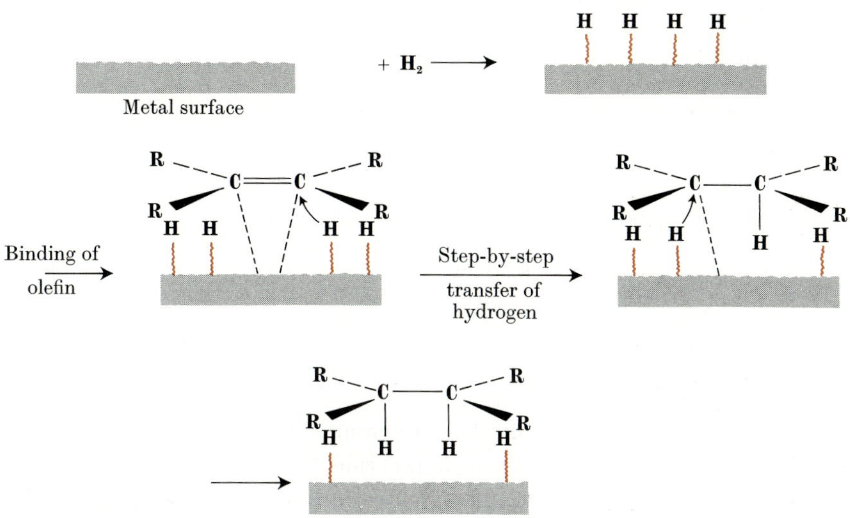

Fig. 13.6 Mechanism of catalytic hydrogenation

The reduction mechanism requires that hydrogen be delivered (probably step by step) to the side of the molecule bound to the metal surface. We might expect, therefore, that a compound which is bulkier on one side than the other will be reduced preferentially from the less hindered side. For example, the bicyclic olefin below, which has an axial methyl group on one face of the molecule, is reduced with palladium to yield principally the product with the *trans* ring fusion:

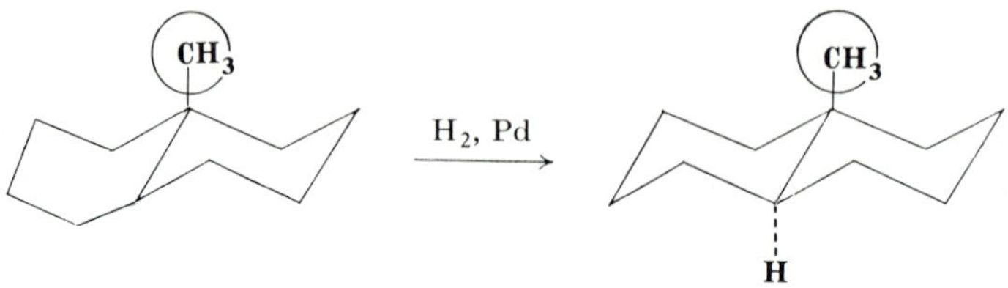

Most of the reduction takes place on the side of the molecule *opposite* the methyl group.

One final point should be made. Conventional hydrogenation can be reversed by using the metal catalyst in absence of hydrogen gas. Two good examples of such a *dehydrogenation* are illustrated in Fig. 13.7.

1)
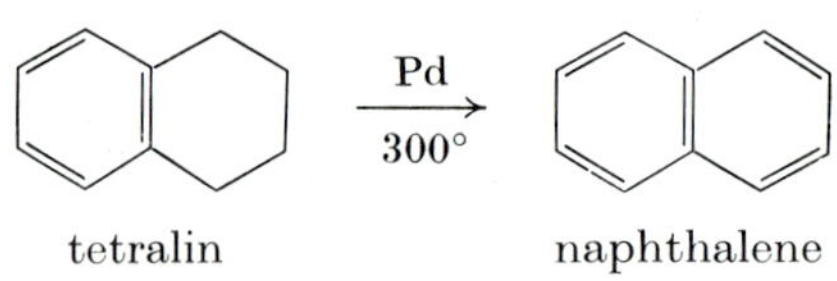

2)
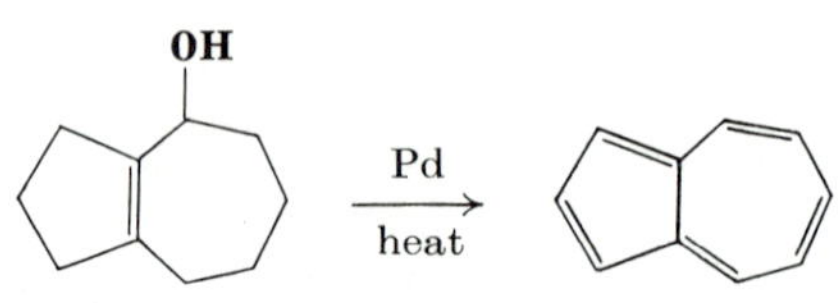

Fig. 13.7 Catalytic dehydrogenation

13.6 CHEMICAL REDUCTION

We have already discussed a number of chemical reductions which use such reagents as lithium aluminum hydride (Section 11.11) and diborane (Section 9.11). These are important reactions and should be reviewed by the reader. In the present section, we will discuss one other chemical reduction, the *Wolf-Kishner reduction*. The Wolf-Kishner reduction provides a way to convert a carbonyl into a methylene group in a single step:

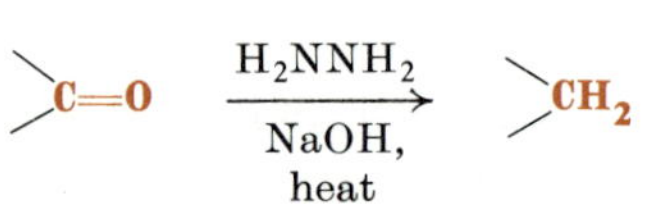

The Wolf-Kishner mechanism is interesting because it illustrates how easy it is to produce reactions that form the stable nitrogen molecule, N_2 (recall the Sandmeyer reaction we discussed in Section 10.11). When the hydrazone of a ketone or aldehyde is treated with strong base, it undergoes a tautomeric equilibrium reaction:

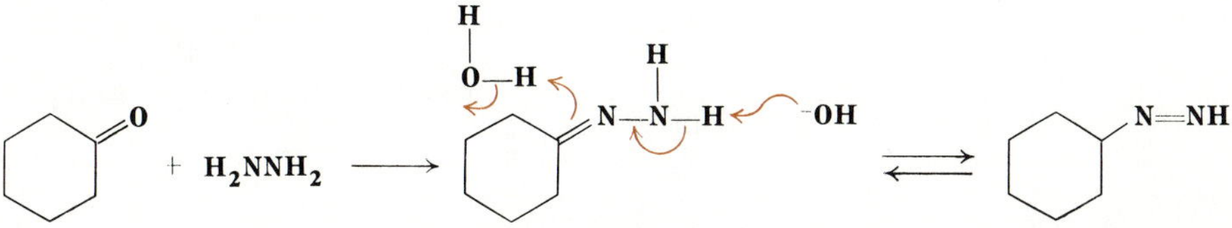

a hydrazone an azo compound

The azo intermediate decomposes to the products, probably through a carbanion intermediate which picks up a proton from water or hydrazine:

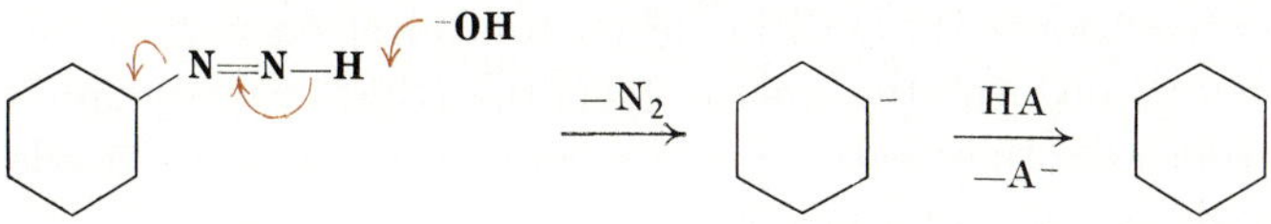

The driving force for this reaction is the ejection of the stable nitrogen molecule.

13.7 DISSOLVING METAL REDUCTIONS

A number of functional groups can be reduced by the combination of a metal and a proton source. The "dissolving metal" reductions include reagent combinations such as sodium in alcohol, lithium in liquid ammonia, and zinc in hydrochloric acid. Among the species reduced by one or more of these reagent combinations are ketones, esters, aromatic rings, and nitro groups. We will cite only one example, the *Birch reduction,* because it is particularly important in the synthesis of steroids and other natural products.

The Birch reduction of an aromatic ring uses lithium metal in a mixture of liquid ammonia and ethanol. For example

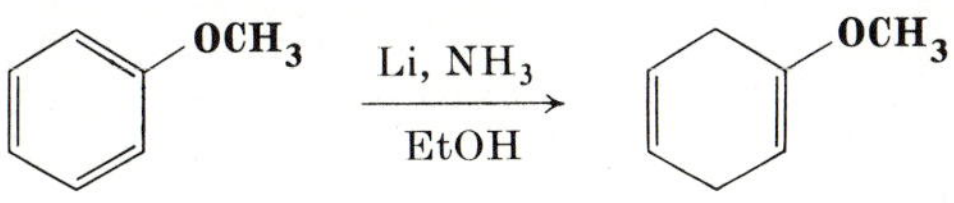

anisole 2,5-dihydroanisole

Reduction occurs by transferring electrons from the metal to the substrate (Fig. 13.8).

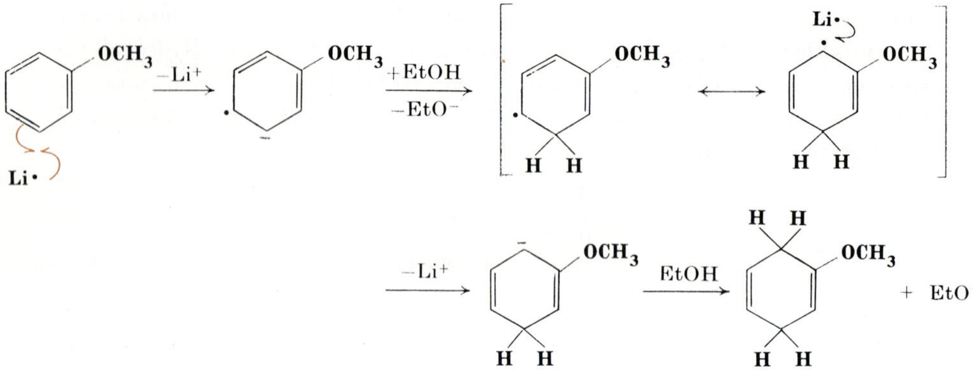

Fig. 13.8 Birch reduction

If one electron is added to the aromatic system, a radical anion is produced which is then protonated prior to the addition of a second electron and proton. Note that the Birch reduction forms the kinetic rather than the thermodynamic product. If the reaction were thermodynamically controlled, the more stable conjugated diene would be the major product.

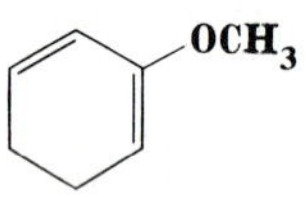

Two other examples of the Birch reduction are given below.

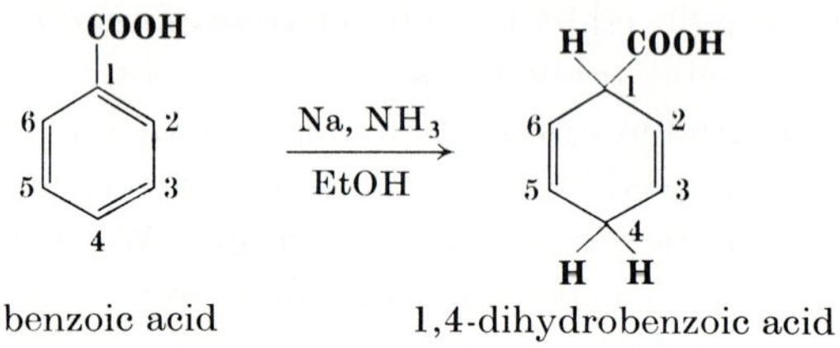

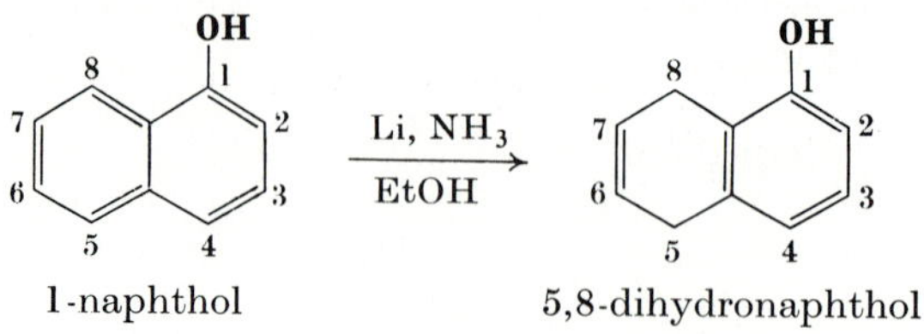

In the Birch reduction of 1-naphthol, only the ring without the hydroxyl group is reduced. This selectivity is expected. Reductions, the addition of electrons, are inhibited by electron-donating substituents.

13.8 ACYLOIN CONDENSATION

Esters undergo a reaction with sodium metal in toluene which results in the formation of α-hydroxyketones or acyloins. This condensation is particularly useful for closing large rings:

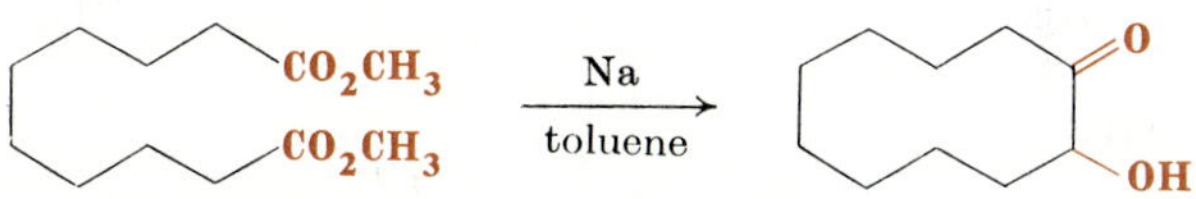

a diester an acyloin

Most other methods for closing rings of ten or more carbons yield large amounts of intermolecular reaction in addition to the desired intramolecular process. In the acyloin condensation, the two ester groups are first adsorbed onto the sodium metal surface, so that the two ends of the molecule do not have to "find" each other in a large three-dimensional space.

The mechanism of the acyloin condensation is shown in Fig. 13.9. Note that, unlike the Birch reduction, no proton donor is available until the reaction mixture is exposed to acid during the work-up.

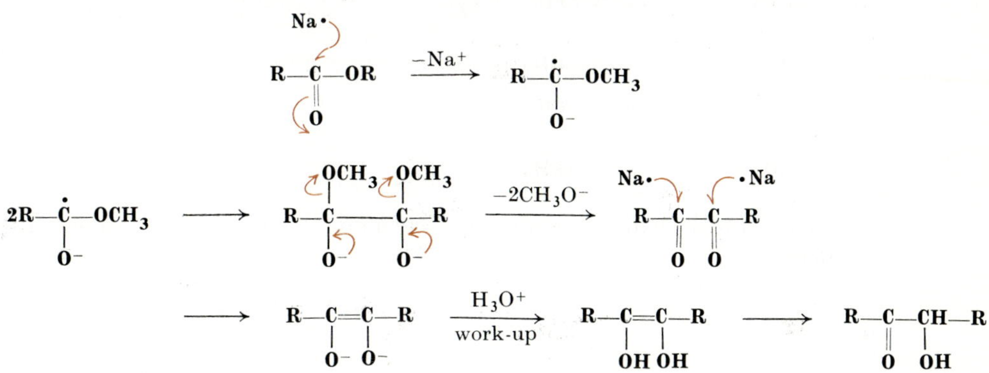

Fig. 13.9 Acyloin reduction

13.9 SUMMARY

It is important to know not only which functionalities will react with a reagent but also which ones will *not* react. Such information regarding certain useful reducing agents is summarized below.

1) Pd/H_2

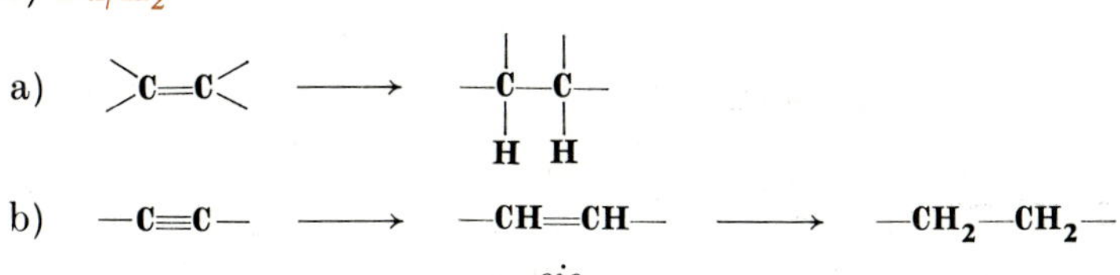

Comment. It is possible to stop this reduction at the olefin stage by using poisoned Pd (Lindlar catalyst—see Section 9.15).

c) $-\overset{O}{\overset{\|}{C}}-Cl \longrightarrow -\overset{O}{\overset{\|}{C}}-H$

d) $-NO_2 \longrightarrow -NH_2$

Comment. Except for acid chlorides, Pd/H_2 will not readily reduce carbonyl compounds.

2) Pt/H_2

a) $>C{=}C< \longrightarrow -\underset{H}{\overset{|}{C}}-\underset{H}{\overset{|}{C}}-$

b) $-C{\equiv}C- \longrightarrow -CH_2-CH_2-$

c) $-\overset{O}{\overset{\|}{C}}- \longrightarrow -\overset{OH}{\overset{|}{C}H}-$

d) $-NO_2 \longrightarrow -NH_2$

Comment. Pt/H_2 will not reduce the carbonyl group of acids, esters, and amides.

3) $LiAlH_4$

a) $-\overset{O}{\overset{\|}{C}}- \longrightarrow -\overset{OH}{\overset{|}{C}H}-$

b) $-CO_2H \longrightarrow -CH_2OH$

c) $-CO_2R \longrightarrow -CH_2OH$

d) $-CONH_2 \longrightarrow -CH_2NH_2$

e) $-CONHR \longrightarrow -CH_2NHR$

f) $-CONR_2 \longrightarrow -CH_2NR_2$

Comment. $LiAlH_4$ will not generally reduce carbon-to-carbon unsaturation.

4) NaBH$_4$

a) 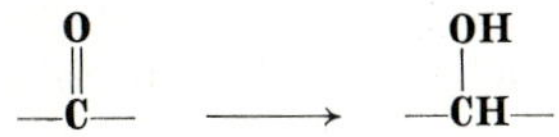

Comment. NaBH$_4$ will not reduce carbon-to-carbon unsaturation, amides, esters, or acids, however it does reduce acid chlorides to primary alcohols.

ADDITIONAL READING

R. Stewart, *Oxidation Mechanism*, Menlo Park, Calif.: Benjamin, 1964.
L. J. Chinn, *Selection of Oxidants in Synthesis*, New York: Marcel Dekker, 1971.

PROBLEMS

1. Name the following compounds. Consult the Nomenclature section in the Appendix if necessary.

a) 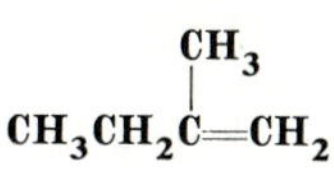 b) 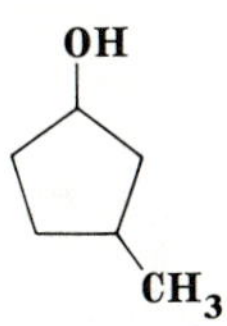 c) 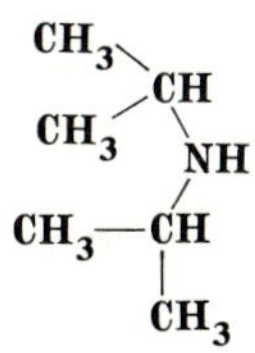 d) 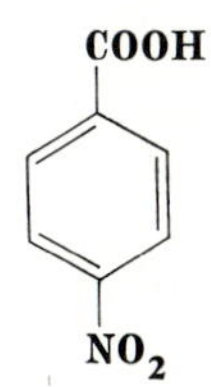

e) $CH_3CH(Cl)(CH_2)_4CH_3$ f) $CH_3C(=O)OCH_2CH_3$ g) $CH_3C(=O)NH_2$ h) $CH_3(CH_2)_3C(=O)CH_3$

i) 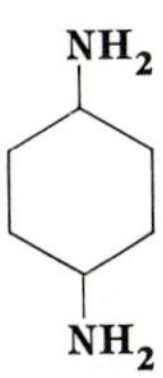 j) 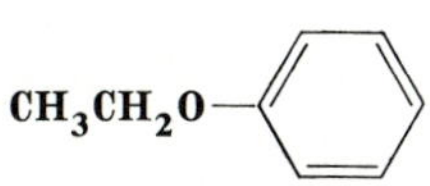 k) 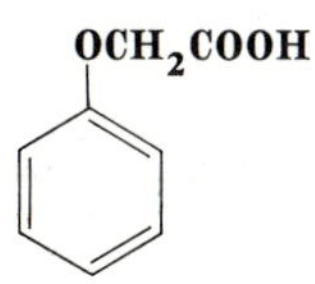 l) 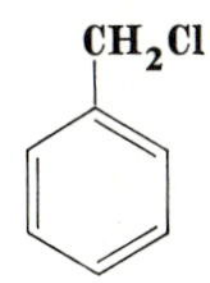

2. What oxidizing or reducing agent would you use to convert:

   a) a double bond into an epoxide?
   b) a double bond into a glycol?
   c) a ketone into a methylene group?
   d) an alkene into an alkane?
   e) a ketone into an alcohol?
   f) a primary alcohol into an acid?

3. For each of the following conversions provide the necessary reagents and the structures of all intermediate compounds. A review of the important reactions discussed in previous chapters will be helpful for these problems.

a) 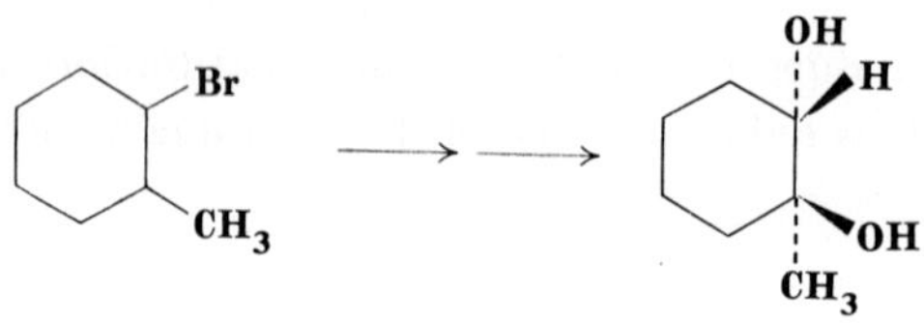

b) 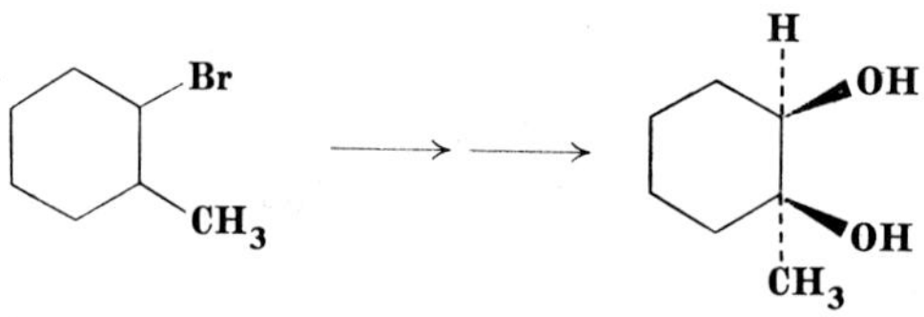

c) 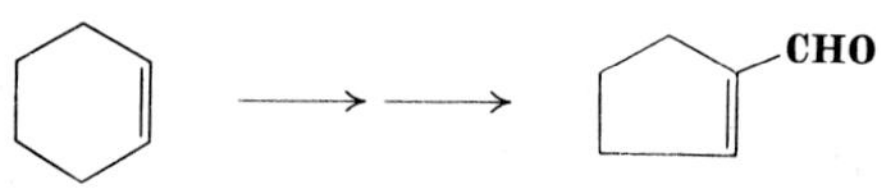

d) 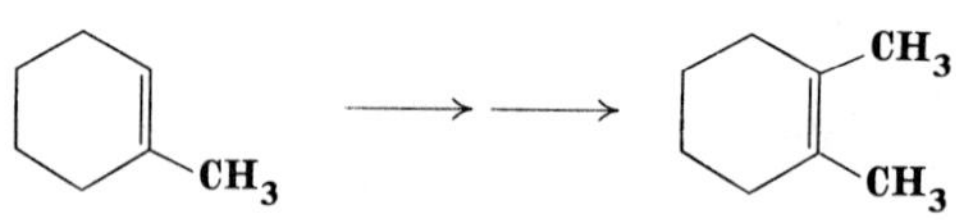

e) 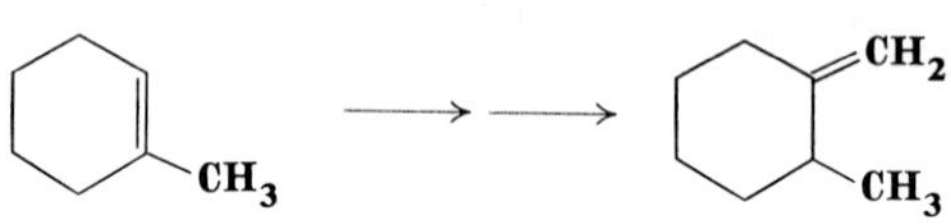

f) 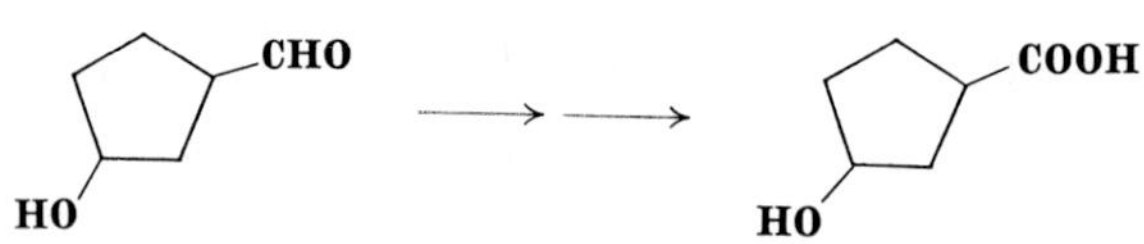

g) 

h) 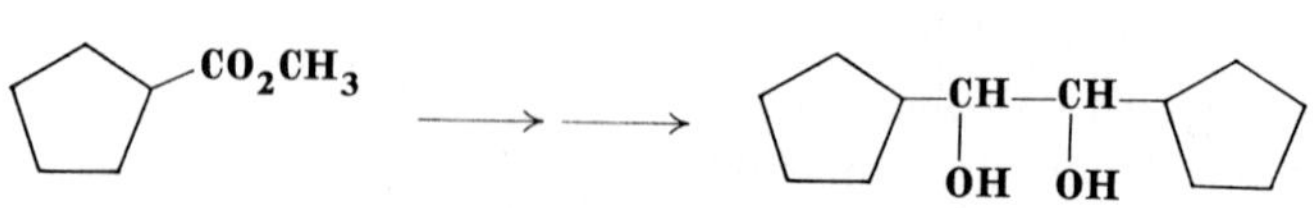

i) $CH_3CH_2CH_2OH \longrightarrow \longrightarrow CH_3CH_2CH_2CH_2NH_2$

j) 

k) 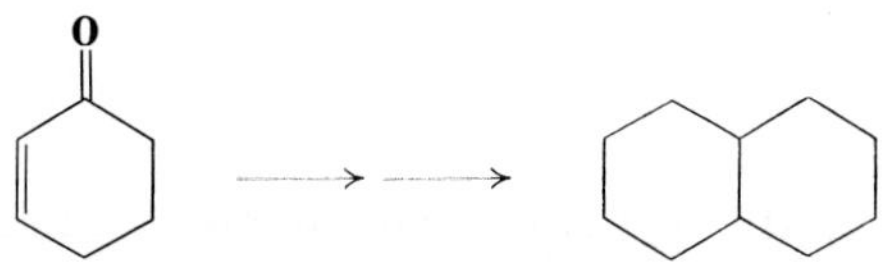

l) 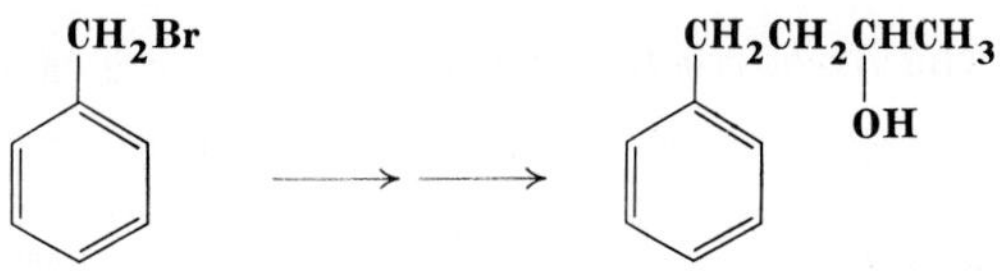

4. Will the *racemic* or the *meso* form of the following glycol be oxidized more rapidly by lead tetraacetate?

$$(CH_3)_2CH-\underset{\displaystyle |}{\overset{\displaystyle OH}{CH}}-\overset{\displaystyle OH}{CH}-CH(CH_3)_2$$

5. Primary halides can be oxidized to aldehydes in good yields using dimethyl sulfoxide (Me_2SO):

$$RCH_2Cl \xrightarrow[\text{2) } Et_3N]{\text{1) } (CH_3)_2\overset{+}{S}-\overset{-}{O}} RCHO + (CH_3)_2S$$

Suggest a reasonable mechanism for this reaction.

6. Suggest a mechanism for the following reaction:

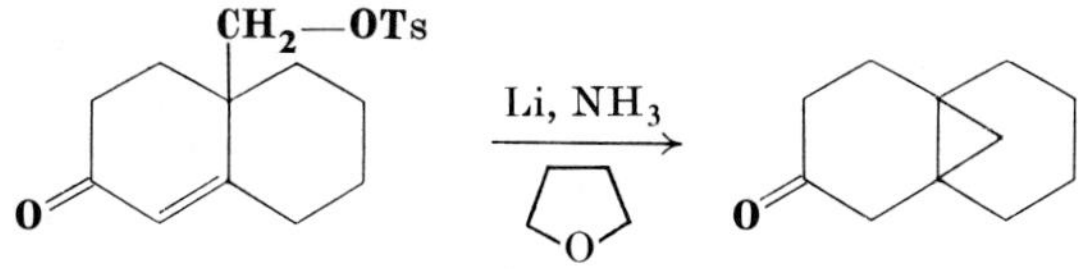

Note: —OTs = tosylate group = 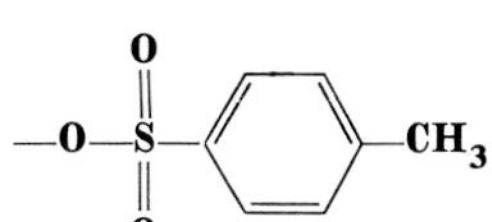

7. The following reactions have not been covered in this book. Predict the products using your knowledge of reaction mechanisms.

a) 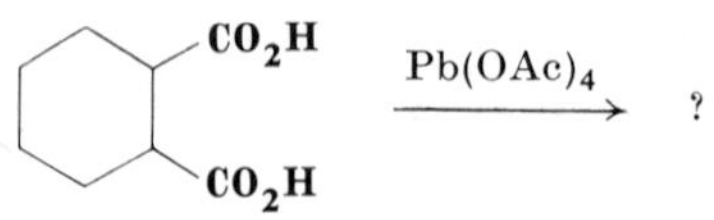

b) 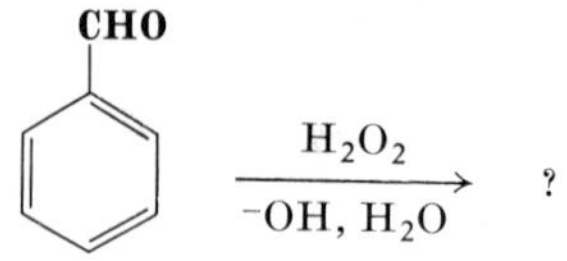

c) $R{-}C{\equiv}CH \xrightarrow[\text{base}]{Cu^{+2}} ?$

d) $(C_6H_5)_3C^+ + CH_3CH_2OH \longrightarrow ?$

8. Write a mechanism for the *Kolbe reaction*, the electrolysis of a carboxylic acid:

$$2RCO_2^- \xrightarrow{\text{anode}} R{-}R + 2CO_2$$

9. The selective oxidation of an unactivated methyl group (such as one bonded to a methylene) is difficult to achieve. Conditions which oxidize the methyl group are likely to tear up the entire molecule. One way to approach this problem is to use an intramolecular oxidizing agent: a group close to the methyl group is used to transform the methyl group into a functionalized carbon. A good example of this method is shown below in the oxidation of an unactivated methyl group of a steroid molecule. Suggest a mechanism for this light-induced free radical reaction.

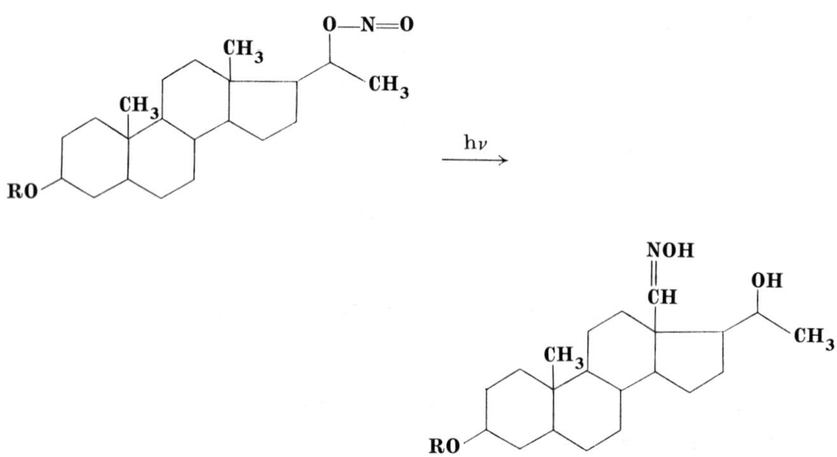

10. Suggest a mechanism for the oxidation of a tertiary amine to an enamine with mercuric acetate:

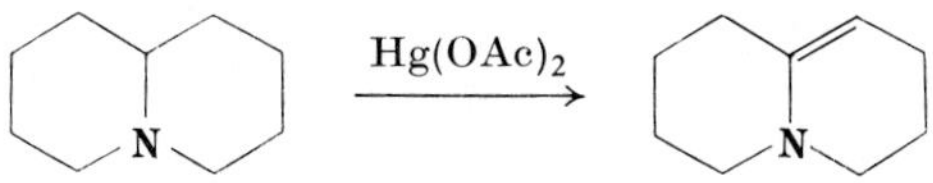

11. What reagents (A through D) would you use to carry out the following reductions?

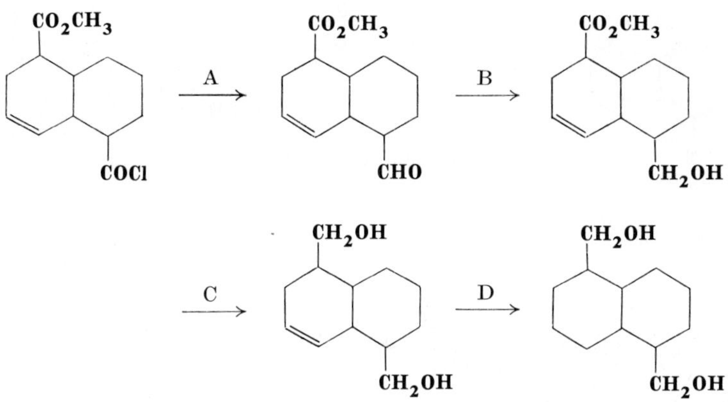

12. Lysergic acid is the parent compound of a series of natural products found in rye grain fungus. The psychological effects of eating spoiled rye have been known since pre-Christian times. Interest in the synthesis of lysergic acid was heightened by the discovery of the vivid (and dangerous) psychedelic states induced by the diethylamide derivative of lysergic acid (LSD).

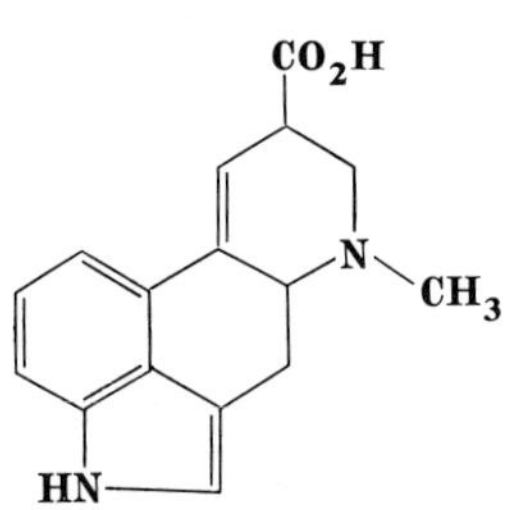

lysergic acid

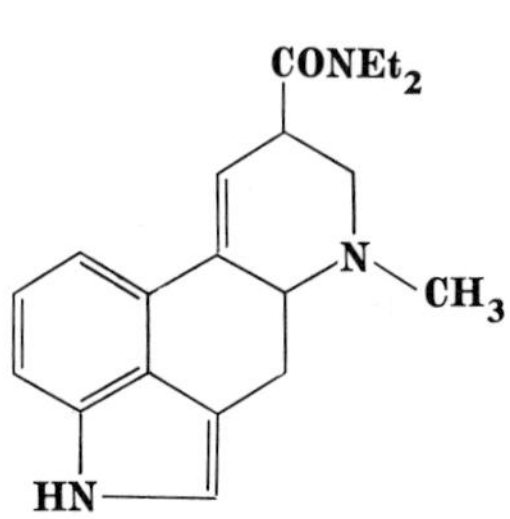

lysergic acid diethylamide (LSD)

The total synthesis of lysergic acid is given below. Study it carefully, review the reactions which you have forgotten, and provide the correct reagents where necessary.

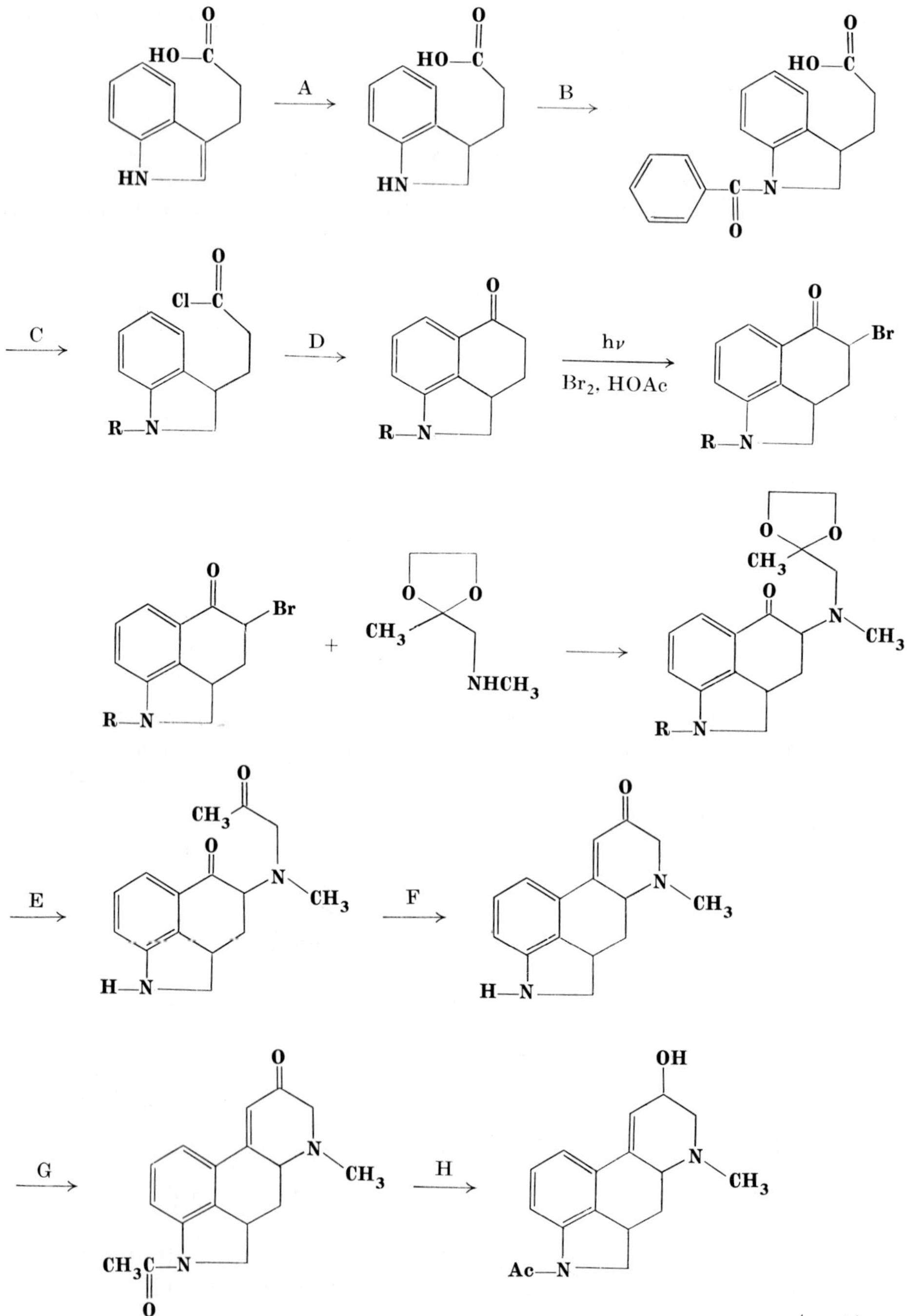

(*cont.*)

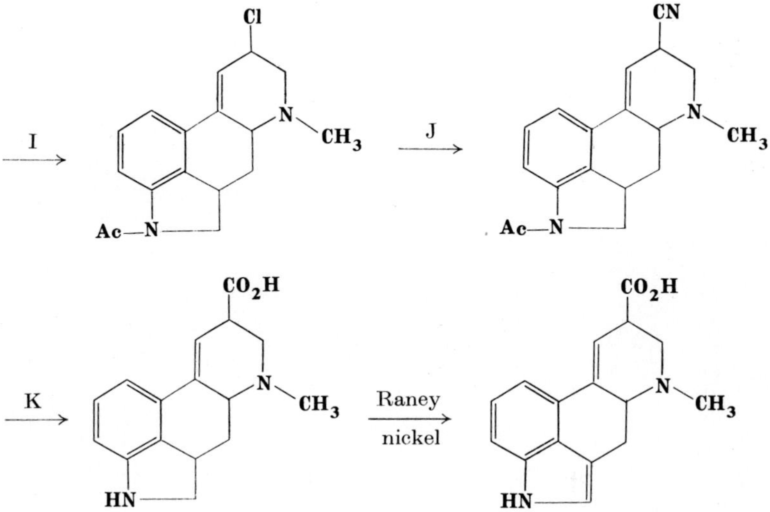

dl-lysergic acid

14
ELECTROCYCLIC REACTIONS

14.1 MOLECULAR ORBITALS

Perhaps the most important single advance in organic chemistry over the past decade has been the development of the Woodward-Hoffmann rules governing the behavior of electrocyclic reactions and related processes. Before we can delve into this subject however, we must have a rudimentary understanding of molecular orbitals.

The π molecular orbital (MO) of ethylene is formed from two p orbitals each containing one electron. Since the number of molecular orbitals always equals the number of original atomic orbitals, ethylene has an empty MO in addition to the one which is filled. This empty molecular orbital is of high energy and is called a π^* MO. We may show this by the following diagram where the arrows represent electrons of differing spin.

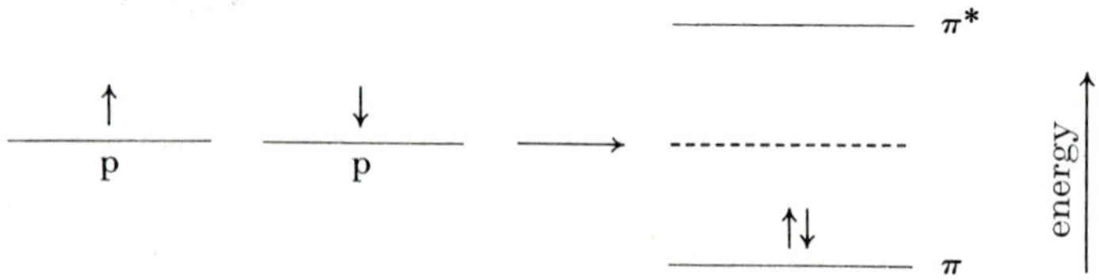

In most of the ethylene chemistry with which we are familiar (*e.g.*, addition of bromine) the π^* MO plays no role. However, it does become important in photochemical reactions. Exposure of ethylene to UV light elevates an electron in a so-called $\pi \rightarrow \pi^*$ transition:

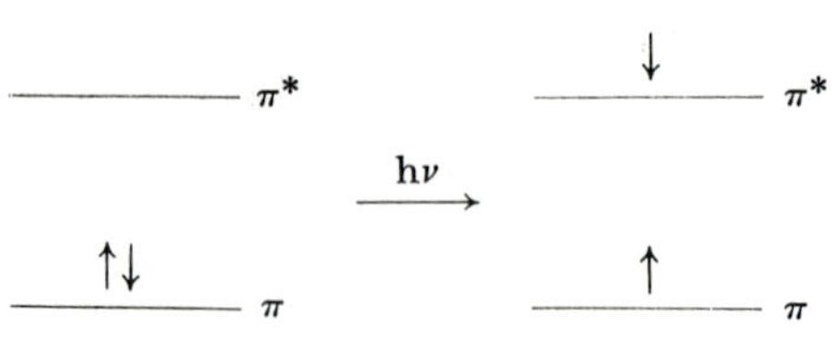

In depicting molecular orbitals, p-orbital lobes are assigned either + or −. This notation is useful in understanding electrocyclic reactions. As the student will learn in physical chemistry, these signs are related to signs of wave functions and have nothing to do with negative or positive charge. The important point is that, according to this convention, lobes of the *same* sign can overlap constructively and assist in the binding of atoms. Lobes of opposite sign do not interact favorably. The two molecular orbitals of ethylene can then be pictured as:

π $\qquad\qquad$ π^*

The π MO is called a *bonding orbital* because lobes of the same sign are adjacent to each other and overlap. On the other hand, in the π^* configuration the adjacent lobes have different signs and contribute nothing to the bonding of the carbons. In fact, the π^* MO is destabilizing and is therefore called *antibonding*.

The π system of butadiene, $CH_2{=}CH{-}CH{=}CH_2$, results from the overlap of four atomic p orbitals which combine to form four molecular orbitals. Two of these are bonding and two are antibonding:

(1)	+ + + + **C—C—C—C** − − − − bonding	(2)	+ + − − **C—C—C—C** − − + + bonding
(3)	+ − − + **C—C—C—C** − + + − antibonding	(4)	+ − + − **C—C—C—C** − + − + antibonding

As one might guess by inspection, the energies of the four orbitals bear the following relationship: $1 < 2 < 3 < 4$. In terms of an energy diagram, this can be represented as:

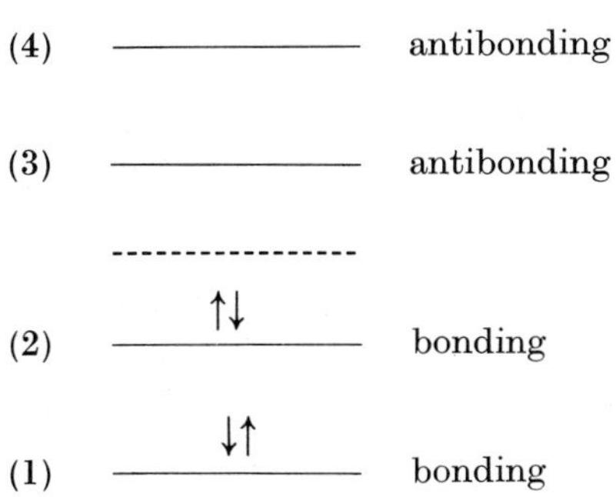

The four π electrons of butadiene have been placed in the two bonding orbitals (molecular orbitals, like atomic orbitals, can hold no more than two electrons). If butadiene is irradiated with UV light, an electron moves from the highest bonding orbital to the lowest antibonding orbital. The excited state looks like:

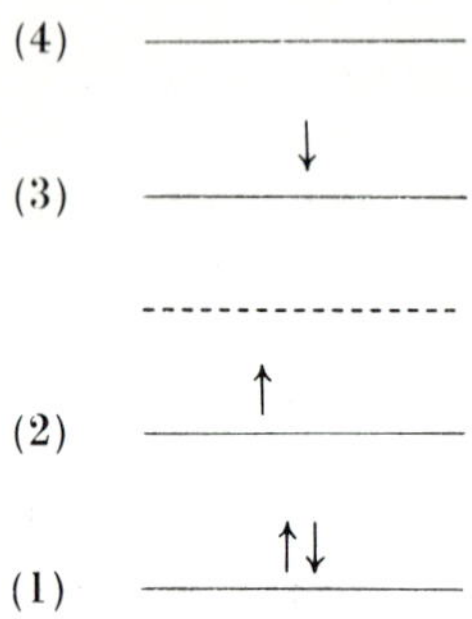

Hexatriene, $CH_2{=}CH{-}CH{=}CH{-}CH{=}CH_2$, has six molecular orbitals:

+ + + + + + C—C—C—C—C—C − − − − − − bonding	+ + + − − − C—C—C—C—C—C − − − + + + bonding	+ + − − + + C—C—C—C—C—C − − + + − − bonding
+ − − + + − C—C—C—C—C—C − + + − − + antibonding	+ − + + − + C—C—C—C—C—C − + − − + − antibonding	+ − + − + − C—C—C—C—C—C − + − + − + antibonding

An easy way of assessing the relative energies of these orbitals is to count nodes. A node is a point midway between two carbons at which a change in sign takes place—the fewer the nodes, the lower the energy. The number of nodes for hexatriene varies from zero (for the lowest-energy bonding orbital) to five (for the highest-energy antibonding orbital).

14.2 ELECTROCYCLIC REACTIONS

With this brief introduction, we are in a position to understand electrocyclic reactions. An electrocyclic reaction involves the formation of a single bond between termini of a system of conjugated double bonds (or the reverse process). A few examples will clarify this definition:

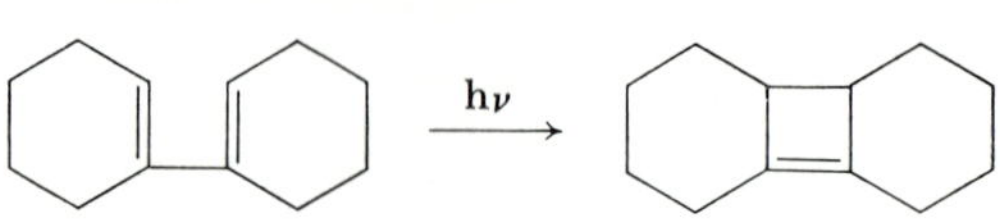

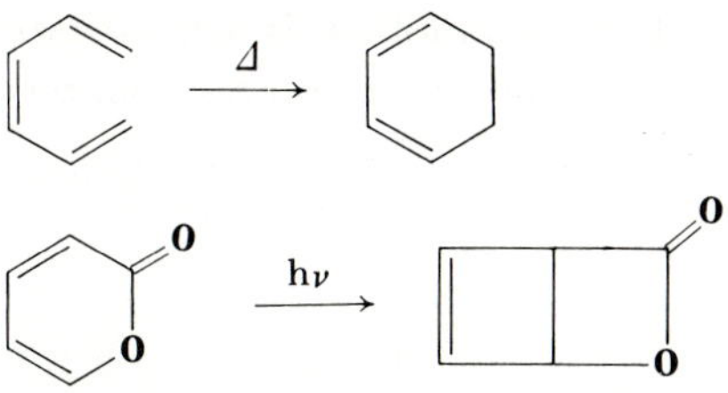

Note that electrocyclic reactions can be induced either photochemically or thermally. In either case, the reactions are believed to be *concerted*: they occur in one step with one transition state and no intermediate of any kind. Conversion of reactant into product *via* "electron pushing" can be written as:

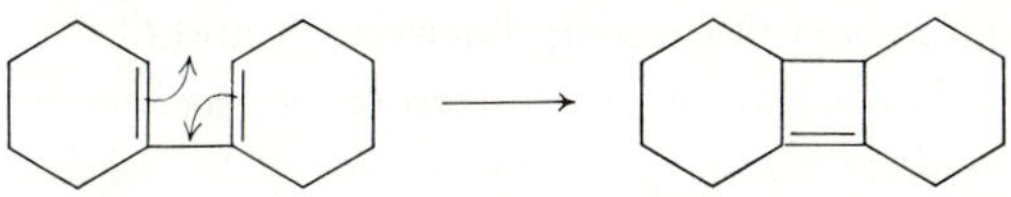

By far the most interesting feature of electrocyclic reactions is their stereochemistry. For example, the thermal cyclization shown below gives only *cis* product. Why should this be so? The simple electron-pushing mechanism does not answer this question.

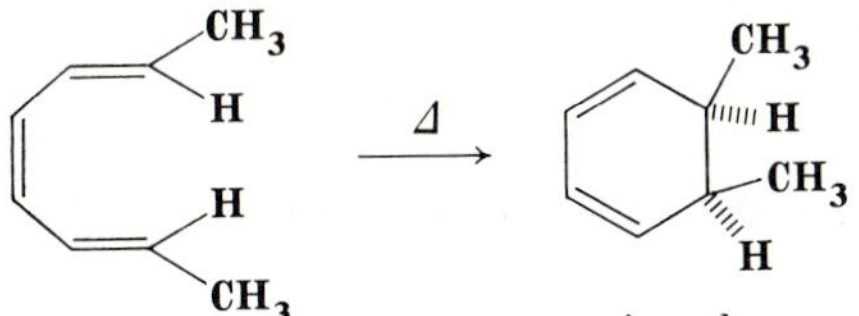

cis only

14.3 STEREOSPECIFICITY

The remainder of this section is devoted to explaining the stereospecificity of electrocyclic reactions. First, we must define two words, *conrotatory* and *disrotatory*. When the terminal carbons of a compound participating in an electrocyclic reaction rotate either both clockwise or both counterclockwise, the reaction is said to be *conrotatory*:

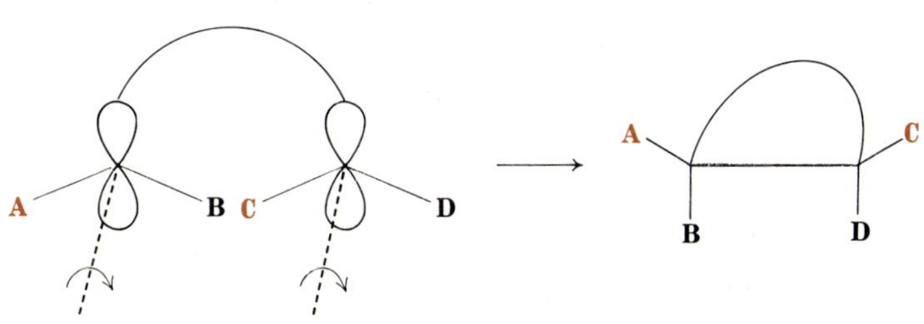

In the example shown above, A and C become *cis* to one another. In a *disrotatory* reaction, one carbon rotates clockwise whereas the other rotates counter-clockwise:

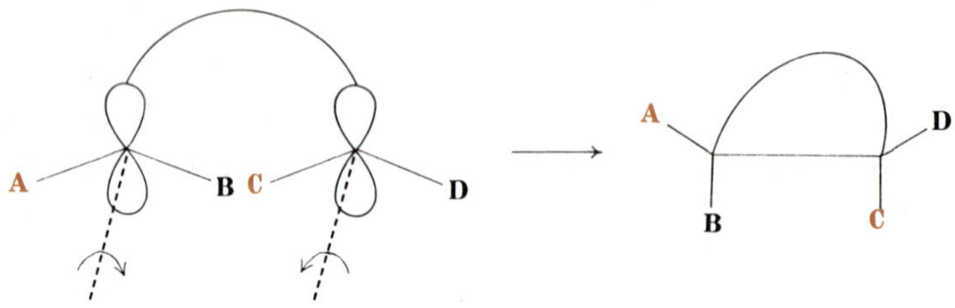

This mode of ring closure leads to a *trans* relationship between A and C.

A few years ago Woodward and Hoffmann proposed that *the stereochemical course of an electrocyclic reaction is determined by the properties of the highest occupied molecular orbital.* This important idea will be better understood after we present a few examples.

Butadiene

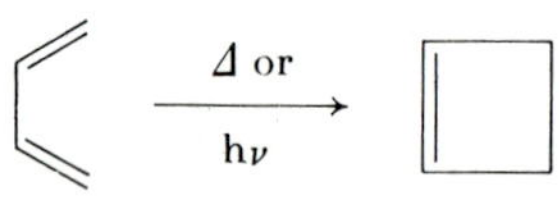

1) *Thermally induced.* The highest occupied butadiene molecular orbital is (see above discussion):

$$\begin{matrix} + & + & - & - \\ \mathbf{C} - & \mathbf{C} - & \mathbf{C} - & \mathbf{C} \\ - & - & + & + \end{matrix}$$

Consequently, formation of the single bond between the termini must be *conrotatory* in order that the + lobe of one terminal carbon overlaps the + lobe of the other:

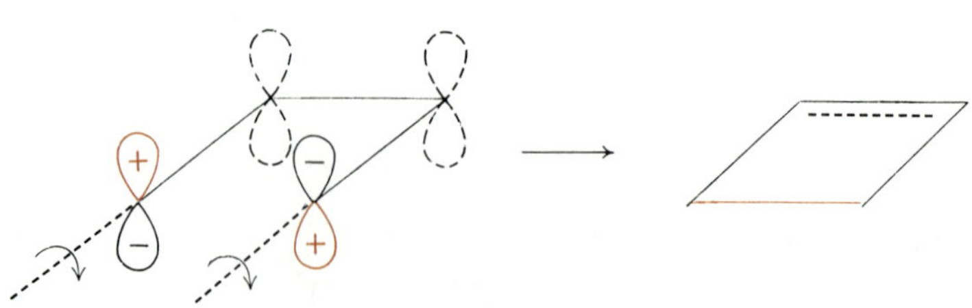

The new σ bond in cyclobutene is formed from the two electrons which were originally in the highest occupied MO. Parenthetically, we may add that the electrons of the new π bond in cyclobutene orginate from the two electrons in the *lowest* MO:

+ + + +
C—C—C—C
− − − −

Thus,

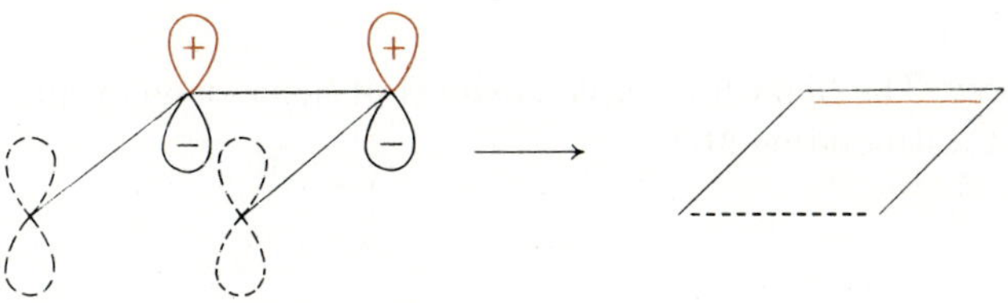

However, for our purposes we will focus only on the new σ bond because this is the site of the interesting stereochemistry. As the rule stated, this stereochemistry at the termini is determined by the highest occupied orbital.

2) *Photochemically induced.* By irradiating butadiene, an electron is elevated to the lower antibonding MO, and the highest occupied orbital becomes:

+ − − +
C—C—C—C
− + + −

Thus the new σ bond must be formed in a disrotatory ring closure:

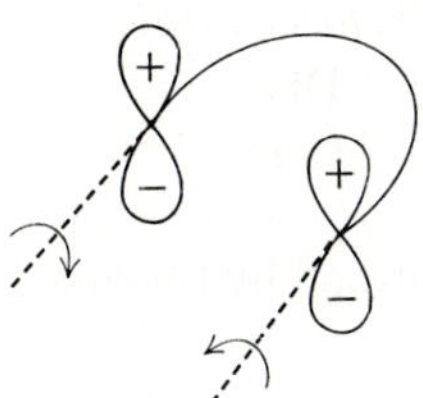

Hexatriene

1) *Thermally Induced.* In the unexcited molecule, the occupied MO of highest energy is:

+ + − − + +
C—C—C—C—C—C
− − + + − −

Thermally induced electrocyclic reactions of trienes therefore display disrotatory behavior:

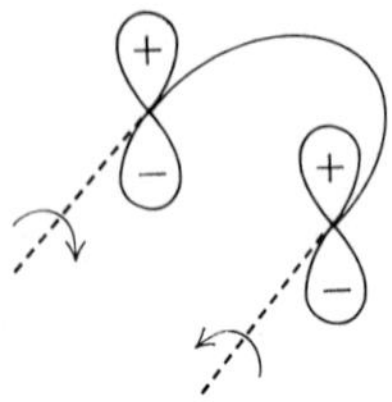

2) *Photochemically induced*. The highest occupied orbital of hexatriene in the excited state is its lowest antibonding MO:

```
+   −   −   +   +   −
C—C—C—C—C—C
−   +   +   −   −   +
```

Cyclization therefore occurs by a conrotatory mode:

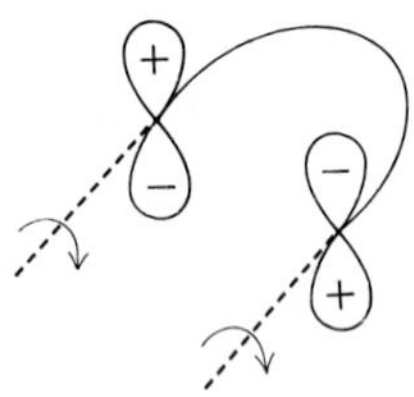

The following table summarizes the conclusions based on the Woodward-Hoffmann concepts:

	Thermal	*Photochemical*
Butadiene	Con	Dis
Hexatriene	Dis	Con

We can now explain the *cis* stereochemistry of the product formed by thermal cyclization of the triene:

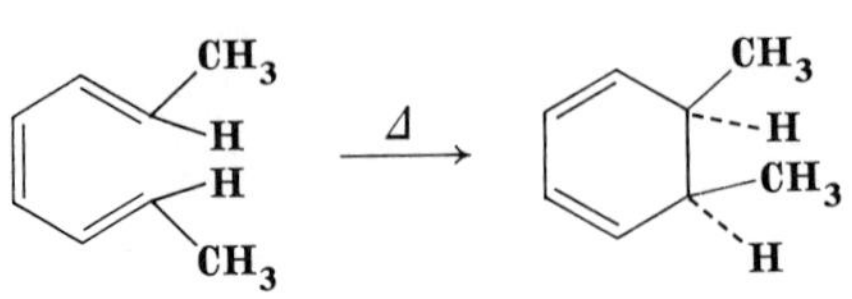

cis

The Woodward-Hoffmann rules predict a disrotatory mode leading to the *cis* product:

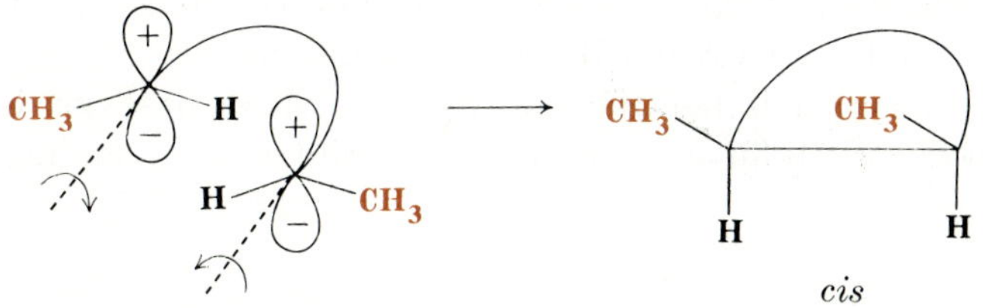

cis

The following examples provide additional evidence that the stereochemistry of the product is consistent with the ideas that we have presented:

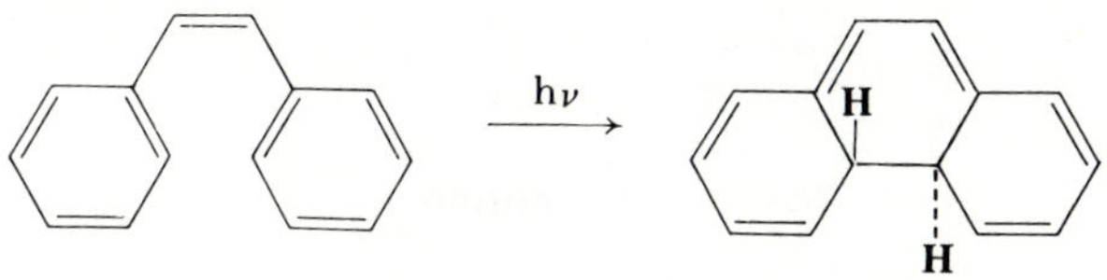

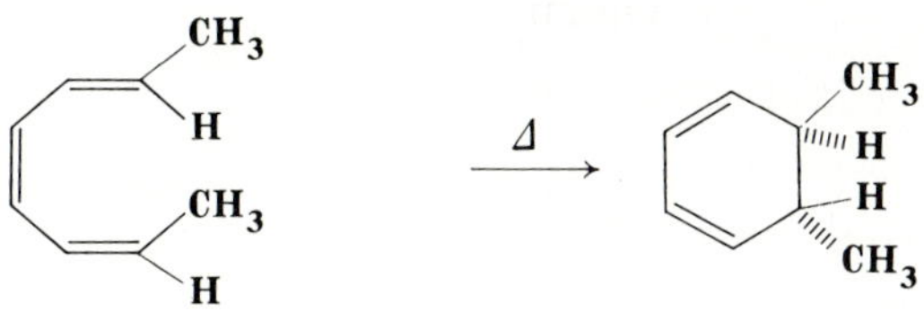

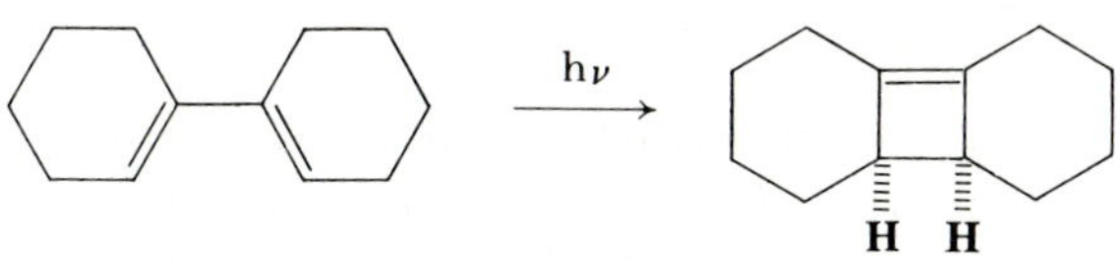

An interesting ambiguity arises in the following thermal cleavage of a disubstituted cyclobutene to a diene (which, like the reverse process, is conrotatory). In theory, this reaction can yield either of *two* products depending on the nature of the conrotatory opening:

1)

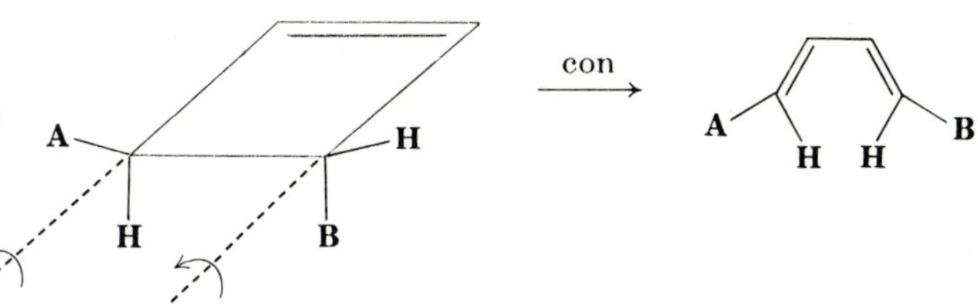

2)

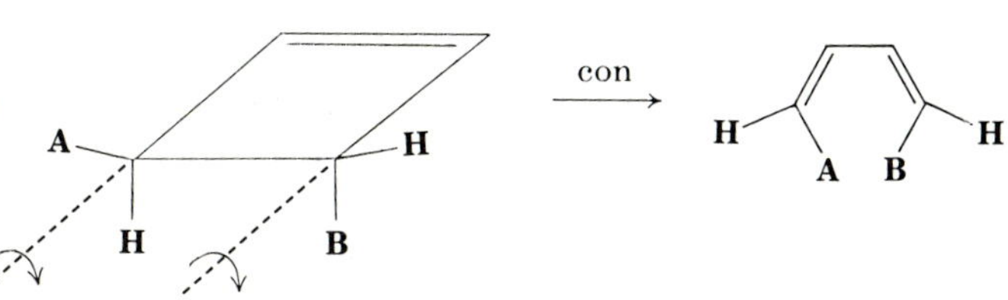

In actual fact, only the first of these possibilities is observed, presumably because steric interaction between A and B inhibits the alternative.

The electrocyclic reaction is not the only process whose behavior is described by the Woodward-Hoffmann concepts. *Cycloadditions* also fall into this category:

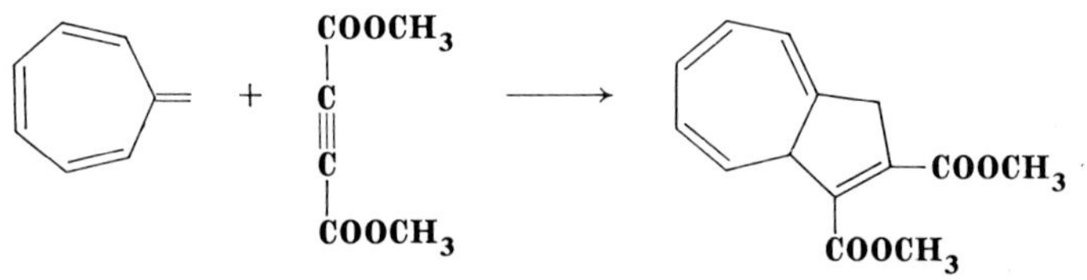

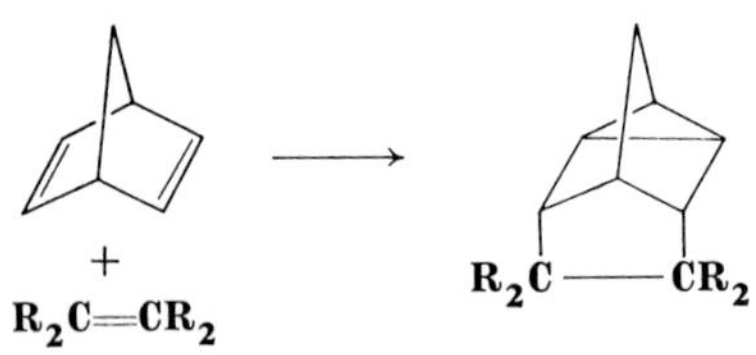

The *sigmatropic rearrangement* is yet another reaction which is dependent on the symmetry of the molecular orbitals. The reaction may be defined as a migration of a σ bond, flanked by one or more double bonds, to a new position within the molecule in an uncatalyzed intramolecular process. The following uncatalyzed rearrangement is a good example:

15
BIOLOGICAL MOLECULES

Much of our knowledge of organic chemistry has resulted from investigations of the composition of naturally occurring substances. These researches have included structure elucidation and efforts at total syntheses. We will discuss a few of these studies in this chapter, in order to illustrate the application of some of the basic principles of organic chemistry.

15.1 GLUCOSE: AN INTRODUCTION TO CARBOHYDRATE CHEMISTRY

We shall approach the chemistry of glucose as a research problem in structure elucidation, assuming that none of the modern spectroscopic techniques are available. Thus, let us suppose that we have a white crystalline substance produced from the hydrolysis of sucrose (cane sugar). Our task is to determine the complete structure of this material (glucose).

First we must establish the purity of our sample, its molecular formula, and its molecular weight. The molecular formula of glucose is $C_6H_{12}O_6$. The general formula, $C_nH_{2n}O_n$, suggests that this material is a *carbohydrate* (hydrate of carbon), a family of compounds which can be obtained from various plant sources. Moreover, the high water solubility of glucose, as well as the number of oxygens in its molecular formula, suggests that hydroxyl functions may be present.

We can determine the number of hydroxyl groups in glucose by allowing it to react with acetic anhydride in the presence of pyridine (acetylation conditions). From this reaction we would isolate a product of molecular formula $C_{16}H_{22}O_{11}$.

$$\underset{\text{glucose}}{C_6H_{12}O_6} \xrightarrow[\text{pyridine}]{(CH_3\overset{\overset{\displaystyle O}{\|}}{C})_2O} \underset{\text{glucosepentaacetate}}{C_{16}H_{22}O_{11}}$$

The change in molecular formula indicates a pentaacetate, since converting an hydroxyl group into an acetate group adds C_2H_2O to each hydroxyl group.

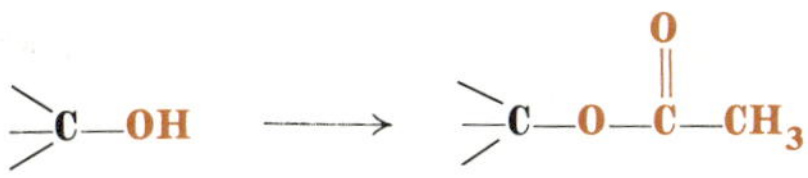

The difference between glucose and its acetate is $5(C_2H_2O)$, and it is therefore likely that glucose has five hydroxyl groups.

The sixth oxygen of glucose is probably part of a ketone or an aldehyde function since glucose is a *reducing* carbohydrate (or sugar): it will reduce an ammoniacal solution of silver nitrate to metallic silver (Tollens' test), or a basic solution of cupric ion to red cuprous oxide (Fehling's test). These reactions indicate the presence of the easily oxidized aldehyde or α-hydroxyketone groupings (see Section 13.3).

We can decide between the ketone and aldehyde alternatives and at the same time establish the carbon skeleton of glucose, by means of the following reaction sequence. Glucose is treated with hydrogen cyanide to form a cyanohydrin, which is then hydrolyzed into a polyhydroxyacid.

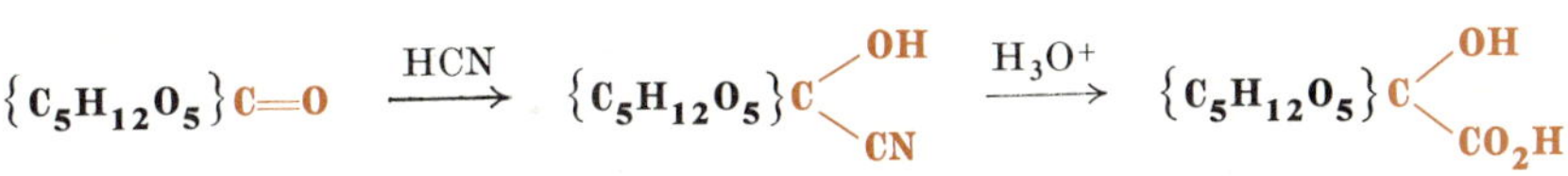

glucose glucose cyanohydrin

We are now left with a glucose derivative whose only functional groups are six hydroxyls and one carboxyl. We can remove all of the hydroxyls, leaving only the carboxyl group, by prolonged heating of glucose with a reducing agent, hydriodic acid:

$$\text{ROH} \xrightarrow{\text{HI}} \text{RI} \xrightarrow[\text{heat}]{\text{HI}} \text{R—H} + \text{I}_2$$

Such treatment would convert hexahydroxyheptanoic acid into a simple aliphatic seven-carbon carboxylic acid. In fact, when the above reaction is carried out with the glucose derivative, *n*-heptanoic acid is obtained. (This is established by comparing the physical properties of the product acid with authentic *n*-heptanoic acid prepared by an unambiguous total synthesis.)

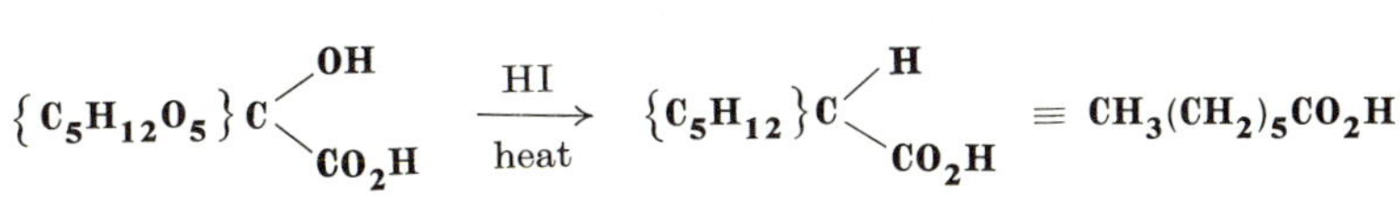

This finding means that the six carbons of glucose must be arranged linearly and that the carbonyl group in glucose must be an aldehyde, not a ketone. If the carbonyl group had been a ketone, a branched chain heptanoic acid would have been produced:

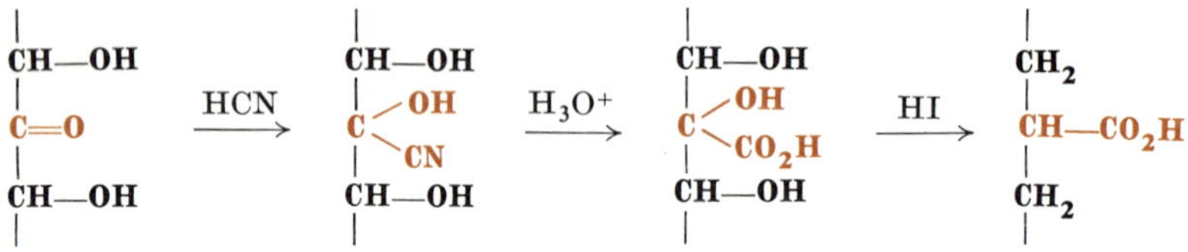

We can now write a gross structure for glucose if we assume that no glucose carbon can bear more than one hydroxyl group. This assumption is based on the knowledge that the gem-diol-to-carbonyl equilibrium lies far to the side of the carbonyl:

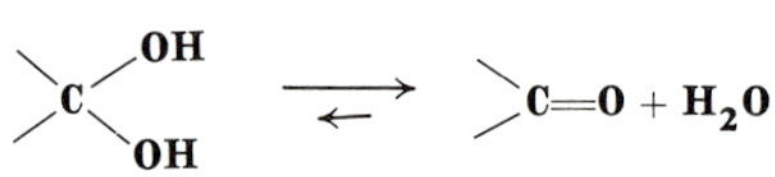

The structure for glucose, then, must be

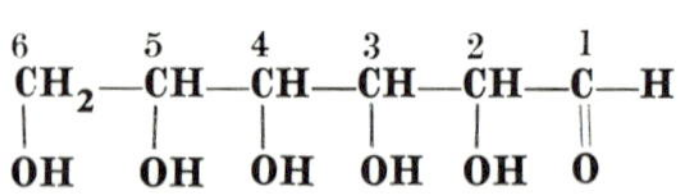

This structure has four asymmetric carbon atoms (C-2, C-3, C-4, and C-5), and therefore the above representation is incomplete; we have yet to show the relative configuration of the asymmetric centers and the absolute configuration of the molecule.

Since the gross structure we have determined contains four asymmetric carbons, there are 16 (2^4) possible optical isomers (or eight pairs of enantiomers). One of the members of each pair is drawn below:

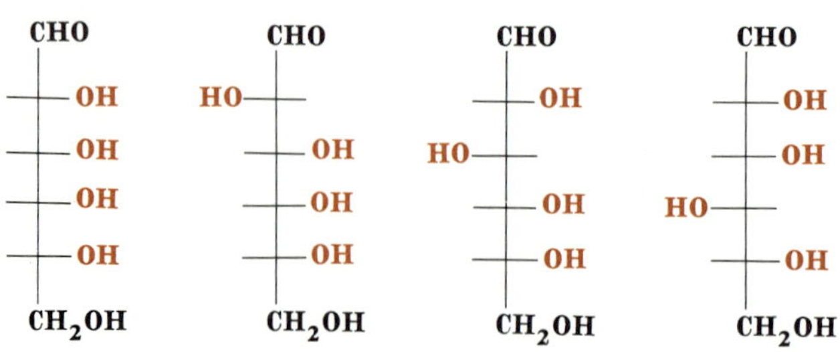

1 2 3 4

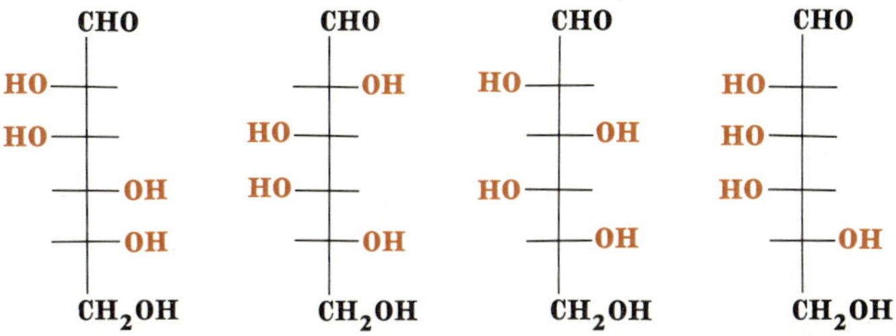

5 6 7 8

Each of these diastereoisomers has a corresponding optical antipode, yielding 16 total possible optical isomers. Glucose is one of these 16 compounds. It is dextrorotatory, which we indicate by calling it (+)-glucose.

We can eliminate some of the possible relative configurations for glucose by oxidizing the sugar with hot dilute nitric acid. This reaction produces a six-carbon diacid, glucosaccharic acid:

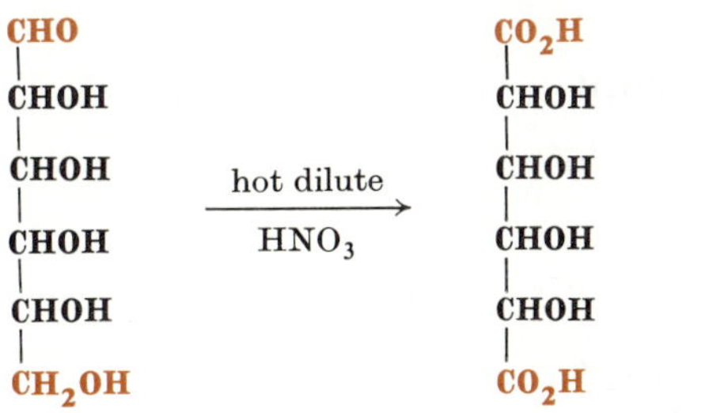

glucose glucosaccharic acid

This diacid is found to be optically active, which rules out configurations **1** and **6** and their enantiomers because they would have given rise to optically inactive (*meso*) diacids (having a plane of symmetry).

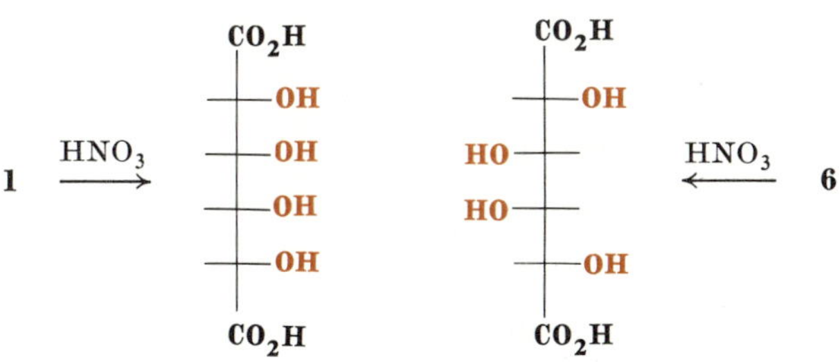

meso *meso*

Three other configurations can be eliminated by degrading the six-carbon aldehydo-sugar (an aldohexose) to a five-carbon sugar (an aldopentose), and then oxidizing to a saccharic acid. This can be effected via a reaction sequence called the *Ruff degradation*, followed by oxidation with hot dilute nitric acid:

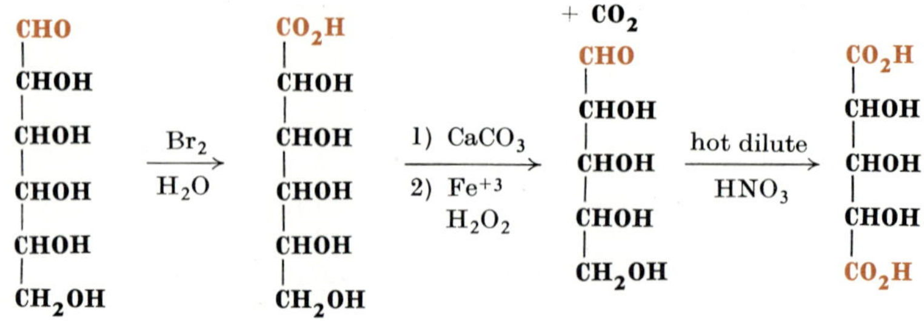

an aldohexose a glyconic acid an aldopentose a 5-carbon saccharic acid

When glucose is subjected to this Ruff degradation, the saccharic acid product is optically active. This eliminates configurations **2**, **4**, and **7** and their enantiomers, since Ruff degradation of these compounds would have produced *meso* five-carbon saccharic acids

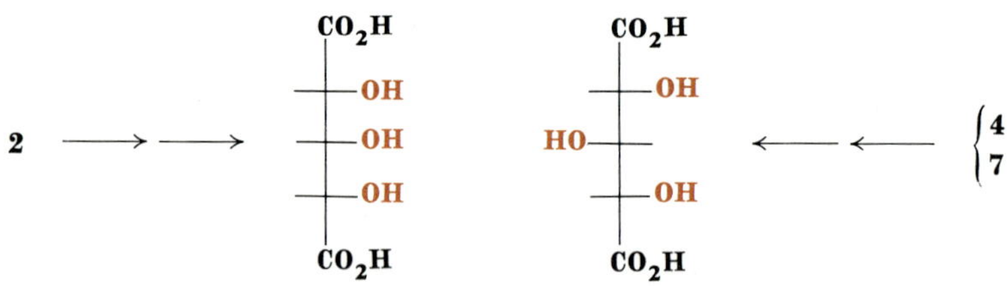

meso *meso*

Glucose must then have one of the relative configurations represented by formulas **3**, **5**, or **8**.

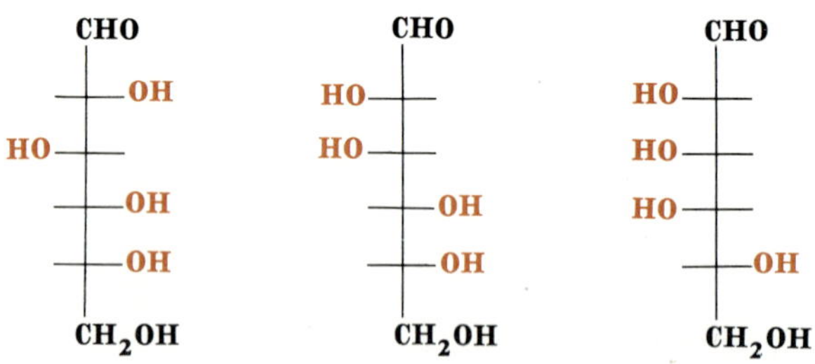

3 **5** **8**

We can distinguish among the three remaining alternatives (**3**, **5**, **8**) by a chain-lengthening procedure. This reaction sequence begins by adding hydrogen cyanide to the sugar's carbonyl group to yield a cyanohydrin, and then hydrolyzing to produce a hydroxy acid. Finally, the hydroxy acid is oxidized with hot dilute nitric acid, yielding a saccharic acid with an additional carbon atom:

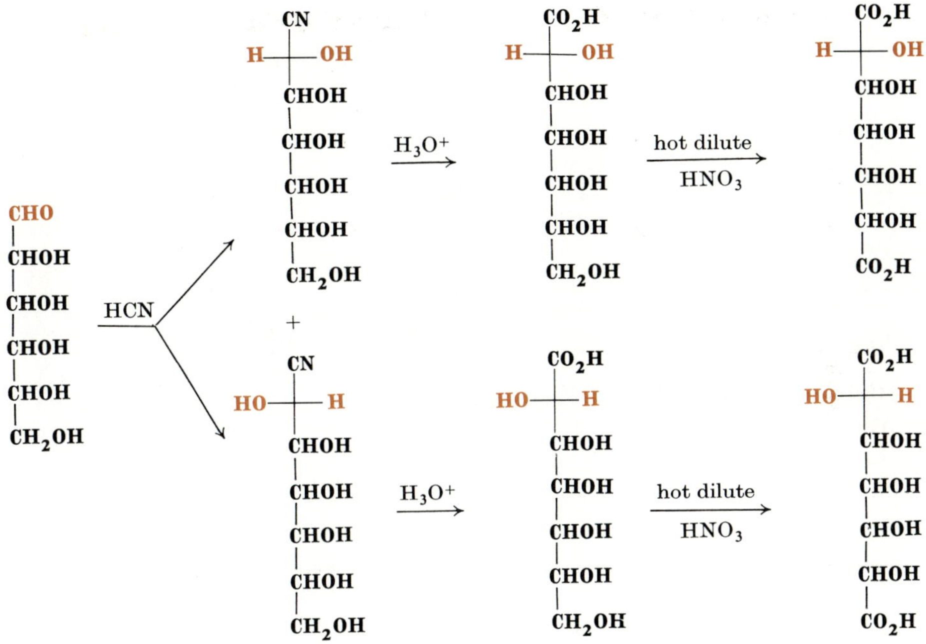

Note that the addition of hydrogen cyanide introduces a new asymmetric center into the molecule, producing two diastereomeric cyanohydrins. This means that *two* seven-carbon saccharic acids will result. When glucose is subjected to these transformations, one of the two saccharic acid products is optically active and the other is not. The relative configurations of formulas **3** and **8** can accommodate this finding, while **5** cannot (**5** would yield *two* optically active saccharic acids):

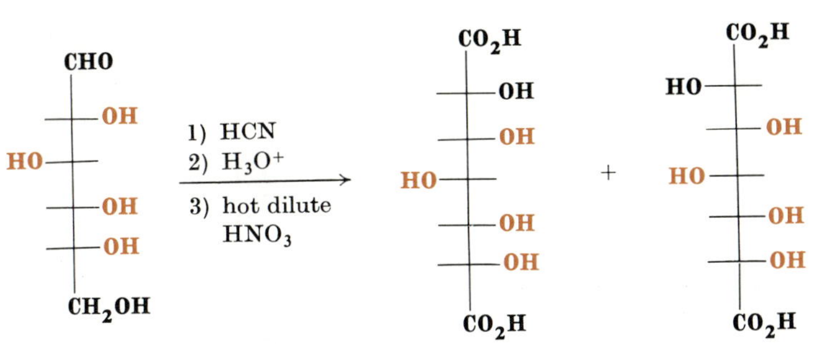

3 *meso* optically active

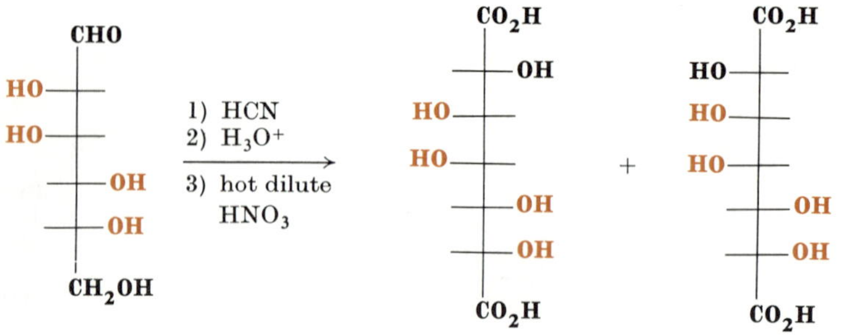

5 — optically active — optically active

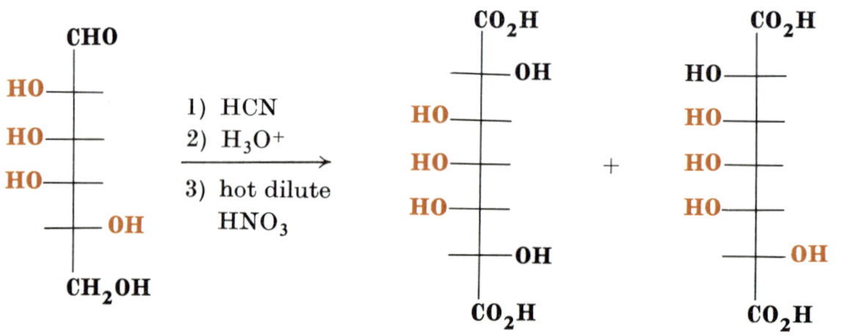

8 — *meso* — optically active

To finally distinguish between configurations **3** and **8** we subject them to two sequential Ruff degradations, followed by nitric acid oxidation of the saccharic acid product. Glucose, on such treatment, yields a *meso* diacid (*meso*-tartaric acid) and must therefore be **3** or its enantiomer:

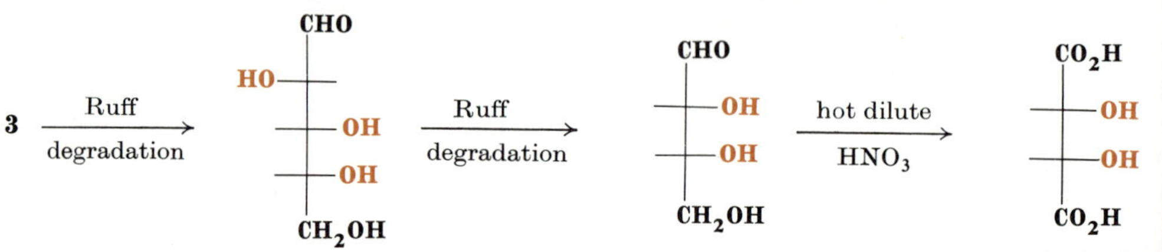

meso-tartaric a

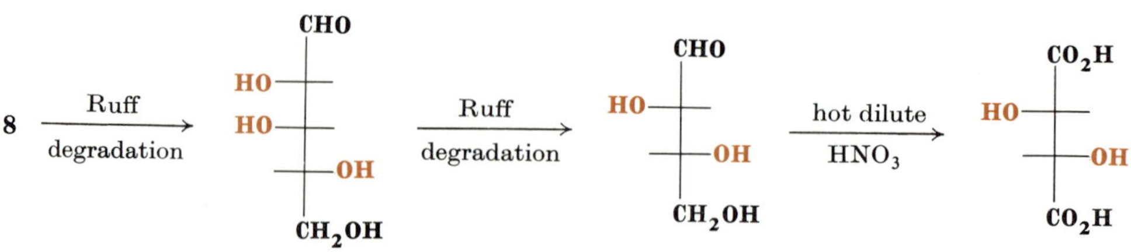

Now we have only to determine the absolute configuration of (+)-glucose, that is, whether it's structure is **3** or the enantiomer, **3′**.

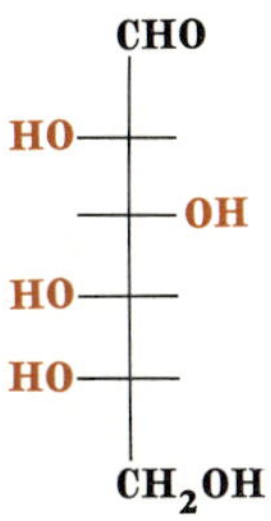

3′

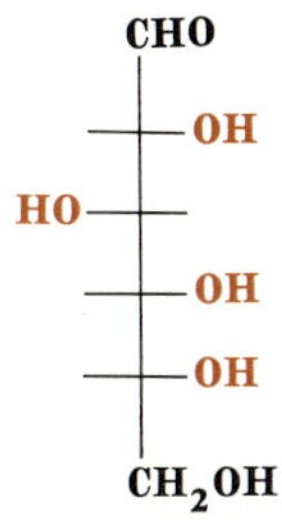

3

To do this we degrade (+)-glucose sequentially via the Ruff degradation to glyceraldehyde, an aldotriose with only one asymmetric center:

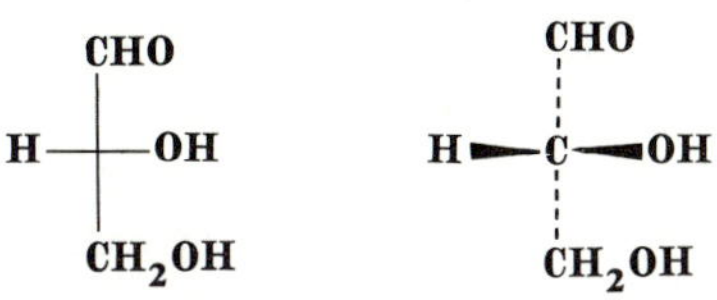

A

(+)-glyceraldehyde

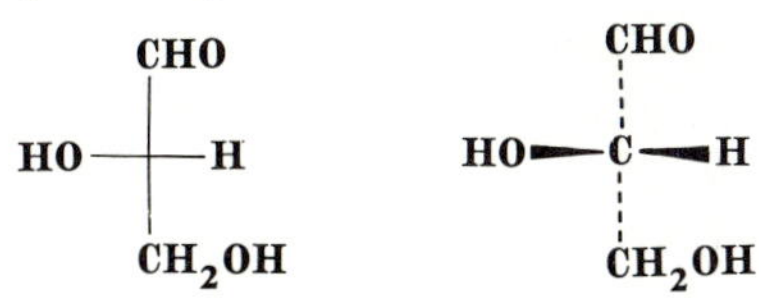

B

(−)-glyceraldehyde

The absolute configurations of (+)-glyceraldehyde and (−)-glyceraldehyde have been determined and are given above; that is, optically active glyceraldehyde that rotates plane polarized light clockwise (to the right) has the structure of formula **A** and optically active glyceraldehyde that rotates plane polarized light counterclockwise has the structure of formula **B**.

The glyceraldehyde obtained by the Ruff degradation of (+)-glucose is dextrorotatory, and thus the absolute configuration in glucose of the asymmetric carbon atom farthest away from the aldehyde function is that shown in **3** and not **3′**. The remaining carbons must be as shown in **3** since their configuration relative to this last asymmetric carbon has been given in the arguments presented above. This will always be true: given the relative configuration of a molecule, once the absolute configuration at any one center is determined, the absolute configuration of the molecule follows. Thus, the structure of (+)-glucose is:

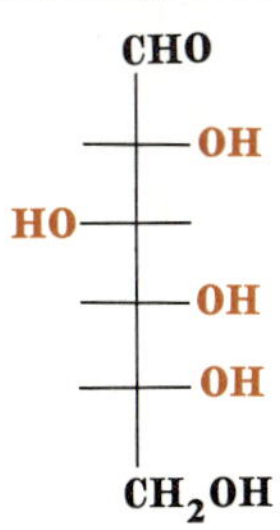

D-(+)-glucose

The name given above is D-(+)-glucose. The capital D denotes that the sugar is configurationally related to (+)-glyceraldehyde, and the (+) indicates the rotation of the sugar.

The structure developed above for D-(+)-glucose is not completely adequate; one further experimental result remains to be rationalized.

Two types of D-(+)-glucose can be isolated depending on how one recrystallizes crude glucose; α-D-(+)-glucose (mp, 146°) is obtained by ordinary recrystallization techniques and β-D-(+)-glucose (mp, 150°) is obtained by recrystallization at elevated temperatures. These are not merely different crystalline forms; their optical activities are different: α-D-(+)-glucose, +112°; β-D-(+)-glucose, +19°. Moreover, the optical activity of a freshly prepared solution of either the α-glucose or the β-glucose will change upon standing until an equilibrium value of +52.7° is reached. If the solution is evaporated, either α-glucose or β-glucose will be recovered, depending on the recrystallization technique employed. This optical behavior is called *mutarotation*.

The structure we have deduced for glucose can be modified to accommodate these facts. Recall that an aldehyde and an alcohol are in equilibrium with the corresponding hemiacetal (see Section 11.6):

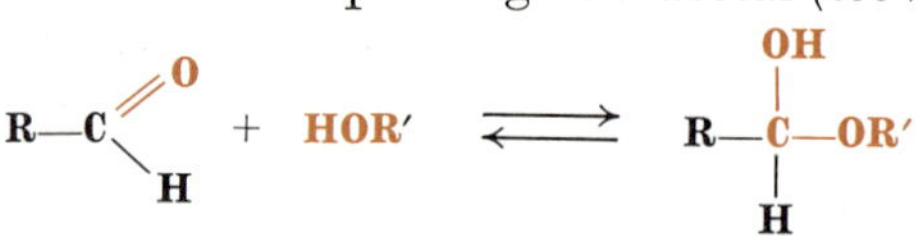

a hemiacetal

The structure we have developed so far for D-(+)-glucose contains both of these functionalities, and an analogous intramolecular equilibrium is therefore possible. Let us postulate that one of the hydroxyls adds across the aldehyde carbonyl to yield a cyclic compound. We shall choose the hydroxyl at the C-5 position so as to yield a six-membered ring, because six-membered rings are often more stable than other ring sizes. (We shall not go into the rigorous proof that this is so in the case of glucose.) The reaction generates a new asymmetric center, and two products result:

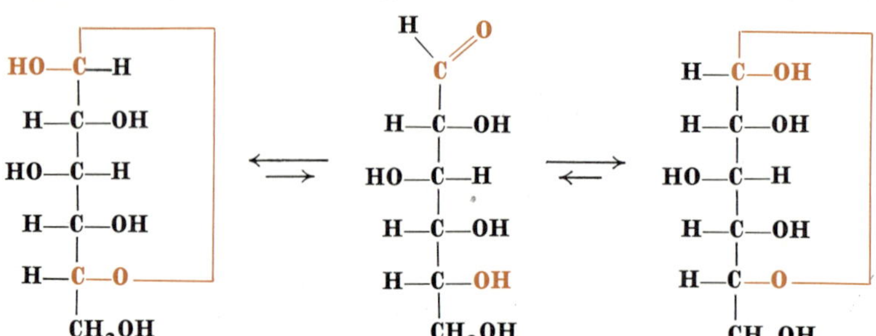

β-D-(+)-glucose D-(+)-glucose α-D-(+)-glucose

We will give the three-dimensional representation of the two hemiacetals later. The important point now is that the hemiacetals are the α-glucose and the β-glucose mentioned above. They are more stable than the open-chain hydroxy aldehyde, and in solution an equilibrium mixture of the two is formed via the open-chain material. Mutarotation is the result of this equilibrium; the final rotation represents an equilibrium mixture of the two diastereoisomers, with the β-form predominating.

These six-membered ring structures are called pyranoses after the pyran ring system to which they are related. We shall see later that similar structures with five-membered rings are also known, and they are called furanoses after the furan ring system.

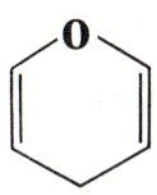

pyran

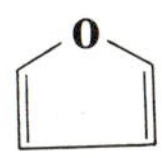

furan

It is instructive to consider the three-dimensional formulations of α-D-(+)-glucose and β-D-(+)-glucose. In order to transform the Fisher projection formulas above to their more informative three-dimensional counterparts, we shall first show them with the asymmetric carbon at 5 rotated to a different conformation (accomplished by two exchanges of substituents, see Sec. 5.6):

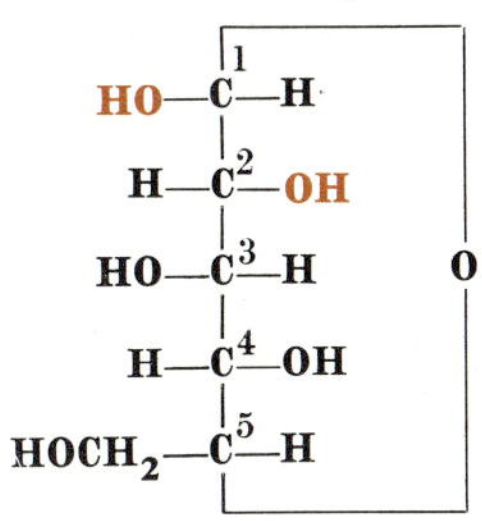

β

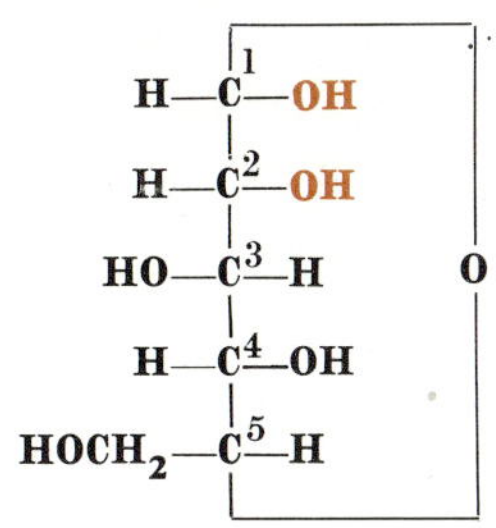

α

These designations make apparent the relative disposition of the substituents on the pyranose ring: in the β-hemiacetal, all the hydrogens are *trans*, whereas in the α-hemiacetal, the hydrogens on C-1 and C-2 (and therefore the hydroxyls as well) are *cis* oriented:

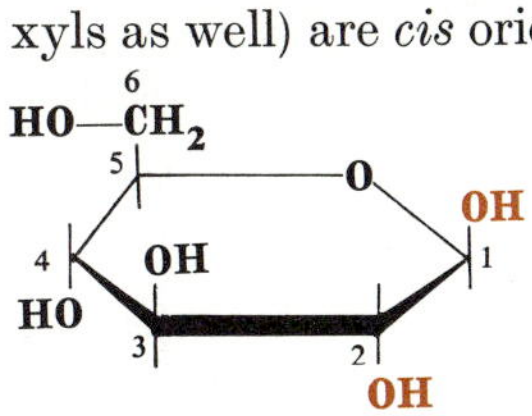

β-D-(+)-glucose

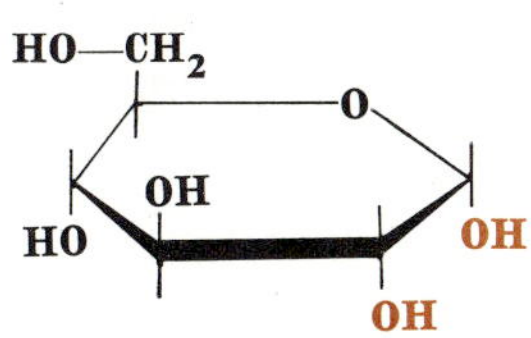

α-D-(+)-glucose

Now, we can show the conformational disposition of the substituents on the ring. Note that all the substituents in the β-isomer can be equatorial, so that we would expect this to be the more stable isomer.

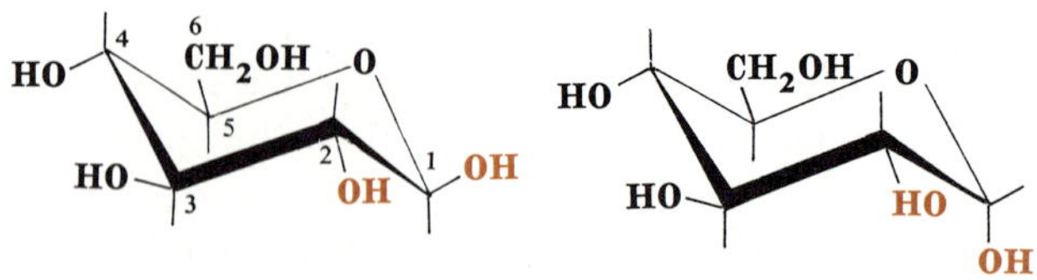

β-D-(+)-glucose α-D-(+)-glucose

In addition to glucose, a second carbohydrate, called fructose, can be isolated from the hydrolysis of sucrose. The structure of fructose can be established quite readily by correlating it with glucose via a reaction called *osazone formation*. An osazone is the product formed from the reaction of an α-hydroxyaldehyde or an α-hydroxyketone with phenylhydrazine:

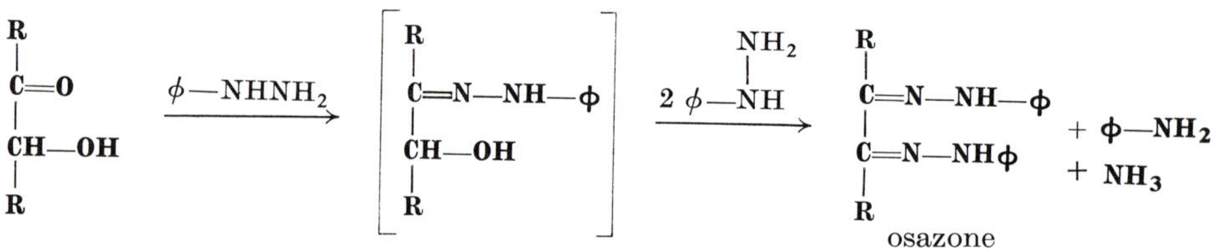

Note that in the second step the alcohol group is oxidized to a carbonyl group prior to reaction with the phenylhydrazine.

A ketose with a carbonyl on the C-2 position forms an osazone in which the oxidation has taken place on C-1.

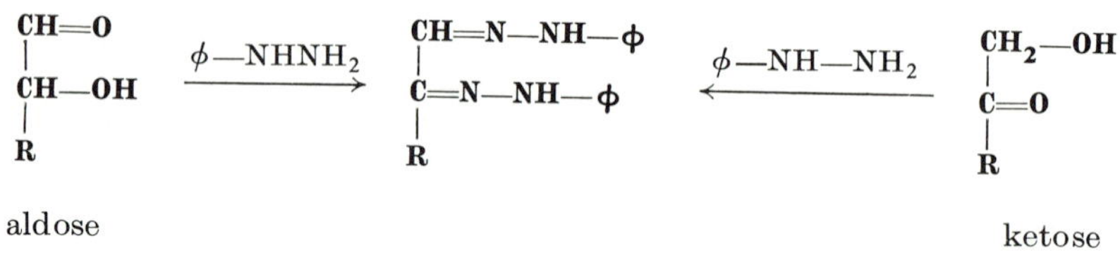

We can establish that fructose is a ketohexose that yields the same osazone as D-(+)-glucose. Because of this, we can conclude that the asymmetric centers at C-3, C-4, and C-5 of fructose and D-(+)-glucose have the same absolute and relative stereochemistry. Fructose, being a ketohexose, has only three asymmetric centers, and its structure is therefore defined:

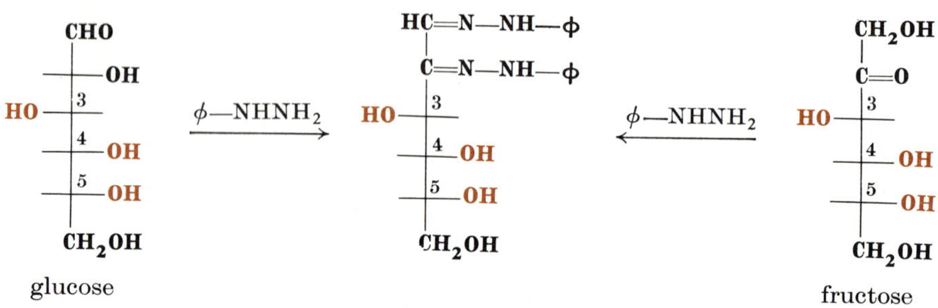

Fructose can exist as a hemiketal similar to α-glucose and β-glucose. In this instance a five-membered furanose ring is formed.

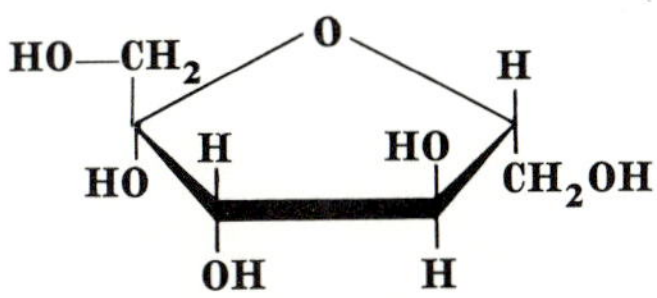

β-D-fructofuranose

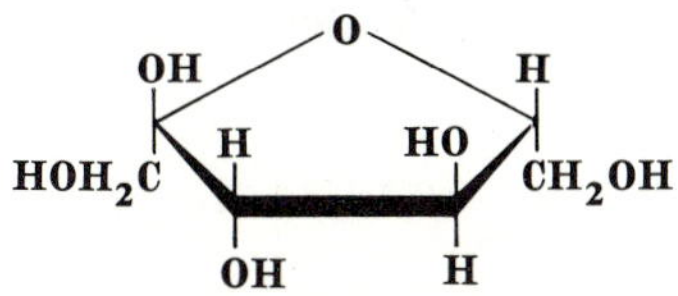

α-D-fructofuranose

15.2 DISACCHARIDES AND POLYSACCHARIDES

Carbohydrates such as glucose and fructose are called monosaccharides. Monosaccharides with other than six carbon atoms are known and can be typified by the aldopentose ribose, a monosaccharide with five carbon atoms:

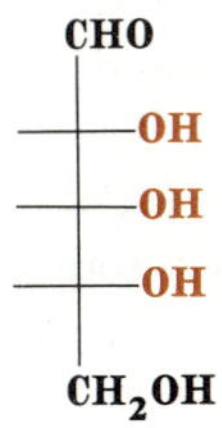

D-(−)-ribose

Carbohydrates consisting of two monosaccharide units joined via acetal or ketal linkages exist in nature. These sugars are called disaccharides. Furthermore, carbohydrates containing many monosaccharide units bound together with acetal or ketal linkages have been isolated. These are known as polysaccharides. Some examples are given below:

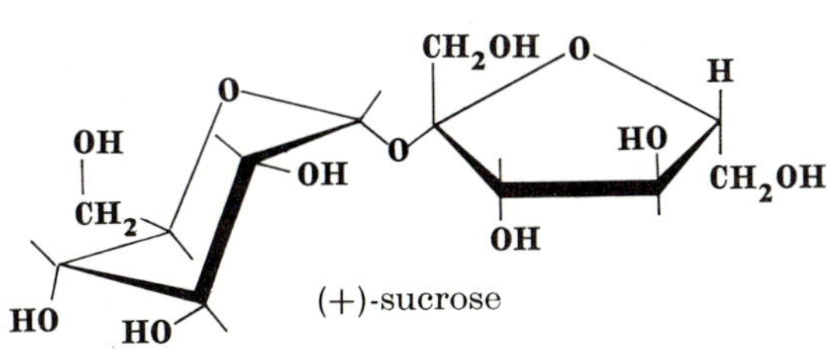

(+)-sucrose

α-D-glucopyranosyl-β-D- fructofuranoside

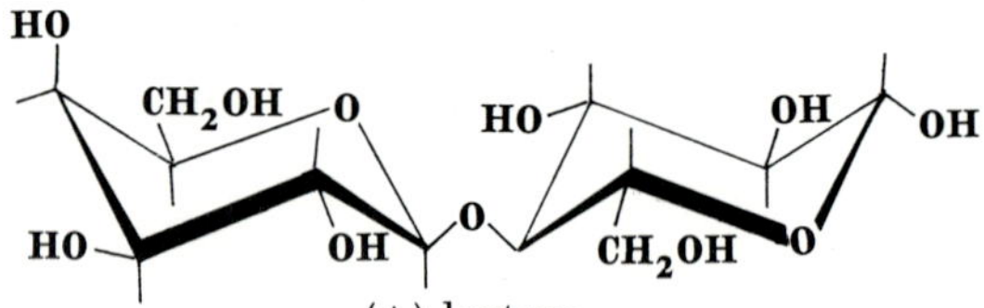

(+)-lactose

4-O-(β-D-galactopyroanosyl)-D-glucopyranose

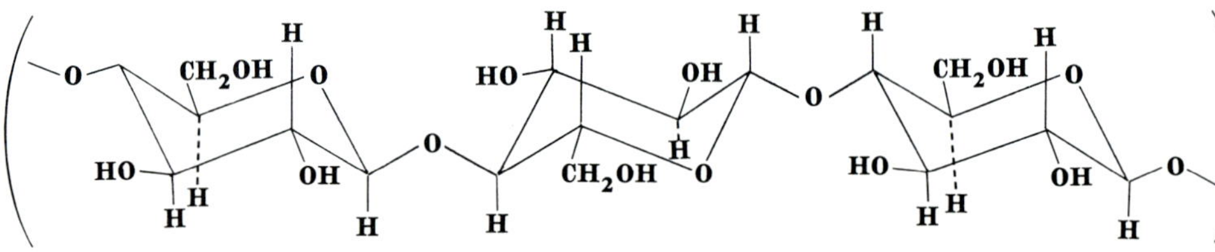

cellulose

Note that in sucrose the carbonyl groups of the glucose and fructose fragments are replaced by an acetal and a ketal, respectively. Therefore, this disaccharide will not reduce Tollens' reagent or Fehling's solution, and this sugar is said to be *nonreducing*. In contrast, the hemiacetal function in the left-hand ring of lactose is in equilibrium with the free aldehyde, and will react with Tollens' reagent and Fehling's solution; lactose is therefore called a *reducing* sugar.

Acid hydrolysis will cleave acetal or ketal bonds and can therefore be used to release the monosaccharide units that compose a disaccharide or a polysaccharide. The elucidation of the stereochemistry and structure of the disaccharides and polysaccharides involves some exceedingly interesting chemistry which is well worth the reader's time; however, it is beyond the scope of this text.

15.3 AMINO ACIDS, PEPTIDES, AND PROTEINS

One of the most important organic chemical constituents of living organisms is the class of compounds called proteins. These substances are found as structural material in such widely diverse forms as skin, hair, muscle, and horn. In addition, all the enzymes which catalyze biological reactions are proteins.

When proteins, or peptides, which are structurally related to proteins, are subjected to hydrolysis with either acid or base, they cleave to produce a variety of amino acids. These amino acids have the general structure shown below,

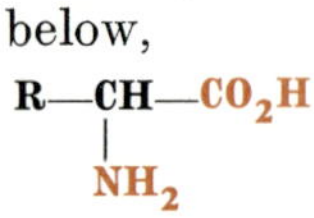

Table 15.1 Representation amino acids

Name	Abbreviation	Formula	$[M]_D^{25°}$ of s-isomer*	Isoelectric point
alanine	Ala	$CH_3CH(NH_2)CO_2H$	+1.6°	6.0
arginine	Arg	$H_2N(HN{=})C{-}NH(CH_2)_3CH(NH_2)CO_2H$	+21.8°	11.2
aspartic acid	Asp	$HO_2CCH_2CH(NH_2)CO_2H$	+6.7°	2.8
cysteine	CySH	$HSCH_2CH(NH_2)CO_2H$	−20.0°	5.1
glycine	Gly	$CH_2(NH_2)CO_2H$		6.0
histidine	His	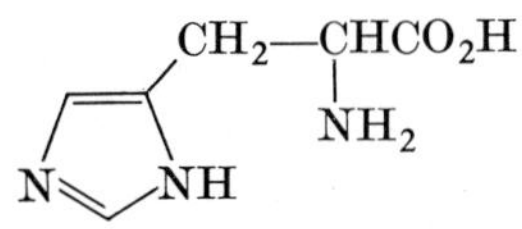	−59.8°	7.5
leucine	Leu	$(CH_3)_2CHCH_2CH(NH_2)CO_2H$	+6.6°	6.0
lysine	Lys	$H_2N(CH_2)_4CH(NH_2)CO_2H$	+19.7°	9.6
phenylalanine	Phe	$C_6H_5CH_2CH(NH_2)CO_2H$	−57.0°	5.5
serine	Ser	$HOCH_2CH(NH_2)CO_2H$	−7.9°	5.7
tryptophan	Try	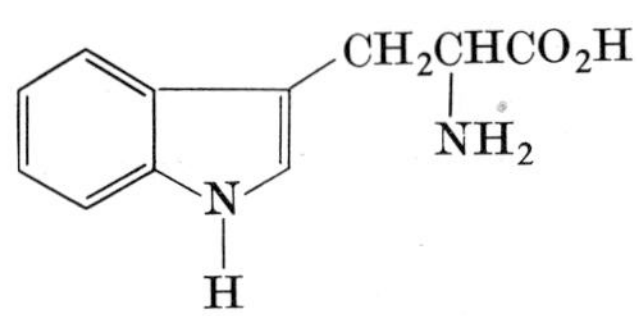	−68.8°	5.9
tyrosine	Tyr	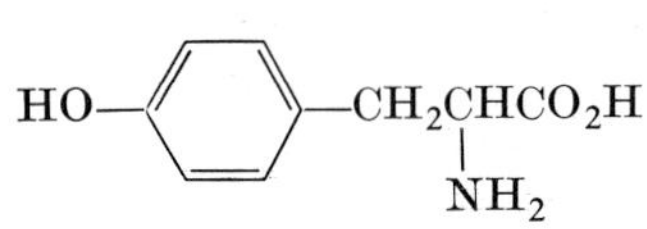	−18.0°	5.7

* $[M]_D^{25°} = \frac{[\alpha]_D^{25°} \times M}{100}$; M = molecular weight, D = 5893 Å.

and they are referred to as α-amino acids. Approximately 20 of these compounds are found as constituents of organisms. Table 15.1 lists the names, structures, and some of the physical properties of representative amino acids.

15.4 THE STRUCTURE AND PROPERTIES OF α-AMINO ACIDS

The general structure we wrote for α-amino acids indicates that when the α-substituent R— is not hydrogen, these compounds will be capable of optical isomerism. In general, the naturally occurring amino acids are all of the same chirality, and belong to the S-series configuration. Using an older form of nomenclature they are referred to as L-amino acids. The L- (for *levo* or left) indicates that the α carbon asymmetric center has the same configuration as L- (S-) glyceraldehyde:

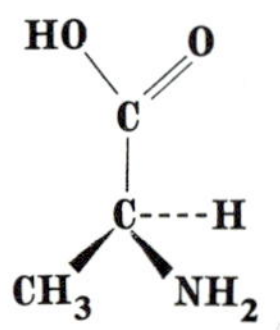

L- (or S-) alanine

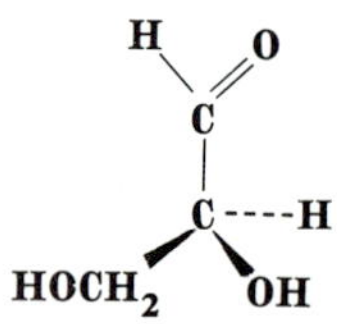

L- (or S-) glyceraldehyde

The general structure for amino acids also suggests that these compounds should be amphoteric, i.e., that they should behave as both acids and bases. For example, using the simplest amino acid, glycine, we would expect the following equilibria in a water solution of this compound:

$$H_3\overset{+}{N}—CH_2CO_2H \rightleftharpoons \left[H_3\overset{+}{N}—CH_2CO_2^- \rightleftharpoons H_2N—CH_2CO_2H\right] + H^+$$

$$\rightleftharpoons H_2NCH_2CO_2^- + 2H^+$$

The position of this equilibrium will, of course, depend on the pH of the solution. The equilibrium between the neutral amino acid and its *zwitterion* (dipolar form), however, does not require external acid or base. This reaction involves only the acid and base groups of the amino acid itself. The physical properties of α-amino acids show that these molecules actually exist as zwitterions since they are generally water soluble and have very high melting or decomposition points (properties of ionic species).

15.5 ANALYSIS OF α-AMINO ACIDS

A number of chemical and physical properties of α-amino acids are characteristic of these compounds. α-Amino acids can be chemically distinguished from most other amines by the *ninhydrin* test. Heating a solution of an α-amino acid and ninhydrin produces an intense blue-violet color. The sequence of reactions which leads to the colored species is shown below:

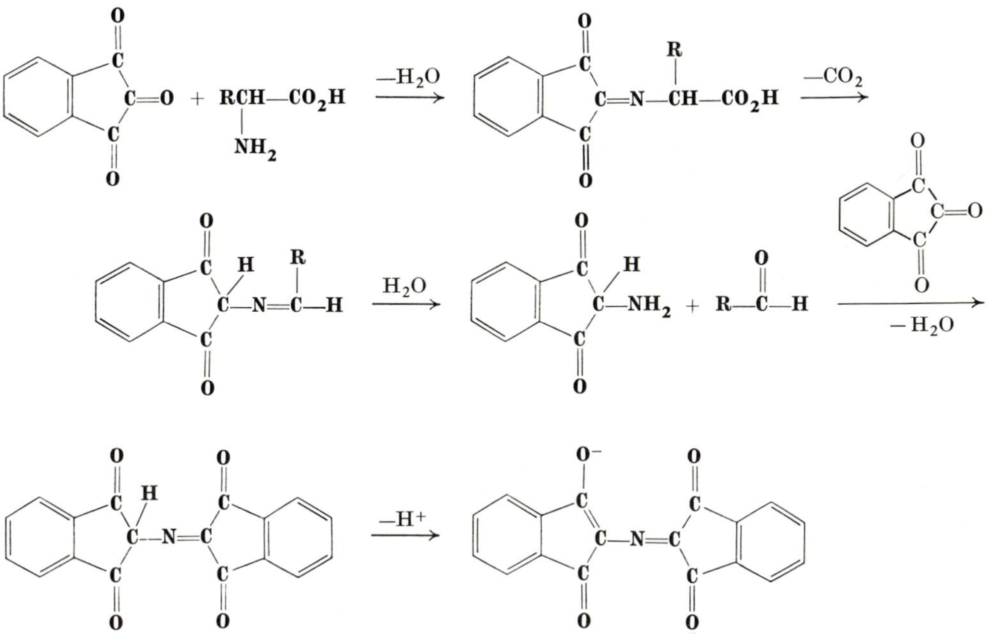

blue species

Amino acids can be individually characterized by their *isoelectric points.* The equilibrium which we wrote previously for glycine in water solution shows the predomination of the conjugate acid of glycine at low pH,

$H_3\overset{+}{N}—CH_2CO_2H$

and the predomination of the conjugate base at high pH,

$H_2N—CH_2CO_2^-$

These ions will migrate to opposite poles in an electric field. At some intermediate pH, however, the electrically neutral zwitterion will dominate the equilibrium, and there will be no migration. This pH is called the isoelectric point, and it has a characteristic value for each amino acid (see Table 15.1).

15.6 SYNTHESIS OF α-AMINO ACIDS

There are many ways to synthesize α-amino acids. We shall examine representative examples of three different approaches to the construction of these compounds. In the first method, the synthesis of an α-halo acid is followed by the displacement of the halogen atom by ammonia. This is illustrated by the preparation of alanine from propionic acid. The initial step in this synthesis, the bromination of propionic acid, is called the *Hell-Volhard-Zelinsky reaction* and is related to the α-bromination of ketones (Section 11.4). The second step is simply a nucleophilic displacement.

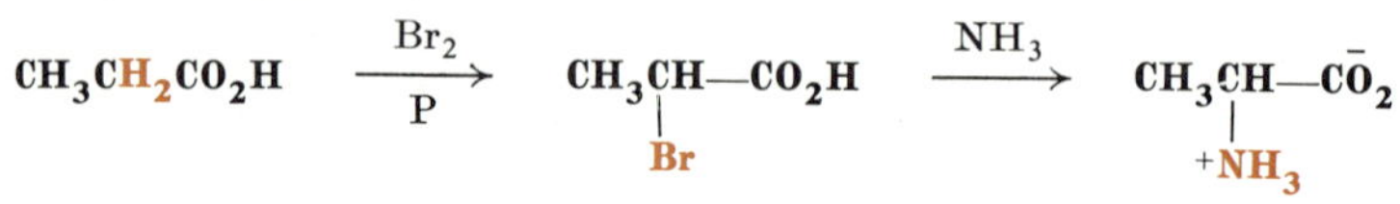

alanine

In the second method, the α-amino carboxyl unit is prepared first, and the rest of the side chain is then added via a displacement or condensation reaction. This is illustrated by the preparation of serine from malonic ester:

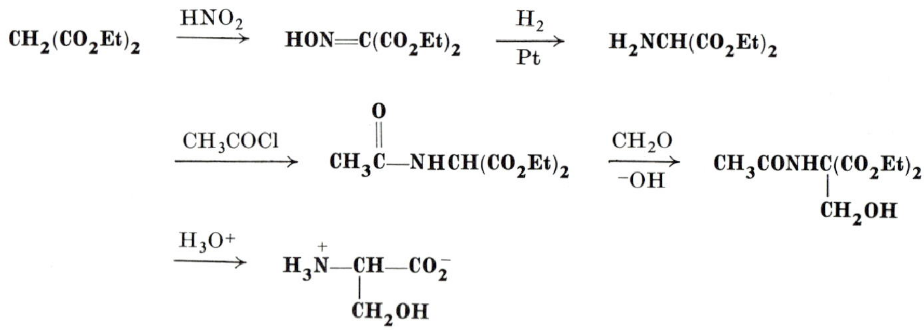

serine

In the third method, both the carboxyl carbon atom and the amino function are added in one step. This procedure, called the *Strecker synthesis*, is illustrated by the synthesis of valine. The key reaction in this sequence, the addition of ammonia and cyanide ion to an aldehyde, is related to the mechanism of cyanohydrin formation (Section 11.7).

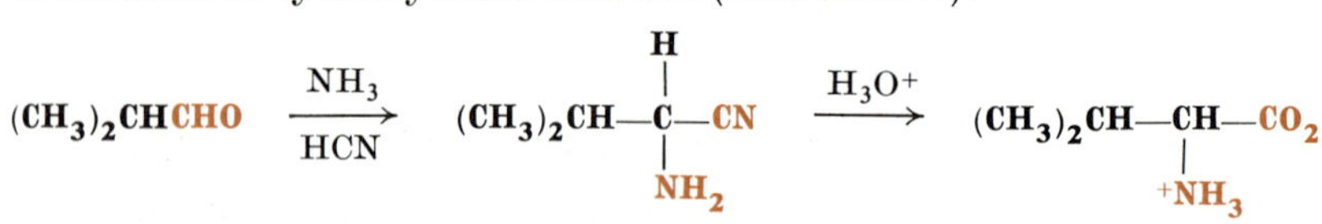

valine

15.7 PEPTIDES AND PROTEINS

Peptides and proteins are composed of several amino acids joined by amide bonds; that is, the carboxy group of one amino acid is linked to the amino group of a second amino acid by an amide, or peptide, bond.

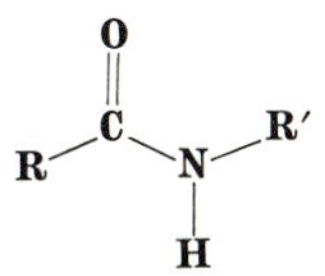

amide or peptide bond

The term, peptide bond, is usually used when discussing the chemistry of peptides and proteins.

The distinction between proteins and peptides is arbitrary. Compounds which have molecular weights greater than 10,000 are generally referred to as proteins. The general structure of a typical peptide is

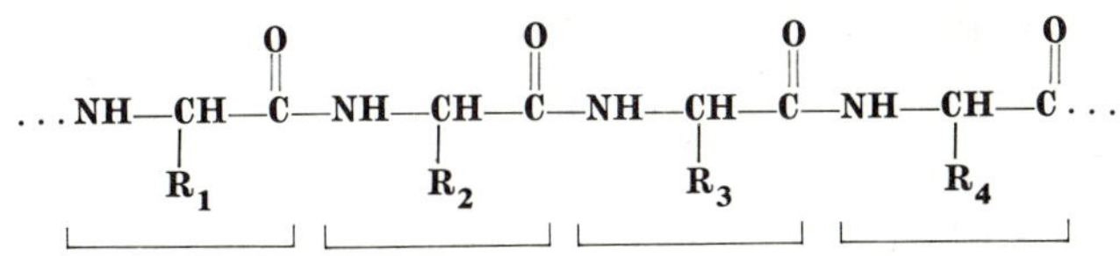

Although proteins also have the characteristic peptide bond structure shown above, the long chains of natural protein molecules tend to have coiled *secondary* structures. These coiled or helical structures are caused by hydrogen bonding and by interactions between nonpolar side chains of the amino acid units. In addition, the protein chain coils are folded into *tertiary* structures by additional hydrogen bonding and by covalent bonding of sulfur atoms of cysteine units. When natural proteins are subjected to physical or chemical "stress," the secondary and tertiary structures are often disrupted. A typical example of these structural disruptions is the change in egg albumin (egg white) when it is beaten or heated. Such a *denaturation* process is often irreversible, and as a result the protein molecules may permanently lose their biological properties (such as their ability to catalyze biological reactions).

15.8 PEPTIDE SYNTHESIS

The synthesis of a peptide from a number of amino acids requires only that the amino function of one amino acid be joined to the carboxyl function of another by an amide bond. However, if we examine the structure of even a simple peptide made from only two amino acids, we will realize how difficult this process is in actual laboratory practice.

Peptides are classified according to the number of amino acids which are linked together. A dipeptide is made from two amino acid molecules (the individual amino acid molecules may be the same, such as two glycines, or different); a tripeptide is made from three amino acid molecules, and so on.

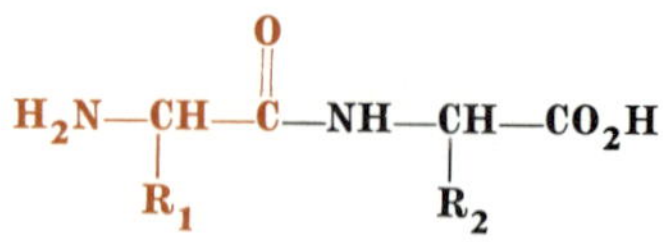

a dipeptide

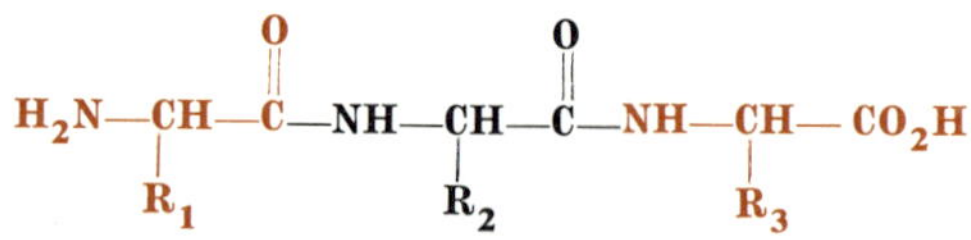

a tripeptide

Let us consider a specific dipeptide made from glycine and phenylalanine. There are two possible isomeric peptides which these compounds might form. In one isomer the amino function of glycine is linked to the carboxyl group of the phenylalanine,

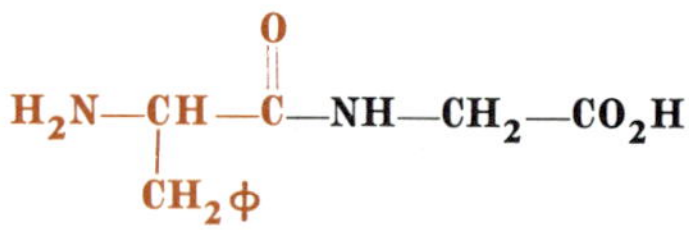

phenylalanylglycine

and in the other isomer the amino group of the phenylalanine is joined to the carboxyl function of the glycine,

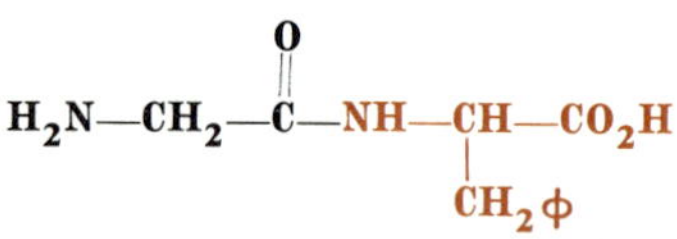

glycylphenylalanine

It is customary to abbreviate the structures of these two peptides as Phe-Gly and Gly-Phe, respectively (the amino acid abbreviated on the left is always the one that has a free amino group, and the one abbreviated on the right has the free carboxyl function).

Now it should be clear that if we attempt our synthesis using a standard method of amide formation, we will encounter a problem. For example, we might try to react an amino group and an acid chloride:

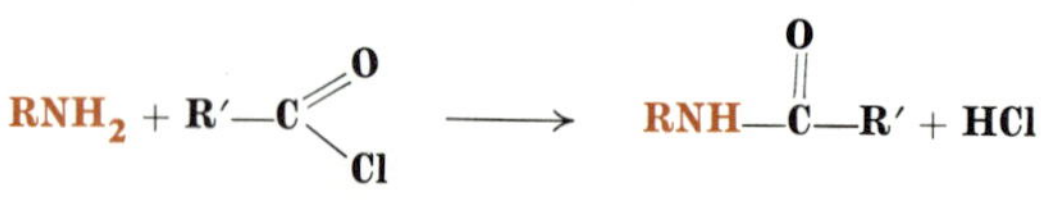

If we try to prepare the acid chloride of glycine, we will be working with a molecule that already contains an amino function. Thus, as soon as one glycine molecule has been converted into its acid chloride it might well react with another molecule of glycine:

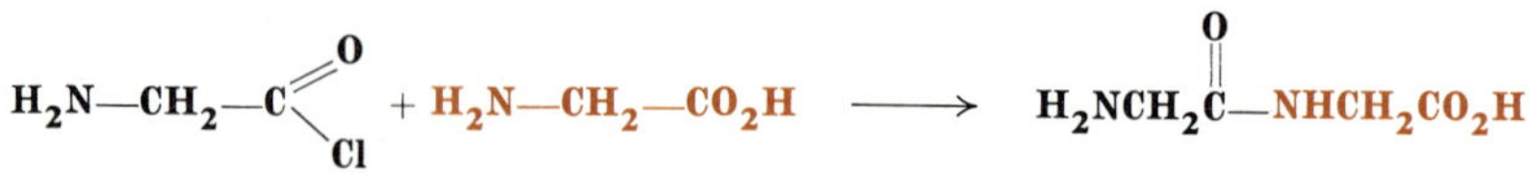

This process might continue until all of the glycine is converted into a polyglycine peptide. Thus, in peptide synthesis the amino function of one of the compounds must be *protected* so that it cannot react with the carboxyl group of another compound of the same structure. One of the most common protecting groups for amino functions is the *carbobenzoxy group* (abbreviated CBO), which is formed by reacting an amino acid with carbobenzoxy chloride:

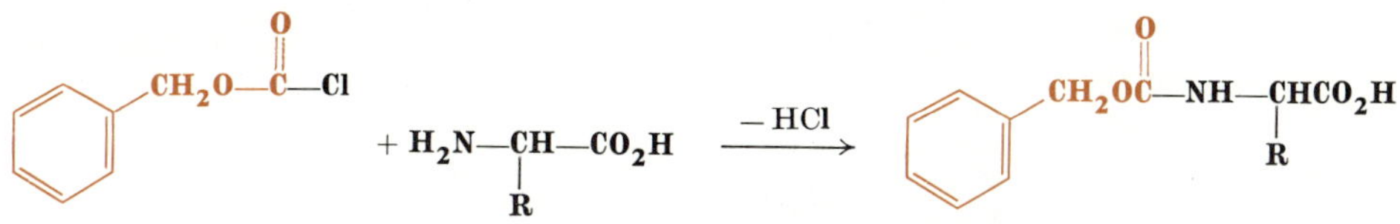

carbobenzoxy chloride amino acid protected amino acid

The carbobenzoxy group is readily removed at the end of a peptide synthesis by cleaving the benzyl carbon-oxygen bond with hydrogen and palladium (hydrogenolysis) or with sodium metal in liquid ammonia:

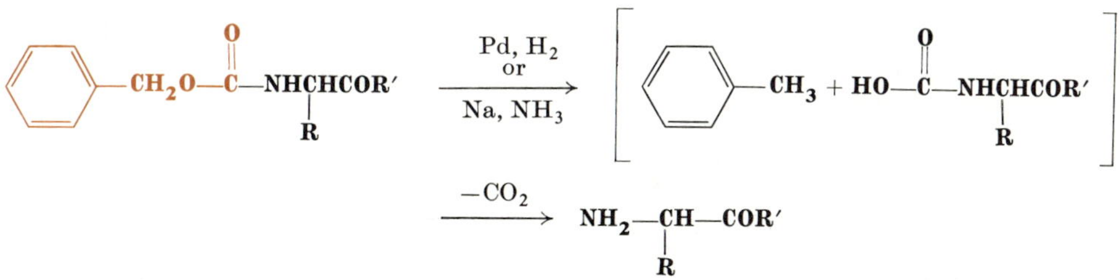

The next step in the synthesis is the "activation" of the carboxyl group of our protected amino acid. An activated carboxyl group is one that has been converted into a derivative which will react readily with an amino group to form an amide. In normal amide formation, the acid chloride function would be an excellent activating group. However, when one attempts to prepare the acid chloride of an optically active amino acid, extensive racemization occurs. A number of other activating groups have, therefore, been devised, of which the cyanomethyl ester is typical:

Let us assume now that the particular dipeptide we wish to make is phenylalanylglycine (Phe-Gly). First we protect the amino group of phenylalanine and activate its carboxyl moiety (Fig. 15.1). Next we must react the amino function of glycine with this derivative. The full sequence of reactions for forming the peptide link and removing the protecting group is shown in Fig. 14.1.

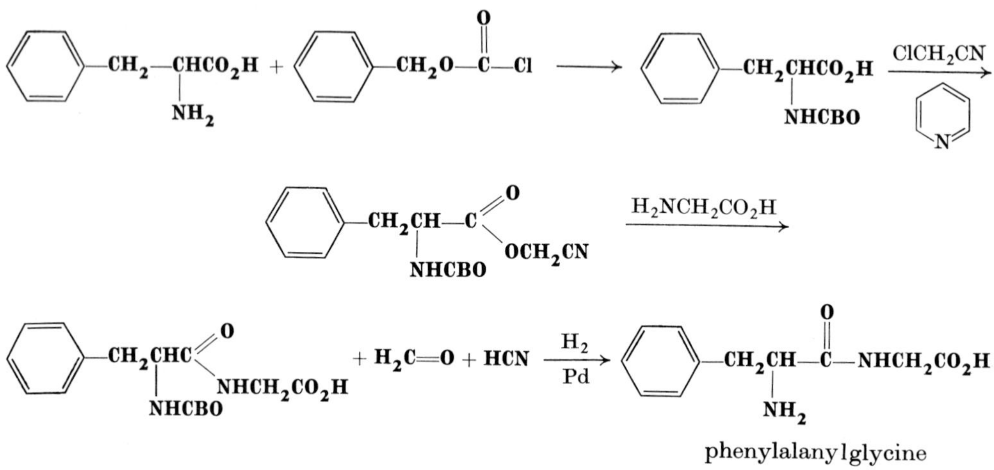

Fig. 15.1 Peptide synthesis using the carbobenzoxy group

To synthesize a longer peptide chain, the same general sequence of reactions is repeated as often as necessary. Thus, the protected dipeptide, shown in Fig. 14.1 before removal of the carbobenzoxy group, would be converted to an activated carboxyl compound and reacted with another amino acid.

Solid phase synthesis. One of the most difficult aspects of long-chain peptide synthesis is the necessity of isolating and purifying each intermediate peptide along the synthetic route. A new method has been developed which avoids these difficulties. Figure 15.2 outlines the steps of this *solid phase synthesis.* First a protected amino acid is reacted with a polymeric resin containing chloromethyl groups (denoted in Fig. 15.2 by Ⓡ—CH_2Cl). The carboxyl groups of the amino acid molecules react with these chloromethyl functions to form ester linkages. Thus the amino acid is now part of the polymeric resin and is insoluble in most common solvents. Consequently, impurities can be easily washed out. Note that the protecting function for the amino group in this example is the *t*-butoxycarbonyl (*t*-BOC) function. It is used here in preference to the carbobenzoxy group because it is more easily hydrolyzed.

Next (step 2), the *t*-BOC group is removed by acid hydrolysis, and the resulting "resin-amino acid" is then coupled (step 3) to a second amino acid (as its *t*-BOC derivative) using the reagent dicyclohexylcarbodiimide (DCC). The mechanism for peptide bond formation using DCC involves once again

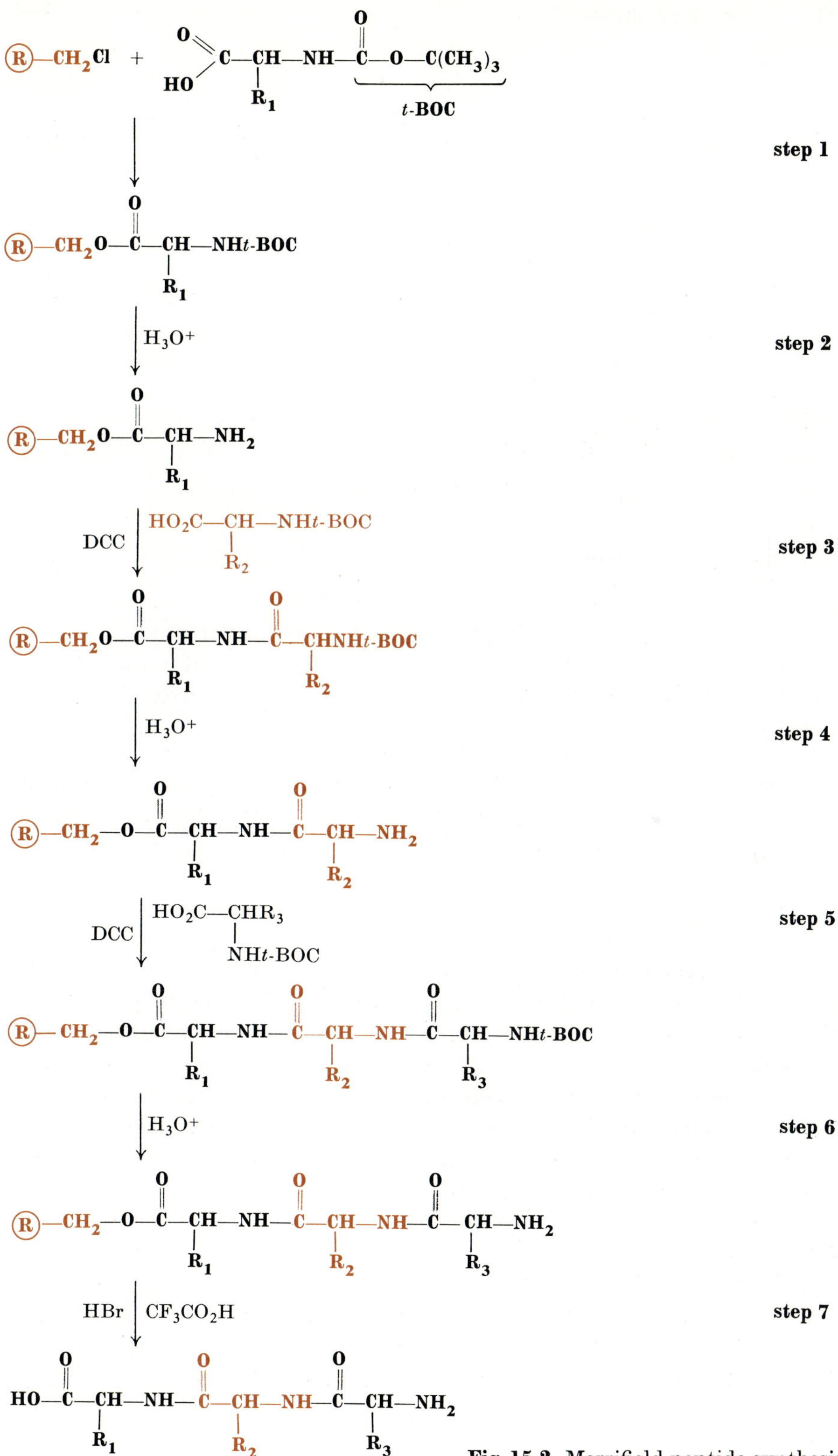

Fig. 15.2 Merrifield peptide synthesis

the formation of an activated carboxyl group, and proceeds along the following course:

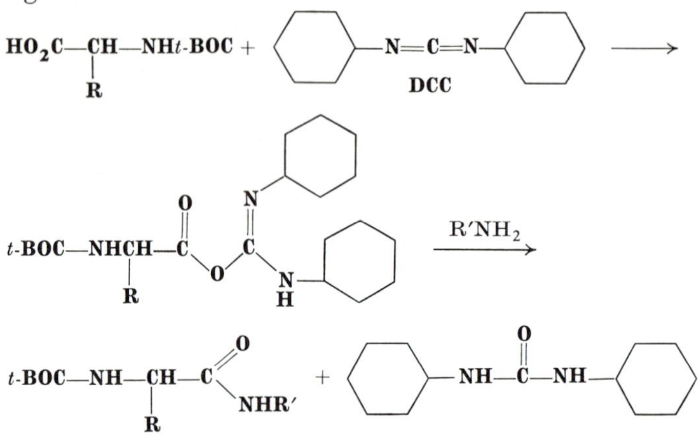

dicyclohexylurea

At this stage we have a dipeptide with the terminal carboxyl group still linked to the resin, and the terminal amino acid group linked to the protecting *t*-butyloxycarbonyl function. In order to increase the length of the peptide chain it is necessary to remove the protecting function, and add a third amino acid (as its *t*-BOC derivative) and DCC (steps 4 and 5).

When sufficient amino acids have been thus linked together (in our case, three of them), the synthetic peptide is removed from the resin by acid cleavage, using hydrogen bromide in trifluoroacetic acid (steps 6 and 7).

15.9 NATURAL PRODUCTS

The organic constituents of living systems can be divided into two major groups: primary metabolites and natural products. The first group includes the proteins, amino acids, carbohydrates, fats, and nucleic acids. The second group, often derived from primary metabolites, include terpenes, steroids, alkaloids, phenolic compounds, and pigments. Many of these compounds are of great chemical interest because of their complex structures, and of great biological importance because of their physiological activity. For example, some important natural products are cortisone (a steroid hormone), quinine (an alkaloid), penicillin (a mold metabolite), and camphor (a terpenoid).

15.10 ACETOGENINS

Various natural products are synthesized in living systems from acetic acid and are known as acetogenins. Principal among these are the terpenoids (including steroids) and phenolic compounds. (Fatty acids are also biosynthesized from acetic acid.)

15.11 TERPENOIDS AND STEROIDS

Terpenoids are naturally occurring substances; some are responsible for the odor, taste, and color of plants, some are found in the resin of various trees, or in the nonhydrolyzable material of both plant and animal tissue. In general, the carbon chains of these compounds appear to be based on a basic "building block" of five carbon atoms. The structures of most terpenoids can in fact be "divided" by the so-called "isoprene rule" into repeating isopentane units joined together "head-to-tail." This arrangement can be seen in the structure of a simple monoterpenoid (two isopentane units) like geraniol.

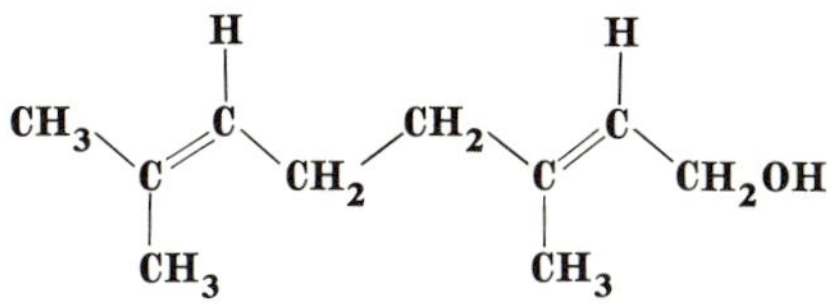

geraniol

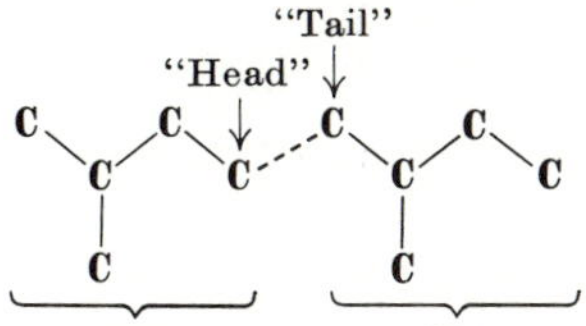

More complex terpenoids result from linking together five additional carbon units to form rings. Examples of this class of terpenoid and the way they fit the "isoprene rule" are illustrated in Fig. 15.3.

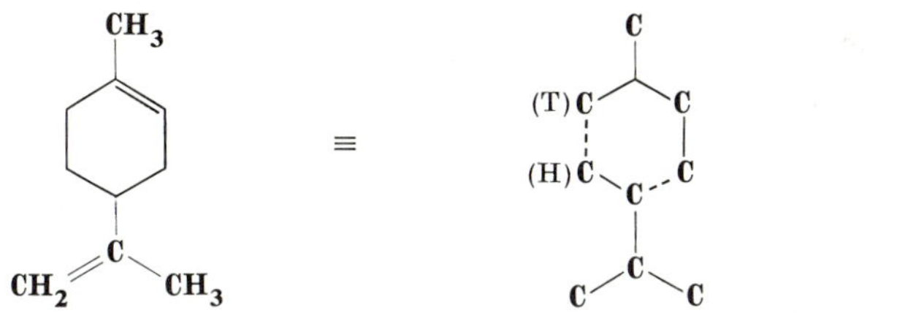

limonene
a monoterpenoid
(two isopentane units)

two isopentane units
joined head (H)-to-tail (T)
plus bond to form ring

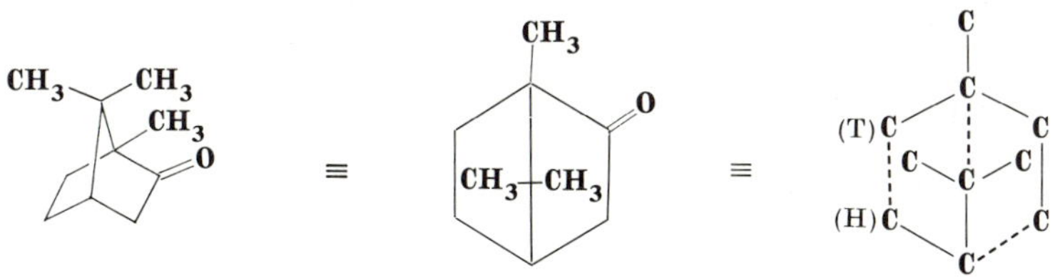

camphor
a monoterpenoid
(two isopentane units)

two isopentane units
joined head-to-tail plus
two bonds to form rings

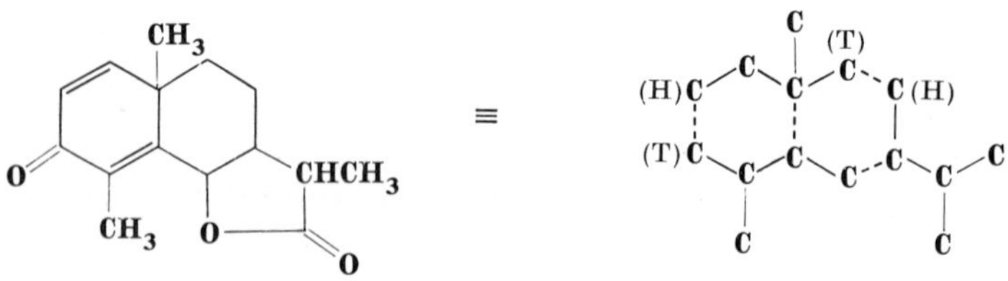

santonin
a sesquiterpenoid
(three isopentane units)

three isopentane units
joined head-to-tail
plus bonds to form rings

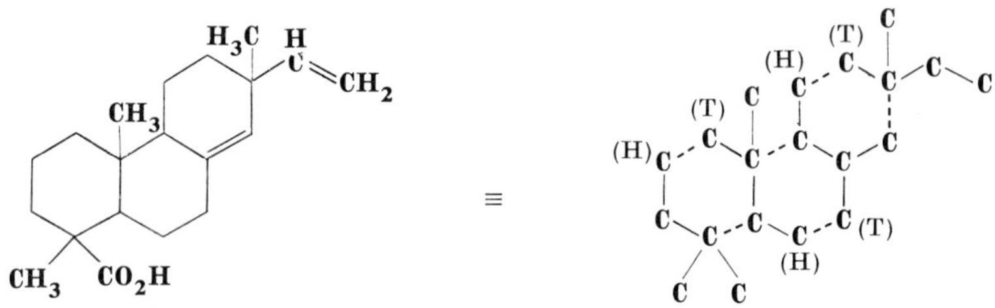

pimaric acid
a diterpenoid
(four isopentane units)

four isopentane units
joined head-to-tail
plus bonds to form rings

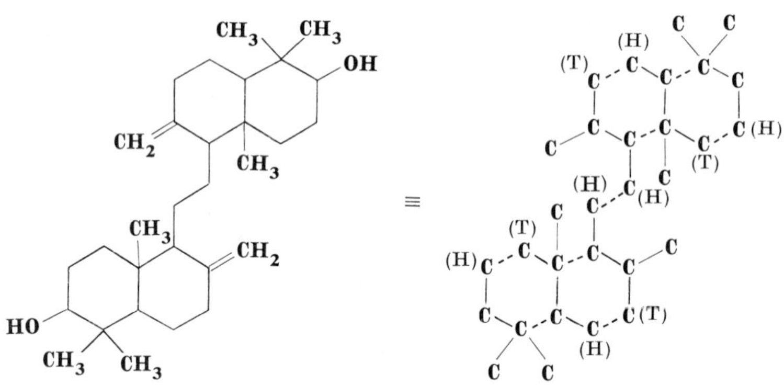

onocerin
a triterpenoid
(six isopentane units)

two units of three isopentane
units each joined head-to-head

Fig. 15.3 Terpenes and isoprene rule

Steroids, many of which are of great biological and medicinal importance, are in a class of rearranged and degraded triterpenoids characterized by the tetracyclic system illustrated in Fig. 15.4. The biological origin of the common steroid cholesterol will be discussed in the following section.

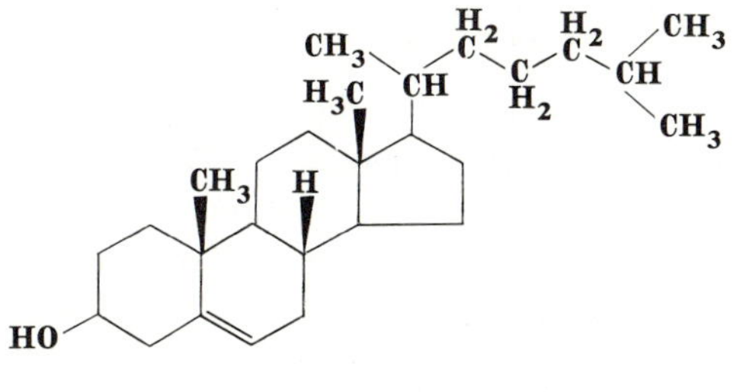

cholesterol

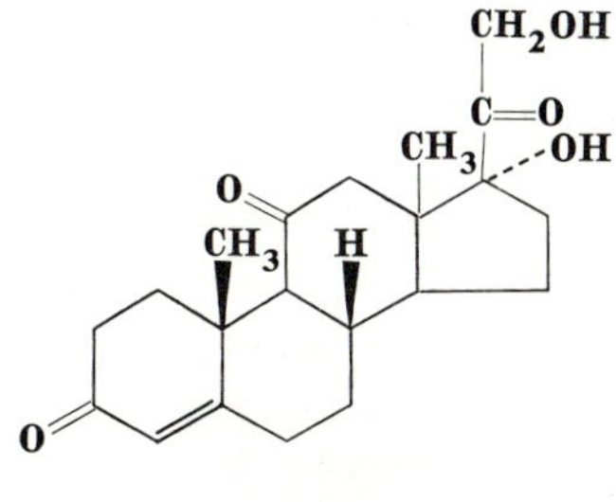

cortisone

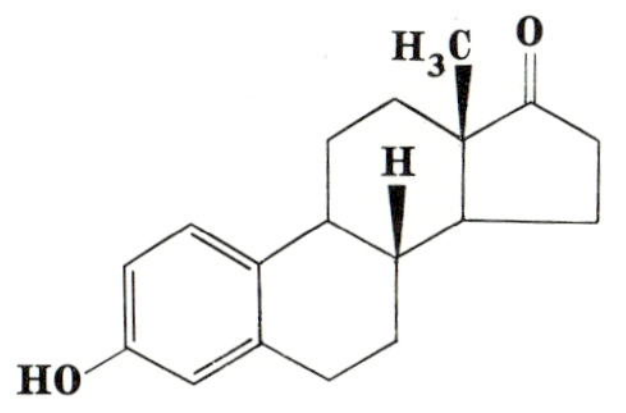

estrone

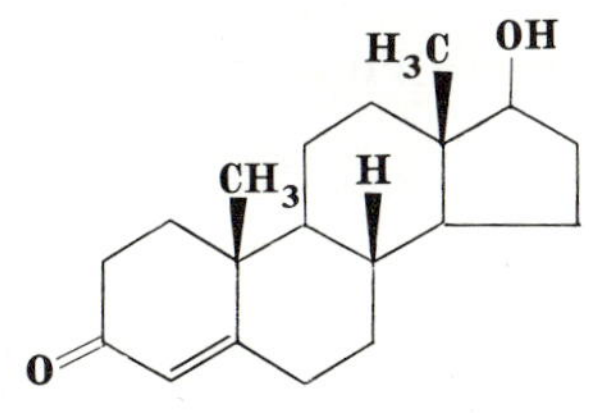

testosterone

Fig. 15.4 Some steroids

15.12 BIOSYNTHESIS

The basic five-carbon-atom building blocks of the terpenoids are formed from acetic acid via a series of condensations and decarboxylations. Figure 15.5, although it omits many of the details of these processes, illustrates that in a

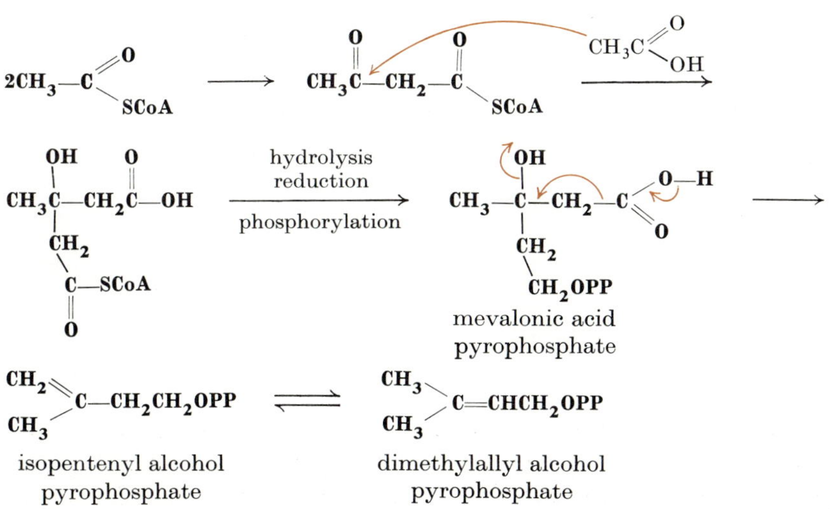

Fig. 15.5 Biosynthesis of the basic isopentane units of terpenoid biosynthesis

living system the reactions which make and break bonds are exactly analogous to those employed in the laboratory.

The reader will notice that all of the carbon-carbon bond-forming steps shown in Fig. 15.5 are variations on the Claisen condensation. Whereas, in the laboratory, the Claisen condensation usually involves alkyl esters, the biological equivalent involves a *thio* ester (the reaction product of an acid and a mercaptan). The particular thio ester employed in this biochemical process is acetylcoenzyme A.

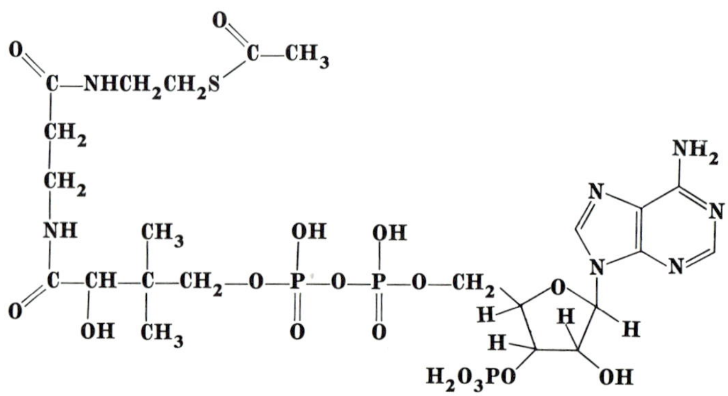

acetyl coenzyme A

Two double-bond isomers, isopentenyl alcohol and dimethylallyl alcohol, result from the sequence shown in Fig. 15.5, and these are further employed in enzymatic systems for the formation of terpenoid chains. The synthesis of

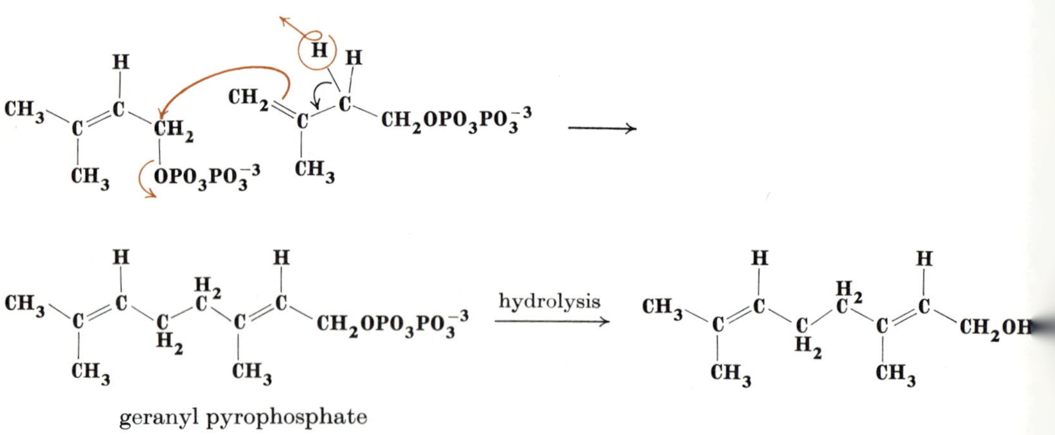

Fig. 15.6 Biosynthesis of geraniol

geraniol is illustrated in Fig. 15.6. Note that the chain-forming step is essentially a substitution reaction in which the electrons of a double bond displace a good leaving group. The latter, as in many similar biochemical displacement reactions, is a pyrophosphate group.

The formation of rings in terpenoid systems, like the chain-building reaction shown in Fig. 15.6, usually involves the displacement of an oxygen function by the π electrons of a double bond. Figure 15.7 illustrates this process for the formation of the citrus terpenoid limonene.

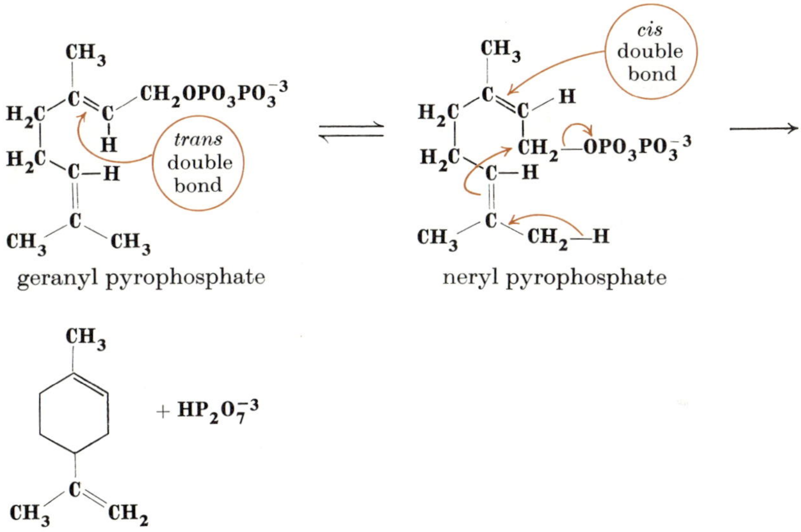

Fig. 15.7 Biosynthesis of limonene

One of the most complex terpenoid biosyntheses which has been elucidated is the synthesis of cholesterol from the thirty-carbon-atom *acyclic* triterpenoid, squalene. The ring forming process in this synthesis (Fig. 15.8) is once again the displacement of an oxygen function (an epoxide) by a double bond. However, in contrast to the production of the simple monoterpenoid limonene shown in Fig. 15.7, cyclization of squalene epoxide is followed by extensive rearrangement and degradation. The rearrangement of the methyl groups as shown in Fig. 15.8 yields lanosterol which is then converted into cholesterol by the loss of three methyl groups, saturation of one double bond, and migration of the other double bond.

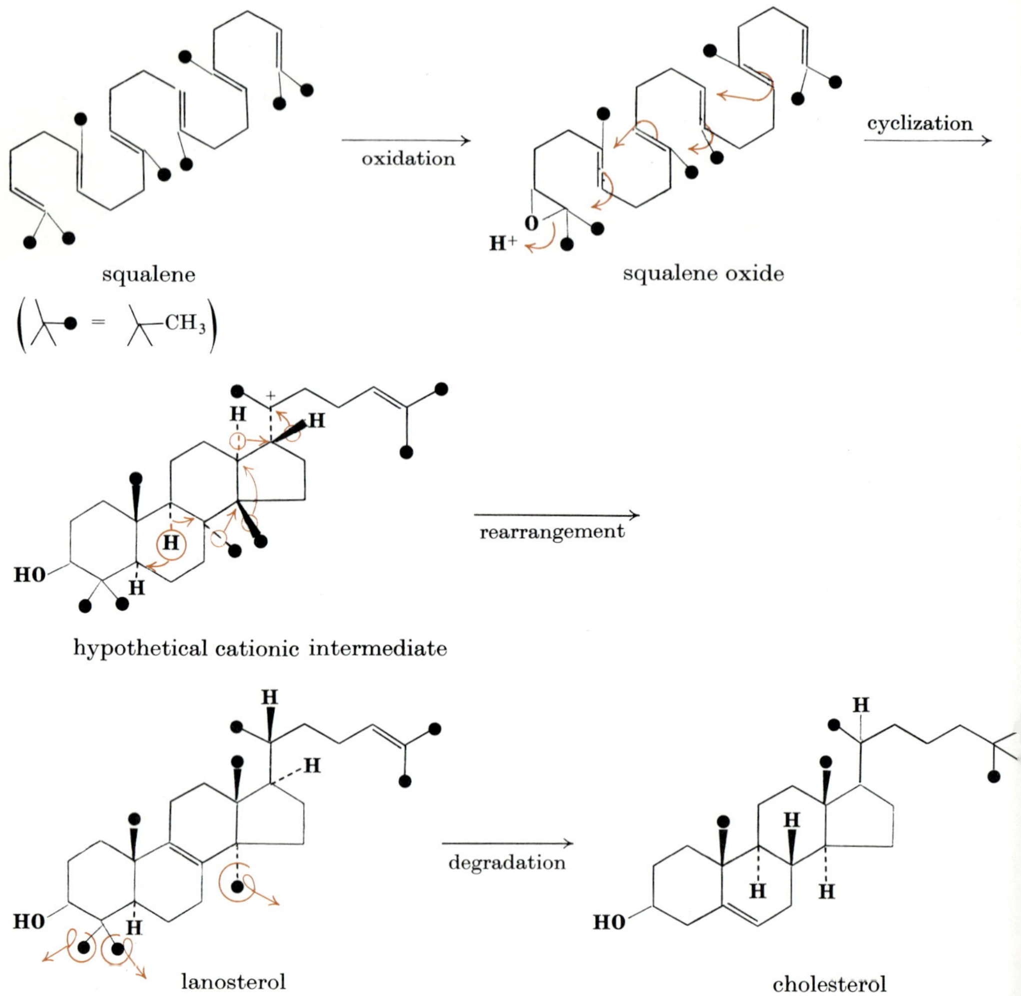

Fig. 15.8 Biosynthesis of cholesterol from squalene

15.13 LABORATORY SYNTHESIS

The creative power of the organic chemist has found its fullest flowering in the synthesis of natural products. Such diverse and complex molecules as chlorophyll, strychnine, tetracycline, and vitamin B_{12} have been prepared from simple precursors. The "art" of synthesis and the application of the principles previously discussed are illustrated in the following description of the synthesis of the terpenoid, sativene (carried out by J. E. McMurry of the University of California, Santa Cruz).

15.14 SATIVENE: STRUCTURE ANALYSIS

The sesquiterpenoid sativene, isolated from *Helminthosporium sativum*, has the following structure:

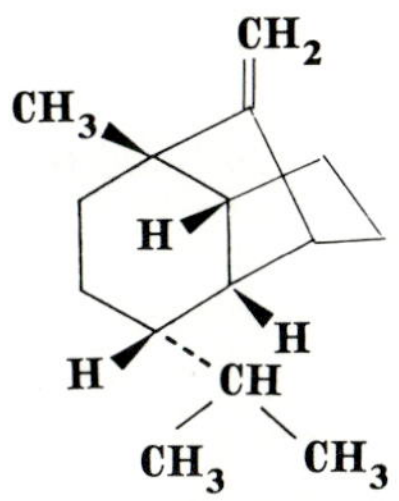

Sativene contains only fifteen carbon atoms, and a single functional group, namely, an *exocyclic* (exterior to a ring) double bond. The fifteen carbon atoms, however, are joined together to form a complex tricyclic system. The synthesis of such a system is a formidable problem. Let us see how in the conception of a synthesis for sativene a *simplification* of the structure led to a rational synthetic plan.

15.15 SIMPLIFYING THE STRUCTURE

As in our earlier synthesis problems, we begin our synthetic planning for sativene with a recognition of the type of structure to be assembled. Then we must breakdown the carbon skeleton into precursor molecules which are simpler to construct than the target structure. In previous problems our analysis of the carbon skeleton has always been closely tied to *specific* methods for forming carbon-carbon bonds. In the sativene problem, however, we will analyze the carbon skeleton in a general way without initial consideration of exact construction methods.

First, imagine the cleavage of a single carbon-carbon bond of sativene and then consider the properties of the resulting structure. To completely analyze the sativene problem we would have to consider the structures obtained from breaking *each* of the seventeen carbon-to-carbon bonds in the molecule. For the sake of brevity we shall limit our discussion to just one of these bonds, the C-4 ,C-8 bond which connects two of the rings to form a third (Fig. 15.9a).

Cleavage of the C-4,C-8 bond generates a structure that is simpler than sativene; it only has two rings (Fig. 15.9b). Furthermore, the resulting structure (Fig. 15.9b) is a member of a common class of bicyclic molecules called decalins.

In decalins, two six-membered rings are fused at adjacent carbon atoms. When we *redraw* the structure in both flat and three-dimensional representations (Figs. 15.9c and 15.9d) it becomes clear that our potential precursor is a decalin.

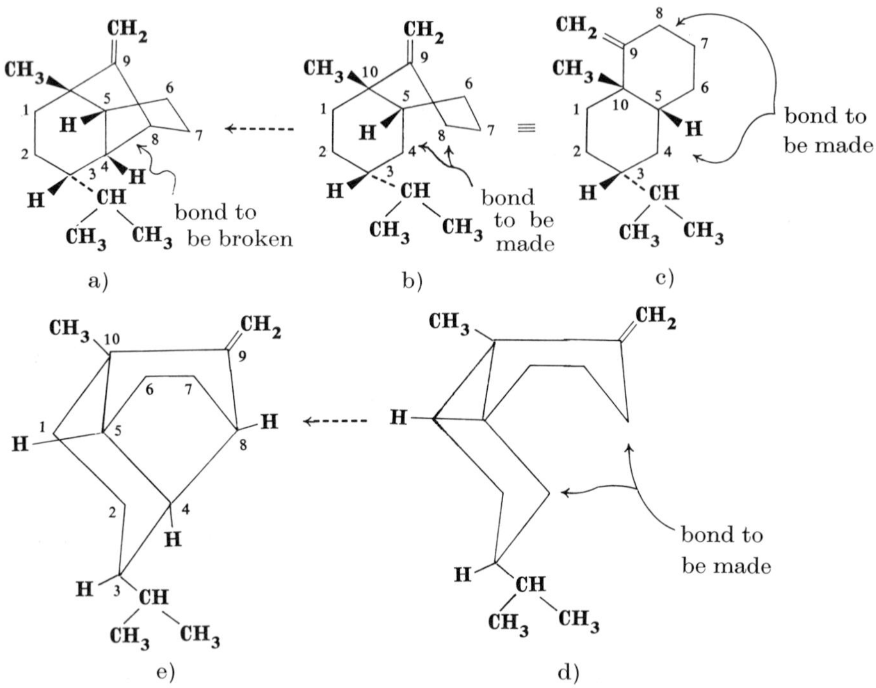

Fig. 15.9 Simplification of the sativene structure to a decalin system—Planar and three-dimensional representations

Now we must consider how the bond we have broken "on paper" might be reconstructed in the laboratory. Bond formation between carbon atoms 4 and 8 requires functionality at both of these sites. The olefin function at C-9 suggests a functionality that could be used. A double bond is a functional equivalent of a carbonyl group. Addition of a Wittig reagent (Section 11.13) to a ketone, or addition of a Grignard reagent followed by dehydration of the resulting alcohol can convert the carbonyl group to an olefin. A ketone, in turn, if it has an α-hydrogen can be used to make a new carbon-carbon bond by alkylation of its enolate ion (Section 11.22):

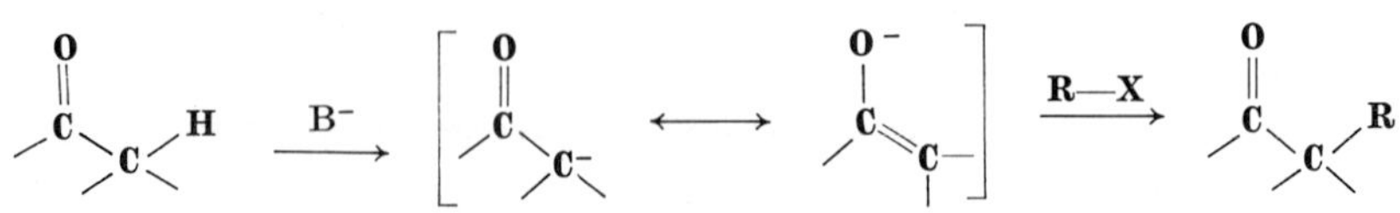

Thus the third ring of sativene may be constructed by a displacement reaction. The functionality that we require in our precursor molecule is a ketone group at C-9 and a leaving group at C-4 (Fig. 15.10).

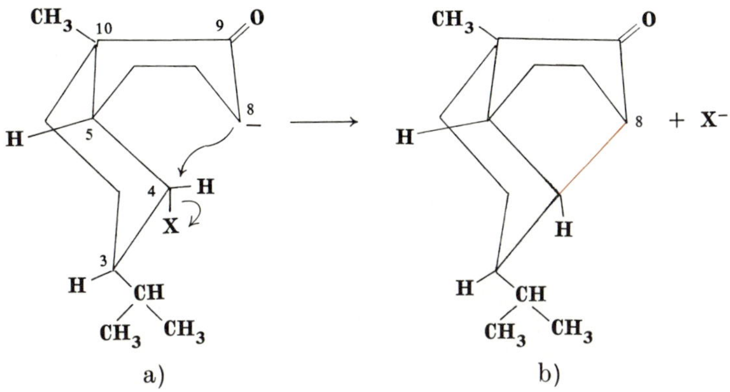

Fig.15.10 Reconstruction of the third ring of sativene

15.16 STEREOCHEMISTRY

The synthetic plan which we are following must accommodate both the stereochemistry of sativene and the steric requirements of the reactions used in its construction. The stereochemistry of sativene is defined by the relationships of the methyl group at C-10, the hydrogens at C-4 and C-5, and the isopropyl group at C-3 (Fig. 15.9e). The first three are *cis* with respect to each other, and *trans* to the isopropyl substituent. The reaction we are using to form the third ring is an S_N2 process. Our synthetic problem is thus reduced to the preparation of a bicyclic ketone with the structure and precise stereochemistry shown in Fig. 15.10a.

15.17 THE BICYCLIC KETONE

When we examine the structure of our precursor ketone (Fig. 15.10a) we see that there are five positions in that molecule that must be subject to our structural and stereochemical control. We must have the ability to introduce into the decalin system an isopropyl group at C-3, a leaving group at C-4, a hydrogen at C-5, a methyl group at C-10 (all with appropriate stereochemistry) and a ketone function at C-9. The Michael-addition aldol condensation–annelation (ring forming) sequence described in Section 11.23 provides a means of accomplishing these aims.

The condensation of methylcyclohexanedione and methyl vinyl ketone (shown in Fig. 15.11) affords a substrate suitable for the introduction of all of

our necessary substituents. To convert this product into the target bicyclic ketone we must
a) reduce the double bond to give a *cis*-decalin system
b) introduce a leaving group at C-4
c) convert the ketone group at C-3 to an isopropyl substituent (Fig. 15.11).
We already have general methods for accomplishing all of these goals.

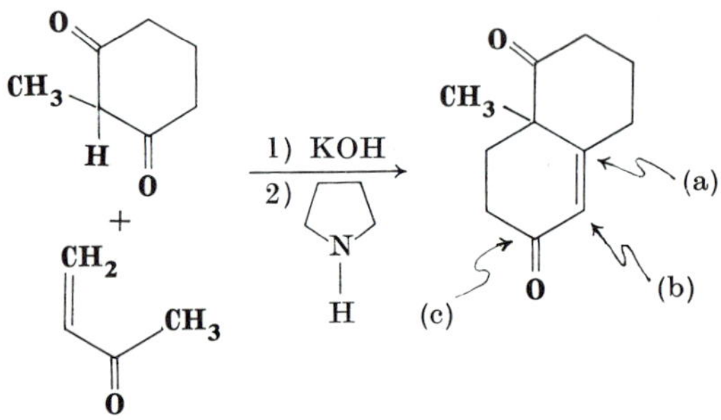

Fig. 15.11 First bicyclic intermediate—Goals for its transformation

The isopropyl group might be added via a Grignard reaction, and both the ring junction hydrogen and the leaving group might be introduced by addition reactions to the carbon-carbon double bond. These processes cannot be carried out in a random manner, however. A specific synthetic plan must ensure that each of these transformations may be carried out without affecting any other parts of the molecule. We must proceed in a stepwise manner making appropriate use of selective reactions and protected functional groups.

15.18 TOTAL SYNTHETIC PLAN

The total synthetic plan for sativene, beginning with the Michael-aldol annelation product (Fig. 15.12*a*), is illustrated in Fig. 15.12. Here we demonstrate how the general ideas which we have discussed for the control of structure and stereochemistry were specifically applied in the laboratory synthesis.

Step 1. Saturated ketones form ketals more readily than do unsaturated ketones. A *selective* reaction may thus be performed on diketone *a* to afford the monoketal *b*. As a consequence, transformations of the unsaturated ketone function may now be accomplished while the saturated ketone group remains *protected*.

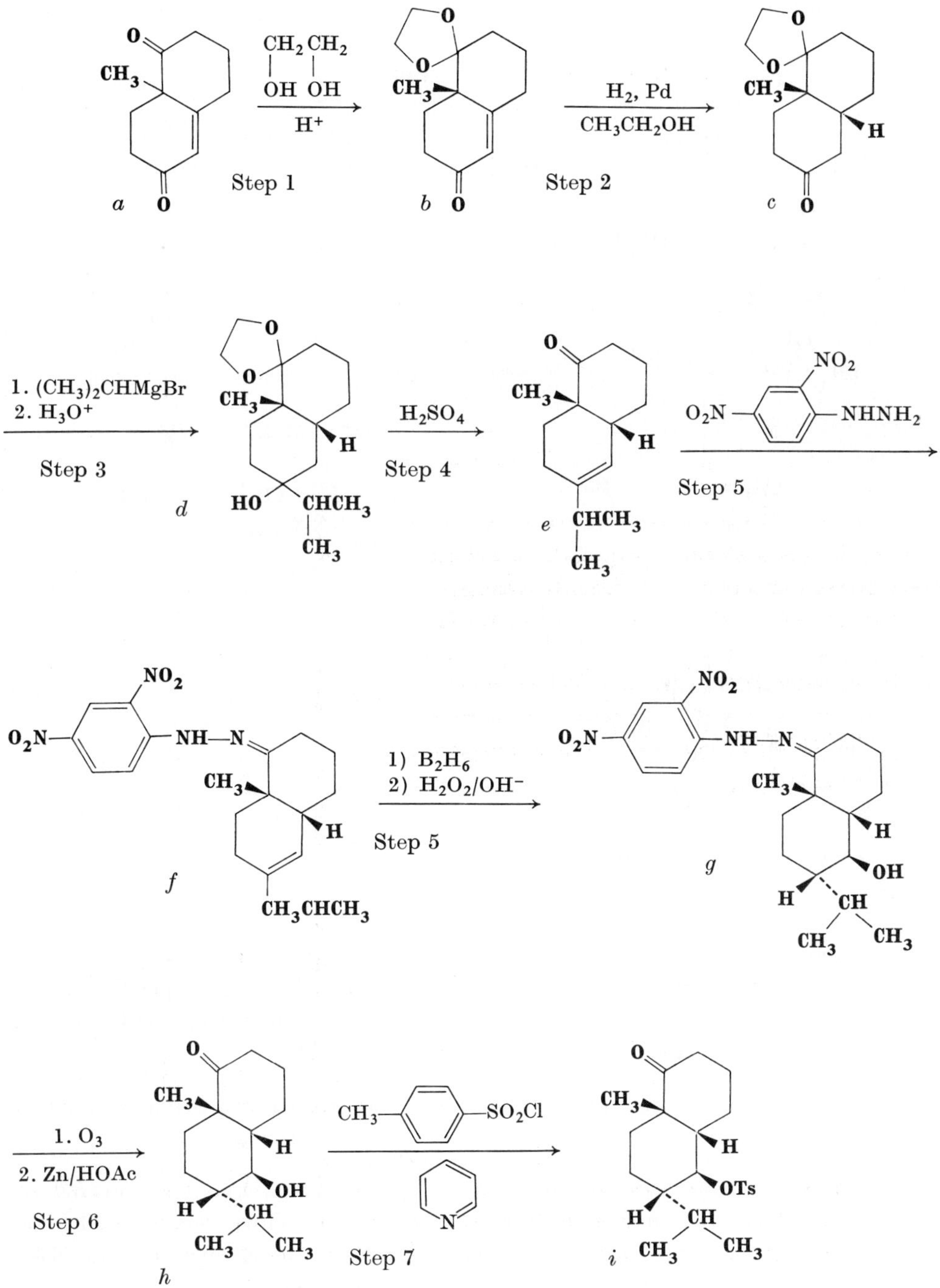

CH_2CH_2 OH OH
H^+
Step 1
H_2, Pd
CH_3CH_2OH
Step 2
a
b
c
1. $(CH_3)_2CHMgBr$
2. H_3O^+
Step 3
d
H_2SO_4
Step 4
e
NO_2
O_2N
$NHNH_2$
Step 5
f
1) B_2H_6
2) H_2O_2/OH^-
Step 5
g
1. O_3
2. Zn/HOAc
Step 6
h
CH_3
SO_2Cl
N
Step 7
i
OTs

Contd.

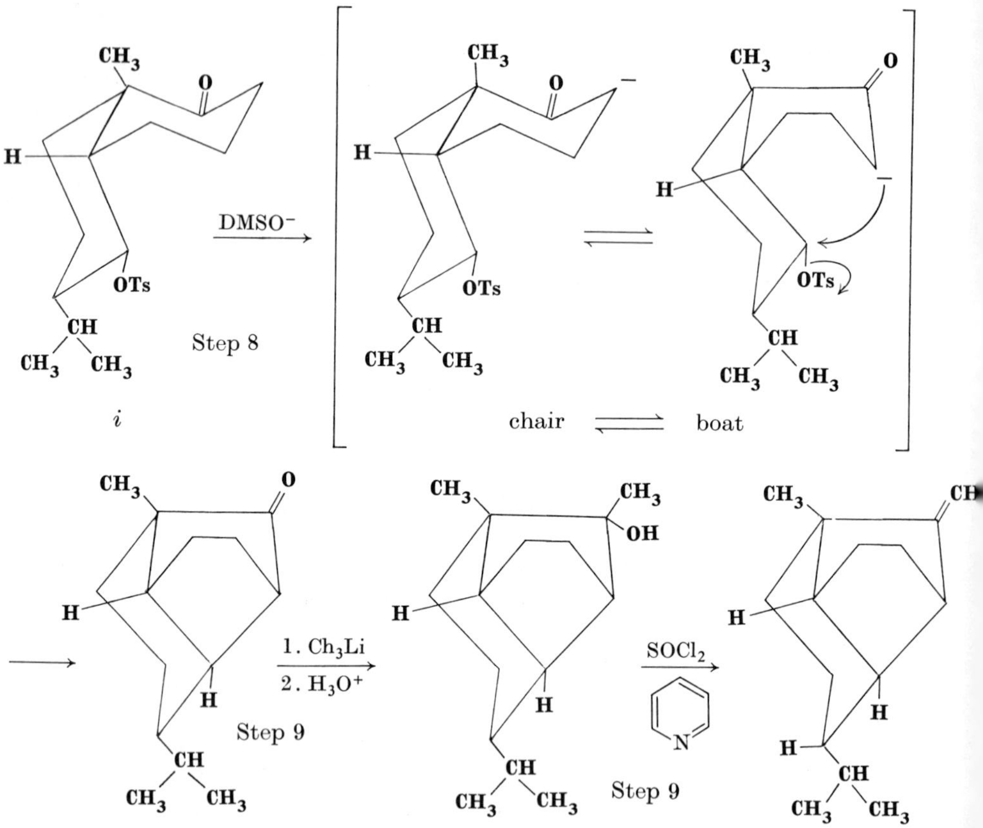

Fig. 15.12 Synthetic plan for sativene

Step 2. The catalytic reduction in neutral solution of the double bond of unsaturated ketones like *b* yields *cis*-decalins. In this stereospecific reaction we establish the required *cis* relationship of the C-1 hydrogen and C-6 methyl groups of our precursor bicyclic ketone (Fig. 15.10a).

Step 3. The isopropyl group of sativene is introduced by a Grignard reaction. Only the *free* ketone group of *c* is affected by the Grignard reagent.

Step 4. When alcohol *d* is dehydrated, a new double bond is introduced and the ketal function is hydrolyzed (Section 11.14). Three isomeric olefins are possible from the dehydration reaction. It is known, however, that in related systems the double-bond is formed in the position shown in structure *e*.

Step 5. Two of the asymmetric centers of the desired bicyclic ketone precursor have already been fixed (Step 2). The remaining two can be introduced in a singie step. A *cis*-decalin has a cup-like shape. One face of the molecule is highly hindered while the other is open to attack by a reagent (Fig. 15.13).

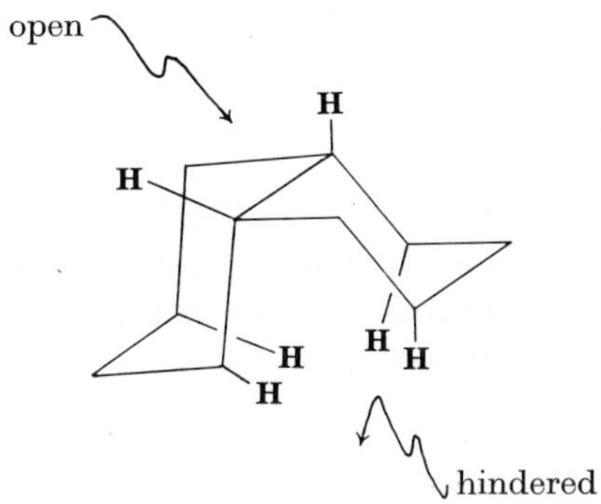

Fig. 15.13 Ease of attack on *cis*-decalin

What we require is a reagent that will supply both the hydrogen at C-3 and the potential leaving group at C-4 from the same side of the molecule (see Fig. 15.10a). 'The 'anti-Markovnikov'' addition of diborane to an olefin (Section 9.11) fits this requirement.

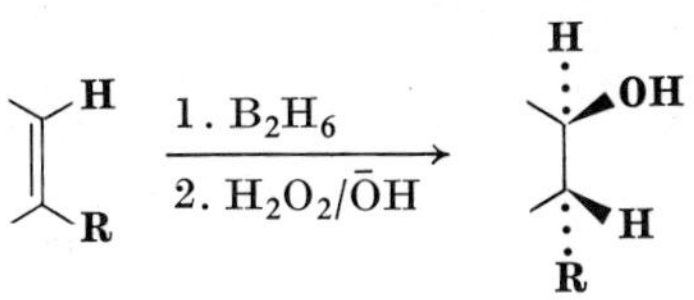

In laboratory practice, however, it was found that direct treatment of *e* with diborane resulted only in reduction of the carbonyl function. It was therefore necessary to again *protect* the ketone group. The protected compound in this case was *f* the 2,4-dinitrophenyl hydrazine derivative of the ketone (Section 11.12).

Step 6. The carbon-nitrogen double bond of *g* may be cleaved with ozone, in a manner analogous to the cleavage of carbon-carbon double bonds to restore the ketone function at C-9, *h*.

Step 7. Compound *i* is the desired precursor ketone (Fig. 15.10a). The leaving group for the intramolecular alkylation reaction is the p-toluenesulfonate or ''tosylate'' group abbreviated "OTs".

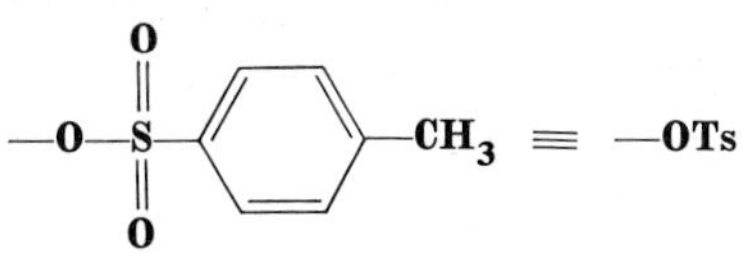

Step 8. The base employed for the ring closure step is the anion of dimethyl sulfoxide (DMSO). It is prepared by the reaction of dimethyl sulfoxide with sodium hydride.

$$CH_3\overset{\overset{\large O}{\|}}{S}—CH_3 + NaH \longrightarrow CH_3\overset{\overset{\large O}{\|}}{S}—CH_2^-Na^+ + H_2$$

Steps 9, 10. The sativene synthesis was completed first by treatment of ketone *j* with methyl lithium which acts in a manner equivalent to a Grignard reagent and second by using thionyl chloride in pyridine, a standard method for the dehydration of tertiary alcohols under basic conditions.

15.19 ALKALOIDS

The alkaloids are plant products which contain a basic nitrogen function. Some typical alkaloids are shown in Fig. 15.14.

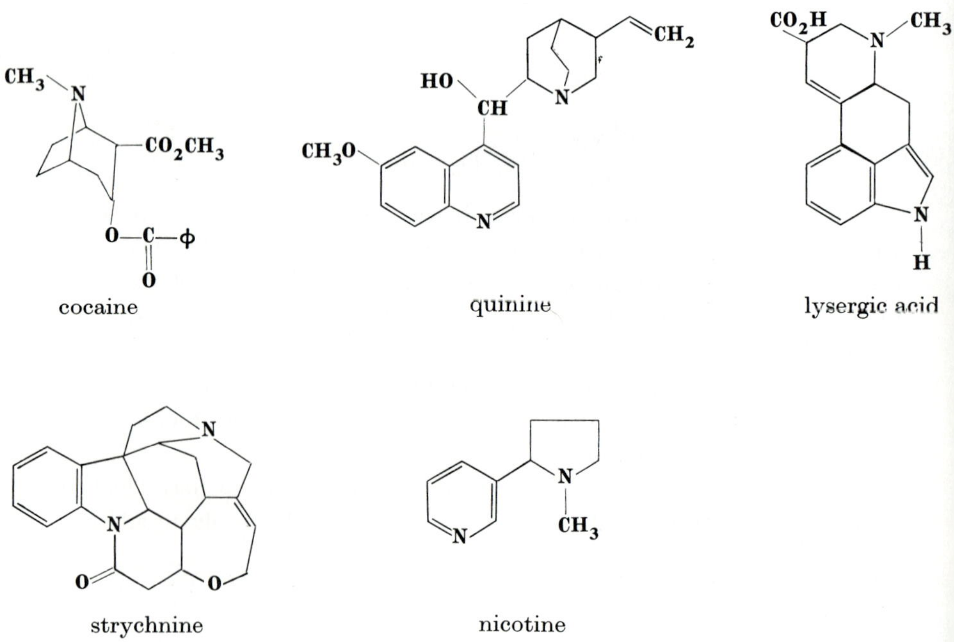

Fig. 15.14 Alkaloids

Most alkaloids appear to be derived biosynthetically from amino acids. Although a general discussion of these biosynthetic pathways is beyond the scope of this text, a formal scheme for the conversion on the amino acid phenylalanine to morphine is shown in Fig. 15.15.

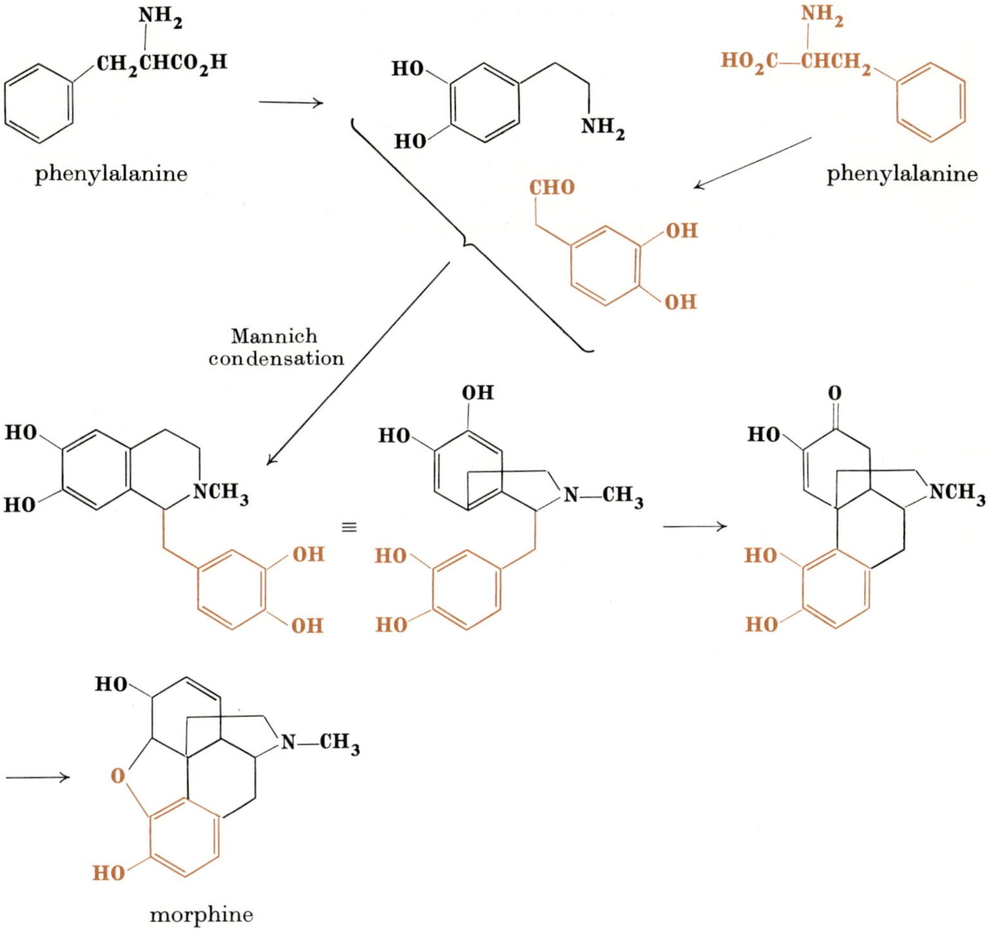

Fig. 15.15 Biosynthesis of morphine

ADDITIONAL READING

M. Goodman and F. Morehouse, *Organic Molecules in Action*, New York: Gordon and Breach, 1973.

J. B. Hendrickson, *The Molecules of Nature*, Menlo Park, Calif.: Benjamin, 1965. Paperback.

E. M. Kosower, *Molecular Biochemistry*, New York: McGraw-Hill, 1962.
L. L. Ingraham, *Biochemical Mechanisms*, New York: Wiley, 1962.
R. E. Ireland, *Organic Synthesis*, Englewood Cliffs. N.J.: Prentice-Hall, 1969. Paperback.
T. A. Geissman, *Workbook in Organic Chemistry*, San Francisco Calif.: W. H. Freeman, 1972. Chapters 4, 5, 6, 11. Paperback.

16
BIOLOGICAL TRANSFORMATIONS

16.1 BIOLOGICAL TRANSFORMATIONS

At the outset of our discussion about Organic Chemistry we stated that the principles which govern the course of a reaction are the same whether that reaction takes place in a test tube or as part of a physiological system. The purpose of this section is to illustrate this statement by utilizing the mechanistic interpretations of organic chemistry developed earlier in this book to rationalize some important biological processes. It is important to remember, however, that enzymatic catalysis is required for all of the biological processes discussed in this chapter.

16.2 GLYCOLYSIS

The energy requirements of highly organized mobile living systems necessitate the intake of biologically degradable nutrients, the breakdown of which results in the release of energy. During the course of these degradations, biological synthetic intermediates are produced which are useful in the biosynthesis of physiologically important factors such as fatty acids, steroids, and the genetic material (DNA). Glycolysis is one such process in which body sugar is degraded to triose phosphates. There are several routes by which this transformation is effected. The most important of these is shown in Fig. 16.1.

We shall not consider the conversion of glycogen to α-D-glucose-1-phosphate (which involves the cleavage of acetal functions) but shall focus on how nature splits a glucose carbon-to-carbon bond into two three-carbon fragments.

In the first step phosphate is removed from the C-1 position of glucose and the 6-hydroxy group is phosphorylated. The mechanism of this conversion is closely related to the carbonyl addition–elimination process described in Sections 3.8, 11.2, and 11.3. For a carboxyl derivative the sequence is called acyl transfer; in the case of phosphoric acid derivatives it is known as phosphate transfer. The two analogous mechanisms are:

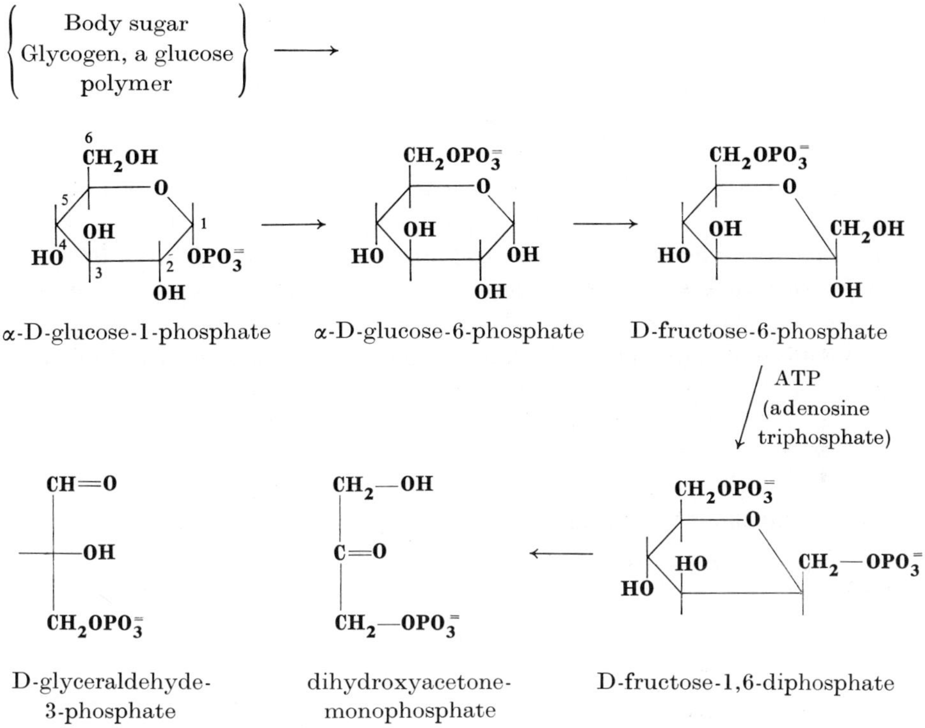

Fig. 16.1 Glycolysis of glucose

acyl transfer:

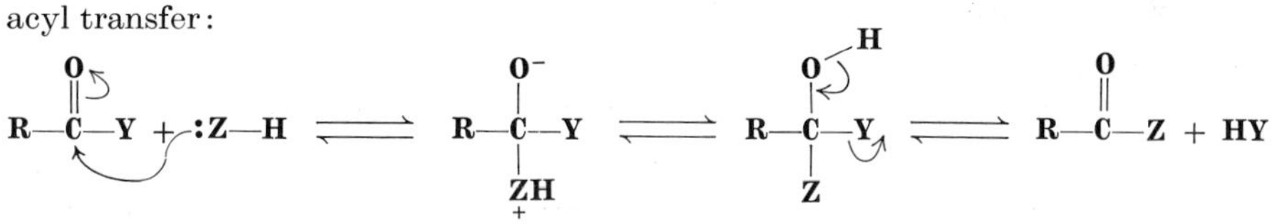

phosphate transfer:

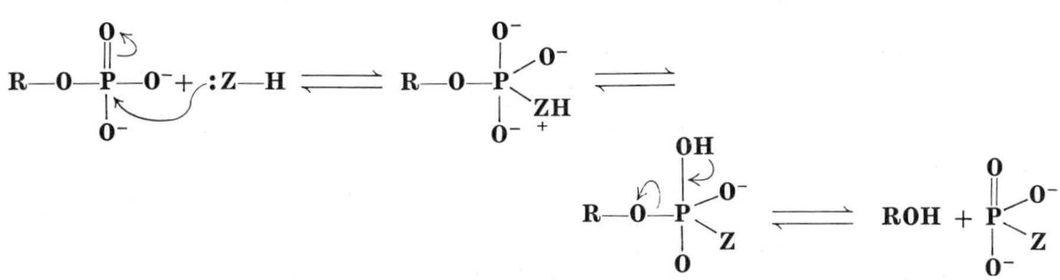

Note that in the first instance the acyl group is transferred from Y to Z and in the second case the phosphoryl group is transferred from R—O to Z. If in the latter example either R or Z is part of an enzyme, the above then becomes a phosphate transfer to and from an enzyme. In the conversion of α-D-glucose-1-phosphate to α-D-glucose-6-phosphate, the enzyme first accepts a phosphate from the α-D-glucose-1-phosphate. (R-glucose, Z = enzyme). Subsequently, the enzyme transfers a phosphate to the 6 position of glucose (R = enzyme, Z = glucose).

The analogy between phosphorylation and acylation may be extended with regard to how nature activates phosphoryl groups. Thus, by substituting chlorine for hydroxyl in the carboxyl, an acyl group is activated (RCOCl is more reactive than RCOOH). Similarly, by substituting a pyrophosphate group for the hydroxyl of phosphoric acid, nature activates the phosphoryl group in the biological phosphorylating agent adenosine triphosphate (ATP).

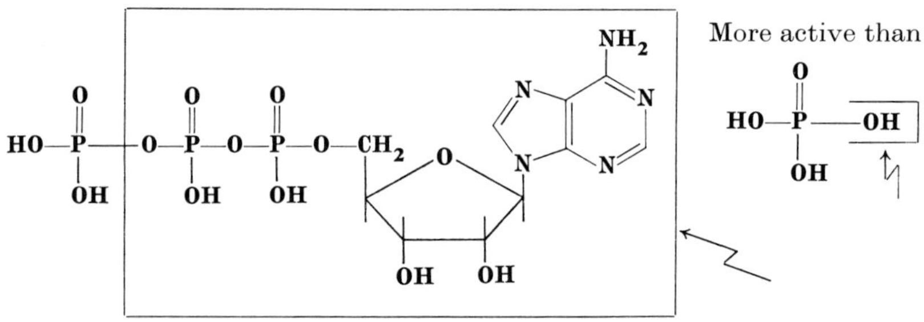

adenosine triphosphate (ATP) pyrophosphate leaving group *vs.* hydroxyl leaving group

Once there is a free hydroxyl group at C-1 of glucose, the following equilibrium can produce D-fructose-6-phosphate:

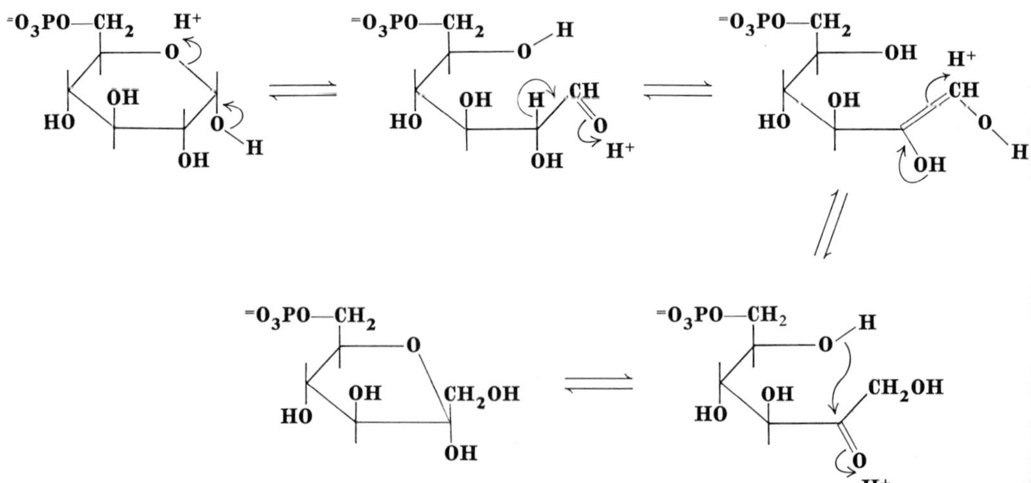

D-Fructose-6-phosphate

D-Fructose-6-phosphate is converted to its 1,6-diphosphate derivative by adenosine triphosphate *via* a phosphate transfer as described previously.

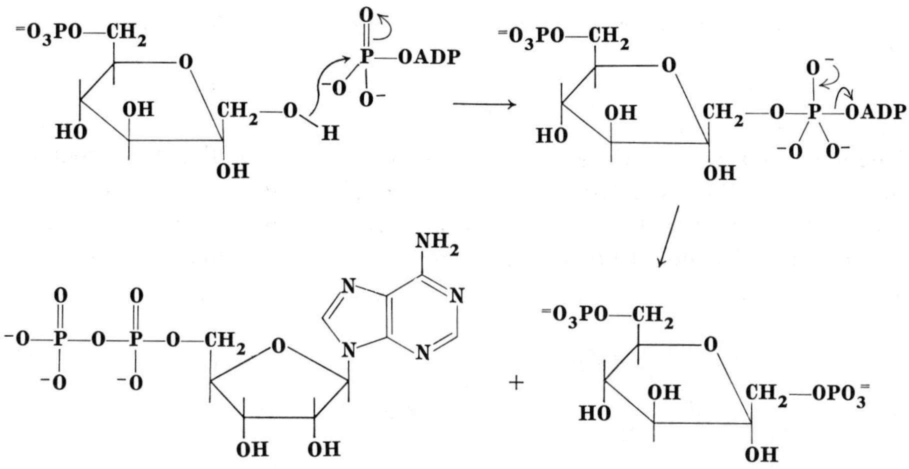

adenosine diphosphate (ADP)

D-fructose-1,6-diphosphate

Carbon-to-carbon bond cleavage is now achieved in a reverse aldol reaction (refer to Section 11.23) as follows:

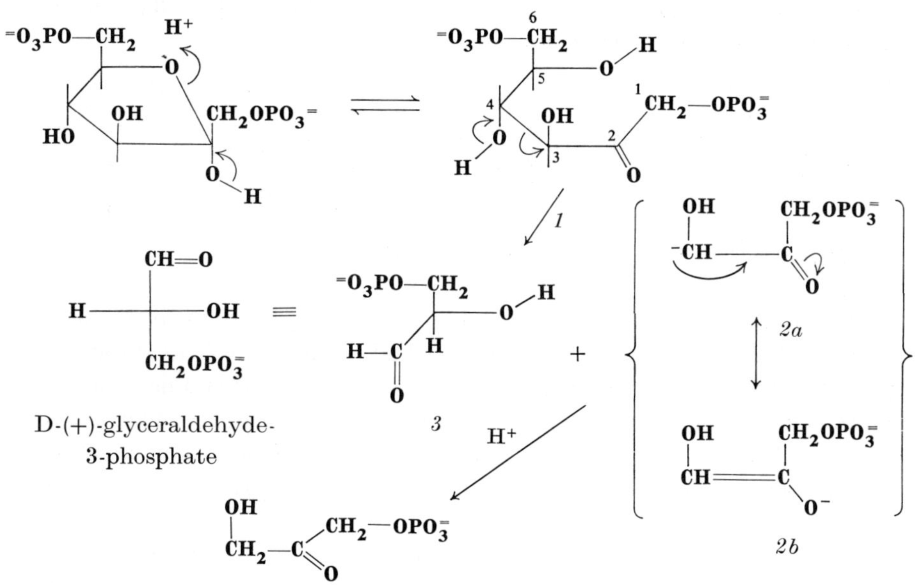

D-(+)-glyceraldehyde-3-phosphate

3

2b

dihydroxyacetone monophosphate

The carbonyl group at the 2 position of the fructose derivative 1 makes possible the particular carbon-to-carbon bond cleavage that occurs (1 to 2a and 3) because bond fission develops anionic character at a site where it can be stabilized by resonance interaction with the carbonyl group (2a ↔ 2b).

16.3 THIAMINE

Thiamine is a biological catalyst which is involved in glycolysis routes alternative to the pathway discussed above. The structure of thiamine is given in Fig. 16.2(a). The catalytic properties we shall consider rest in the thiazole ring. For the sake of simplicity, we shall represent thiamine as shown in Fig. 16.2(b).

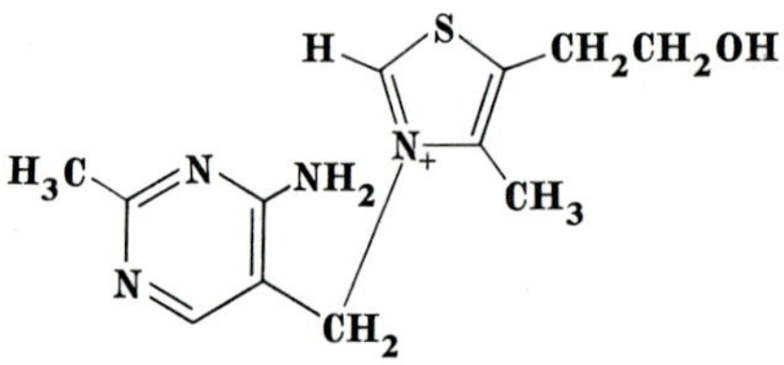

Fig. 16.2a Thiamine

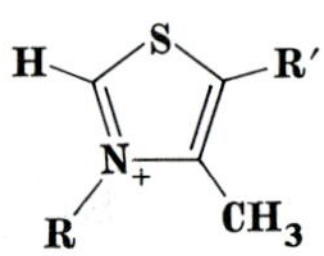

Fig. 16.2b Thiazole ring

The uniqueness of thiamine is due to the ability of the thiazole ring to stabilize anionic character both directly on the ring (Fig. 16.3a) and on a carbon substituting the ring (Fig. 16.3b).

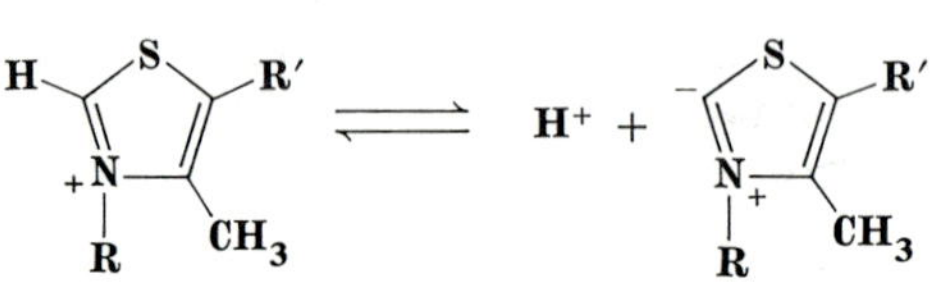

Fig. 16.3a

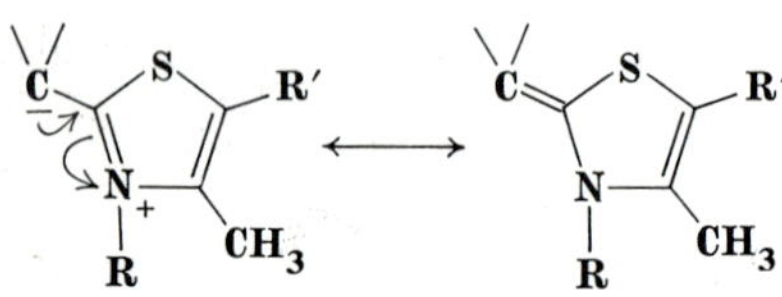

Fig. 16.3b

In Fig. 16.3(a), the anion is stabilized inductively by the positively charged nitrogen and by d-orbital overlap with the adjacent sulfur. This stabilization provides the impetus for the indicated dissociation of thiamine to afford a proton and thiamine anion. In the second case (Fig. 16.3b) the anion is resonance stabilized as indicated. An example of how nature uses thiamine catalysis in the biosynthesis of acetylcoenzyme A (see Section 15.12) from coenzyme A and pyruvate is given in Fig. 16.4.

In Fig. 16.4 (step 1), thiamine anion adds to the pyruvate carbonyl to afford the indicated intermediate (refer to Chapter 11). Decarboxylation occurs in step 2. The driving force for decarboxylation is provided by the thiamine moiety which stabilizes the resulting anion by resonance, A ↔ B.

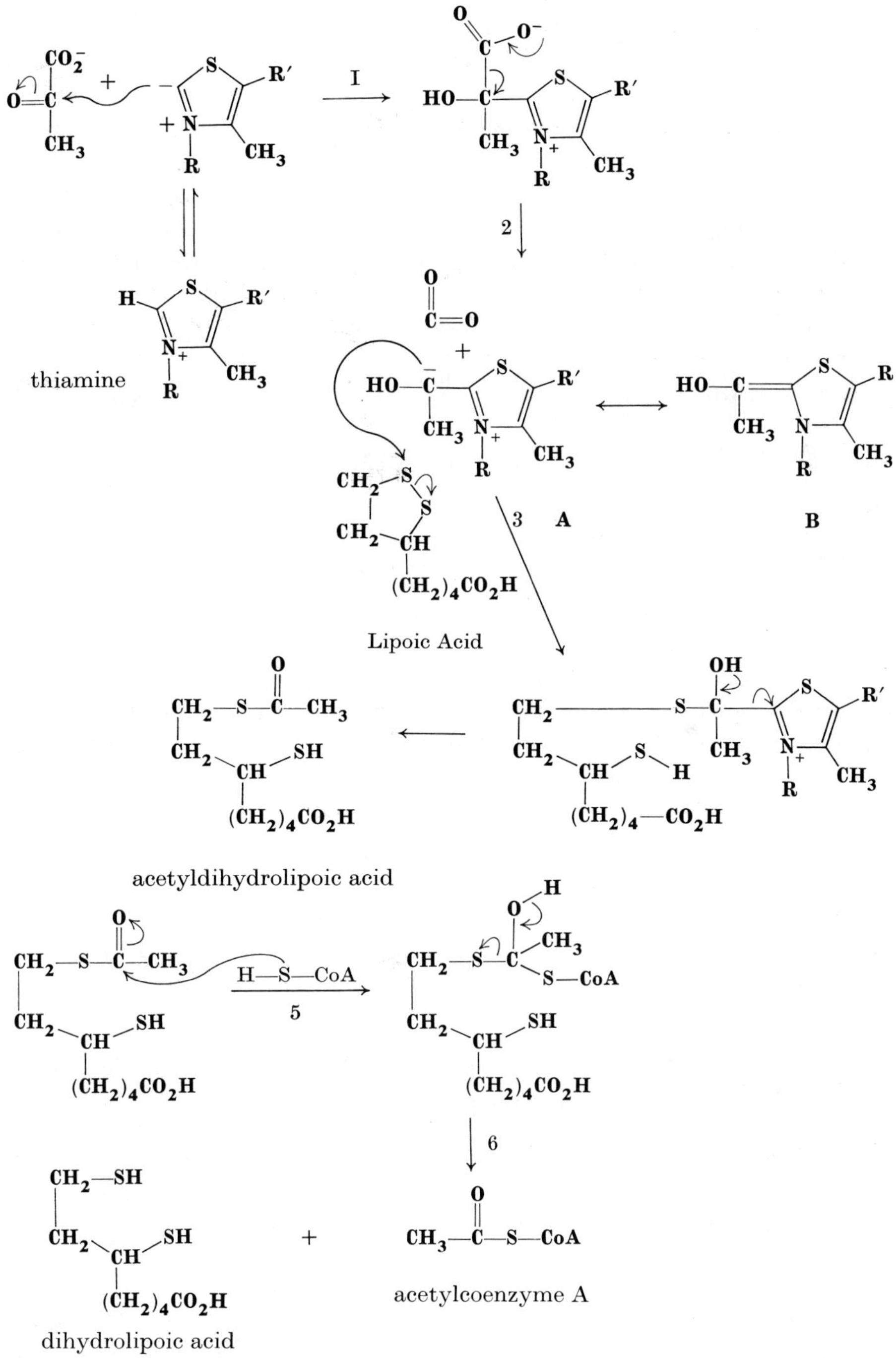

Fig. 16.4 Biosynthesis of acetylcoenzyme A

In step 3, anion A displaces the disulfide bond of lipoic acid in a substitution reaction. The substitution product then decomposes (step 4) to produce acetyldihydrolipoic acid and the anion of thiamine. Notice that this sequence has produced a thioester, the same functionality found in acetylcoenzyme A. Thioester interchange (steps 5 and 6, see also acyl transfer under glycolysis) results in acetylcoenzyme A.

The byproduct of this reaction sequence, dihydrolipoic acid, is reoxidized back to lipoic acid by the biological oxidizing agent nicotinamide adenine dinucleotide (NAD).

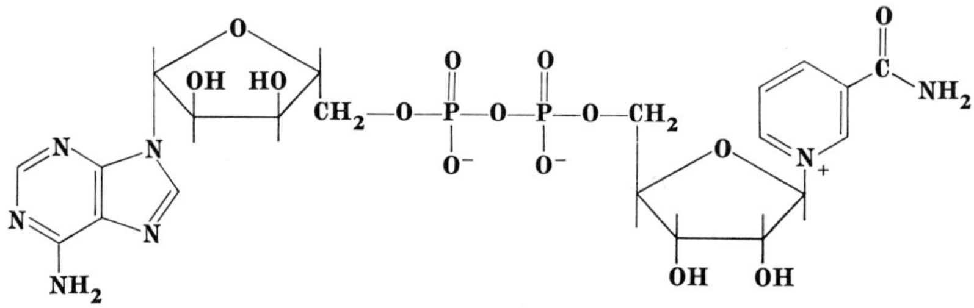

nicotinamide adenine dinucleotide (NAD)

This agent serves as an electron acceptor and in its reduced form as an electron donor. The part of the molecule responsible for this activity is the pyridine ring. The reaction may be represented as follows:

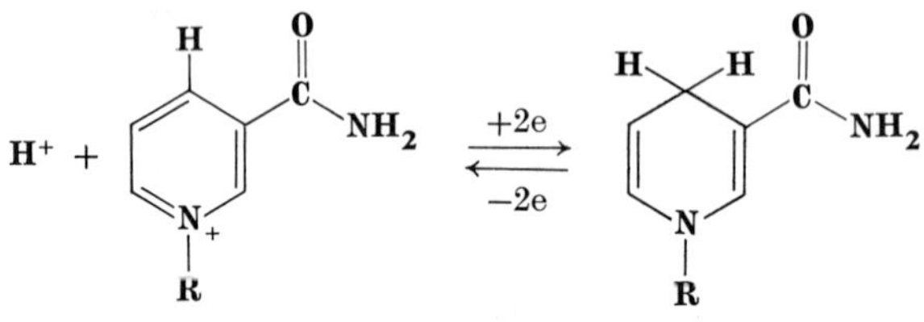

NAD NADH

In the particular oxidation of dihydrolipoic acid to lipoic acid, a possible mechanism for the process would be:

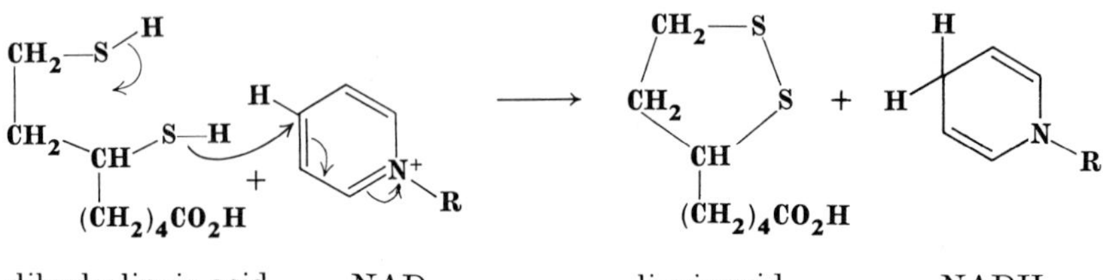

dihydrolipoic acid NAD lipoic acid NADH

16.4 TRICARBOXYLIC ACID CYCLE

The sequence of reactions known as the tricarboxylic acid cycle constitutes one part of aerobic metabolism. During the course of this series of transformations, pyruvic acid is oxidized to carbon dioxide and NAD is converted

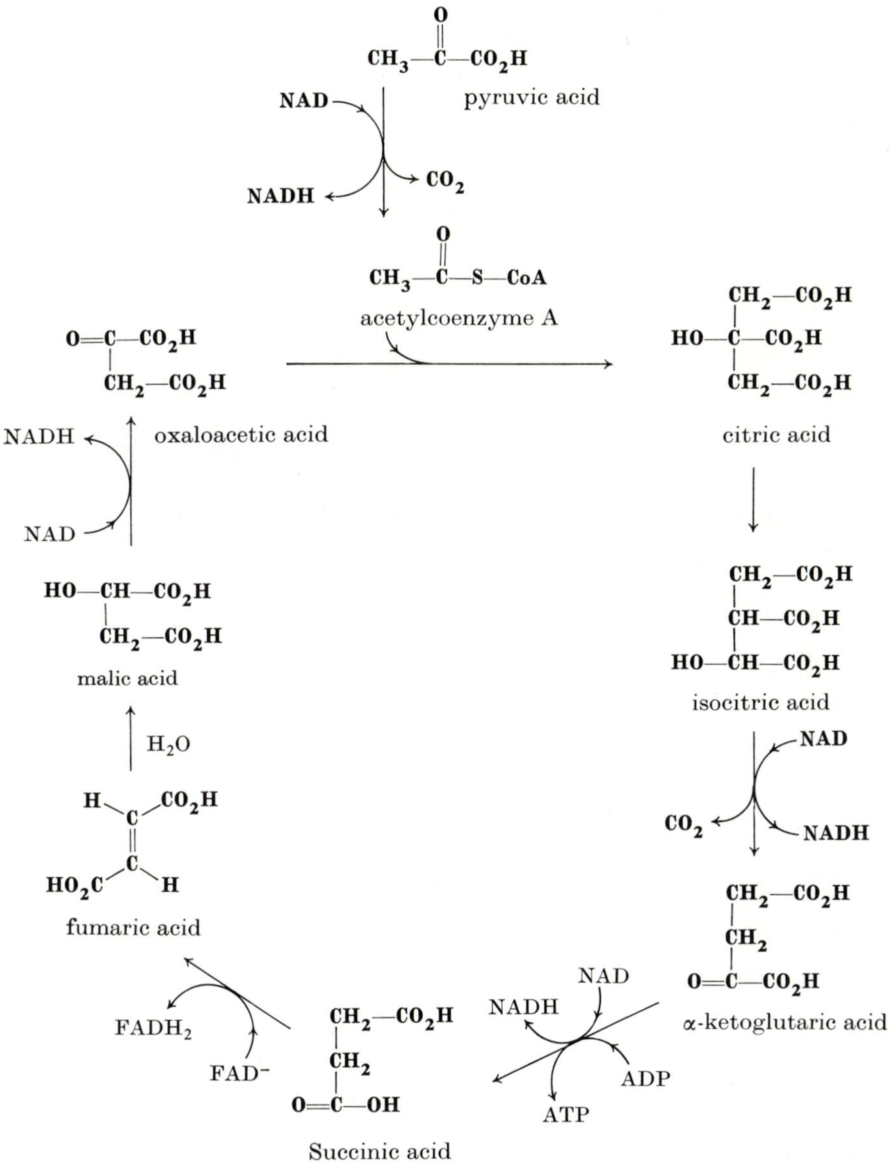

Fig. 16.5 The tricarboxylic acid cycle

to NADH. Further, a number of substances are produced which serve as useful nutrients in the biosyntheses of other biological factors. The cycle of reactions is given in Fig. 16.5.

Earlier we discussed the conversion of pyruvic acid into acetylcoenzyme A. The reaction of acetylcoenzyme A with oxaloacetic acid to yield citric acid is similar to an aldol addition to a carbonyl group (see Section 11.23).

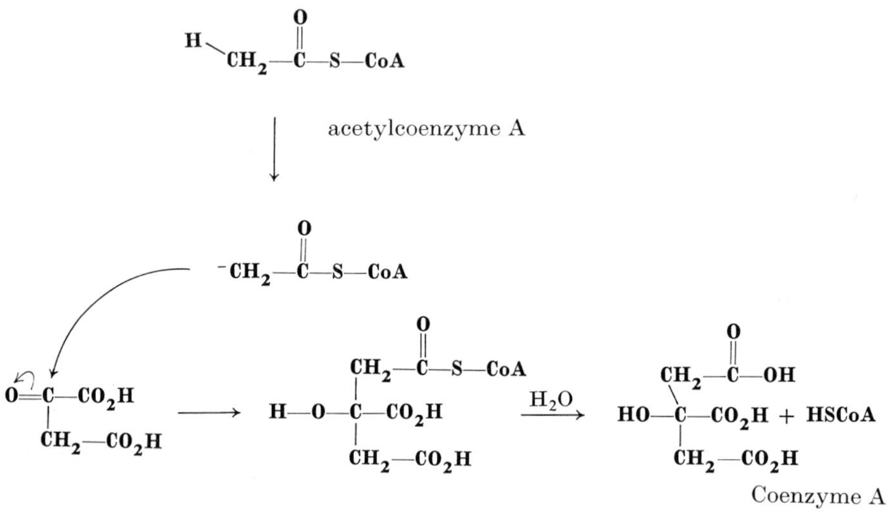

oxaloacetic acid citric acid

Note that anion formation in acetylcoenzyme A is made possible by resonance stabilization of the anion by the thioester carbonyl group.

Citric acid is isomerized to isocitric acid *via* an elimination and addition process which is enzymatically catalyzed.

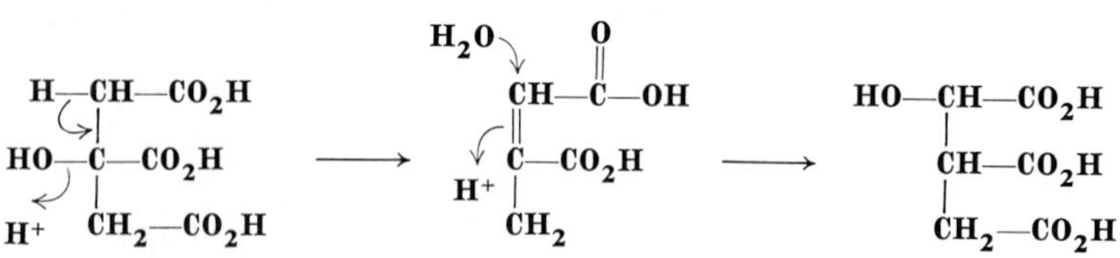

citric acid *cis* aconitic acid isocitric acid

Oxidation of isocitric acid by NAD then produces NADH and an intermediate ketotriacid. The latter decarboxylates to afford α-ketoglutaric acid:

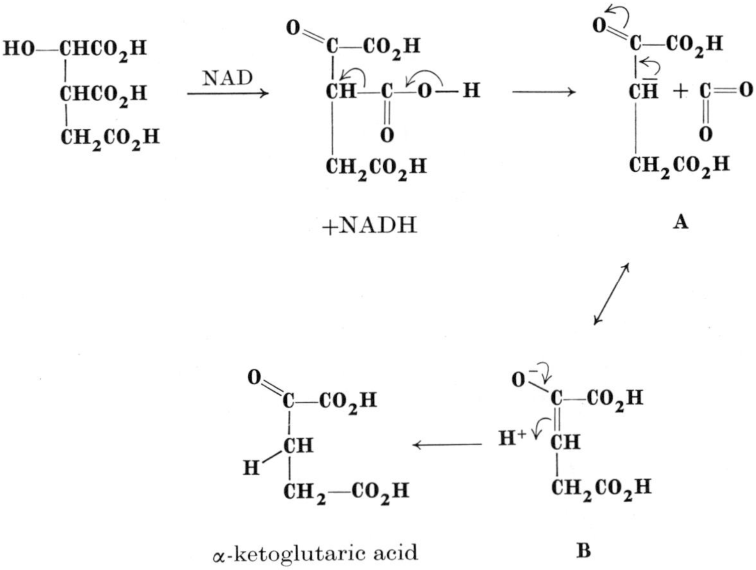

This decarboxylation of the keto-triacid is facilitated by the proximity of the ketone group to the carboxyl group that is lost. Thus, the anion, A, that is produced by the loss of carbon dioxide can be stabilized by the indicated resonance (A ↔ B).

α-Ketoglutaric acid is now transformed to succinoylcoenzyme A by a reaction sequence similar to that discussed for the thiamine-catalyzed conversion of pyruvic acid into acetylcoenzyme A. (The student should work out the mechanism for this.)

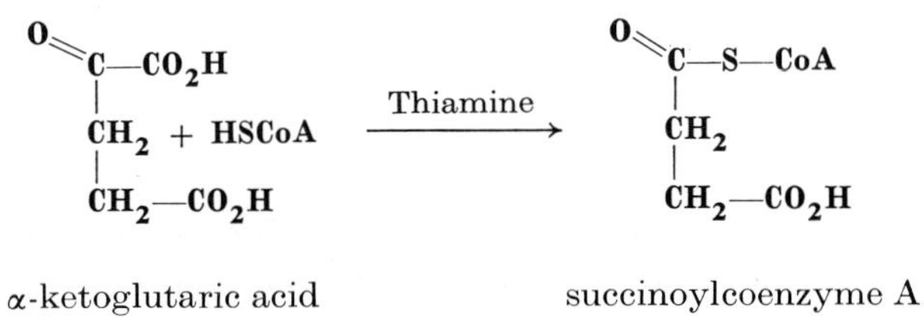

Succinoylcoenzyme A is converted into succinic acid. This is not a simple hydrolysis of the thioester. During the course of this reaction adenosinediphosphate (ADP) and phosphate combine to afford adenosinetriphosphate (ATP). We shall not discuss the mechanism of this transformation but encourage speculation by the student.

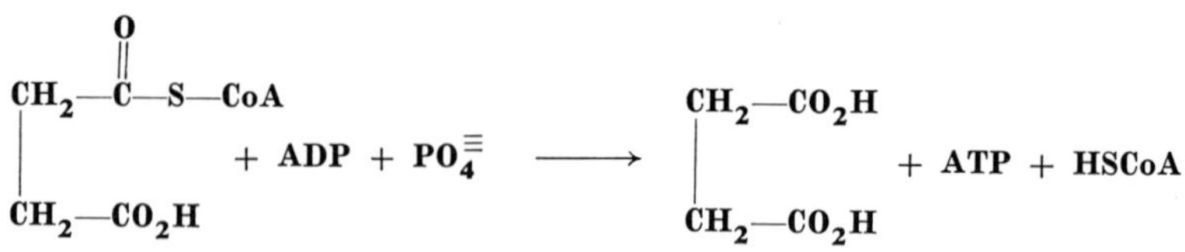

succinoylcoenzyme A succinic acid

An oxidation (dehydrogenation) of succinic acid by flavin adenine dinucleotide (FAD) produces fumaric acid. FAD serves as a biological oxidizing agent in much the same manner as does NAD (see Section 16.3). As succinic acid is oxidized to fumaric acid, FAD is reduced to $FADH_2$:

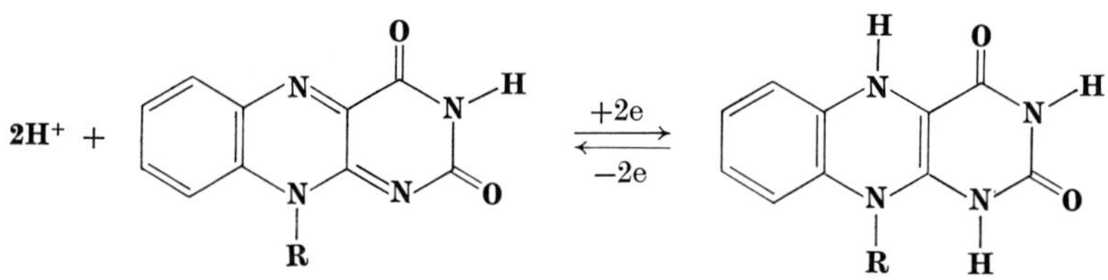

FAD $FADH_2$

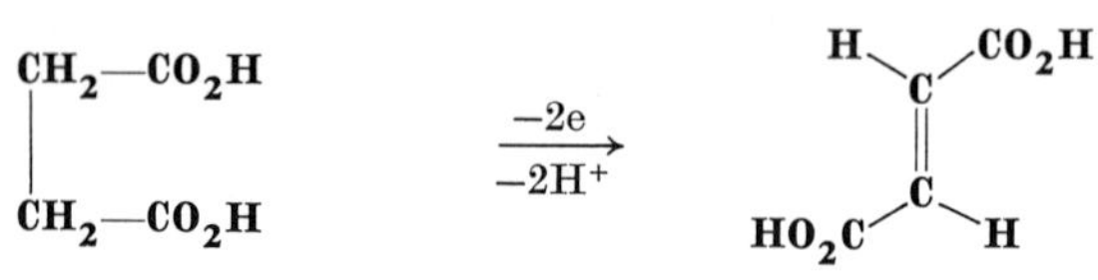

succinic acid fumaric acid

The cycle is completed with two final steps, the hydration of fumaric acid to malic acid and the oxidation of malic acid by NAD to oxaloacetic acid:

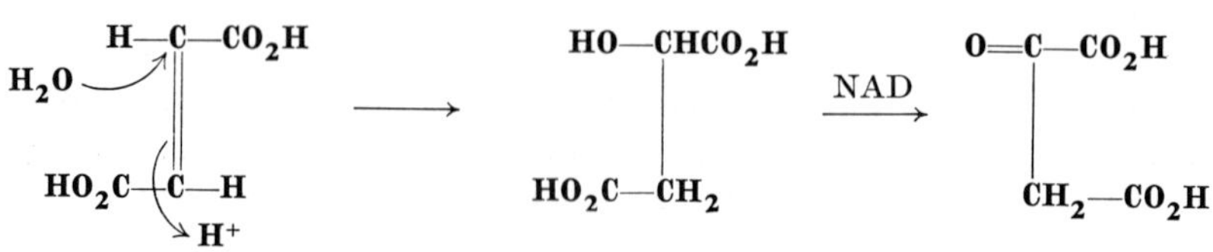

fumaric acid malic acid oxaloacetic acid

16.5 VITAMIN B_6—TRANSAMINATION

The term vitamin B_6 actually includes three agents which are interconvertible in biological systems: pyridoxine, pyridoxal, and pyridoxamine.

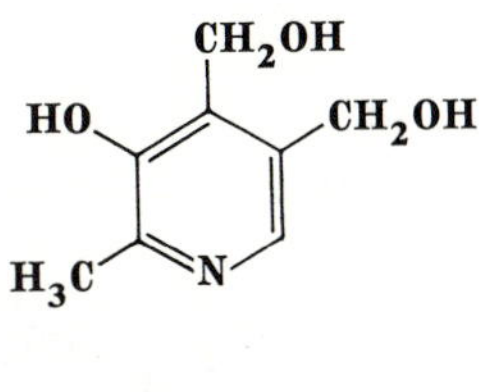

pyridoxine

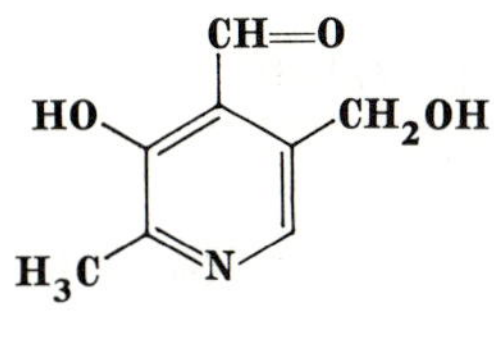

pyridoxal

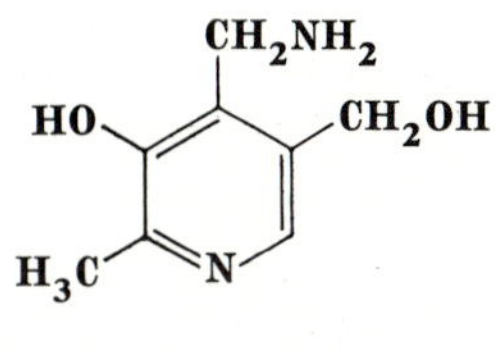

pyridoxamine

These agents as their phosphate esters serve as *cofactors* in amino acid chemistry; that is, they act in concert with enzymes to bring about biochemical transformations. One such reaction is the conversion of alanine to pyruvic acid.

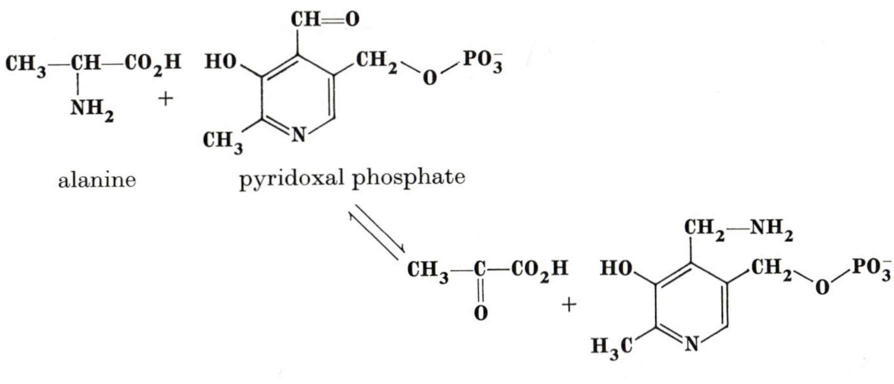

alanine pyridoxal phosphate

pyruvic acid pyridoxamine phosphate

By means of this transfer of the amino group (transamination), amino acids can provide intermediates for the tricarboxylic acid cycle, or conversely, they can be derived from reactants of the tricarboxylic acid cycle.

The mechanism of transamination relates to the addition-elimination process for carbonyl groups discussed in Section 11.12 and is illustrated in Fig. 16.6. The feature of this process not considered earlier (Section 11.12) is the tautomeric equilibrium between imine I and imine II shown in Fig. 16.6. In this reaction protonation of the imine carbon and loss of a proton from the amine carbon results in imine isomerization.

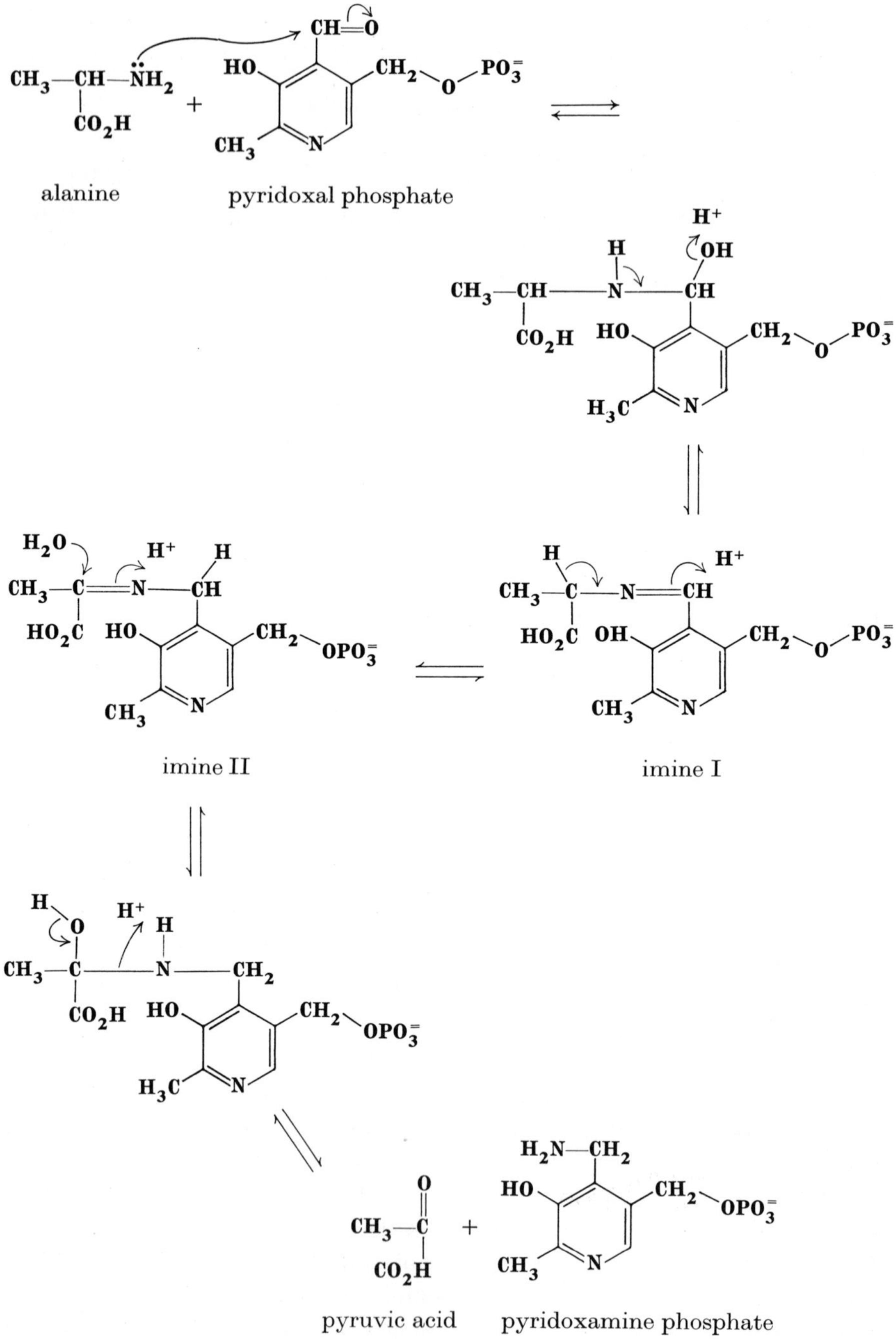

Fig. 16.6 Conversion of alanine to pyruvic acid

This mobile tautomerism in the pyridoximine system is utilized in the biosynthesis of serine from glycine (Fig. 16.7). As before, the first step in the

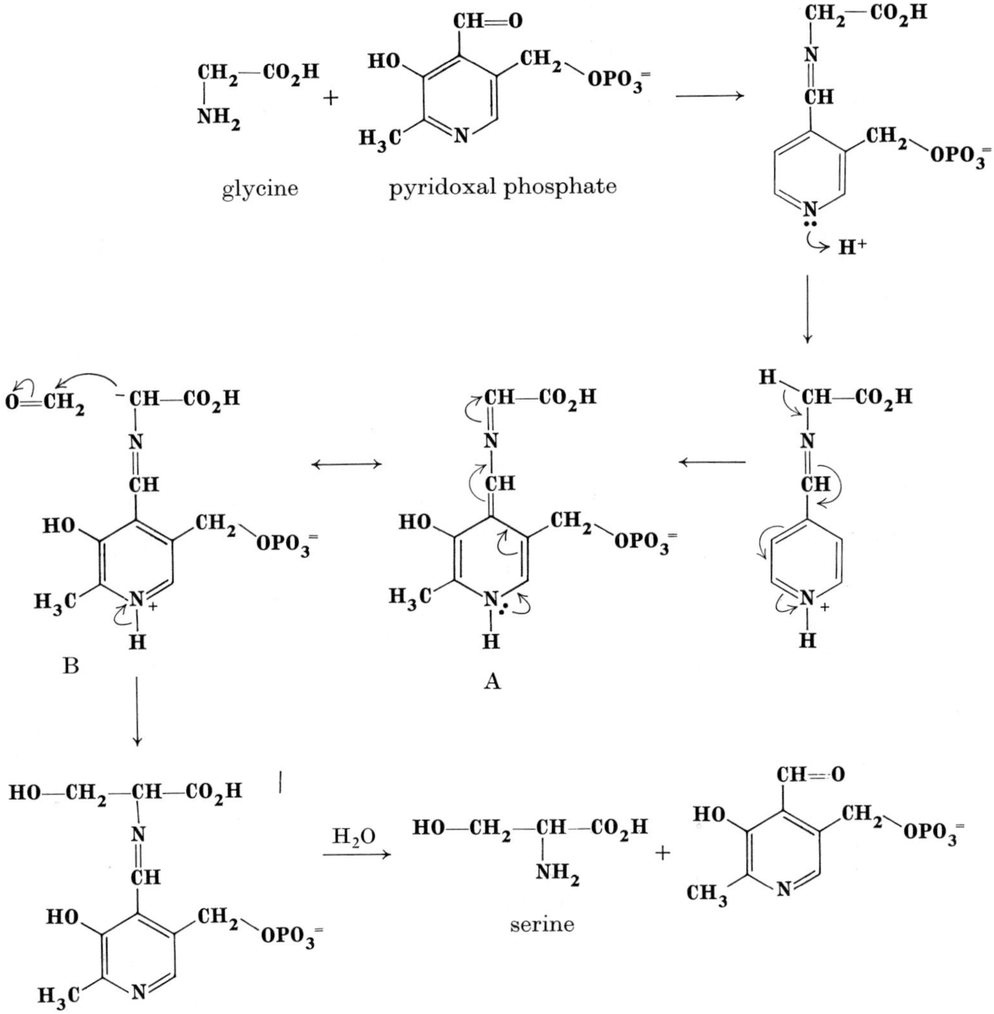

Fig. 16.7 Biosynthesis of serine

sequence is imine formation between the amino acid and pyridoxal phosphate. A tautomerization of this imine occurs through the pyridoxal ring to yield the enamine A. Resonance contributor B shows that anionic character is developed at the carbon adjacent to the carboxyl group. Aldol type addition of this anion to formaldehyde produces the imine of serine which, upon hydrolysis, gives serine and pyridoxal phosphate.

16.6 MECHANISM OF CHYMOTRYPSIN ACTION

One of the great chemical problems today is the elucidation of the mechanism of enzyme action. Probably no enzyme has attracted as much attention as chymotrypsin, a substance secreted by the pancreas for the purpose of catalyzing the breakdown of ingested protein:

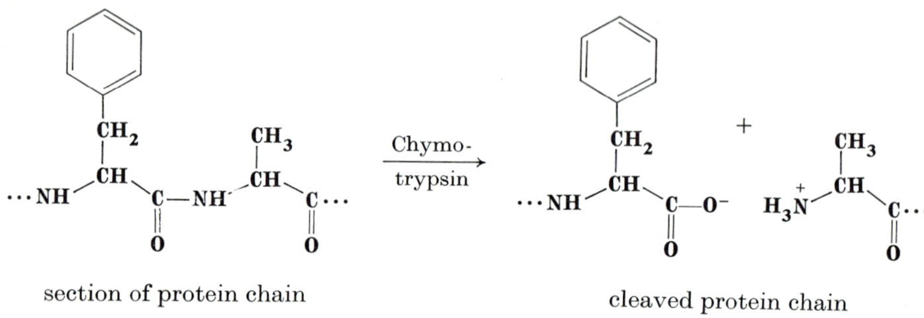

Chymotrypsin performs this feat at neutral pH and 37° with remarkable facility. If a chemist wanted to duplicate the rate of protein hydrolysis in a test tube, he would have to resort to concentrated acid or base at elevated temperatures. The question of how chymotrypsin catalyzes reactions of its substrates is therefore intriguing.

Chymotrypsin, like all enzymes, is a protein. It has 241 amino acid subunits, but for the purposes of understanding the mechanism only three, serine, histidine, and aspartic acid, need be discussed in detail.

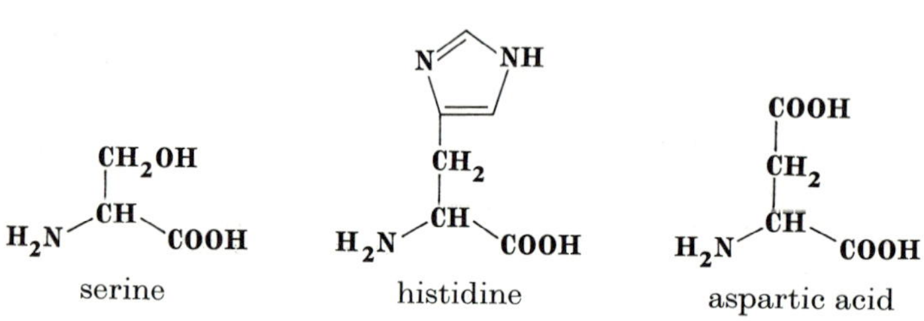

Chymotrypsin does not exist in solution in the form of long extended chains. Instead, the chains are coiled and twisted into a complicated but well-defined three-dimensional structure. The three-dimensional conformation is stabilized in part by attractive interactions between the amino acid side chains and in part by hydrogen bonding between peptide linkages. The three critical amino acid units mentioned above are by no means adjacent to each other in the amino acid sequence. However, entwining of the protein brings them into close proximity to one another. The region where the three amino acids congregate is called the *active site*, and it is here that chemical reaction occurs between substrate and enzyme. Compared to the size of the rest of the molecule, the active site is very small. Most of the other 238 amino acid units are present mainly to ensure that these three chemically important amino acids are properly oriented in space to carry out their catalytic function.

For a rapid reaction to occur between enzyme and substrate, the substrate must be correctly positioned at the active site relative to the enzyme's catalytic groups. Enzymes secure such an orientation by binding the substrate at the active site prior to reaction. The resulting enzyme-substrate complex is held together by weak attractive forces between the substrate and certain amino acid side chains. Enzymatic reactions can therefore be described by a simple equation (where E, S, ES, and P are enzyme, substrate, complex, and product, respectively):

$$E + S \overset{K}{\rightleftharpoons} ES \xrightarrow{k} E + P$$

Acetamide is hydrolyzed very slowly by chymotrypsin because acetamide binds poorly to the active site. An aromatic group greatly assists the binding of a substrate; thus the amide of phenylalanine, for example, hydrolyzes much more rapidly.

How do we know that histidine, serine, and aspartic acid participate in a catalyzed hydrolysis at the active site? One line of evidence supporting the involvement of the imidazole ring of histidine entails chemical alteration of the ring. When chymotrypsin is treated with the alkylating agent shown below, enzymatic activity ceases. Analysis of the amino acid content of the chemically-treated enzyme shows that one of the histidines has been destroyed.

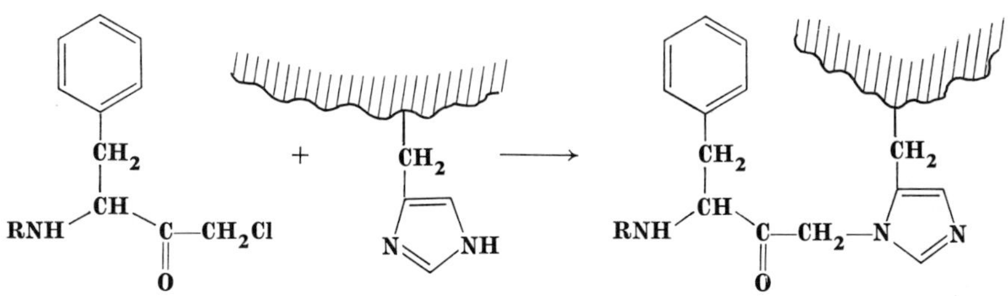

Apparently, histidine is an essential entity at the active site, and alkylation destroys its ability to function. Kinetic data on the hydrolyses of amino acid esters (which are also catalyzed by chymotrypsin) support the involvement of histidine. Thus, the rate of chymotrypsin-catalyzed hydrolysis of an ethyl ester derivative of tryptophan shows a sigmoidal dependence on pH (as in a titration curve):

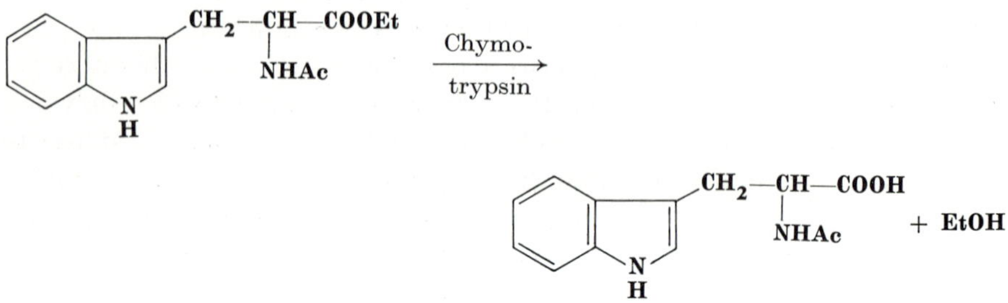

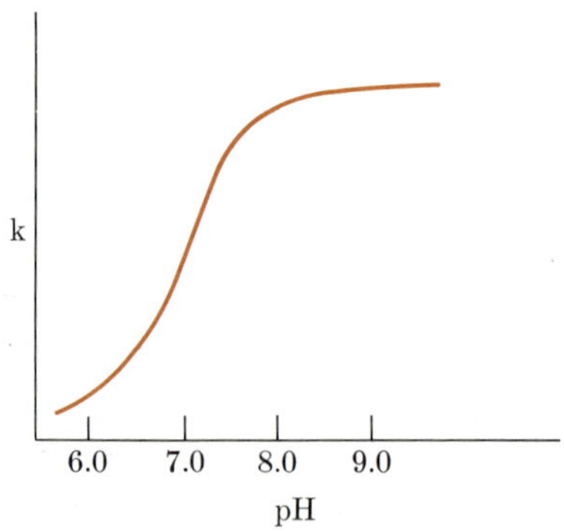

The flex point of the rate curve appears at pH = 7.0. Since no amino acid side chain other than the imidazole of histidine has a pK_a near 7, one may conclude that an unprotonated histidine is required for catalysis.

The evidence for serine is also convincing. Chymotrypsin and diisopropylphosphofluoridate (DFP) react rapidly with a 1:1 stoichiometry to yield an inert enzyme. Only one of the chymotrypsin's serines is derivatized by this reaction:

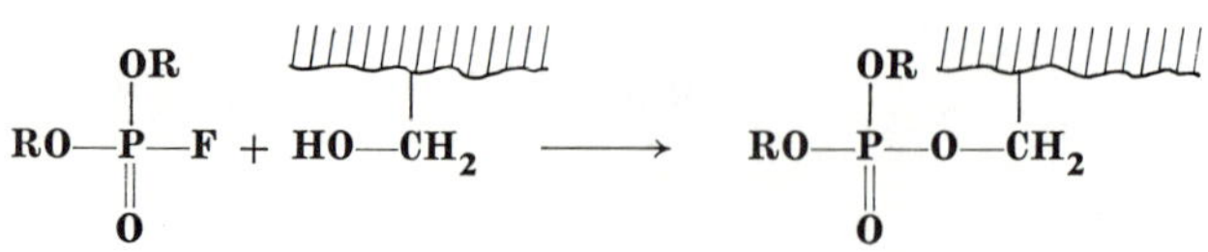

Enzymatic activity can be restored by removing the blocking group with a good nucleophile. DFP does not react readily with a denatured enzyme (an enzyme whose three-dimensional conformation has been disturbed by heat or the presence of urea). These results, along with those from many other experiments, prove the presence of serine at the active site.

Both the complete amino acid sequence of chymotrypsin and the three-dimensional (tertiary) structure of the protein are known as a result of an X-ray diffraction study. The protein resembles a spherical pile of thread. The active site is a shallow depression in the surface containing serine, histidine, and aspartic acid. These amino acids are adjacent to a nonpolar pocket where substrates are thought to bind to the enzyme. There is good agreement between the conclusions derived from X-ray data and from chemical experiments regarding the nature of the active site:

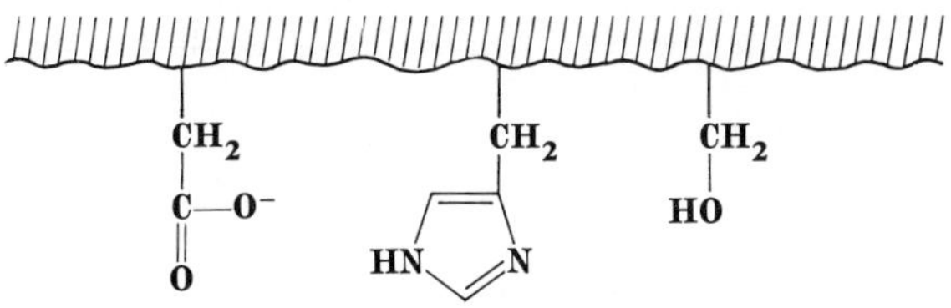

active site of chymotrypsin

Having established *which* groups participate in the catalysis, we must now consider *how* they effect catalysis. One of the most important findings in this regard is formation of an *acyl-enzyme intermediate*:

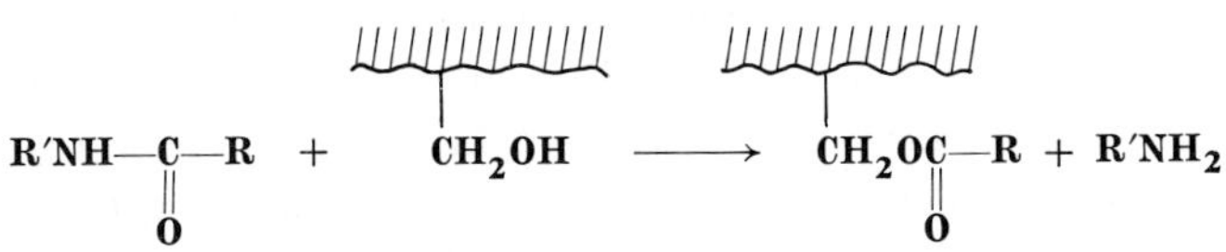

Hydrolysis of the acyl-enzyme intermediate produces the carboxylic acid and regenerates the enzyme:

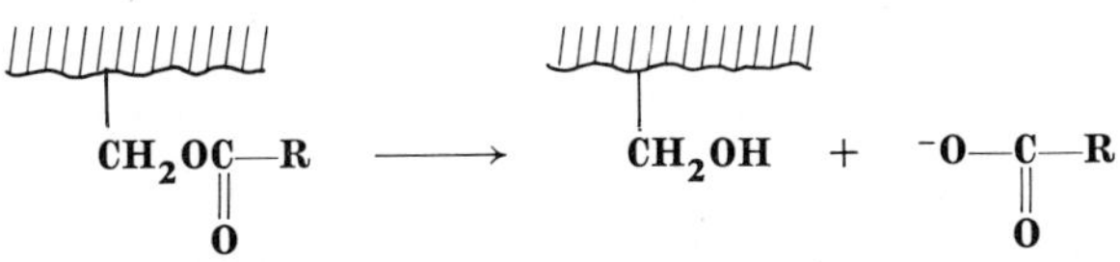

The net reaction is hydrolysis of the amide:

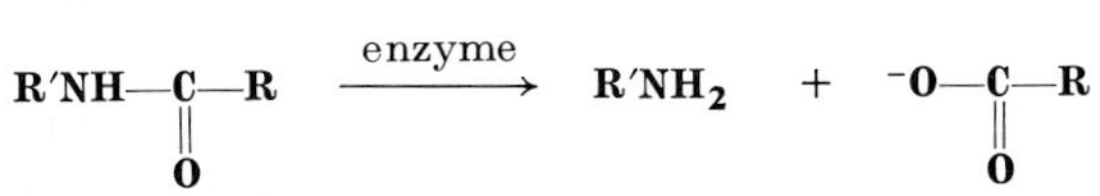

The following experiment is one of many which supports enzyme acylation during the catalyzed hydrolysis. When chymotrypsin is added to a solution of excess N-acetyltryptophan *p*-nitrophenyl ester (shown below), a sudden "burst" of yellow *p*-nitrophenol is observed spectrophotometrically (400 nm in the UV). This is followed by a much slower formation of additional *p*-nitrophenol. A plausible explanation is that chymotrypsin is rapidly and totally acylated by the excess substrate, causing a quick release of *p*-nitrophenol. As the acyl-enzyme slowly deacylates, free enzyme is liberated which subsequently reacts with substrate to give more *p*-nitrophenol. Of course, this latter reaction can proceed no faster than the relatively slow hydrolysis of the acyl-enzyme. Obviously, the experiment says nothing about *where* the enzyme is acylated. However, numerous other data (including the UV spectrum of the acyl-enzyme) point strongly toward an acylated serine. At one time it was thought that the imidazole ring of histidine is acylated, but this idea is no longer accepted.

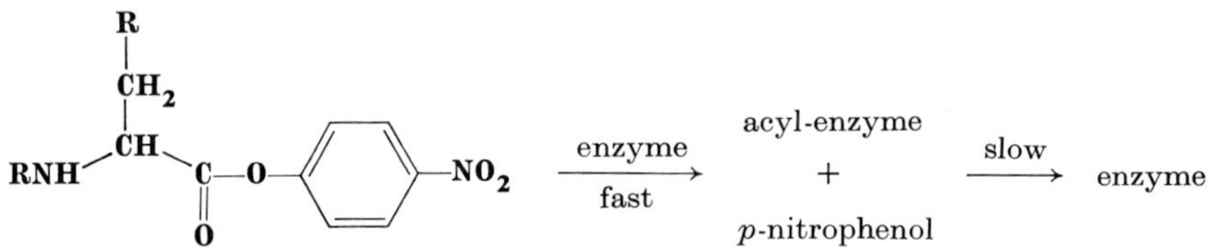

The most important question which we have not yet answered concerns the role of histidine and aspartate. Since an alcohol by itself reacts extremely slowly with amides, the imidazole ring must accelerate formation of the acyl-enzyme. The mechanism of catalysis currently believed correct is shown in Fig. 16.8. Imidazole removes a proton from the serine hydroxyl group as the hydroxyl attacks the amide carbonyl. Proton removal would greatly accelerate the attack, judging for example, from the greater nucleophilicity of methoxide ion relative to methanol. The steady-state tetrahedral intermediate then collapses to the acyl-enzyme by ejecting amine. The amine accepts a proton from the protonated imidazole ring so that the amine can depart as RNH_2, not as RNH^-. Hydrolysis of the acyl-enzyme intermediate occurs by much the same mechanism as its formation, the major difference being that water (not serine) is the nucleophilic agent.

The role of aspartate has not been established. Perhaps it merely holds the imidazole ring (*via* hydrogen bonding) in the correct position for interacting with the serine hydroxyl group, or it may enhance the basic properties of the imidazole.

Although the mechanism in Fig. 16.7 has not been proved, a considerable variety of data is consistent with it. For example, chymotrypsin-catalyzed reactions display a solvent isotope effect; *i.e.*, the reactions are *faster* in water than in deuterium oxide (D_2O). Such behavior is typical of reactions in which a bond to a hydrogen (or deuterium) is broken in a rate-determining step. The mechanism in 16.8 is consistent with the observed solvent isotope effect since a bond between the serine proton and oxygen is cleaved. In deuterium oxide, the serine hydroxyl exchanges the proton for a deuterium, which then transfers less easily to the imidazole in the first step of the mechanism.

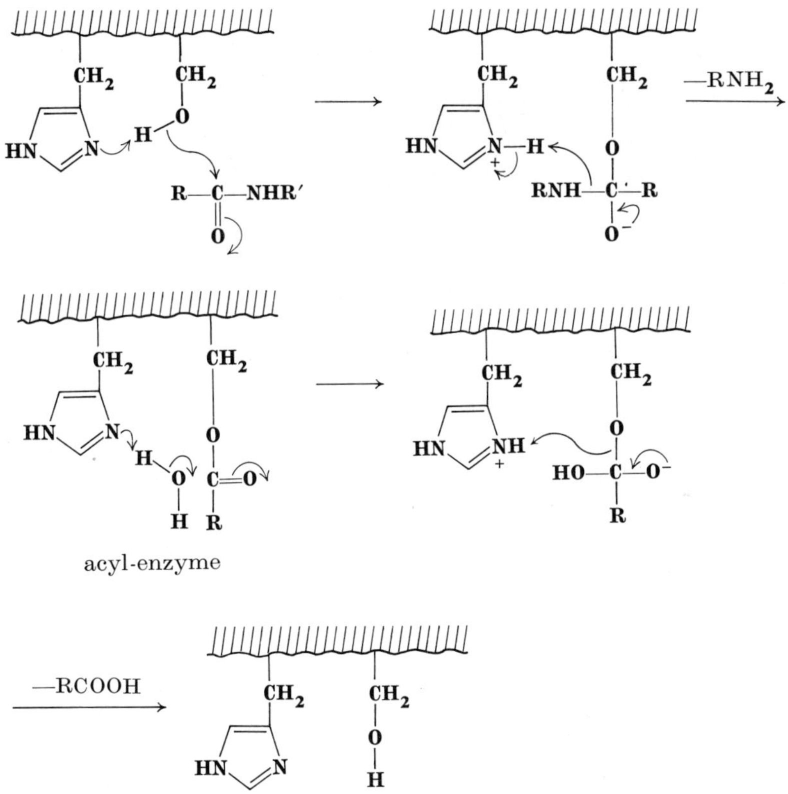

Fig. 16.8 The mechanism of chymotrypsin action

One important approach to the study of enzyme mechanisms not previously mentioned entails the synthesis of enzyme "models": small organic molecules which simulate in some fashion the action of an enzyme. For example, model catalysts which bind substrates prior to catalysis have been synthesized. Thus far no one has succeeded in constructing a model which catalyzes the hydrolysis of amides with anywhere near the efficiency of chymotrypsin.

APPENDIX
NOMENCLATURE

This section will introduce the reader to the principles of chemical nomenclature. The rules given here will allow the naming of only simple structures. For a more complete treatment of the rules of nomenclature, the reader is referred to any of the self-teaching texts on this topic, such as *Organic Nomenclature: A Programmed Introduction*, by J. G. Traynham, Prentice-Hall, Englewood Cliffs, N.J., 1966.

I. ALKANES

A. Saturated hydrocarbons

Names of the straight-chain hydrocarbons through C_{10}

CH_4	methane
$CH_3—CH_3$	ethane
$CH_3—CH_2—CH_3$	propane
$CH_3—CH_2—CH_2—CH_3$	*n*-butane
$CH_3—CH_2—CH_2—CH_2—CH_3$	*n*-pentane
$CH_3—CH_2—CH_2—CH_2—CH_2—CH_3$	*n*-hexane
$CH_3—CH_2—CH_2—CH_2—CH_2—CH_2—CH_3$	*n*-heptane
$CH_3—CH_2—CH_2—CH_2—CH_2—CH_2—CH_2—CH_3$	*n*-octane
$CH_3—CH_2—CH_2—CH_2—CH_2—CH_2—CH_2—CH_2—CH_3$	*n*-nonane
$CH_3—CH_2—CH_2—CH_2—CH_2—CH_2—CH_2—CH_2—CH_2—CH_3$	*n*-decane

"Heptane" refers to any saturated hydrocarbon of seven carbons. If the hydrocarbon is a straight chain with no branching, the prefix *n* is used. Cyclic hydrocarbons are named after the number of carbons in the ring and given the prefix "cyclo":

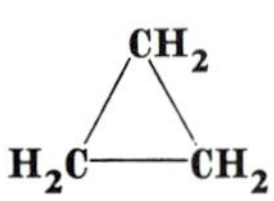

cyclopropane

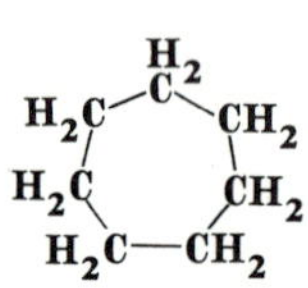

cycloheptane

B. Common names of some alkyl and alkenyl groups

Names of hydrocarbon groups used as substituents

CH_3—	methyl
CH_3CH_2—	ethyl
$CH_3CH_2CH_2$—	*n*-propyl
$CH_3CH(CH_3)$—	isopropyl (*i*-propyl)
$CH_3CH_2CH_2CH_2$—	*n*-butyl
$CH_3CH_2CH(—)CH_3$	secondary butyl (*s*-butyl)
$CH_3CH(CH_3)CH_2$—	isobutyl (*i*-butyl)
$CH_3C(CH_3)_2$—	tertiary butyl (*t*-butyl)
$CH_3CH_2CH_2CH_2CH_2$—	*n*-amyl
$CH_3CH(CH_3)CH_2CH_2$—	isoamyl (*i*-amyl)
$CH_3C(CH_3)(—)CH_2CH_3$	tertiary amyl (*t*-amyl)
$CH_3C(CH_3)_2CH_2$—	neopentyl
CH_2=CH—	vinyl
CH_3CH=	ethylidene
CH_2=$CHCH_2$—	allyl
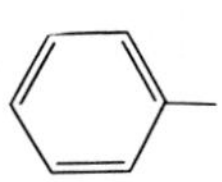	phenyl

C. Geneva or IUPAC system of nomenclature

1) A compound is named as a derivative of the *longest* carbon chain in the molecule. If there are two chains of equal length, choose that chain which has the greatest number of substituents. The name of a saturated hydrocarbon must always end in "ane".
2) The carbons of the chain selected above are numbered consecutively in such a way that when the numbers of the carbons bearing substituents are added together, the smallest possible sum is achieved.

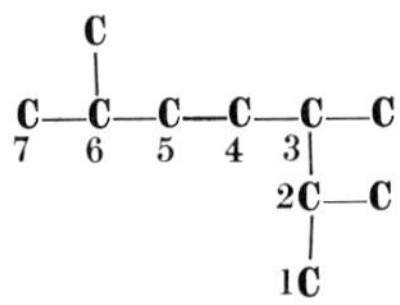

preferred over

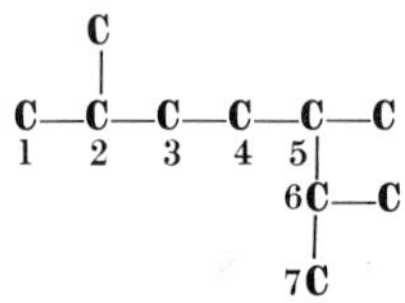

substituents on 2, 3, and 6 substituents on 2, 5, and 6

3) Alkyl groups attached to the main chain are identified by their common names (methyl, *s*-butyl, etc.), and they are located by citing the number of the carbon atom to which the group is attached. In case several groups of the same type are present, a prefix indicating the number of such groups is used (di, tri, tetra, penta, and hexa for one, two, three, four, five, and six groups, respectively). Thus, the above compound is named 2,3,6-trimethylheptane. Note that the numbers are separated from each other by commas, and the last number is joined to the name by a hyphen. Halogen substituents are designated fluoro, chloro, bromo, and iodo:

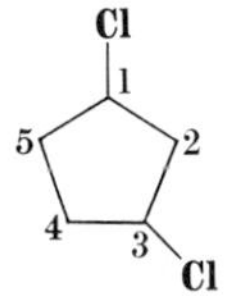

1,3-dichlorocyclopentane

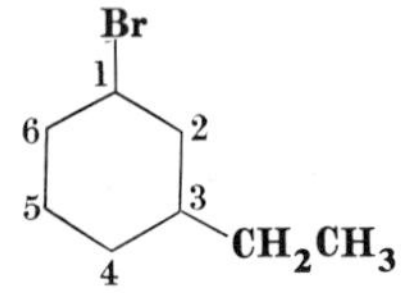

1-bromo-3-ethylcyclohexane

In the cyclohexane derivative, the bromo is mentioned first because alphabetically "bromo" comes before "ethyl."
4) If a substituent does not have a simple common name, a subnumber must be assigned to the carbons in the substituent, starting with the carbon directly attached to the main chain as number 1. The substituent is then named as though it were a compound, except that the name ends in "yl". This entire name is inserted in parentheses following the number of the carbon in the main chain to which the substituent is attached.

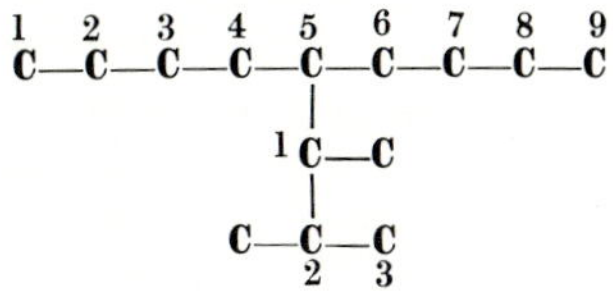

5-(1,2-dimethylpropyl)nonane

5) The prefixes *cis* and *trans* are used when it is desirable to specify the stereochemistry of cyclic compounds.

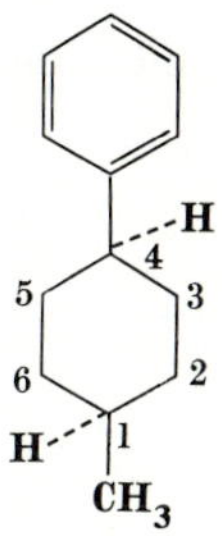

cis-1-methyl-4-phenylcyclohexane

II. ALKENES AND ALKYNES

An unsaturated compound is named as a derivative of the longest carbon-carbon chain containing the unsaturated function. The rules given above for naming alkanes also apply to unsaturated hydrocarbons. The following additional rules should be noted.

1) The parent chain is numbered to designate the unsaturation with the *lowest* possible numbers. The suffix "ene" is used for olefins and "yne" for acetylenes.

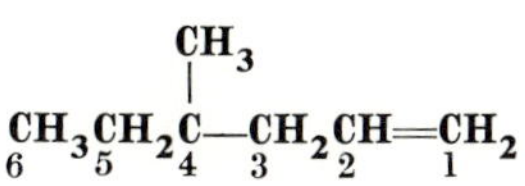

4-methyl-1-hexene

4 3 2 1
$CH_3CH_2CH{=}CH_2$

1-butene

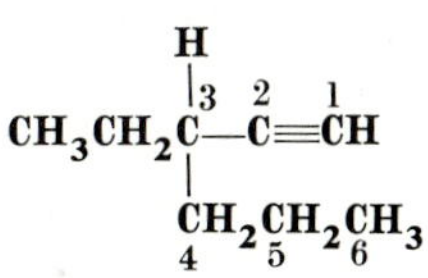

3-ethyl-1-hexyne

The position of the unsaturated center is always indicated by citing the *lower* of the two numbers of the carbons between which the unsaturated center is situated:

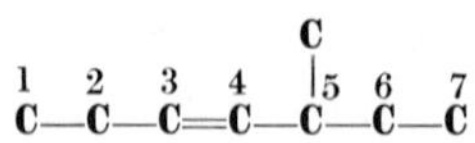

5-methyl-3-heptene
(not 5-methyl-4-heptene
or 3-methyl-4-heptene)

2) The stereochemistry of substituents on the double bond is designated by the prefixes *cis* and *trans*:

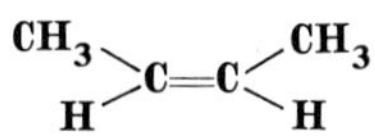

cis-2-butene

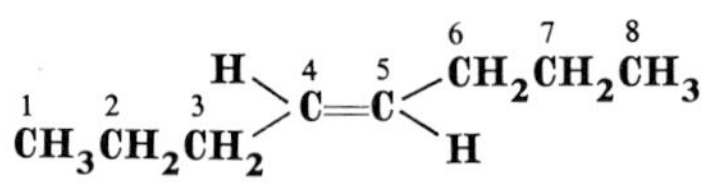

trans-4-octene

The *cis-trans* nomenclature is satisfactory for disubstituted olefins, but for trisubstituted and tetrasubstituted alkenes the "EZ" convention is used. The rules are as follows:

a) Assign priority (as in the R and S convention, given in Section 5.5) to the two groups on one of the olefinic carbons; repeat for the neighboring olefinic carbon.
b) If the two higher-priority groups are on the same side (*cis*), the compound is designated Z (for German *zusammen*). If the two higher-priority groups are *trans*, the compound is designated E (for German *entgegen*).

Two examples are shown below.

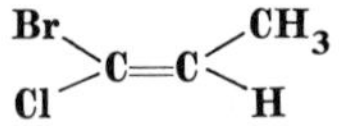

Z-1-bromo-1-chloropropene

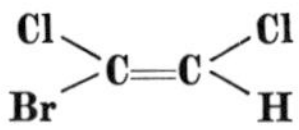

E-1-bromo-1,2-dichloroethene

3) Substituted acetylenes may also be named as derivatives of acetylene by denoting the substituents on the carbon-carbon triple bond:

$CH_3C{\equiv}CH$

methylacetylene

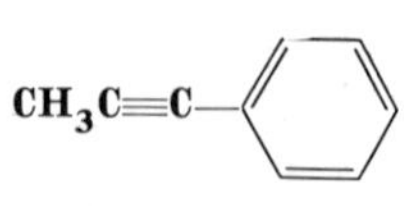

methylphenylacetylene

III. ALCOHOLS

1) The alcohol is named as a derivative of the longest carbon-carbon chain containing the hydroxyl function. This chain is numbered so that the carbon bearing the alcohol group has the lowest possible number. Substituents are designated as discussed above, and the name bears the suffix "ol":

CH_3CH_2OH

ethanol

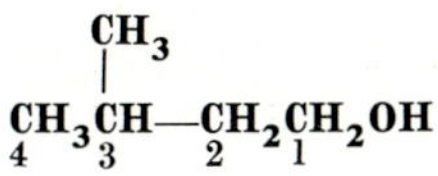

3-methyl-1-butanol

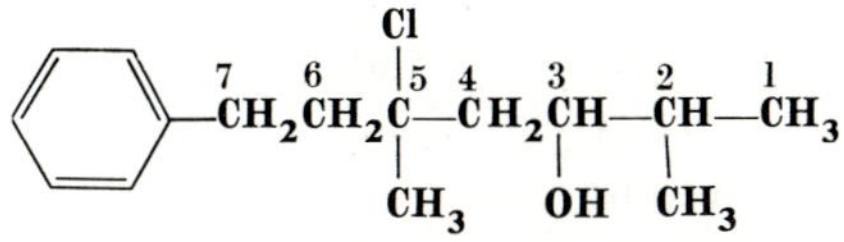

2,5-dimethyl-5-chloro-7-phenyl-3-heptanol

Derived system. The grouping —C—OH is known as the carbinol grouping. Alcohols may be named as substituted derivatives of this grouping by using the common names of alkyl and alkenyl groups, as given in Section IB, to denote the substituents of the carbinol group:

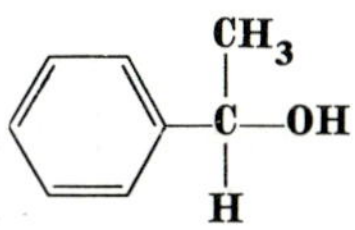

methylphenylcarbinol

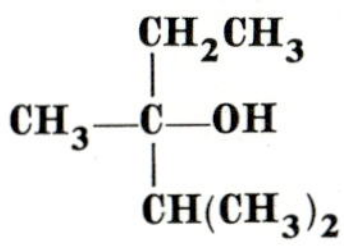

methylethylisopropylcarbinol

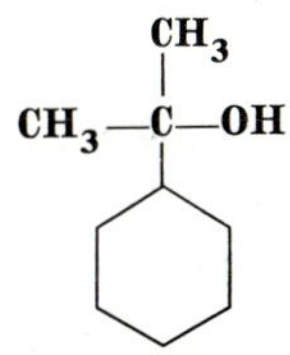

dimethylcyclohexylcarbinol

2) The alcohol or OH substituent is called a "hydroxy" group:

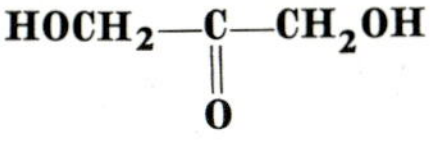

dihydroxyacetone

IV. ALDEHYDES

1) Aldehydes are named after the longest hydrocarbon chain containing the aldehyde function. The chain is always numbered, with the aldehyde group carbon as 1. Substituents and their location are given in the usual manner. The name bears the suffix "al."

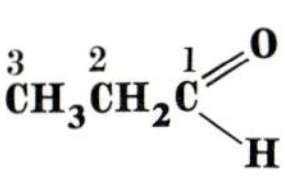

propanal

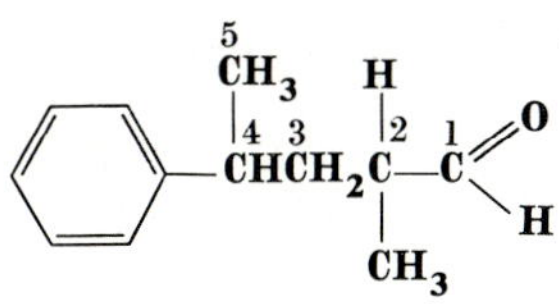

2-methyl-4-phenylpentanal

If, in addition to the aldehyde group, there is a double bond, the compound is named after the longest olefin chain containing the group.

$$\overset{5}{CH_3}\overset{4}{C}H\overset{3}{C}H_2\overset{2}{C}H{=}\overset{1}{C}HCHO$$
$$|$$
$$CH_3$$

5-methyl-2-hexenal

2) Aldehydes are also given "common" names based on the common names of carboxylic acids:

$HCOOH$

formic acid

CH_3COOH

acetic acid

CH_3CH_2COOH

propionic acid

$CH_3CH_2CH_2COOH$

butyric acid

$$CH_3CHCOOH$$
$$|$$
$$CH_3$$

isobutyric acid

$CH_3(CH_2)_3COOH$

valeric acid

To name an aldehyde, the suffix "ic" is dropped from the name of the acid and "aldehyde" is added in its place:

CH_3CH_2CHO

propionaldehyde

The location of a substituent is indicated by the Greek letters α, β, and γ (corresponding to the 2, 3, and 4 positions, respectively):

$$\overset{\gamma}{CH_3}\overset{\beta}{CH_2}\overset{\alpha}{CH}{-}CHO$$
$$|$$
$$Cl$$

α-chlorobutyraldehyde
(or 2-chlorobutanal)

$$HO\overset{\beta}{CH_2}\overset{\alpha}{CH_2}CHO$$

β-hydroxypropionaldehyde
(or 3-hydroxypropanal)

V. KETONES

1) Ketones are named after the longest carbon-carbon chain containing the ketone carbonyl group. The chain is numbered so that the ketone carbonyl carbon is assigned the lowest possible number. The substituents and their location are indicated in the usual manner. The name bears the suffix "one."

$$\underset{1}{CH_3}{-}\underset{2}{\overset{\overset{O}{\|}}{C}}{-}\underset{3}{CH_2}\underset{4}{CH_3}$$

2-butanone

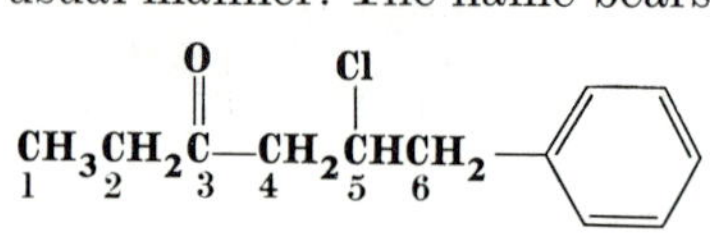

5-chloro-6-phenyl-3-hexanone

2) Ketones may also be named by denoting the two substituents on the carbonyl group using the names given in Section I.B for these substituents.

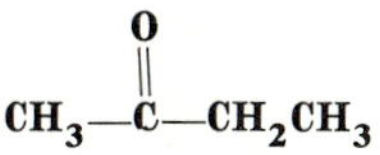

methyl ethyl ketone

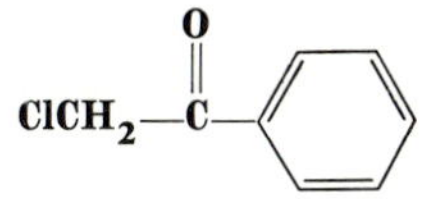

chloromethyl phenyl ketone

3) Still another designation for the location of substituents relative to the carbonyl group, which is similar to that given in Section IV.B for aldehydes, uses α, β, and γ. The carbons attached directly to the carbonyl carbon are designated as the alpha positions, the next carbons, the beta positions, and so on. Primed Greek letters (α', β', etc.) are used when both sides of the carbonyl group are substituted.

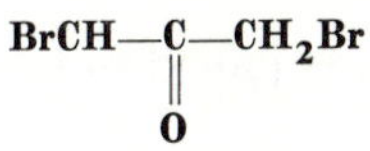

α,α'-dibromoacetone

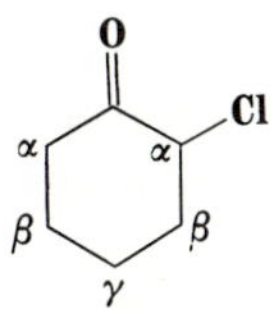

α-chlorocyclohexanone
or 2-chlorocyclohexanone

VI. AMINES

Amines are normally named after the substituents on the amine nitrogen and given the suffix "amine."

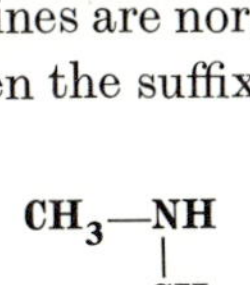

dimethylamine

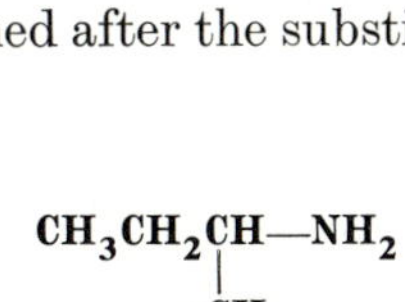

s-butylamine

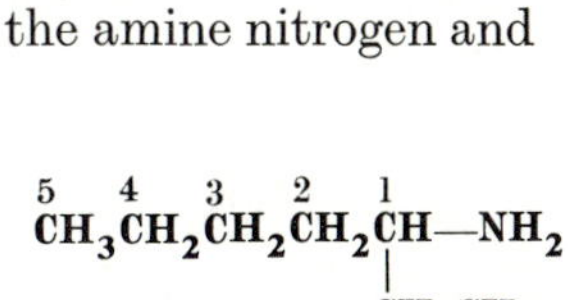

1-ethylpentylamine

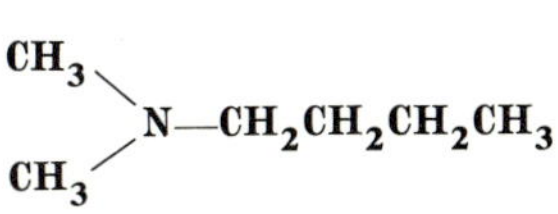

N,N-dimethyl-*n*-butylamine

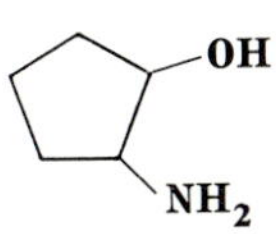

2-aminocyclopentanol
or 2-hydroxylcyclopentylamine

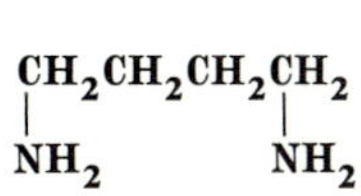

1,4-butadiamine

The letter N indicates that a substituent is on the nitrogen. If the substituent is an amine group, the name "amino" is used. Primary, secondary, and tertiary amines have one, two, and three groups, respectively, on the nitrogen. Thus, dimethylamine is called a secondary amine.

VII. ETHERS

Ethers are named in a manner similar to amines except that the suffix "ether" is used.

$CH_3—O—CH_3$

dimethyl ether

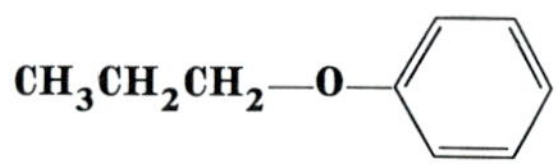

n-propyl phenyl ether

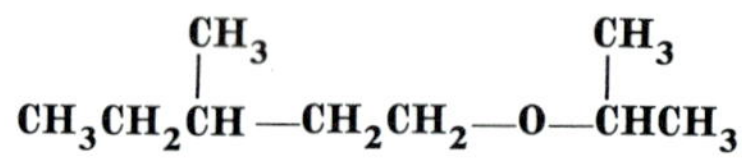

3-methylpentylisopropyl ether

Note that "ether" is a separate word, unlike "amine," which is joined to the word which designates the substituents (see Section VI).

VIII. CARBOXYLIC ACIDS

The common names of some acids are given in Section IV.2. Carboxylic acids are named after the longest carbon-carbon chain containing the acid function; the "e" of the alkane is replaced by "oic."

$CH_3CH_2CH_2COOH$

butanoic acid
(or butyric acid)

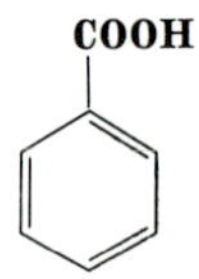

benzoic acid

The chain is always numbered so that the carboxyl group has position 1.

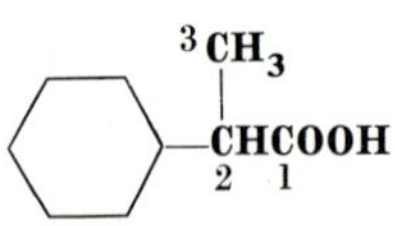

2-cyclohexylpropanoic acid

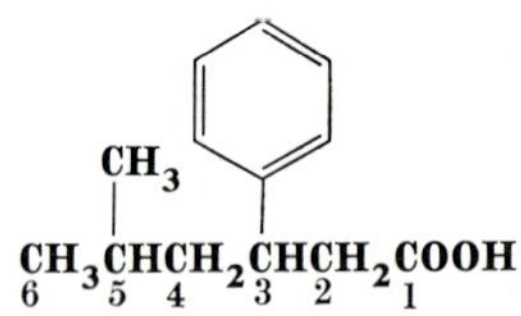

5-methyl-3-phenylhexanoic acid

IX. ESTERS

Esters are named after the alcohol and acid from which they may be prepared. The alcohol portion is mentioned first, followed by the acid name which includes the suffix "oate" in place of the "ic."

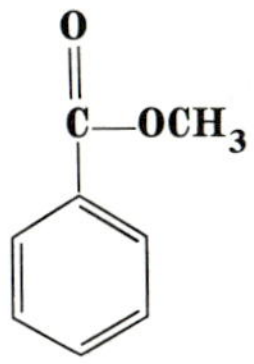

methyl benzoate

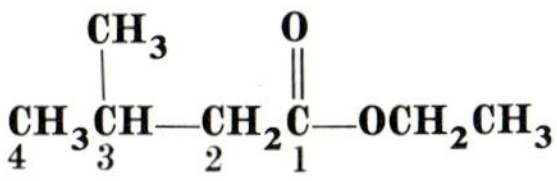

ethyl 3-methylbutanoate

X. AMIDES

Amides are named by first citing the groups that substitute the amide nitrogen, denoting this substitution with an N, and then naming the acid portion of the amide and adding the suffix "amide."

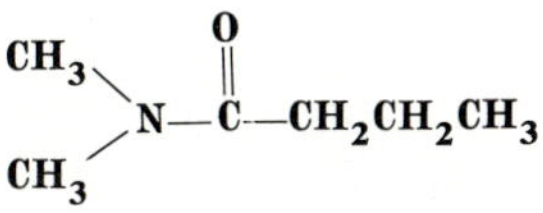

N,N-dimethylbutanamide
or N,N-dimethylbutyramide

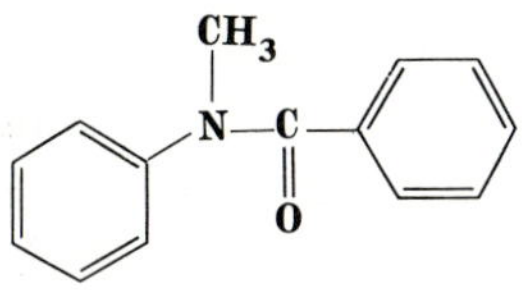

N-methyl-N-phenylbenzamide

XI. AROMATICS

1) Substituted benzenes are named by numbering the carbons on the ring from one to six, and naming and locating the substituents in the usual manner.

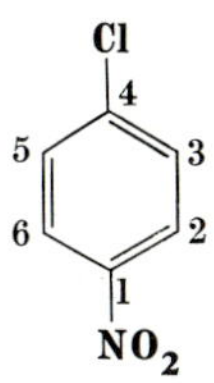

4-chloronitrobenzene
(or 4-nitrochlorobenzene)

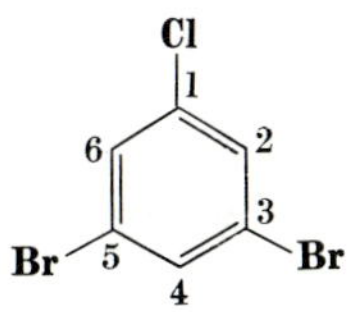

3,5-dibromochlorobenzene

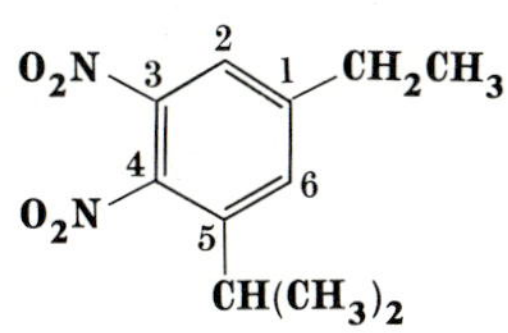

3,4-dinitro-
5-isopropylethylbenzene

2) The letters *o*, *m*, and *p* (for *ortho*, *meta*, and *para*) may be used to denote 1-2, 1-3, and 1-4 disubstitution, respectively.

p-dichlorobenzene

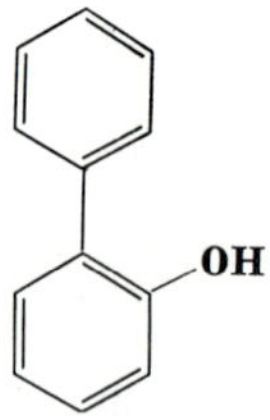

o-hydroxydiphenyl

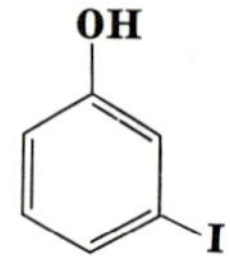

m-iodophenol

XII. NAMES OF MISCELLANEOUS SUBSTITUENTS OR "RADICALS"

$H_2N—$	amino
$HO—$	hydroxy
$CH_3C(=O)—$	acetyl
$CH_3C(=O)—NH—$	acetamido
$HOOC—$	carboxy
$CH_3O—C(=O)—$	carbomethoxy
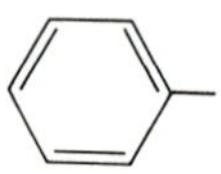	phenyl
$CH_2{=}CH—$	vinyl
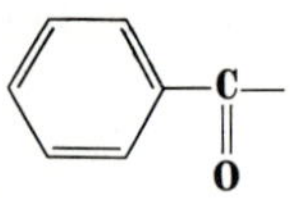	benzoyl
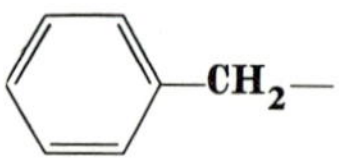	benzyl
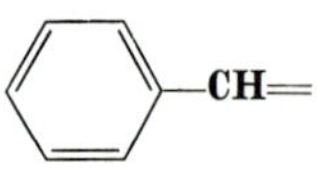	benzal
$N{\equiv}C—$	cyano
F—, Cl—, Br—, I—	fluoro, chloro, bromo, iodo

HS—	mercapto
O_2N—	nitro
CH_3O—	methoxy
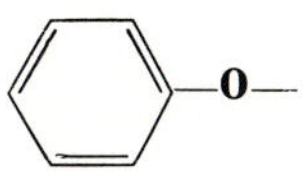	phenoxy
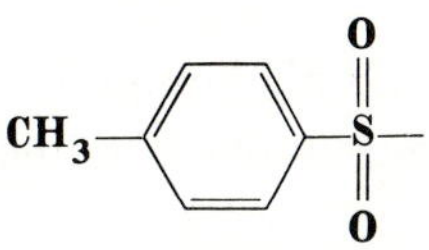	tosyl (toluenesulfonyl)

XIII. THE FOLLOWING COMPOUNDS AND THEIR NOMENCLATURE ILLUSTRATE THE ABOVE RULES

1) $C_2H_5CH(OH)CH(CH_3)_2$

2-methyl-3-pentanol

The longest chain is five carbons long; hence the compound is a pentane derivative. In order to locate the methyl and hydroxy groups the chain is numbered from right to left so that the numbers remain small. The suffix "ol" indicates the compound is an alcohol, and the "3" fixes the position of the hydroxy group.

2) $(CH_3)_2CHCH(OH)CH{=}CH_2$

4-methyl-1-penten-3-ol

The longest straight chain has five carbons and contains a double bond, hence the "penten." The number 1 locates the position of the double bond. Since the compound is an alcohol, the compound is a pentenol; the number 3 locates the position of the hydroxy group. The 3 is placed between the "penten" and the "ol," rather than before "penten," to avoid confusion regarding the location of the double bond.

3)

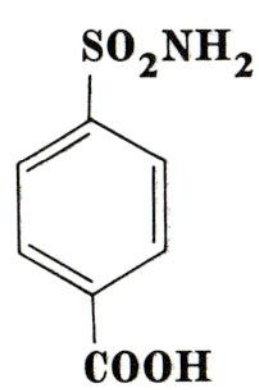

4-carboxybenzenesulfonamide

This compound could also be called *p*-sulfamylbenzoic acid.

4) 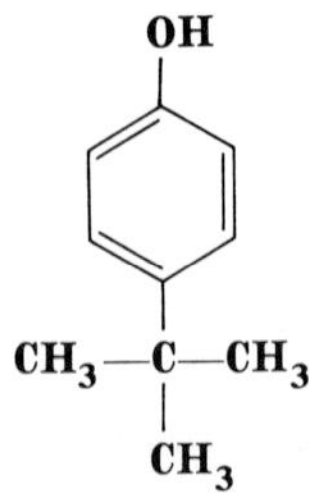

4-*t*-butyphenol

This compound is named as a derivative of phenol. One could also name the compound 1-hydroxy-4-*t*-butylbenzene, although the shorter name is much preferred.

5) $Cl(CH_2)_3COOCH_3$

methyl 4-chlorobutyrate

As explained in Section IX, esters are named from the corresponding acid (in this case, butyric acid). The "ic" becomes "ate," hence the "butyrate." Since the carbonyl carbon is numbered 1, the chloro group is on carbon-4. The word "methyl" indicates that the compound is a methyl ester (*i.e.*, a derivative of methyl alcohol).

6) 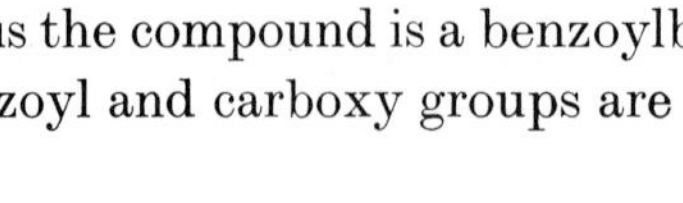

4-benzoylbenzoic acid

In Section XII it is noted that the C_6H_5CO— group is called the benzoyl group. Thus the compound is a benzoylbenzoic acid. The number 4 indicates that the benzoyl and carboxy groups are *para* to one another.

7) $CH_3CH(OH)CH_2NH_2$

1-amino-2-propanol

Another conceivable name would be 2-hydroxy-1-propylamine.

8) $CH_3(CH_2)_3CO(CH_2)_3CH_3$

5-nonanone

The "e" of nonane is dropped and replaced by the ketone suffix "one." The compound could also be called di-*n*-butyl ketone.

9) 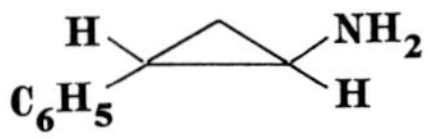

trans-2-phenylcyclopropylamine

"Cyclopropyl" means a 3-membered ring. The prefix *trans* indicates that the amino group and the phenyl group are on opposite sides of the ring.

10) $(C_6H_5CH_2)_3N$

tribenzylamine

This tertiary amine possesses three benzyl groups.

11) 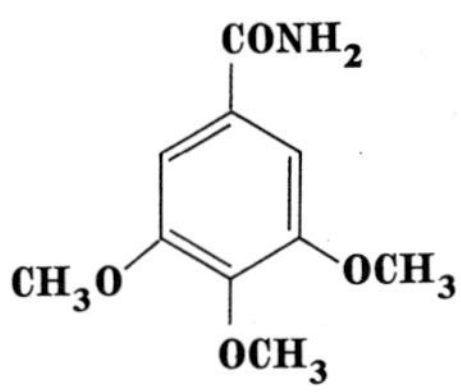

3,4,5-trimethoxybenzamide

Amides (which are carboxylic acid derivatives) are named by dropping the "ic" from the acid and adding "amide." Thus, "benzoic" becomes "benzamide". The rest of the name is self-explanatory.

12) 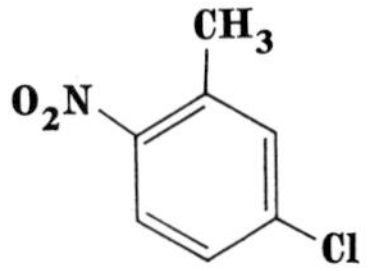

3-chloro-6-nitrotoluene

The compound is named as a toluene derivative in preference to 1-methyl-3-chloro-6-nitrobenzene since the former is somewhat shorter.

13) 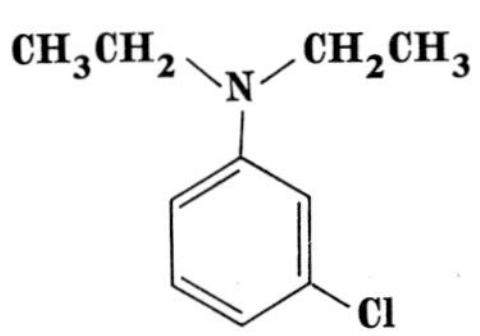

3-chloro-N,N-diethylaniline

The "N,N" indicates that the two ethyl groups are on the nitrogen and not on the benzene ring, (see section I for contrasting use of lower case n).

14)

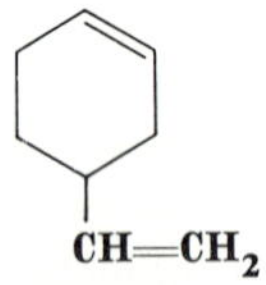

4-vinyl-1-cyclohexene

In numbering the cyclic olefin, carbons 1 and 2 incorporate the double bond. This places the vinyl group on carbon-4.

ADDITIONAL READING

J. G. Traynham, *Organic Nomenclature, A Programmed Introduction,* Englewood Cliffs, N.J.: Prentice-Hall, 1966. Paperback.

ANSWERS TO PROBLEMS

CHAPTER 1

1. No. Lithium has a single electron in the second shell which it readily donates to achieve a filled shell configuration. Therefore lithium is not electronegative (*i.e.*, electron attracting).
2. Since the two hydrogen atoms have identical electronegativities, there will be little tendency for either atom to take *both* electrons (to form $H:^-$) leaving the other with none (H^+). By sharing the pair of electrons, both hydrogens can attain filled shells.
3. A covalent bond involves the sharing of a pair of electrons by two atoms. The pair of electrons helps "bond" the atoms together by reducing the repulsion between the two positive nuclei. The covalent bond is represented by a dash.
4. Two.
5. a) 4, 5, and 6.

   b)

   c) 0, 1, and 2.

6.

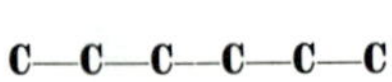

n-hexane

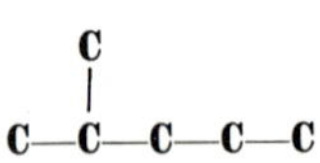

2-methylpentane

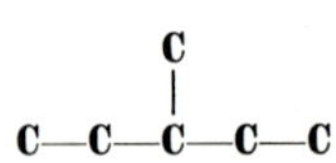

3-methylpentane

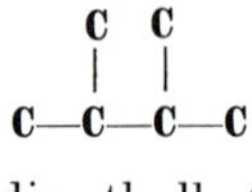

2,3-dimethylbutane

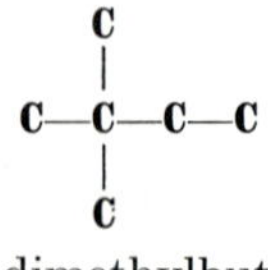

2,2-dimethylbutane

7. 120°. 10.5°.

8.

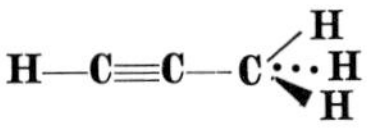

9. If the carbon were planar, one would expect two isomers of CH_2Cl_2:

10. The acetylenic group is normally linear, and we would therefore expect cyclopentyne to be highly strained and extremely unstable.

11.

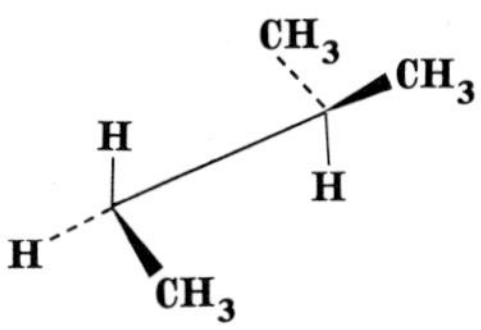

12. a) 10 σ, 0π b) 7σ, 3π c) 10σ, 1π

13. The carbons of the carbon-chlorine bonds of ethyl chloride, vinyl chloride, and chloroacetylene are sp^3, sp^2, and sp hybridized, respectively. The carbon-chlorine bond distances decrease in the manner shown, because the greater the percentage of s character the smaller the orbital.

14. a) Same b) Isomers c) Same d) Isomers

15.

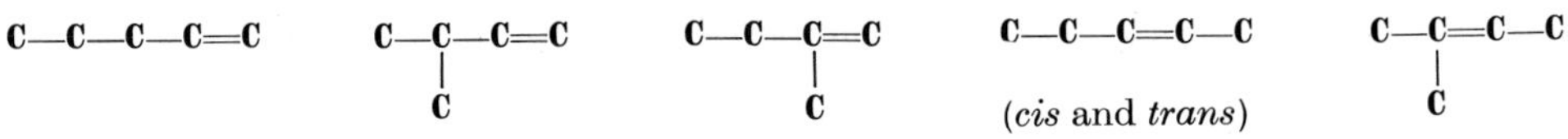

16. Alkanes: C_nH_{2n+2}, Alkenes: C_nH_{2n}, Alkynes: C_nH_{2n-2}
17. a) One b) One c) Two
18. An sp hybridized carbon can bear a negative charge because the electrons are in an sp orbital which has 50% s character and which is relatively close to the positive nucleus.
19. Two

20. a)

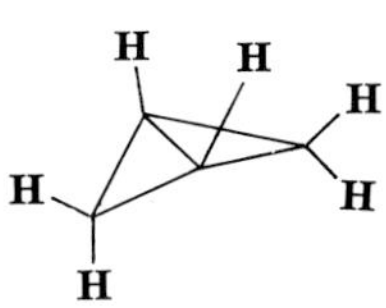

b)

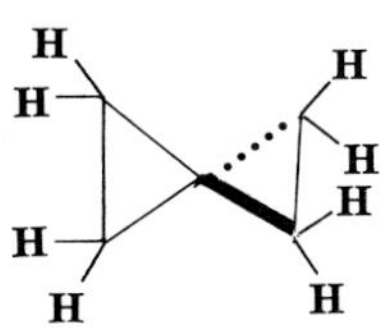

21.

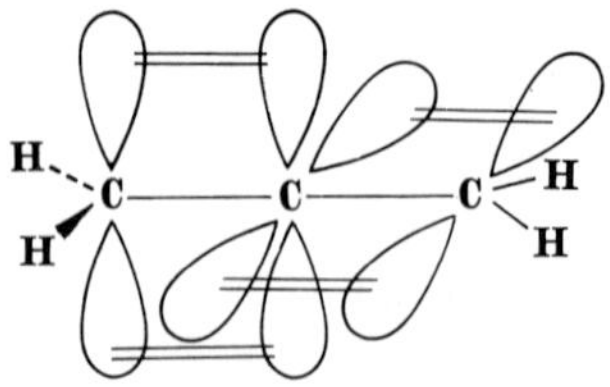

22.

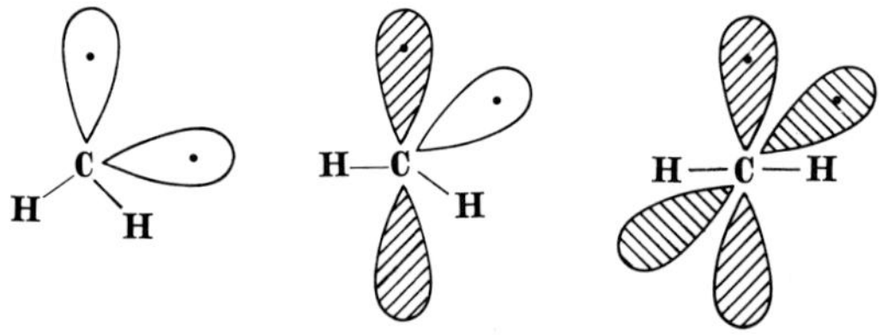

sp^3 sp^2 sp

23.

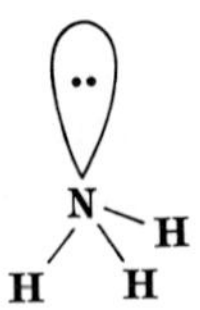

The bond formed between the ammonia nitrogen and the proton is comprised of a pair of electrons donated solely by the nitrogen.

24.

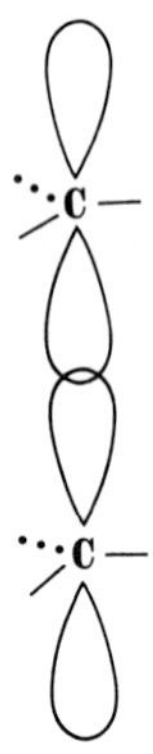

25.

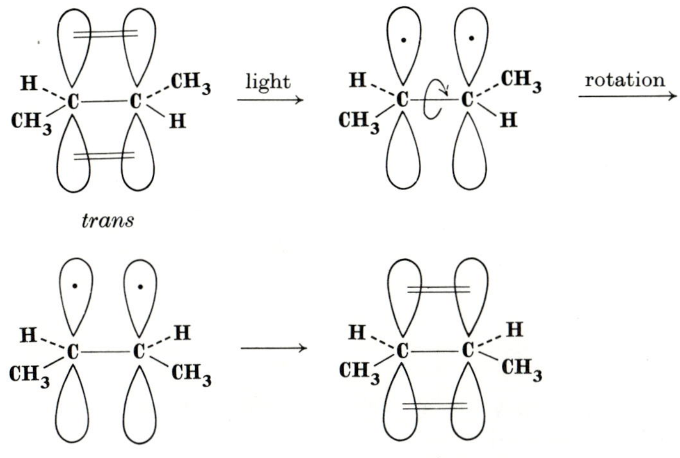

cis

The light energy breaks the π bond (*i.e.*, the overlap between the p orbitals), permitting free rotation about the central carbon-carbon bond. Reformation of the π bond gives the *cis* material.

CHAPTER 2

1. $H\ddot{\underset{\cdot\cdot}{O}}:^-$ $\quad B = 2,\ C = 6.$

2. $H_3N:$ $\quad B = 6,\ C = 2.$

3. $\overset{+\delta}{H}—\overset{-\delta}{F}$ $\quad H_3\overset{+\delta}{C}—\overset{-\delta}{Cl}$ $\quad H\overset{-\delta}{O}—\overset{+\delta}{Cl}$

4. a) Not very important because charge is created and because the cationic carbon lacks a complete shell. Negative charge is borne by carbon which is not electronegative.
   b) Totally unimportant contributor. Resonance contributors involve redistribution of π electrons and unshared pairs of electrons, not σ electrons.
   c) Impossible resonance contributor because hydrogen cannot have four electrons in its bonding shell.
5. The central carbon bears a partial positive charge (and is hence attacked by hydroxide ion) because of

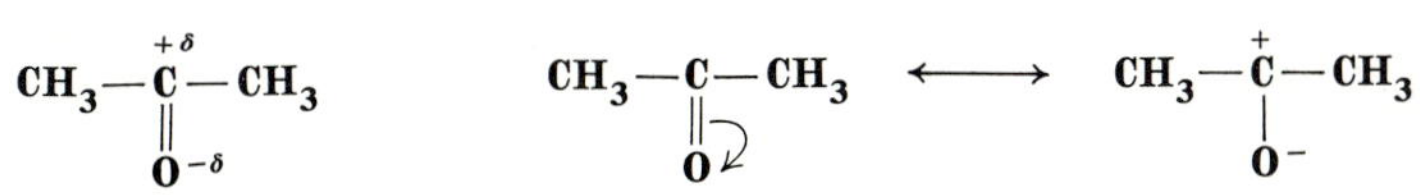

a) inductive effects $\qquad$ b) resonance effects

6. a) 

b) 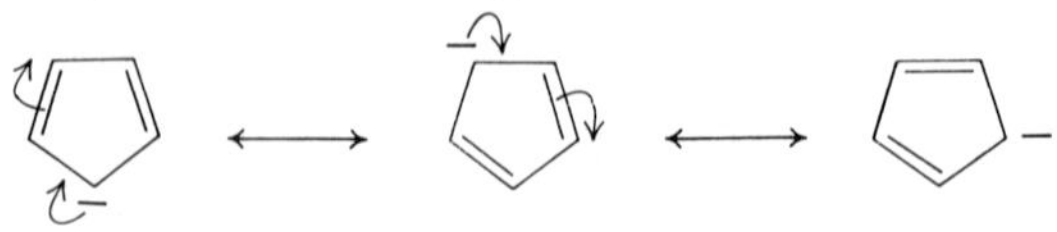

c) 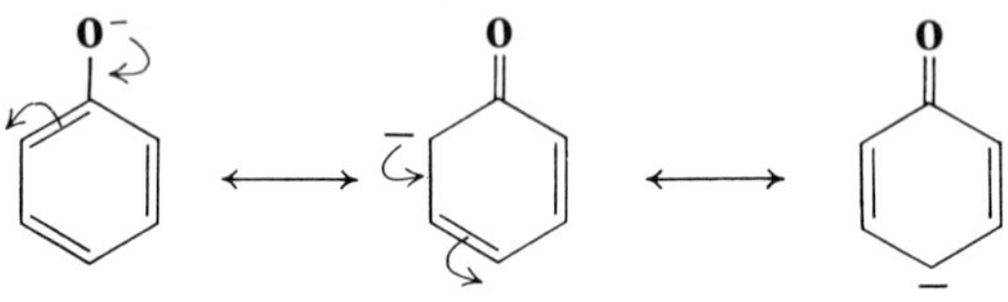

7. a) Positive b) Positive c) Negative d) Neutral e) Negative
f) Negative g) Neutral

8. a) 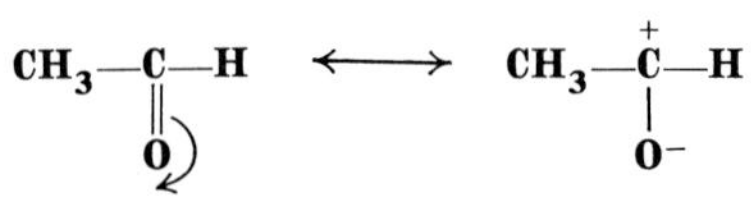

b) 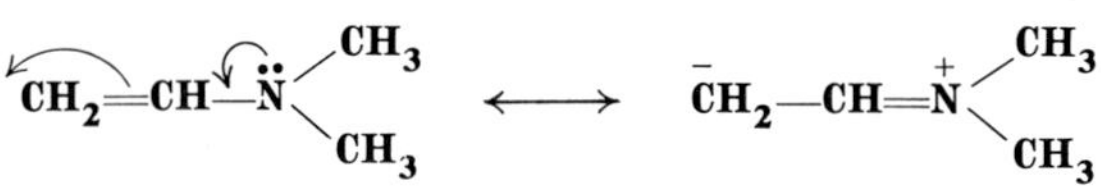

c) 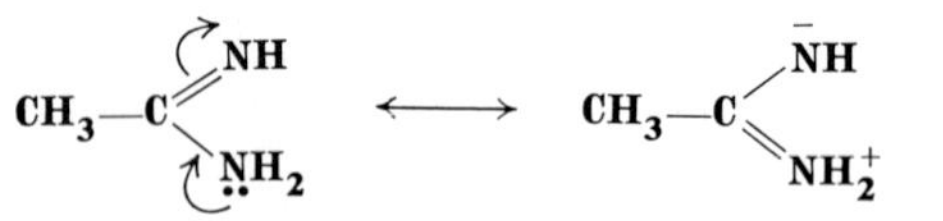

d) 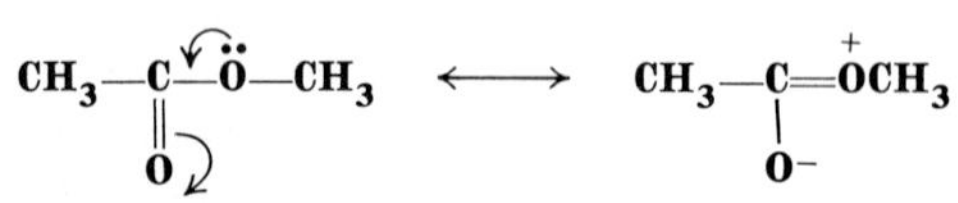

e) 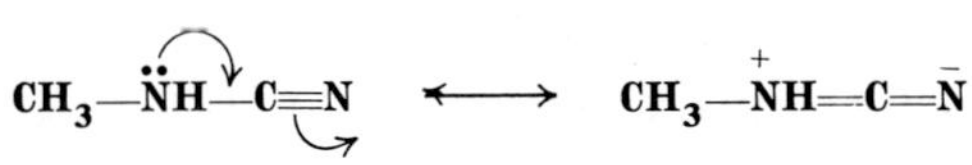

9. A dipolar resonance contributor diminishes the dipole moment which arises from inductive polarization of electrons in the direction of the chlorine.

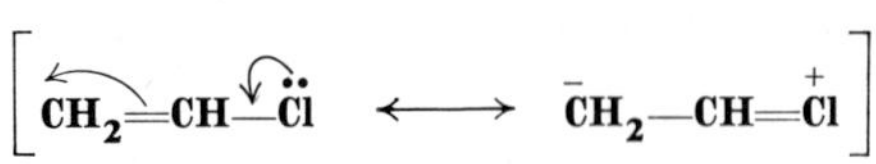

10. Neutral. Two pair

11. a) Two b) One c) One d) One e) One f) One g) Three
h) One

12. NCO^- is a hybrid of two resonance contributors:

13. There are five identical resonance contributors:

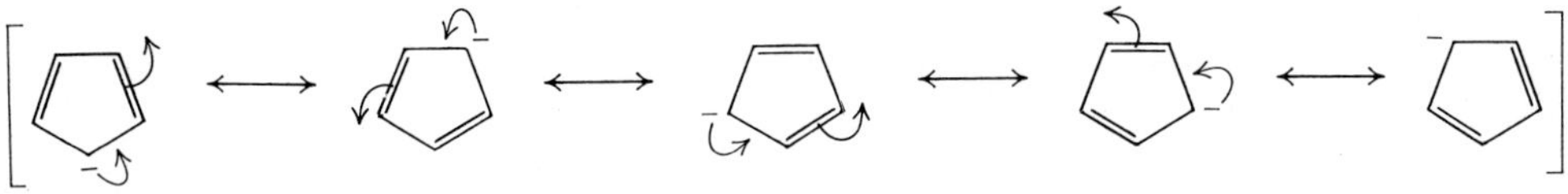

Therefore, each carbon bears $\frac{1}{5}$ negative charge in the resonance hybrid.

14. a) Different b) Contributors c) Different d) Contributors
e) Contributors f) Different

15. The benzene ring stabilizes the carbonium ion by allowing the positive charge to be delocalized over several atoms.

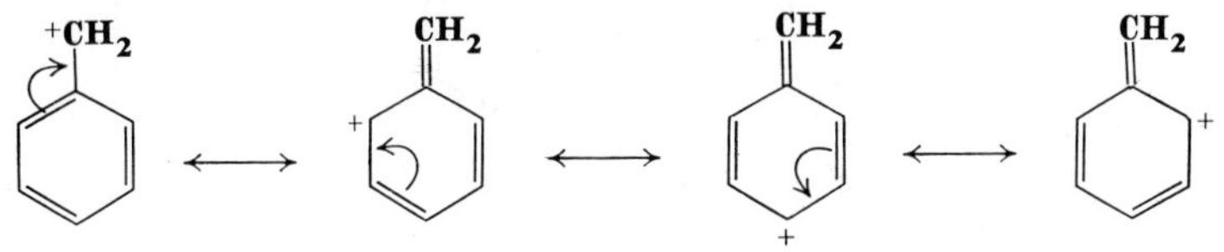

16.

a) 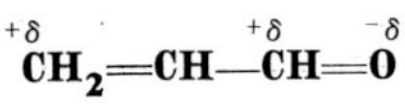 b) 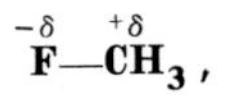

c) 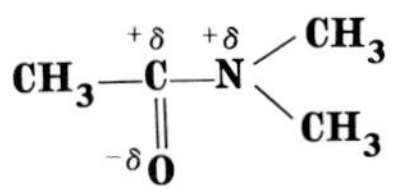 d) 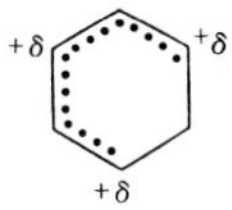 e) 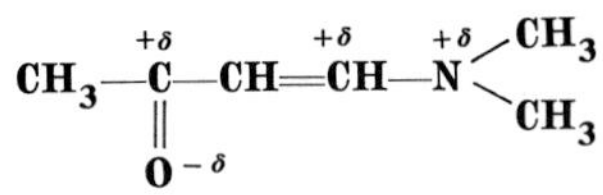

17. The central nitrogen-carbon bond has considerable double-bond character because of a dipolar resonance contributor.

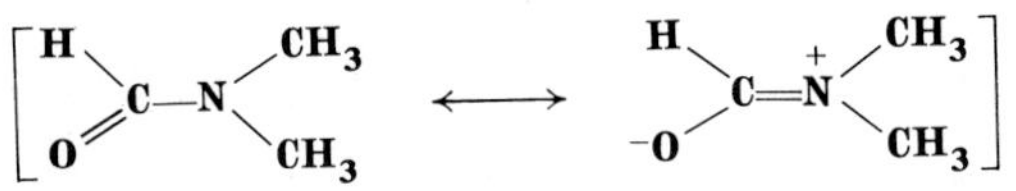

18. The two free electrons on the pyrrole nitrogen are delocalized into the ring, making them less available for coordination with a proton:

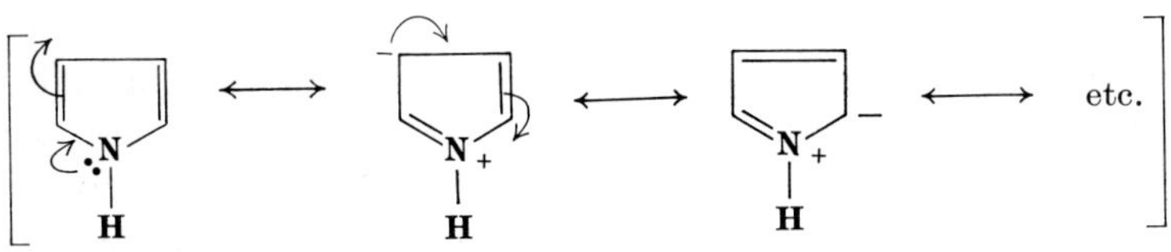

19. The resonance contributor is unreasonable because bond angles are severely distorted. The acetylenic portion, for example, is not linear. Moreover, the bond energies of $C\equiv C$ and $C=C$ versus $3C-C$ and $C=C$ reflect the improbability of the contributor.

20. An sp^2 carbon is slightly more electronegative than an sp^3 carbon because the former has more s character.

21.

a) 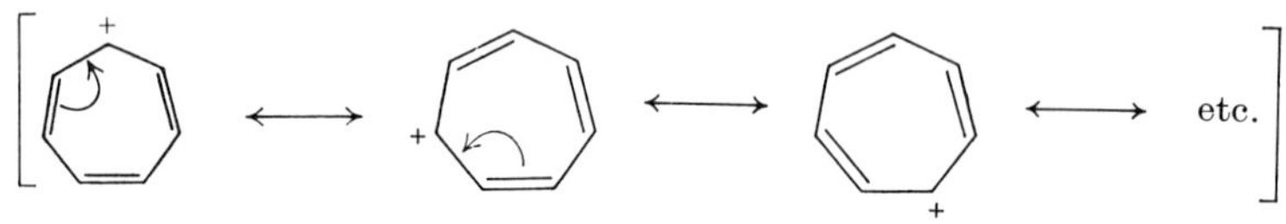

Seven equivalent forms are possible. The charge is +1/7 on each carbon.

b) 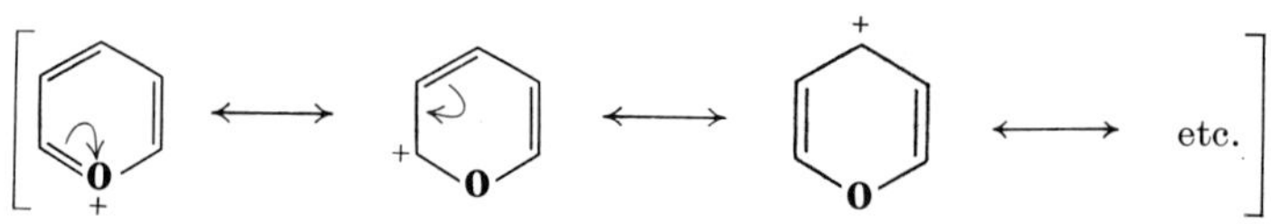

c) 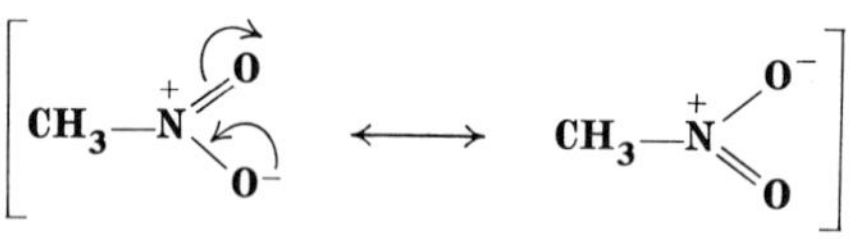

d) 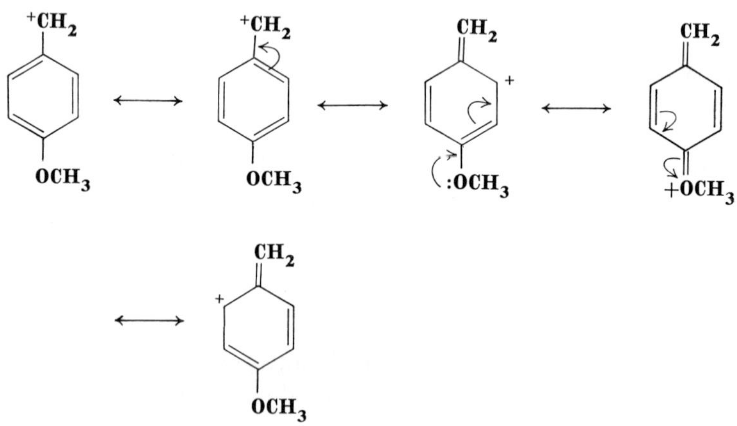

Note: The fourth resonance contributor is actually rather important, despite the fact that electronegative oxygen bears a positive charge, because *each* atom of this contributor has a full outer shell of electrons. This contrasts with the other contributors in which a cationic carbon has only six electrons.

e)

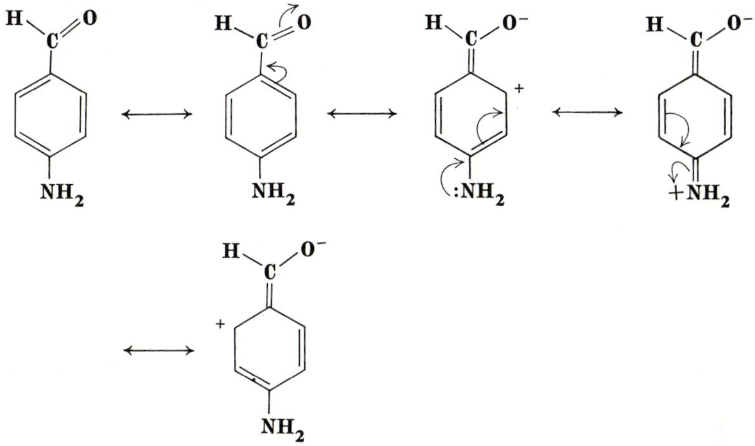

Note: For a reason similar to that given in d above, the third resonance contributor is important relative to the third and fifth contributors.

CHAPTER 3

a) Mg, $CH_3CH_2CH_3$

b) 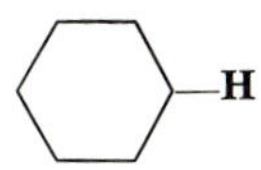

c) $CH_2{=}CH{-}CH_3$

d) KOH

e) 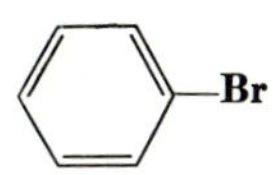

f) CrO_3

g) $CH_2{=}O$, H_3O^+

h) CrO_3

i) CH_3CH_2Br, ^-OH

j) $(CH_3)_3C{-}O^-\ Na^+$

k) CO_2, H_3O^+

l) 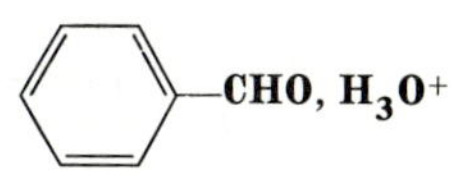

m) $CH_3CH_2CH_2CHCH_3$ (Br on the CH)

n) $CH_3CH_2CH{=}CHCH_3$

o) $LiAlH_4$

2.

a) 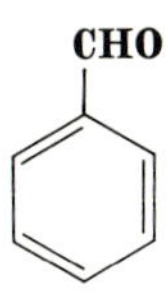

b) 

c) CH_3CHCH_3 (OH on the CH)

d) 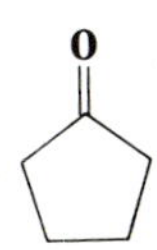

e) 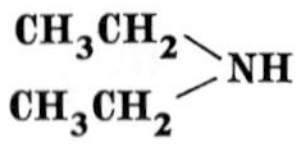

f) $O{=}C{=}O$

g) 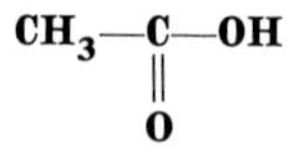

h) $(CH_3CH_2CH_2)_2O$

i) CH_3NH_2

j) 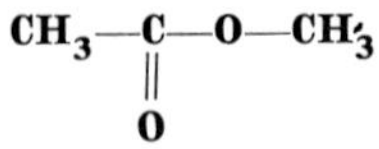

3. a) $LiAlH_4$; HBr
   b) Br_2/Fe; Mg
   c) CH_3MgBr; Na; CH_3Br

4.

a) 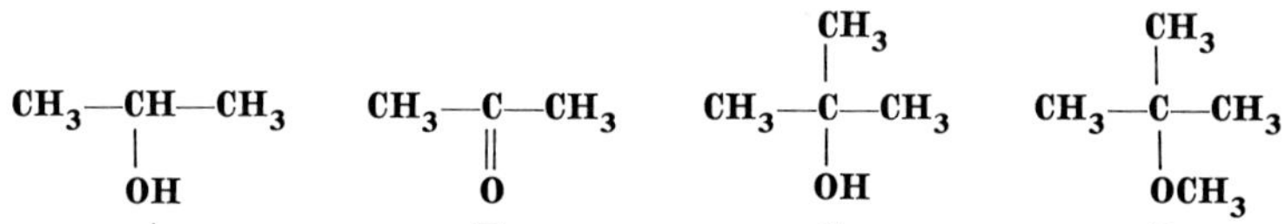

A B C D

b) 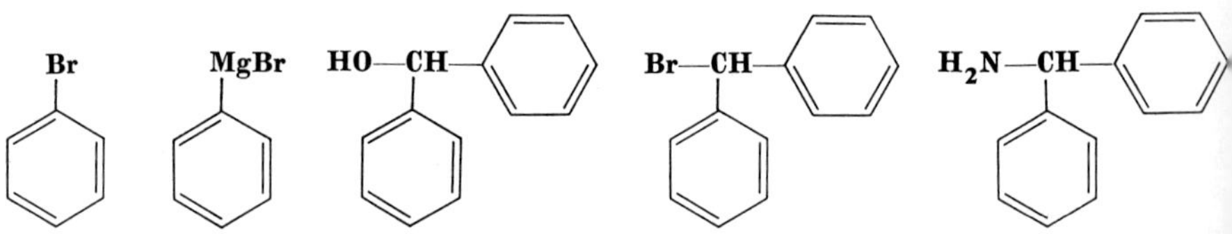

A B C D E

5.

a) 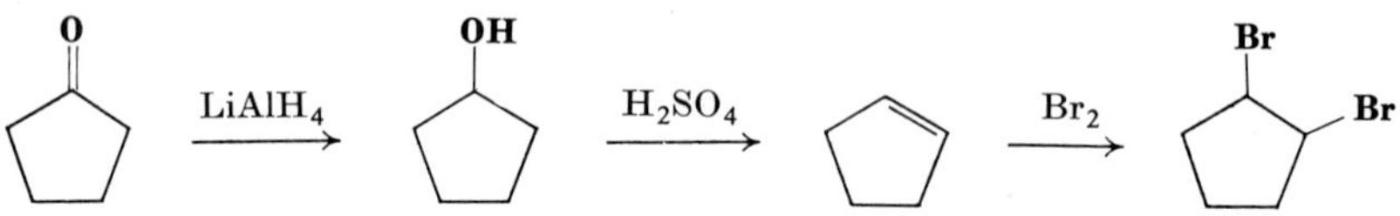

b) 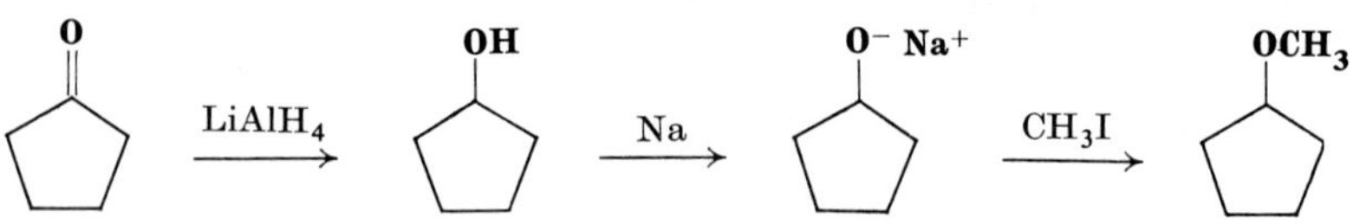

c) 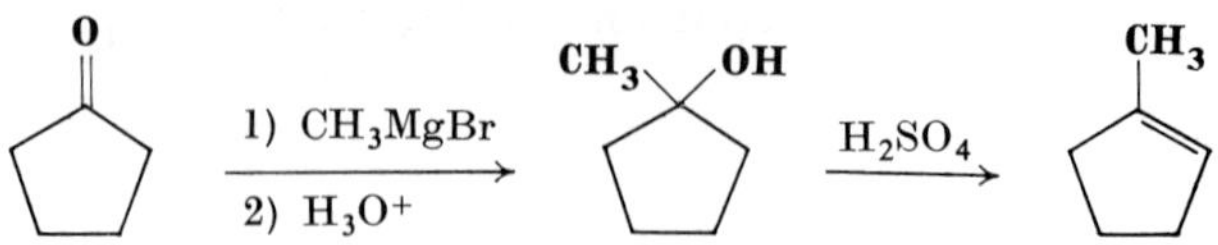

d) 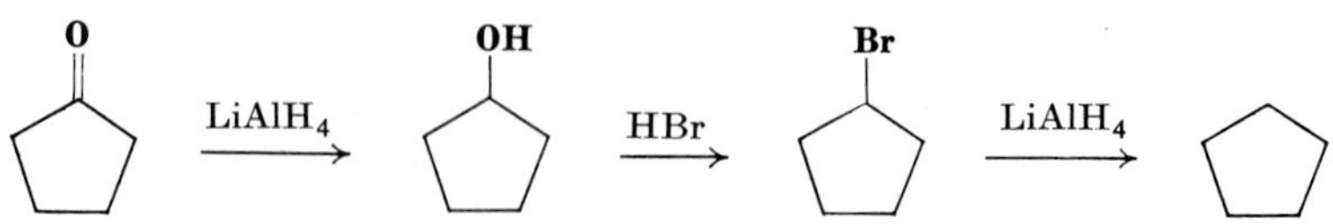

e) 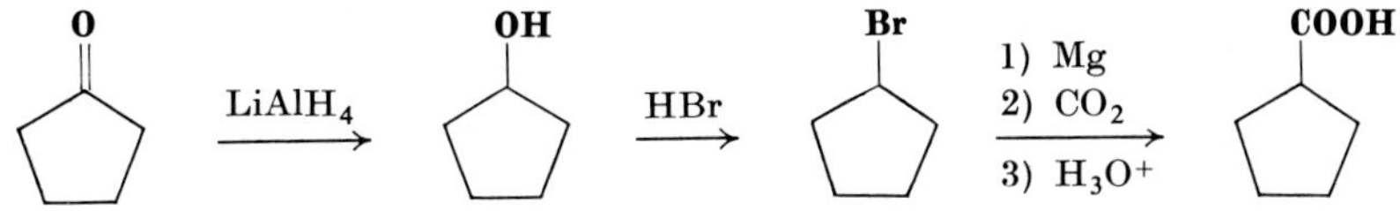
O
LiAlH$_4$
OH
HBr
Br
1) Mg
2) CO$_2$
3) H$_3$O$^+$
COOH

6.

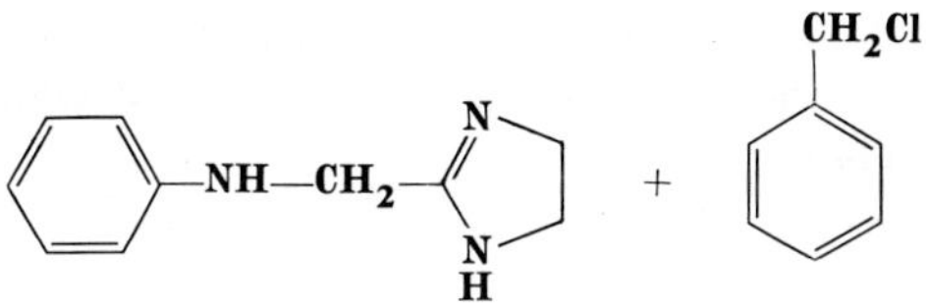
NH—CH$_2$
N
N
H
+
CH$_2$Cl

7.

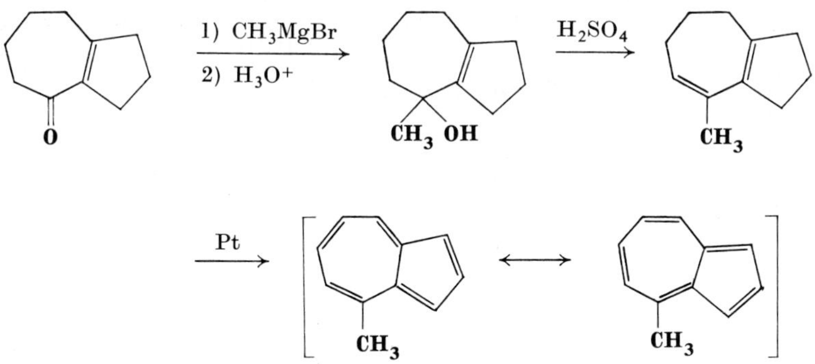
O
1) CH$_3$MgBr
2) H$_3$O$^+$
CH$_3$ OH
H$_2$SO$_4$
CH$_3$
Pt
CH$_3$
CH$_3$

8.

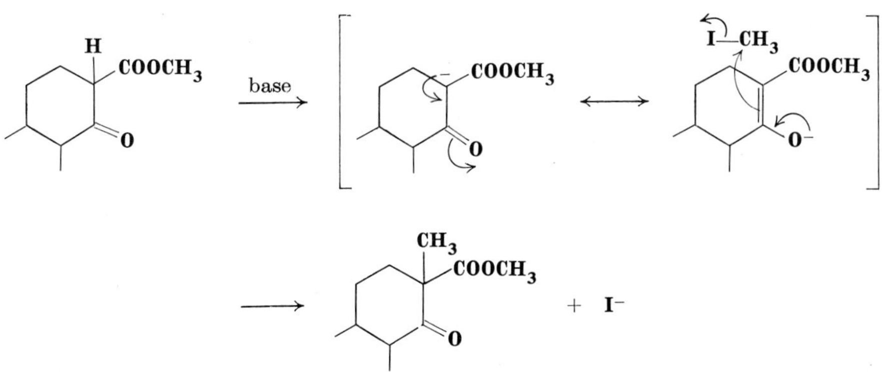
H
COOCH$_3$
O
base
COOCH$_3$
O
I—CH$_3$
COOCH$_3$
O$^-$
CH$_3$
COOCH$_3$
O
+ I$^-$

9.

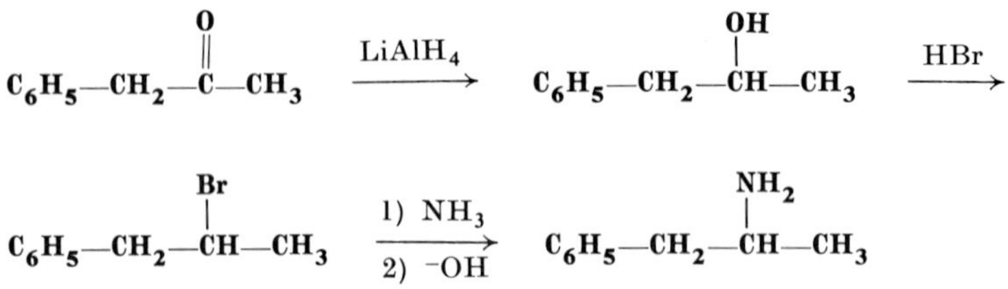

10. Addition of water to trichloroacetaldehyde reduces the unfavorable electrostatic interaction between the two carbon atoms, both of which have large partial positive charges:

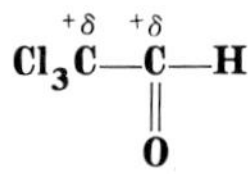

11.

a) $CH_3CH_2CH_2Br \xrightarrow{Mg} CH_3CH_2CH_2MgBr \xrightarrow[2)\ H_3O^+]{1)\ CH_2O} CH_3CH_2CH_2CH_2OH$

$\xrightarrow[2)\ CH_3I]{1)\ Na} CH_3CH_2CH_2CH_2OCH_3$

b)

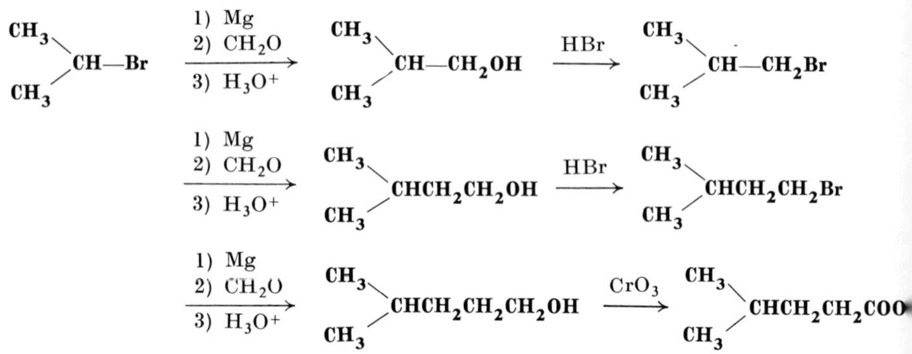

12.

a)

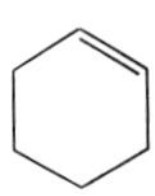

b)

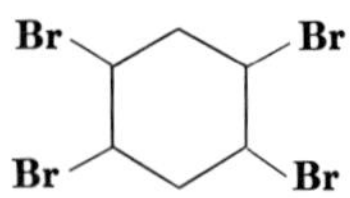

c) $CH_3CH_2CH_2CH{=}CHCH_3$

d)

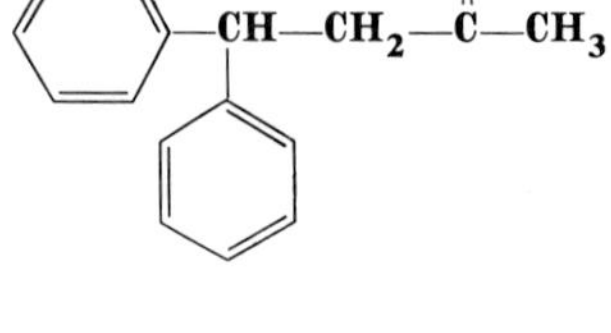

13.

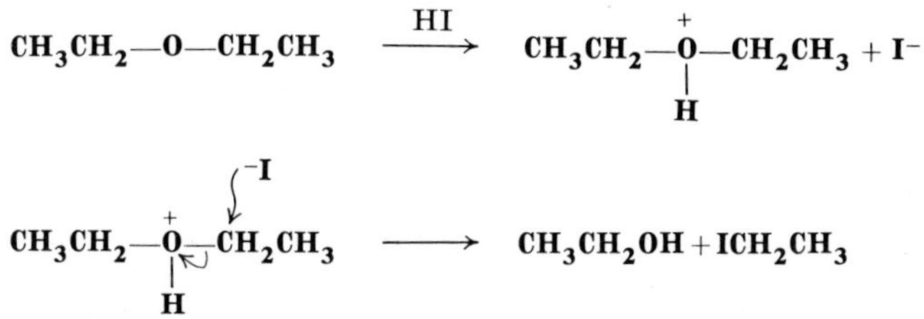

14. The solutions to these problems are by no means unique. Compare your answers with those of your colleagues and with those given below.

a) 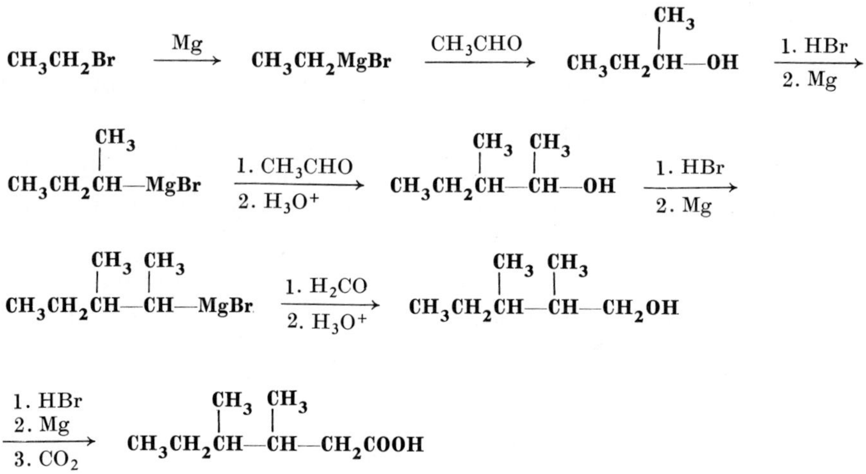

b) 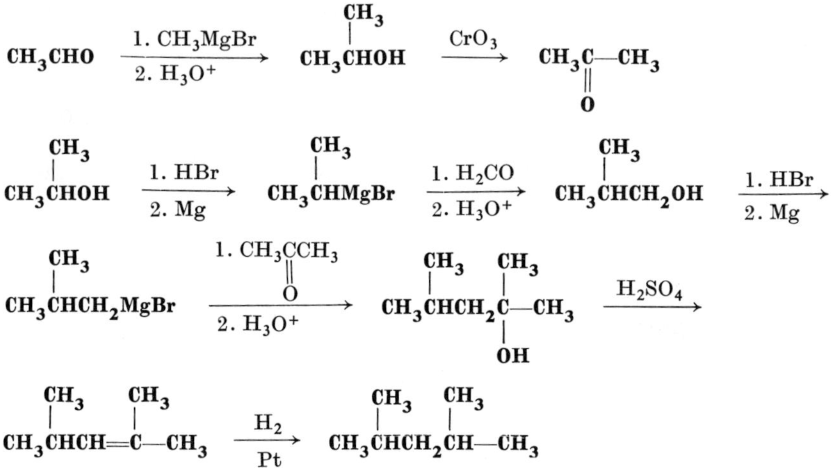

c) $CH_3CH_2Br \xrightarrow{Mg} CH_3CH_2MgBr \xrightarrow[2.\ H_3O^+]{1.\ H_2CO} CH_3CH_2CH_2OH \xrightarrow[2.\ Mg]{1.\ HBr}$

$CH_3CH_2CH_2MgBr \xrightarrow[2.\ H_3O^+]{1.\ CH_3CHO} CH_3CH_2CH_2CH(OH){-}CH_3 \xrightarrow{H_2SO_4}$

$CH_3CH_2CH{=}CH{-}CH_3$

Note: Dehydration of an alcohol (last step in 3c) leads predominantly to the olefin with the *largest* number of alkyl groups.

CHAPTER 4

1. The acid of pK_a 4.8 is the stronger. The values of K differ by a factor of ten.
2. a) Weaker b) Stronger
3. One cannot tell how the equilibrium will be affected without knowing whether HA or A^- is more stabilized. If they are equally stabilized, the equilibrium will be unchanged.
4. In simplified terms, the acetate ion has its negative charge distributed by resonance over two electronegative oxygens, whereas the phenolate ion has its negative charge on only one oxygen and on three nonelectronegative carbons.
5. 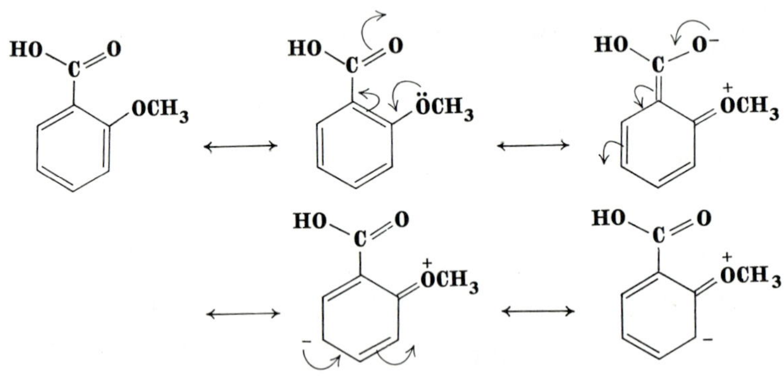
6. With a strong acid, the proton readily departs from the conjugate base. Conversely, the conjugate base will not readily accept a proton. In other words, it is a poor base. HCl is a good example.
7. At one pH unit above the pK_a, 90% of the material exists as the conjugate base. Thus the concentration of phenolate is 0.09 M.

8. The first of the following contributors is the most important because oxygen is much more electronegative than carbon.

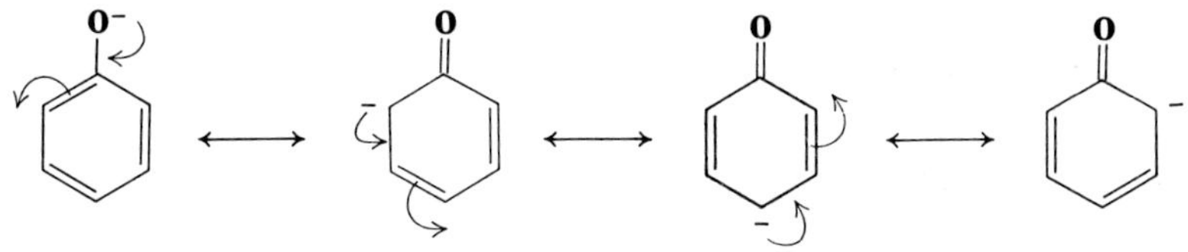

9. The rate of a reaction depends only on the activation energy. If the activation energy of an endothermic reaction is less than that of an exothermic reaction, the former will be faster.

10. All the statements are true.

11.

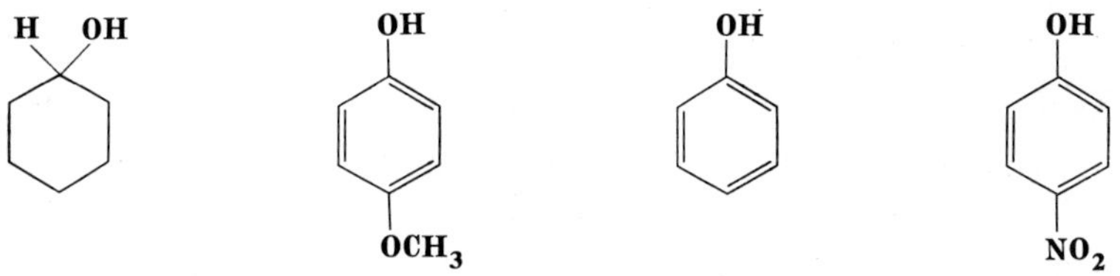

12. Protonation of the aromatic amine on the nitrogen avoids disruption of the aromatic ring system. Protonation of the vinyl amine on the carbon leads to a resonance-stabilized cation, unlike protonation on the nitrogen.

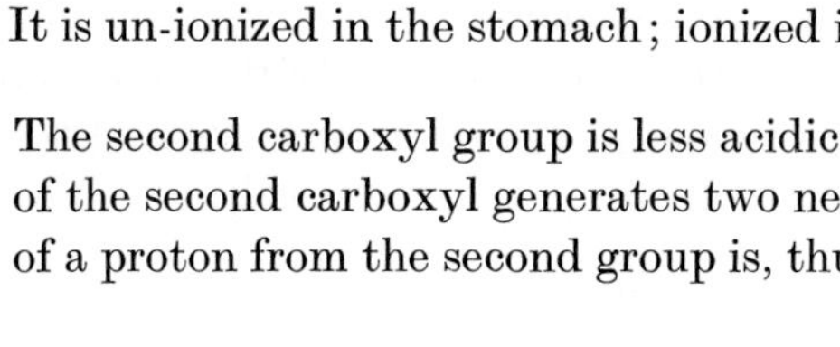

13. It is un-ionized in the stomach; ionized in the small intestines.

14. The second carboxyl group is less acidic than the first because ionization of the second carboxyl generates two neighboring negative charges. Loss of a proton from the second group is, thus, electrostatically impeded.

15. The pK_a of acetic acid is greater in benzene than in water. The ionization equilibrium, drawn below, is driven to the left in benzene, because this nonpolar solvent is less capable than water of solvating the charged species.

$$CH_3COOH \rightleftharpoons CH_3COO^- + H^+$$

16. $$K_a = \frac{[H^+][A^-]}{[HA]}$$

$$\log K_a = \log [H^+] + \log \frac{[A^-]}{[HA]}$$

$$-\log K_a = -\log [H^+] - \log \frac{[A^-]}{[HA]}$$

$$pK_a = pH - \log \frac{[A^-]}{[HA]}$$

17. a) Histidine, neutral. Serine, neutral. Lysine, positive. Aspartic acid, negative. Arginine, positive.

b) Arginine is more basic than lysine because the conjugate acid of the former is highly resonance stabilized:

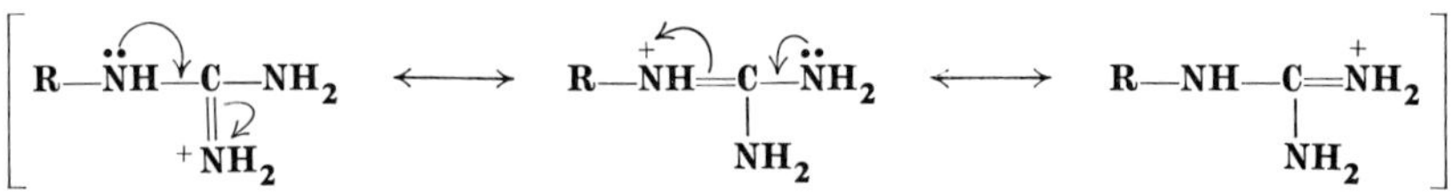

c) Histidine is protonated in the manner shown because this leads to a resonance-stabilized cation.

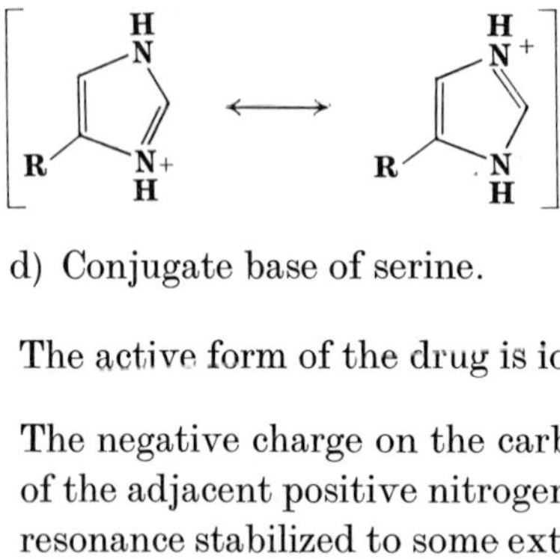

d) Conjugate base of serine.

18. The active form of the drug is ionic.

19. The negative charge on the carbon is stabilized by inductive withdrawal of the adjacent positive nitrogen. The conjugate base of thiamine may be resonance stabilized to some extent, although the neutral contributor has a divalent carbon with only six electrons in the outer shell:

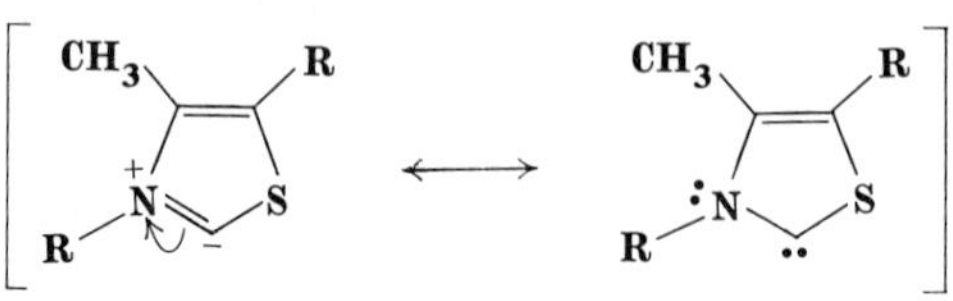

Finally, the carbon orbital with the unshared pair of electrons might overlap with an empty d orbital of sulfur.

20. The ring nitrogen of 4-aminopyridine is a stronger base than the corresponding atom in 3-aminopyridine because only the conjugate acid of the former can have the positive charge delocalized over *both* nitrogens:

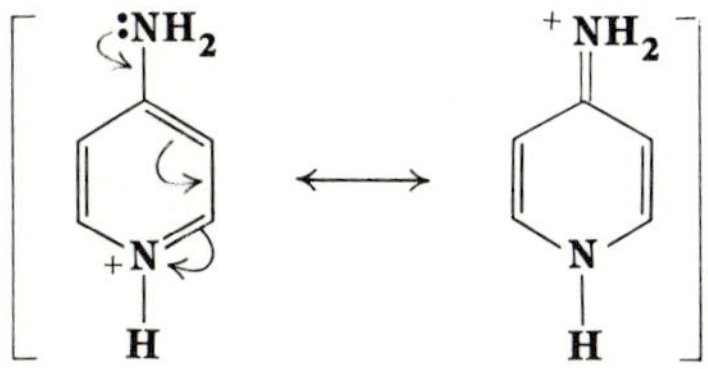

The —NH_2 group of 4-aminopyridine is less basic than the corresponding group of 3-aminopyridine because the free pair of electrons of the former can be delocalized in the manner shown:

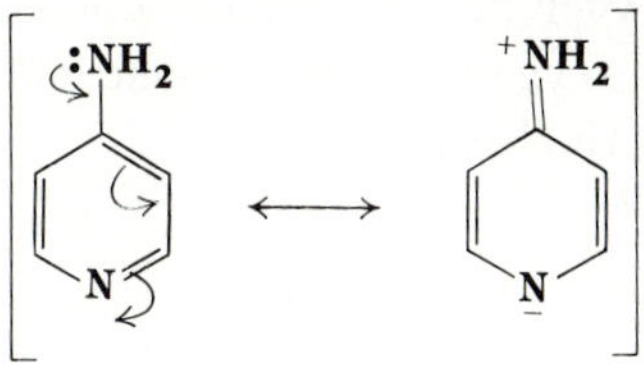

The electronegative ring nitrogen can accept the negative charge, while the 3-isomer cannot. Consequently, the latter has a more basic —NH_2 group.

21.

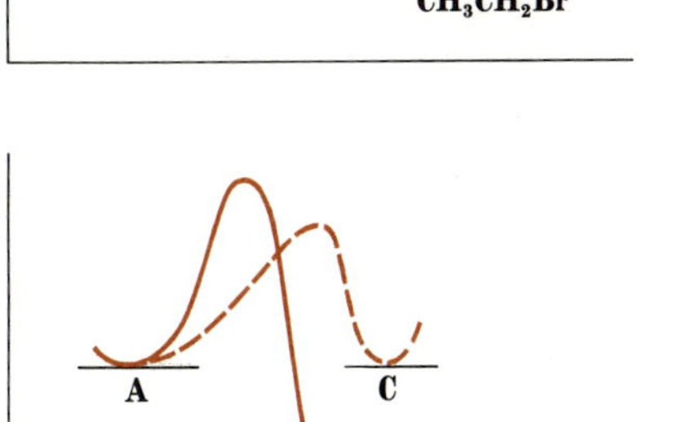

22.

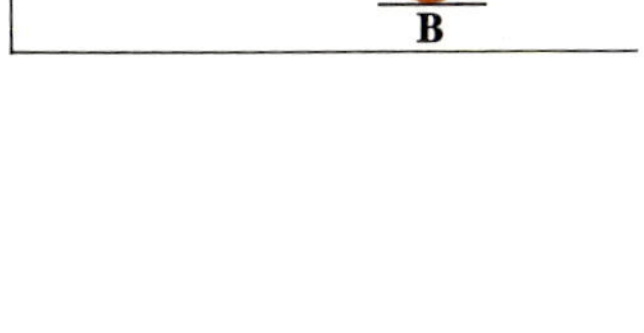

23.

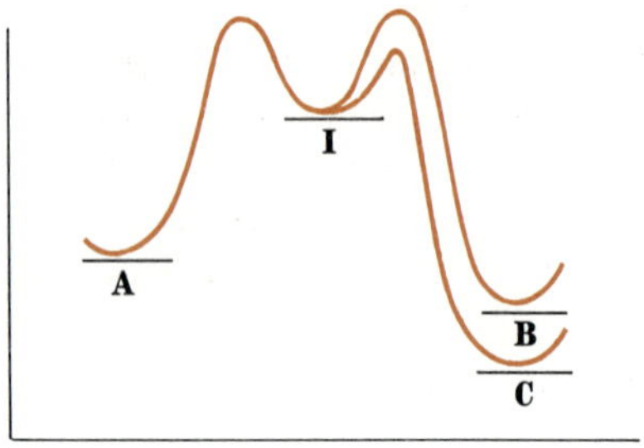

24. The two oxygens of acetyl peroxide have identical electronegativity, and therefore the oxygen-oxygen bond cleaves symmetrically, with each oxygen getting one electron. Since the oxygen in acetic acid is much more electronegative than the hydrogen, the oxygen-hydrogen bond cleaves heterolytically with the oxygen taking both electrons.

25.

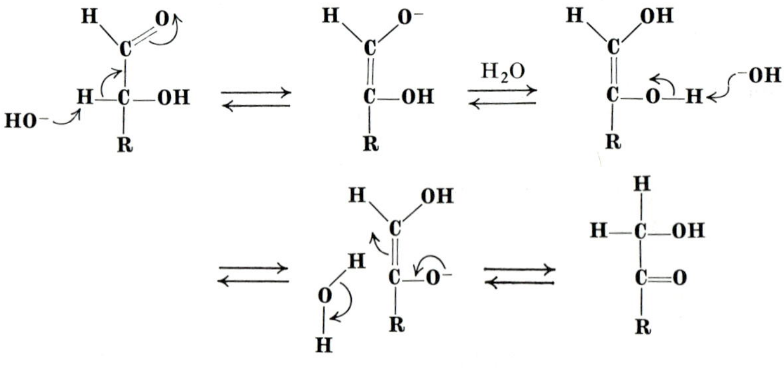

26.

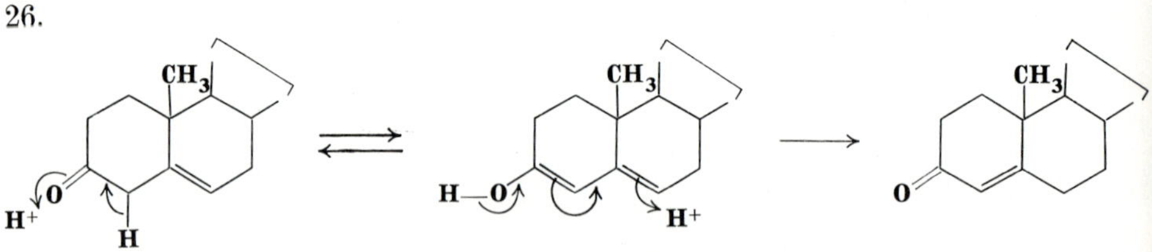

27.

a) 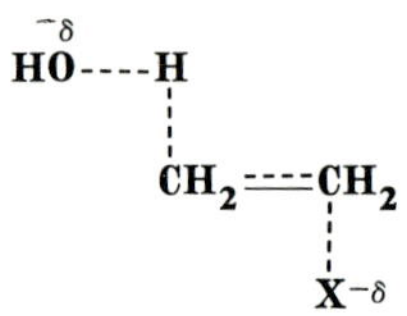

b) 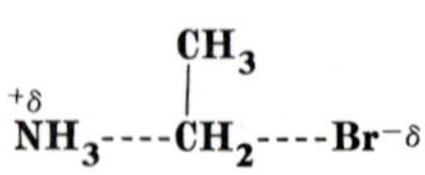

28. The mixture of three compounds is dissolved in benzene, and the solution is shaken with aqueous acid. The aqueous layer is separated, made basic, and the resulting insoluble material (1-naphthylamine) is taken up in benzene. The benzene is removed, leaving pure naphthylamine. The original benzene solution, now containing only two compounds, is shaken with aqueous base. The water layer is acidified and extracted with benzene. Evaporation of the benzene yields *p*-methylphenol. At this point, the original benzene solution contains only the decanol, which can be obtained by removing the solvent.

CHAPTER 5

1. a) inactive b) inactive c) inactive d) inactive e) active f) inactive

2. The material is racemic (50%+, 50%−).

3. One cannot determine the direction of rotation from the absolute configuration, nor can one determine the absolute configuration from the sign of rotation.

4.

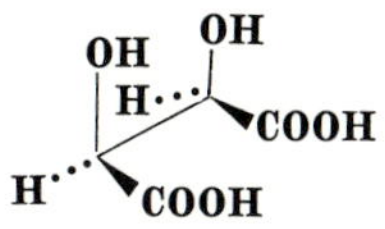

5.

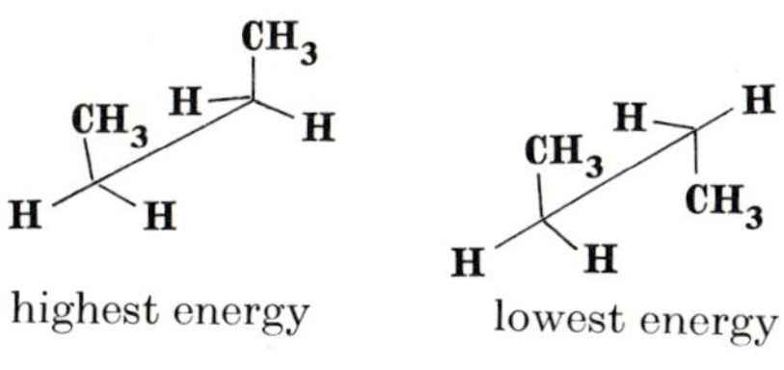

The highest energy conformer has a particularly unfavorable nonbonded interaction between the eclipsed methyl groups that is not present in the lowest energy conformer.

6. All the statements are true.

7.

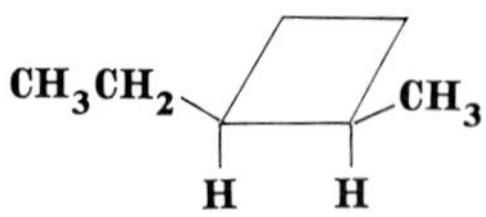

A

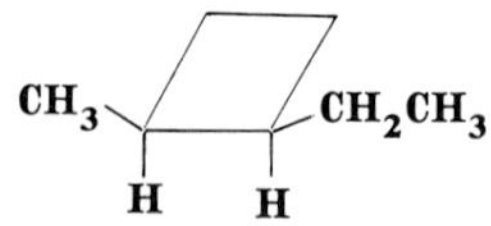

B

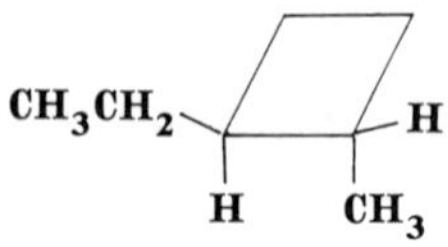

C

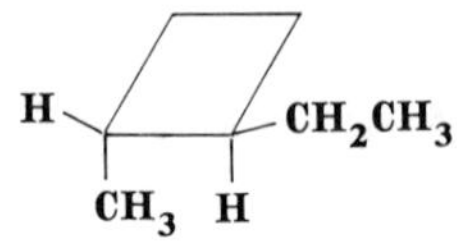

D

Enantiomers: *A* and *B*; *C* and *D*.
Diastereomers: *A* and *C*; *A* and *D*; *B* and *C*; *B* and *D*.

8. a) Dissymmetric, 1, R b) Dissymmetric, 2, S,S c) Symmetric (*meso*), 2, 1-R, 2-S d) Symmetric, 0, e) Dissymmetric, 2, 1-S 2-R f) Symmetric, 0 g) Dissymmetric, 2, 2-R, 3-S h) Dissymmetric, 1, S i) Dissymmetric, 1, R j) Dissymmetric, 5, 1-R, 2-R, 3-R, 4-S, 5-R

9. a) 4 b) None c) 1 dl pair, 1 (*meso*) d) 1 (*meso*) e) 2 f) 256

10.

a) 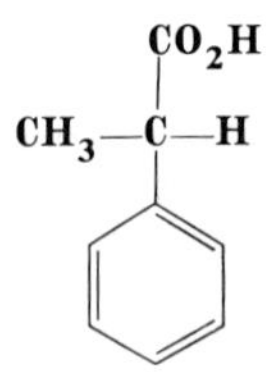

S

b) 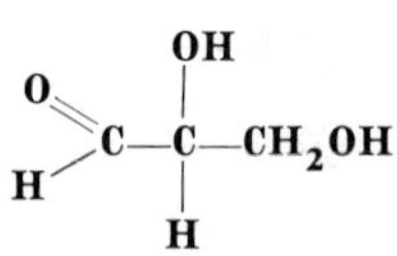

S

c) 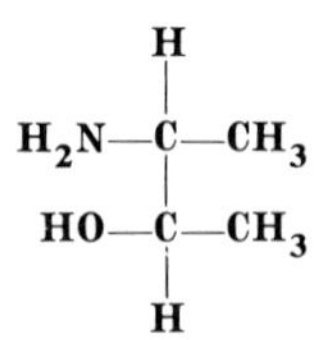

S
R

11. a) Enantiomers b) Diastereoisomers c) Identical

12.

a) 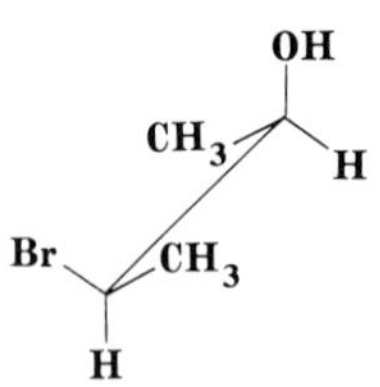

b) 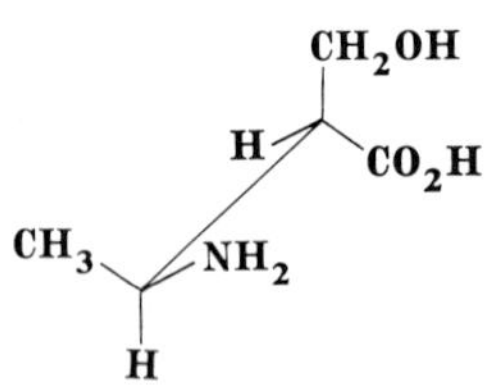

c) 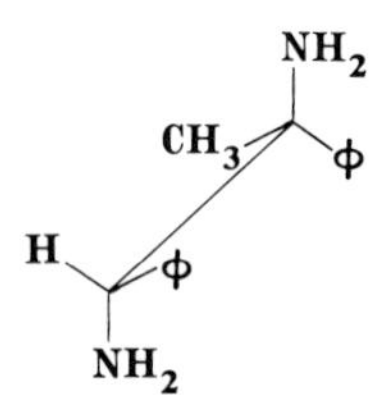

13. Only the isomer drawn below and its mirror image are possible.

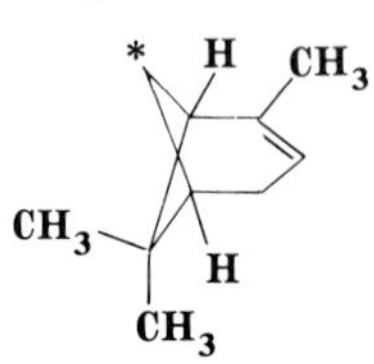

The two bonds which connect the methylene bridge (i.e., the starred carbon atom) must both be on the *same* side of the molecule. This restriction limits the number of optical isomers to two.

14.

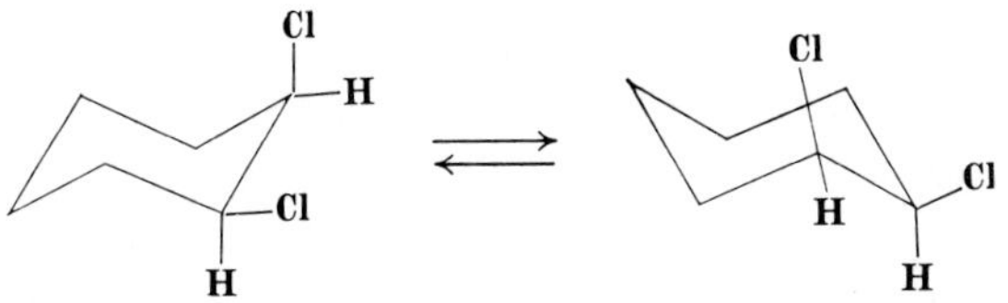

a) No b) Mirror images c) At normal temperatures these conformational isomers are in rapid equilibrium.

15.

a)

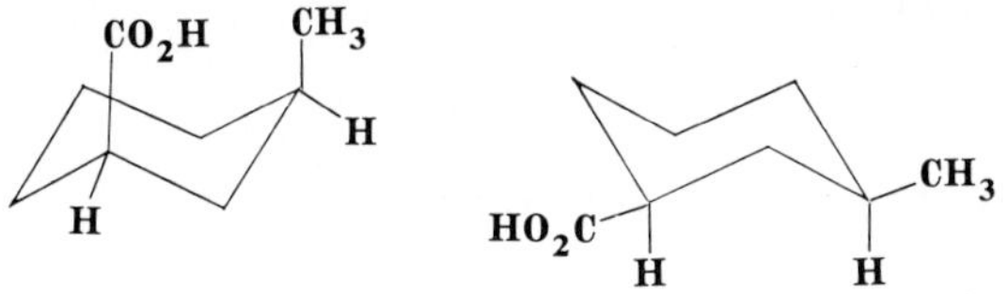

b)

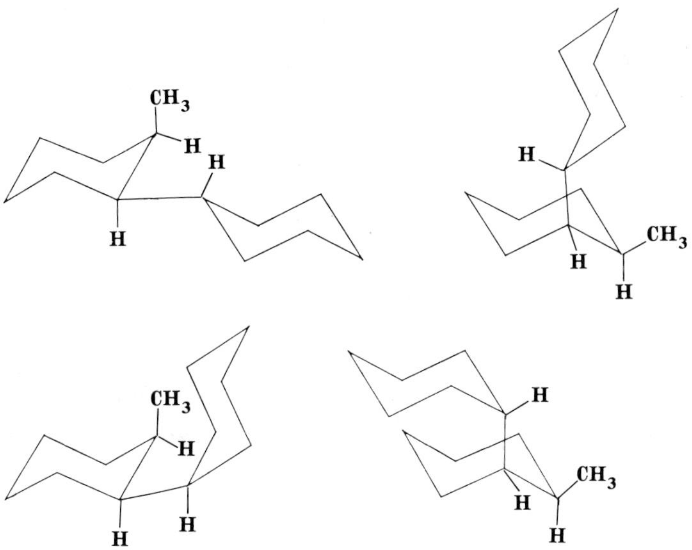

c)

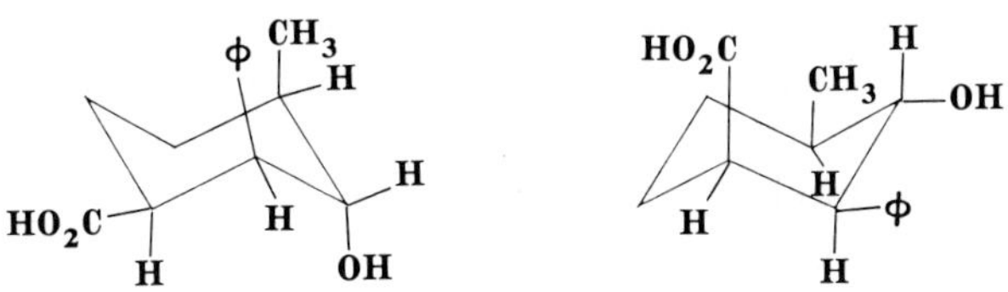

16. The *cis* compound has three 1,3-diaxial interactions which are absent in the *trans* compound.

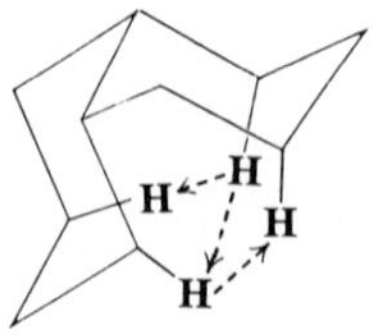

cis

17. a) The dl-mixture cannot be separated since the optical antipodes have the same chemical and physical properties. Compound Z has two asymmetric centers. Since Y has the R configuration, the two forms of Z must be R,R and S,R. These are diastereoisomers and they have different chemical and physical properties.

b)

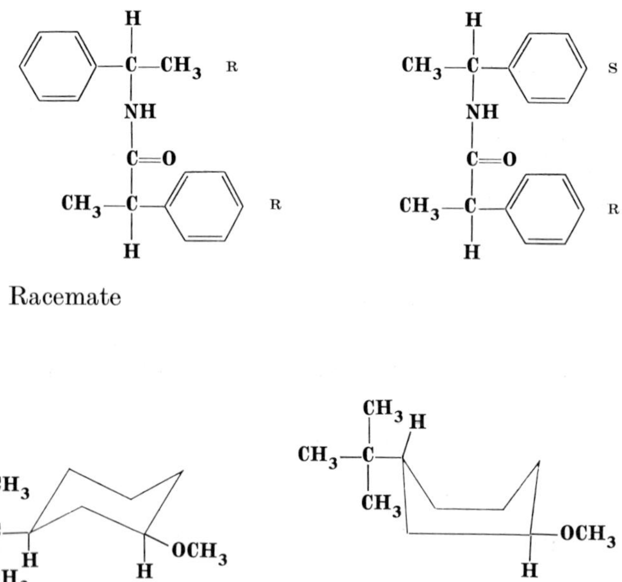

c) Racemate

18.

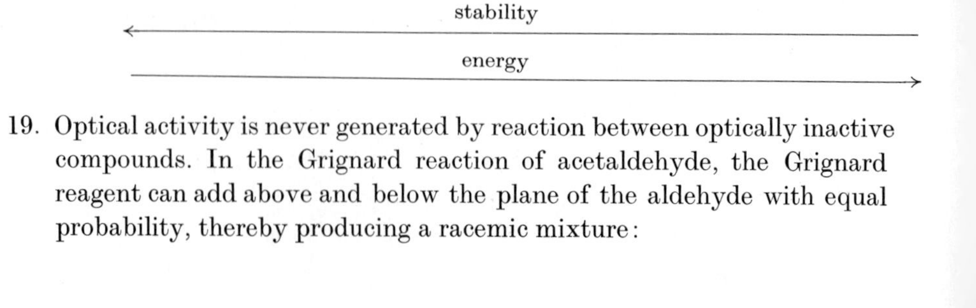

19. Optical activity is never generated by reaction between optically inactive compounds. In the Grignard reaction of acetaldehyde, the Grignard reagent can add above and below the plane of the aldehyde with equal probability, thereby producing a racemic mixture:

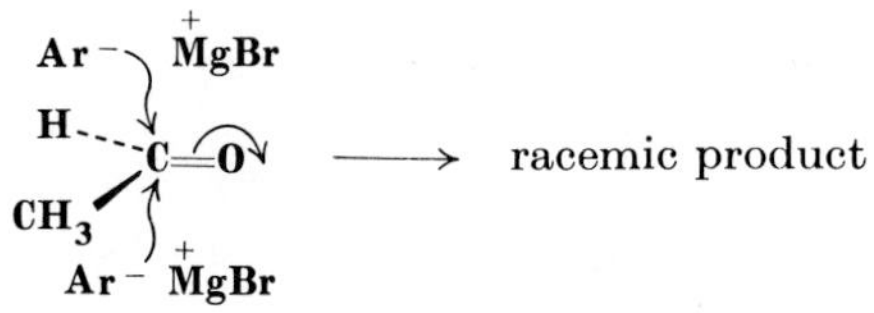

20. The ketone enolizes in strong base; the enolate anion is symmetric and protonates to form inactive ketone:

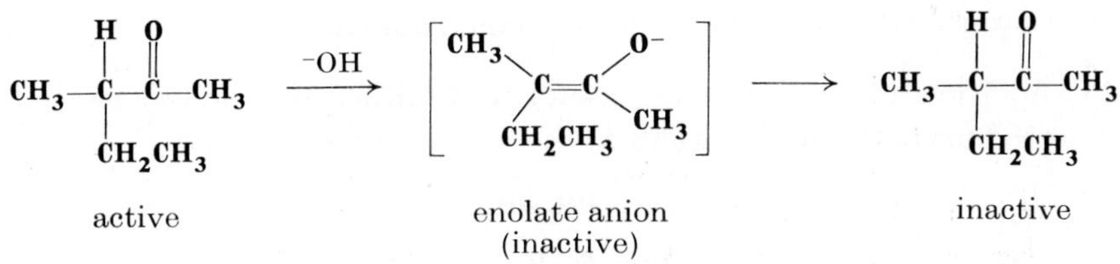

21. Let

k_a = the rate constant for reaction of pure axial OH
k_e = the rate constant for reaction of pure equatorial OH
N_a = fraction of cyclohexanol molecules with axial OH
N_e = fraction of cyclohexanol molecules with equatorial OH
k = the rate constant for reaction of cyclohexanol

$$k = N_a k_a + N_e k_e$$
$$1.24 = N_a \times 2.97 + N_e \times 1.00$$
$$1.24 = N_a \times 2.97 + (1 - N_a) \times 1.00$$
$$N_a = 0.12 \quad \text{and} \quad N_e = 0.88$$

OH groups in cyclohexanol are 12% axial and 88% equatorial.

22. Yes. The two mirror images below are nonsuperimposable.

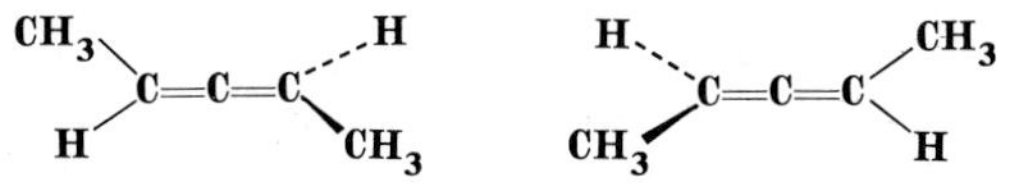

Note that because of the central sp hybridized carbon, the allene is not planar.

23. The desired "antiparallel" arrangement is achieved only in the less stable, diaxial conformation. The hydrogen on the carbon bearing the methyl group does not have the proper geometry for elimination, whereas the other one does.

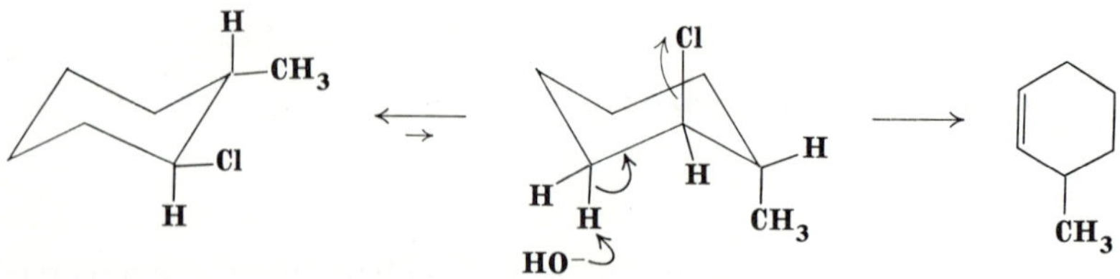

24. They are not optically active. The two alcohols are diastereoisomers (with different chemical and physical properties and different energy), and they are not necessarily formed in equal amounts.

25. Addition to the aldehyde occurs from either side of the molecule to produce both configurations at the new asymmetric center.

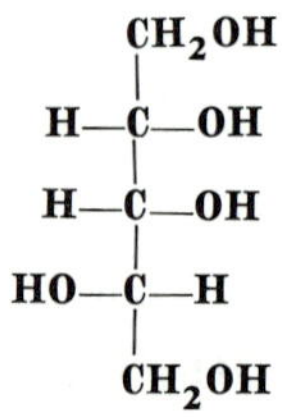

optically active

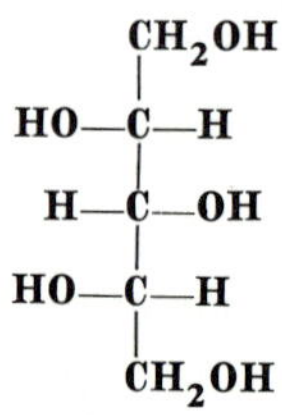

optically inactive
(*meso*)

26. Yes.

CHAPTER 6

1. An aldehyde.
2. Acetone, CH_3COCH_3.
3. 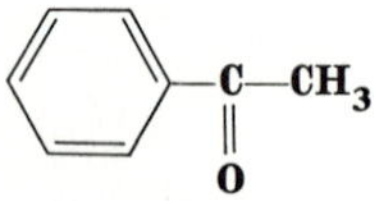

4. a) 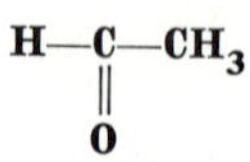 b) 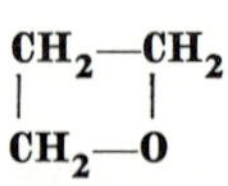 c) 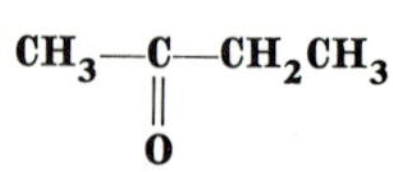

d) 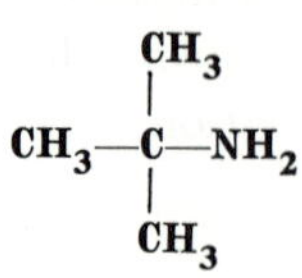 e) 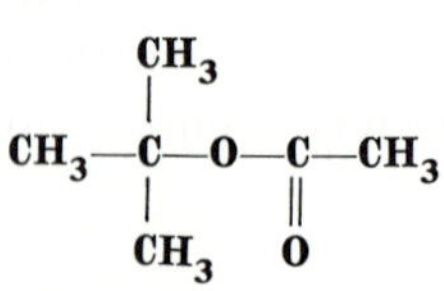 f) 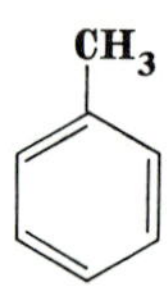

g) 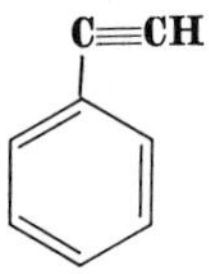

h) 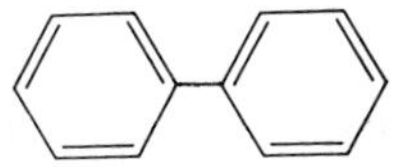

i) $ClCH_2CH_2CH_2Cl$

5.

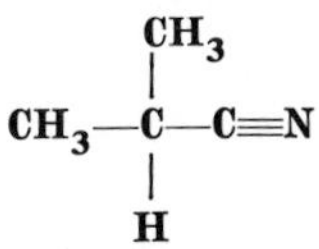

6. Either *ortho*, *meta*, or *para*-methyl phenol.

7.

a) 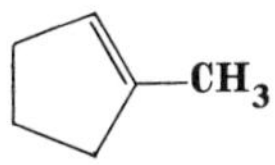

b) 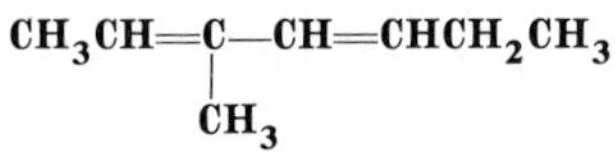 or 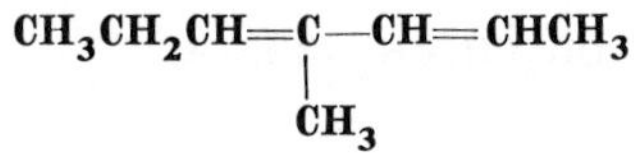

c) 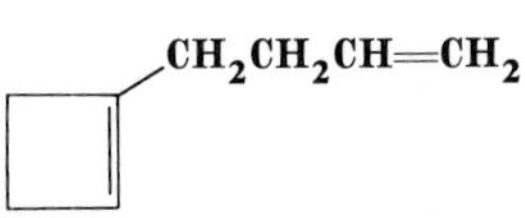 or 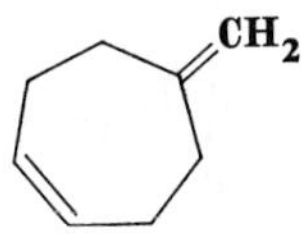

d) 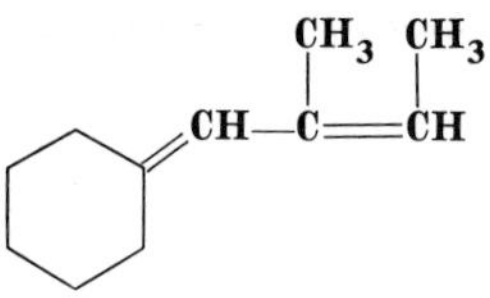 or 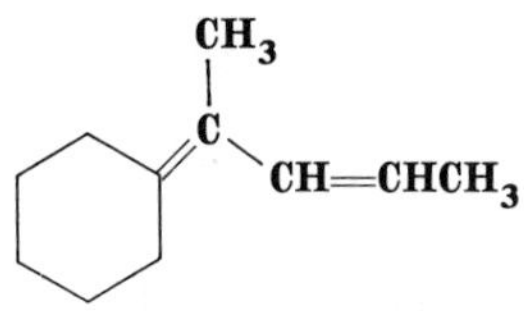

8.

a) CH_3CHO; $>C{=}O$ stretching, $\sim 1730\ cm^{-1}$

b) $(CH_3)_2CHNH_2$; N—H stretching, $\sim 3400\ cm^{-1}$

c) $CH_3CH_2CH{=}CH_2$; $C{=}C$ stretching, $\sim 1650\ cm^{-1}$

d) $CH_3CH_2C{\equiv}CH$; $\equiv C{-}H$ stretching, $\sim 3300\ cm^{-1}$

e) CH_3CO_2H; $>C{=}O$ stretching, $\sim 1725\ cm^{-1}$

f) CH_3CH_2CHO; $-\underset{O}{\overset{\|}{C}}-H$ stretching, $\sim 2720\ cm^{-1}$

9.

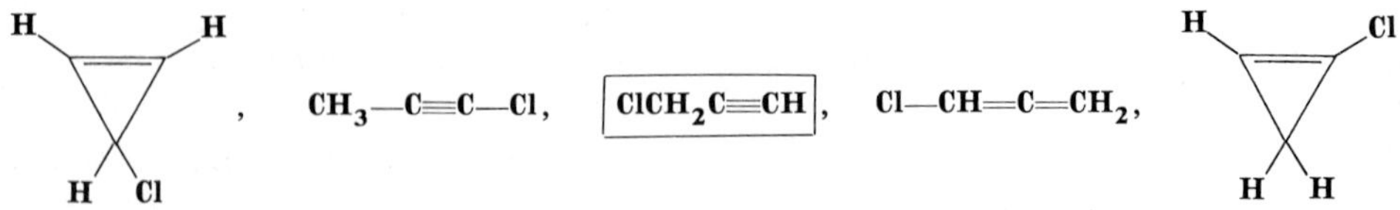

10. a) One b) Four c) Two d) Two e) Three f) Two g) Three
h) Four i) Five

11.

a) Compound A: 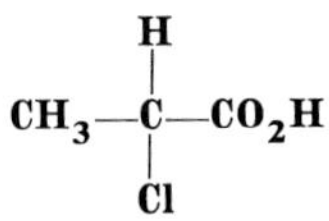 b) Compound B: 

c) Compound C: $CH_3OCH_2CH_2CN$

12.

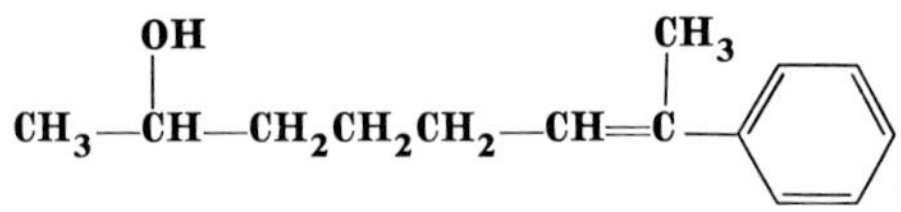

A

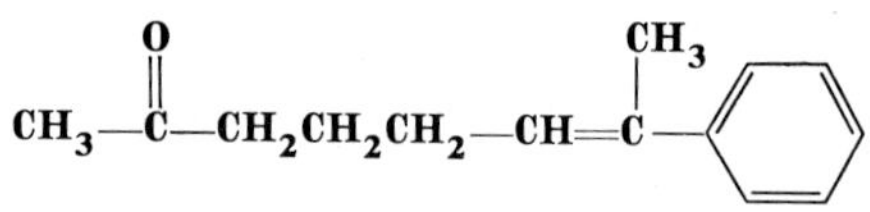

B

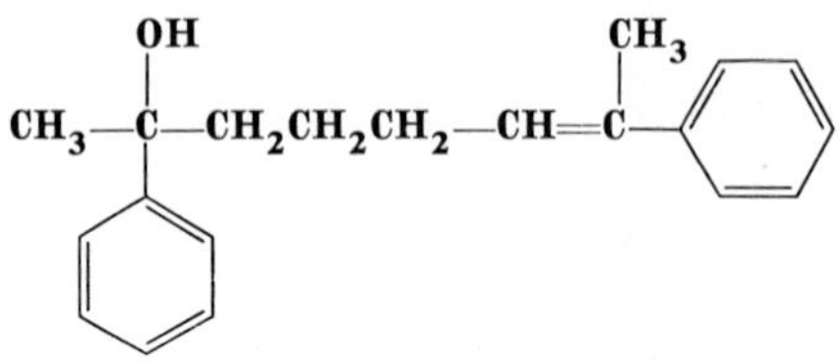

C

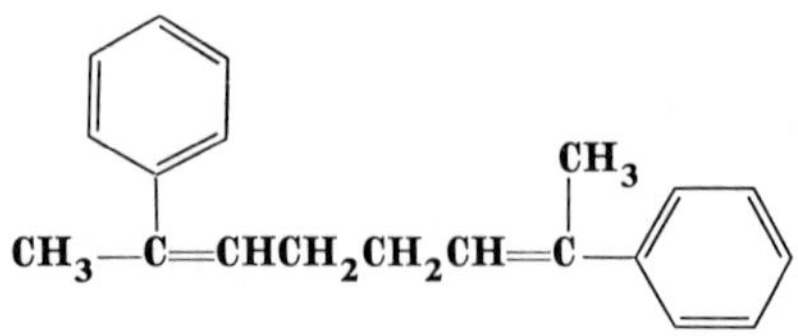

D

13. I,3; II,5; III,4; IV,1; V,2

14. a) The amide carbonyl group has more single-bond character than the ketone carbonyl group because of an important resonance contributor:

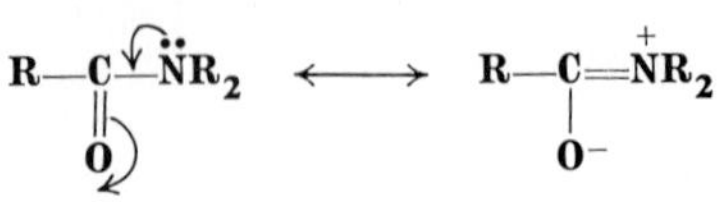

Since a single bond is easier to stretch than a double bond, it will take less energy to stretch the amide carbonyl than to stretch the ketone carbonyl. Hence, the absorption frequency of an amide is lower than that of a ketone.

b) The oxygen of $ROCH_3$ inductively withdraws electrons away from the methyl group. This diminished electron density around the methyl protons of $ROCH_3$ means that there is less "local opposition" to the applied magnetic field. Hence, the methyl protons of $ROCH_3$ resonate at lower applied fields than do the methyl protons of RCH_3. Stated in another way, the methyl protons of $ROCH_3$ are less shielded by valence electrons than are the methyl protons of RCH_3.

CHAPTER 7

1.

a) 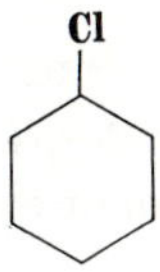

b) 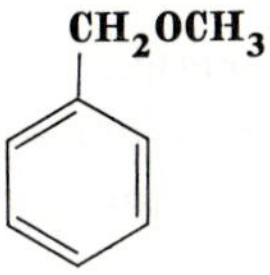

c) 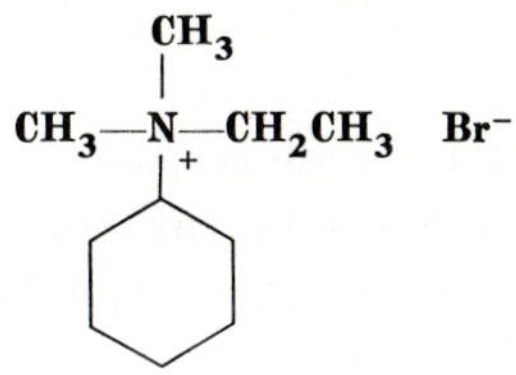

d) 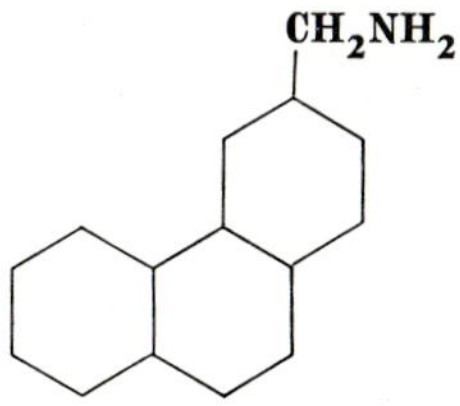

e) 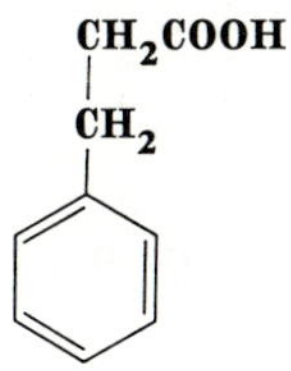

f) 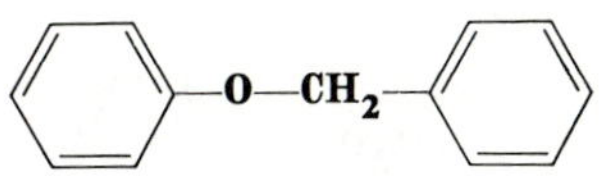

g) 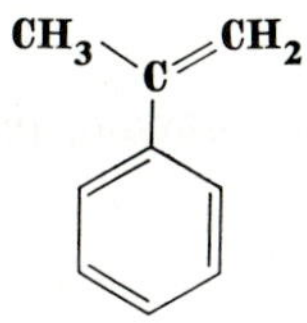

h) 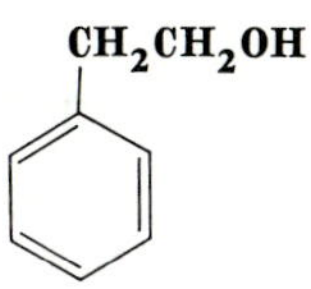

i) 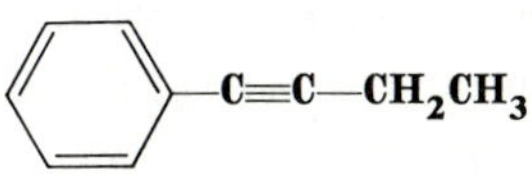

2. a) more b) carbonium c) elimination d) tertiary e) backside f) malonic g) epoxide h) Gabriel i) ammonia j) two k) polarizable l) $SOCl_2$ m) bromobenzene

3.

a) H_2SO_4 b) $CH_3CH_2O^- \ Na^+$ c) $(CH_3CH_2)_2C{=}CHCH_3$ d) $CH_2{=}CHCH_2CH_2C{\equiv}CH$

e) 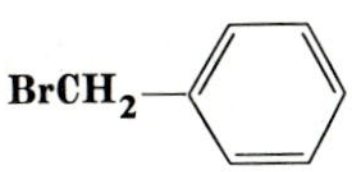

f) $H_2 + Pt$ g) 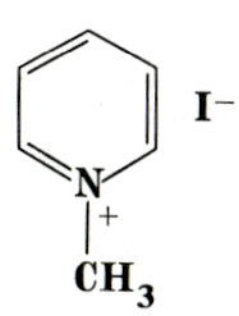

h) $SOCl_2$

4.

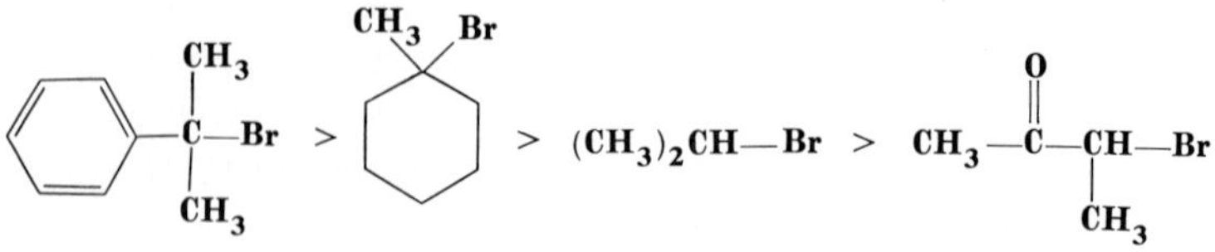

5. S_N1 reactions proceed faster in methanol than in benzene because methanol has a higher dielectric constant and is better able to solvate the intermediate carbonium ion.

6. The bicyclic bromide resists the S_N1 reaction because the ring system prevents the resulting carbonium ion from achieving the desired planarity. It resists the S_N2 reaction because the nucleophile cannot reach the back side of the carbon.

7.

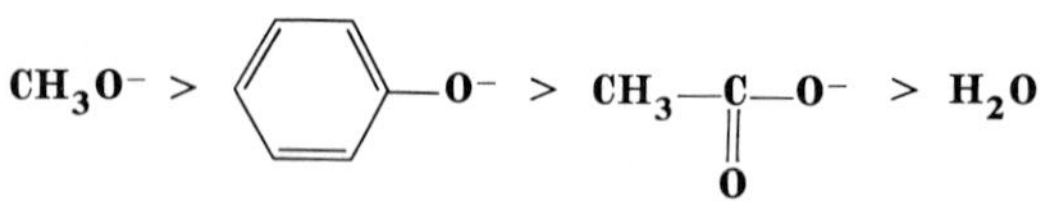

8. The enzymatic reaction, involving a group E on the enzyme, proceeds by means of *two* S_N2 displacements:

$$E + R - X \rightarrow E - R$$
$$E - R + Y \rightarrow E + R - Y$$

(retained configuration)

Another possible mechanism involves carbonium ion formation *via* an S_N1 reaction with the backside of the molecule blocked by the enzyme. Thus, approach of the nucleophile is possible only from the front side.

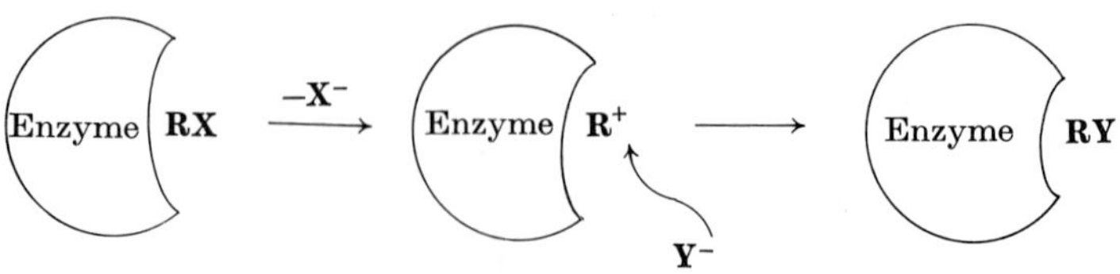

9.

a) 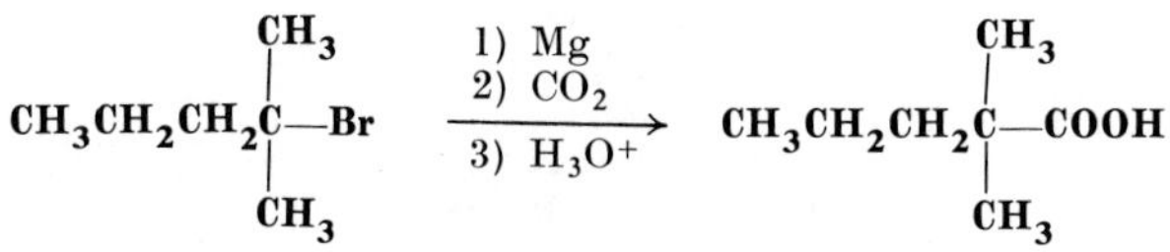

b) 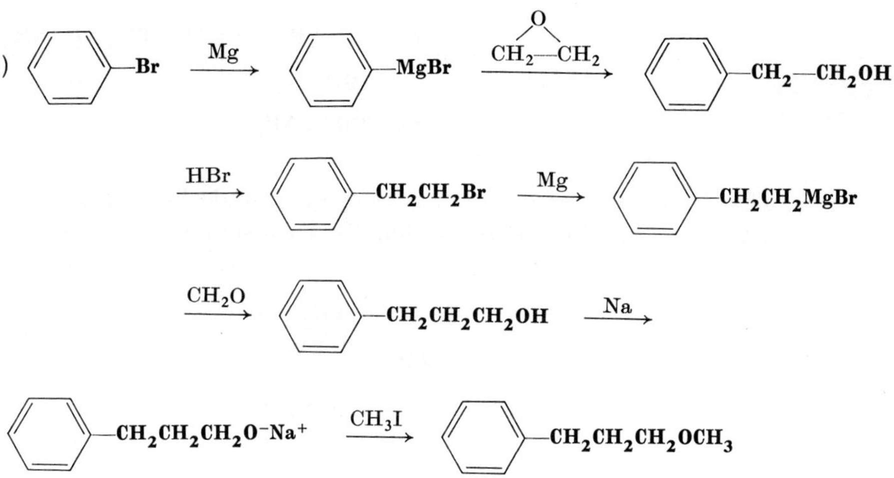

c) 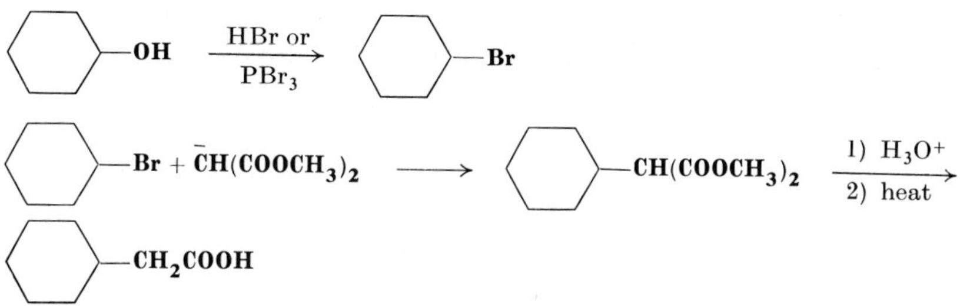

d) 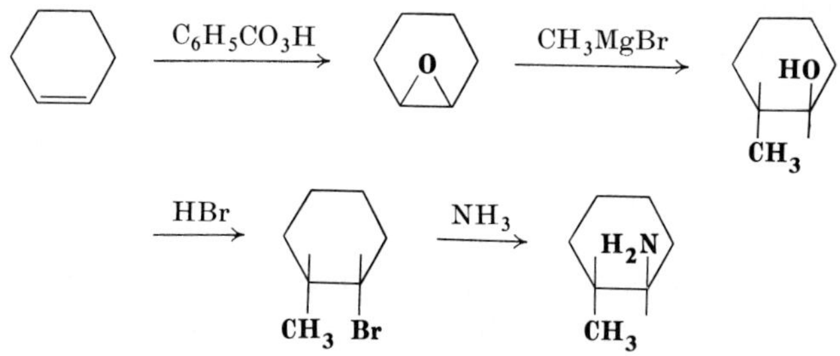

10.

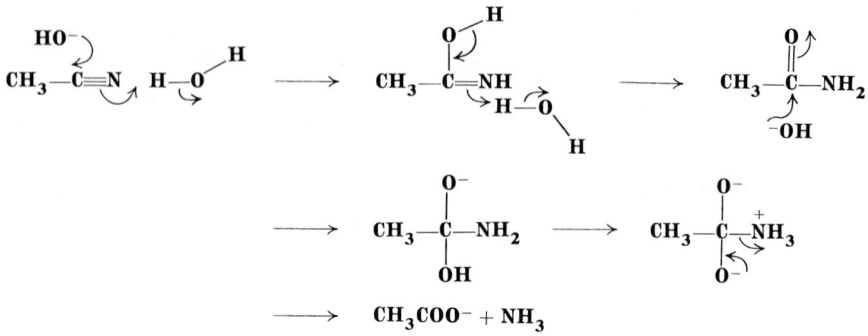

11. The alkylating species is not a primary chloride but a reactive three-membered ring formed by intramolecular displacement of the chloride by nitrogen:

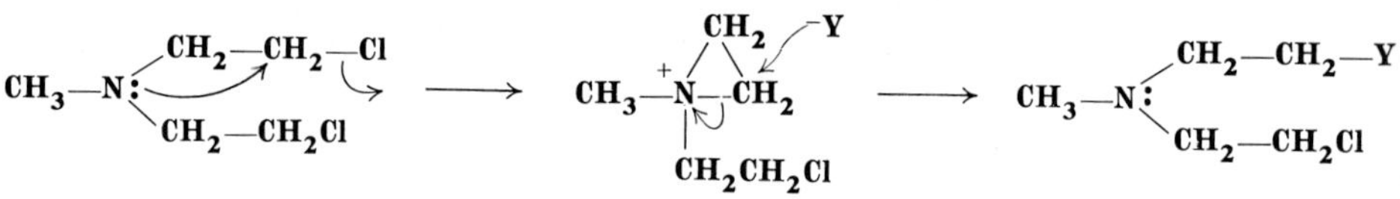

12. In the first system, the hydroxyl group is in the correct position for an intramolecular back-side attack, thereby forming the epoxide. In the second reaction, intramolecular back-side substitution is not possible, and the alkyl bromide simply reacts with hydroxide ion.

13. Rate of product formation

$= k_3[R^+]$

Rate of change in concentration of R^+

$= k_1[RBr] - k_2[R^+][Br^-] - k_3[R^+] = 0$

Thus

$$[R^+] = \frac{k_1[RBr]}{k_3 + k_2[Br^-]}$$

rate of product formation

$$= \frac{k_1 k_3[RX]}{k_3 + k_2[Br^-]}$$

The progress of the reaction can be followed by titrimetrically determining the release of acid with standard base as a function of time.

14. Two reactions occur in $0.1M$ sodium acetate in acetic acid:

a) $R_2CHI + CH_3COO^- \longrightarrow R_2CHOCCH_3$ (C=O)

(inversion)

$$\text{rate} = k_2[R_2CHI][CH_3COO^-]$$

b) $R_2CHI \longrightarrow R_2CH^+ \xrightarrow{HOAc} R_2CHOCCH_3$ (C=O)

(racemization)

$$\text{rate} = k_1[R_2CHI]$$

At $1M$ acetate, the first reaction (whose rate depends on acetate) becomes the predominant one, and all the product is inverted.

15. Pyridine was used. The second step of the first sequence must go with inversion.

16.

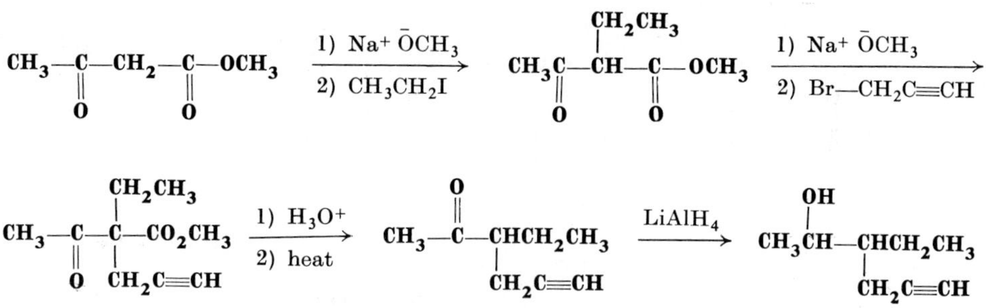

17.

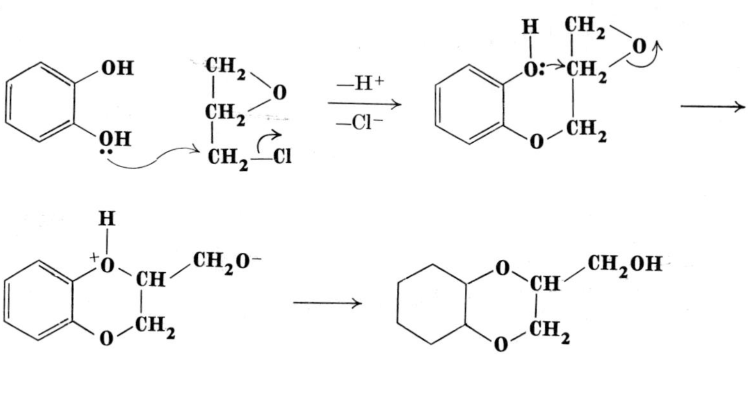

CHAPTER 8

1. A) $LiAlH_4$ B) CrO_3 C) H_2SO_4 D) HBr E) KOH F) Mg G) CH_3CHO H) H_2SO_4 or $POCl_3$ I) CH_3MgBr J) Na, CH_3I K) H_2SO_4 L) Br_2 M) Zn

2. a) pyrolysis b) Saytzeff c) dehydration d) coplanar e) amine oxide f) E1 g) *trans* h) vicinal i) acidic j) rate determining k) *trans* l) E1

3. a) 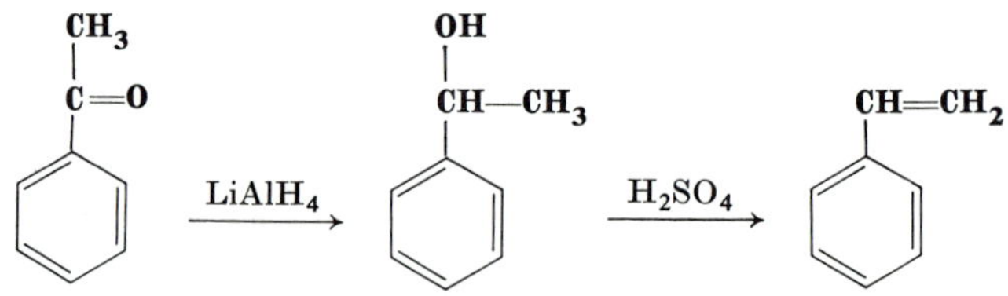

b) 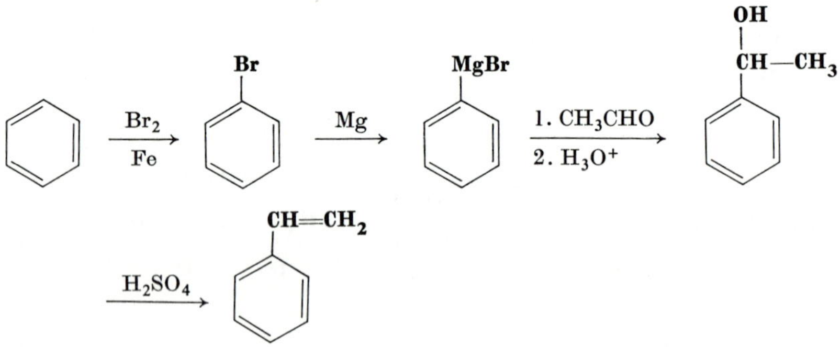

c) 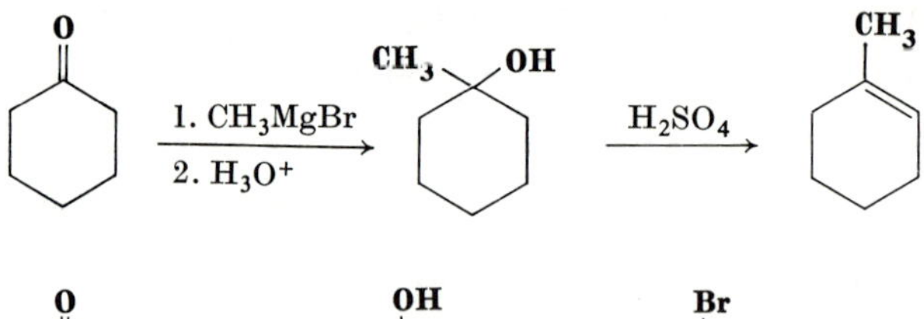

d) 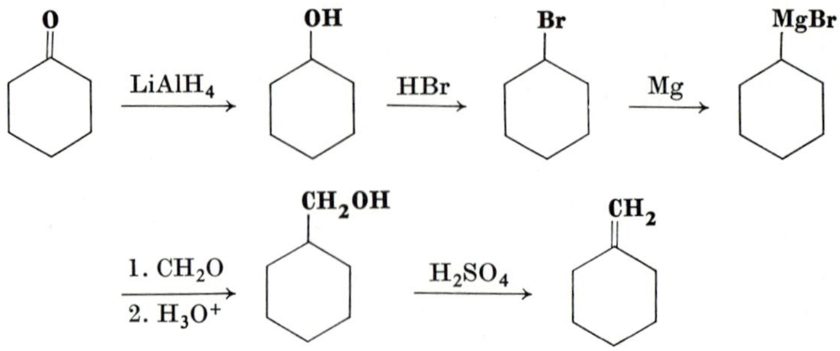

e)

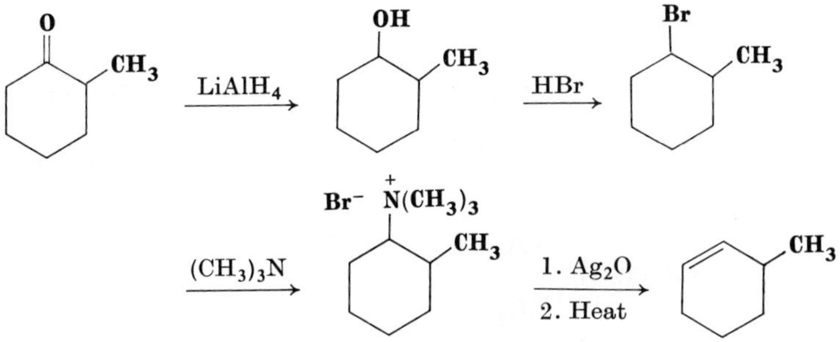
O
CH3
LiAlH4
OH
CH3
HBr
Br
CH3
(CH3)3N
Br⁻ N(CH3)3
CH3
1. Ag2O
2. Heat
CH3

f)

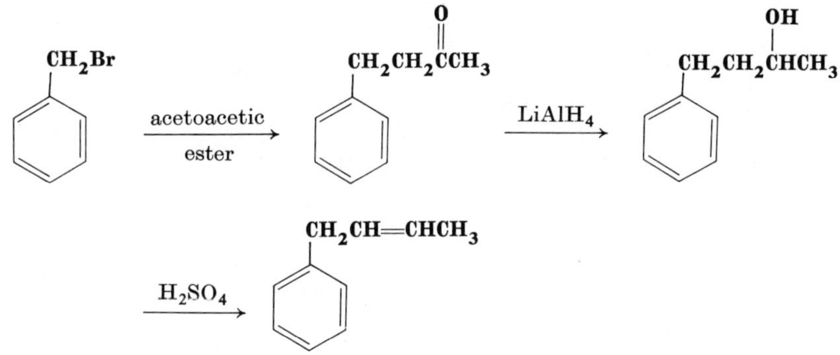
CH2Br
acetoacetic
ester
O
CH2CH2CCH3
LiAlH4
OH
CH2CH2CHCH3
H2SO4
CH2CH=CHCH3

g)

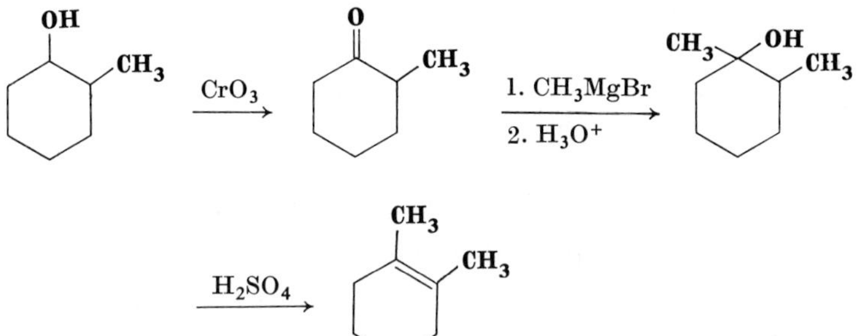
OH
CH3
CrO3
O
CH3
1. CH3MgBr
2. H3O+
CH3
OH
CH3
H2SO4
CH3
CH3

h)

1.

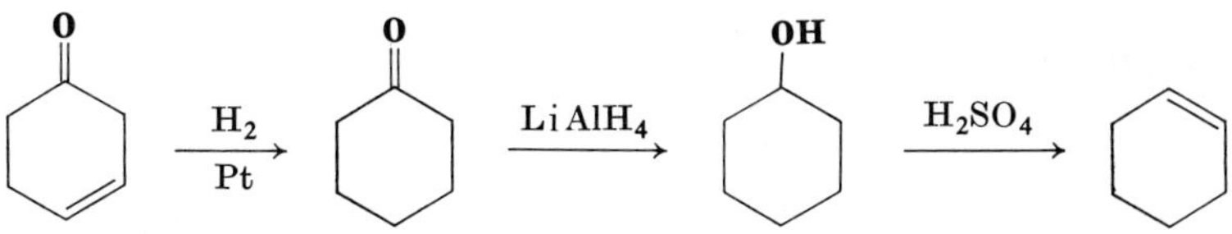
O
H2
Pt
O
Li AlH4
OH
H2SO4

2.

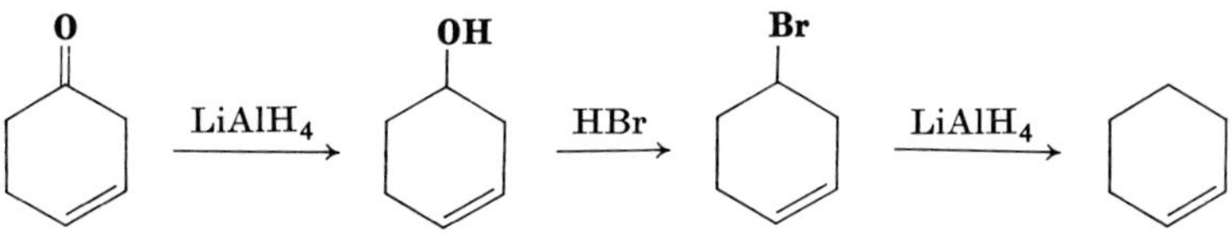
O
LiAlH4
OH
HBr
Br
LiAlH4

4.

a) 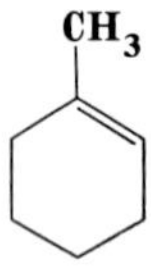

b) 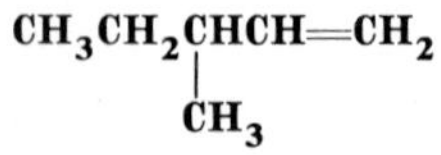

c) 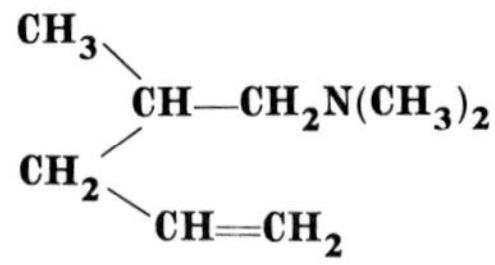

d) 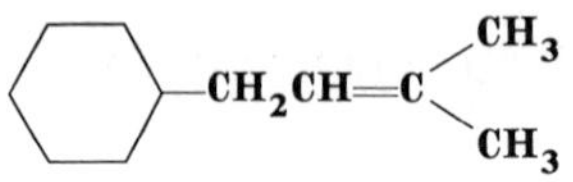

e) 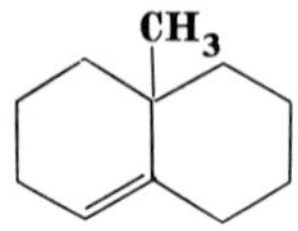

5.

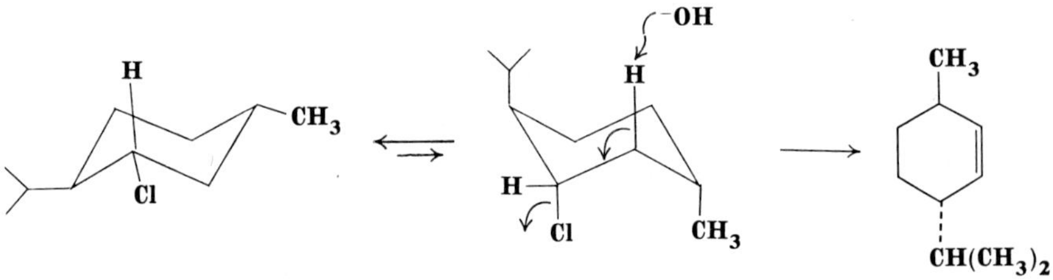

A B

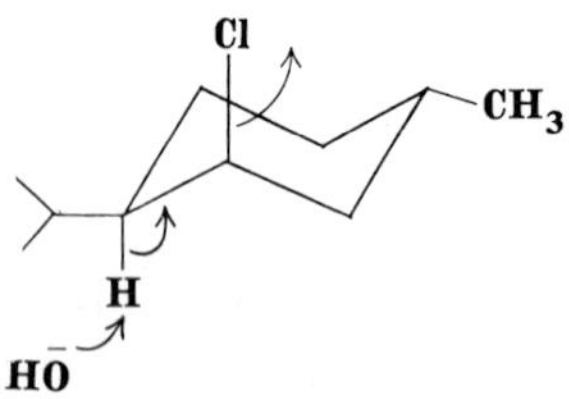

C

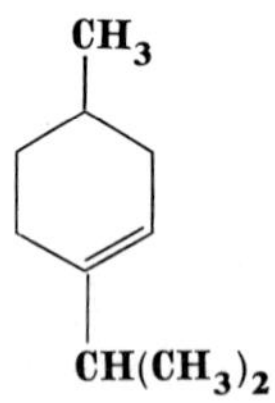

D

In the first reaction, the *trans*-coplanar requirement overrides the Saytzeff rule. In the second reaction, there is a choice between two axial hydrogens, and the hydrogen is removed which gives the Saytzeff product.

6.

a) 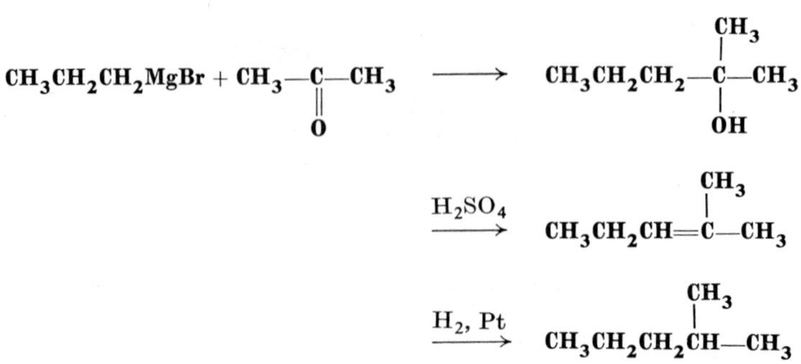

b) 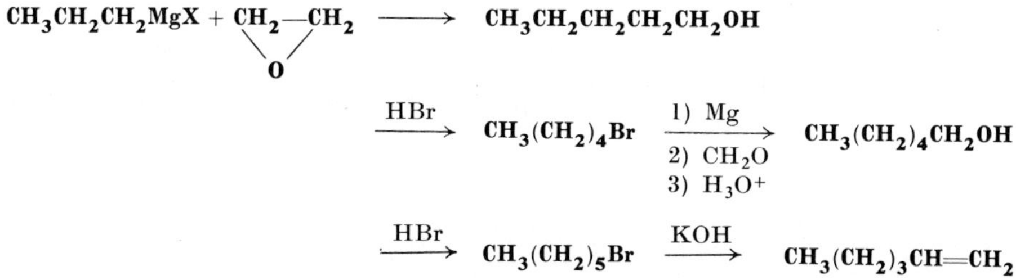

c) 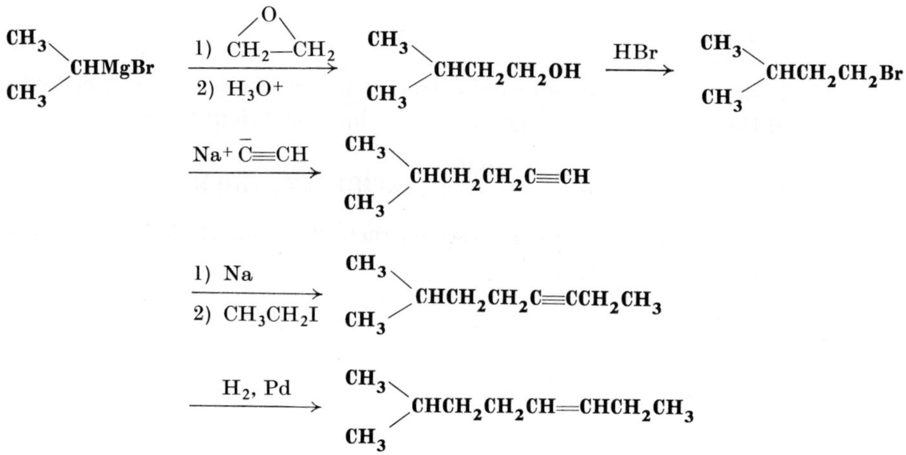

d) 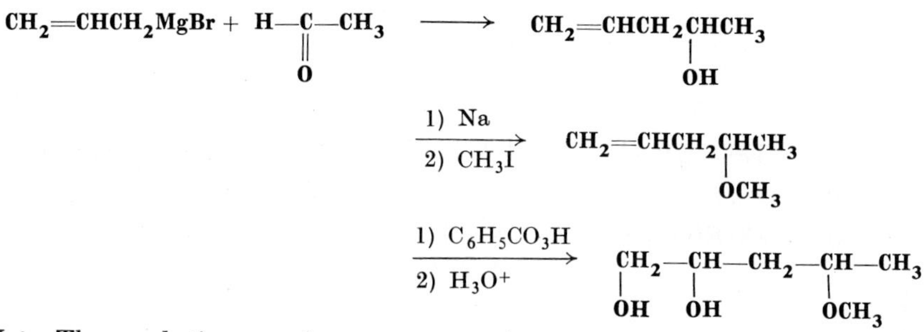

Note: These solutions are by no means unique.

7. 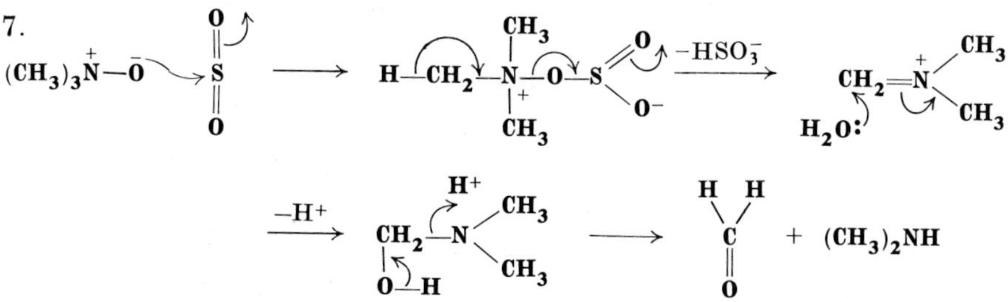

8.

a) 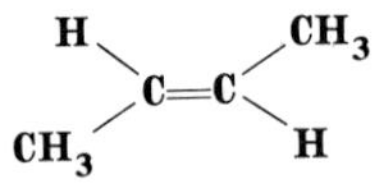

trans

b) 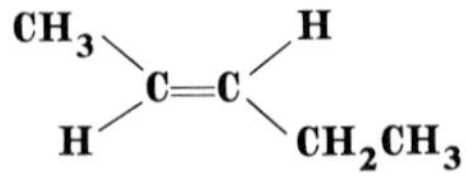

trans

c) 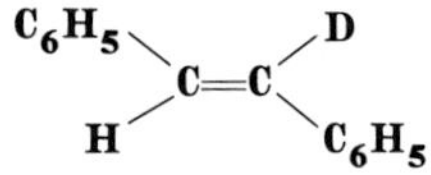

d) 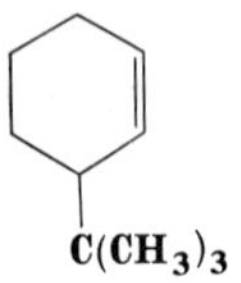

e) 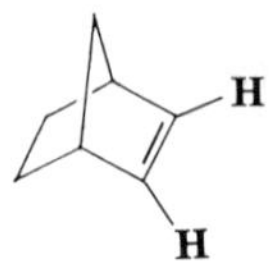

9. If the first mechanism were correct, then one would have expected some of the recovered starting material to have lost deuterium:

$$\phi\text{—}\bar{C}D\text{—}CH_2\text{—}\overset{+}{N}(CH_3)_3 \xrightarrow{EtOH} \phi CHD\text{—}CH_2\text{—}\overset{+}{N}(CH_3)_3$$

If the second mechanism were correct, the trimethylamine would have contained deuterium.

10.

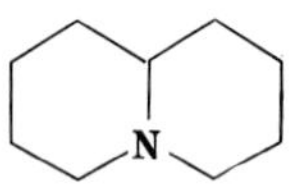 among others

11. rate of product formation

$$= k[R_2\bar{C}\text{—}CXR_2]$$

$$K = \frac{[R_2\bar{C}\text{—}CXR_2]}{[EtO^-][R_2CH\text{—}CXR_2]}$$

Thus

$$\text{rate} = kK[EtO^-][R_2CH\text{—}CXR_2]$$

12.

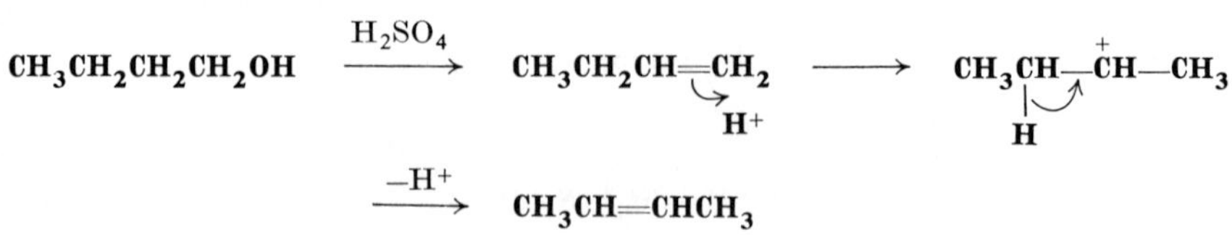

This side reaction is of no concern in the dehydration of cyclohexanol because double-bond isomerization leads to no change in the product (cyclohexene). Double-bond migration in the dehydration product of the other alcohol is unlikely because it would lead to a less substituted and less stable double bond.

13.

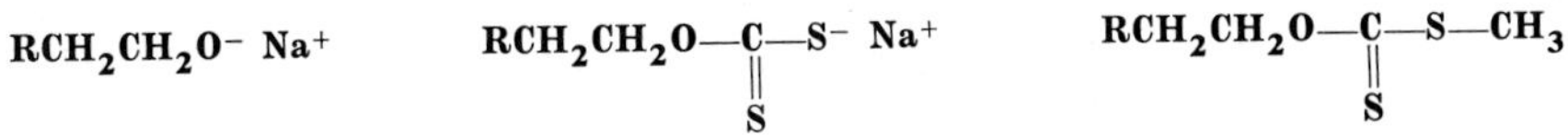

A B C

14. Perform the reaction with aromatic halide labeled with ^{14}C as shown:

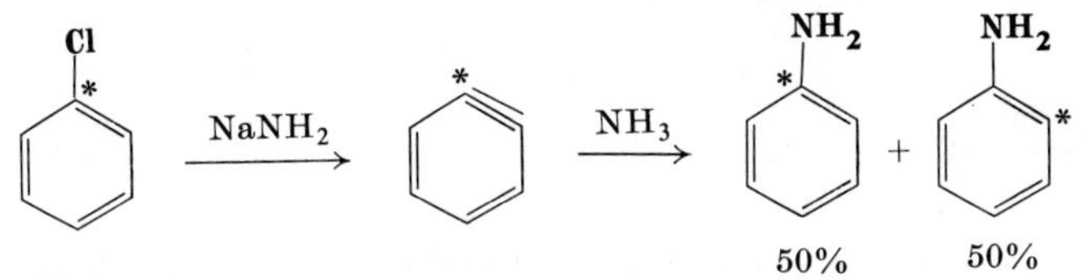

Product from an $\bar{S}_N1$ or $\bar{S}_N2$ reaction:

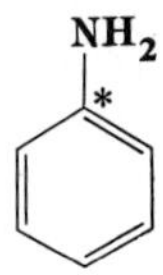

15.

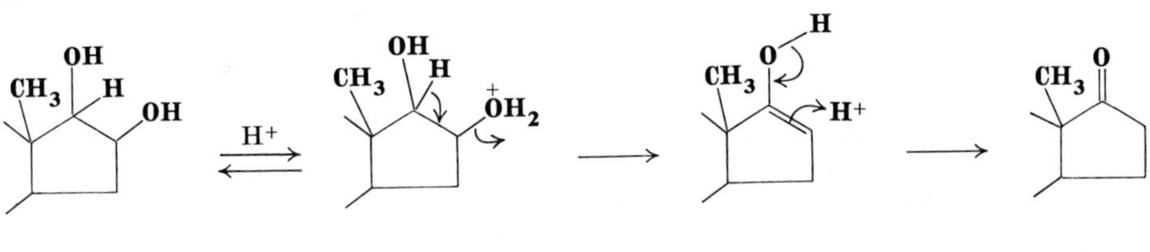

16.

a)

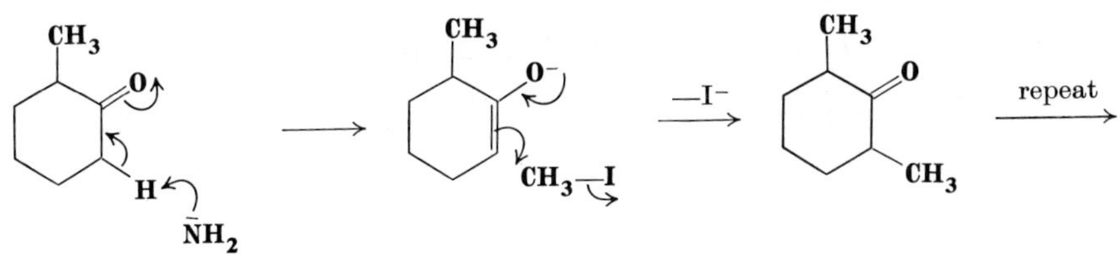

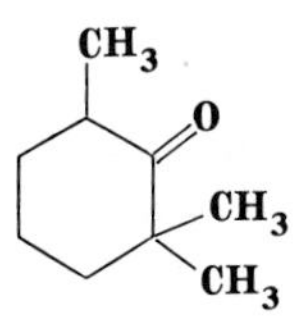

b) 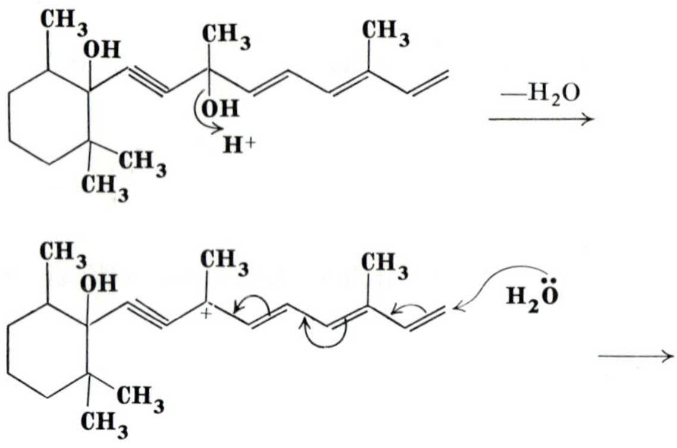

5

c) E1. The carbonium ion is tertiary and highly resonance stabilized.

d) The primary alcohol is much less sterically hindered than the tertiary alcohol, and it reacts preferentially.

17. a) E2 mechanism

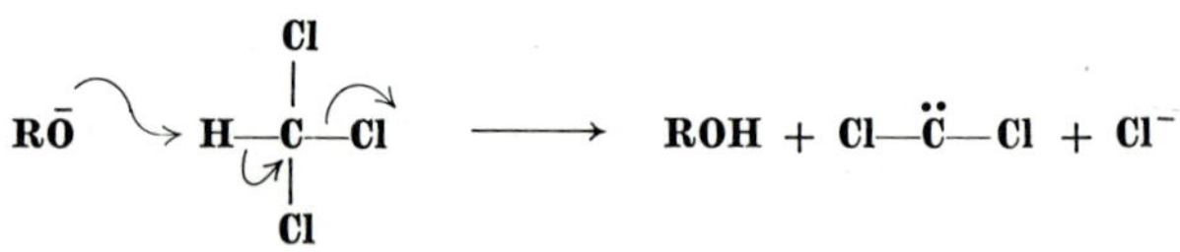

b) E1cb mechanism

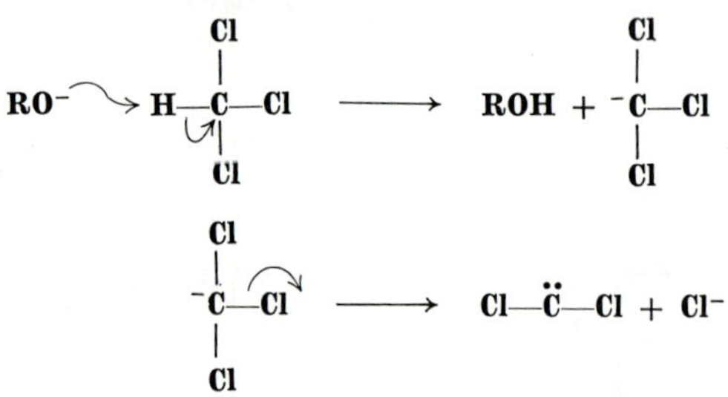

CHAPTER 9

1. a) dibromide b) alcohol c) cyclic d) Michael e) Markovnikov f) glycol g) acetone h) Lindlar i) NBS j) *trans* k) ion pair l) dienophile m) kinetic; thermodynamic n) peracid

2.

a) 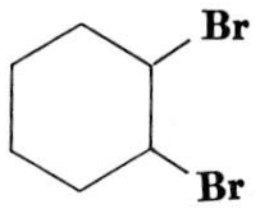 b) 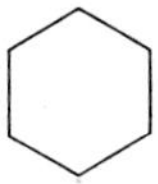 c) NR d) 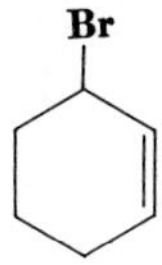

e) 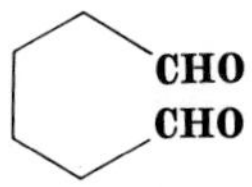 f) NR g) 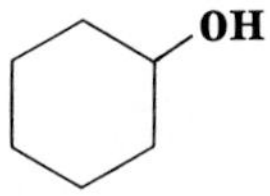 h) 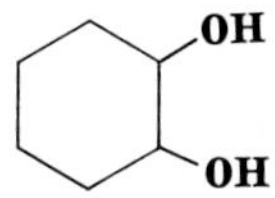

i) 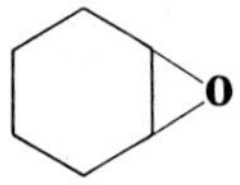 k) NR l) NR

3.

$CH_2{=}CH{-}C({=}O){-}CH_3$ $\quad$ $(CH_3)_2N{-}CH_2{-}CH_2{-}C({=}O){-}CH_3$

4.

a) $CH_2{=}CH{-}C({=}O){-}CH_3$ b) 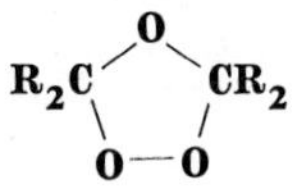 c) $CH_3{-}CH(OH){-}CH(OH){-}CH_3$

d) $CH_3{-}C({=}O){-}O{-}OH$ e) 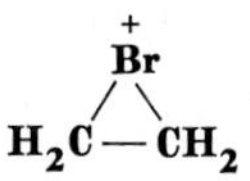 f) HO—OH g) 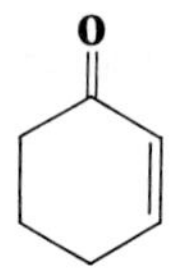

h) HOCl i) 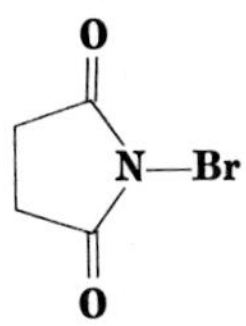 j) 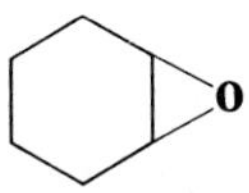

5.

a) $CH_3{-}C(CH_3)(Br){-}CH_2CH_2CH_3$ b) $KMnO_4$ c) O_3, $Zn + H_2O$ d) $CH_3CH(NHCH_3)CH_2C({=}O)CH_3$

e) BH_3, $H_2O_2 + NaOH$ f) H_2, Pd g) 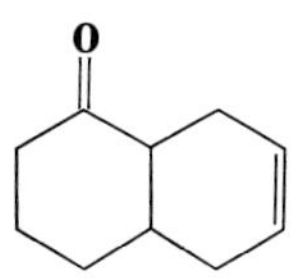 h) $HBr + H_2O_2$

i) 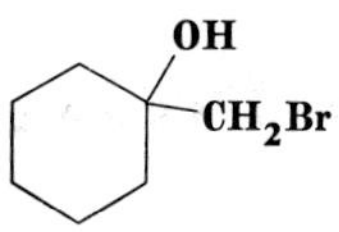 j) N-bromosuccinimide

6. Hydration of a triple bond leads to an enol which then tautomerizes to the carbonyl compound:

$$R{-}C{\equiv}C{-}R \xrightarrow[H_3O^+]{HgSO_4} \left[R{-}\underset{\underset{OH}{|}}{C}{=}CHR\right] \longrightarrow R{-}\underset{\underset{O}{\|}}{C}{-}CH_2R$$

7.

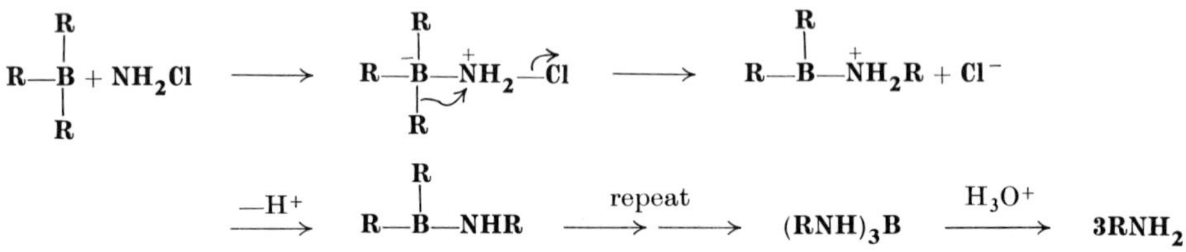

8.

a) $C_6H_5{-}CHBr{-}CHBr{-}C_6H_5 \xrightarrow{Zn} C_6H_5CH{=}CHC_6H_5 \xrightarrow[2)\ Zn,\ H_2O]{1)\ O_3} 2C_6H_5CH$

b) $CH_3OOC{-}CH_2{-}COOCH_3 + CH_2{=}CHCH_2Br \xrightarrow{Na^+\ \bar{O}CH_3} CH_3OOC{-}\underset{\underset{CH_2CH=CH_2}{|}}{CH}{-}COOCH_3$

$$\xrightarrow[2)\ heat]{1)\ H_3O^+} HOOC{-}CH_2CH_2CH{=}CH_2$$

$$\xrightarrow{C_6H_5CO_3H} HOOC{-}CH_2CH_2CH{-}CH_2 \text{ (epoxide, with O bridging the last two carbons)}$$

c)

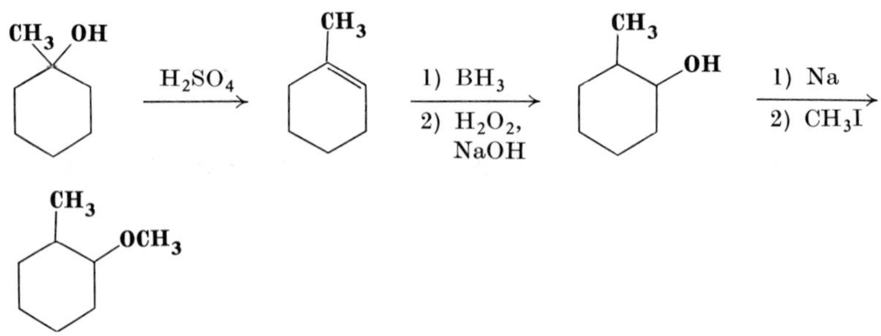

d)

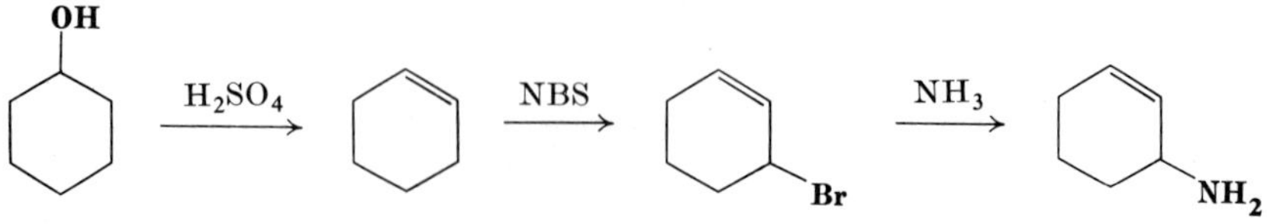

e) 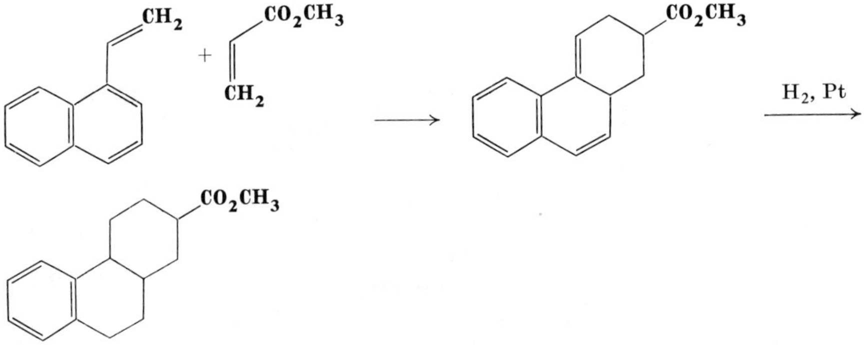

f) 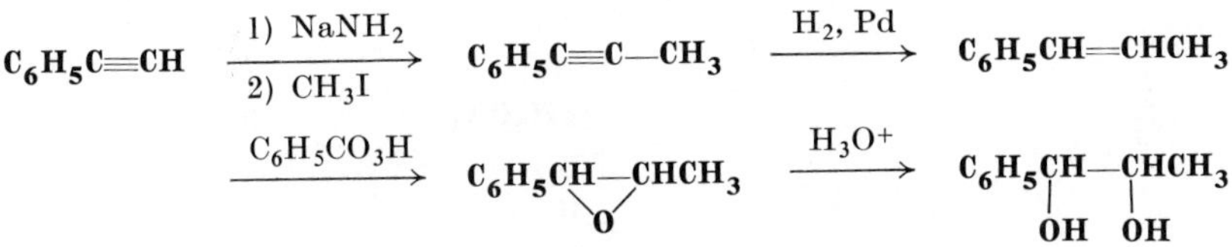

g) 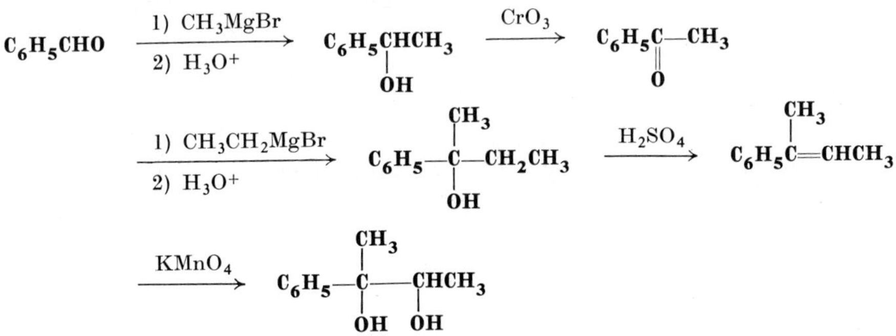

9. 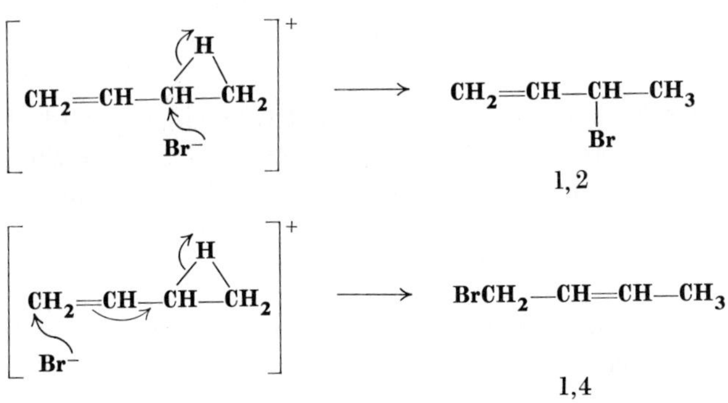

1,4 material is the more stable (the thermodynamic product). The 1,2 isomer forms faster, however, because nucleophilic attack by bromide occurs preferentially at the center with the lower electron density.

10.

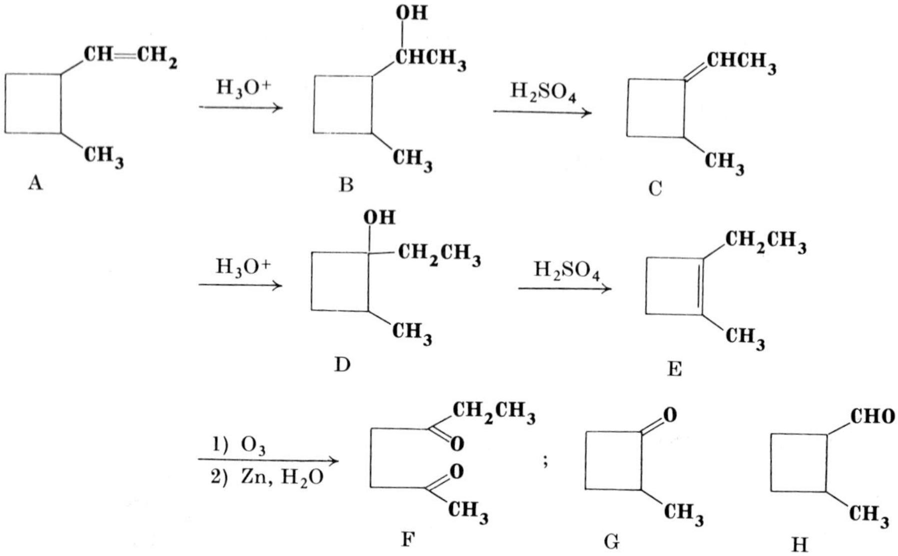

11.

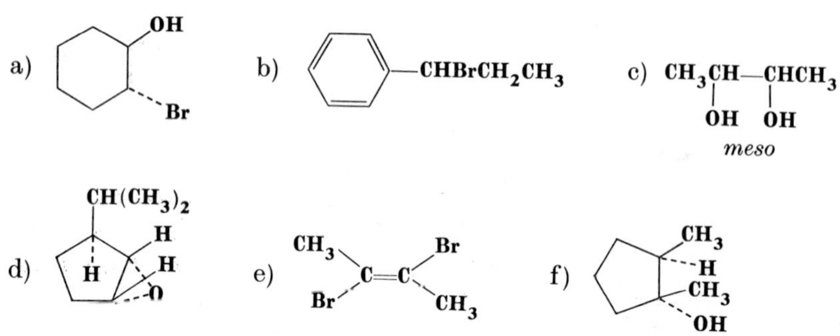

12. Bromine addition leads to more *trans* product than chlorine addition because bromine is better able to bear a positive charge in the cyclic intermediate. The cyclic intermediate is responsible for exclusive *trans* addition. Intermediate 3 (Fig. 9.1) is relatively more important in the case of chlorine.

13. No. The olefin would have picked up deuterium because the first step of the mechanism is reversible.

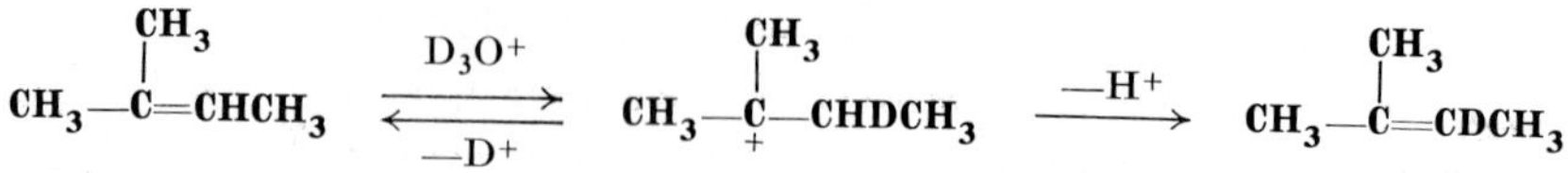

14.

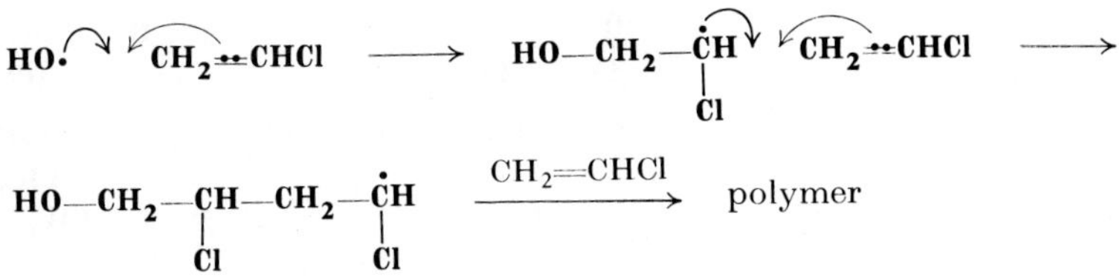

Add a small amount of a bifunctional monomer:

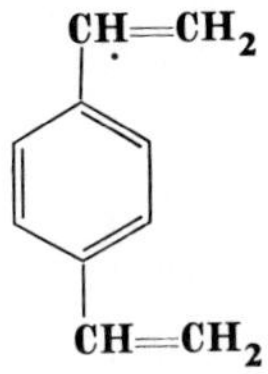

15. The product from the nonenzymatic synthesis of monodeuterated malic acid is A:

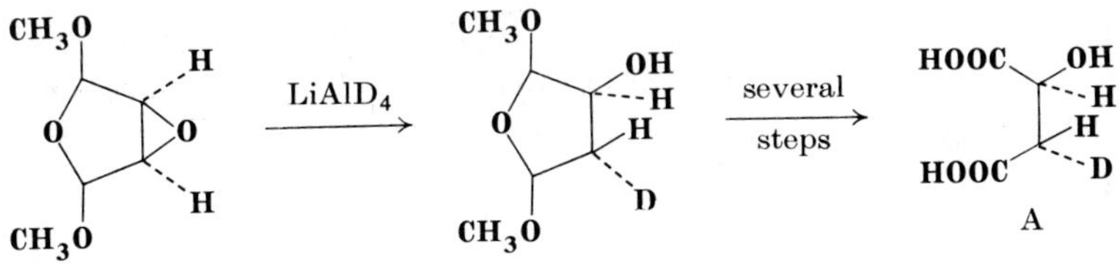

Since the enzymatic product is distinctly different, its structure must be B:

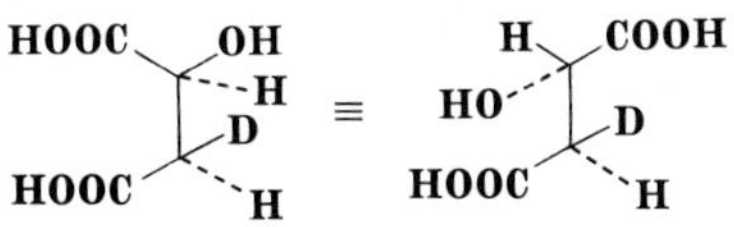

B

Clearly, the water (OH and D) have undergone *trans* addition.

16.

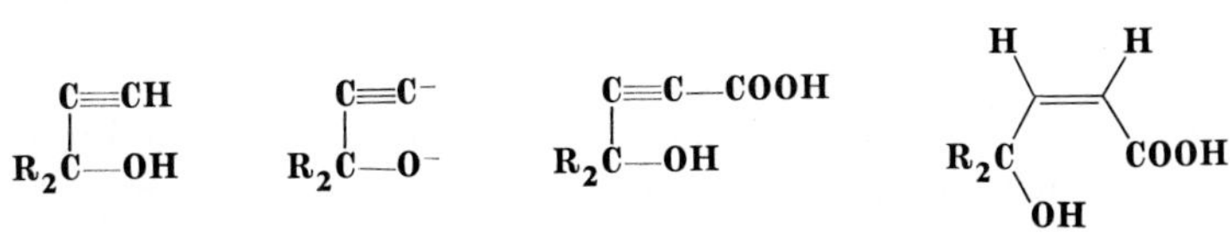

A B C D

17.

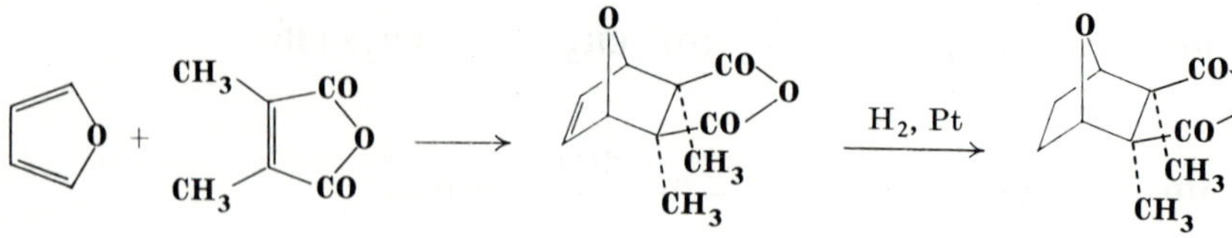

18.

$$(CH_3)_2C{=}CH_2 + H^+ \longrightarrow (CH_3)_2\overset{+}{C}{-}CH_3$$

$$(CH_3)_2\overset{+}{C}{-}CH_3 \quad CH_2{=}C(CH_3)_2 \longrightarrow (CH_3)_3CCH_2\overset{+}{C}(CH_3)_2$$

$$\xrightarrow{-H^+} (CH_3)_3CCH{=}C(CH_3)_2 + (CH_3)_3CCH_2C(CH_3){=}CH_2$$

19. b) The reducing agent, $LiAlH_4$, must approach the two carbonyl groups from the outside. Delivery of the hydride ions from the outside leads to hydroxyl groups pointing inwardly.

c) Yes.

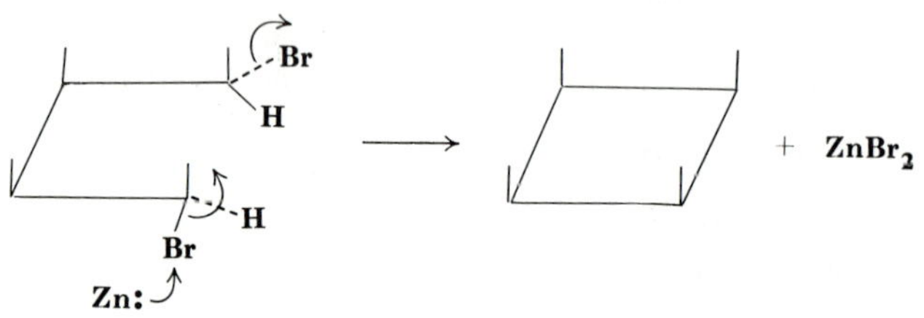

CHAPTER 10

1. Benzene is a highly resonance-stabilized system. Addition of bromine to benzene would destroy the aromatic stability; substitution to form bromobenzene retains the aromatic character of the compound. Simple addition to alkenes involves no loss of resonance energy.

2.

a) 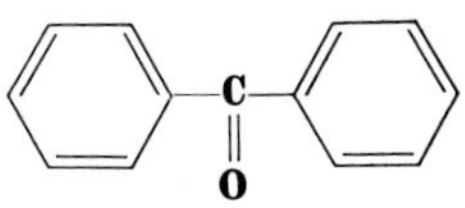

b) 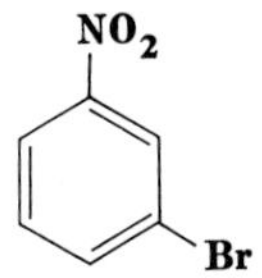

c) 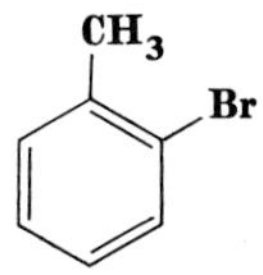 and 

d) 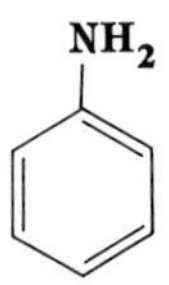

e) 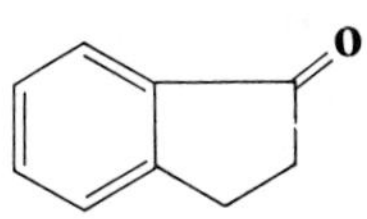

f) 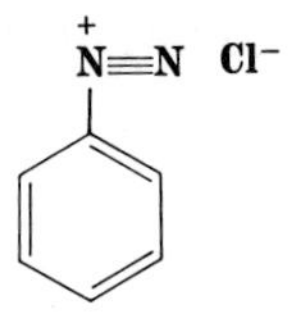

g) 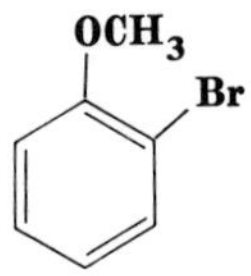 and 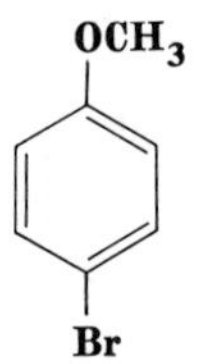

h) 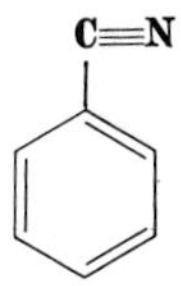

3. a) *o*-nitrotoluene and *p*-nitrotoluene
   b) acetanilide
   c) phenyl magnesium bromide
   d) benzenesulfonyl chloride
   e) benzenediazonium chloride
   f) *m*-nitrobenzene
   g) 1-bromonaphthalene
   h) benzyl chloride

4.

a) $—NO_2$

b) 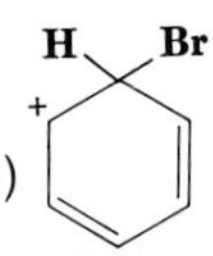

c) $FeBr_3$ or $AlCl_3$

d) 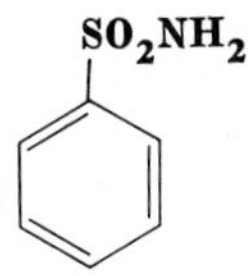

e) $CH_3\overset{\displaystyle O}{\overset{\|}{C}}—O—\overset{\displaystyle O}{\overset{\|}{C}}—CH_3$

f) 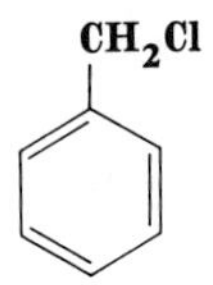

g) $—CH_3$

h) 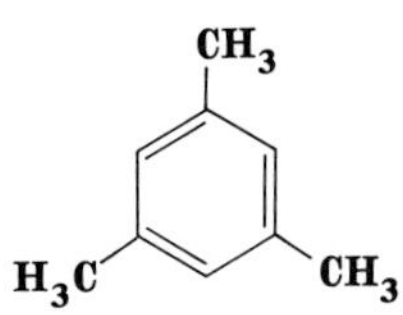

i) 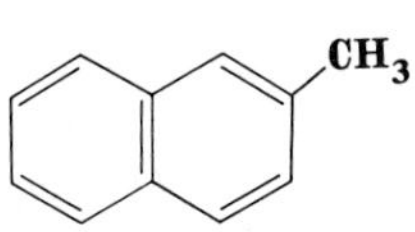

j) $CH_3\overset{+}{C}{=}O$

k) 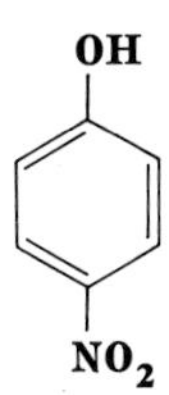

l) 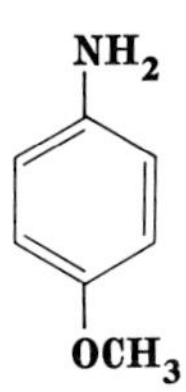

5.

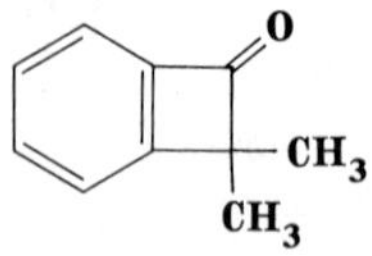

6. A: CH_3COCl; B: HNO_3, H_2SO_4; C: H_3O^+;
D: HNO_2, H_3PO_2; E: H_2, Pt; F: HNO_2, H_3O^+

7.

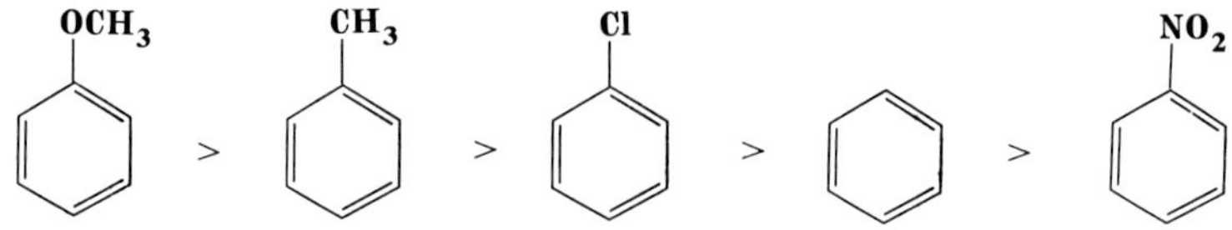

8.

a)

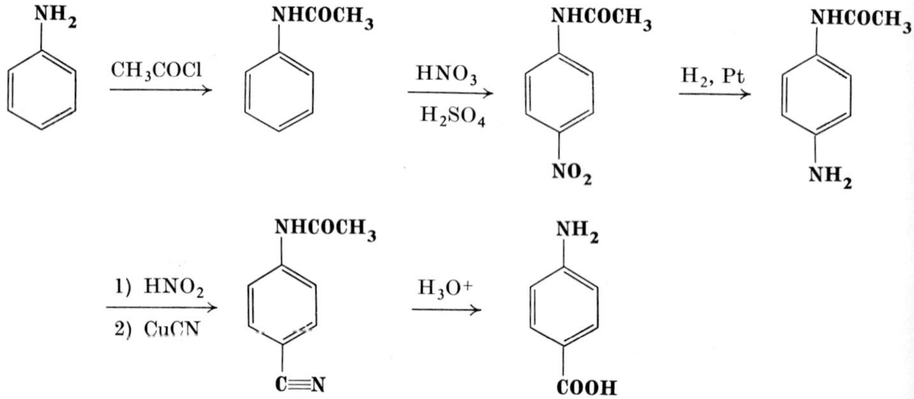

b)

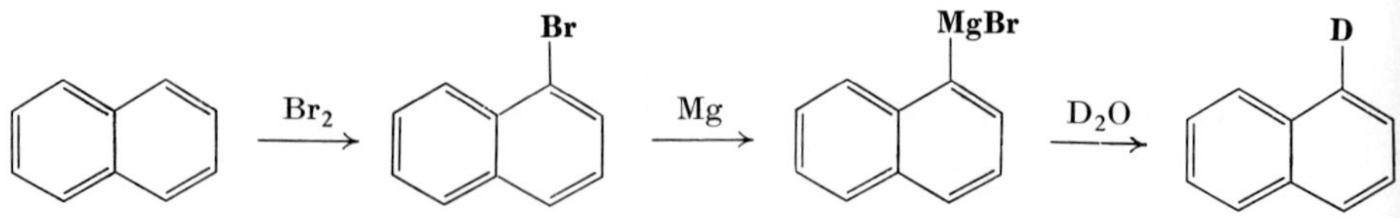

c)

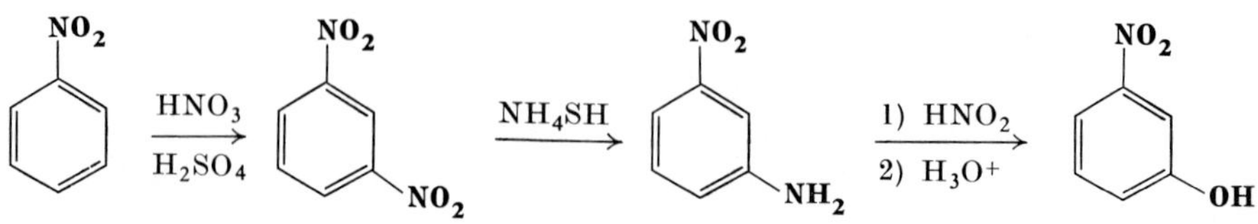

d)

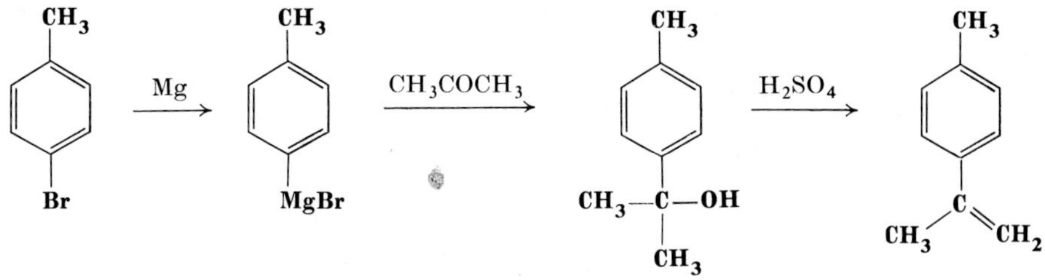
CH3
Br
Mg
MgBr
CH3COCH3
CH3—C—OH
CH3
H2SO4
C
CH2

e)

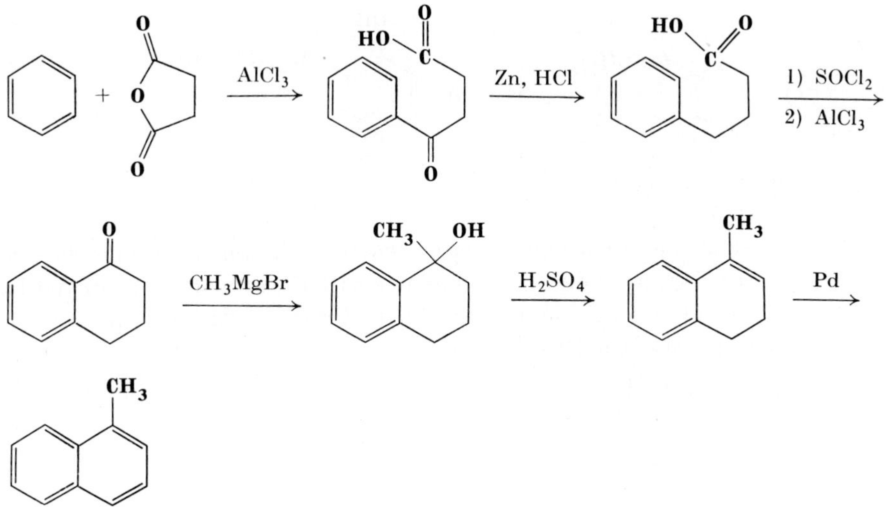
AlCl3
HO
Zn, HCl
1) SOCl2
2) AlCl3
CH3MgBr
OH
H2SO4
Pd

f)

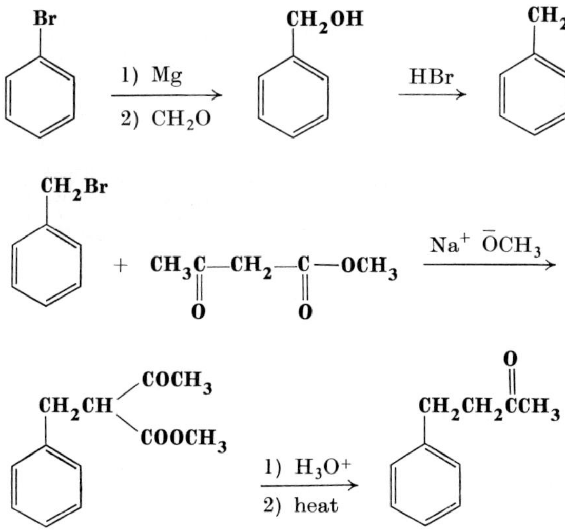
Br
1) Mg
2) CH2O
CH2OH
HBr
CH2Br
Na+ OCH3
CH3C—CH2—C—OCH3
COCH3
CH2CH
COOCH3
1) H3O+
2) heat
CH2CH2CCH3

9.

a) 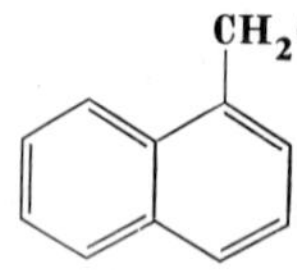

b) 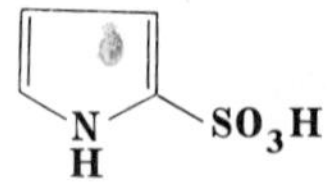

c) 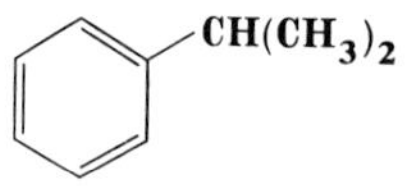

d) 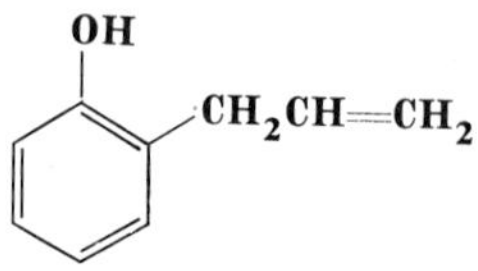

10. At higher concentrations, intermolecular acylation predominates. That is, the acyl group of one molecule reacts with the aromatic ring of another. The rate of this intermolecular Friedel-Crafts acylation is proportional to the square of the RCOCl concentration:

rate $= k_2[\mathrm{RCOCl}]^2$

11.

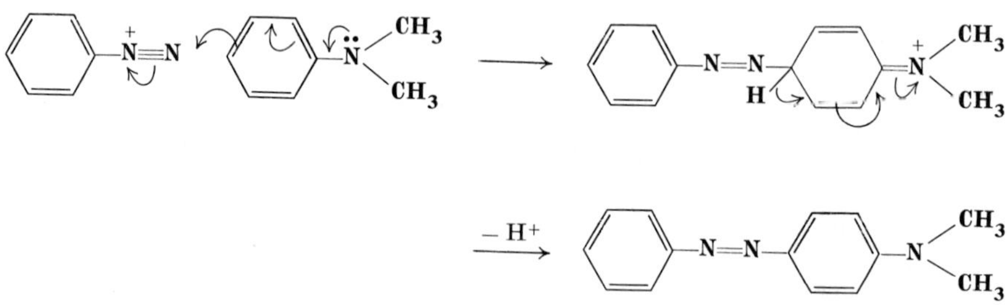

12. A: CH_3COCl; B: CH_3I, NaOH; C: H_3O^+;

D: 1) HONO, 2) CuI + HI; E: Mg; F: 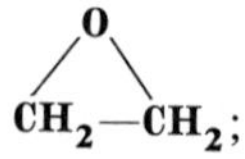

G: HBr; H: 1) $CH_2(COOCH_3)_2 + \overset{+}{Na}\overset{-}{O}CH_3$, 2) H_3O^+ + heat;
I: 1) $SOCl_2$, 2) $AlCl_3$

13.

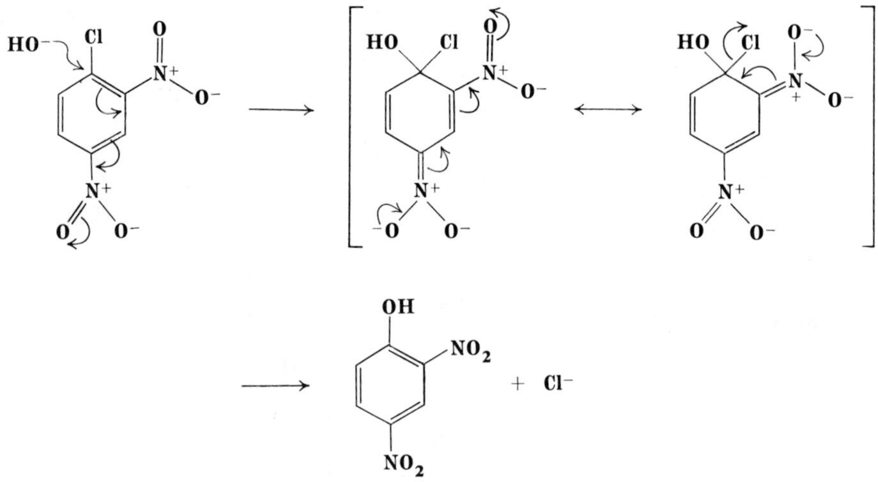

14.

a)

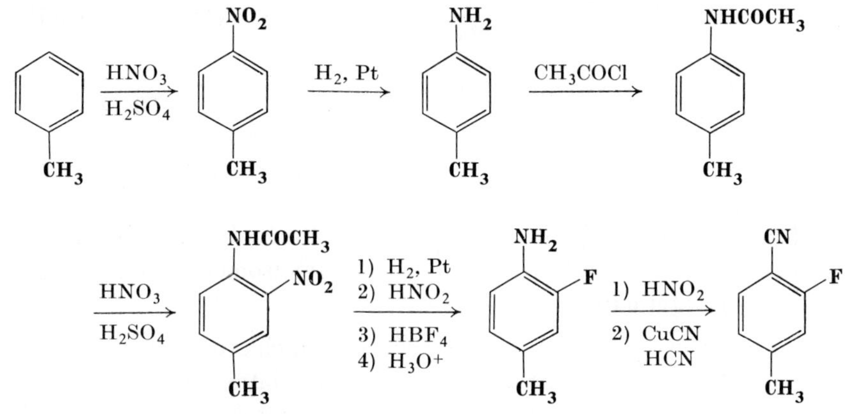

b)

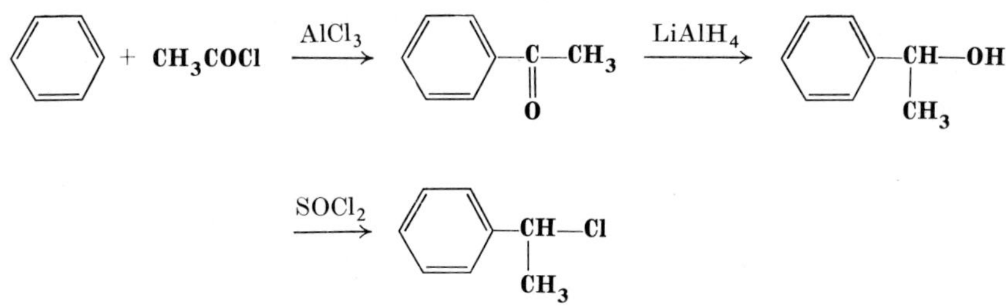

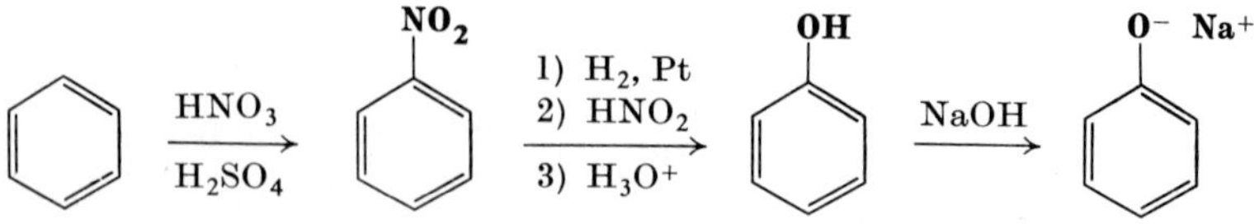

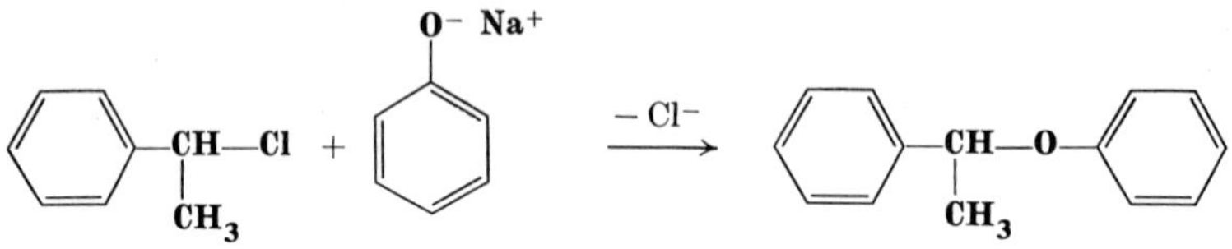

c) 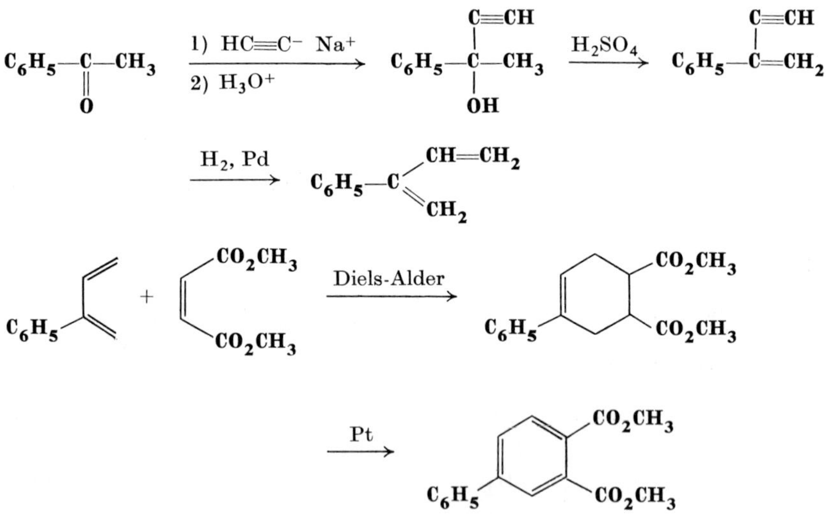

15. a) At 100° the process becomes more thermodynamically controlled and thus the sterically more stable para isomer dominates to an even greater extent.

b) Fluorine is more electronegative than chlorine. Therefore, the transition state leading to the *ortho* addition intermediate for fluorobenzene is destabilized, compared to that for chlorobenzene, by inductive withdrawal.

c) The nitro group is a better electron-withdrawing group than the carboxyl group. Nitration occurs at the least deactivated position.

16.

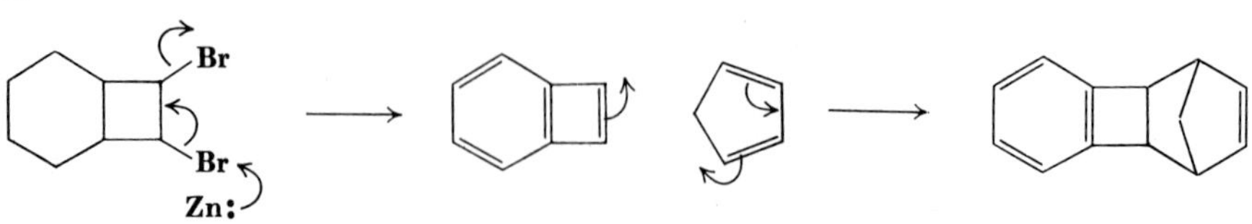

17.

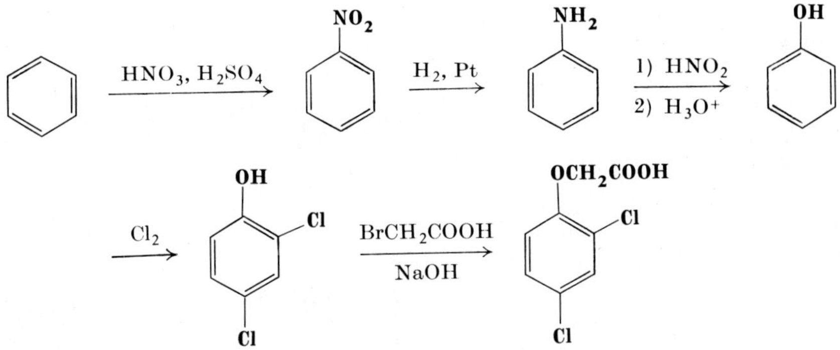

18. A: HNO_2; B: $Cu\overset{*}{C}N$, HCN; C: H_3O^+; D: H_2, Pt;
E: Mg; F: $\overset{*}{C}O_2$; G: $SOCl_2$; H: $AlCl_3$; I: Zn, HCl; J: Pt

19.

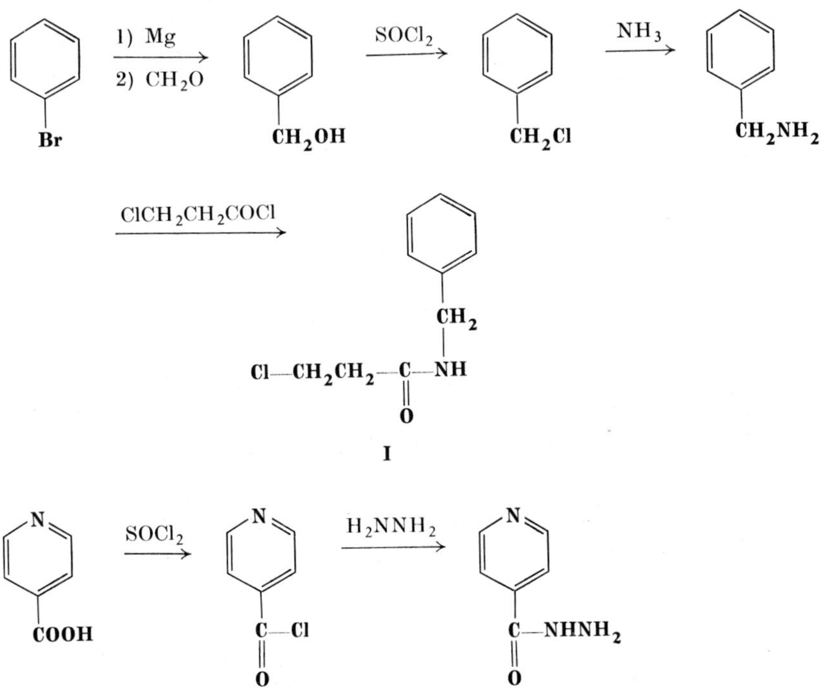

I + II ⟶ nialamide

20.

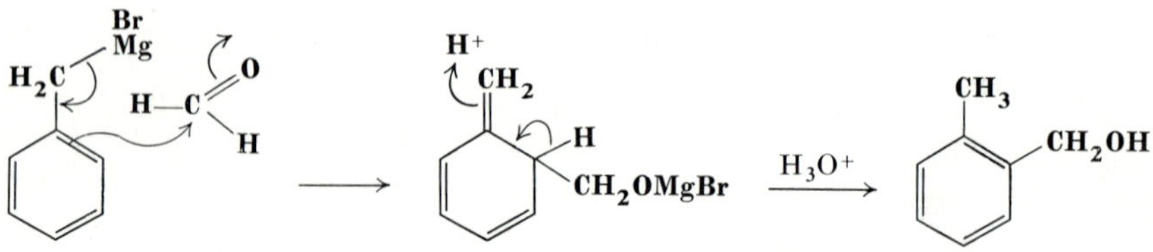

21.

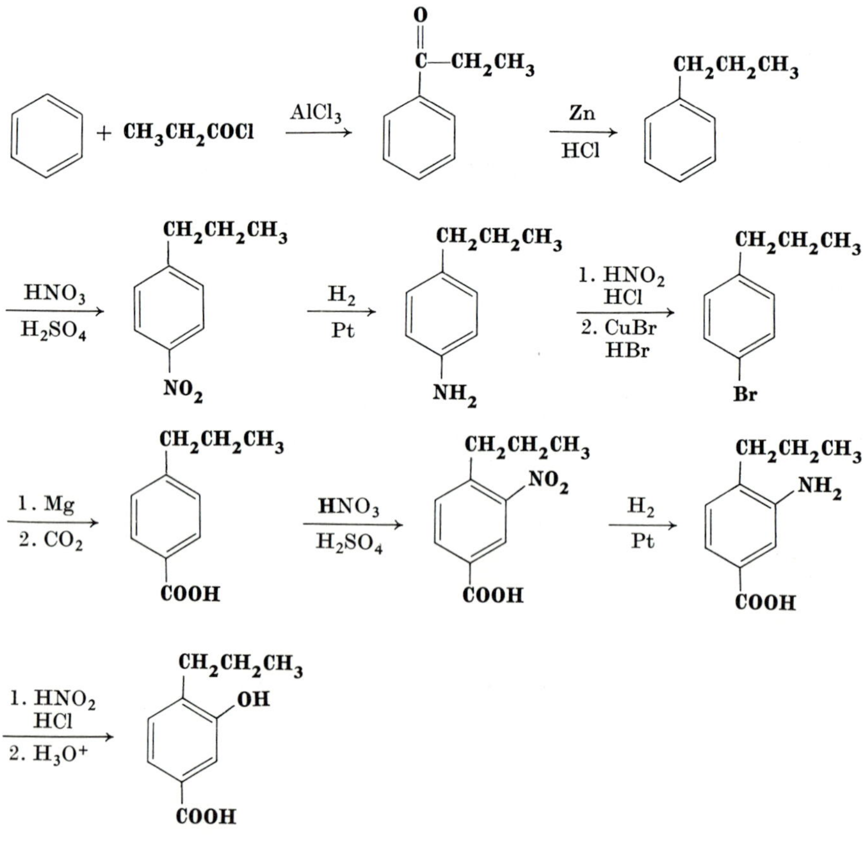

CHAPTER 11

1. Study list
2. a) ^{-}CN b) hydroxylamine (H_2NOH) c) alcohols d) ylides such as $(C_6H_5)_3\overset{+}{P}CH_2^-$ e) primary amines f) enolate of acetic anhydride
3. a) Wittig b) Dieckmann c) Mannich d) Grignard e) Claisen

4. a) NH_3 b) CH_3OH c) $LiAlH_4$ d) $LiAlH_4$ e) HBr f) NH_3
g) Mg; CO_2 h) $LiAlH_4$ i) $HOCH_2CH_2OH$, H^+ j) H_3O^+
k) $LiAl[O\text{-}t\text{-}Bu]_3H$ l) CH_3CdCl m) I_2, NaOH n) $Ph_3P^+\text{-}O^-$
o) H_2O p) $SOCl_2$

5.

a) 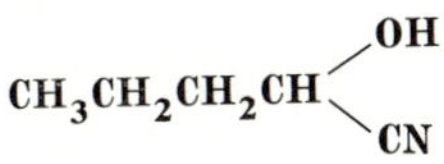

b) 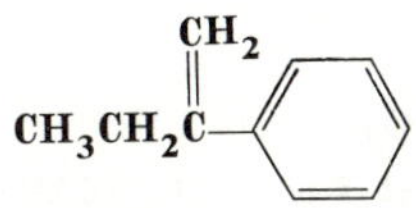

c) 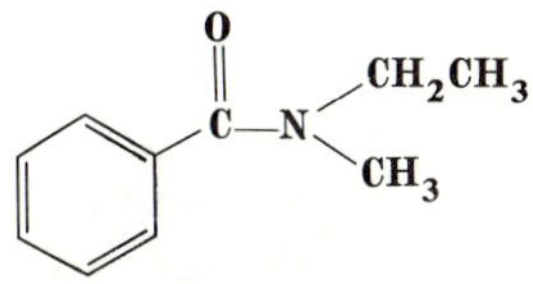

d) 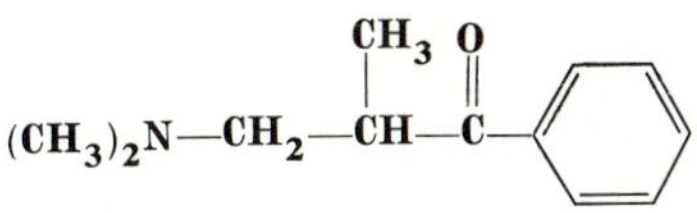

e) 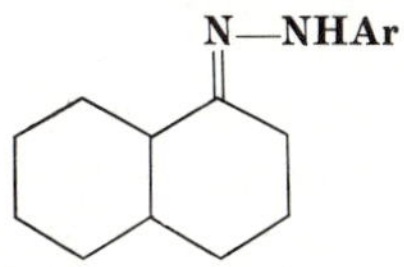

f) 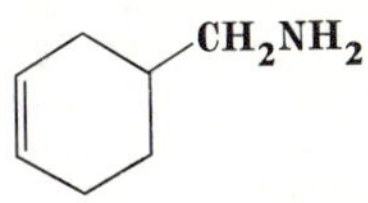

g) 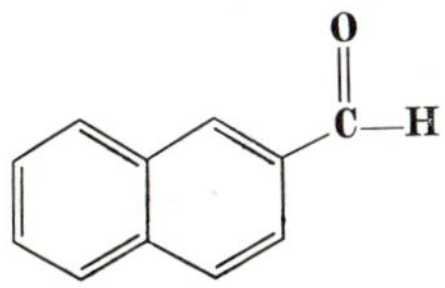

h) 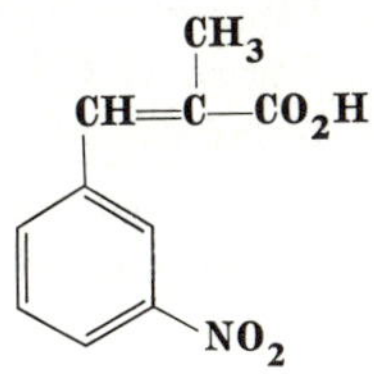

i) 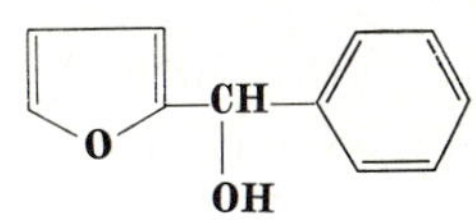

j) 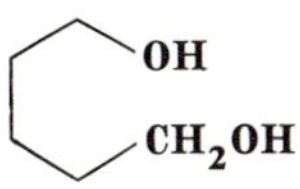

6. a) A: CH_3MgBr; B: H_2SO_4; C: BH_3, $H_2O_2 + {}^-OH$;
D. CrO_3;

E: $\underset{OH}{CH_2}—\underset{OH}{CH_2}$, H^+

b) A: $LiAlH_4$; B: $SOCl_2$; C: NaCN; H_3O^+;
D: $SOCl_2$; E: $(CH_3)_2Cd$

c) A: O_3, Zn, H_2O; B: Na^+ ^-OEt; C: $\overset{-}{C}H_2—\overset{+}{P}(C_6H_5)_3$

d) A: CH_3COCl, $AlCl_3$; B: I_2, NaOH; C: CH_3OH, HCl
D: $LiAlH_4$; E: Na, CH_3I

e) A: $Na^+\ ^-OEt$; B: $BrCH_2COOEt$; C: H_3O^+, heat

f) A: $CH_3-\overset{\overset{\displaystyle O}{\|}}{C}-CH_3$; B: $CH_2(COOEt)_2$; $Na^+\ ^-OEt$;

C: $Na^+\ \bar{O}Et$; D: H_3O^+, heat

7. A: $NaBH_4$; B: H_2SO_4; C: Mg; D: CH_2O; E: HBr;
F: Mg; G: O_3, Zn, H_2O; H: $LiAlH_4$;
I: $CH_3C_6H_4SO_2Cl$; J: NaBr; K: $(C_6H_5)_3P$;
L. $NaNH_2$; M: H_3O^+

8. One cannot prepare an organomagnesium derivative of an alkyl halide which has a group that reacts with Grignard reagents. Thus, it is not possible to prepare the Grignard of a bromoester. Utilization of the organozinc derivative in the Reformatsky reaction is a clever way around this difficulty. Organozinc compounds are not very reactive, and they therefore add to ketones but not to esters.

9.

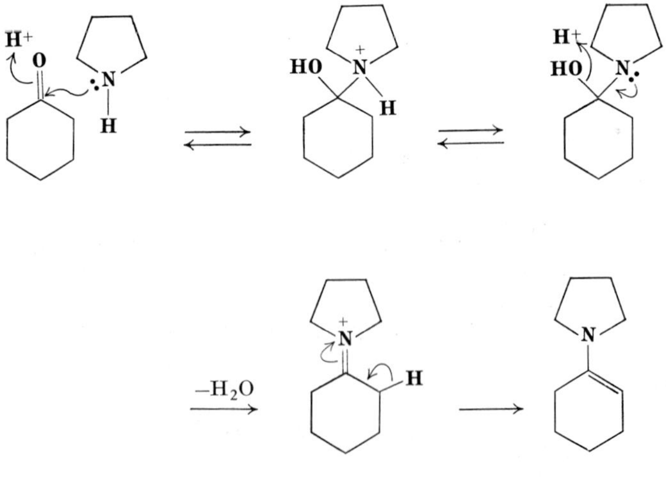

10.

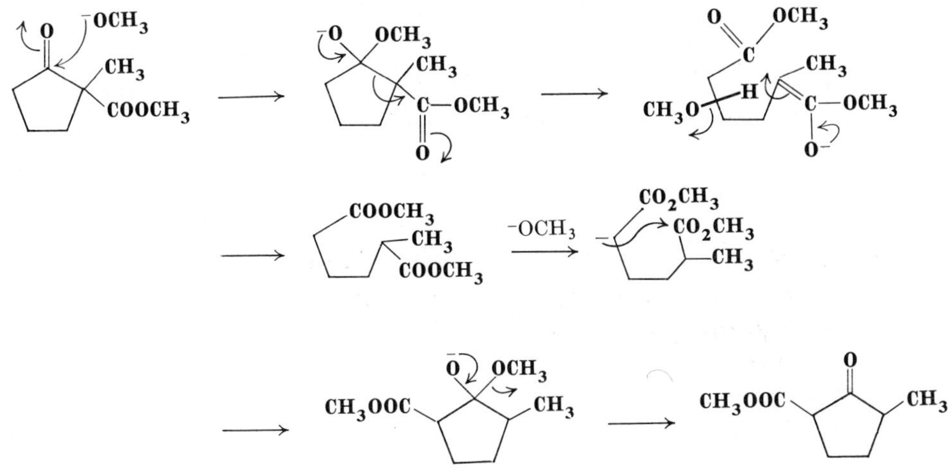

11.

a)

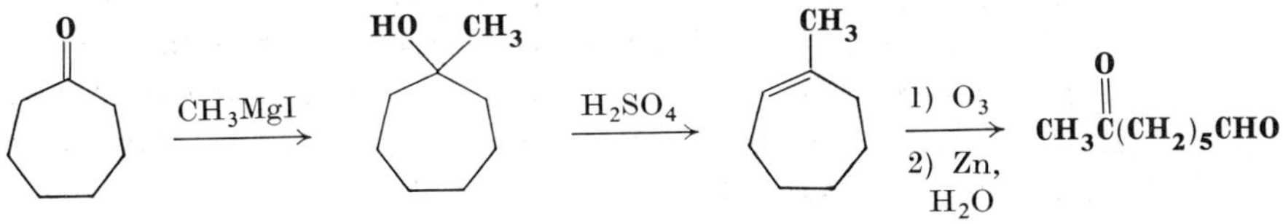

$\xrightarrow[\text{Na}^+\ {}^-\text{OEt}]{CH_2(CO_2Et)_2}$ $CH_3\overset{O}{\overset{\|}{C}}(CH_2)_5CH{=}C(CO_2Et)_2$

$\xrightarrow[\text{2) heat}]{\text{1) } H_3O^+}$ $CH_3\overset{O}{\overset{\|}{C}}(CH_2)_5CH{=}CHCOOH$

b). $HO(CH_2)_4CHO$ $\xrightarrow[\text{Na}^+\ {}^-\text{OEt}]{CH_2(CO_2Et)_2}$ $HO(CH_2)_4CH{=}C(CO_2Et)_2$

$\xrightarrow[\text{2) heat}]{\text{1) } H_3O^+}$ $HO(CH_2)_4CH{=}CHCOOH$ $\xrightarrow{PBr_3}$ $Br(CH_2)_4CH{=}CHCOOH$

$\xrightarrow[2\text{Na}^+\ {}^-\text{OEt}]{CH_3COCH_2CO_2Et}$ $CH_3CO\overset{CO_2Et}{\overset{|}{C}H}(CH_2)_4CH{=}CHCOOH$

$\xrightarrow[\text{2) heat}]{\text{1) } H_3O^+}$ $CH_3\overset{O}{\overset{\|}{C}}(CH_2)_5CH{=}CHCOOH$

12. A: $LiAlH_4$; B: CH_3SO_2Cl; C: $NaCN$; D: H_3O^+; E: H_2, Pt; F: $LiAlH_4$; G: CH_3SO_2Cl; H: CrO_3; I: NaH; J: H_2NNH_2, KOH

a) A hydroxyl group cannot be displaced in an S_N2 reaction. It must first be converted into a good leaving group.

b)

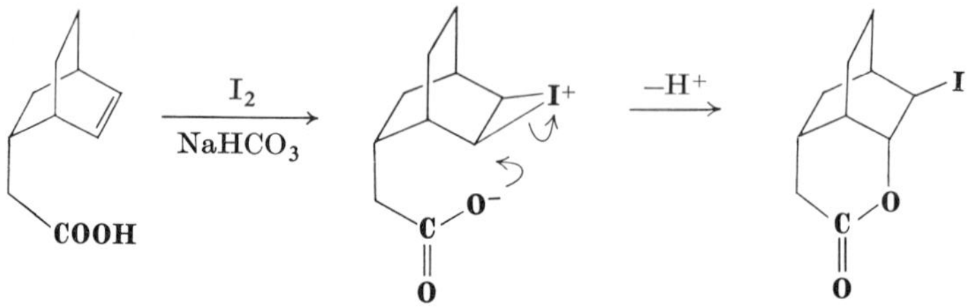

c) The primary alcohol is much less sterically hindered, and therefore more reactive, than the secondary alcohol. The primary alcohol can therefore be selectively converted to a derivative.

13.

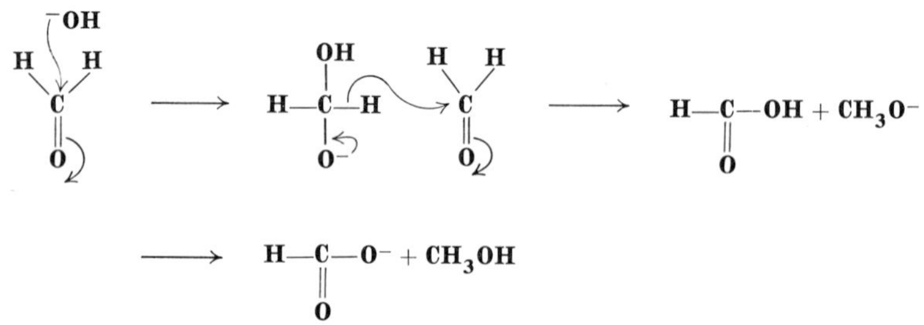

14.

a)

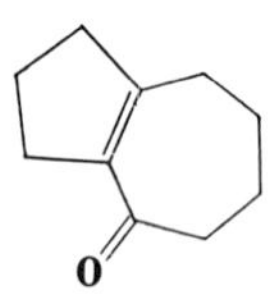

b) $C_6H_5COCH_2COOH$

c)

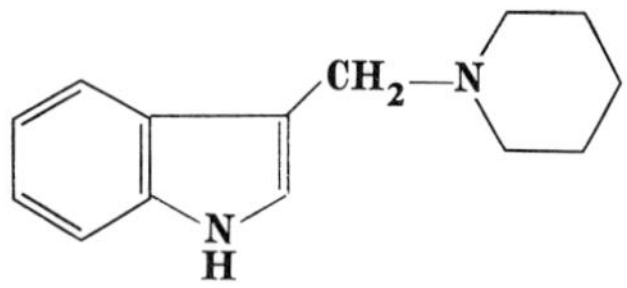

d)

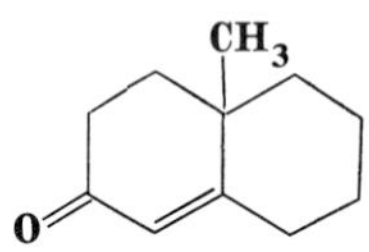

e)

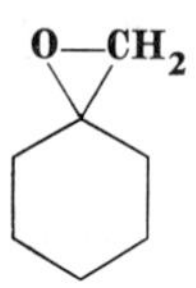

15.

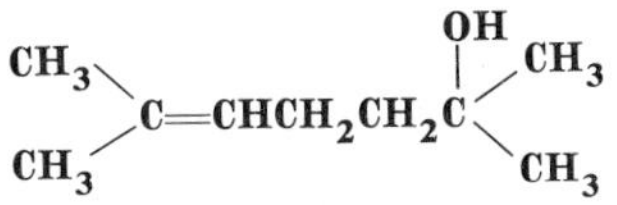

I

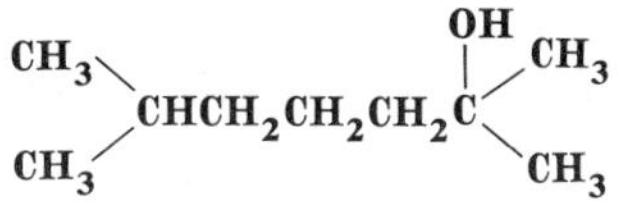

II

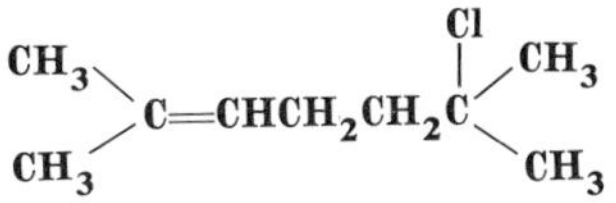

III

IV

16.

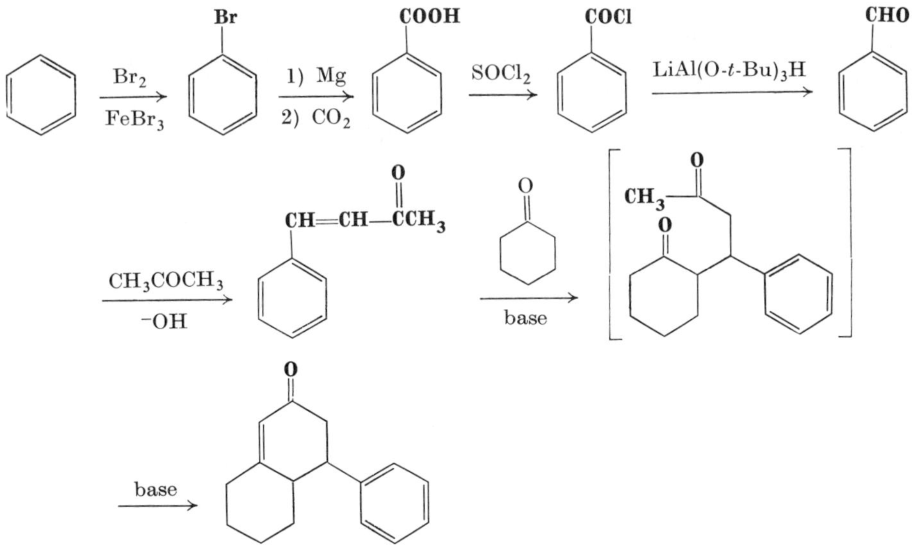

17.

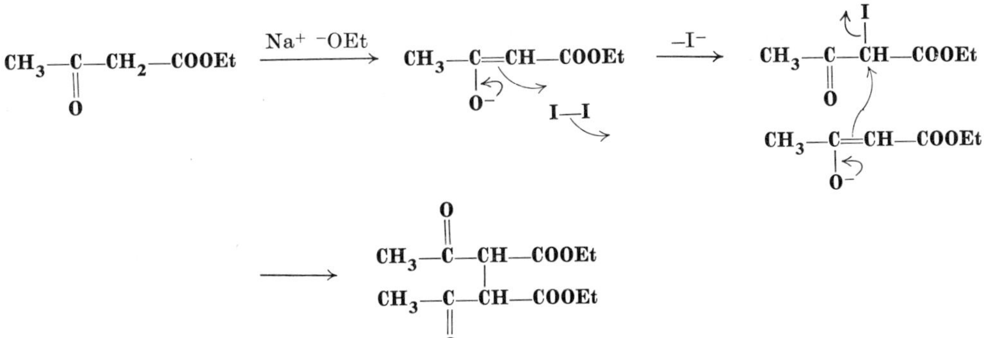

18.

a) $\mathbf{RBr + Mg \longrightarrow RMgBr \xrightarrow[2)\ H_3O^+]{1)\ CO_2} RCOOH}$

b) $\mathbf{RBr + CH_2(CO_2Et)_2 \xrightarrow{Na^+\ \bar{O}Et} RCH(CO_2Et)_2 \xrightarrow[heat]{H_3O^+} RCH_2COOH}$

c) $\mathbf{RCOOH \xrightarrow{SOCl_2} RCOCl \xrightarrow{LiAl(O\text{-}t\text{-}Bu)_3H} RCHO}$

d) $\mathbf{RCOCH_3 \xrightarrow[NaOH]{I_2} RCOOH}$

e) 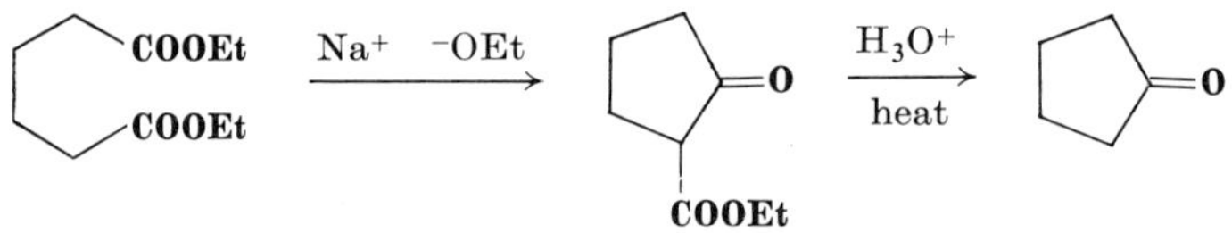

f) $\mathbf{RCH{=}CH_2 \xrightarrow[2)\ H_2O_2,\ ^-OH]{1)\ BH_3} RCH_2CH_2OH}$

g) $\mathbf{RCOOH \xrightarrow{SOCl_2} RCOCl \xrightarrow{(CH_3)_2Cd} RCOCH_3}$

h) $\mathbf{RCOOH \xrightarrow{SOCl_2} RCOCl \xrightarrow{NH_3} RCONH_2 \xrightarrow{LiAlH_4} RCH_2NH_2}$

i) 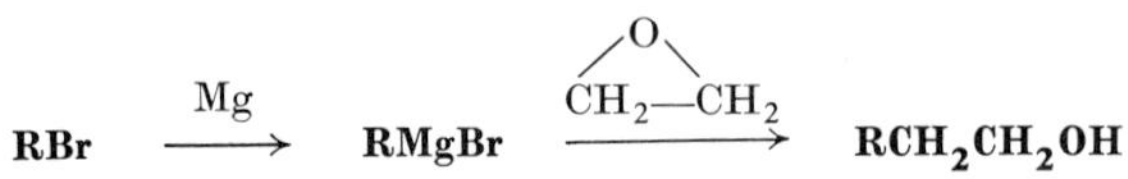

j) $\mathbf{R_2C{=}O \xrightarrow{(C_6H_5)_3\overset{+}{P}-\overset{-}{C}H_2} R_2C{=}CH_2}$

k) $\mathbf{RCH{=}CHCH_3 \xrightarrow{NBS} RCH{=}CHCH_2Br}$

l) 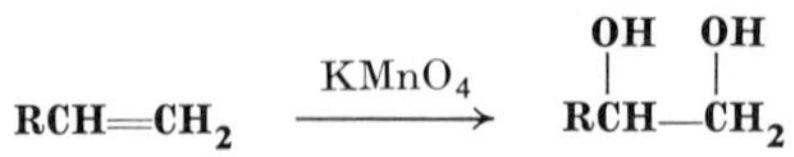

m) 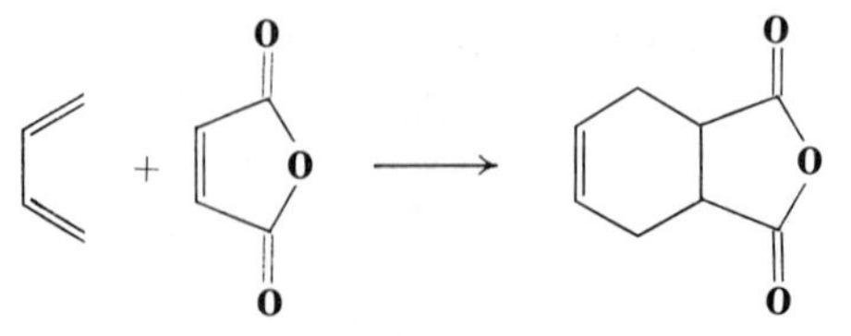

19.

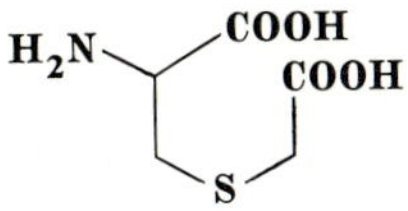

A

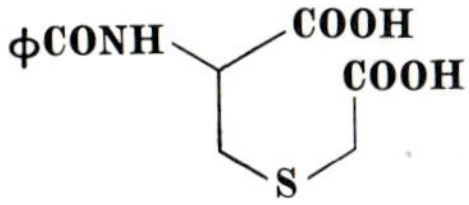

B

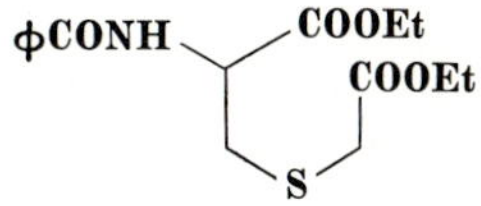

C

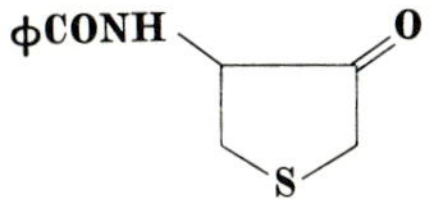

D

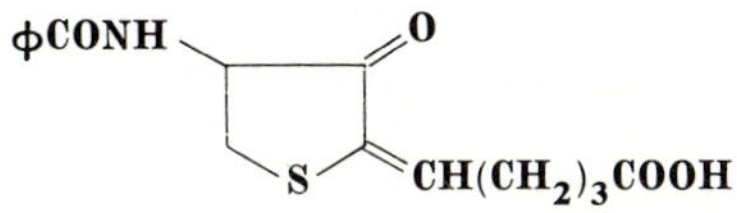

E

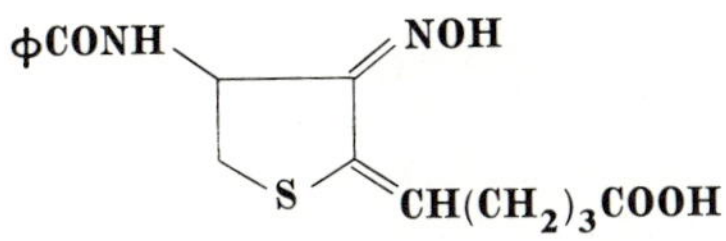

F

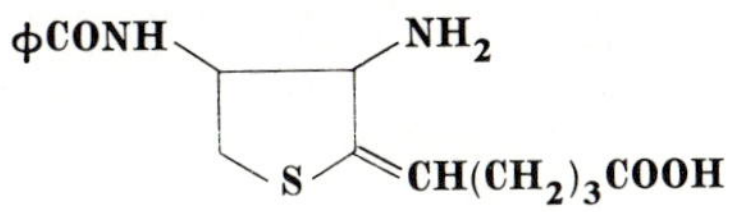

G

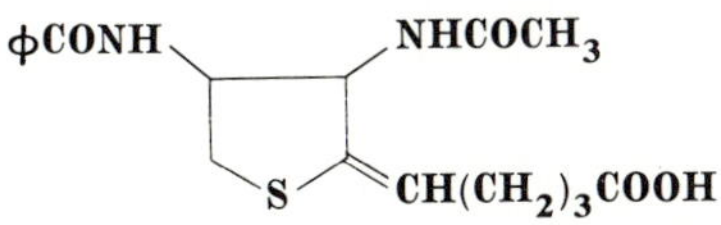

H

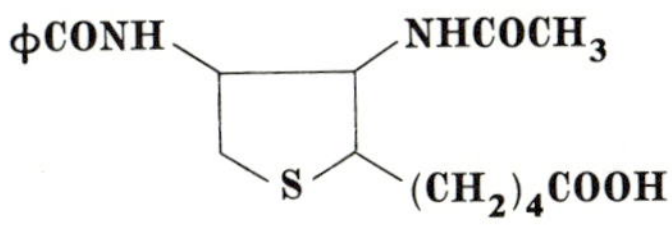

I

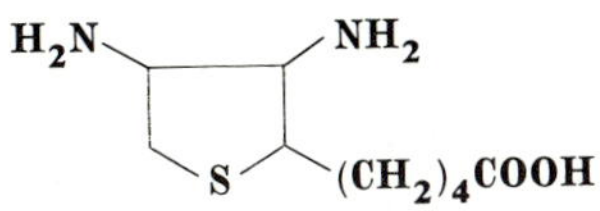

J

CHAPTER 12

1. a) N b) C c) O d) N
2. a) Wagner-Meerwein and Beckmann.
   b) The group which is *trans* to the departing water migrates.
   c) To form the amide anion and make it a better nucleophile toward Br_2.
   d) It is bonded to another electron-attracting oxygen.
   e) No, such a reaction would not occur because a tertiary carbonium ion is much more stable than a primary carbonium ion.
   f) No, it has one less carbon.
   g) Benzoic acid.

3.

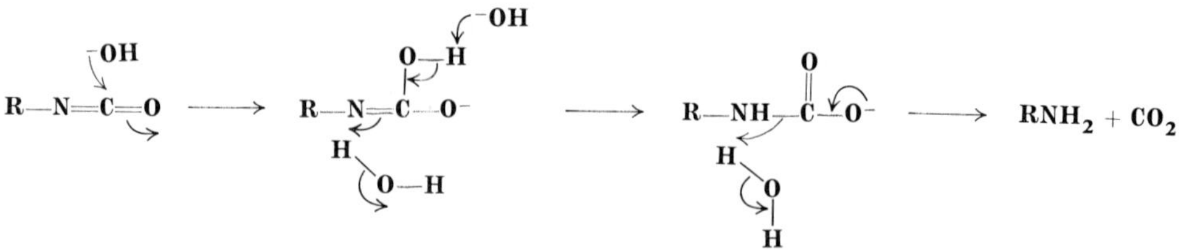

The reaction would produce a carbamate ($RNHCOOCH_3$).

4.

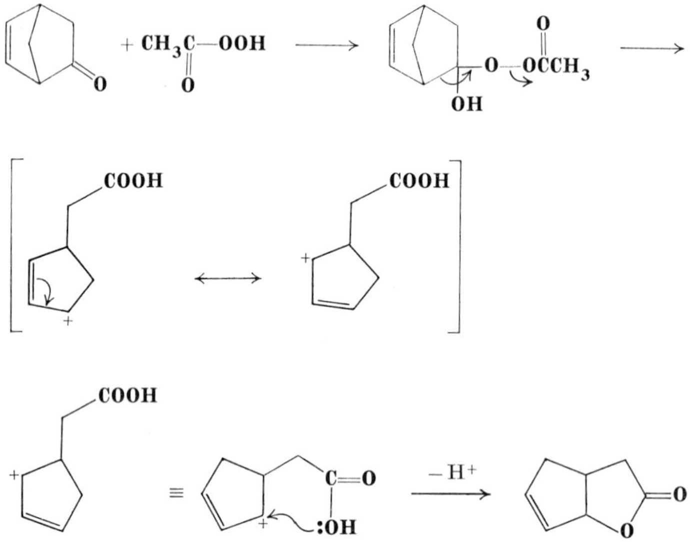

5.

a)

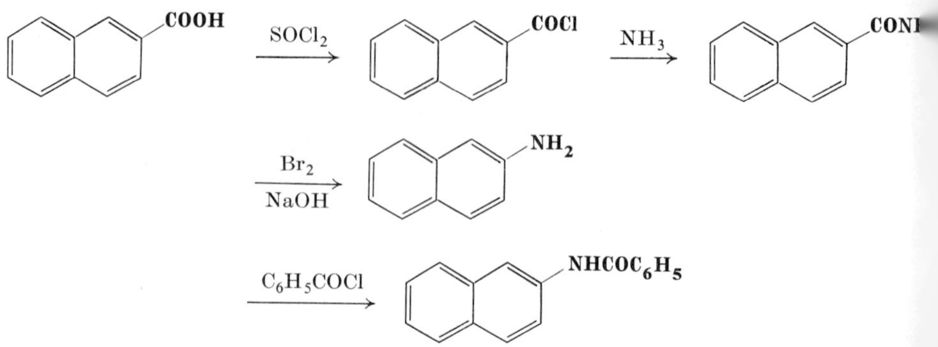

b) 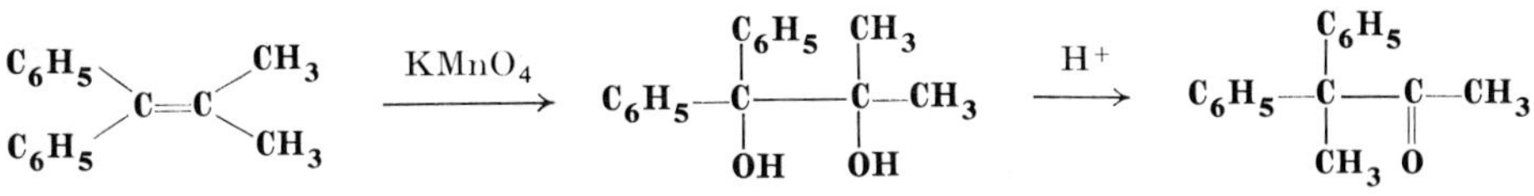
KMnO4
H+

c) 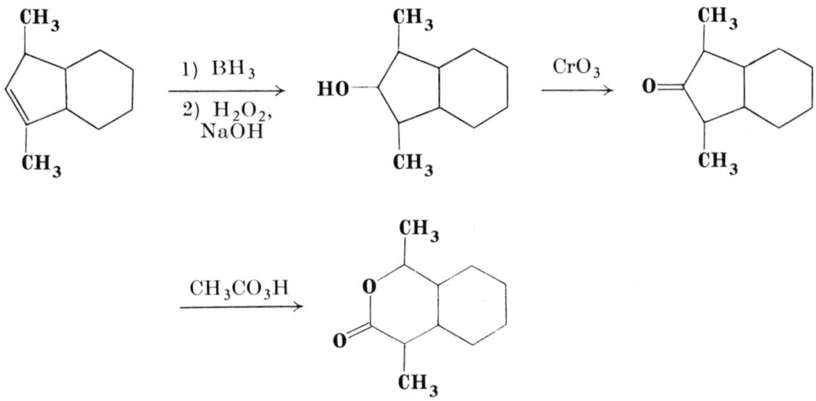
1) BH3
2) H2O2, NaOH
CrO3
CH3CO3H

d) 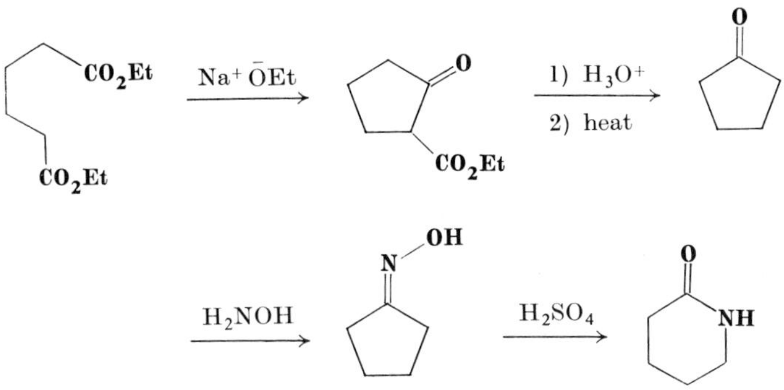
Na+ ŌEt
1) H3O+
2) heat
H2NOH
H2SO4

6.

a) 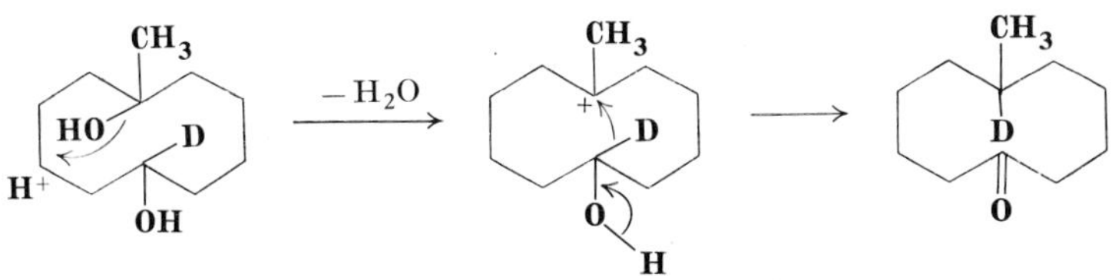
– H2O

b)

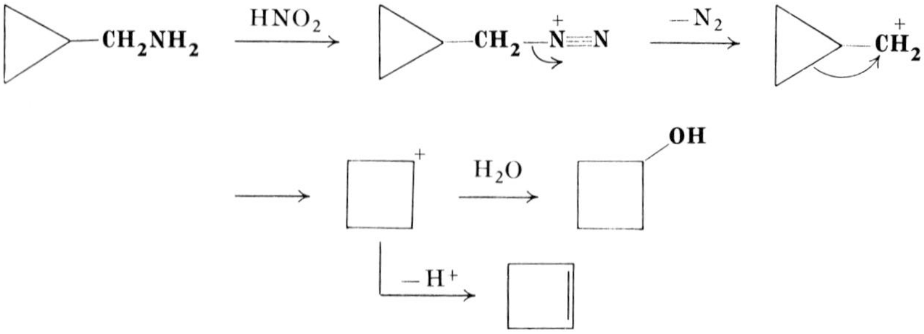
CH₂NH₂
HNO₂
CH₂
N≡N
−N₂
CH₂
H₂O
OH
−H⁺

c)

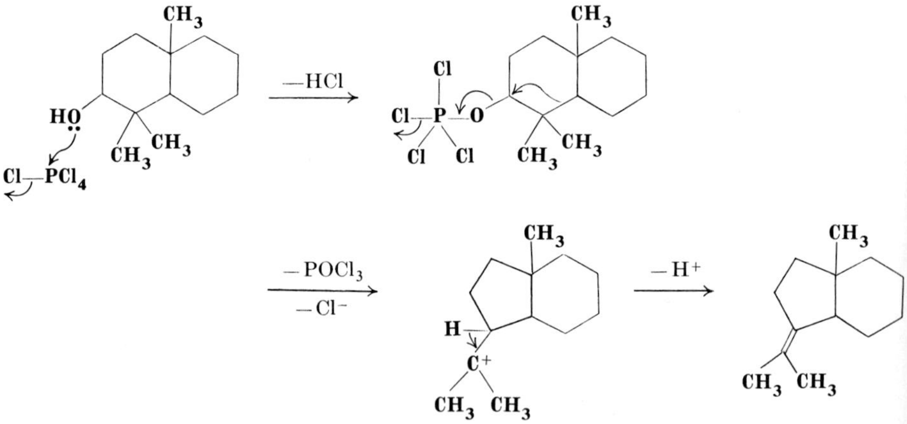
CH₃
HO
Cl
PCl₄
—HCl
P
O
−POCl₃
−Cl⁻
H
C⁺
−H⁺

d)

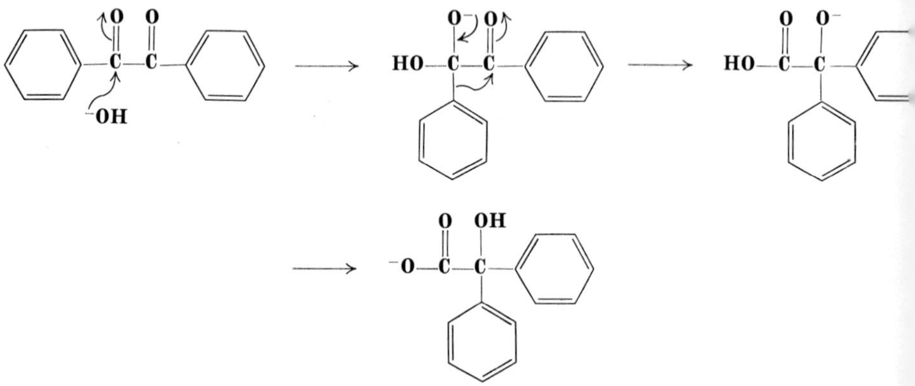
O
C
⁻OH
HO
O⁻

e)

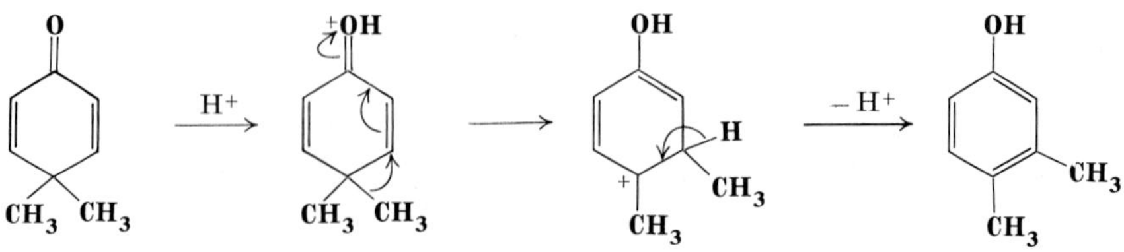
O
CH₃ CH₃
H⁺
+OH
OH
H
CH₃
−H⁺

7.

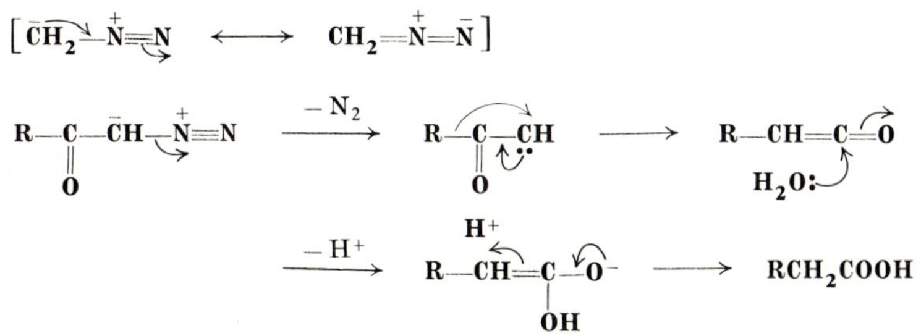

8. When the rearrangement is performed with the labeled material shown below, the product is found to be labeled only in the carboxyl group. This is consistent with the second mechanism but not with the first mechanism.

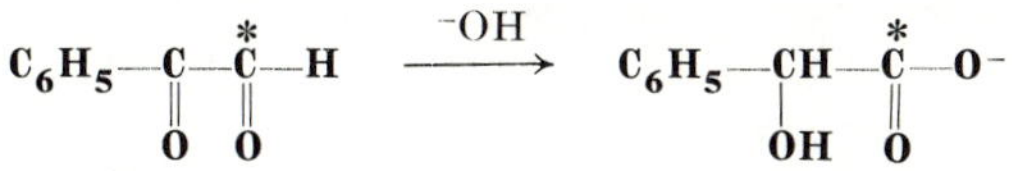

9. The reaction takes place with retention of configuration. The migrating alkyl group does not become entirely free from the rest of the molecule. If it did, the product would be racemic.

10.

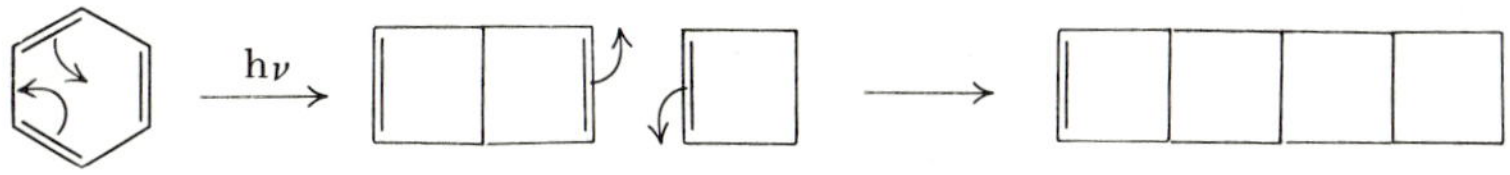

CHAPTER 13

1. a) 2-methyl-1-butene
   b) 3-methylcyclopentanol
   c) diisopropylamine
   d) *p*-nitrobenzoic acid
   e) 2-chloroheptane
   f) ethyl acetate
   g) acetamide
   h) 2-hexanone
   i) 1,4-diaminocyclohexane
   j) ethyl phenyl ether
   k) phenoxyacetic acid
   l) benzyl chloride

2. a) perbenzoic acid b) $KMnO_4$ c) H_2NNH_2, NaOH d) Pd, H_2
e) $LiAlH_4$ f) CrO_3

3.

a) 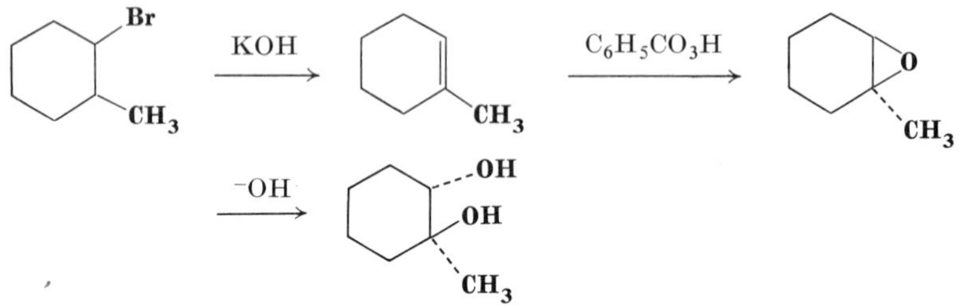

b) 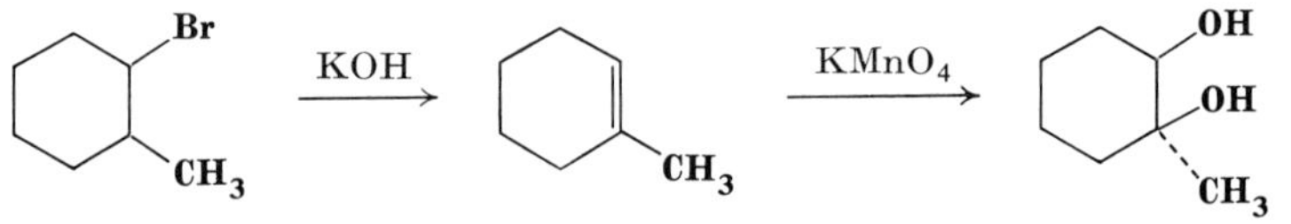

c) 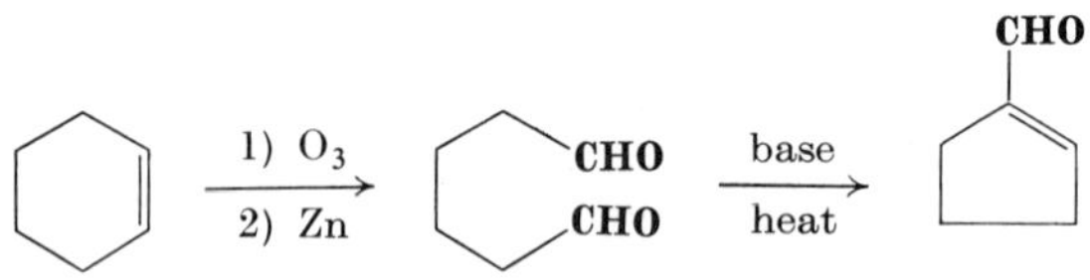

d) 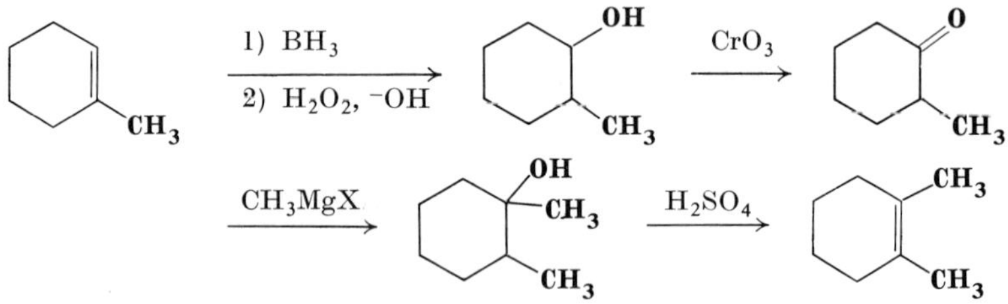

e) 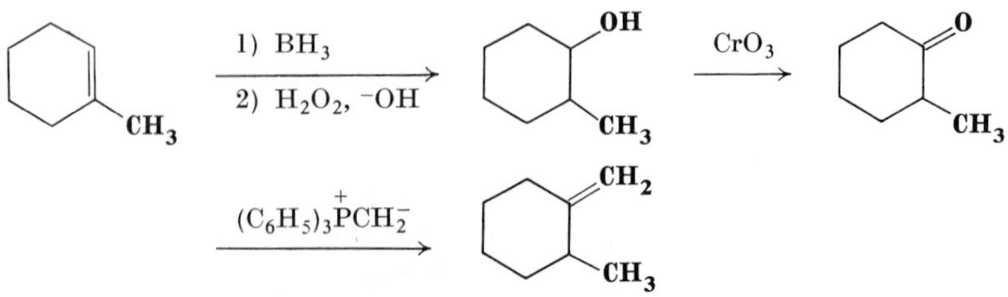

f) 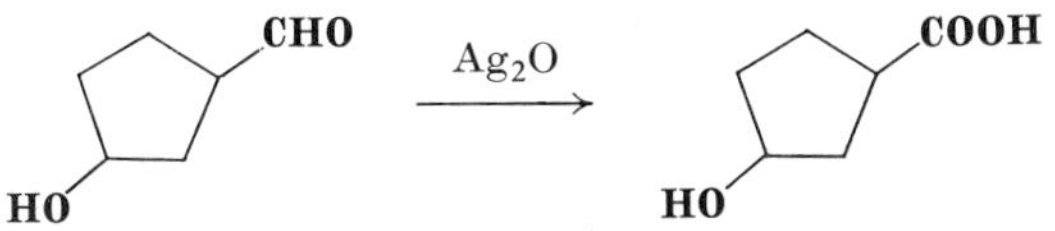

g) 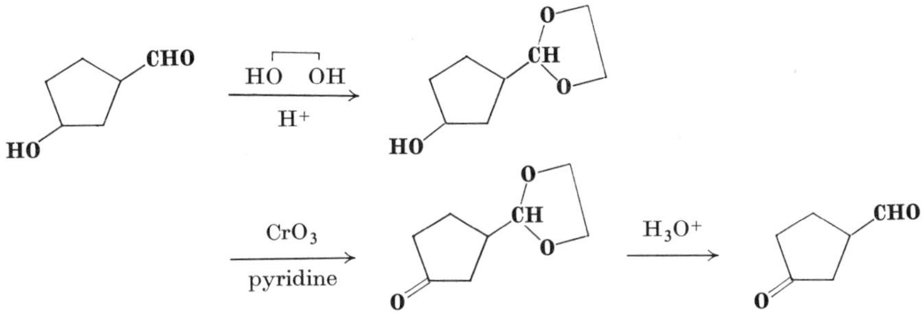

h) 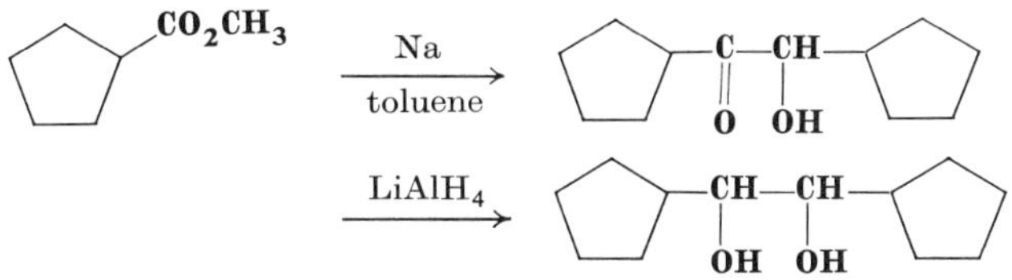

i) $CH_3CH_2CH_2OH \xrightarrow{HBr} CH_3CH_2CH_2Br \xrightarrow{NaCN} CH_3CH_2CH_2CN$

$\xrightarrow{H_2,\ Pt} CH_3(CH_2)_3NH_2$

j) 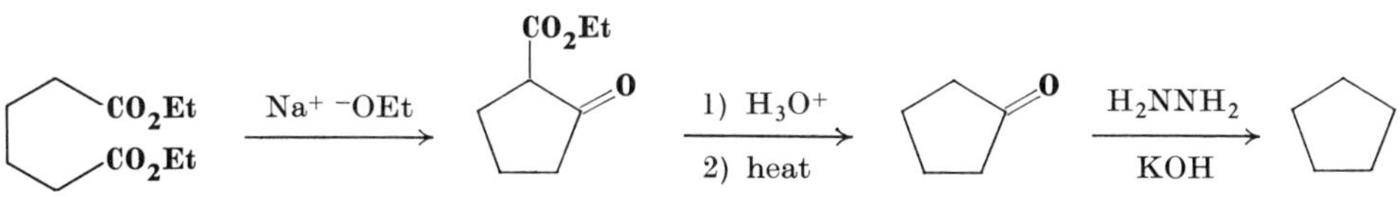

k) 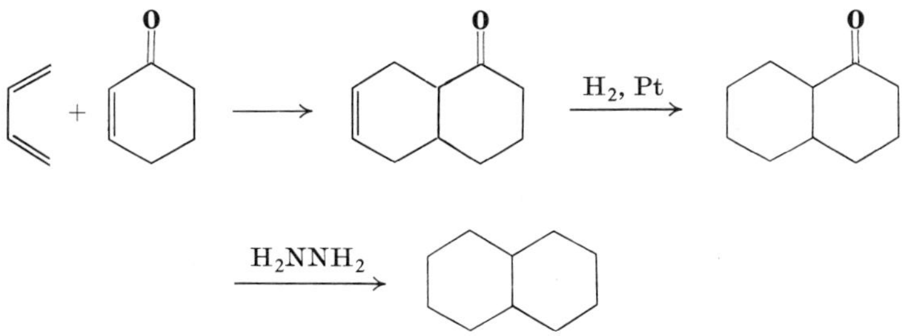

1) 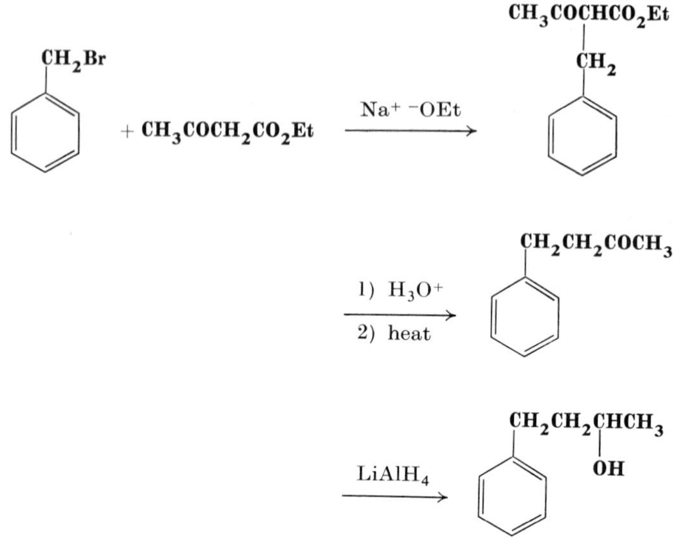

4. Racemic.

5.

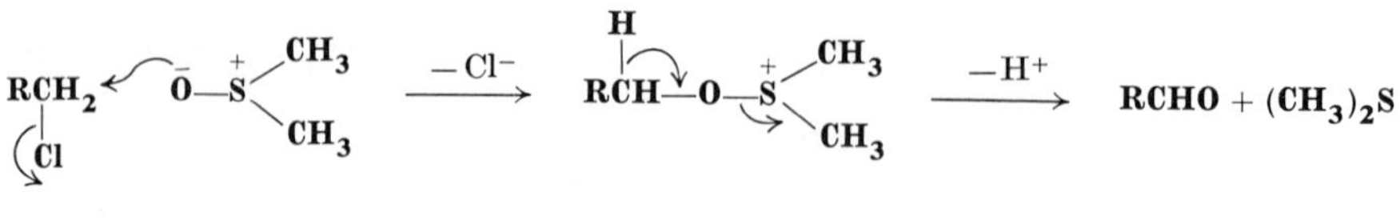

6.

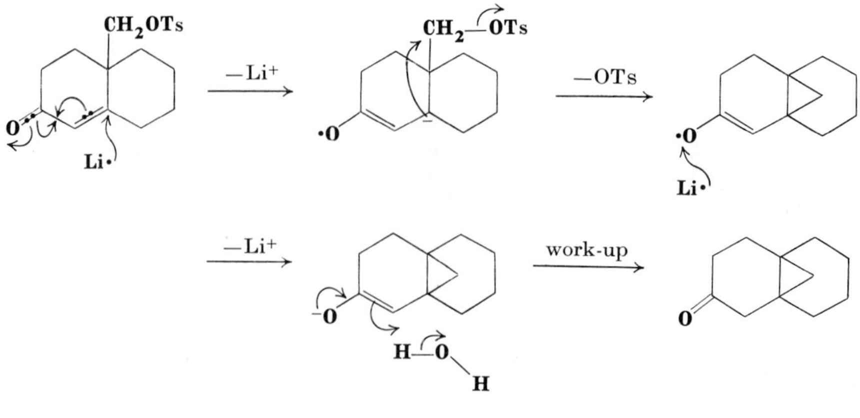

7.

a) 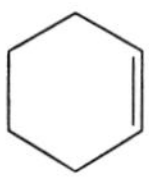 + $2CO_2$ b) 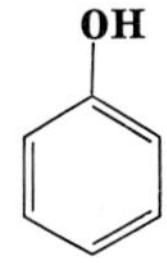

c) $RC{\equiv}C{-}C{\equiv}CR + Cu^+$ d) $(C_6H_5)_3CH + CH_3CHO$

8.

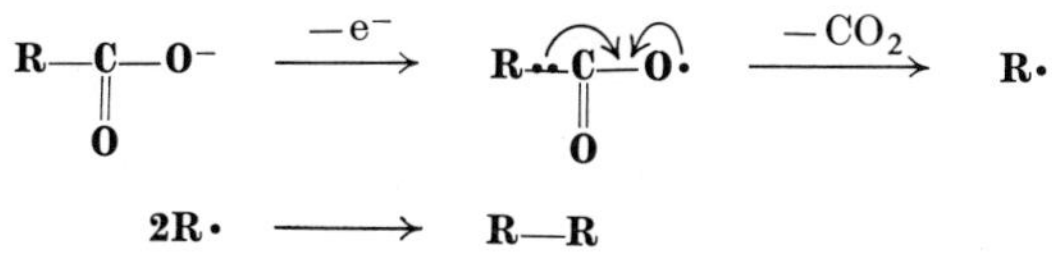

9.

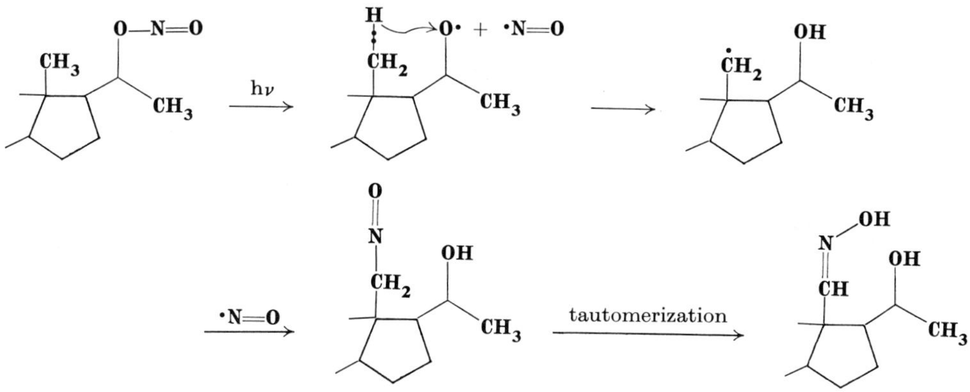

10.

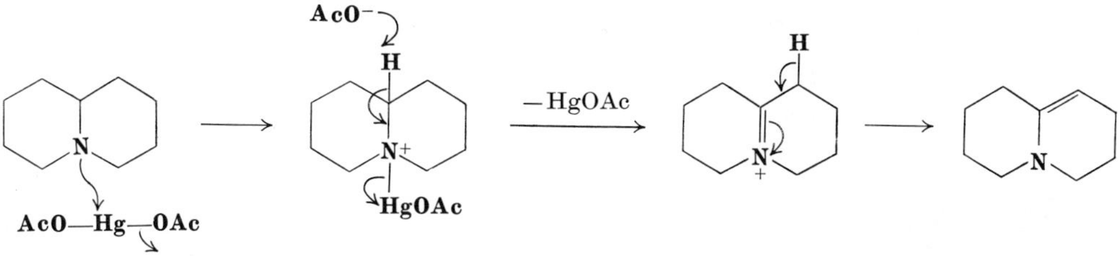

11. A: $LiAl(O\text{-}t\text{-}Bu)_3H$ B: $NaBH_4$ C: $LiAlH_4$ D: H_2, Pt

12.

A: H_2, Pt; B: —COCl; C: $SOCl_2$; D: $AlCl_3$; E: H_3O^+;

F: NaOH; G: $(CH_3CO)_2O$; H: $NaBH_4$; I: $SOCl_2$; J: NaCN;

K: H_3O^+

INDEX